Methods in Enzymology

Volume 243

INORGANIC MICROBIAL SULFUR METABOLISM

METHODS IN ENZYMOLOGY

EDITORS-IN-CHIEF

John N. Abelson Melvin I. Simon

DIVISION OF BIOLOGY
CALIFORNIA INSTITUTE OF TECHNOLOGY
PASADENA, CALIFORNIA

FOUNDING EDITORS

Sidney P. Colowick and Nathan O. Kaplan

Methods in Enzymology

Volume 243

Inorganic Microbial Sulfur Metabolism

EDITED BY

Harry D. Peck, Jr.

Jean LeGall

DEPARTMENT OF BIOCHEMISTRY
SCHOOL OF CHEMICAL SCIENCES
THE UNIVERSITY OF GEORGIA
BOYD GRADUATE STUDIES RESEARCH CENTER
ATHENS, GEORGIA

ACADEMIC PRESS

San Diego New York Boston London Sydney Tokyo Toronto

This book is printed on acid-free paper. ∞

Academic Press, Inc.
A Division of Harcourt Brace & Company
525 B Street, Suite 1900, San Diego, California 92101-4495

United Kingdom Edition published by
Academic Press Limited
24-28 Oval Road, London NW1 7DX

International Standard Serial Number: 0076-6879

International Standard Book Number: 0-12-182144-7

PRINTED IN THE UNITED STATES OF AMERICA
94 95 96 97 98 99 MM 9 8 7 6 5 4 3 2 1

Table of Contents

Section I. Sulfate Transport

Section II. Dissimilatory Sulfate Reduction

A. Dehydrogenases: Hydrogenases from Sulfate-Reducing Bacteria

B. Electron Carrier Proteins from Sulfate-Reducing Bacteria

C. Terminal Reductases of Sulfate-Reducing Bacteria

D. Molecular Biology

Section III. Dissimilatory Sulfur Reduction

Section IV. Oxidation of Reduced Sulfur Compounds

Section V. Metabolism of Polythionates

Section VI. Special Techniques

Contributors to Volume 243

Article numbers are in parentheses following the names of contributors.
Affiliations listed are current.

J. M. Akagi (17), *Department of Microbiology, University of Kansas, 7042 Haworth Hall, Lawrence, Kansas 66045-2103*

Belarmino A. S. Barata (4), *Departamento de Quimica, Faculdade de Ciências, Universidade de Lisboa, Lisboa and ITQB, Oeiras, Portugal*

Larry L. Barton (7), *Department of Biology, University of New Mexico, Albuquerque, New Mexico 87131*

D. H. Boxer (26), *Department of Biochemistry, The University of Dundee, Dundee, Tayside DD1 4HN, United Kingdom*

Mireille Bruschi (10), *Unité de Bioénergetique et Ingéniérie des Proteines, Centre National de la Recherche Scientifique, 31 Chemin Joseph Aiguier, 13402 Marseille Cedex 20, France*

Richard Cammack (5), *Division of Life Sciences, King's College, London W8 7AH, United Kingdom*

Cristina Costa (21), *Departamento de Quimica, Faculdade de Ciências e Tecnologia, Universidade Nova de Lisboa, 2825 Monte de Caparica, Portugal*

Isabel B. Coutinho (9), *Instituto de Tecnologia Química e Biológica, Universidade Nova de Lisboa, 2780 Oeiras, Portugal*

Heribert Cypionka (1), *Institut für Chemie und Biologie des Meeres, Universität Oldenburg, D-26111 Oldenburg, Germany*

Christiane Dahl (23, 28), *Institut für Mikrobiologie und Biotechnologie, Rheinische Friedrich-Wilhelms-Universität Bonn, 53115 Bonn, Germany*

Daniel V. DerVartanian (18), *Department of Biochemistry, Life Sciences Building, The University of Georgia, Athens, Georgia 30602*

H. L. Drake (17), *BITÖK, Universität Bayreuth, D-8580 Bayreuth, Germany*

Richard C. Ebersole (41), *Central Research and Development, E. I. DuPont De Nemours and Company, Inc., Wilmington, Delaware 19880*

Paula Fareleira (39), *Instituto de Tecnologia Química e Biológica, Universidade Nova de Lisboa, 2780 Oeiras, Portugal*

Guy D. Fauque (24, 25), *Centre d'Océanologie de Marseille, URA CNRS 41, Campus de Luminy, Case 901, 13288 Marseille Cedex 9, France*

Victor M. Fernandez Lopez (5), *Instituto de Catalisis, Consejo Superior de Investigaciones Científicas, Universidad Aútonoma Cantoblanco, 28049 Madrid, Spain*

Diane Gevertz (17), *Agouron Institute, La Jolla, California 92037*

Theo A. Hansen (2, 3), *Department of Microbiology, University of Groningen, 9751 NN Haren, The Netherlands*

E. Claude Hatchikian (5, 19), *Unité de Bioénergetiques et Ingéniérie des Proteines Centre National de la Recherche Scientifique, 13402 Marseille Cedex 20, France*

Dirk Heering (13), *Department of Biochemistry, Agricultural University, 6703 HA, Wageningen, The Netherlands*

Charles M. H. Hensgens (2), *Department of Microbiology, University of Groningen, 9751 NN Haren, The Netherlands*

Yoshiki Higuchi (11), *Department of Life Sciences, Himeji Institute of Technology, Kamigori, Hyogo 678-12, Japan*

Boi Hanh Huynh (38), *Department of Physics, Emory University, Atlanta, Georgia 30322*

JOHN W. INGLEDEW (26), *Department of Biological Sciences, St. Andrews University, St. Andrews, Fife KY16 9AL, United Kingdom*

DON P. KELLY (35, 36, 37), *Department of Science Education, University of Warwick, Coventry CV4 7AL, United Kingdom*

JAE-HO KIM (17), *Department of Biology, Kyungsung University, Pusan 608-736, South Korea*

OLIVER KLIMMEK (25), *Institut für Microbiologie–Biozentrum Niderusel, J. W. Goethe-Universität Marie Curie Strasse 9, 60439 Frankfurt am Main, Germany*

ACHIM KRÖGER (25), *Institut für Microbiologie, J. W. Goethe-Universität, Frankfurt am Main, W-6000 Frankfurt am Main, Germany*

JORGE LAMPREIA (16), *Departamento de Quimica, Faculdade de Ciências e Tecnologia, Universidade Nova de Lisboa, 2825 Monte de Caparica, Portugal*

JEAN LEGALL (39), *Department of Biochemistry, University of Georgia, Athens, Georgia 30602*

ANA R. LINO (20), *Centro de Tecnologia Quimica e Biológica, 2780 Oeiras, Portugal*

MING-CHEH LIU (21), *Department of Biochemistry, University of Texas Health Center at Tyler, Tyler, Texas 75710*

ANJOS L. MACEDO (12), *Departamento de Quimica, Faculdade de Ciências e Tecnologia, Universidade Nova de Lisboa, 2825 Monte de Caparica, Portugal*

T. E. MEYER (30, 31), *Department of Biochemistry, University of Arizona, Tucson, Arizona 85721*

ISABEL MOURA (15, 20, 21), *Departamento de Quimica, Faculdade de Ciências e Tecnologia, Universidade Nova de Lisboa, 2825 Monte de Caparica, Portugal*

JOSÉ J. G. MOURA (4, 12, 16), *Departamento de Quimica, Faculdade de Ciências e Tecnologia, Universidade Nova de Lisboa, 2825 Monte de Caparica, Portugal*

J. MARTIN ODOM (41), *Central Research and Development, E. I. DuPont De Nemours and Company, Inc., Wilmington, Delaware 19880*

P. NUNO PALMA (12), *Departamento de Quimica, Faculdade de Ciências e Tecnologia, Universidade Nova de Lisboa, 2825 Monte de Caparica, Portugal*

DAULAT S. PATIL (6), *Department of Biochemistry, University of Georgia, Athens, Georgia 30602*

SJAAK PEELEN (13), *Department of Biochemistry, Agricultural University, 6703 HA Wageningen, The Netherlands*

ALICE S. PEREIRA (16), *Departamento de Quimica, Faculdade de Ciências e Tecnologia, Universidade Nova de Lisboa, 2825 Monte de Caparica, Portugal*

NATARAJAN RAVI (15), *Department of Physics, Emory University, Atlanta, Georgia 30322*

HELENA SANTOS (39), *Instituto de Tecnologia Química e Biológica, Universidade Nova de Lisboa, 2780 Oeiras, Portugal*

LARRY C. SIEKER (14), *Department of Biological Structure, SM-20, University of Washington, Seattle, Washington 98195*

NORBERT SPEICH (23), *Institut für Humangenetik, Philipps-Universität Marburg, 35037 Marburg, Germany*

JACK P. W. G STOKKERMANS (22), *Laboratory for Monoclonal–Antibody Technology, Agricultural University, Binnenhaven, Wageningen, 6709 PD, The Netherlands*

ISAMU SUZUKI (32, 33), *Department of Microbiology, University of Manitoba, Winnipeg, Manitoba, Canada R3T 2N2*

PEDRO TAVARES (15), *Departamento de Quimica, Faculdade de Ciências e Tecnologia, Universidade Nova de Lisboa, 2825 Monte de Caparica, Portugal*

BARRIE F. TAYLOR (27), *Rosenstiel School of Marine and Atmospheric Science, University of Miami, Miami, Florida 33149-1098*

HANS G. TRÜPER (23, 28, 29), *Institut für Mikrobiologie und Biotechnologie, Rheinische Friedrich-Wilhelms-Universität Bonn, 53115 Bonn, Germany*

WILLEM VAN BERKEL (13), *Department of Biochemistry, Agricultural University, 6703 HA Wageningen, The Netherlands*

WALTER M.A.M. VAN DONGEN (22), *Department of Biochemistry, Agricultural University, 6703 HA Wageningen, The Netherlands*

WILLY A. M. VAN DEN BERG (22), *Department of Biochemistry, Agricultural University, 6703 HA Wageningen, The Netherlands*

JACQUES VERVOORT (13), *Department of Biochemistry, Agricultural University, 6703 HA Wageningen, The Netherlands*

GERRIT VOORDOUW (11), *Department of Biological Sciences, The University of Calgary, Calgary, Alberta, Canada T2N 1N4*

JOHN E. WAMPLER (40), *Computational Center for Molecular Structure and Design, Department of Biochemistry, University of Georgia, Athens, Georgia 30602*

ANN P. WOOD (35, 36, 37), *Division of Life Sciences, King's College London, London W8 7AH, United Kingdom*

ANTÓNIO V. XAVIER (9, 39), *Instituto de Tecnologia Química e Biológica, Universidade Nova de Lisboa, 2780 Oeiras, Portugal*

TATSUHIKO YAGI (8, 11), *Department of Chemistry, Shizuoka University, Shizuoka 422, Japan*

TATEO YAMANAKA (34), *Department of Industrial Chemistry, College of Science and Technology, Nihon University, Kanda-Surugadai 1-5, Chiyoda-ku, Tokyo 101, Japan*

Preface

One of the major aims of biochemical and microbiological studies of sulfur metabolism is to relate the functional chemistry and biochemistry of the metabolism and conversions of sulfur reduction as carried out by the sulfate-reducing bacteria and the assimilatory or biosynthetic sulfate reduction for the biosynthesis of cysteine. A most important concept in this regard is the existence of aerobic assimilatory and anaerobic dissimilatory and sulfur cycles. (The enzymes of the assimilatory pathway are reviewed in Volume 143 of *Methods of Enzymology*.) Among various oxidative and reductive phases of these cycles, which do not include nonenzymatic catalysis, microbial conversion has become essential in linking and completing the sulfur cycles.

HARRY D. PECK, JR.

METHODS IN ENZYMOLOGY

Volume I. Preparation and Assay of Enzymes
Edited by Sidney P. Colowick and Nathan O. Kaplan

Volume II. Preparation and Assay of Enzymes
Edited by Sidney P. Colowick and Nathan O. Kaplan

Volume III. Preparation and Assay of Substrates
Edited by Sidney P. Colowick and Nathan O. Kaplan

Volume IV. Special Techniques for the Enzymologist
Edited by Sidney P. Colowick and Nathan O. Kaplan

Volume V. Preparation and Assay of Enzymes
Edited by Sidney P. Colowick and Nathan O. Kaplan

Volume VI. Preparation and Assay of Enzymes (*Continued*)
Preparation and Assay of Substrates
Special Techniques
Edited by Sidney P. Colowick and Nathan O. Kaplan

Volume VII. Cumulative Subject Index
Edited by Sidney P. Colowick and Nathan O. Kaplan

Volume VIII. Complex Carbohydrates
Edited by Elizabeth F. Neufeld and Victor Ginsburg

Volume IX. Carbohydrate Metabolism
Edited by Willis A. Wood

Volume X. Oxidation and Phosphorylation
Edited by Ronald W. Estabrook and Maynard E. Pullman

Volume XI. Enzyme Structure
Edited by C. H. W. Hirs

Volume XII. Nucleic Acids (Parts A and B)
Edited by Lawrence Grossman and Kivie Moldave

Volume XIII. Citric Acid Cycle
Edited by J. M. Lowenstein

Volume XIV. Lipids
Edited by J. M. Lowenstein

Volume XV. Steroids and Terpenoids
Edited by Raymond B. Clayton

Volume XVI. Fast Reactions
Edited by Kenneth Kustin

VOLUME XVII. Metabolism of Amino Acids and Amines (Parts A and B)
Edited by HERBERT TABOR AND CELIA WHITE TABOR

VOLUME XVIII. Vitamins and Coenzymes (Parts A, B, and C)
Edited by DONALD B. MCCORMICK AND LEMUEL D. WRIGHT

VOLUME XIX. Proteolytic Enzymes
Edited by GERTRUDE E. PERLMANN AND LASZLO LORAND

VOLUME XX. Nucleic Acids and Protein Synthesis (Part C)
Edited by KIVIE MOLDAVE AND LAWRENCE GROSSMAN

VOLUME XXI. Nucleic Acids (Part D)
Edited by LAWRENCE GROSSMAN AND KIVIE MOLDAVE

VOLUME XXII. Enzyme Purification and Related Techniques
Edited by WILLIAM B. JAKOBY

VOLUME XXIII. Photosynthesis (Part A)
Edited by ANTHONY SAN PIETRO

VOLUME XXIV. Photosynthesis and Nitrogen Fixation (Part B)
Edited by ANTHONY SAN PIETRO

VOLUME XXV. Enzyme Structure (Part B)
Edited by C. H. W. HIRS AND SERGE N. TIMASHEFF

VOLUME XXVI. Enzyme Structure (Part C)
Edited by C. H. W. HIRS AND SERGE N. TIMASHEFF

VOLUME XXVII. Enzyme Structure (Part D)
Edited by C. H. W. HIRS AND SERGE N. TIMASHEFF

VOLUME XXVIII. Complex Carbohydrates (Part B)
Edited by VICTOR GINSBURG

VOLUME XXIX. Nucleic Acids and Protein Synthesis (Part E)
Edited by LAWRENCE GROSSMAN AND KIVIE MOLDAVE

VOLUME XXX. Nucleic Acids and Protein Synthesis (Part F)
Edited by KIVIE MOLDAVE AND LAWRENCE GROSSMAN

VOLUME XXXI. Biomembranes (Part A)
Edited by SIDNEY FLEISCHER AND LESTER PACKER

VOLUME XXXII. Biomembranes (Part B)
Edited by SIDNEY FLEISCHER AND LESTER PACKER

VOLUME XXXIII. Cumulative Subject Index Volumes I–XXX
Edited by MARTHA G. DENNIS AND EDWARD A. DENNIS

VOLUME XXXIV. Affinity Techniques (Enzyme Purification: Part B)
Edited by WILLIAM B. JAKOBY AND MEIR WILCHEK

VOLUME XXXV. Lipids (Part B)
Edited by JOHN M. LOWENSTEIN

VOLUME XXXVI. Hormone Action (Part A: Steroid Hormones)
Edited by BERT W. O'MALLEY AND JOEL G. HARDMAN

VOLUME XXXVII. Hormone Action (Part B: Peptide Hormones)
Edited by BERT W. O'MALLEY AND JOEL G. HARDMAN

VOLUME XXXVIII. Hormone Action (Part C: Cyclic Nucleotides)
Edited by JOEL G. HARDMAN AND BERT W. O'MALLEY

VOLUME XXXIX. Hormone Action (Part D: Isolated Cells, Tissues, and Organ Systems)
Edited by JOEL G. HARDMAN AND BERT W. O'MALLEY

VOLUME XL. Hormone Action (Part E: Nuclear Structure and Function)
Edited by BERT W. O'MALLEY AND JOEL G. HARDMAN

VOLUME XLI. Carbohydrate Metabolism (Part B)
Edited by W. A. WOOD

VOLUME XLII. Carbohydrate Metabolism (Part C)
Edited by W. A. WOOD

VOLUME XLIII. Antibiotics
Edited by JOHN H. HASH

VOLUME XLIV. Immobilized Enzymes
Edited by KLAUS MOSBACH

VOLUME XLV. Proteolytic Enzymes (Part B)
Edited by LASZLO LORAND

VOLUME XLVI. Affinity Labeling
Edited by WILLIAM B. JAKOBY AND MEIR WILCHEK

VOLUME XLVII. Enzyme Structure (Part E)
Edited by C. H. W. HIRS AND SERGE N. TIMASHEFF

VOLUME XLVIII. Enzyme Structure (Part F)
Edited by C. H. W. HIRS AND SERGE N. TIMASHEFF

VOLUME XLIX. Enzyme Structure (Part G)
Edited by C. H. W. HIRS AND SERGE N. TIMASHEFF

VOLUME L. Complex Carbohydrates (Part C)
Edited by VICTOR GINSBURG

VOLUME LI. Purine and Pyrimidine Nucleotide Metabolism
Edited by PATRICIA A. HOFFEE AND MARY ELLEN JONES

VOLUME LII. Biomembranes (Part C: Biological Oxidations)
Edited by SIDNEY FLEISCHER AND LESTER PACKER

VOLUME LIII. Biomembranes (Part D: Biological Oxidations)
Edited by SIDNEY FLEISCHER AND LESTER PACKER

VOLUME LIV. Biomembranes (Part E: Biological Oxidations)
Edited by SIDNEY FLEISCHER AND LESTER PACKER

VOLUME LV. Biomembranes (Part F: Bioenergetics)
Edited by SIDNEY FLEISCHER AND LESTER PACKER

VOLUME LVI. Biomembranes (Part G: Bioenergetics)
Edited by SIDNEY FLEISCHER AND LESTER PACKER

VOLUME LVII. Bioluminescence and Chemiluminescence
Edited by MARLENE A. DELUCA

VOLUME LVIII. Cell Culture
Edited by WILLIAM B. JAKOBY AND IRA PASTAN

VOLUME LIX. Nucleic Acids and Protein Synthesis (Part G)
Edited by KIVIE MOLDAVE AND LAWRENCE GROSSMAN

VOLUME LX. Nucleic Acids and Protein Synthesis (Part H)
Edited by KIVIE MOLDAVE AND LAWRENCE GROSSMAN

VOLUME 61. Enzyme Structure (Part H)
Edited by C. H. W. HIRS AND SERGE N. TIMASHEFF

VOLUME 62. Vitamins and Coenzymes (Part D)
Edited by DONALD B. MCCORMICK AND LEMUEL D. WRIGHT

VOLUME 63. Enzyme Kinetics and Mechanism (Part A: Initial Rate and Inhibitor Methods)
Edited by DANIEL L. PURICH

VOLUME 64. Enzyme Kinetics and Mechanism (Part B: Isotopic Probes and Complex Enzyme Systems)
Edited by DANIEL L. PURICH

VOLUME 65. Nucleic Acids (Part I)
Edited by LAWRENCE GROSSMAN AND KIVIE MOLDAVE

VOLUME 66. Vitamins and Coenzymes (Part E)
Edited by DONALD B. MCCORMICK AND LEMUEL D. WRIGHT

VOLUME 67. Vitamins and Coenzymes (Part F)
Edited by DONALD B. MCCORMICK AND LEMUEL D. WRIGHT

VOLUME 68. Recombinant DNA
Edited by RAY WU

VOLUME 69. Photosynthesis and Nitrogen Fixation (Part C)
Edited by ANTHONY SAN PIETRO

VOLUME 70. Immunochemical Techniques (Part A)
Edited by HELEN VAN VUNAKIS AND JOHN J. LANGONE

VOLUME 71. Lipids (Part C)
Edited by JOHN M. LOWENSTEIN

VOLUME 72. Lipids (Part D)
Edited by JOHN M. LOWENSTEIN

VOLUME 73. Immunochemical Techniques (Part B)
Edited by JOHN J. LANGONE AND HELEN VAN VUNAKIS

VOLUME 74. Immunochemical Techniques (Part C)
Edited by JOHN J. LANGONE AND HELEN VAN VUNAKIS

VOLUME 75. Cumulative Subject Index Volumes XXXI, XXXII, XXXIV–LX
Edited by EDWARD A. DENNIS AND MARTHA G. DENNIS

VOLUME 76. Hemoglobins
Edited by ERALDO ANTONINI, LUIGI ROSSI-BERNARDI, AND EMILIA CHIANCONE

VOLUME 77. Detoxication and Drug Metabolism
Edited by WILLIAM B. JAKOBY

VOLUME 78. Interferons (Part A)
Edited by SIDNEY PESTKA

VOLUME 79. Interferons (Part B)
Edited by SIDNEY PESTKA

VOLUME 80. Proteolytic Enzymes (Part C)
Edited by LASZLO LORAND

VOLUME 81. Biomembranes (Part H: Visual Pigments and Purple Membranes, I)
Edited by LESTER PACKER

VOLUME 82. Structural and Contractile Proteins (Part A: Extracellular Matrix)
Edited by LEON W. CUNNINGHAM AND DIXIE W. FREDERIKSEN

VOLUME 83. Complex Carbohydrates (Part D)
Edited by VICTOR GINSBURG

VOLUME 84. Immunochemical Techniques (Part D: Selected Immunoassays)
Edited by JOHN J. LANGONE AND HELEN VAN VUNAKIS

VOLUME 85. Structural and Contractile Proteins (Part B: The Contractile Apparatus and the Cytoskeleton)
Edited by DIXIE W. FREDERIKSEN AND LEON W. CUNNINGHAM

VOLUME 86. Prostaglandins and Arachidonate Metabolites
Edited by WILLIAM E. M. LANDS AND WILLIAM L. SMITH

VOLUME 87. Enzyme Kinetics and Mechanism (Part C: Intermediates, Stereochemistry, and Rate Studies)
Edited by DANIEL L. PURICH

VOLUME 88. Biomembranes (Part I: Visual Pigments and Purple Membranes, II)
Edited by LESTER PACKER

VOLUME 89. Carbohydrate Metabolism (Part D)
Edited by WILLIS A. WOOD

VOLUME 90. Carbohydrate Metabolism (Part E)
Edited by WILLIS A. WOOD

VOLUME 91. Enzyme Structure (Part I)
Edited by C. H. W. HIRS AND SERGE N. TIMASHEFF

VOLUME 92. Immunochemical Techniques (Part E: Monoclonal Antibodies and General Immunoassay Methods)
Edited by JOHN J. LANGONE AND HELEN VAN VUNAKIS

VOLUME 93. Immunochemical Techniques (Part F: Conventional Antibodies, Fc Receptors, and Cytotoxicity)
Edited by JOHN J. LANGONE AND HELEN VAN VUNAKIS

VOLUME 94. Polyamines
Edited by HERBERT TABOR AND CELIA WHITE TABOR

VOLUME 95. Cumulative Subject Index Volumes 61–74, 76–80
Edited by EDWARD A. DENNIS AND MARTHA G. DENNIS

VOLUME 96. Biomembranes [Part J: Membrane Biogenesis: Assembly and Targeting (General Methods; Eukaryotes)]
Edited by SIDNEY FLEISCHER AND BECCA FLEISCHER

VOLUME 97. Biomembranes [Part K: Membrane Biogenesis: Assembly and Targeting (Prokaryotes, Mitochondria, and Chloroplasts)]
Edited by SIDNEY FLEISCHER AND BECCA FLEISCHER

VOLUME 98. Biomembranes (Part L: Membrane Biogenesis: Processing and Recycling)
Edited by SIDNEY FLEISCHER AND BECCA FLEISCHER

VOLUME 99. Hormone Action (Part F: Protein Kinases)
Edited by JACKIE D. CORBIN AND JOEL G. HARDMAN

VOLUME 100. Recombinant DNA (Part B)
Edited by RAY WU, LAWRENCE GROSSMAN, AND KIVIE MOLDAVE

VOLUME 101. Recombinant DNA (Part C)
Edited by RAY WU, LAWRENCE GROSSMAN, AND KIVIE MOLDAVE

VOLUME 102. Hormone Action (Part G: Calmodulin and Calcium-Binding Proteins)
Edited by ANTHONY R. MEANS AND BERT W. O'MALLEY

VOLUME 103. Hormone Action (Part H: Neuroendocrine Peptides)
Edited by P. MICHAEL CONN

VOLUME 104. Enzyme Purification and Related Techniques (Part C)
Edited by WILLIAM B. JAKOBY

VOLUME 105. Oxygen Radicals in Biological Systems
Edited by LESTER PACKER

VOLUME 106. Posttranslational Modifications (Part A)
Edited by FINN WOLD AND KIVIE MOLDAVE

VOLUME 107. Posttranslational Modifications (Part B)
Edited by FINN WOLD AND KIVIE MOLDAVE

VOLUME 108. Immunochemical Techniques (Part G: Separation and Characterization of Lymphoid Cells)
Edited by GIOVANNI DI SABATO, JOHN J. LANGONE, AND HELEN VAN VUNAKIS

VOLUME 109. Hormone Action (Part I: Peptide Hormones)
Edited by LUTZ BIRNBAUMER AND BERT W. O'MALLEY

VOLUME 110. Steroids and Isoprenoids (Part A)
Edited by JOHN H. LAW AND HANS C. RILLING

VOLUME 111. Steroids and Isoprenoids (Part B)
Edited by JOHN H. LAW AND HANS C. RILLING

VOLUME 112. Drug and Enzyme Targeting (Part A)
Edited by KENNETH J. WIDDER AND RALPH GREEN

VOLUME 113. Glutamate, Glutamine, Glutathione, and Related Compounds
Edited by ALTON MEISTER

VOLUME 114. Diffraction Methods for Biological Macromolecules (Part A)
Edited by HAROLD W. WYCKOFF, C. H. W. HIRS, AND SERGE N. TIMASHEFF

VOLUME 115. Diffraction Methods for Biological Macromolecules (Part B)
Edited by HAROLD W. WYCKOFF, C. H. W. HIRS, AND SERGE N. TIMASHEFF

VOLUME 116. Immunochemical Techniques (Part H: Effectors and Mediators of Lymphoid Cell Functions)
Edited by GIOVANNI DI SABATO, JOHN J. LANGONE, AND HELEN VAN VUNAKIS

VOLUME 117. Enzyme Structure (Part J)
Edited by C. H. W. HIRS AND SERGE N. TIMASHEFF

VOLUME 118. Plant Molecular Biology
Edited by ARTHUR WEISSBACH AND HERBERT WEISSBACH

VOLUME 119. Interferons (Part C)
Edited by SIDNEY PESTKA

VOLUME 120. Cumulative Subject Index Volumes 81–94, 96–101

VOLUME 121. Immunochemical Techniques (Part I: Hybridoma Technology and Monoclonal Antibodies)
Edited by JOHN J. LANGONE AND HELEN VAN VUNAKIS

VOLUME 122. Vitamins and Coenzymes (Part G)
Edited by FRANK CHYTIL AND DONALD B. MCCORMICK

VOLUME 123. Vitamins and Coenzymes (Part H)
Edited by FRANK CHYTIL AND DONALD B. MCCORMICK

VOLUME 124. Hormone Action (Part J: Neuroendocrine Peptides)
Edited by P. MICHAEL CONN

VOLUME 125. Biomembranes (Part M: Transport in Bacteria, Mitochondria, and Chloroplasts: General Approaches and Transport Systems)
Edited by SIDNEY FLEISCHER AND BECCA FLEISCHER

VOLUME 126. Biomembranes (Part N: Transport in Bacteria, Mitochondria, and Chloroplasts: Protonmotive Force)
Edited by SIDNEY FLEISCHER AND BECCA FLEISCHER

VOLUME 127. Biomembranes (Part O: Protons and Water: Structure and Translocation)
Edited by LESTER PACKER

VOLUME 128. Plasma Lipoproteins (Part A: Preparation, Structure, and Molecular Biology)
Edited by JERE P. SEGREST AND JOHN J. ALBERS

VOLUME 129. Plasma Lipoproteins (Part B: Characterization, Cell Biology, and Metabolism)
Edited by JOHN J. ALBERS AND JERE P. SEGREST

VOLUME 130. Enzyme Structure (Part K)
Edited by C. H. W. HIRS AND SERGE N. TIMASHEFF

VOLUME 131. Enzyme Structure (Part L)
Edited by C. H. W. HIRS AND SERGE N. TIMASHEFF

VOLUME 132. Immunochemical Techniques (Part J: Phagocytosis and Cell-Mediated Cytotoxicity)
Edited by GIOVANNI DI SABATO AND JOHANNES EVERSE

VOLUME 133. Bioluminescence and Chemiluminescence (Part B)
Edited by MARLENE DELUCA AND WILLIAM D. MCELROY

VOLUME 134. Structural and Contractile Proteins (Part C: The Contractile Apparatus and the Cytoskeleton)
Edited by RICHARD B. VALLEE

VOLUME 135. Immobilized Enzymes and Cells (Part B)
Edited by KLAUS MOSBACH

VOLUME 136. Immobilized Enzymes and Cells (Part C)
Edited by KLAUS MOSBACH

VOLUME 137. Immobilized Enzymes and Cells (Part D)
Edited by KLAUS MOSBACH

VOLUME 138. Complex Carbohydrates (Part E)
Edited by VICTOR GINSBURG

VOLUME 139. Cellular Regulators (Part A: Calcium- and Calmodulin-Binding Proteins)
Edited by ANTHONY R. MEANS AND P. MICHAEL CONN

VOLUME 140. Cumulative Subject Index Volumes 102–119, 121–134

VOLUME 141. Cellular Regulators (Part B: Calcium and Lipids)
Edited by P. MICHAEL CONN AND ANTHONY R. MEANS

VOLUME 142. Metabolism of Aromatic Amino Acids and Amines
Edited by SEYMOUR KAUFMAN

VOLUME 143. Sulfur and Sulfur Amino Acids
Edited by WILLIAM B. JAKOBY AND OWEN GRIFFITH

VOLUME 144. Structural and Contractile Proteins (Part D: Extracellular Matrix)
Edited by LEON W. CUNNINGHAM

VOLUME 145. Structural and Contractile Proteins (Part E: Extracellular Matrix)
Edited by LEON W. CUNNINGHAM

VOLUME 146. Peptide Growth Factors (Part A)
Edited by DAVID BARNES AND DAVID A. SIRBASKU

VOLUME 147. Peptide Growth Factors (Part B)
Edited by DAVID BARNES AND DAVID A. SIRBASKU

VOLUME 148. Plant Cell Membranes
Edited by LESTER PACKER AND ROLAND DOUCE

VOLUME 149. Drug and Enzyme Targeting (Part B)
Edited by RALPH GREEN AND KENNETH J. WIDDER

VOLUME 150. Immunochemical Techniques (Part K: *In Vitro* Models of B and T Cell Functions and Lymphoid Cell Receptors)
Edited by GIOVANNI DI SABATO

VOLUME 151. Molecular Genetics of Mammalian Cells
Edited by MICHAEL M. GOTTESMAN

VOLUME 152. Guide to Molecular Cloning Techniques
Edited by SHELBY L. BERGER AND ALAN R. KIMMEL

VOLUME 153. Recombinant DNA (Part D)
Edited by RAY WU AND LAWRENCE GROSSMAN

VOLUME 154. Recombinant DNA (Part E)
Edited by RAY WU AND LAWRENCE GROSSMAN

VOLUME 155. Recombinant DNA (Part F)
Edited by RAY WU

VOLUME 156. Biomembranes (Part P: ATP-Driven Pumps and Related Transport: The Na,K-Pump)
Edited by SIDNEY FLEISCHER AND BECCA FLEISCHER

VOLUME 157. Biomembranes (Part Q: ATP-Driven Pumps and Related Transport: Calcium, Proton, and Potassium Pumps)
Edited by SIDNEY FLEISCHER AND BECCA FLEISCHER

VOLUME 158. Metalloproteins (Part A)
Edited by JAMES F. RIORDAN AND BERT L. VALLEE

VOLUME 159. Initiation and Termination of Cyclic Nucleotide Action
Edited by JACKIE D. CORBIN AND ROGER A. JOHNSON

VOLUME 160. Biomass (Part A: Cellulose and Hemicellulose)
Edited by WILLIS A. WOOD AND SCOTT T. KELLOGG

VOLUME 161. Biomass (Part B: Lignin, Pectin, and Chitin)
Edited by WILLIS A. WOOD AND SCOTT T. KELLOGG

VOLUME 162. Immunochemical Techniques (Part L: Chemotaxis and Inflammation)
Edited by GIOVANNI DI SABATO

VOLUME 163. Immunochemical Techniques (Part M: Chemotaxis and Inflammation)
Edited by GIOVANNI DI SABATO

VOLUME 164. Ribosomes
Edited by HARRY F. NOLLER, JR., AND KIVIE MOLDAVE

VOLUME 165. Microbial Toxins: Tools for Enzymology
Edited by SIDNEY HARSHMAN

VOLUME 166. Branched-Chain Amino Acids
Edited by ROBERT HARRIS AND JOHN R. SOKATCH

VOLUME 167. Cyanobacteria
Edited by LESTER PACKER AND ALEXANDER N. GLAZER

VOLUME 168. Hormone Action (Part K: Neuroendocrine Peptides)
Edited by P. MICHAEL CONN

VOLUME 169. Platelets: Receptors, Adhesion, Secretion (Part A)
Edited by JACEK HAWIGER

VOLUME 170. Nucleosomes
Edited by PAUL M. WASSARMAN AND ROGER D. KORNBERG

VOLUME 171. Biomembranes (Part R: Transport Theory: Cells and Model Membranes)
Edited by SIDNEY FLEISCHER AND BECCA FLEISCHER

VOLUME 172. Biomembranes (Part S: Transport: Membrane Isolation and Characterization)
Edited by SIDNEY FLEISCHER AND BECCA FLEISCHER

VOLUME 173. Biomembranes [Part T: Cellular and Subcellular Transport: Eukaryotic (Nonepithelial) Cells]
Edited by SIDNEY FLEISCHER AND BECCA FLEISCHER

VOLUME 174. Biomembranes [Part U: Cellular and Subcellular Transport: Eukaryotic (Nonepithelial) Cells]
Edited by SIDNEY FLEISCHER AND BECCA FLEISCHER

VOLUME 175. Cumulative Subject Index Volumes 135–139, 141–167

VOLUME 176. Nuclear Magnetic Resonance (Part A: Spectral Techniques and Dynamics)
Edited by NORMAN J. OPPENHEIMER AND THOMAS L. JAMES

VOLUME 177. Nuclear Magnetic Resonance (Part B: Structure and Mechanism)
Edited by NORMAN J. OPPENHEIMER AND THOMAS L. JAMES

VOLUME 178. Antibodies, Antigens, and Molecular Mimicry
Edited by JOHN J. LANGONE

VOLUME 179. Complex Carbohydrates (Part F)
Edited by VICTOR GINSBURG

VOLUME 180. RNA Processing (Part A: General Methods)
Edited by JAMES E. DAHLBERG AND JOHN N. ABELSON

VOLUME 181. RNA Processing (Part B: Specific Methods)
Edited by JAMES E. DAHLBERG AND JOHN N. ABELSON

VOLUME 182. Guide to Protein Purification
Edited by MURRAY P. DEUTSCHER

VOLUME 183. Molecular Evolution: Computer Analysis of Protein and Nucleic Acid Sequences
Edited by RUSSELL F. DOOLITTLE

VOLUME 184. Avidin–Biotin Technology
Edited by MEIR WILCHEK AND EDWARD A. BAYER

VOLUME 185. Gene Expression Technology
Edited by DAVID V. GOEDDEL

VOLUME 186. Oxygen Radicals in Biological Systems (Part B: Oxygen Radicals and Antioxidants)
Edited by LESTER PACKER AND ALEXANDER N. GLAZER

VOLUME 187. Arachidonate Related Lipid Mediators
Edited by ROBERT C. MURPHY AND FRANK A. FITZPATRICK

VOLUME 188. Hydrocarbons and Methylotrophy
Edited by MARY E. LIDSTROM

VOLUME 189. Retinoids (Part A: Molecular and Metabolic Aspects)
Edited by LESTER PACKER

VOLUME 190. Retinoids (Part B: Cell Differentiation and Clinical Applications)
Edited by LESTER PACKER

VOLUME 191. Biomembranes (Part V: Cellular and Subcellular Transport: Epithelial Cells)
Edited by SIDNEY FLEISCHER AND BECCA FLEISCHER

VOLUME 192. Biomembranes (Part W: Cellular and Subcellular Transport: Epithelial Cells)
Edited by SIDNEY FLEISCHER AND BECCA FLEISCHER

VOLUME 193. Mass Spectrometry
Edited by JAMES A. MCCLOSKEY

VOLUME 194. Guide to Yeast Genetics and Molecular Biology
Edited by CHRISTINE GUTHRIE AND GERALD R. FINK

VOLUME 195. Adenylyl Cyclase, G Proteins, and Guanylyl Cyclase
Edited by ROGER A. JOHNSON AND JACKIE D. CORBIN

VOLUME 196. Molecular Motors and the Cytoskeleton
Edited by RICHARD B. VALLEE

VOLUME 197. Phospholipases
Edited by EDWARD A. DENNIS

VOLUME 198. Peptide Growth Factors (Part C)
Edited by DAVID BARNES, J. P. MATHER, AND GORDON H. SATO

VOLUME 199. Cumulative Subject Index Volumes 168–174, 176–194

VOLUME 200. Protein Phosphorylation (Part A: Protein Kinases: Assays, Purification, Antibodies, Functional Analysis, Cloning, and Expression)
Edited by TONY HUNTER AND BARTHOLOMEW M. SEFTON

VOLUME 201. Protein Phosphorylation (Part B: Analysis of Protein Phosphorylation, Protein Kinase Inhibitors, and Protein Phosphatases)
Edited by TONY HUNTER AND BARTHOLOMEW M. SEFTON

VOLUME 202. Molecular Design and Modeling: Concepts and Applications (Part A: Proteins, Peptides, and Enzymes)
Edited by JOHN J. LANGONE

VOLUME 203. Molecular Design and Modeling: Concepts and Applications (Part B: Antibodies and Antigens, Nucleic Acids, Polysaccharides, and Drugs)
Edited by JOHN J. LANGONE

VOLUME 204. Bacterial Genetic Systems
Edited by JEFFREY H. MILLER

VOLUME 205. Metallobiochemistry (Part B: Metallothionein and Related Molecules)
Edited by JAMES F. RIORDAN AND BERT L. VALLEE

VOLUME 206. Cytochrome P450
Edited by MICHAEL R. WATERMAN AND ERIC F. JOHNSON

VOLUME 207. Ion Channels
Edited by BERNARDO RUDY AND LINDA E. IVERSON

VOLUME 208. Protein–DNA Interactions
Edited by ROBERT T. SAUER

VOLUME 209. Phospholipid Biosynthesis
Edited by EDWARD A. DENNIS AND DENNIS E. VANCE

VOLUME 210. Numerical Computer Methods
Edited by LUDWIG BRAND AND MICHAEL L. JOHNSON

VOLUME 211. DNA Structures (Part A: Synthesis and Physical Analysis of DNA)
Edited by DAVID M. J. LILLEY AND JAMES E. DAHLBERG

VOLUME 212. DNA Structures (Part B: Chemical and Electrophoretic Analysis of DNA)
Edited by DAVID M. J. LILLEY AND JAMES E. DAHLBERG

VOLUME 213. Carotenoids (Part A: Chemistry, Separation, Quantitation, and Antioxidation)
Edited by LESTER PACKER

VOLUME 214. Carotenoids (Part B: Metabolism, Genetics, and Biosynthesis)
Edited by LESTER PACKER

VOLUME 215. Platelets: Receptors, Adhesion, Secretion (Part B)
Edited by JACEK J. HAWIGER

VOLUME 216. Recombinant DNA (Part G)
Edited by RAY WU

VOLUME 217. Recombinant DNA (Part H)
Edited by RAY WU

VOLUME 218. Recombinant DNA (Part I)
Edited by RAY WU

VOLUME 219. Reconstitution of Intracellular Transport
Edited by JAMES E. ROTHMAN

VOLUME 220. Membrane Fusion Techniques (Part A)
Edited by NEJAT DÜZGÜNEŞ

VOLUME 221. Membrane Fusion Techniques (Part B)
Edited by NEJAT DÜZGÜNEŞ

VOLUME 222. Proteolytic Enzymes in Coagulation, Fibrinolysis, and Complement Activation (Part A: Mammalian Blood Coagulation Factors and Inhibitors)
Edited by LASZLO LORAND AND KENNETH G. MANN

VOLUME 223. Proteolytic Enzymes in Coagulation, Fibrinolysis, and Complement Activation (Part B: Complement Activation, Fibrinolysis, and Nonmammalian Blood Coagulation Factors)
Edited by LASZLO LORAND AND KENNETH G. MANN

VOLUME 224. Molecular Evolution: Producing the Biochemical Data
Edited by ELIZABETH ANNE ZIMMER, THOMAS J. WHITE, REBECCA L. CANN, AND ALLAN C. WILSON

VOLUME 225. Guide to Techniques in Mouse Development
Edited by PAUL M. WASSARMAN AND MELVIN L. DEPAMPHILIS

VOLUME 226. Metallobiochemistry (Part C: Spectroscopic and Physical Methods for Probing Metal Ion Environments in Metalloenzymes and Metalloproteins)
Edited by JAMES F. RIORDAN AND BERT L. VALLEE

VOLUME 227. Metallobiochemistry (Part D: Physical and Spectroscopic Methods for Probing Metal Ion Environments in Metalloproteins)
Edited by JAMES F. RIORDAN AND BERT L. VALLEE

VOLUME 228. Aqueous Two-Phase Systems
Edited by HARRY WALTER AND GÖTE JOHANSSON

VOLUME 229. Cumulative Subject Index Volumes 195–198, 200–227 (in preparation)

VOLUME 230. Guide to Techniques in Glycobiology
Edited by WILLIAM J. LENNARZ AND GERALD W. HART

VOLUME 231. Hemoglobins (Part B: Biochemical and Analytical Methods)
Edited by JOHANNES EVERSE, KIM D. VANDEGRIFF AND ROBERT M. WINSLOW

VOLUME 232. Hemoglobins (Part C: Biophysical Methods)
Edited by JOHANNES EVERSE, KIM D. VANDEGRIFF AND ROBERT M. WINSLOW

VOLUME 233. Oxygen Radicals in Biological Systems (Part C)
Edited by LESTER PACKER

VOLUME 234. Oxygen Radicals in Biological Systems (Part D)
Edited by LESTER PACKER

VOLUME 235. Bacterial Pathogenesis (Part A: Identification and Regulation of Virulence Factors)
Edited by VIRGINIA L. CLARK AND PATRIK M. BAVOIL

VOLUME 236. Bacterial Pathogenesis (Part B: Integration of Pathogenic Bacteria with Host Cells)
Edited by VIRGINIA L. CLARK AND PATRIK M. BAVOIL

VOLUME 237. Heterotrimeric G Proteins
Edited by RAVI IYENGAR

VOLUME 238. Heterotrimeric G-Protein Effectors
Edited by RAVI IYENGAR

VOLUME 239. Nuclear Magnetic Resonance (Part C)
Edited by THOMAS L. JAMES AND NORMAN J. OPPENHEIMER

VOLUME 240. Numerical Computer Methods (Part B)
Edited by MICHAEL L. JOHNSON AND LUDWIG BRAND

VOLUME 241. Retroviral Proteases
Edited by LAWRENCE C. KUO AND JULES A. SHAFER

VOLUME 242. Neoglycoconjugates (Part A)
Edited by Y. C. LEE AND REIKO T. LEE

VOLUME 243. Inorganic Microbial Sulfur Metabolism
Edited by HARRY D. PECK, JR., AND JEAN LEGALL

VOLUME 244. Proteolytic Enzymes: Serine and Cysteine Peptidases
Edited by ALAN J. BARRETT

VOLUME 245. Extracellular Matrix Components (in preparation)
Edited by E. RUOSLAHTI AND E. ENGVALL

VOLUME 246. Biochemical Spectroscopy (in preparation)
Edited by KENNETH SAUER

VOLUME 247. Neoglycoconjugates (Part B: Biomedical Applications)
Edited by Y. C. LEE AND REIKO T. LEE

VOLUME 248. Proteolytic Enzymes: Aspartic and Metallo Peptidases (in preparation)
Edited by ALAN J. BARRETT

VOLUME 249. Enzyme Kinetics and Mechanism (Part D) (in preparation)
Edited by DANIEL L. PURICH

VOLUME 250. Lipid Modifications of Proteins (in preparation)
Edited by PATRICK J. CASEY AND JANICE E. BUSS

VOLUME 251. Biothiols, Part A (in preparation)
Edited by LESTER PACKER

VOLUME 252. Biothiols, Part B (in preparation)
Edited by Lester Packer

VOLUME 253. Adhesion of Microbial Pathogens (in preparation)
Edited by Ron J. Doyle and Stzhak Ofek

Section I

Sulfate Transport

[1] Sulfate Transport

By HERIBERT CYPIONKA

Introduction

Sulfate must be activated by means of ATP prior to reduction. Because ATP is available only inside the cell, uptake is the precondition of assimilatory and dissimilatory sulfate metabolism. Owing to the different roles of assimilatory and dissimilatory sulfate metabolism, the features of the sulfate transport systems differ. Sulfur usually constitutes less than 1% of the dry mass of living organisms. Therefore assimilatory sulfate reduction, as carried out by many bacteria, fungi, and algae, is quantitatively of minor importance. The performance of an assimilatory sulfate uptake system requires effective uptake of sulfate in order to secure biosynthesis. High affinity is more advantageous than high rates of uptake or a low energy requirement for the process. Accordingly, in many bacteria that carry out assimilatory sulfate reduction, sulfate uptake involves primary transport systems, by which the uptake of sulfate is driven by ATP hydrolysis. Such a unidirectional mechanism protects the cell from loss of intracellular sulfate.

In bacteria performing dissimilatory sulfate reduction the energetic aspects prevail. An amount of sulfate at least 10 times the cell dry mass is reduced per doubling of the cell. Sulfate transport systems must allow uptake at high rates. High affinity is required, especially in fresh-water environments, which typically reveal low sulfate concentrations. The low free energy change of sulfate respiration forces the cells to use economic mechanisms for sulfate transport. For this, secondary transport systems, in which transport is driven by other, already existing gradients across the membrane, are appropriate. We know from previous studies that sulfate is symported with protons in fresh-water sulfate reducers and with sodium ions in marine strains. In both groups high-accumulating transport systems are expressed, depending on the sulfate availability during growth, whereas low-accumulating transport systems are expressed constitutively. Accumulation of sulfate is additionally regulated at the activity level. At high sulfate concentration the high-accumulating transport is switched off.

This chapter is about sulfate transport in microorganisms. Uptake experiments with whole cells of dissimilatory sulfate-reducing bacteria are described. Techniques used for studies with vesicles are described in this series by Turner.[1] An important aspect of sulfate transport, namely its export by bacteria that use reduced sulfur compounds as electron donors for respiration processes (colorless sulfur bacteria), fermentation (sulfate-reducing bacteria that disproportionate sulfite or thiosulfate to sulfate and sulfide), or photosynthesis (purple and green sulfur bacteria), is not considered here because it has not yet been studied in detail. The methods described here are also applicable to the study of transport of other inorganic sulfur compounds, such as thiosulfate or (with different tracers) other anions such as phosphate.

Sulfur Sources during Growth

Because assimilatory and dissimilatory sulfate transport systems are regulated on the genetic level, some consideration must be given to the sulfur compounds added to the growth medium. L-Methionine and thiosulfate, and in some species also cysteine, as sulfur sources have been reported to repress assimilatory sulfate transport systems, whereas the activity was present after growth with djenkolic acid (*S,S*′-methylenebis-L-cysteine).[2,3] Washing the cells and a subsequent preincubation period in sulfur-free medium might favor derepression of sulfate transport activity.

With dissimilatory sulfate-reducing bacteria the highest accumulation factors are obtained after continuous growth in chemostats[4] under sulfate limitation.[5]

Test for Periplasmic Sulfate Transport Systems

Many organisms carrying out assimilatory sulfate reduction have so-called periplasmic transport systems. In these, a periplasmic protein specifically binds one molecule of sulfate prior to the transport step, which is driven by ATP hydrolysis. The periplasmic protein can be washed away from the cell by a mild osmotic shock, resulting in loss of sulfate transport activity.

Other characteristics of a periplasmic transport system are sensitivity to the ATPase inhibitor *N,N*′-dicyclohexylcarbodiimide (DCCD); lack of reversibility (because the transport system exerts only sulfate uptake)[6];

[1] R. J. Turner, this series, Vol. 191, p. 479.

[2] A. B. Pardee, L. S. Prestidge, M. B. Whipple, and J. Dreyfuss, *J. Biol. Chem.* **241,** 3962 (1966).

[3] J. W. Tweedie and I. H. Segel, *Biochim. Biophys. Acta* **196,** 95 (1970).

[4] H. Cypionka and N. Pfennig, *Arch. Microbiol.* **143,** 366 (1986).

[5] J. Stahlmann, R. Warthmann, and H. Cypionka, *Arch. Microbiol.* **155,** 554 (1991).

[6] H. W. Van Veen, T. Abee, J. J. Kortstee, W. N. Konings, and A. J. B. Zehnder, *J. Bacteriol.* **175,** 200 (1993).

and rapid binding of small amounts of sulfate, independent of the external sulfate concentration, with a low saturation constant (less than 10 μM).

For the osmotic shock,[7] washed cells are suspended in diluted buffer with ethylenediaminetetraacetic acid (EDTA; 1 mM) and 0.3 to 0.75 M sucrose or mannitol. After 10 min the cells are centrifuged and diluted in cold water followed by stirring for an additional 10 min. The supernatant obtained after a further centrifugation step will contain the periplasmic binding protein.

Shocked cells with a periplasmic sulfate transport system will lose most of their sulfate transport activity by this procedure, but their viability should not be affected. The sulfate bound to the periplasmic protein can be separated from the solution by dialysis or ultrafiltration. *Salmonella typhimurium* has been reported to contain up to 10,000 copies of the binding protein per cell.[2]

Harvest and Storage of Dissimilatory Sulfate-Reducing Bacteria

Although sulfate-reducing bacteria are capable of aerobic respiration at low O_2 tensions[8–10] they are sensitive to molecular oxygen, especially at increased concentrations and in the presence of H_2S.[11] Furthermore, during preparation of the cells sulfide is easily oxidized to products interfering with sulfate accumulation. Therefore, H_2S should be removed by sparging the culture with CO_2 for 15 min prior to harvesting the cells. Instead of pure CO_2, a mixture of N_2 (or another inert gas) and CO_2 (80 : 20, v/v) may be used; pure N_2, however, would result in an alkalinization of the medium by loss of CO_2 and H_2S. In alkaline milieu, H_2S is transformed to HS^- (pK 6.9), which is not volatile. Most sulfate-reducing bacteria are not immediately affected by O_2, if harvesting is carried out quickly, and if the head space of the centrifuge tubes is gassed with inert gas. For sensitive strains, however, strictly anoxic preparation might be necessary.

In our laboratory, harvested cells are suspended at cell densities of 10 to 100 mg dry mass/ml in buffer (e.g., 50 mM morpholinepropanesulfonic acid [MOPS]), KCl (150 mM), or, for marine strains, in salt solutions containing NaCl, KCl, and $MgCl_2$, in concentrations depending on the experiments to be carried out. The gas phase is N_2 or H_2. The latter provides an electron donor for the removal of residual O_2. The best temperature for storage of more than a few hours is room temperature. Overnight

[7] H. C. Neu and L. A. Heppel, *J. Biol. Chem.* **240,** 3685 (1965).
[8] W. Dilling and H. Cypionka, *FEMS Microbiol. Lett.* **71,** 123 (1990).
[9] S. Dannenberg, M. Kroder, W. Dilling, and H. Cypionka, *Arch. Microbiol.* **158,** 93 (1992).
[10] C. Marschall, P. Frenzel, and H. Cypionka, *Arch. Microbiol.* **159,** 168 (1993).
[11] H. Cypionka, F. Widdel, and N. Pfennig, *FEMS Microbiol. Ecol.* **31,** 39 (1985).

storage at 8° is possible, while on ice the activity of the cells decreases more rapidly.

Sodium Ion-Dependent Transport in Marine Sulfate Reducers

In principle, the same procedures for transport studies may be used with fresh-water and marine strains. With marine strains, however, the presence of sufficient Na^+ must be considered. Although marine sulfate reducers pump out Na^+ and keep a low intracellular sodium ion level, some strains are deenergized if they are prepared in the presence of less than 5 m*M* Na^+. Lithium ions might replace Na^+ to some extent.[5,12]

Multielectrode Device for Simultaneous Monitoring of Concentrations of H^+, H_2S, and O_2 (or H_2), and of Changes in Redox Potential

A multielectrode device, by which various parameters relevant for sulfate transport and sulfide metabolism can be followed simultaneously, has been developed in our laboratory and is described briefly here. This setup has provided valuable information during experiments on sulfate transport,[5,12–14] disproportionation of sulfur compounds,[15] oxidation of sulfide by sulfate-reducing bacteria[8,9] or phototrophic bacteria,[16] and on H_2 formation by phototrophic bacteria.[17]

A 3-ml glass vessel above an O_2 (or, with reversed polarity, H_2) electrode (CB1-D or DW1H_2; Hansatech, Norfolk, United Kingdom) is closed by a rubber stopper through which another three electrodes and a hollow needle are fitted (Fig. 1). The assay is temperature controlled, stirred magnetically, and can be illuminated for use with phototrophic sulfur bacteria. The chamber is completely filled with buffer or, when pH changes are to be measured, with nonbuffered salt solutions as described above. The hollow needle allows additions by microliter syringes and also functions as overflow outlet for the added volumes.

The Clark-type O_2 electrode at the bottom of the reaction chamber is especially suited for measurements in sulfide-containing suspensions. The stirring bar is located directly on the platinum cathode. The ring-shaped silver/silver chloride anode is separated from the assay by means of a salt

[12] R. Warthmann and H. Cypionka, *Arch. Microbiol.* **154,** 144 (1990).
[13] H. Cypionka, *Arch. Microbiol.* **148,** 144 (1987).
[14] H. Cypionka, *Arch. Microbiol.* **152,** 237 (1989).
[15] M. Krämer and H. Cypionka, *Arch. Microbiol.* **151,** 232 (1989).
[16] J. Overmann, H. Cypionka, and N. Pfennig, *Limnol. Oceanogr.* **37,** 150 (1992).
[17] R. Warthmann, N. Pfennig, and H. Cypionka, *Appl. Microbiol. Biotechnol.* **39,** 358 (1993).

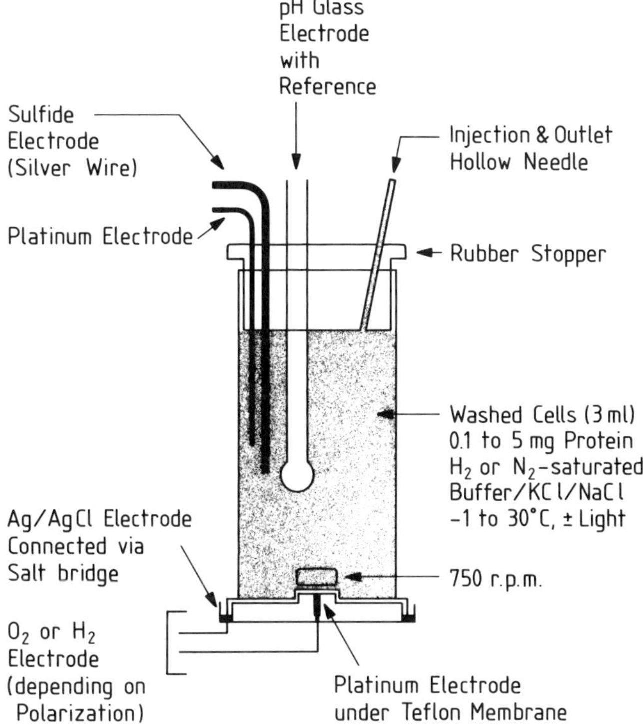

FIG. 1. Multielectrode setup for simultaneous measurement of sulfide, pH, E_h, and O_2 or H_2.

bridge. This consists of a cigarette paper in half-saturated KCl beneath the O_2-permeable membrane. This arrangement avoids any disturbing effects of H_2S on the silver/silver chloride electrode. By changing the polarity of anode and cathode the O_2 electrode can also be used as a sulfide-insensitive H_2 electrode.[17]

A small combined pH electrode (4.5-mm diameter, LoT 406-M4; Ingold, Steinbach, Germany) is used to follow pH changes. The pH signal is also required in order to calculate the sulfide concentration, because the sulfide electrode reacts in a pH-sensitive manner. The reference of the pH electrode is also utilized as a reference for the sulfide electrode and the platinum electrode.

The sulfide electrode is made from silver wire (1-mm diameter). Coating with silver sulfide is obtained by applying a voltage of 1 V in sulfide-containing medium. This electrode reacts proportional to the logarithm of the sulfide activity.[18-20] The detection limit is below 1 μM sulfide (i.e., $H_2S + HS^-$; free S^{2-} does not exist in aqueous solutions because the second pK is at 17 to 19[21]) when the electrode is preincubated anoxically in sulfide-containing solution.

The third electrode used in the multielectrode device is a platinum electrode made from a platinum wire of 0.8-mm diameter, which is coated with platinum black (fine granulate of platinum particles) by applying a voltage of −5 V in a solution of hexachloroplatinate to increase its stabilty. This electrode responds to changes of the redox potential and of the H_2 concentration. However, owing to the presence of changing concentrations of H_2S, H_2, O_2, and H^+, only qualitative interpretation is possible. Additional information on the redox potential is easily obtained by the redox indicator resazurin (redox potential about −45 mV[22], which is used at a concentration of 20 μM after it has been proved that it does not interfere with the processes studied.

The analog signals of the four different electrodes are amplified by pH/mV meters (model pH 91; WTW, Weilheim, Germany) and converted by means of an A/D converter card (PGA DAS-8; Keithley, Germering, Germany). A Turbo Pascal program run on a personal computer (PC 386 or PC 486) allows numerical and graphical data processing and transfer to standard software.

Calibration of the electrodes is achieved by small pulses (10 to 50 nmol) of HCl and Na_2S from O_2-free 1 mM solutions, or with O_2-saturated (or H_2-saturated) water with known concentrations.[23]

To calculate the sulfide concentration a correction factor (k) describing the pH sensitivity of the sulfide signal is determined by comparing the changes of the sulfide and pH electrode readings (R_{pS} and R_{pH}, respectively) in response to a small H^+ pulse that does not influence the sulfide concentration.

$$k = \Delta R_{pS}/\Delta R_{pH}$$

This H^+ pulse is also used to quantify proton uptake or production during sulfate uptake or reduction. The actual concentration of sulfide

[18] R. A. Berner, *Geochim. Cosmochim. Acta* **27,** 563 (1963).
[19] H. Guterman, S. Ben-Yaakov, and A. Abeliovich, *Anal. Chem.* **55,** 1731 (1983).
[20] H. Cypionka, *J. Microbiol. Methods* **5,** 1 (1986).
[21] R. J. Myers, *J. Chem. Educat.* **63,** 687 (1986).
[22] R. S. Twigg, *Nature (London)* **155,** 401 (1945).
[23] E. Wilhelm, R. Battino, and R. J. Wilcock, *Chem. Rev.* **77,** 219 (1977).

(c_s, the sum of H_2S and HS^-) is calculated from a calibration pulse (c_p) according to

$$c_s = \frac{c_p 10^{(\Delta R_{pS} - k\Delta R_{pH})/\mathrm{sl}_{pS}}}{(V)10^{[(\Delta R_{pS} - k\Delta R_{pH})/\mathrm{sl}_{pS}]} - 1}$$

where c_s is the concentration of sulfide (H_2S plus HS^-) (μM), c_p is the sulfide calibration pulse (nmol), ΔR_{pS} is the change in the sulfide signal (mV), ΔR_{pH} is the change in the pH signal (mV), k is the correction factor $\Delta R_{pS}/\Delta R_{pH}$, sl_{pS} is the slope of the sulfide electrode per 10-fold increase in the sulfide concentration (~ −33.5 mV), and V is the reaction volume (ml).

Figure 2 shows pH and sulfide traces during reduction of various sulfur compounds by *Desulfovibrio desulfuricans*. Proton uptake (due to transport and due to sulfide formation) and sulfide formation show different kinetics, and depend on the electron acceptor during growth.

Selective Inhibition of Dissimilatory Sulfate Reduction

Generally, assimilatory sulfate transport is easier to follow than dissimilatory sulfate transport, as the uptake rates are lower, primary transport systems (if involved) are unidirectional, and the label is fixed in the cells. Dissimilatory sulfate reduction, however, results in the formation of volatile H_2S, which is immediately set free (Fig. 2). Therefore, for transport studies dissimilatory sulfate reduction must be inhibited, if possible without interfering with the driving forces for sulfate accumulation.

For this purpose two methods have been applied in our laboratory: the first is to chill the cell suspension to about 0 to −1°. Under these conditions, sulfate is taken up immediately after its addition, whereas the onset of sulfate reduction and H_2S release is delayed for up to 5 min. However, the delay is observed only after the first sulfate pulse and must be proved by means of a sulfide electrode as described above.

The low temperature (although not unusual for the natural habitats of sulfate-reducing bacteria) could change the state of the membrane and thus affect the driving forces for sulfate uptake. Therefore the second method to prevent sulfate reduction is molecular oxygen. Exposure to air prevents sulfate reduction. Cells of several strains incubated under air for 5 to 10 min revealed higher sulfate accumulation than after chilling to 0°.[5,12] It must, however, be considered that, depending on the availability of endogenous or exogenous electron donors, aerobic respiration might occur and influence the driving forces for sulfate transport, and that prolonged exposure to air might be toxic for the cells.

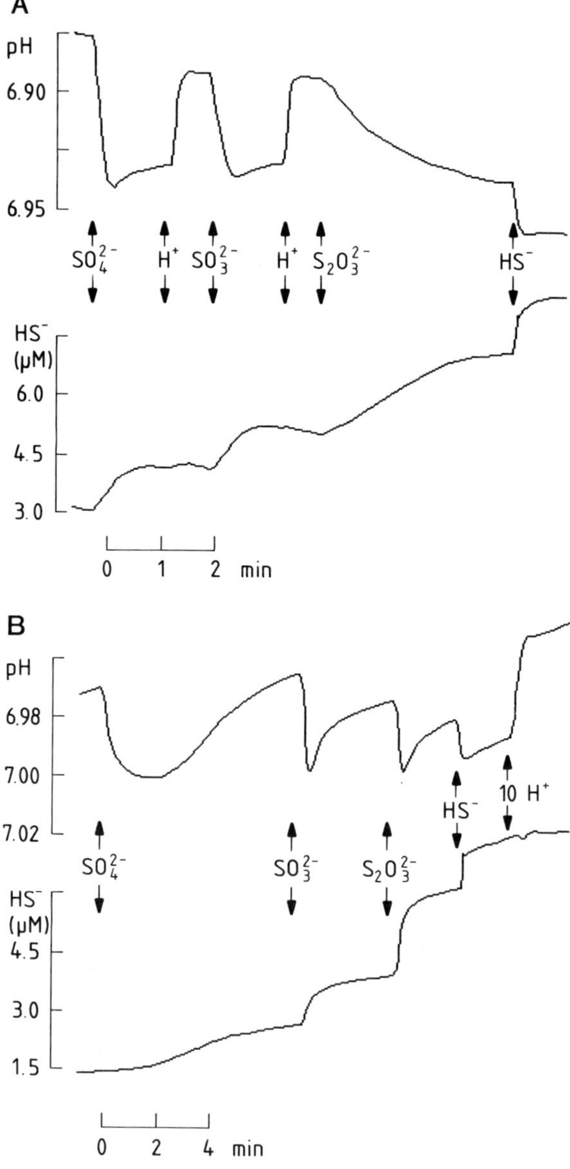

FIG. 2. pH traces and sulfide production during uptake and reduction of sulfate, sulfite, and thiosulfate by *Desulfovibrio desulfuricans*. (A) Washed cells (0.16 mg of protein/ml) pregrown under sulfate limitation. (B) Washed cells (0.22 mg of protein/ml) pregrown with thiosulfate at limiting concentration. The cells were incubated in H_2-saturated 150 m*M* KCl at (A) 20° or (B) 30°. At the times indicated by the arrows, 5-nmol additions of the sulfur compounds or HCl were made [except the last pulse in (B), 10 nmol of HCl].

Measurement of Sulfate Transport

Photometric Study of Plasmolysis

A simple but elegant method to qualitatively follow uptake of a solute added to a cell suspension is a plasmolysis and reswelling experiment, in which changes in light scattering are followed.[24] Concentrated solutes (5 *M*) are added to cell suspensions in diluted buffer with an optical density of about 1, to give a final concentration of 0.2 *M*. Shrinkage of the cells due to plasmolysis causes in increase in light scattering, whereas a subsequent decrease in optical density indicates uptake of the solute. By comparing various solutes and by use of ionophores (in order to destroy the membrane potential), uptake mechanisms can be studied. These[25] and similar[26] experiments have been used to demonstrate that *Desulfovibrio* species are not simply permeable to sulfate ions, even if high concentrations are applied and if the membrane potential, which might repel the two charges of sulfate, is destroyed.

Following pH Changes Coupled to Sulfate Transport

If transport experiments are carried out in weakly buffered cell suspensions the pH electrode allows the proton symport to be followed during sulfate uptake. To monitor eventual sulfide production a multielectrode setup as described above is required for those measurements. The H^+/SO_4^{2-} ratio is calculated by comparison with equimolar calibration pulses of 1 m*M* HCl (Fig. 3). The pH measurement is generally suited to discriminate between symport with sodium ions as occurring in marine strains (Fig. 3A) or with protons as in fresh-water strains (Fig. 3B). The method does, however, underestimate the stoichiometry of electrogenic uptake in symport with protons.[13,14] Detailed analysis of the driving forces for sulfate accumulation has revealed that high sulfate accumulation by fresh-water strains requires three protons per sulfate,[27] whereas disappearance of only two protons was detected by the pH electrode. Obviously, electrogenic proton uptake is compensated by the membrane potential before it is detected by the pH electrode.

[24] P. Mitchell and J. Moyle, *Eur. J. Biochem.* **9,** 149 (1969).

[25] A. Varma, P. Schönheit, and R. K. Thauer, *Arch. Microbiol.* **136,** 69 (1983).

[26] D. Littlewood and J. R. Postgate, *J. Gen. Microbiol.* **16,** 596 (1957).

[27] B. Kreke and H. Cypionka, *Arch. Microbiol.* **158,** 183 (1992).

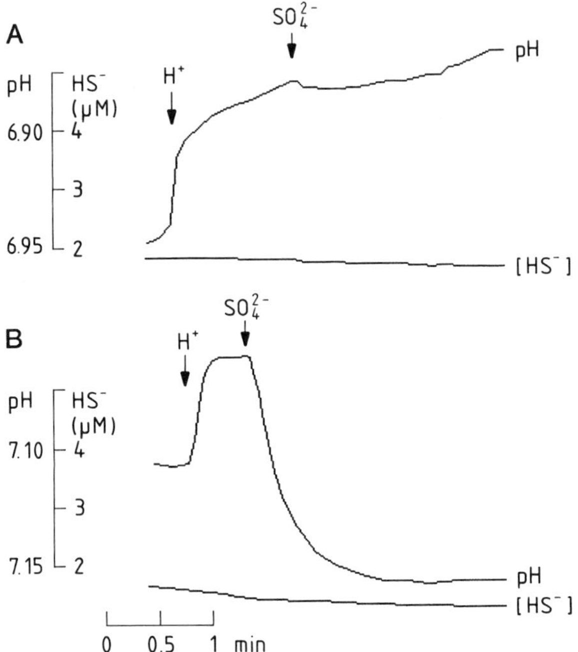

FIG. 3. Sulfate transport experiments with (A) the marine species *Desulfococcus multivorans* and (B) the fresh-water species *Desulfobulbus propionicus*. The cells were incubated in nonbuffered salt solutions containing (A) 350 m*M* NaCl, 150 m*M* KCl, and 14 m*M* $MgCl_2$, and (B) 150 m*M* KCl. Sulfate reduction was prevented by air; the sulfide traces prove that there was no H_2S formation during the experiment. At the times indicated by the arrows, 10 nmol of HCl or radiolabeled sulfate was added. After 3 min more than 80% of the sulfate had been taken up in both experiments.

Use of Labeled Sulfate

Most conveniently, transport of sulfate is measured by means of radioactively labeled sulfate. The half-life time of ^{35}S is about 88 days and allows use of a batch for periods of up to 1 year, after which the initial activity is decreased by about 95%. The energy of ^{35}S β radiation is similar to that of ^{14}C. Therefore ^{35}S may be measured by the ^{14}C program in a scintillation counter. The required activities are relatively low, when sulfate is accumulated by the cells. For a cell suspension of dissimilatory sulfate reducers grown under sulfate limitation the addition of 5 μM sulfate, resulting in an activity of 100 Bq [i.e., 6000 disintegrations per minute (dpm)] per milliliter may be sufficient. In a typical experiment, after 2 min 90% of the label is accumulated by the cells, while the concentration outside is decreased 10-fold.

Silicone Oil Centrifugation Technique

The silicone oil centrifugation technique is used to separate cells from a suspension without addition of a stop solution or washing of the cells. This method is suited for steady state measurements and has also been applied for determination of the cell volume, membrane potential, and pH gradient across the membrane.[25] It is, however, not suited for the study of rapid kinetics.

After addition of labeled sulfate, a defined sample volume (0.5 to 1 ml) is taken by means of a syringe and carefully layered on 0.2 to 0.4 ml of silicone oil in the tip of an Eppendorf tube. Centrifugation for 2 min at 10,000 rpm in a desktop centrifuge drives the cells through the oil phase to the bottom of the tube, while the supernatant remains above the oil phase.

A noncentrifuged sample is used to indicate the zero time concentration, and the supernatant and cells are used for sulfate transport analysis. The cells are most simply obtained by cutting off the tip of the Eppendorf tube after the supernatant and oil phase have been removed by means of a syringe.

For liquid scintillation counting various cocktails have been used successfully [Lumagel (Baker, Deventer, the Netherlands) or Emulsifier-Safe (Packard, Downers Grove, IL)].

The appropriate density of the oil phase (1.02 to 1.04 g/ml) must be determined for each strain and medium, and is adjusted by mixing silicone oils of different densities (e.g., DC200 and AR200; Fluka, Buchs, Switzerland). Not all silicone oils are mixable; we also obtained good results by diluting silicone oil (DC704; Serva, Heidelberg, Germany) with *n*-hexadecane.[24] Difficulties were encountered with some marine strains; because of the higher buoyancy of the medium the amount of water in the cell pellet is increased.

Rapid Filtration Technique

The disadvantage of the silicone oil method is that at least 30 sec is required for cells and supernatant to be separated by centrifugation.

However, the rapid filtration technique, as described by Turner,[1] includes a twofold application of a cold stop solution. This changes the transmembrane sulfate gradients and gave poor results with sulfate-reducing bacteria in our hands. It has, however, been used successfully with membrane vesicles of *Paracoccus denitrificans*.[28]

[28] J. N. Burnell, P. John, and F. R. Whatley, *Biochem. J.* **150,** 527 (1975).

Calculation of Number of Cations Symported

If sulfate is transported by secondary transport systems, that is, if it is driven by existing transmembraneous gradients (proton motive force, membrane potential, gradient of the symported cation), an equilibrium between sulfate accumulation and the driving forces can be calculated for the final steady state. A general equation for accumulation of an anion in steady state equilibrium with the membrane potential and the gradient of the symported cation is

$$\log(c_i/c_o) = -(m + n)\Delta\psi/z + n \log(x_o/x_i)$$

where c_i, c_o, x_o, and x_i are the concentrations of the anion (c) and the symported cation (x) inside (i) or outside (o) the cell (μM), m is the charge of the anion, n is the number of symported cations, $\Delta\psi$ is the membrane potential, inside relative to outside (V), and $z = 2.3RT/F \approx 0.06$ V.

Resolved for n, this gives

$$n = [\log(c_i/c_o) + m\Delta\psi/z]/[\log(x_o/x_i) - \Delta\psi/z]$$

The stoichiometry of the transport systems can be calculated only if the membrane potential and the gradient of the symported cation are known.[25]

Acknowledgments

I wish to thank my co-workers Rolf Warthmann, Jochem Stahlmann, Bernd Kreke, and Christoph Marschall, who have refined many of the methods used in our laboratory. Our research on sulfate transport has been supported by grants from the Deutsche Forschungsgemeinschaft and the Bundesministerium für Forschung und Technologie.

Section II

Dissimilatory Sulfate Reduction

A. Dehydrogenases; Hydrogenases from Sulfate-Reducing Bacteria
Articles 2 through 7

B. Electron Carrier Proteins from Sulfate-Reducing Bacteria
Articles 8 through 15

C. Terminal Reductases of Sulfate-Reducing Bacteria
Articles 16 through 21

D. Molecular Biology
Articles 22 through 25

[2] NAD-Dependent Alcohol Dehydrogenase from *Desulfovibrio gigas*

By Theo A. Hansen and Charles M. H. Hensgens

$$R—CH_2OH + NAD^+ \rightarrow R—CHO + NADH + H^+$$

Desulfovibrio gigas NCIB 9332 and several other *Desulfovibrio* strains employ an NAD-dependent alcohol dehydrogenase for the dehydrogenation of short primary alcohols (especially ethanol and propanol) to the corresponding aldehydes.[1,2] Thus far, only the purification of the alcohol dehydrogenase from *Desulfovibrio gigas* has been reported.[3]

Assay Methods

Principle

Alcohol dehydrogenase can be assayed by measurement of the change in absorbance at 340 nm as a result of the formation (forward reaction) or disappearance of NADH (aldehyde reductase activity). In view of the oxygen lability of the enzyme, the assays are preferably carried out under N_2, although with sufficiently active preparations acceptable results can be obtained in aerobic assays.

Reagents

Alcohol dehydrogenase assay
- Tris-HCl (pH 9.0), 100 m*M*
- NAD^+, 50 m*M*
- Ethanol (or other alcohol), 100 m*M*

Aldehyde reductase assay
- Tris-HCl (pH 7.5), 100 m*M*
- NADH, 2 m*M*
- Acetaldehyde (or other aldehyde), 100 m*M*

Procedure. The reaction mixture for the forward reaction (in a 1-ml quartz cuvette; 1-cm light path) contains 0.5 ml of Tris buffer (pH 9.0), 0.1 ml of an appropriate alcohol, 0.1 ml of NAD^+, and water (0.29 ml);

[1] D. R. Kremer, H. E. Nienhuis-Kuiper, and T. A. Hansen, *Arch. Microbiol.* **150,** 552 (1988).
[2] A. S. Ouattara, N. Cuzin, A. S. Traore, and J.-L. Garcia, *Arch. Microbiol.* **158,** 218 (1992).
[3] C. M. H. Hensgens, J. Vonck, J. Van Beeumen, E. F. J. van Bruggen, and T. A. Hansen, *J. Bacteriol.* **175,** 2859 (1993).

the assay system is made anaerobic by bubbling with N_2 for at least 3 min (in the spectrophotometer at 30°) and closed with a butyl rubber stopper. The reaction is started by adding an appropriate volume of enzyme-containing solution (usually 10 μl; if necessary the enzyme solution is diluted) with a microsyringe. The reaction is monitored for at least 3 min; controls without alcohol or enzyme yield no or negligible activities.

For the reverse reaction (aldehyde reductase) the reaction mixture contains 0.5 ml of Tris buffer (pH 7.5), 20 μl of aldehyde solution, 0.1 ml of NADH, and 0.37 ml of water. The subsequent procedure is as for the dehydrogenase assay.

Unit of Enzyme Activity. One unit is defined as the amount of enzyme that catalyzes the reduction of 1 μmol of NAD^+ per minute. Specific activity is expressed as units per milligram of protein. Protein is determined by the method of Bradford[4] with bovine serum albumin as a standard.

Purification Procedure

Growth of Organism

Desulfovibrio gigas NCIB 9332 is grown in 20-liter cultures (in glass vessels) in a bicarbonate-buffered medium with ethanol (up to 40 m*M*) as a substrate at 30°. The vessels are closed with butyl rubber stoppers (with holes), equipped with butyl rubber tubing and side-arm flasks to allow sterile additions. See Ref. 5 for general techniques employed in the cultivation of sulfate-reducing bacteria. The basal medium consists of (final concentrations per liter) 1.0 g of NaCl, 0.4 g of $MgCl_2 \cdot 6H_2O$, 0.1 g of $CaCl_2 \cdot 2H_2O$, 4.0 g of Na_2SO_4, 0.25 g of NH_4Cl, 0.2 g of KH_2PO_4, 0.5 g of KCl, 0.2 g of yeast extract (Difco, Detroit, MI), 1 ml of a trace elements solution, 100 n*M* sodium selenite, and 100 n*M* sodium tungstate. The trace elements solution contains (per liter) 5 g of ethylenediaminetetraacetic acid (EDTA), 12.5 ml of 7.7 *M* HCl, 2 g of $FeSO_4 \cdot 7H_2O$, 100 mg of $ZnSO_4 \cdot 7H_2O$, 30 mg of $MnCl_2 \cdot 4H_2O$, 300 mg of H_3BO_3, 200 mg of $CoCl_2 \cdot 6H_2O$, 10 mg of $CuCl_2 \cdot 2H_2O$, 20 mg of $NiCl_2 \cdot 6H_2O$, and 30 mg of $Na_2MoO_4 \cdot 2H_2O$. The autoclaved basal medium is completed by the following sterile additions (per liter): 1 ml of a filter-sterilized vitamin solution [contains (per liter) 20 mg of biotin, 200 mg of niacin, 100 mg of *p*-aminobenzoic acid, 200 mg of thiamin, 100 mg of pantothenic acid, 500

[4] M. M. Bradford, *Anal. Biochem.* **72,** 248 (1976).

[5] F. Widdel and F. Bak, *in* "The Prokaryotes" (A. Balows, H. G. Trüper, M. Dworkin, W. Harder, and K.-H. Schleifer, eds.), 2nd Ed., p. 3352. Springer-Verlag, New York, 1991.

mg of pyridoxamine, 100 mg of cobalamin, and 100 mg of riboflavin], 20 ml of filter-sterilized 2 *M* ethanol, 30 ml of 1 *M* $NaHCO_3$, and 8 ml of 1 *M* HCl; then the head space above the medium is made anaerobic by flushing with N_2 and 4 ml of 0.5 *M* sodium sulfide is added. The pH should be 7.3. Inoculation is made from a similarly grown culture (40 ml/liter).

After 48 to 60 hr, the cells are harvested by Pellicon cassette filtration (HVLP 0.5-μm Durapore filter; Millipore Corp., Bedford, MA) and centrifugation (20 min at 8000 *g* and 4°), washed three times with 50 m*M* Tris-HCl (pH 8.0), suspended in a small volume of 50 m*M* Tris (pH 8.0) containing 2 m*M* dithiothreitol (DTT), and stored under N_2 at −20°. Frozen cells should be used within 1 week to prevent a considerable loss of activity.

Step 1: Preparation of Cell Extract. Cells are disrupted by passage through a French pressure cell operating at 106 MPa; the pressure cell is filled in an anaerobic glove box (approximately 90% N_2 and 10% H_2) and the effluent is collected in a serum bottle prefilled with oxygen-free N_2. Cell debris is removed by anaerobic centrifugation (30 min at 48,000 *g* and 4°). The supernatant is decanted in an anaerobic glove box and used as the cell extract for enzyme purification or stored at −20° under N_2 until further use. A 20-liter culture grown with 40 m*M* ethanol yields approximately 35 ml of cell extract, with a protein content of 15 to 20 mg/ml. Chromatography steps described below are carried out in an anaerobic glove box and should be completed within 36 hr.

Step 2: Anion-Exchange Chromatography. Cell extract (800 mg in 50 ml; see Table I) is applied to a Q-Sepharose Fast Flow (Pharmacia, Piscataway, NJ) column (27 × 2.6 cm; cooled to 10° and preequilibrated with 20 m*M* Tris-HCl [pH 8.0] containing 2 m*M* DTT); alcohol dehydrogenase is eluted after 73 ml of buffer. Fractions containing alcohol dehydro-

TABLE I
PURIFICATION OF ALCOHOL DEHYDROGENASE FROM *Desulfovibrio gigas*[a]

Fraction	Protein (mg)	Total units[b]	Specific activity (units/mg)	Yield (%)	Purification (-fold)
Step 1. Extract	800	504	0.63	100	1
Step 2. Q-Sepharose pool	16.0	219.5	13.8	43.5	21.8
Step 3. Phenyl-Sepharose pool	4.48	142.5	31.8	28	50.5

[a] Reproduced from C. M. H. Hensgens, J. Vonck, J. Van Beeumen, E. F. J. van Bruggen, and T. A. Hansen, *J. Bacteriol.* **175,** 2859 (1993).

[b] Measured with ethanol as a substrate.

genase are pooled (84 ml) and prepared for hydrophobic interaction chromatography by adding NaCl to a concentration of 1 *M*.

Step 3: *Hydrophobic Interaction Chromatography*. The preparation of step 2 is applied to a phenyl-Sepharose CL-4B (Pharmacia) column (15 × 1.6 cm; preequilibrated with 1 *M* NaCl in 20 m*M* Tris-HCl [pH 7.0] containing 2 m*M* DTT). The column is eluted with linear gradients of 1 to 0 *M* NaCl (100 ml; at 1 ml/min) and 0 to 87% (w/w) glycerol (200 ml; during the chromatography the flow rate is adjusted to approximately 0.25 ml/min). Both gradients are prepared in 20 m*M* Tris-HCl (pH 7.0) containing 2 m*M* DTT. The alcohol dehydrogenase is eluted between 54 and 78% glycerol in a volume of 56 ml.

A typical purification is summarized in Table I. When necessary, glycerol can be removed by passing the enzyme solution over a Blue Sepharose CL-6B (Pharmacia) column (15 × 1.6 cm) that has been equilibrated with 20 m*M* Tris-HCl (pH 7.0) containing 2 m*M* DTT. The column is washed with two bed volumes of buffer; the enzyme is eluted with 5 m*M* NADH in the same buffer.

Properties

Purity. As judged by gel filtration on Superose 6 HR (Pharmacia), sodium dodecyl sulfate-polyacrylamide gel electrophoresis (SDS–PAGE), and N-terminal analysis the enzyme preparation purified by the procedure described above is pure.

Molecular Weight and Structure. SDS–PAGE (12.5% acrylamide gels) with marker proteins yields a molecular weight of 43,000. Electron microscopy showed that the enzyme is composed of two stacked pentameric rings and thus is a homodecamer of M_r 430,000. On the basis of the sequence of the first 42 N-terminal amino acids the *D. gigas* alcohol dehydrogenase was recognized as belonging to the same family of alcohol dehydrogenases as ADH2 from *Zymomonas mobilis*, methanol dehydrogenase from *Bacillus methanolicus* C1, and *Saccharomyces cerevisiae* ADH4.

Stability. The purified enzyme is rapidly inactivated by exposure to air. At 0° in 0.1 *M* Tris-HCl (pH 9.0 or 7.0), 80% of the activity is lost on incubation under air for 30 and 90 min, respectively. Under N_2 in 20 m*M* Tris-HCl (pH 7.0) with 2 m*M* DTT, more than 90% of the activity is lost after incubation at 4 or −20° for 6 days. The general instability of the enzyme makes it necessary to use the enzyme during the first 2 days after the purification (anaerobic storage at 4°).

Enzyme Inhibitors. The enzyme is strongly inhibited by 2 m*M* *N*-ethylmaleimide, 2 m*M* Cu^{2+}, and 0.04 m*M* Zn^{2+}.

Specifity. The enzyme is NAD(H) dependent; activities with NAD(P) are less than 2% of the values with NAD(H). The enzyme rapidly catalyzes the oxidation of ethanol and propanol, is less active toward 1-butanol, 1,3-propanediol, 1,2-propanediol, and glycerol, and only weakly active toward methanol and 2-propanol (a few percent of the activity with ethanol). The reduction of acetaldehyde was 4.9-fold faster than the oxidation of ethanol in the standard assays used.

Michaelis Constants. The apparent K_m values are 0.15 mM for ethanol and 0.28 mM for 1-propanol (under the standard dehydrogenase assay conditions), and 1.8 mM for acetaldehyde (aldehyde reductase assay conditions).

[3] NAD(P)-Independent Lactate Dehydrogenase from Sulfate-Reducing Prokaryotes

By THEO A. HANSEN

$$\text{Lactate} + \text{acceptor} \longrightarrow \text{pyruvate} + \text{reduced acceptor}$$

Lactate is one of the most widely used substrates for the cultivation of dissimilatory sulfate reducers. Lactate dehydrogenases (LDHs) have been demonstrated in sulfate-reducing bacteria[1–4] and archaea[5] but none of the enzymes has been purified to homogeneity; all of the enzymes are $NAD(P)^+$ independent and are assayed with artificial acceptors. The enzyme activities are usually tightly associated with the membrane fraction except for the D(−)-LDH of *Desulfovibrio vulgaris* Miyazaki, of which a considerable part of the activity was reported to be soluble.[2] The membrane-bound LDHs were reported to be localized on the cytoplasmic aspect of the membrane.[3,6] Requirement for a reducing environment (inclusion of thiol compounds in buffers) or oxygen sensitivity was reported for the L(+)-LDHs[3,6] but not for the D(−)-LDHs.[2]

[1] M. H. Czechowski mand H. W. Rossmore, *Dev. Ind. Microbiol.* **21,** 349 (1980).
[2] M. Ogata, K. Arihara, and T. Yagi, *J. Biochem.* (*Tokyo*) **89,** 1423 (1981).
[3] A. J. M. Stams and T. A. Hansen, *FEMS Microbiol. Lett.* **13,** 389 (1982).
[4] I. P. Pankhania, A. M. Spormann, W. A. Hamilton, and R. K. Thauer, *Arch. Microbiol.* **150,** 26 (1988).
[5] D. Möller-Zinkhan, G. Börner, and R. K. Thauer, *Arch. Microbiol.* **152,** 362 (1989).
[6] H. D. Peck, Jr., and J. LeGall, *Philos. Trans. R. Soc. London B* **298,** 443 (1982).

Assay Method

Principle

The assay is based on the reduction of dichlorophenol-indophenol in the presence of L(+)-lactate, which is monitored by the change in absorbance at 600 nm (extinction coefficient, 22 mM^{-1} cm^{-1}). Because of the oxygen sensitivity of the enzyme the assay is carried out anaerobically.

Reagents

Potassium phosphate buffer (pH 7.8), 100 m*M*
2,6-Dichlorophenol-indophenol, 0.7 m*M*
L(+)-Lithium lactate, 100 m*M*

Procedure. The reaction mixture contains 0.8 ml of phosphate buffer, and 0.1 ml of dichlorophenol-indophenol. The reaction mixture (in a 1-ml cuvette at 30° in a spectrophotometer) is made anaerobic by bubbling with N_2 for at least 3 min and closed with a butyl rubber stopper. Crude enzyme (cell extract or membrane fraction) is injected with a microsyringe (1 to 10 μl containing 20 to 200 μg of protein). The reaction is started by the injection of 0.1 ml of lactate solution (kept under N_2 in a screw-cap tube with a butyl rubber septum).

Unit of Enzyme Activity. One unit is defined as the amount of enzyme catalyzing the oxidation of 1 μmol of L(+)-lactate per minute. Specific activity is expressed as units per milligram protein.

Growth of Organism. *Desulfovibrio baculatus* HL21 DSM 2555 (this rod-shaped, desulfoviridin-negative strain was designated as *Desulfovibrio desulfuricans* by Stams and Hansen,[3] is maintained as *D. baculatus* by the German Collection of Microorganisms (Braunschweig, Germany), and may need transfer to the genus *Desulfomicrobium*[7]) or another sulfate-reducing bacterium such as *Desulfovibrio vulgaris* Marburg DSM 2119 is grown at its optimum temperature (usually 30 to 35°) in a bicarbonate-buffered medium with 20 m*M* L(+)-lactate. See Ref. 8 for the composition of the basal medium and preparation of the complete medium. The substrate is added from a separately sterilized solution of 1 *M* L(+)-lactate or DL-lactate. At the end of the exponential growth phase (usually after 48 to 72 hr), cells are harvested by centrifugation (20 min at 8000 *g* and 4°), washed twice in 100 m*M* Tris-HCl (pH 7.8) containing 2 m*M* dithiothreitol (DTT), suspended in a small volume of buffer, and stored under N_2 at −80° or used immediately for the preparation of a cell extract.

[7] M. Vainshtein, H. Hippe, and R. M. Kroppenstedt, *Syst. Appl. Microbiol.* **15,** 554 (1992).
[8] T. A. Hansen and C. M. H. Hensgens, this volume [2].

Preparation of Cell Extracts and Cell Fractionation. Cell extracts can be prepared by passage through a French pressure cell (see Ref. 8); an alternative method consists of ultrasonic disintegration (under a flow of N_2) of approximately 3 ml of a cell suspension (20 to 40 mg of protein per milliliter) in 100 m*M* Tris-HCl (pH 7.8) containing 1 m*M* DTT, followed by anaerobic centrifugation at 10,000 *g* (20 min at 4°). The supernatant is stored under oxygen-free N_2 in a 20-ml screw-cap tube with a butyl rubber septum (Bellco, Vineland, NY) at 0°. Membrane fractions can be prepared by ultracentrifugation of the cell extract (130,000 *g* for 1.5 hr at 4°).

Properties

Extracts of *D. baculatus* HL21 contain 0.27 to 042 U of L(+)-LDH per milligram protein.[3] A value of 0.16 U/mg protein (measured with phenazinemethosulfate as electron acceptor) was reported for *D. vulgaris* Marsburg DSM 2119.[4] If care is taken to prevent contact of the extract and membranes with air, virtually all of the L(+)-LDH activity can be recovered in the membrane fraction of *D. baculatus*. The activity in *D. baculatus* extracts is extremely unstable under air; 0.15% O_2 in the gas phase above the extract (kept at 0°) led to a loss of over 50% within 15 min. Extracts stored under O_2-free N_2 lost 50% of their LDH activity in approximately 60 hr.[3] Incubation of an oxygen-inactivated extract under N_2 for 1 to 24 hr with 1 m*M* dithiothreitol leads to a 64 to 95% restoration of activity.

Substrate Specificity. The membrane-bound lactate dehydrogenase activity from L(+)-lactate-grown cells of *D. baculatus* is active toward L(+)-lactate; with D(−)-lactate only a trace of activity is observed. Approximately the same activity was observed with 2-hydroxybutyrate as with L(+)-lactate.

Electron Acceptor Specificity. The activity of *D. baculatus* L(+)-LDH can be assayed with a variety of electron acceptors, including dichlorophenol-indophenol (70 μ*M*), 3-(4′,5′-dimethylthiazol-2-yl)-2,4-diphenyltetrazoliumbromide (2 m*M*), potassium ferricyanide (2 m*M*), phenazinemethosulfate (60 μ*M*), and methylene blue (50 μ*M*). The enzyme from *Archaeoglobus fulgidus* can be assayed with 2,3-dimethyl-1,4-naphthoquinone (0.2 m*M*).[5]

[4] Aldehyde Oxidoreductases and Other Molybdenum-Containing Enzymes

By JOSE J. G. MOURA and BELARMINO A. S. BARATA

Introduction: General Survey

Molybdenum (Mo) is a relevant transition metal in biological systems. Two groups of molybdo-containing proteins have been described. The first group comprises the nitrogenase enzyme, in which molybdenum is associated with iron in a complex cluster-type structure (FeMoco factor) that contains [4Fe-3S] and [1Mo-3Fe-3S] clusters bridged by three nonprotein ligands.[1,2] Two X-ray crystal structures of nitrogenase molybdenum–iron protein, isolated from *Azotobacter vinelandii*[1,2] and *Clostridium pasteurianum*,[3] were solved. Enzymes in a second group contain molybdenum coordinated to an organic structural component (pterin), designated as Molybdenum cofactor (molybdopterin), associated in general with iron-sulfur centers and/or a flavin or heme center, as in the case for molybdenum oxotransferases.[4] No three-dimensional structural data are yet available for this enzyme group.

The detailed knowledge of the structure and catalytic mechanism of molybdenum-containing enzymes has considerably increased during the last decade and, in particular, the biochemistry of aldehyde oxidoreductases has been extensively studied.[4–9] The participation of tungsten in related systems is less explored and was shown to be present in formate dehydrogenase from methanogenes[10] and thermophiles,[11] in aldehyde fer-

[1] J. Kim and D. C. Rees, *Science* **257,** 1677 (1992).

[2] J. Kim and D. C. Rees, *Nature (London)* **360,** 553 (1992).

[3] J. Kim and D. C. Rees, *Biochemistry* **32,** 7104 (1993).

[4] R. C. Bray, *in* "The Enzymes" (P. D. Boyer, ed.), Vol. 12, p. 299. Academic Press, New York, 1975.

[5] R. C. Bray, *Adv. Enzymol.* **51,** 107 (1980).

[6] R. C. Bray, *in* "Biological Magnetic Resonance" (L. J. Berliner and J. Reuben, eds.), Vol. 2, p. 45. Plenum, New York, 1980.

[7] R. Hille and V. Massey, *in* "Molybdenum Enzymes—Metal Ions in Biology" (T. G. Spiro, ed.), Vol. 7, p. 443. Wiley (Interscience), New York, 1985.

[8] M. Coughland, *in* "Molybdenum and Molybdenum Containing Enzymes" (M. Coughland, ed.), p. 119. Pergamon Press, Oxford, England, 1980.

[9] R. C. Bray, *Q. Rev. Biophys.* **21,** 299 (1988).

[10] J. B. Jones and T. C. Stadman, *J. Biol. Chem.* **256,** 656 (1981).

[11] L. G. Ljungdahl, *Annu. Rev. Microbiol.* **40,** 415 (1986).

redoxin oxidoreductase from hyperthermophiles,[12] and to be a component of carboxylic acid reductase[13] and of formylmethanofuran dehydrogenase,[14] with analogies to the molybdopterin-containing enzymes.

Enzymes related to the second group have been isolated and characterized from sulfate-reducing bacteria (strict anaerobes).[15–22] They represent unique situations for the presence of such a group of enzymes in the prokaryotic world, because most often these proteins are studied in eukaryotes.

The molybdenum iron-sulfur protein isolated from *Desulfovibrio gigas* is analogous in some ways to the molybdenum hydroxylases and was shown to have aldehyde oxidoreductase activity, but does not contain a flavin moiety.[15,16,18] This enzyme is used here as a case study. Molybdenum-containing proteins similar to the *D. gigas* enzyme are also found in *Desulfovibrio desulfuricans* ATCC 27774, Berre Sol, and Berre Eau (Ref. 21 and J. J. G. Moura, R. Duarte, and B. A. S. Barata, unpublished results, 1994). Unique and different molybdenum proteins were isolated from *Desulfovibrio africanus* Benghazi[23] and *Desulfovibrio salexigens* British Guiana.[24] The visible spectrum is distinct from that of the *D. gigas* enzyme and information concerning the iron-sulfur center types,

[12] M. W. W. Adams, *Annu. Rev. Microbiol.* **47,** 627 (1993); S. Mukund and M. W. W. Adams, *J. Biol. Chem.* **266,** 14208 (1991); S. Mukund and M. W. W. Adams, *J. Biol. Chem.* **265,** 11508 (1990).

[13] H. White and H. Simon, *Arch. Microbiol.* **158,** 81 (1992).

[14] R. A. Schimtz, S. J. P. Albracht, and R. K. Thauer, *Eur. J. Biochem.* **209,** 1013 (1992); and R. A. Schimtz, M. Richter, D. Linder, and R. K. Thauer, *Eur. J. Biochem.* **207,** 559 (1990).

[15] J. J. G. Moura, A. V. Xavier, M. Bruschi, J. LeGall, D. O. Hall, and R. Cammack, *Biochem. Biophys. Res. Commun.* **72,** 782 (1976).

[16] J. J. G. Moura, A. V. Xavier, R. Cammack, D. O. Hall, M. Bruschi, and J. LeGall, *Biochem. J.* **173,** 419 (1978).

[17] S. P. Cramer, J. J. G. Moura, A. V. Xavier, and J. LeGall, *J. Inorg. Biochem.* **20,** 275 (1984).

[18] N. Turner, B. Barata, R. C. Bray, J. Deistung, J. LeGall, and J. J. G. Moura, *Biochem. J.* **243,** 755 (1987).

[19] R. C. Bray, N. A. Turner, J. LeGall, B. A. S. Barata, and J. J. G. Moura, *Biochem. J.* **280,** 817 (1991).

[20] B. A. S. Barata, J. Liang, I. Moura, I., J. LeGall, J. J. G. Moura, and B. H. Huynh, *Eur. J. Biochem.* **204,** 773 (1992).

[21] B. A. S. Barata, Ph.D. Thesis, Faculdade de Ciências da Universidade de Lisboa, Lisboa, Portugal (1992).

[22] J. J. G. Moura, A. V. Xavier, J. LeGall, and J. M. P. Cabral, *J. Less-Common Met.* **4,** 555 (1977).

[23] E. C. Hatchikian and M. Bruschi, *Biochem. Biophys. Res. Commun.* **86,** 725 (1979).

[24] M. Czechowski, G. Fauque, N. Galliano, B. Dimon, I. Moura, J. J. G. Moura, A. V. Xavier, B. A. S. Barata, A. Lino, and J. LeGall, *J. Ind. Microbiol.* **1,** 1 (1986).

the coordination, and the redox properties of the molybdenum site are not available. In addition, their physiological role is unknown.

Desulfovibrio gigas Aldehyde Oxidoreductase

Desulfovibrio gigas molybdenum-containing protein was isolated on the basis of its optical absorption characteristics and molybdenum was first detected by electron paramagnetic resonance (EPR).[15,24a] The generation of Mo(V) EPR signals in the presence of aldehydes, reminiscent of those observed in molybdenum hydroxylases, suggested the search for its physiological role. This is an example of how spectroscopy can lead to the discovery of a biological function. A wealth of spectroscopic data are available for *D. gigas* aldehyde oxidoreductase (AOR).[15–22] The molecular cloning and sequence analysis of the gene encoding *D. gigas* AOR have been completed.[24b] Crystals were obtained, diffracting to 2.4 Å, and an electron density map is currently being interpreted.[25a,b]

Physicochemical Characterization

Purification. The purification method has been described in detail.[15,16] All operations are carried out at 4°, starting from a *D. gigas* NCIB 9332 cell-free crude extract.[26] A more straightforward purification scheme involving fewer steps and using DEAE-52, DEAE-BioGel, HTP columns, and an additional high-performance liquid chromatography (HPLC) anionic chromatographic step was introduced in order to improve the protein purifica-

[24a] EPR has been used as a valuable tool for revealing the different aspects of the Mo(V) active site in molybdenum-containing hydroxylases. The exhaustive work of R. C. Bray and co-workers with xanthine oxidase and related enzymes established the nomenclature currently used. The catalytically active form of the enzyme contains a sulfido ligand at the molybdenum site. The conversion to an oxo ligand (desulfo form) results in the loss of catalytic activity. Molybdenum-containing hydroxylases are mixtures of inactive desulfo forms (that originate slow EPR-type signals after a long reduction time with dithionite, 15–30 min) and active species (yielding rapid-type signals generated in the presence of substrates or short chemical reduction, within the turnover time scale). Kinetics and thermodynamics of the reduction of Mo(VI) to Mo(V) and Mo(IV) differ markedly within these species. EPR rapid-type 2 Mo(V) signal is defined as a catalytically competent species that shows two strongly coupled protons that can be exchanged. Type 1 Mo(V) EPR signal has one strong and one weakly coupled proton.

[24b] V. Thoener, O. L. Flores, A. Neves, B. Devreese, J. J. Van Beeumen, R. Huber, M. J. Romãs, J. LeGall, J. J. G. Moura, and C. Rodrigues-Pousada, *Eur. J. Biochem.* **220,** 901 (1994).

[25a] Romão and Huber, in preparation (1994).

[25b] M. J. Romão, B. A. S. Barata, M. Archer, K. Lobeck, I. Moura, M. A. Carrondo, J. LeGall, F. Lottspeich, R. Huber, and J. J. G. Moura, *Eur. J. Biochem.* **215,** 729 (1993).

[26] J. LeGall, G. Mazza, and N. Dragoni, this series, Vol. 6, p. 819.

TABLE I
PURIFICATION OF ALDEHYDE OXIDOREDUCTASE ACTIVITY IN *Desulfovibrio gigas*[a,b]

Purification step	Total protein (mg)	Specific activity[c] (U/mg)	Total units (U)
Step 1	5744.3	0.04	231.4
DEAE-52	2314.8	0.10	230.3
DEAE-52	633.4	0.18	113.9
DEAE-BioGel	210.5	0.52	114.1
HTP	118.1	0.54	63.9
HPLC	53.1	1.17	62.0

[a] Adapted from Ref. 27.
[b] From 1500 g of wet cells.
[c] 1 unit = 1 μmol of DCPIP reduced per minute. $[\text{DCPIP}]_{\text{final}}$ = 35 μM, [benzaldehyde] = 100 μM.

tion in the final steps.[27] A purification outline is given in Table I. Invariant spectral features and stable activity are found if the protein is stored at low temperature (liquid nitrogen, in phosphate buffer).

Molecular Mass, N-Terminal, and Metal Content. Desulfovibrio gigas AOR is a dimer composed of two identical subunits of about 100 kDa [single band detected by electrophoresis in the presence of sodium dodecyl sulfate (SDS) and 2-mercaptoethanol, after a boiling step]. The molecular mass determined by calibrated gel exclusion is around 200 kDa for the native protein in 0.3 *M* NaCl and 0.1 *M* Tris-HCl, pH 7.0.[25] Previous results showed a molecular mass of about 120 kDa by gel filtration using 1 *M* NaCl and 0.02 *M* phosphate buffer.[15,16] Sedimentation equilibria performed in formic acid indicate a bimodal sedimentation distribution that can be explained by molecular masses of 96 and 180 kDa.[25] These results are supported by genetic analysis indicating that the gene spans 2718 bp of genomic DNA and codes for 906 amino acid residues.[27a]

The amino acid sequence of the first 29 residues of the NH_2-terminal portion of *D. gigas* AOR was identified as

```
1               5                   10                  15
Met Ile Gln Lys Val Ile Thr Val Asn Gly Ile Glu Gln Asn Leu Phe Val Asp Ala
20                  25
Glu Ala Leu Leu Ser Asp Val Leu Arg Gln
```

Molybdenum was quantified by neutron activation analysis.[22] Only iron and molybdenum were detected by plasma emission analysis.[20] The

[27] B. A. S. Barata, J. LeGall, and J. J. G. Moura, *Biochemistry* **32,** 11559 (1993).
[27a] Thoener, Flores, Neves, and R.-Pousada, in preparation.

average values found are 3.5 mol of iron and 0.53 mol of molybdenum monomeric unit, indicating an Fe : Mo ratio of 7 : 1. These values strongly indicate the presence of 4 mol of iron per subunit (arranged as two [2Fe-2S] centers; see spectroscopic studies, below) and suggest that some of the protein molecules may be devoid of molybdenum, a feature also detected in other hydroxylases[28] (demolybdenum forms). Another possibility is that the molybdenum site is shared by the two identical subunits. The resolution of the three-dimensional structure of the protein is anticipated with great interest.

Aldehyde Oxidoreductase Activity

The obvious analogies between the EPR spectra of *D. gigas* AOR and the molybdenum-containing hydroxylases spurred the search for the enzymatic activity[18] (see below) and detailed study of the catalytic properties of the *D. gigas* molybdenum [Fe-S] protein, in particular its apparent kinetic parameters (K_m and V_{max}) toward different aldehydes and other substrates typically utilized by the molybdenum hydroxylase group.[18,27]

The absence of a flavin group suggests the association with a dehydrogenase activity using 2,6-dichlorophenol-indophenol (DCPIP) as electron acceptor (aldehyde–DCPIP–oxidoreductase activity). The data initially determined for a limited range of substrates[18] were extended to a large number of aldehydes and related compounds measured under optimized conditions.[27]

Activity Measurements. Activity is determined aerobically at 25° by measuring the rate of DCPIP reduction at 600 nm ($\varepsilon = 21\ mM^{-1}\ cm^{-1}$) in a 1-cm optical path spectrophotometer quartz cell, containing the following assay mixture: 100 m*M* Tris-HCl buffer (pH = 7.6), 40 μM DCPIP, and 100 μM aldehyde to which is added 20 μl of enzyme solution. Under these experimental conditions 1 enzymatic unit corresponds to 1 μmol of DCPIP reduced per minute and the specific activity is the number of units per milligram of enzyme.[18,27]

A pH profile of the enzyme activity is determined in different buffer systems (at the same ionic strength, 200 m*M*). The buffer systems used—citrate, citrate/phosphate, phosphate, Tris-HCl, Tris-glycine, glycine-NaOH—cover a pH range of 4–11.[27]

For the Michaelis–Menten parameter determinations, the added volumes of the aldehydes (acetaldehyde, propionaldehyde, benzaldehyde, and salicylaldehyde) from different stock solutions are varied concomitantly with the buffer volume, so that the final value will always be 1 ml.

[28] A. M. Ventom, J. Deistung, and R. C. Bray, *Biochem. J.* **255,** 949 (1988).

The proportion, dye to protein, is constant. Substrate is the last component to be added after a 2.5-min equilibration of the protein with the electron acceptor.[27]

Electron Donors and Acceptors, Kinetic Parameters, and Substrate Specificity. The DCPIP-dependent aldehyde oxidoreductase activity was studied in detail using a wide range of aldehydes and analogs. A steady state kinetic analysis (K_m and V_{max}) was made for acetaldehydes, propionaldehydes, benzaldehydes, and salicylaldehydes in excess DCPIP concentration.[18,27] As an example, Fig. 1 superimposes three independent data sets obtained with benzaldehyde and a simple Michaelis–Menten model shown to be applicable as a first kinetic approach. Apparent kinetic parameters for different aldehydes are collected in Table II.

Xanthine, purine, allopurinol, and N^1-methylnicotinamide (NMN) could not be utilized as enzyme substrates. DCPIP and ferricyanide were shown to be capable of cycling the electronic flow, whereas other cation and anion dyes do not. O_2 and $NAD(P)^+$ were not active in this process. The enzyme showed an optimal pH activity profile around 7.8. High concentrations of all the reducing substrates inhibited the activity; methanol and cyanide also inhibited the reaction.[27]

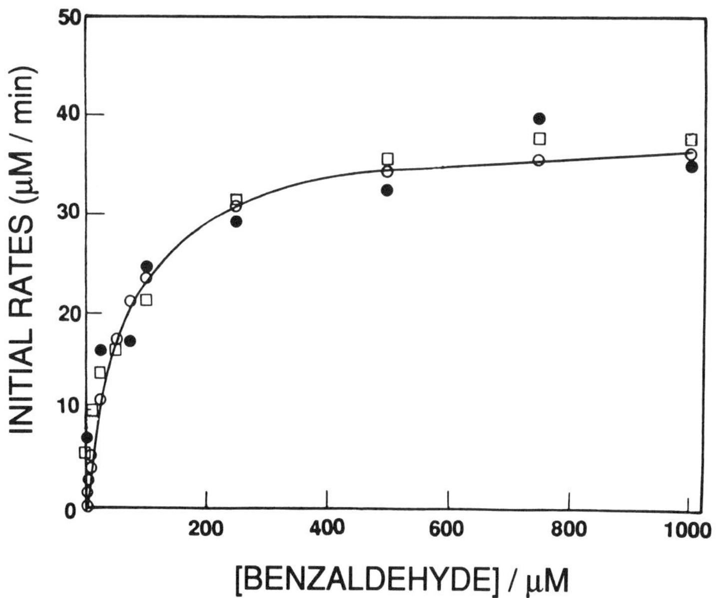

FIG. 1. Michaelis–Menten kinetic curves for different batches of *D. gigas* AOR [determined as indicated in text (Activity Measurements)].

TABLE II[a]
APPARENT KINETIC PARAMETERS[b] OF *Desulfovibrio gigas* ALDEHYDE OXIDOREDUCTASE FOR DIFFERENT REDUCING SUBSTRATES USING 2,6-DICHLOROPHENOL-INDOPHENOL AS ELECTRON ACCEPTOR

Aldehyde	K_m (μM)	V_{max} (μM min^{-1})	[IS][c] (μM)
Acetaldehyde	13	31	500
Propionaldehyde	15	12	250
Benzaldehyde	52	39	1000
Salicylaldehyde	0.25	4.0	500

[a] Adapted from Ref. 27.
[b] Apparent kinetic parameters, valid for DCPIP concentration of 35 m*M*. Extention of these values for a wide range of DCPIP values using benzaldehyde as a substrate is presented in Ref. 27.
[c] [IS], Typical inhibitory substrate concentration.

The enzyme specificity toward different substrates is indicated in Table III and compared with xanthine oxidase (XO) performance.

Spectroscopic Studies

A set of Mo(V) EPR signals detected in *D. gigas* AOR shows close analogies with the molybdenum-containing hydroxylases. Mössbauer and X- and Q-band EPR spectroscopic studies,[19,20] complementing ultraviolet–visible (UV–vis), circular dichroic (CD), and EPR data,[15,16] revealed a [2Fe-2S] arrangement of the iron-sulfur cores.

Optical Absorption and Circular Dichroism Spectra

The visible absorption spectrum of the protein is similar to that observed for the deflavo forms of xanthine and aldehyde oxidases[4,5,15,29,30] and reminiscent of the one observed for plant-type ferredoxins. Figure 2 shows a typical spectrum of a pure native preparation. The extinction coefficients and optical ratios at characteristic wavelengths are indicated in the caption. Anaerobic addition of benzaldehyde decreases the optical absorption, measured at 462 nm, by approximately 8%, after a 1-min reaction, and by 11% after 1 hr (compared with the maximal bleaching obtained with dithionite reduction).

The circular dichroism spectrum is quite intense and was one of the first distinct fingerprints obtained for the presence of [2Fe-2S] centers,

[29] H. Komai, V. Massey, and G. Palmer, *J. Biol. Chem.* **244,** 1692 (1969).
[30] K. V. Rajagopalan and P. Handler, *J. Biol. Chem.* **239,** 2022 (1964).

TABLE III
SUBSTRATE SPECIFICITY COMPARISON BETWEEN *Desulfovibrio gigas* ALDEHYDE OXIDOREDUCTASE AND XANTHINE OXIDASE[a–c]

Substrate	*D. gigas* AOR	XO
Acetaldehyde	100	21
Propionaldehyde	35	10
Benzaldehyde	92	11
Salicylaldehyde	19	7
Xanthine	0.5	103
Hipoxanthine	0	52
1-Methylxanthine	0	170
Allopurinol	0	120
Purine	0	55

[a] Adapted from Ref. 27.
[b] Acetaldehyde/DCPIP oxidoreductase activity is taken as a reference value.
[c] Assay conditions: [enzyme], 450 n*M*; [substrate], 1 m*M*; [DCPIP], 35 μM; [Tris-HCl], 20 m*M*; pH 7.8; T = 25°.

being similar to the characteristic spectra of plant ferredoxins and xanthine oxidase.[31]

Molybdenum Center

Electron Paramagnetic Resonance Studies. The first Mo(V) EPR signals detected were of the resting- and slow-type EPR signals reminiscent of those observed for the desulfo form of xanthine oxidase (inactive)[16–18] (see Ref. 24a for nomenclature of the EPR signals[4–7,9]). In addition, it was later demonstrated that a new Mo(V) EPR signal, centered at g_{av} = 1.9742, could be obtained by brief reduction of the protein with small amounts of dithionite or by brief reaction with aldehydes like those observed for the active form of molybdenum hydroxylase enzymes.[18] These rapid-type EPR signals have been shown to be physiologically significant, as they develop within the enzyme turnover time scale.

Different EPR signals can be generated and observed in *D. gigas* AOR that reflect the reactivity and coordination versatility of the molybdenum

[31] K. Garbett, R. D. Gilard, P. F. Knowles, and J. E. Stangroom, *Nature* (*London*) **215,** 824 (1967).

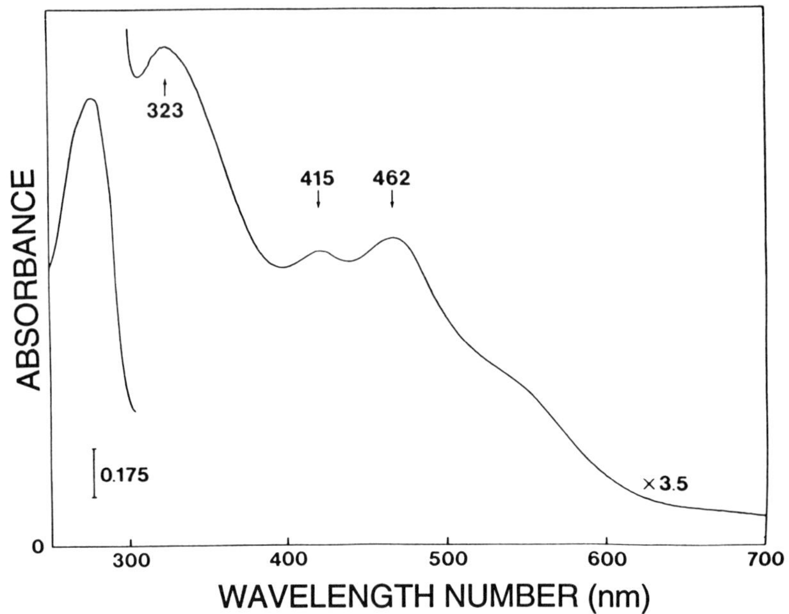

FIG. 2. Optical absorption spectra of *D. gigas* AOR (10 μM protein) in 20 mM Tris-HCl, pH 7.6. The absorbance ratios at different wavelengths are $A_{280}/A_{323} = 3.17$, $A_{280}/A_{415} = 5.25$, and $A_{280}/A_{462} = 5.05$. Extinction coefficient at 462 nm = 24 mM cm^{-1}.

site and enables placement of the enzyme among the molybdenum hydroxylases (see Fig. 3).

Resting Electron Paramagnetic Resonance Signal. Untreated *D. gigas* AOR (as isolated) gives EPR signals presumed to be due to Mo(V) and the exact form varies somewhat depending on preparation. They do not modify after exchange with D_2O and have a low intensity (1–7%). These signals have not been extensively studied.

Slow Electron Paramagnetic Resonance Signal. Long treatment of *D. gigas* AOR with dithionite (15–30 min) makes the resting signal decline and eventually replaced by another EPR-active species.[15–17] By analogy to xanthine oxidase this signal is referred to as the slow signal. The signal was developed in ^1H_2O and ^2H_2O, indicating the interaction of the molybdenum site with an exchangeable proton. The EPR spectra were recorded in X- and Q-bands and simulated with the same set of parameters (except the linewidth), indicating that a single Mo(V) species is involved.[19] The signal has spectral parameters similar to those of the slow signal detected in the desulfo forms of various molybdenum-containing hydroxy-

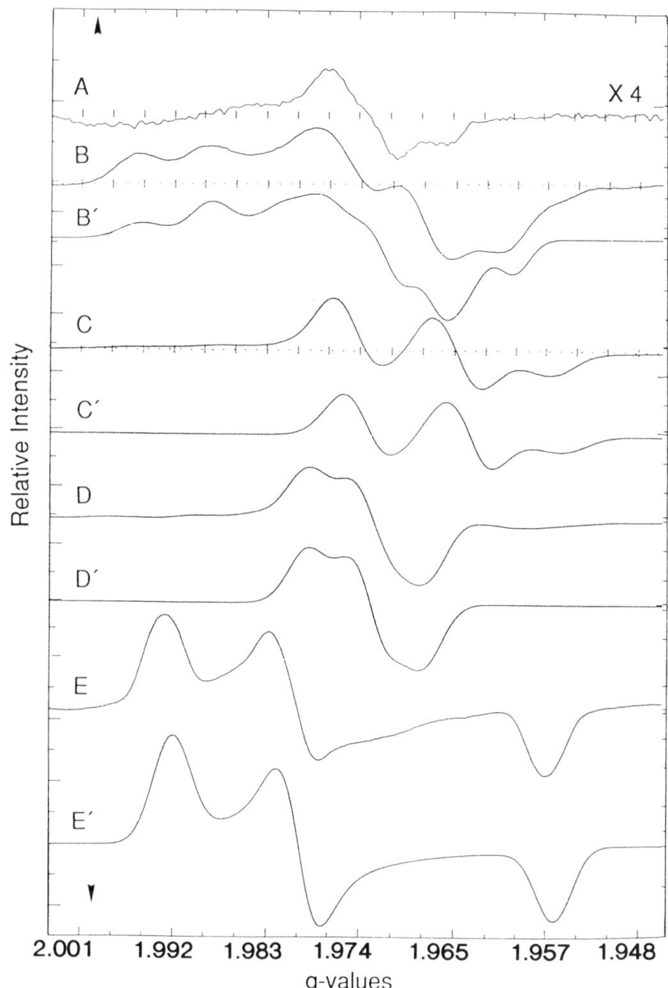

FIG. 3. Collection of EPR spectra of the different faces of the Mo(V) site in *D. gigas* AOR. Spectra obtained at 110 K (for experimental details see Refs. 16, 18, 19, and 21). (A) Resting, (B) rapid type 2, (C) slow, (D) desulfo inhibited, and (E) inhibited signals. Spectra designated with a prime are simulations of the experimental data. Details on spectral parameters can be found in Table IV.

lases (see Table IV and Extended X-Ray Absorption Fine Structure, below).

Rapid Electron Paramagnetic Resonance Signal. It was possible to generate a new Mo(V) EPR signal (rapid type 2) centered at $g_{av} = 1.9750$

TABLE IV
COMPARISON OF Mo(V) ELECTRON PARAMAGNETIC RESONANCE SIGNALS DETECTED IN MOLYBDENUM HYDROXYLASES[a]

Enzyme	Signal	g_1	g_2	g_3	^1H hyperfine interactions (mT) A_1	A_2	A_3
D. gigas AOR	Slow (dith)[b]	1.9705	1.9680	1.9580	1.68	1.60	1.44
AOR	Slow (dith)	1.9720	1.9668	1.9556	1.60	1.60	1.65
XO	Slow (dith)	1.9719	1.9671	1.9551	1.50	1.48	1.52
D. gigas AOR	Rapid 2 (dith)	1.9882	1.9702	1.9643	1.15	1.66	1.26
					1.14	0.63	0.93
	Rapid 2 (Ald)	1.9895	1.9715	1.9640	1.4	1.3	0.7
					0.35	0.5	1.3
AOR	Rapid 2	1.9895	1.9700	1.9622	1.38	1.50	1.50
					1.13	0.65	0.71
XO	Rapid 2 (P-3-A)	1.9861	1.9695	1.9623	1.35	1.57	1.45
					1.35	0.74	0.99
	Rapid 1	1.9906	1.9707	1.9654	1.20	1.30	1.30
					0.47	0.0	0.0
D. gigas AOR	Desulfo inhibited	1.9787	1.9725	1.9673			
AOR	Desulfo inhibited	1.980	1.973	1.964			
XO	Desulfo inhibited	1.9784	1.9707	1.9641			
D. gigas AOR	Inhibited	1.9915	1.9795	1.9555	nd	nd	nd
AOR	Inhibited	1.994	nd	1.952	0.65	nd	0.65
XO	Inhibited	1.9911	1.9772	1.9513	0.50	0.34	0.54

[a] Adapted from Ref. 18 and the following references: S. Gutteridge, S. J. Tanner, and R. C. Bray, *Biochem. J.* **175,** 887 (1978); R. C. Bray, *in* "Flavins and Flavoproteins" (R. C. Bray, P. C. Engels, and S. G. Mayhew, eds.), p. 707. de Gruyter, Berlin, 1984; M. J. Barber, M. P. Coughland, K. V. Rajagopalan, and L. M. Siegel, *Biochemistry* **21,** 3561 (1982); R. C. Bray, G. N. George, S. Gutteridge, L. Norlander, J. G. Stell, and C. Stubley, *Biochem. J.* **203,** 263 (1982).

[b] P-3-A, Pyridine-3-aldehyde; nd, not determined; dith, dithionite; Ald, aldehyde.

analogous to those observed for xanthine and aldehyde oxidases.[18] The rapid signals were not only obtained after a brief chemical reduction (with dithionite) but also in the presence of salicylaldehyde,[18] suggesting that other substrates might also produce rapid EPR signals. These signals were visualized after spectral subtraction, because they were obtained in a complex mixture of different Mo(V) EPR-active species. The spectral parameters were extracted by spectral simulation. These active species were also generated in the presence of benzaldehydes and acetaldehydes.[27] Hyperfine interactions with two strongly coupled (exchangeable) protons were observed (see Table IV). The rapid EPR signals, classified as type 2, are similar and analogous to the ones detected in aldehyde oxidase and in xanthine oxidase reacted with pyridine-3-aldehyde. Rapid type 1 EPR signals were also not detected in aldehyde oxidase but are the most common in XO.

Desulfo-Inhibited Electron Paramagnetic Resonance Signals. Reaction of the enzyme with a high concentration of ethylene glycol (100 m*M* to 5 *M*) for an extended period of time (30–120 min) in a sample in which the slow signal develops, results in the appearance of a new Mo(V) species designated as desulfo inhibited.[21] The kinetics of slow/desulfo-inhibited forms of conversion can be followed by EPR. The signal develops in the presence of formaldehyde, methanol, and glycerol and persists on air oxidation. This form is interesting for the study of Mo(V) in the absence of magnetic interactions with other centers (paramagnetic iron-sulfur clusters).

Inhibited Electron Paramagnetic Resonance Signal. The inhibited species is detected in *D. gigas* AOR, mixed with rapid signals, and clearly originated from a functional form.[21] Its detection is relevant for the comparison of the enzyme with those of the hydroxylase group.

A general comparison of the EPR data of the most relevant Mo(V) EPR signals detected in *D. gigas* AOR with other molybdenum-containing hydroxylases is presented in Table IV.

Extended X-Ray Absorption Fine Structure. The molybdenum extended X-ray absorption fine structure (EXAFS) of the *D. gigas* AOR was examined using fluorescence detection and synchroton radiation.[17] In the oxidized form the molybdenum environment is found to contain two terminal oxo groups and two long (2.47-Å) Mo—S bonds. Evidence was also found for an oxygen or nitrogen ligand at 1.90 Å. Addition of dithionite to the oxidized enzyme results in loss of a terminal oxo group, perhaps due to protonation. In addition, a 0.1-Å contraction in the Mo—S bond lengths is observed. The behavior of both oxidized and dithionite-treated forms is similar to that observed previously with desulfoxanthine oxidase and resembles sulfite oxidase and nitrate reductase (with two terminal oxo groups).

Iron-Sulfur Centers

Electron Paramagnetic Resonance. EPR studies (complemented with Mössbauer studies, see below) reveal the presence of two types of [2Fe-2S] cores, named Fe-S I and Fe-S II centers.[16,18,19] X- and Q-band EPR studies established the presence of two EPR signals in approximately equal amounts (Fe-S I and Fe-S II centers), as determined by computer simulation using the same set of parameters at the two different frequencies and double intergrations performed at different powers and temperatures. The Fe-S I center is observed at 77 K (g values at 2.021, 1.938, and 1.919; average, 1.959). The Fe-S II center is observable only below 65 K (g values at 2.057, 1.970, and 1.900; average, 1.976).

Variation of amplitude of Fe-S I as a function of redox potential was observed to be complex (inflexions at −260 and −440 mV)[16] and led to the postulation of the presence of a third iron-sulfur center (Fe-S IB); this was proved to be wrong.[19,20] The width of the g_1 feature of Fe-S I increased with decreasing temperature and a splitting was detected consistent with magnetic coupling of different paramagnetic centers in the molecule. Interaction between Fe-S I and Fe-S II was first invoked,[16] but most probably can be explained as a result of the interaction between Fe-S I with the Mo(V) center.[19] Fe-S II has a midpoint potential of −285 mV. Long equilibration was necessary to observe the slow-type molybdenum signals with midpoint redox potentials at −415 mV [Mo(VI)/Mo(V)] and −530 mV [Mo(V)/Mo(IV)].[16]

The overall data are consistent with the presence of only two iron-sulfur centers (see Mössbauer, below).

Table V compares the EPR spectral data on the iron–sulfur centers of different molybdenum-containing hydroxylases.

Mössbauer. An important advance in these studies was the possibility of isolating the enzyme from ^{57}Fe-enriched media with obvious interest in an iron-sulfur center site label, the enhanced sensitivity of the Mössbauer studies (an advantage with respect to mammalian systems; see below),

TABLE V
COMPARISON OF ELECTRON PARAMAGNETIC RESONANCE g-VALUES OF Fe-S CENTERS IN PROKARYOTIC AND EUKARYOTIC MOLYBDENUM HYDROXYLASES[a,b]

Enzyme	g_1	g_2	g_3
Fe-S center I			
D. gigas	2.021	1.938	1.919
Eukaryotes	2.021 ± 0.004	1.933 ± 0.005	1.907 ± 0.01
Volinella alcalescens	2.026	1.939	1.925
Clostridium acidiurici	2.034	1.945	1.918
Fe-S center II			
D. gigas	2.057	1.979	1.900
Eukaryotes	2.103 ± 0.005	2.002 ± 0.006	1.912 ± 0.01
Volinella alcalescens	nd	nd	nd
Clostridium acidiurici	2.075	1.924	1.871

[a] Adapted from Ref. 19 and references therein.
[b] Values of eukaryotic molybdenum hydroxylases refer to six different sources.[19]

and the possibility of directly measuring the substrate binding. Spectra of the enzyme in its oxidized, partially reduced, and benzaldehyde-reacted and fully reduced states were recorded at different temperatures and applied magnetic fields.[20] All the iron atoms in *D. gigas* AOR are organized as [2Fe-2S] centers. In the oxidized enzyme, the clusters are diamagnetic and exhibit a single quadrupole doublet with parameters ($\Delta E_Q = 0.62 \pm 0.02$ mm/sec and $\delta = 0.27 \pm 0.01$ mm/sec) typical of the +2 state. Mössbauer spectra of the reduced clusters also show the characteristic values of the +1 state and could be explained by a spin-coupling model proposed for the [2Fe-2S] cluster, where the high-spin ferrous site ($S = 2$) is antiferromagnetically coupled to a high-spin ferric site ($S = 5/2$) to form an $S = 1/2$ system. Two ferrous sites with different ΔE_Q (3.42 and 2.93 mm/sec) at 85K are observed, indicating the presence of two types of [2Fe-2S] clusters in the *D. gigas* enzyme. From this observation, together with the metal content and EPR data, it is concluded that, similarly to other molybdenum-containing hydroxylases, the *D. gigas* AOR also contains two spectroscopically distinguishable [2Fe-2S] centers.

A Mössbauer study of the protein reacted with benzaldehyde shows partial reduction of the iron–sulfur centers, indicating the involvement of the clusters in the process of substrate oxidation and rapid intramolecular electron transfer from the molybdenum to the iron-sulfur sites.

Besides the larger ΔE_Q values observed for the ferrous sites in reduced [2Fe-2S] clusters, the hyperfine parameters obtained for the Fe-S clusters in the *D. gigas* AOR are similar to those of the [2Fe-2S] centers in ferredoxins.[32] The only Mössbauer study of molybdenum hydroxylases prior to this work was performed on a nonenriched sample of milk xanthine oxidase,[33] and an unusually large ΔE_Q (3.2 mm/sec at 175 K) is also observed for the ferrous site of one of the clusters.

Molybdenum Cofactor Extrusion

The molybdenum cofactor was liberated from *D. gigas* AOR, and under appropriate conditions was transferred quantitatively to apo nitrate reductase in extracts of *Neurospora crassa* (*nit-1* mutant) to yield active nitrate reductase. This procedure, followed by assay of the nitrate reductase activity, forms the basis for the standard assay of the molybdenum cofactor, so well documented for xanthine oxidase.[34,35] The data on recon-

[32] W. R. Dunham, A. J. Bearden, I. T. Salmeen, G. Palmer, R. H. Sands, W. H. Orme-Johnson, and Beinert, *Biochim. Biophys. Acta* **253,** 134 (1971); E. Muenck, P. G. Debruner, J. C. M. Tsibris, and I. C. Gunsalus, *Biochemistry* **11,** 855 (1972).

[33] R. Hille, W. R. Hagen, and W. R. Dunham, *J. Biol. Chem.* **260,** 10569 (1985).

[34] T. R. Hawkes and R. C. Bray, *Biochem. J.* **219,** 481 (1984).

[35] T. R. Hawkes and R. C. Bray, *Biochem. J.* **222,** 587 (1984).

stitution of nitrate reductase activity clearly indicate that the molybdenum cofactor is liberated from *D. gigas* AOR after treatment with dimethyl sulfoxide (DMSO), and after complementation with apo nitrate reductase achieved.[18] On the basis of molybdenum content, the activity observed for reconstitution with molybdenum cofactor of *D. gigas* was lower (25%) than the values observed for molybdenum cofactor of XO (21 ± 1 nmol NO^{2-} formed/min/pg-atom Mo). This difference is not fully understood, and clearly a more complete characterization of the pterin cofactor present in *D. gigas* AOR is required.

Experimental Procedure. The method is described in Refs. 34 and 35. All operations, except the incubation of the complementation mixture, are performed under strict anaerobic conditions (oxygen less than 1 ppm). The protein, after dilution with 25 m*M* potassium phosphate buffer (pH 7.4) containing 1 m*M* EDTA, is denatured by the addition to an excess of DMSO containing 1–2 m*M* sodium dithionite. Alternative denaturing procedures use SDS or *N*-methylformamide. Anaerobic complementation for 24 hr at 3.5° with the apo nitrate reductase of partially purified extracts of the *nit-1* mutant of *N. crassa* yields active nitrate reductase. Complementation is carried out in the presence of Na_2MoO_4 (10 m*M*).

Functionality of Aldehyde Oxidoreductase

The EPR and Mössbauer data analysis, together with the metal determination and the effect of aldehydes in the chromophore absorption of the protein, indicated that AOR samples are a mixture of active, inactive, and demolibdo forms. Electron paramagnetic resonance indicates functional (30%) and desulfo forms (70%) coexist.[19] Therefore, to define the functionality of the enzyme [in the absence of a flavin group it is not possible to determine the activity-to-flavin ratio (AFR factor) as done for the example using xanthine oxidase[36]], a set of complementary measurements is required. This is also an inherent limitation of the EXAFS measurements when dealing with mixtures of species and a failure to detect the active form of the enzyme if present in low concentration, being difficult to find evidence for a short Mo-sulfido bond ligand.[17,18]

Genetic Advances

Cloning and expression were directed in a first stage to the determination of primary amino acid sequences. Molecular cloning and sequence analysis of the gene encoding the molybdenum-containing protein of *D. gigas* are now underway. A 4023-bp genomic *Pst*I fragment of *D. gigas*

[36] L. I. Hart, M. McGartoll, H. R. Chapman, and R. C. Bray, *Biochem. J.* **116,** 851 (1970).

harboring the AOR gene was isolated, and encodes 906 amino acids. The procedure begins from the known N-terminal sequence. The protein sequence shows some homology to xanthine dehydrogenase from different organisms.[25b,27a]

Preliminary X-Ray Studies

Desulfovibrio gigas AOR was crystallized at 4°, pH 7.2, using 2-propanol and $MgCl_2$ as precipitants.[25] The crystals diffract beyond 3.0-Å resolution and belong to space group P_{6122} or its enantiomorph, with cell dimensions $a = b = 144.5$ Å and $c = 163.2$ Å. The structural analysis by crystallographic methods is in progress and some heavy derivatives have been found and analyzed. Preliminary phase information from multiple-isomorphous replacement data has allowed the calculation of an electron-density map, which confirms the packing of the protein as a dimer of about 2 × 100 kDa. Successful results on crystallization of *D. desulfuricans* ATCC 27774 AOR have been obtained (M. J. Romão and R. Huber, personal communication, 1994).

Physiological Studies: From Aldehydes to Molecular Hydrogen

Desulfovibrio gigas AOR was shown to be part of an electron transfer chain comprising 4 different soluble proteins from the same organism, with a total of 11 discrete redox centers, which is capable of linking the oxidation of aldehydes to the reduction of protons,[27] as indicated:

$$\begin{array}{c} \text{aldehyde} \\ \downarrow \\ \text{carboxylic acid} \end{array} \longrightarrow \text{AOR} \longrightarrow \text{flavodoxin} \longrightarrow \text{cytochrome } c_3 \text{ (26 kDa)} \longrightarrow \text{hydrogenase} \longrightarrow \begin{array}{c} 2H^+ \\ \downarrow \\ H_2 \end{array}$$

The proposed physiological electron transfer chain is entirely cytoplasmic. The *in vitro* reconstitution of the electron transfer chain from molecular hydrogen to sulfite,[37] one of the bacteria respiratory substrates, has been demonstrated. These two reconstituted electron transfer pathways complete an electron transfer scheme that is compatible with the hydrogen cycling hypothesis.[38]

[37] L. Chen, M.-Y. Liu, and J. LeGall, *Arch. Biochem. Biophys.* **303,** 44 (1993).

[38] J. M. Odom and H. D. Peck, Jr., *FEMS Lett.* **12,** 47 (1981); J. M. Odom and H. D. Peck, Jr., *J. Bacteriol.* **147,** 161 (1981).

Other Molybdenum-Containing Proteins

Proteins homologous to *D. gigas* AOR have been isolated and preliminarily characterized from *D. desulfuricans* ATCC 27774, Berre Eau, and Berre Sol (Ref. 21 and J. J. G. Moura and B. A. S. Barata, unpublished results, 1994). The molecular mass ranges from 110 to 130 kDa (monomer). The UV–vis spectra have analogies with the former enzyme with maxima around 280, 300, 420, and 460 nm, and a shoulder at 520 nm. Mo(V) EPR signals (slow and rapid type 2) were detected, as well as two sets of Fe-S centers [for *D. desulfuricans* ATCC AOR *g* values around 2.07 and 1.89 (center II) and 2.01, 1.93 and 1.92 (center I).[21]] Apparent kinetic parameters were determined for *D. desulfuricans* ATCC AOR (acetaldehyde, K_m 24 μM and V_{max} 43 μM/min; benzaldehyde, K_m 41 μM and V_{max} 28 μM/min).[21]

Different molybdenum proteins have been isolated from other sulfate reducers, and it is difficult at this stage to classify them. A molybdenum-containing iron–sulfur protein has been isolated from *D. africanus* Benghazi.[23] The protein appears to be a complex protein of high molecular mass (112 kDa) composed of 10 subunits (each subunit contains about 106 amino acids and a molecular mass of 11.5 kDa, with 2 cysteines) and 5–6 molybdenum atoms and 20 iron atoms and labile sulfur. The spectrum shows peaks around 615 (48.4), 410 (64.4), 325 (141), and 280 nm. N-Terminal sequence was determined for up to 26 residues. From *D. salexigens,* a similar protein was isolated, characterized by its blue-gray color with bands at 612, 410, and 275 nm, and a shoulder at 319 nm, having a molecular mass of 110 kDa (13-kDa subunits).[24] At the moment no information is available on the molybdenum site and the iron–sulfur center arrangement, but preliminary EPR data seem to indicate that they are of a new type.

Conclusions: Physiological Significance

A classification of molybdenum-containing proteins, based on amino acid sequence data and spectroscopic properties, has been completed.[39] A high degree of similarity, both in the amino acid composition and prosthetic group composition, is found in the molybdenum-containing enzymes isolated from eukaryotic organisms, including xanthine oxidase/dehydrogenase and aldehyde oxidase. Great diversity, however, is observed for the molybdenum hydroxylases found in prokaryotic systems. In particular, significant differences are found both in the content and spectroscopic

[39] J. C. Wootton, R. E. Nicolson, J. M. Cock, D. E. Walters, J. F. Bruke, W. A. Doyle, and R. C. Bray, *Biochim. Biophys. Acta* **1057,** 157 (1991).

properties of the iron–sulfur centers.[19,40,41] The presence of two types of [2Fe-2S] clusters in the *D. gigas* AOR and the different Mo(V) forms detected by EPR (see Tables IV and V) have established a close relationship between the *D. gigas* enzyme and the molybdenum hydroxylases isolated from eukaryotic systems. An important difference is the absence of FAD. The amino acid sequence associated with the binding of Fe-S I was identified in eukaryotic enzymes by analogy with plant-type [2Fe-2S] ferredoxin sequence.[39] The high degree of homology between Fe-S I in eukaryotic and prokaryotic enzymes anticipates that these sequences must be conserved in these proteins. On the other hand, the binding site of Fe-S II seems to be less well defined.[19]

The presence of aldehyde- and carboxylic acid-metabolizing enzymes raises questions concerning novel metabolic pathways and represents a unique situation for extrapolating data from the prokaryotic to the eukaryotic world, where most often these proteins are studied. However, the question of the physiological significance of the molybdenum protein in *D. gigas* [and other sulfate-reducing bacteria (SRB)] other than as a scavenger of environmental aldehydes remains. Some sulfate-reducing organisms can use aldehydes as energy sources.[42] The existence of both ethanol dehydrogenase and aldehyde oxidase in SRB has been previously demonstrated,[43] as has the ability of *D. gigas* to utilize internal polyglucose.[2,43] Glyceraldehyde, another substrate used by *D. gigas* AOR, is of special interest because of the observation of Stams *et al.*[43] confirming earlier results by Thomas[44] that *D. gigas* accumulates large amounts of polyglucose under normal growth conditions. This molecule has been viewed as a storage polymer because it can be degraded into acetate and hydrogen in the absence of sulfate and as an energy source for sulfate-reducing bacteria. *In vivo* nuclear magnetic resonance (NMR) studies indicate that nucleotide triphosphates are formed in the presence of oxygen from polyglucose degradation, and in the absence of oxygen the importance of glyceraldehyde was suggested.[45] Observations by Mukund and Adams[12] showed that a tungsten-containing aldehyde ferredoxin oxidoreductase from *Pyrococcus furiosus* could be involved in a glycolytic pathway

[40] R. Wagner, R. Cammack, and J. R. Anderson, *Biochim. Biophys. Acta* **791,** 63 (1984).

[41] H. Dalton, D. J. Lowe, R. T. Pawlik, and R. C. Bray, *Biochem. J.* **153,** 287 (1976).

[42] F. Widdel, *in* "Biology of Anerobic Microorganisms" (A. J. B. Zehnder, ed.), p. 469. (1988).

[43] F. J. M. Stams, M. Veenhuis, G. H. Weenk, and T. A. Hansen, *Arch. Microbiol.* **136,** 54 (1983).

[44] P. Thomas, *J. Microsc.* **13,** 349 (1972).

[45] M. H. Santos, P. Fareleira, A. V. Xavier, L. Chen, M.-Y. Liu, and J. LeGall, *Biochem. Biophys. Res. Commun.* **195,** 551 (1993).

through which glucose is converted into acetate. The enzyme converts glyceraldehyde to glycerate with ferredoxin-linked hydrogen production through hydrogenase. *Desulfovibrio gigas* aldehyde oxidoreductase could play a similar role and be involved in the degradation of glyceraldehyde, a possible intermediate in endogeneous polyglucose utilization. However, the electron transfer chain present in *D. gigas* is far more complicated than in *P. furiosus,* which requires only ferredoxin as intermediate electron carrier, as discussed earlier.

Desulfovibrio gigas is in a special situation because it lacks an uptake system for glucose and/or hexokinase, preventing it from growing directly on glucose,[43] but it probably derives glyceraldehyde (plus pyruvate) from internal polyglucose. This variation of the Entner–Doudoroff pathway[12] that was proposed to be operative in *P. furiosus* could well exist in *D. gigas,* for all the activities of the enzymes necessary to phosphorylate glycerate [obtained from the NAD(P)$^+$-independent molybdenum-containing aldehyde oxidoreductase reaction] and to transform it into pyruvate (3-phosphoglycerate mutase, enolase, and pyruvate kinase) are present in significant levels in the cell-free extracts of lactate-grown bacteria.[46]

Optimization, stimulation, and coupling of this chain to pyridine nucleotides or to nucleotide phosphates is currently under progress. In contrast with developments in the field of the molybdenum-containing nitrogenases,[1–3] the lack of fine structural information on the molybdenum-containing hydroxylating enzymes, especially with regard to the relative spatial arrangement of the metal sites and the nature of the molybdenum cofactor, has hampered the detailed understanding of the mechanisms operating in these enzymes. Structure determination of *D. gigas* AOR, now in progress, will help in the understanding of most of these questions.

[46] D. R. Kremer and T. A. Hensen, *Arch. Microbiol.* **147,** 249 (1987).

[5] Nickel-Iron Hydrogenase

By RICHARD CAMMACK, VICTOR M. FERNANDEZ, and E. CLAUDE HATCHIKIAN

Introduction

The [Ni-Fe] hydrogenases are the most commonly found hydrogenases in sulfate-reducing bacteria.[1-3] Even in those bacteria that contain hydrogenases of more than one type, one of the hydrogenases is generally of the [Ni-Fe] type. In species such as *Desulfovibrio gigas*, in which the [Ni-Fe] hydrogenase is the only hydrogenase found, this type of hydrogenase appears to be capable of both hydrogen production and consumption under different circumstances. H_2 production occurs during fermentation of pyruvate, and is involved in interspecies hydrogen transfer, while hydrogen uptake is used during growth on H_2 plus sulfate. The [Ni-Fe] hydrogenases, when isolated, are found to catalyze both hydrogen production and uptake, with low-potential multiheme cytochromes such as cytochrome c_3 acting as either electron donors or acceptors, depending on their oxidation state.[4] The enzymes can also use artificial dyes such as methyl viologen as electron donors or acceptors. The [Ni-Fe] hydrogenases contain a nickel center, which is believed to be the site at which hydrogen binds, and iron-sulfur clusters that probably serve as secondary electron carriers.[5-7]

In this chapter we first describe the isolation of [Ni-Fe] hydrogenase and its physical properties. Next, we detail various methods for estimation of its activity. The apparent activity of the [Ni-Fe] hydrogenases is vari-

[1] J. J. G. Moura, M. Teixeira, I. Moura, and J. LeGall, *in* "Bioinorganic Chemistry of Nickel" (J. R. Lancaster, Jr., ed.), p. 191. VCH Publ., Deerfield Beach, Florida, 1988.

[2] G. Fauque, H. D. Peck, Jr., J. J. G. Moura, B. H. Huynh, Y. Berlier, D. V. DerVartanian, M. Teixeira, A. E. Przybyla, P. A. Lespinat, I. Moura, and J. LeGall, *FEMS Microbiol. Rev.* **54,** 299 (1988).

[3] E. C. Hatchikian, V. M. Fernandez, and R. Cammack, *in* "Microbiology and Biochemistry of Strict Anaerobes Involved in Interspecies Hydrogen Transfer" (J.-P. Bélaich, M. Bruschi, and J.-L. Garcia, eds.), p. 53. Plenum, New York and London, 1990.

[4] H. G. R. Bell, J. LeGall and H. D. Peck, Jr., *J. Bacteriol.* **120,** 994 (1974).

[5] R. Cammack, V. M. Fernandez, and K. Schneider, *in* "Bioinorganic Chemistry of Nickel" (J. R. Lancaster, Jr., ed.), p. 167. VCH Publ., Deerfield Beach, Florida, 1988.

[6] M. Teixeira, I. Moura, A. V. Xavier, J. J. G. Moura, J. LeGall, D. V. DerVartanian, H. D. Peck, Jr., and B. H. Huynh, *J. Biol. Chem.* **264,** 16435 (1989).

[7] R. Cammack, *in* "Bioinorganic Catalysis" (J. Reedijk, ed.), p. 189. Dekker, New York, 1993.

able, depending on the way in which they are prepared; the enzyme may be reversibly deactivated and reactivated. To understand the catalytic and spectroscopic properties of the enzymes, it is important to understand this behavior. The changes in activity are interpreted in terms of (at least) three different forms of the enzyme, which differ in the state of the nickel center.[8] Methods are described for the preparation of the enzyme in these three forms. The spectroscopic properties of the three forms are summarized, with emphasis on electron paramagnetic resonance (EPR) spectra, which are particularly diagnostic of the state of the various centers in the enzyme.

Other types of hydrogenases, the [Ni-Fe-Se] and [Fe] hydrogenases, are described in other chapters of this volume.[9,10] The assay methods are applicable to all these hydrogenases but some properties such as activation phenomena are different. The description here is based principally on the widely studied hydrogenase from *D. gigas*.[11] This enzymc is found in the periplasmic space of the cells and appears to be loosely associated with the membrane. Other [Ni-Fe] hydrogenases such as that from *Desulfomicrobium baculatum* Norway 4, are more tightly membrane bound and require detergents for solubilization.[12] The enzymes from other species differ somewhat in their activity, stability, and donor/acceptor specificity.

Growth Conditions and Extraction of Hydrogenase

Desulfovibrio gigas (NCIB 9332) is grown at 37° in a basic lactate–sulfate medium.[13] The stock cultures are kept in Hungate tubes under an atmosphere of argon. For large-scale cultures, the bacteria are grown in a 300-liter fermenter and cells are harvested at the stationary phase after 44 hr of growth.

The purification is carried out at 4° under aerobic conditions and, unless otherwise stated, the buffers used are Tris(hydroxymethyl)aminomethane-HCl (Tris-HCl) buffer and potassium phosphate buffer at pH 7.6. Hydrogenase is monitored in the extracts by its enzymatic activity and its absorbance ratio ($A_{400\text{ nm}}/A_{280\text{ nm}}$) in the final steps of purification.

[8] V. M. Fernandez, E. C. Hatchikian, and R. Cammack, *Biochim. Biophys. Acta* **832,** 69 (1985).

[9] D. S. Patil, this volume [6].

[10] B. H. Huynh, this volume [38].

[11] E. C. Hatchikian, M. Bruschi, and J. LeGall, *Biochem. Biophys. Res. Commun.* **82,** 451 (1978).

[12] W. V. Lalla-Maharajh, D. O. Hall, R. Cammack, and J. LeGall, *Biochem. J.* **209,** 445 (1983).

[13] E. C. Hatchikian, this volume [19].

The periplasmic location of *D. gigas* [Ni-Fe] hydrogenase[4,11,14,15] allows the extraction of the enzyme from whole cells by washing them with slightly alkaline buffer. Freshly harvested cells (260 g wet weight) are rapidly frozen at −20°. After storage for 2–3 days at this temperature, the cells are thawed and carefully suspended in 280 ml of 40 m*M* Tris-HCl buffer (pH 7.6) containing 5 m*M* ethylenediaminetetraacetic acid (EDTA) and the bacterial suspension is homogenized with a stirring rod as previously reported for the extraction of cytochrome c_3 (M_r 13,000).[16] The suspension is then centrifuged at 35,000 *g* for 30 min and the reddish-brown supernatant is collected. The wash fraction is subsequently centrifuged at 140,000 *g* for 2 hr to remove sulfide present in the extract and the supernatant is used as the source of the enzyme. This procedure allows the extraction of approximately 35% of the total hydrogenase activity present in the cells. The yield of *D. gigas* [Ni-Fe] hydrogenase extraction is clearly lower than that obtained with the periplasmic [Fe] hydrogenase from *Desulfovibrio vulgaris* (98%).[17] The extraction of the enzyme requires slightly alkaline buffer, but no improvement is obtained with pH values higher than 7.6 as well as with higher EDTA concentration. Freezing and thawing of cells are required to obtain maximum extraction. In these extraction conditions, some lysis of cells estimated at about 8% from absorbance by bisulfite reductase (desulfoviridin) at 628 nm,[9] is observed.

Purification of the Hydrogenase of *Desulfovibrio gigas*

The purification procedure is summarized in Table I.

Silica Gel Fractionation

A settled volume of silica gel [Merck (Rahway, NJ) type 60 GF_{254}], equal to 40 ml and equilibrated in 10 m*M* Tris-HCl, is added to the brownish-red wash fraction previously obtained and the mixture is stirred for 1 hr at 4°. This suspension is subsequently poured into a column and the unadsorbed protein fraction containing hydrogenase is collected. During this step, a part of cytochrome c_3 (M_r 13,000) present in the extract is tightly bound to silica gel and separated from hydrogenase.

[14] J. M. Odom and H. D. Peck, Jr., *J. Bacteriol.* **147,** 161 (1981).

[15] V. Nivière, A. Bernadac, N. Forget, V. M. Fernandez, and E. C. Hatchikian, *Arch. Microbiol.* **155,** 579 (1991).

[16] J. LeGall, G. Mazza, and N. Dragoni, *Biochim. Biophys. Acta* **99,** 385 (1965).

[17] H. Van der Western, S. G. Mayhew, and C. Veeger, *FEBS Lett.* **86,** 122 (1978).

TABLE I
TYPICAL PURIFICATION SCHEME OF NICKEL-IRON HYDROGENASE FROM *Desulfovibrio gigas*

Step	Volume, (ml)	Total protein (mg)	Total activity (units)	Specific activity (units/mg)	Recovery (%)
1. Wash fraction[a]	320	2,530	15,200	6.0	100.0
2. Silica gel fractionation	330	2,380	13,800	5.8	90.8
3. First DEAE-cellulose column chromatography	340	1,365	13,100	9.6	86.2
4. Second DEAE-cellulose column chromatography	85	433	11,600	26.8	76.3
5. Ultrogel AcA 44 gel filtration	65	79	9,800	124.1	64.5
6. Hydroxylapatite column chromatography	42	50	8,500	170	55.8
7. Preparative HPLC on TSK DEAE-5PW column	22	40	7,200	180	47.4

[a] From 260 g of cells (wet weight).

First DEAE-Cellulose Column Chromatography

The hydrogenase-containing fraction is loaded onto a DEAE-cellulose [Whatman (Clifton, NJ) DE-52] column (3 × 6 cm) previously equilibrated with 40 m*M* Tris-HCl, pH 7.6. The column is washed with 500 ml of 40 m*M* Tris-HCl and the unadsorbed proteins containing hydrogenase activity are collected in a volume of 340 ml.

Second DEAE-Cellulose Column Chromatography

The active fraction is then adsorbed on a DEAE-cellulose column (4 × 20 cm) equilibrated with 40 m*M* Tris-HCl. The column is washed with 100 ml of the same buffer and the enzyme is eluted by means of a discontinuous gradient (600 ml) of Tris-HCl (50 to 300 m*M*, in 50 m*M* steps). The hydrogenase is eluted from the column with 150 m*M* Tris-HCl and collected in a volume of 85 ml.

Ultrogel AcA 44 Gel Filtration

The previous hydrogenase-containing fraction is concentrated to 35 ml by ultrafiltration using a PM-30 membrane (Amicon, Danvers, MA) and filtered through an Ultrogel AcA 44 (IBF) column (5 × 100 cm) equilibrated with 10 m*M* Tris-HCl. The flow rate of the column is 35 ml/hr and the protein is collected in 8-ml fractions. Most of the brownish band of hydrogenase is separated from cytochrome c_3 (M_r 13,000) at this step.

Hydroxylapatite Column Cromatography

The subsequent hydrogenase fraction is adsorbed onto a hydroxylapatite [Bio-Rad (Richmond, CA) BioGel HTP] column (2.6 × 8 cm) equilibrated with 10 m*M* Tris-HCl. The slight amount of cytochrome c_3 (M_r 13,000) still present in the preparation is eluted using a linear gradient from 10 m*M* Tris-HCl to 50 m*M* potassium phosphate, pH 7.6. The volume of each buffer is 200 ml. Hydrogenase is then eluted from the column with 100 m*M* potassium phosphate. At this stage, hydrogenase is 97% pure and exhibits an absorbance ratio $A_{400}/A_{280} = 0.25$. Further purification is achieved using an ion-exchange preparative high-performance liquid chromatography (HPLC) column.

Preparative HPLC on TSK DEAE-5PW Column

Fractions of approximately 25 mg of protein, pooled from the hydroxylapatite column, are diluted with 2 vol of distilled water and applied to an UltroPac TSK DEAE-5PW column (LKB, Bromma, Sweden) (2.15 × 15 cm), equilibrated with 50 m*M* Tris-HCl. The flow rate is 1.5 ml/min throughout the whole procedure. The column is washed with the same buffer for 30 min and a linear gradient from 50 m*M* Tris-HCl to 150 m*M* NaCl, 50 m*M* Tris-HCl is applied over 150 min. The enzyme is eluted at 75 m*M* NaCl, 50 m*M* Tris-HCl from the preparative HPLC column. At this stage, the hydrogenase preparation exhibits an absorbance ratio A_{400}/A_{280} of 0.26 and the specific activity is close to 180 units/mg, using the manometric assay described below. Note that the activities given during the purification are for the unactivated enzyme (see below p. 57). The enzyme is judged to be pure by both polyacrylamide gel electrophoresis (PAGE) and analytical ultracentrifugation and it is suitable for crystallization assays. The yield is usually 30–50 mg.

Crystallization

The crystallization procedure is taken from that described by Nivière *et al.*[18] For crystallization experiments, the protein solution obtained from the preparative HPLC column is concentrated through centricon-30 (Amicon) with M_r 30,000 cutoff and the protein is subsequently dialyzed by diafiltration against 50 m*M* cacodylate buffer, pH 6.5. The final protein content is 6–8%. Two crystal forms are obtained both at pH 6.5, using the hanging drop method.[19] Droplets (10 μl) of protein solution are placed

[18] V. Nivière, E. C. Hatchikian, C. Cambillau, and M. Frey, *J. Mol. Biol.* **195,** 969 (1987).
[19] A. McPherson, *Methods Biochem. Anal.* **23,** 249 (1976).

on silicon-treated coverslips and vapor equilibrated with 1 ml of reservoir solution.

Form A. Light brown needle-like crystals (0.1 × 0.1 × 1.0 mm) are obtained in hanging drops after several days to 1 week at 4° with 230 m*M* cacodylate, 6–9% (w/v) polyethylene glycol 6000 in precipitant reservoirs, and 10-μl droplets made from 5 μl of the protein solution (40 mg/ml, 50 m*M* cacodylate) filtered through 0.22-μm pore size filters (Millipore, Bedford, MA) with 5 μl of the reservoir solution. These crystals are orthorhombic and diffract to only 6-Å resolution. The cell dimensions and probable space group are determined to be $C222_1$ or $C222$, with $a = 125.0$ Å, $b = 200.0$ Å, and $c = 136$ Å.

Form B. A protein solution (40 mg/ml) is prepared in 40% saturated ammonium sulfate, buffered with 50 m*M* cacodylate at pH 6.5, and subsequently filtered through 0.22-μm pore size filters. Light-brown, platelike crystals (0.6 × 0.4 × 0.10 mm) appear after 2 weeks at 4° with 50 m*M* cacodylate, 53 to 57% saturated ammonium sulfate in precipitant reservoirs and 10-μl droplets of the protein solution. The crystals are monoclinic, with space group $C2$, and cell parameters $a = 257.0$ Å, $b = 184.7$ Å, $c = 148.3$ Å, and $b = 101.3°$. The diffraction pattern extends beyond 3-Å resolution when a synchrotron radiation source is used. These crystals are stable for 48 hr in the X-ray beam of a rotating anode operating at 5 kW.

Isotope Substitution Methods

For some spectroscopic investigations it is necessary to produce enzyme enriched in isotopes such as ^{61}Ni or ^{57}Fe. Because it is not possible to remove these elements and replace them *in vitro*, with retention of enzyme activity, it is necessary to grow the organism in a medium enriched in that isotope. The method used here is for enrichment with ^{61}Ni, which was necessary for unequivocal identification of EPR signal from nickel.[20–22] Hydrogenase is isolated from cells grown on the basic lactate-sulfate medium supplemented with ^{61}Ni. The content of natural nickel (isotopes ^{58}Ni and ^{60}Ni) in the medium is approximately 1 μM as measured by atomic absorption. The residual natural nickel isotopes originate from impurities present in the components of the medium, including mainly potassium phosphate, sodium lactate, and yeast extract. The medium is supplemented with 550 μg/liter of 94.91% enriched ^{61}Ni (AERE Harwell,

[20] J. R. Lancaster, *Science* **216,** 1324 (1982).

[21] S. P. J. Albracht, E. G. Graf, and R. K. Thauer, *FEBS Lett.* **140,** 311 (1982).

[22] J. J. G. Moura, I. Moura, B. H. Huynh, H.-J. Krüger, M. Teixeira, R. C. DuVarney, D. V. DerVartanian, A. V. Xavier, H. D. Peck, Jr., and J. LeGall, *Biochem. Biophys. Res. Commun.* **108,** 1388 (1982).

Oxfordshire, United Kingdom). This corresponds to a nickel concentration of 9 μM and the enrichment in the culture medium is estimated to be 83%, as judged by the EPR spectrum. The bacteria are grown in 20-liter flasks and the purification of ^{61}Ni-enriched hydrogenase is carried out from approximately 25 g of bacteria (wet weight). The purification procedure is modified as follows: after thawing, the cells are washed twice with 30 ml of 40 mM Tris-HCl buffer (pH 7.6) containing 5 mM EDTA and the suspension is centrifuged as reported above. The purification of hydrogenase can be achieved in three steps using only step 3 of the purification procedure (first DEAE-cellulose column, 1.8 × 3.5 cm), step 5 (Ultrogel AcA 44 column, 2.5 × 90 cm), and step 6 (hydroxylapatite column, 1.8 × 5 cm) as described previously. This procedure gives 10 mg of pure ^{61}Ni-enriched hydrogenase with a specific activity of 190 units/mg, using the manometric H_2 evolution assay described below. Hydrogenase isolated from cells grown in ^{61}Ni-enriched medium contains 1.05 g-atom of nickel and 11.2 g-atom of iron per molecule, respectively.

Physical and Chemical Properties

Hydrogenase from *D. gigas*, pure by electrophoresis criteria, exhibits the properties listed in Table II.[11,23–28] The molecular weight of hydrogenase determined by analytical ultracentrifugation at equilibrium sedimentation,[29] was estimated to be 89,500. Sodium dodecyl sulfate (SDS) gel electrophoresis reveals that the enzyme contains two subunits of approx M_r 28,000 and 62,000, respectively. These values fit well with the molecular weight of the subunits (28,420 and 61,477) calculated from their primary structure derived from the nucleotide sequence.[28]

The enzyme contains 1 nickel atom, 11 iron atoms, and 11 sulfide groups per molecule. Electron paramagnetic resonance analysis and Möss-

[23] R. Cammack, D. S. Patil, R. Aguirre, and E. C. Hatchikian, *FEBS Lett.* **142,** 289 (1982).

[24] J. LeGall, P. O. Ljungdahl, I. Moura, H. D. Peck, Jr., A. V. Xavier, J. J. G. Moura, M. Teixeira, B. H. Huynh, and D. V. DerVartanian, *Biochem. Biophys. Res. Commun.* **106,** 610 (1982).

[25] M. Teixeira, I. Moura, A. V. Xavier, B. H. Huynh, D. V. DerVartanian, H. D. Peck, Jr., J. LeGall, and J. J. G. Moura, *J. Biol. Chem.* **260,** 8942 (1985).

[26] V. M. Fernandez, E. C. Hatchikian, D. S. Patil, and R. Cammack, *Biochim. Biophys. Acta* **883,** 145 (1986).

[27] B. H. Huynh, D. S. Patil, I. Moura, M. Teixeira, J. J. G. Moura, D. V. DerVartanian, M. H. Czechowski, B. C. Prickril, H. D. Peck, Jr., and J. LeGall, *J. Biol. Chem.* **262,** 795 (1987).

[28] G. Voordouw, N. K. Menon, J. LeGall, E.-S. Choi, H. D. Peck, Jr., and A. E. Przybyla, *J. Bacteriol.* **171,** 2894 (1989).

[29] D. A. Yphantis, *Biochemistry* **3,** 297 (1964).

TABLE II
PHYSICAL AND CHEMICAL PROPERTIES OF *Desulfovibrio gigas* NICKEL-IRON HYDROGENASE

Property	Value
Molecular weight	89,500
Subunits	2 (61,477 and 28,420)
p*I*	6.0
Nickel (atoms/mol)	1
Iron (atoms/mol)	11
Acid-labile sulfide (mol/mol)	11
Half-cystine (mol/mol)	21
A_{400} (mM^{-1} cm^{-1})	
Oxidized form	46.5
Reduced form	31.52
Iron-sulfur clusters	2 × $[4Fe\text{-}4S]^{2+/1+}$ 1 × $[3Fe\text{-}4S]^{1+/0}$

bauer studies indicate that the iron atoms and acid-labile sulfide present in the protein are assembled in two $[4Fe\text{-}4S]^{2+/1+}$ and one $[3Fe\text{-}4S]^{1+/0}$ cluster per molecule.[6,27] The protein exhibits a p*I* of 6.0, in agreement with its amino acid composition, which shows a preponderance of acidic amino acids.[11,28] It contains 21 cysteine residues, which is higher than the number required to chelate the three iron-sulfur clusters.

The optical absorption spectrum of oxidized *D. gigas* hydrogenase is typical of a nonheme iron protein, with a broad absorption peak around 400 nm and a slight shoulder in the 310-nm region. The extinction coefficients of the oxidized enzyme at 400 and 280 nm are 46.5 and 178 mM^{-1} cm^{-1}, respectively. Reduction of the enzyme under H_2 results in a decrease of A_{400} of about 33%.

Assay Methods

Four general methods may be used for the assay of hydrogenases, based on their ability to catalyze the following reactions.

1. Evolution of H_2:

$$D_{red} + 2H^+ \rightarrow D_{ox} + H_2 \qquad (1)$$

where the electron donor D is a low-potential compound such as cytochrome c_3 or methyl viologen.

2. Oxidation of H_2:

$$H_2 + A_{ox} \rightarrow 2H^+ + A_{red} \qquad (2)$$

where the electron acceptor A may be either a low-potential compound such as cytochrome c_3 or methyl viologen, or a higher potential compound such as methylene blue or 2,6-dichloro-indophenol (DCIP).

3. Deuterium or tritium exchange reactions with H^+, in the absence of electron donors or acceptors. The following is an example with deuterium:

$$^2H_2 + 2\,^1H_2O \rightarrow {}^1H_2 + 2\,^1H^2HO \quad (3)$$

and

$$^2H_2 + {}^1H_2O \rightarrow {}^1H^2H + {}^1H^2HO \quad (4)$$

The proportion of these two reactions depends on the type of hydrogenase, and other conditions such as pH.[2,30]

4. Conversion of *para* to *ortho* H_2[31]:

$$pH_2(\uparrow\downarrow) + H_2O \rightarrow oH_2(\uparrow\uparrow) + H_2O \quad (5)$$

A previous chapter in this series[32] describes the principles of these reactions and describes detailed procedures for the assay of hydrogenases, based on (1) manometric measurement of hydrogen evolution or uptake, (2) counting, with a gas ionization chamber, of the radioactivity of tritium evolved by the exchange reaction of hydrogen with tritiated water, (3) mass spectrometric analysis of masses 2, 3, and 4 evolved from the exchange reaction between normal hydrogen with deuterated water, and (4) measurement of the change in thermal conductivity that accompanies the interconversion of *para* and *ortho* H_2 gas.

Another method is based on the analysis by gas chromatography of hydrogen evolved by hydrogenase from solutions of dithionite-reduced methyl viologen.[33] All these methods are limited by being based on discrete sampling. The methods we describe here for the study of [Ni-Fe] hydrogenases from *Desulfovibrio* species, with the exception of the hydrogen/tritium exchange reaction, use continuous monitoring. They have proved useful in the characterization of the different states of activity of this group of hydrogenases. The assays will be grouped according to the instrumental technique utilized.

Amperometric Assays (Hydrogen Electrode)

Since its introduction in 1956 by Peck and Guest,[34] the measurement of hydrogen from dithionite-reduced methyl viologen has been a pop-

[30] Y. M. Berlier, G. D. Fauque, J. LeGall, P. A. Lespinat, and H. D. Peck, *FEBS Lett.* **221,** 241 (1987).

[31] Y. Berlier, P. A. Lespinat, and B. Dimon, *Anal. Biochem.* **188,** 427 (1990).

[32] A. I. Krasna, this series, Vol. 53, p. 296.

[33] K. K. Rao, L. Rosa and D. O. Hall, *Biochem. Biophys. Res. Commun.* **68,** 21 (1976).

[34] H. D. Peck, Jr., and H. Gest, *J. Bacteriol.* **71,** 70 (1956).

ular method of hydrogenase assay. The rate of H_2 production has been estimated manometrically in a Warburg apparatus[32,35] or with a gas chromatograph.[33] The amperometric determination of hydrogen with a Clark-type electrode[36] has made it possible to measure the rate continuously.

A description of the functioning of the electrode, and conditioning procedures to obtain more stable responses, have been published in this series.[37] An inexpensive circuit for the measurement of hydrogen with a commercially available Clark-type electrode is described in the paper by Sweet *et al.*[38]

We have used this technique to gain insight into the activation–deactivation processes of [Ni-Fe] hydrogenase from *D. gigas* and other *Desulfovibrio* species. Important information may be obtained from the early phase of the curve of H_2 evolution.[39] For this purpose, we have used a thermostatted reaction vessel similar to that described by Wang,[37] modified with a lateral port connected to a nitrogen line and with a magnetic stirrer bar included. Into this cell a Clark-type electrode (type 4004; Yellow Springs Instruments) was fitted. Two different methods were used, depending on the way of reduction of methyl viologen.

Assay with Dithionite-Reduced Methyl Viologen. Dithionite is a sufficiently strong reducing agent for H_2 production, and is capable of reducing the metal centers in the enzyme. However, the reduction rate is too slow for a convenient assay of enzyme activity, hence the need for electron-transferring mediators such as methyl viologen.

A typical experiment is run as follows: 5.5 ml of the appropriate buffer, containing 1 m*M* methyl viologen, is placed in the vessel and deaerated by passing N_2 through the lateral port. After 5 min of bubbling with vigorous stirring, both the N_2 flow and agitation are stopped and the glass stopper is inserted, taking care to avoid any residual N_2 bubbles. Through the capillary channel that traverses the glass male stopper an aliquot of a sodium dithionite solution (final concentration 15 m*M*, prepared under nitrogen) is added, to reduce the methyl viologen. The stirrer is turned on and the reaction started by addition of the enzyme. The evolved H_2 is measured with the Clark electrode.

It has been observed that the response of the electrode to the hydrogen

[35] L. E. Mortenson, this series, Vol. 53, p. 286.

[36] R. T. Wang, F. P. Healey, and J. Myers, *Plant Physiol.* **48,** 108 (1971).

[37] R. T. Wang, this series, Vol. 69, p. 409.

[38] W. J. Sweet, J. P. Houchins, P. R. Rosen, and D. J. Arp, *Anal. Biochem.* **107,** 337 (1980).

[39] V. M. Fernandez, R. Aguirre, and E. C. Hatchikian, *Biochim. Biophys. Acta* **790,** 1 (1984).

concentration decreases in a stepwise manner during a series of experiments. Therefore after each hydrogenase assay the electrode is recalibrated by addition of H_2-saturated water.

Assay with Electrochemically Reduced Methyl Viologen. A serious drawback of the use of dithionite as a reductant for methyl viologen in this assay is the pH dependence of its reduction potential, in contrast to the pH independence of the oxidized/reduced methyl viologen (MV^{2+}/MV^{+}) potential. As a consequence the concentration of MV^{+} in equilibrium with a fixed concentration of dithionite is pH dependent; dithionite is inefficient in the reduction of viologen at pH values below 6.5. Electrochemical reduction does not suffer from this limitation and is an efficient and clean alternative to dithionite. A schematic diagram of an electrochemical cell[40] that incorporates a Clark-type electrode for polarographic measurement of H_2 is given in Fig. 1. With this cell the electrochemical reduction is carried out *in situ*, and hydrogenase activity can be monitored continuously in a wide range of pH and redox potentials. This design allows the MV^{2+} produced by the reaction to be recycled, keeping the concentration of MV^{+} constant during the assay. This point is important because the usual concentration of MV^{+} in most hydrogenase assays is 1 m*M*, a value far below the saturating concentration. The cell, similar to that described above, incorporates three electrodes required for the electrochemical reduction of MV^{2+}. The electrical bridges are made of 3% agar in 0.5 m*M* KCl; vycor tips (Corning Glass 6, Princeton Applied Research, N.J.) are a convenient alternative. The reference electrode is a saturated calomel electrode (Radiometer), and a platinum wire serves as the auxiliary electrode. As a working electrode a gold minigrid (Buckbee-Mears, St. Paul, MN) was chosen because of its high specific area. Moreover, the large overpotential of gold for H^{+} reduction allows the reduction of MV^{2+} over a large pH range without simultaneous H_2 evolution. The gold minigrid is welded to a gold wire that traverses the glass male stopper by the capillary channel. The redox potential is controlled with a potentiostat (Wenking LB 75). The hydrogen electrode and control are as described in the previous section.

The buffer containing MV^{2+} is degassed as described above. After closing the cell, viologen reduction is started and allowed to continue until the current is negligible (about 20 μA). At an applied potential of -505 mV vs. standard hydrogen electrode (SHE), the reduction of 1 m*M* methyl viologen is accomplished in about 30 min. After checking that the rest potential coincides with the imposed potential, the enzymatic reaction is started by addition of the hydrogenase. Hydrogen evolution is measured with the Clark electrode as before.

[40] V. M. Fernandez, *Anal. Biochem.* **130,** 54 (1983).

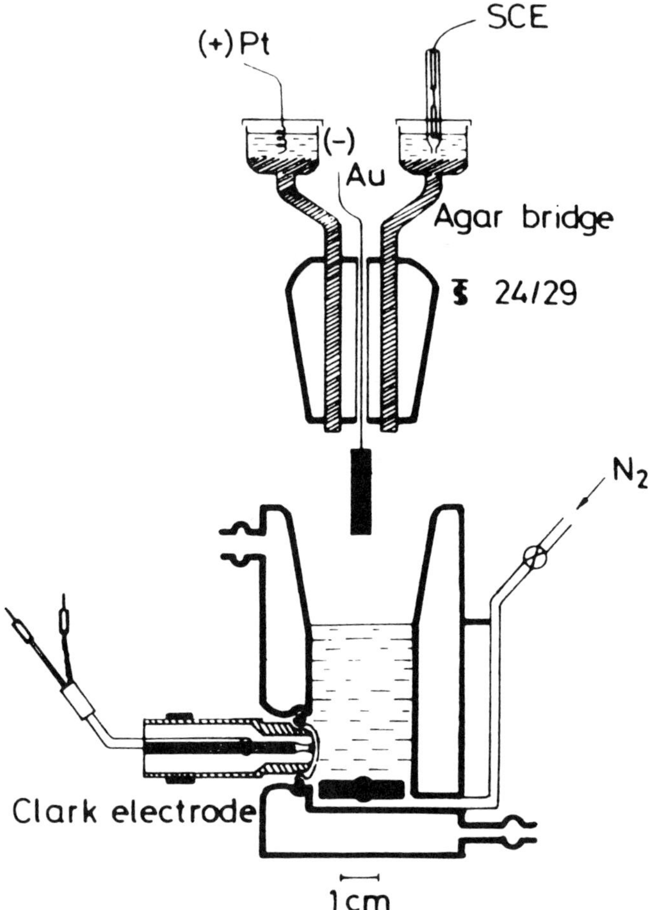

FIG. 1. Electrochemical cell showing the Clark electrode for H_2 measurements and the three electrodes required for electrochemical reduction of MV^{2+}. The taper is a standard 24/29 glass joint. (Reproduced, with permission, from Ref. 40.)

Mass spectrometric assays

Deuterium Exchange Reaction. In this assay, the evolution of masses $m/z = 2$ ($^1H_2$), 3 ($^1H^2H$), and 4 ($^2H_2$) in reactions (3) and (4) is followed. The method is based on the analysis of dissolved gases admitted to a mass spectrometer via a membrane inlet, as first described by Hoch and Kok.[41]

[41] G. Hoch and B. Kok, *Arch. Biochem. Biophys.* **101,** 160 (1963).

The inlet membrane system has been described for the measurement of oxygen exchange in algae.[42]

Nickel-iron hydrogenases after reductive activation are sensitive to trace amounts of oxygen and precautions are required for good anaerobiosis. The gases employed are deoxygenated by passage over a BASF catalyst (R3-11) (BASF, Actiengesellschaft, Ludwigshafen, Germany) heated at 130°. A thermostatted reaction chamber especially designed for anaerobic conditions is manufactured by Secia Industry (Manosque, France). The volume of the reaction mixture is adjustable between 3 and 10 ml by displacement of a plunger (15-mm diameter and 100-mm height). The bottom of the vessel consists of a stainless steel grid covered with a 12.5-μm-thick Teflon membrane (Du Pont, Wilmington, DE). Details of the construction are provided by Jouanneau *et al.*[43] During the assay the reaction vessel is kept anaerobic by flushing argon through an external circuit. Routinely 10 ml of buffer is sparged with a gas mixture of 20% $^2H_2$, 80% argon, and stirred with a magnetic stirrer until the 32-amu signal declines fully. Next, the plunger is lowered to eliminate the gas phase, taking care to avoid all bubbles. The reaction is started by addition of enzyme by syringe through a rubber septum. If strict anaerobiosis is required, a trace of dithionite solution is added before the enzyme. Alternatively an oxygen scavenger system of glucose and glucose oxidase may be used; in this case, to avoid foaming, glucose oxidase is added after the system is closed, and 5 min before hydrogenase is added.

The formation of $^1H^2H$ and $^1H_2$ is followed on-line by mass spectrometry. The dissolved gases diffuse through the Teflon membrane to a vacuum line, connected to the ion source of a mass spectrometer (Masstorr 200 DX quadrupole from VG Quadrupoles, Ltd., Middlewich, Cheshire, United Kingdom). The vacuum in the line is maintained with a rotary/oil vapor diffusion pump. The gases are passed through a liquid nitrogen cold trap, where water vapor is condensed, into the ion source of the spectrometer. Several masses may be monitored simultaneously with a peak-jumping system controlled by a microcomputer. To estimate the initial rates of formation of $^1H^2H$ and H_2 the system is calibrated with saturated solutions of $^1H^2H$ and H_2 in ^1H_2O at 25°. At a partial pressure of 10^5 Pa the concentration of $^2H_2$ in water is 0.811 m*M* and that of $^1H_2$ is 0.758 m*M*[44]; the concentration of $^1H^2H$ is assumed to be 0.785 m*M*.

Hydrogen Evolution Reaction. The setup just described can be used to monitor the evolution of hydrogen from reduced methyl viologen by

[42] R. Radmer and O. Ollinger, this series, Vol. 69, p. 547.

[43] Y. Jouanneau, B. C. Kelley, Y. Berlier, P. A. Lespinat, and P. M. Vignais, *J. Bacteriol.* **143,** 628 (1980).

[44] J. Muccitelli and W. Y. Wen, *J. Solution Chem.* **7,** 257 (1978).

TABLE III
ELECTRON ACCEPTORS OF NICKEL-IRON HYDROGENASES

Electron acceptor	$Em_{7,NHE}$ (mV)	Wavelength (nm) Oxidized	Wavelength (nm) Reduced	ε (mM^{-1} cm^{-1})
Cytochrome c_3 (*D. gigas*)	−195, −295, −315, −330		552	111
Methyl viologen	−446		604	13.9
Benzyl viologen	−358		604	8.4
Methylene blue	11	600		32.8
Phenazine methosulfate	80	387		23.8
DCIP	217	600		20.6
Ferricyanide, $K_3Fe(CN)_6$	430	420		1.04

looking at the mass $m/z = 2$. The assay procedure is the same as that described for the exchange reaction, except that after bubbling the buffer with argon, a solution of sodium dithionite is added to reduce MV^{2+}. After closing the cell the reaction is started by the injection of the enzyme, and the formation of $^1H_2$ is monitored. Alternatively, at acid pH values where dithionite does not reduce MV^+ effectively, electrochemically reduced methyl viologen may be used. In this case, MV is reduced in an electrochemical cell with a mercury cathode as described by Thorneley.[45] In this cell 60 ml of 1 m*M* methyl viologen can be reduced in about 2 hr. Aliquots of MV^+ can be removed from the cell with a glass syringe through a rubber septum, and transferred to the anaerobic vessel of the mass spectrometer that has previously been gassed with argon. The plunger is then lowered, the magnetic stirrer turned on, and the reaction started by addition of enzyme.

Spectrophotometric Assays

Reduction of Electron Acceptors with H_2. Besides cytochrome c_3, which has been suggested to be the natural substrate for [Ni-Fe] hydrogenases of *Desulfovibrio* species, some organic dyes are reduced with H_2 by these enzymes. The reduction of these compounds by H_2 may be followed spectrophotometrically. Their relevant properties are described in Table III.

The procedure to be described here is for the measurement of the rate of reduction of methyl viologen:

$$H_2 + 2MV^{2+} \rightarrow 2H^+ + 2MV^+ \tag{6}$$

[45] R. N. F. Thorneley, *Biochim. Biophys. Acta* **333,** 487 (1974).

The reaction mixture, in a cuvette with a serum stopper, containing 200 μmol of the appropriate buffer, 2 μmol of methyl viologen, 20 μmol of glucose, 3 μl of ethanol, and 1000 units of catalase, in a final volume of 2.0 ml, is bubbled with hydrogen gas through a fine-syringe needle for 20 min. Then 20 μl of a 1-mg/ml solution of glucose oxidase is added and the cell is incubated at 30°. The reaction is started by addition of hydrogenase and the absorbance at 604 nm is followed. For studies of enzyme activation it is necessary to assay concentrated enzyme, such as is used for EPR spectroscopy; the enzyme is not diluted before introducing it into the assay mixture. In this case the volume of the assay medium in the cuvette is increased to 6.0 ml and a small volume (≥0.05 μl) is transferred with a plunger-in-needle microsyringe (total capacity, 0.5 μl; SGE, Ltd, Milton Keynes, U.K.).

In this assay, nonactivated hydrogenase shows an induction period. This is suppressed if, before adding the enzyme, small amounts of dithionite solution are added to produce an absorbance at 604 nm of about 0.1–0.2. This serves to remove residual traces of oxygen, and to reduce the hydrogenase at the start of the reaction. For assays with alternative electron acceptors (Table III), typical concentrations used are 0.1 m*M* DCIP, methylene blue, or phenazine methosulfate, or 4 μ*M* cytochrome c_3.

Manometric Assay

Manometric measurement of hydrogen evolution has previously been described in this series.[32] Here we detail the method used routinely in our protein purifications.[11] The reaction is carried out in 15-ml Warburg flasks under argon at 30°. Tris-HCl buffer at pH 8.0 (50 m*M*) is used in the assay and in the preparation of all the reagents, which are degassed and flushed with argon. All the reagents except the enzyme are introduced in the main compartment of the Warburg flasks, which are flushed with argon. The reaction is started by tipping in the enzyme from the side arm. The 3 ml of reaction mixture contains 1 m*M* methyl viologen (3 μmol) and 15 m*M* sodium dithionite (45 μmol). The rate of hydrogen evolution is directly proportional to enzyme concentration when the amount of hydrogenase used is chosen so as to give values in the range of 10 to 25 μl of gas evolved per 10 minutes.

Radioassays

An assay for hydrogen exchange with tritiated water, based on the analysis of tritium in the gas phase with a gas ionization chamber, has been described.[32] The procedure described here is for the reverse reaction, for exchange of tritium gas into normal water:

$$^3H^1H + {}^1H_2O \rightarrow {}^1H_2\ {}^3H^1HO \quad (7)$$

For this assay the tritium-labeled hydrogen gas is conveniently produced by the reaction of lithium metal with tritiated water. The reaction mixture, in a 10-ml septum-sealed vial, contains 400 μmol of the appropriate buffer, 40 μmol of glucose, 10 μl of ethanol, and 1000 units of catalase in a final volume of 4.0 ml. The vial is flushed with argon for 5 min before 20 μl of a 1-mg/ml solution of glucose oxidase is added. Then 300 μl of tritium-labeled hydrogen gas is introduced with a gas-tight syringe (Hamilton Co., Bonaduz, Switzerland). After equilibration in a shaking water bath at 30°, duplicate 50-μl samples of the liquid phase are taken for liquid scintillation counting to provide a background reading. The reaction is started by the addition of the enzyme, and duplicate 50-μl samples are taken at intervals for counting. Samples are added to 5 ml of scintillant (Scintillator 299; Packard Instrument Co., Meriden, CT), and counted in the tritium channel of a liquid scintillation counter. Results are corrected for the decrease, due to sampling, in the amount of the enzyme present in the reaction vessel.

Tritium gas is prepared by controlled addition of lithium metal to tritiated water (1 mCi/ml; Amersham, Arlington Heights, IL) in a sealed and evacuated vessel connected to a vacuum gauge (Fig. 2). In the limb at the bottom of the cell, 1–2 ml of tritiated water is subjected to five freezing/vacuum/thawing cycles. By dipping the lithium-containing basket into the tritiated water, tritium gas is evolved and the reaction is allowed to continue until the desired pressure is reached.

The specific radioactivity of the tritium-labeled gas is determined by reacting a known volume with air over an incandescent platinum wire to produce tritiated water. This is done in a sealed glass vessel. The vapor produced is condensed in a side arm containing a known volume of water (stirred continuously), which is immersed in ice. The reaction is allowed to proceed for 30 min, and the water is cooled further by dipping the limb in liquid nitrogen, to condense any tritiated water vapor. Samples of the water are removed for liquid scintillation counting as previously described.

Catalytic Properties

One unit of enzyme activity is defined as the amount of hydrogenase catalyzing the evolution of 1 μmol of H_2 per minute under the assay conditions.

The natural redox partner of [Ni-Fe] hydrogenase in *D. gigas* is thought to be the tetraheme cytochrome c_3 (M_r 13,000). Both proteins are located

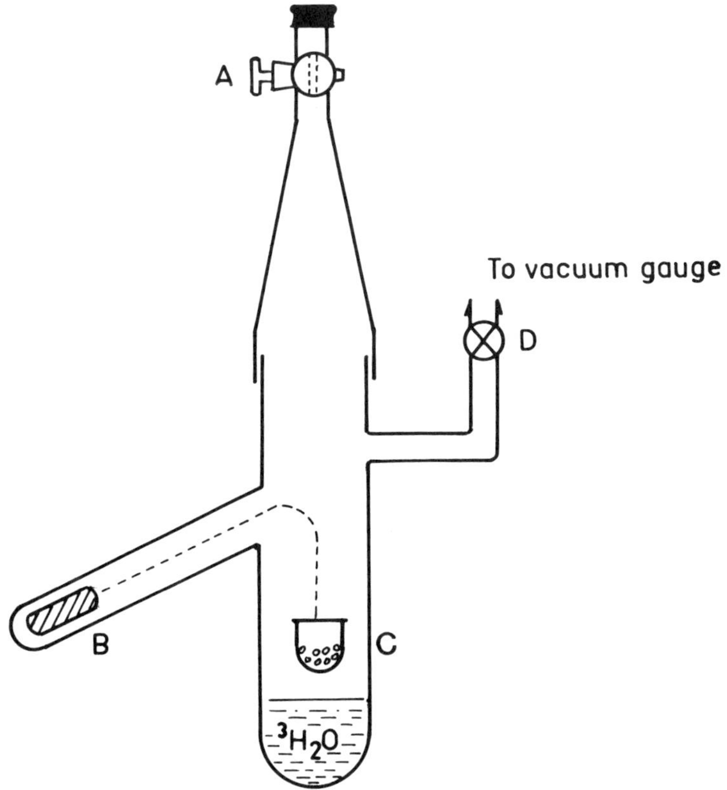

FIG. 2. Apparatus for generating tritium-labeled H_2 gas. The end of stopcock A fits a serum stopper and the cell is connected to a manifold through a hypodermic needle. A gold chain passes through the pierced end of a magnetic bar situated in the lateral arm (B); the other end of the chain is fixed to a perforated glass container (C), in which lithium metal (about 50 mg) is placed. The bar can be displaced in compartment B with the help of a magnet outside. Samples of the gas are removed with a syringe through the septum and central hole of stopcock A.

in the periplasm. With dithionite as reductant, the enzyme catalyzes H_2 formation in the presence of cytochrome c_3 (M_r 13,000). The K_m for reduced cytochrome c_3 (M_r 13,000) is 0.4 μM at pH 5.5.[46] In the absence of electron carriers, hydrogenase produces H_2 from dithionite but at a rate 30 times lower than with methyl viologen as carrier. Nickel-iron hydrogenase reduces cytochrome c_3 (M_r 13,000) under hydrogen but is inactive toward ferredoxin, flavodoxin, or rubredoxin in the absence of

[46] G. R. Bell, J.-P. Lee, H. D. Peck, Jr., and J. LeGall, *Biochimie* **60,** 315 (1978).

cytochrome c_3 (M_r 13,000).[46,47] The K_m of the reactivated *D. gigas* hydrogenase for oxidized cytochrome c_3 (M_r 13,000) is 7.4 μM at pH 7.6.[24] The second-order rate constant of electron transfer between [Ni-Fe] hydrogenase and cytochrome c_3 (M_r 13,000) is $k = 6.5 \times 10^7\ M^{-1}\ \sec^{-1}$ at 25° and pH 7.6.[48]

Activity States of Nickel-Iron Hydrogenases

For studies of the activity or spectroscopic properties it is necessary to prepare the enzyme such that the nickel and iron-sulfur clusters are in defined states of activation and oxidation/reduction. In this section, some methods for preparation of such samples are described. It is first necessary to understand the various types of changes that the metal centers can undergo, and their spectroscopic properties.

A difficulty in estimating the purity and yield of [Ni-Fe] hydrogenase preparations is the fact that the activity depends considerably on the history of the sample. Nickel-iron hydrogenases from several *Desulfovibrio* species isolated under normal aerobic conditions do not display activity in the hydrogen isotope exchange assay[49] even after prolonged flushing with oxygen-free argon or incubation with an oxygen scavenger system consisting of glucose and glucose oxidase. The same preparations show activity, although with variable kinetics, when assayed for H_2 evolution[39] or H_2 uptake with viologens.[49,50] All of these three activities are slowly but greatly stimulated by reductive treatments.[8,39,50,51] The apparently complex changes in activity of hydrogenases such as those from *D. gigas* have been interpreted in terms of interconversion between three states, designated the *unready*, *ready*, and *active* states. This interpretation may be explained by reference to Fig. 3. At present the reasons for the differences in activity are not known, but the procedures for interconversion between the various states have been investigated.[8]

The conversion of the [Ni-Fe] hydrogenase from *D. gigas* to the active state is an exceptionally slow process (half-life of activation at 4° about 30 hr) that is strongly temperature dependent, with a high activation energy of 88 kJ mol^{-1}.[8] The rate of the activation process was found to be virtually independent of the concentration of protein, and of the presence of thiore-

[47] M. R'zaigui, E. C. Hatchikian, and D. Benlian, *Biochem. Biophys. Res. Commun.* **92,** 1258 (1980).

[48] V. Nivière, E. C. Hatchikian, P. Bianco, and J. Haladjian, *Biochim. Biophys. Acta* **935,** 34 (1988).

[49] Y. M. Berlier, G. Fauque, P. A. Lespinat, and J. LeGall, *FEBS Lett.* **140,** 185 (1982).

[50] T. Lissolo, S. Pulvin, and D. Thomas, *J.Biol. Chem.* **259,** 11725 (1984).

[51] V. M. Fernandez, K. K. Rao, M. A. Fernandez, and R. Cammack, *Biochimie* **68,** 43 (1986).

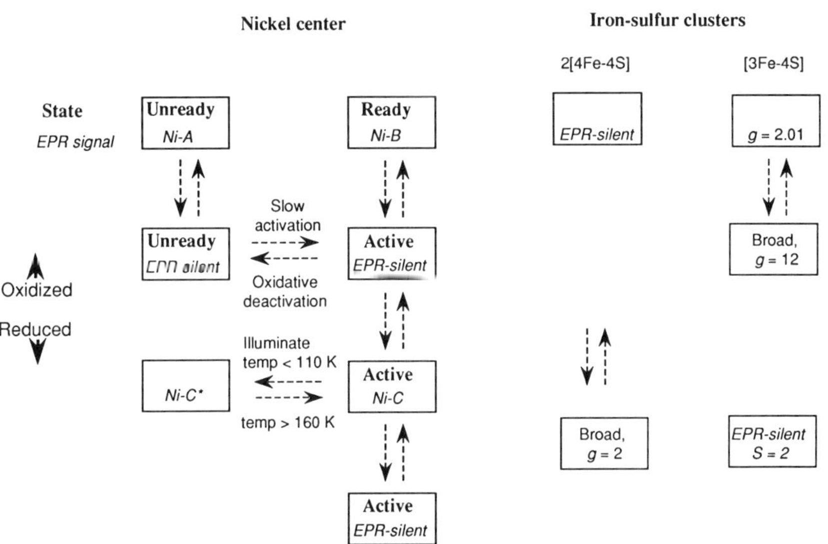

FIG. 3. States of the nickel and iron–sulfur clusters in a hydrogenase such as that from *D. gigas*. *Top*: Most oxidized states. *Bottom*: Most reduced.

doxin, Ni(II), Fe(II), sulfide, ascorbate or dithiothreitol, and only slightly affected by pH in the range 6–9.[8] The conditions required for reductive activation were found to be (1) the absence of oxygen, and (2) the presence of a reductant. Methods for effective deoxygenation of gas streams are described in another section of this chapter (p. 55). When anaerobic conditions are to be maintained for long incubations we have used the flask shown in Fig. 4. The enzyme solution is kept in the small lateral arm (1 ml) of the septum-sealed flask. After flushing the flask with argon, an oxygen scavenger solution (1 m*M* MV^+–15 m*M* dithionite–100 m*M* Tris-HCl buffer, pH 7.6) is placed in the bottom of the flask and gently stirred. After 15 min of stirring, the hydrogenase solution is introduced into the lateral arm with a syringe through the septum. When required, the whole system is flushed with hydrogen for 30 min.

The extent of activation of *D. gigas* hydrogenase has been found to be dependent on the applied redox potential ($E^{\circ}_{\mathrm{pH}\,7} = -310\ \mathrm{mV_{SHE}}$).[50] Partial activation has been obtained by incubation under different partial pressures of hydrogen[40] or with dithiothreitol (5 m*M*) plus indigotetrasulfonate[8] (because this electron mediator is no longer commercially available, indigodisulfonate is an effective alternative). Saturation with hydrogen, or addition of dithionite solution (1 m*M*), provides reductive conditions strong enough to achieve full activation of [Ni-Fe] hydrogenases. The

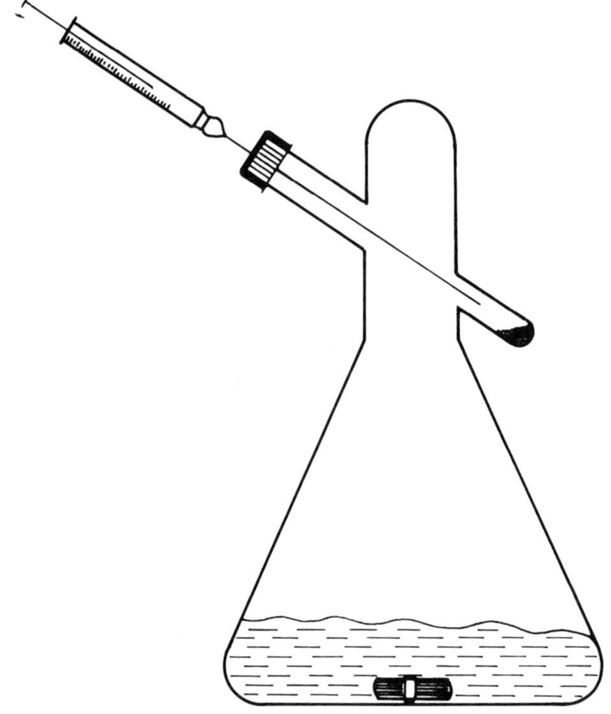

FIG. 4. Anaerobic flask for sample treatments.

presence of low-potential electron carriers, such as reduced cytochrome c_3 or methyl viologen, facilitates the onset of the activation process, but does not affect the final activity. We have found activation with dithionite to be reliable, because anaerobic conditions are more easily maintained. However, incubation under hydrogen is advantageous when further treatments of the enzyme are sought, because hydrogen can be removed easily and substituted by nitrogen or argon.

The ready and unready states of [Ni-Fe] hydrogenases are produced by oxidative treatments of the activated enzyme.[8] In our experience it is not possible to isolate the enzyme purely in one of these forms, but the type of oxidative treatment influences the major form that is produced. The treatments are carried out at room temperature in small conical tubes with septum stoppers. The unready state is produced by oxidation of the active state by displacement of hydrogen with oxygen or air, and allowing the gas to diffuse into the solution over the course of several hours at room temperature. The ready state can be prepared either by displacement of hydrogen with vacuum/argon cycles or by anaerobic addition of a high-potential acceptor, DCIP (1 m*M*), at pH 7.8.

Catalytic Properties of Activity States

The unready, ready, and active states of [Ni-Fe] hydrogenases are easily differentiated by their pattern of reactivity in the different assays of hydrogenase.

Hydrogen Uptake Assay

The unready form is active with acceptors of low redox potential, such as methyl viologen, but inactive with high-potential acceptors such as methylene blue or DCIP. The activity in the spectrophotometric assay with MV^{2+} is about 10% of the active form. When high- and low-potential acceptors are present simultaneously, the activity with methyl viologen is suppressed.

The ready form is active with MV^{2+}, displaying sigmoid kinetics, with an induction period but without the lag phase. The maximum slope of the sigmoidal curve is higher than that of the unready form, and similar to that of the active state.

The active form reduces all acceptors, without lag phase and induction period, even if both low- and high-potential acceptors are simultaneously present in the assay. In this case, the high-potential acceptor is reduced first. Figure 5 summarizes the pattern of H_2 uptake activity of the three states of *D. baculatum* Norway 4, membrane-bound hydrogenase, with low- and high-potential acceptors.[51]

H_2 Evolution Assay

The unready form shows an induction period in the H_2 evolution assay. The activity is typically one-third of the active state. The ready and active forms show identical activity and cannot be differentiated with this assay.

$^2H_2$ or $^3H_2$ Exchange Assay

Both the unready and ready forms are inactive in the $^2H_2$ or $^3H_2$ exchange assay. However, if dithionite is already present in the assay, enzyme in the ready state displays an activity similar to the active state. For the active form, the assay is linear from the beginning of the reaction.

Electron Paramagnetic Resonance Spectroscopy and Redox States

Whereas the optical absorption spectra of [Ni-Fe] hydrogenases are broad and featureless, their EPR spectra are particularly informative about

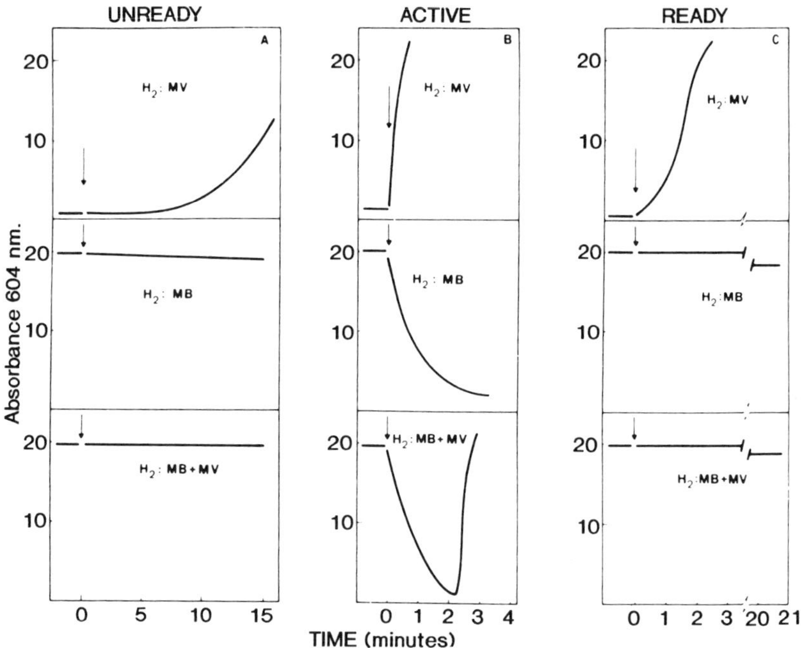

FIG. 5. Activity of different forms of *Desulfomicrobium baculatum* Norway 4, membrane-bound hydrogenase. (A) Unready state, enzyme as isolated, flushed with argon; (B) active state after 6 hr under hydrogen; (C) ready state, active enzyme after anaerobic addition of 1 m*M* DCIP. Spectrophotometric assay with hydrogen uptake and either 1 m*M* methyl viologen (MV; low-potential acceptor), 0.1 m*M* methylene blue (MB; high-potential acceptor), or 1 m*M* methyl viologen plus 0.1 m*M* methylene blue as electron acceptors. All assays were run at 30° in 0.1 *M* Tris-glycine buffer, pH 8.5. At the time indicated by the arrow, 5 μl of a 0.15-mg/ml solution of protein was added to the cuvette. (Reprinted, with permission, from Ref. 18.)

the states of the nickel and iron-sulfur centers, and the magnetic interactions between them. The EPR spectrum of a sample depends on the proportions of the various paramagnetic centers in the protein in different oxidation and redox states, and the temperature of measurement. The EPR spectra of nickel are different for the ready, unready, and active states, which all have characteristic midpoint potentials. Furthermore, the midpoint potentials of the nickel and [4Fe-4S] clusters are pH dependent, so that the state of the enzyme depends on the pH, the applied redox potential, and the time for which the redox potential has been applied. Values for the midpoint potentials of the redox centers are given for the [Ni-Fe] hydrogenase of *D. gigas*, in Table IV.

The EPR spectra of nickel may readily be detected in samples at

TABLE IV
MIDPOINT POTENTIALS OF *Desulfovibrio gigas* HYDROGENASE[a]

Redox event	Midpoint potential (mV)
[3Fe-4S] cluster[23]	−35
Nickel (Ni-A signal)[23]	−150
Nickel (Ni-C/EPR-silent)[52]	−270
Nickel (EPR-silent/Ni-C)[52]	−390
[4Fe-4S] clusters[6]	−290, −340
Reductive activation[50]	−310
Oxidative deactivation[b]	−133
Catalytic activity (H_2 evolution)[39]	−360

[a] Potentials are expressed relative to the standard hydrogen electrode.

[b] R. M. Mege and C. Bourdillon, *J. Biol. Chem.* **260,** 14701 (1985).

temperatures from liquid helium temperature (4.2 K) up to room temperature. Concentrations of 10 μM or greater are required at 4.2–77 K. At 4.2 K the spectra are strongly saturated, and low microwave power must be used, unless the relaxation is enhanced by interaction with reduced [4Fe-4S] clusters.[52] The EPR spectra of the iron-sulfur clusters are detected only at liquid helium temperatures; the spectra broaden out above 30 K for the [3Fe-4S] clusters, and above 10 K for the [4Fe-4S] clusters. Procedures for measurement of EPR spectra at low temperatures have been described.[53,54]

The redox potential of a hydrogenase preparation may be adjusted by titration with oxidants and reductants in the presence of dye mediators, under argon or nitrogen atmosphere, as described by Dutton.[55] The potential may be measured with a gold or platinum electrode. A vessel for titration of small volumes has previously been described.[54] As the potential falls to within 150 mV of the potential of the H^+/H_2 couple (−414 mV at pH 7), the equilibrium concentration of hydrogen gas becomes significant and the potential is unstable as hydrogen gas is produced and diffuses out of the solution. For lower potentials it is necessary to have hydrogen present in the gas phase, and the potential may be controlled by the proportion of hydrogen in the gas mixture. A suitable gas mixing apparatus may either be obtained commercially (e.g., Signal Instrument Co, Camber-

[52] R. Cammack, D. S. Patil, E. C. Hatchikian, and V. M. Fernandez, *Biochim. Biophys. Acta* **912,** 98 (1987).

[53] H. Beinert, W. H. Orme-Johnson, and G. Palmer, this series, Vol. 54, p. 111.

[54] R. Cammack and C. E. Cooper, this series, Vol. 227, p. 353.

[55] P. L. Dutton, this series, Vol. 54, p. 411.

ley, Surrey, U.K.), or constructed from flow meters and flow control valves, coupled together with all-metal tubing.

The oxidation, reduction, and activation and deactivation pathways for the nickel site in hydrogenases may be interpreted in terms of Fig. 3.[7] A disadvantage of EPR is that in general it can detect only those oxidation states of a metal center that have an odd number of unpaired electrons. During progressive reduction steps the nickel site is in turn EPR detectable (having an odd number of unpaired electrons) and EPR silent. Therefore, in order to obtain a complete picture of the state of the metal centers in a sample of the protein, including the EPR-silent states, quantitative monitoring of the EPR signals is necessary.

The oxidized [Ni-Fe] hydrogenases give a mixture of EPR signals, Ni-A and Ni-B (Fig. 6, spectra a and b), which correlate with the relative amounts of the ready and unready enzyme.[26] To quantify these signals it is necessary to measure the spectrum at 100 K or higher temperature, owing to broad signals from the oxidized [3Fe-4S] clusters.

Samples of *D. gigas* hydrogenase that are stored in the frozen state tend to be mainly the Ni-A form, whereas for the hydrogenase of *Desulfovibrio fructosovorans* the Ni-B state tends to predominate. The redox state giving rise to the Ni-C EPR signal (Fig. 6, spectrum c) represents the active enzyme in an intermediate state of reduction.[25,52] On reduction with hydrogen the signal is small. Evacuation, or purging with an inert gas such as argon (in the absence of oxygen), oxidizes the enzyme by displacing hydrogen, and the Ni-C signal increases. The Ni-C signal is also seen in [Ni-Fe-Se] hydrogenases,[56] although the Ni-A and Ni-B signals are not.

The iron-sulfur clusters respond to the oxidation–reduction potential, independent of the state of the nickel. The [3Fe-4S] cluster gives an extremely prominent EPR signal at $g = 2.01$ in the oxidized state. In the reduced state it gives rise to a broad signal around $g = 12$, which, however, disappears when the [4Fe-4S] clusters are reduced. The reduced [4Fe-4S] clusters yield another broad signal around $g = 2$.

The EPR signals of the nickel are influenced by spin–spin interactions with the paramagnetic states of the [4Fe-4S] clusters. This gives rise to a characteristic splitting of the spectrum of Ni-C when measured at temperatures below 7 K (Fig. 6, spectrum d). At higher temperatures, above 50 K, the Ni-C signal is unchanged in shape but shows more rapid electron spin relaxation. It is possible to obtain the enzyme in various forms, with or without these interactions, by judicious adjustment of the redox potential, pH, and activation states of the enzyme. For example,

[56] S. H. He, M. Teixeira, J. LeGall, D. S. Patil, I. Moura, J. J. G. Moura, D. V. DerVartanian, B. H. Huynh, and H. D. Peck, *J. Biol. Chem.* **264,** 2678 (1989).

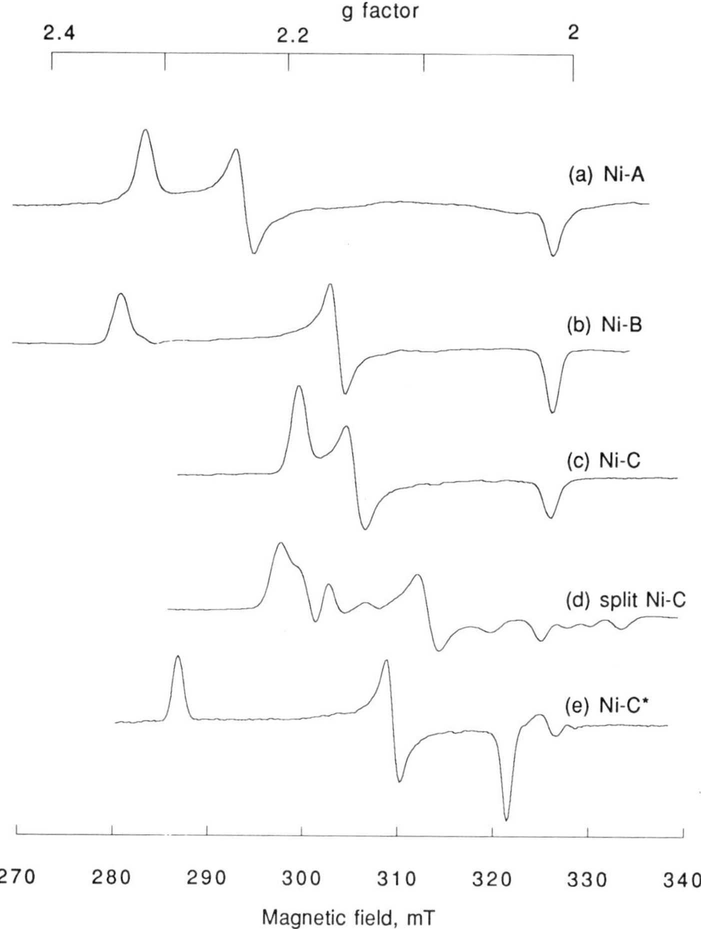

FIG. 6. Representative EPR spectra of nickel from [Ni-Fe] hydrogenases; (a) Ni-A (measured at a temperature of 92 K); (b) Ni-B (92 K); (c) Ni-C (60 K); (d) Ni-C, split by interaction with the reduced iron-sulfur clusters (4.2 K); (e) Ni-C*, induced by illumination of Ni-C (30 K). (a and b) From *D. fructosovorans* hydrogenase; (c–e) from *D. gigas* hydrogenase. For clarity, spectral subtraction has been used to remove small amounts of the Ni-A signal from (a), and of the Ni-B signal from (b). Conditions of measurement: microwave power, 20 mW; frequency, 9.2 GHz; modulation amplitude, 1 mT.

to obtain the enzyme with the nickel in the Ni-C state, but the [4Fe-4S] clusters nonmagnetic, the enzyme is carefully reduced at pH 6, to a potential around −250 mV. To obtain the enzyme with reduced [4Fe-4S] clusters, free from nickel EPR signals, the enzyme in the unready state is reduced for a few minutes with dithionite, at 0°. To show the maximum interaction between Ni-C and [4Fe-4S] clusters, the enzyme is fully activated and reduced to a potential around −350 mV at pH 6; the lower pH optimizes the Ni-C EPR signal.

Hydrogenase in the form giving the Ni-C EPR signal is light sensitive at low temperatures. It is converted by a photochemical reaction to another form with a different EPR signal designated Ni-C*[52,57] (Fig. 6, spectrum e). This form may be produced by illumination of a sample of the enzyme with a photographic spot lamp for 15 min, in a quartz tube immersed in liquid nitrogen. The spectrum reverts to the original form on annealing for a few minutes at temperatures above 150 K.

Acknowledgments

The skilled assistance of Nicole Forget is gratefully acknowledged. We thank our numerous colleagues who have contributed to this work and provided helpful advice. The work is supported by the U.K. Science and Engineering Research Council, the CNRS, and the Royal Society.

[57] A. Chapman, R. Cammack, E. C. Hatchikian, J. McCracken, and J. Peisach, *FEBS Lett.* **242,** 134 (1988).

[6] Nickel-Iron-Selenium Hydrogenase

By Daulat S. Patil

Introduction

The enzyme hydrogenase catalyzes the reversible oxidation of molecular hydrogen and plays a vital role in anaerobic metabolism. A number of reviews summarize the metabolic role and characterization of hydrogenase.[1–7] It is a key enzyme for both hydrogen uptake and hydrogen evolu-

[1] H. D. Peck, Jr., *in* "The Sulfate-Reducing Bacteria: Contemporary Perspectives" (J. M. Odom and R. Singleton, Jr., eds.), p. 41. Springer-Verlag, New York, 1992.

[2] G. Fauque, H. D. Peck, Jr., J. J. G. Moura, B. H. Huynh, Y. Berlier, D. V. DerVartanian, M. Teixeira, A. E. Przybyla, P. A. Lespinat, I. Moura, and J. Le Gall, *FEMS Microbiol. Rev.* **54,** 299 (1988).

[3] M. W. W. Adams, *Biochim. Biophys. Acta* **1020,** 115 (1990).

[4] A. E. Przybyla, J. Robbins, N. Menon, and H. D. Peck, Jr., *FEMS Microbiol. Rev.* **88,** 109 (1992).

[5] L.-F. Wu and M. A. Mandrand, *FEMS Microbiol. Rev.* **104,** 243 (1993).

tion in sulfate-reducing anaerobic bacteria (*Desulfovibrio*), in fermentative anaerobes (*Clostridium*), in facultative anaerobes (*Escherichia coli*), and in aerobic nitrogen-fixing bacteria in symbiosis with *Rhizobia*, photosynthetic bacteria, and cyanobacteria.

The purpose of this chapter is to describe briefly the laboratory-scale growth of bacteria for enzyme purification and to give an account of the [Ni-Fe-Se] hydrogenases in the sulfate-reducing bacteria with reference to experimental methods used for investigation. Experimental techniques of anaerobic preparation of enzyme samples and potentiometric redox titration in conjunction with spectroscopy are also described.

On the basis of metal composition, immunological cross-reactivity, and gene structure, three distinct categories of hydrogenases have been identified in the genus *Desulfovibrio*[8,9]: (1) iron hydrogenases containing two [4Fe-4S] centers in addition to a specialized center thought to be a type of [4Fe-4S] center, (2) nickel-iron hydrogenases, the most common, containing a nickel center coordinated to sulfur, two [4Fe-4S] centers, and a [3Fe-*x*S] center, and (3) nickel-iron-selenium hydrogenases containing a nickel center coordinated to a selenium as well as two [4Fe-4S] centers. The nickel center in hydrogenase is additionally coordinated to nitrogen and/or oxygen in the first ligand sphere, and is thought to be the locus at which hydrogen binds and [Fe-S] clusters function as secondary electron carriers. The biochemical role of selenium is not clearly known or understood. Selenium biochemistry has been reviewed.[10]

Growth of *Desulfovibrio* Cells and Purification of Hydrogenase

Cultivation and growth of the sulfate-reducing bacteria in pure culture has been discussed by Postgate.[11] Purification strategies and procedures for electron-transfer components from the sulfate-reducing bacteria have been discussed by Le Gall and Forget.[12] The general procedure for laboratory cultivation of *Desulfovibrio* bacteria is illustrated with reference to the

[6] B. Friedrich and C. G. Friedrich, *in* "Autotrophic Microbiology and One-Carbon Metabolism" (G. A. Codd, L. Dijkhuizen, and F. R. Tabita, eds.), p. 55. Kluwer Academic Publ., Dordrecht, The Netherlands, 1990.

[7] R. Cammack, V. M. Fernandez, and E. C. Hatchikian, this volume [5].

[8] B. C. Prickril, S.-H. He, C. Li, N. Menon, E.-S. Choi, A. E. Przybyla, D. V. DerVartanian, H. D. Peck, Jr., G. Fauque, J. Le Gall, M. Teixeira, I. Moura, J. J. G. Moura, D. Patil, and B. H. Huynh, *Biochem. Biophys. Res. Commun.* **149,** 369 (1987).

[9] G. Voordouw, *Adv. Inorg. Chem.* **38,** 397 (1992).

[10] T. C. Stadtman, *Annu. Rev. Biochem.* **59,** 111 (1990).

[11] J. R. Postgate, The Sulphate-Reducing Bacteria," 2nd Ed., Cambridge Univ. Press, Cambridge, 1979.

[12] J. Le Gall and N. Forget, this series, Vol. 53, p. 613.

incorporation of isotopes such as ^{57}Fe, ^{61}Ni, ^{77}Se, and ^{15}N. Metalloenzymes isolated by the isotope-supplement method greatly help spectroscopic analysis of the metal centers, especially in electron paramagnetic resonance (EPR), electron nuclear double resonance (ENDOR), and Mössbauer studies.[13a–c] Pure isotopes are available commercially and can be used directly as additional supplements in the growth medium. The ^{57}Fe-enriched cells are grown in lactate–sulfate medium as described previously.[13] Commercially available ^{57}Fe metal (95% enrichment), 200 mg, is brought into solution in a minimum volume (about 10 ml) of concentrated sulfuric acid. Addition of a small amount of concentrated hydrochloric acid with gentle heating on a water bath enhances the completion of solution, which is then neutralized and sterilized in an autoclave for 15 min. The clear sterilized aliquot is then added to the growth medium (40 liters). The addition of the isotope aliquot can be done in portions at different steps of culturing procedure to achieve maximum assimilation of the element.

The growth of bacteria prior to large-scale fermentor growth involves stepwise culturing of 10-fold incremental volumes of culture with 24 hr of incubation at 37° at each step. For a 400-liter fermentor batch, the culturing is started with 0.4 ml of the pure stock culture in sulfate–lactate medium, inoculated into a sterile medium of 40 ml, and incubated for 24 hr at 37°. This in turn is inoculated into 400 ml of medium in the next step and, in the final step, 40 liters of active bacterial growth medium is transferred and ready for use in the fermentor. The cells are generally harvested after 24 hr of growth at 37° in a 400-liter fermentor that is kept anaerobic under a constant stream of an inert gas. During the growth, samples are monitored by microscopic examination to check for contamination and by measurement of optical density to follow growth curve and cell population. The wet cells are collected rapidly by centrifugation and suspended in about 100 ml of lysis buffer (Tris-HCl, 50 m*M*, containing 1 m*M* protease inhibitors) and stored frozen at −70°C. A typical fermentor batch of 400 liters yields 400–500 g of cells (wet weight).

[13a] S. H. He, M. Teixeira, J. Le Gall, D. S. Patil, I. Moura, J. J. G. Moura, D. V. DerVartanian, B. H. Huynh, and H. D. Peck, Jr., *J. Biol. Chem.* **264,** 2678 (1989).

[13b] S. H. Bell, D. P. E. Dickson, R. Rieder, R. Cammack, D. S. Patil, D. O. Hall, and K. K. Rao, *Eur. J. Biochem.* **145,** 645 (1984).

[13c] R. Cammack, D. S. Patil, E. C. Hatchikian, and V. M. Fernandez, *Biochim. Biophys. Acta* **912,** 98 (1987).

[13] B. H. Huynh, D. S. Patil, I. Moura, M. Teixeira, J. J. G. Moura, D. V. DerVartanian, M. H. Czechowski, B. C. Prickril, H. D. Peck, Jr., and J. Le Gall, *J. Biol. Chem.* **262,** 795 (1987).

Composition of Medium: pH 7.2

The growth medium consists of the following components (per liter): sodium lactate (60%), 12.5 ml; NH_4Cl, 2 g; $MgSO_4 \cdot 7H_2O$, 2 g; K_2HPO_4, 0.59 g; Na_2SO_4, 4 g; $FeSO_4 \cdot 7H_2O$, 10 mg; $CaCl_2$, 0.2 g; yeast extract, 1 g; cysteine-HCl, 0.25 g; sodium sulfide, 0.25 g; modified Wolfes mineral elixir, 1 ml. One liter of the Wolfes mineral elixir contains the following: nitrilotriacetic acid (pH to 6.5 with KOH), 1.5 g; $MgSO_4 \cdot 7H_2O$, 3 g; $MnSO_4 \cdot H_2O$, 0.5 g; NaCl, 1.0 g; $FeSO_4 \cdot 7H_2O$, 1 g; $CoCl_2$ or $CoSO_4$, 0.1 g; $NiCl_2 \cdot 6H_2O$, 0.1 g; $CaCl_2$, 0.1 g; $ZnSO_4 \cdot 7H_2O$, 0.1 g; $CuSO_4 \cdot 5H_2O$, 0.01 g; aluminum potassium sulfate, 0.01 g; boric acid, 0.01 g; Na_2MoO_4, 0.01 g; Na_2SeO_3, 0.001 g.

Purification of Hydrogenase

The purification procedures for the [Ni-Fe-Se] hydrogenases from *Desulfomicrobium (Dsm.) baculatum* (Norway 4),[14] *Dsm. baculatum,*[15,16] and *Desulfovibrio salexigens*[17] have been described previously. These protocols involve an application of a combination of ion-exchange and gel-filtration chromatography and tend to be lengthy and cumbersome. A typical purification from 500 g of frozen cells involving four to five steps may require a continuous operation for a period of up to 2 weeks. This is a disadvantage, because the enzyme may suffer loss of specific activity and the formation of apoprotein may occur owing to time-consuming dialyses and diaflo concentrations and chromatographic steps. However, by using a high-performance liquid chromatography (HPLC) column in the final step, purification time can be shortened.[16]

The following is a description of an alternate protocol for the purification of hydrogenase.[18] This has a clear advantage in terms of economy of time and effort and, more importantly, the rapidity of the protocol minimizes chances of loss of specific activity, of proteolysis, and of apoprotein formation. A typical purification at 4° from 200 g of frozen cells can be accomplished in about 96 hr, with enzyme >99% pure, on a silver stain

[14] R. Rieder, R. Cammack, and D. O. Hall, *Eur. J. Biochem.* **145,** 631 (1984).

[15] M. Teixeira, G. Fauque, I. Moura, P. A. Lespinat, Y. Berlier, B. Prickril, H. D. Peck, Jr., A. V. Xavier, J. Le Gall, and J. J. G. Moura, *Eur. J. Biochem.* **167,** 47 (1987).

[16] S. H. He, M. Teixeira, J. Le Gall, D. S. Patil, I. Moura, J. J. G. Moura, D. V. DerVartanian, B. H. Huynh, and H. D. Peck, Jr. *J. Biol. Chem.* **264,** 2678 (1989).

[17] M. Teixeira, I. Moura, G. Fauque, M. Czechowski, Y. Berlier, P. A. Lespinat, J. Le Gall, A. V. Xavier, and J. J. G. Moura, *Biochimie* **68,** 75 (1986).

[18] N. K. Menon, J. Robbins, M. DerVartanian, D. Patil, H. D. Peck, Jr., A. L. Menon, R. L. Robson, and A. E. Przybyla, *FEBS Lett.* **331,** 91 (1993).

gel [sodium dodecyl sulfate–polyacrylamide gel electrophoresis (SDS–PAGE)]. The protocol involves the following steps.

1. Clarification of broken cell extract/cell wash by ultracentrifugation (40,000 rpm, 1 hr): The frozen cells are allowed to thaw in the cold. About 100 ml of Tris-HCl buffer, pH 7.6, is added. The cell suspension is centrifuged for 15 min (8000 rpm), and supernatant extract is used for recovering periplasmic hydrogenase after subjecting it to ultracentrifugation. The cell pellet is preserved or used for the recovery of membrane-bound hydrogenase after treatment with detergent such as 1% Triton.

2. Anion-exchange chromatography column (25 × 2.5 cm): The packing material used consists of cross-linked polystyrene/divinylbenzene particles (Poros II, PerSeptive Biosystems, Cambridge, MA). A variety of packing materials is commercially available.

The concentrated cell wash/extract is loaded on the column and washed with three or four bed volumes of 50 m*M* Tris-HCl buffer (pH 7.6) to elute unbound proteins. A linear gradient of the Tris-HCl buffer (50–500 m*M*) is then applied at a flow rate of 4 ml/min. The fractions eluting around 100 m*M* are collected and assayed for hydrogenase activity. The pool of hydrogenase fractions is then concentrated using a Diaflo YM30 membrane (Amicon, Danvers, MA) and washed four or five times with 50 m*M* Tris-HCl, pH 7.6, and finally concentrated to a volume of about 2 ml to be applied to the next column.

3. Preparative electrophoresis by continuous elution: The concentrated hydrogenase pool from the ion-exchange chromatographic column is loaded on an 8% polyacrylamide gel column of about 10 × 3 cm for continuous-elution electrophoresis. During the run, proteins are electrophoresed vertically through a cylindrical sieving gel. The individual bands migrate off the base of the gel and pass directly into an elution chamber equipped with a thin frit. A dialysis membrane (M_r 6000 cutoff) underneath the elution frit traps proteins within the chamber. Elution buffer (0.1 *M* Tris, pH 7.6) enters the chamber around the perimeter of a special gasket, is drawn radially inward to an elution tube in the center of the cooling core, and flows out for collection [via a peristaltic pump, ultraviolet (UV) monitor and fraction collector]. The apparatus for preparative electrophoresis is commercially available and is illustrated in the manufacturer manual (model 491; Bio-Rad, Richmond, CA).

4. Hydroxylapatite column (20 × 1.5 cm): Purification step 4 may be employed only if the purity of hydrogenase from the previous step indicates the presence of some impurity as judged by silver stain gel. The enzyme is loaded on the column and eluted with a 50 to 250 m*M* gradient of potassium phosphate buffer, pH 7.5, at a flow rate of 1 ml/min. The

hydrogenase elutes around a buffer concentration of 100 m*M*. The pure fractions of hydrogenase are pooled and concentrated by using a Diaflo YM30 membrane. The purified enzyme is preserved at −70°.

The problem of proteolysis, especially during the initial steps of purification, will be encountered; however, in spite of the formation of degraded enzyme molecules due to proteolysis and the presence of apoprotein, the enzyme can retain biological activity. This situation may lead to erroneous conclusions about the size and structure of the enzyme molecule as well as its specific activity. Addition of protease inhibitors such as EDTA and ethylene glycol-bis(β-aminoethyl ether)-*N*,*N*,*N*′,*N*′-tetraacetic acid (EGTA) to the cell extract often helps to minimize or arrest proteolysis. In this respect, working with a small-scale batch (<100 g) of cells is advantageous especially when one is searching/reporting for the presence of new proteins, which requires reproducibility to ensure consistent purification characteristics.

A useful source of information is Vol. 182 of this series, *Guide to Protein Purification,* which deals with various aspects of purification and related techniques to help an experimentor make judicious choice of a purification strategy. For large-scale (>500 g of cells) purification of hydrogenase, the widely used protocol employing ion-exchange and gel-filtration chromatography as described by Le Gall and Peck and coworkers[14–17] is the most suitable method especially because various other metalloproteins can be recovered in addition to hydrogenase. A clear separation of various *Desulfovibrio* proteins into separate, colored bands on the columns helps one to search visually for unidentified new proteins.

Measurement of Protein Concentration and Homogeneity

Different methods for the determination of protein concentration for hydrogenase have been employed.[19] The significance of an accurate protein concentration is that it influences the [spin] per molecule of a paramagnetic EPR species under study, which in turn can affect Mössbauer spectroscopic simulation.[19a,b] The magnitude of protein concentration determination also influences the calculation of the number of iron atoms per molecule, which in turn can influence the assignment of [Fe-S] cluster type, especially at the preliminary stage of characterization. It can affect the assignment of molecular weight and extinction coefficient of metalloenzymes. Therefore, it is of utmost importance that a careful protein determi-

[19] D. S. Patil, J. J. G. Moura, S. H. He, M. Teixeira, B. C. Prickril, D. V. DerVartanian, H. D. Peck, Jr., J. Le Gall, and B. H. Huynh, *J. Biol. Chem.* **263,** 18732 (1988).

[19a] E. Münck, K. K. Surerus, and M. P. Hendrich, this series, Vol. 227, p. 463.

[19b] E. Münck, this series, Vol. 53, p. 346.

nation be performed by using different methods employing various protein standards and by cross-checking the results. It is essential to use freshly prepared reagents, to carefully weigh out pure, dry crystals of reference protein standard [bovine serum albumin (BSA)], calibrate and check the baseline performance of the spectrophotometer, and perform five or more measurements to express the result as a statistical mean and standard deviation.

Methods of protein determination have been described previously in this series.[20] The procedures widely employed are the Lowry, biuret, and Bradford procedures. These methods require a pure reference standard, such as crystalline BSA. Methods are available that do not depend on a reference standard, such as the Kjeldahl method,[21] in which the nitrogen of proteins is liberated as NH_3, which in turn is titrated directly against a standard acid. Another method that does not require a secondary reference standard is the Dumas method,[22] which employs thermal oxidation of the protein in the presence of CuO. The product of oxidation, resulting from protein degradation is nitrogen gas, which can be measured by volume under standard temperature and pressure.

The widely used Lowry method is sensitive and quick to perform, but tends to overestimate metalloproteins and shows a nonlinear standard curve. The Bradford method, which uses the dye Coomassie blue to bind the protein noncovalently, is sensitive and rapid but the protein response to noncovalent binding can fluctuate widely. The biuret (microbiuret) method, in which the peptide bands form a complex with Cu^{2+}, involves the use of unstable reagents, and A_{545} measurement is only moderately sensitive to reaction time compared to the A_{650} and A_{750} measurements used in the Lowry method. The most convenient, fast, and routinely performed method still remains the measurement of UV absorption λ = 260/280, which is best for pure proteins because they can be recovered for reuse.

The biochemical homogeneity of an enzyme can be examined by subjecting it to the silver stain SDS–electrophoresis test. The developed gel should not show any trace of additional bands apart from the subunit bands that indicate a homogeneous preparation. Such a preparation should retain its specific activity and display its correct molecular weight as determined by gel-filtration chromatography and ultracentrifugation method. It is essential that purified hydrogenase be homogeneous and free from apoprotein or trace metal impurity, especially because these

[20] C. M. Stoscheck, this series, Vol. 182, p. 50.

[21] J. Kjeldahl, *Anal. Chem.* **22,** 366 (1883); F. C. Koch and T. L. McMeekin, *J. Am. Chem. Soc.* **42,** 2066 (1924).

[22] W. Kristen, *in* "Comprehensive Analytical Chemistry" (C. L. Wilson and D. W. Wilson, eds.), Vol. 1b, p. 494. Elsevier, Amsterdam, 1960.

can cause complications in the analysis of spectroscopic data of metal centers. Münck and co-workers have drawn attention to the distinction between the biochemical and physical heterogeneity of metalloprotein.[23]

Assay of Hydrogenase and Biochemical Characterization

Methods of assay for hydrogenase have been discussed by Krasna previously.[24] In this volume, Cammack and co-workers (see [5]) have presented an incisive discussion of various methods for estimation of hydrogenase activity. All hydrogenases contain different types of [Fe-S] clusters coordinated, usually, to cysteine residues of the protein in conserved patterns. Methods of identifying [Fe-S] clusters have been described previously.[25–28]

Examples of Nickel–Iron–Selenium Hydrogenases in *Desulfomicrobium* and *Desulfovibrio*

Desulfomicrobium baculatum (Norway 4) Hydrogenase

An unusual soluble hydrogenase, isolated from *Dsm. baculatum* (Norway 4),[14] contains selenium in amounts equal to nickel. Moreover, the enzyme does not have a [3Fe-*x*S] center and only a rudimentary, broadened EPR signal due to nickel. The enzyme contains eight iron, eight inorganic sulfur, one nickel, and one selenium atom. It is substantially different from the membrane-bound hydrogenase from the same organism, for which the presence of selenium has not been reported.[29] The two enzymes differ in their subunit molecular weights and EPR spectra. More importantly, the differences in amino acid compositions support the view that the soluble and membrane-bound hydrogenases are different gene products rather than conformational variants of the same protein. Because the two hydrogenases were separately isolated from the same strain of *Dsm. baculatum* (Norway 4), the genes for both enzymes must be present in the organism, which under different growth conditions expresses two enzymes with different properties. Mössbauer, EPR, and magnetic circu-

[23] E. Münck, K. K. Surerus, and M. P. Hendrich, this series, Vol. 227, p. 463.

[24] A. I. Krasna, this series, Vol. 53, p. 296.

[25] H. Beinert, this series, Vol. 54, p. 435.

[26] W. H. Orme-Johnson and R. H. Holm, this series, Vol. 53, p. 268.

[27] R. Cammack, *Adv. Inorg. Chem.* **38,** 281 (1992).

[28] J. C. Rabinowitz, this series, Vol. 53, p. 275.

[29] W. V. Lalla-Maharajh, D. O. Hall, R. Cammack, K. K. Rao, and J. LeGall, *Biochem. J.* **209,** 445 (1983).

lar dichroism (MCD) spectroscopic studies have established that this enzyme contains two [4Fe-4S] clusters.[13b] Mössbauer data, however, suggest that the two [4Fe-4S] clusters are inequivalent in that one cluster is readily reducible whereas the other remains oxidized, and this has slightly different average Mössbauer parameters compared to those of the native oxidized sample. It was argued that the possiblity of each molecule containing one reducible and one nonreducible cluster was unlikely; instead it was assumed that the enzyme preparations probably contained a fraction of molecules that were rendered inactive due to loss of nickel metal atoms, probably a consequence of the lengthy purification procedure.

Desulfovibrio salexigens Hydrogenase

Purification of and D_2/H^+ exchange at pH 7.6 in a [Ni–Fe–Se] cluster from a halophilic sulfate reducer *D. salexigens* have been reported.[17] The soluble enzyme, composed of two subunits (62 and 36 kDa), contains 1 nickel, 12–15 iron, and 1 selenium atom per enzyme molecule. As was the case with *Dsm. baculatum* (Norway 4) soluble hydrogenase, this enzyme in the native state shows no EPR resonance. A sample poised at −380 mV shows a weak rhombic EPR signal (g = 2.22, 2.16, and 2.0) that, by analogy with the *Desulfovibrio gigas* hydrogenase, is assigned to nickel. The reduced enzyme (below −300 mV) shows weak EPR signals (g_y = 1.94) that are assigned to iron-sulfur centers. Furthermore, from temperature dependence and power saturation measurements, these iron-sulfur centers have been diffentiated into two types: center I has faster relaxation properties than center II. The hydrogenase activity was monitored at pH 7.6 by the D_2/H^+ exchange reaction and showed a H_2/HD ratio higher than 1, which in comparison was significantly higher than that for *D. gigas* hydrogenase. The H_2/HD ratios are used to differentiate between heterolytic and homolytic cleavage of the hydrogen molecule. It was argued that the different exchange kinetics of hydrogen-binding sites could reflect differences either at the active center or at the hydride-binding site. A correlation between the presence of selenium in hydrogenase and D_2/H^+ exchange characteristic was suggested.

Desulfomicrobuim baculatum Hydrogenase

Nickel–iron–selenium hydrogenases were isolated from different cell locations (periplasmic space, cytoplasm, and membrane-bound regions) of *Dsm. baculatum*.[15] The hydrogenases, each composed of two subunits, were differentiated spectroscopically. All three hydrogenases, however, had identical molecular weights of 100,000. Electrophoretic analysis indicated that the small subunit molecular weights were "identical" but that the large subunit molecular weights varied: periplasmic, cytoplasmic, and membrane-bound enzymes had molecular weights of 49,000, 54,000, and 62,000, respectively. Native preparations showed a weak $g = 2.0$ EPR signal. The periplasmic hydrogenase showed a rhombic EPR signal ($g = 2.20, 2.06, 2.0$) that was assigned to Ni(III). The membrane-bound hydrogenase showed two components: $g = 2.34, 2.16, 2.0$ and $g = 2.33, 2.24, 2.0$. On reduction all hydrogenases showed identical EPR signals due to reduced [Fe-S] centers, which were identified as belonging to two groups: center I, $g = 2.03, 1.89, 1.86$ and center II, $g = 2.06, 1.95, 1.88$, which saturated easily at low temperatures (10 K). A rhombic $g = 2.20, 2.16, 2.0$ nickel species attained maximum intensity in a sample poised at −320 mV [vs standard hydrogen electrode (SHE)]. Additional complex EPR signals were also observed during the course of reduction; spin quantitation of these EPR signals was not attempted, probably because of their complexities. The soluble hydrogenase could be readily activated whereas the membrane-bound form required incubation with hydrogen to attain the active state.

A study of the activation and deactivation of membrane-bound hydrogenase from *Dsm. baculatum* (Norway 4)[29] showed that the enzyme had low activity as isolated by the hydrogen-methyl viologen reductase assay and no activity as isolated by the hydrogen-methylene blue reductase assay. This process was interpreted as conversion of the inactive unready state of the enzyme to the active state. This behavior was akin to that of the membrane-bound [Ni–Fe–Se] hydrogenase from *Dsm. baculatum;* however, for the *Dsm. baculatum* (Norway 4) membrane-bound hydrogenase the presence of selenium has not been reported.[29]

D_2/H^+ exchange reactions for the three [Ni–Fe–Se] hydrogenases of *Dsm. baculatum*[30] showed H_2/HD ratios significantly higher than those observed for [Ni–Fe] hydrogenases. Because antibodies to the [Ni–Fe–Se] hydrogenase of *Dsm. baculatum* (Norway 4) cross-reacted with the *Dsm. baculatum* periplasmic hydrogenase, the two strains, although differently named, were considered "identical." It should be noted that many cross reacting hydrogenases from various bacteria are not structurally identical.[31]

Cloning and sequencing of the genes coding for the large (58 kDa) and

[30] P. A. Lespinat, Y. Berlier, G. Fauque, M. H. Czechowski, B. Dimon, and J. LeGall, *Biochimie* **68,** 55 (1986).

[31] K. L. Kovács, L. Seefeldt, G. Tigyi, C. M. Doyle, L. E. Mortenson, and D. J. Arp, *J. Bacteriol.,* **171,** 430 (1989).

small (31 kDa) subunits of the periplasmic [Ni–Fe–Se] hydrogenase from *Dsm. baculatum* have been reported.[32] These findings showed that the genes were arranged in an operon in the small subunit gene that coded for a 32-amino acid leader sequence preceding the large subunit gene and no iron-sulfur coordination sites could be observed. Also, no significant homology between the [Ni–Fe–Se] hydrogenase and the [Fe] hydrogenase from *Desulfovibrio vulgaris* existed, which clearly indicated that the two types of hydrogenases were structurally different. It was also concluded, especially because the [Ni–Fe–Se] hydrogenase gene did not contain a UAG codon, that a different nucleotide sequence coded for selenocysteine, and that the selenium was present in a form other than selenocysteine or was added posttranslationally. A search to locate a second gene homologous to the one encoding the small subunit of the periplasmic [Ni–Fe–Se] hydrogenase was undertaken by using the small subunit gene as a probe to hybridize *Dsm. baculatum* genomic DNA; the second gene was shown to be absent, which indicated that the membrane and cytoplasmic [Ni–Fe–Se] hydrogenases were encoded by different genes. Subsequent work made use of the codon usage method to reanalyze the nucleotide sequence encoding the periplasmic [Ni–Fe–Se] hydrogenase from *Dsm. baculatum.*[33] The corrected nucleotide sequence encoding the small and large subunit genes was determined. This study showed that the [Ni–Fe–Se] and the [Ni–Fe] hydrogenase sequences shared about 35% homology and that both were different from that of the [Fe] hydrogenase of *D. vulgaris,* in which the mature small subunit lacked cysteine residues, whereas the large subunit amino terminus had eight cysteines coordinated to two [4Fe-4S] centers as in [8Fe-8S] ferredoxins. In the small subunit of the [Ni–Fe–Se] as well as the [Ni–Fe] hydrogenase, the 10 conserved cysteines were adequate to coordinate two [4Fe-4S] centers. The arrangement of these cysteines was different from that found in the amino terminus of the large subunit of the [Fe] hydrogenase and it was not homologous to [8Fe-8S] ferredoxin by comparison of the relevant nucleic acid sequences and by analogy with the findings that selenium was present as selenocysteine in *E. coli* formate dehydrogenase, which is encoded by a TGA codon. Also, the TGA codon was replaced by a TGC (cysteine) codon in the *Methanobacterium formicicum* formate dehydrogenase, which lacked selenium. From these observations and from the EXAFS studies[36] it was concluded, contrary to the earlier inference,[32] that selenium

[32] N. K. Menon, H. D. Peck, Jr., J. Le Gall, and A. E. Przybyla, *J. Bacteriol.* **169,** 5401 (1987). Correction, *J. Bacteriol.* **170,** 4429 (1988).

[33] G. Voordouw, N. K. Menon, J. Le Gall, E. S. Choi, H. D. Peck, Jr., and A. E. Przybyla, *J. Bacteriol.* **171,** 2894 (1989).

[36] M. K. Eidsness, R. A. Scott, B. C. Prickril, D. V. DerVartanian, J. Le Gall, I. Moura, J. J. G. Moura, and H. D. Peck, Jr., *Proc. Natl. Acad. Sci. U.S.A.* **86,** 147 (1989).

as selenocysteine indeed formed a ligand to nickel in the [Ni–Fe–Se] hydrogenase of *Dsm. baculatum*. The occurrence of selenocysteine in *Methanococcus vannielii* hydrogenase[34] and the presence of a TGA codon for selenocysteine in mammalian glutathione peroxidase[35] added further support.

The evidence in favor of selenium coordination to the nickel site of [Ni–Fe–Se] *Dsm. baculatum* hydrogenase came from the results of nickel and selenium extended X-ray absorption fine structure spectroscopy (EXAFS) measurements.[36] From the nickel, selenium K edge data, an approximate coordination structure for the nickel site was defined as penta- or hexacoordinate with (N, O) and (S, Cl) ligands and an Ni-Se distance of 2.44 Å. Although a hexacoordinate, pseudooctahedral nickel site was considered probable, the nickel EXAFS curve-fitting results in combination with nickel EPR resonances pointed to the structure of nickel being surrounded by a distorted octahedral symmetry.

The growth of *Dsm. baculatum* on a medium supplemented with the isotope ^{77}Se has been described.[16] The [Ni–Fe–^{77}Se] hydrogenase was studied using EPR spectroscopy. The nickel signal (g = 2.23 EPR signal) in the hydrogen-reduced enzyme showed a clear broadening compared with the unenriched enzyme. The isotope effect, ^{77}Se (I = 1/2), due to the interaction of nuclear and electron spins, showed clearly the g_x = 2.23 and g_y = 2.17 resonances had appreciably broadened (see Fig. 1). The isotope broadening effect was identical for both the high (2 Se/mol) and low (1 Se/mol) enzyme. Theoretical EPR spectral simulations with resonances at g_x = 2.228, g_y = 2.174, and g_z = 2.01 and respective linewidths of 1.95, 2.20, and 2.00 mT were matched with the experimental spectra of the unenriched and the ^{77}Se-enriched enzymes. Hyperfine coupling constants that best described the observed broadening were found to be A_x = 1.0 mT and A_y = 1.8 mT. This experimental observation indicated that the unpaired electron was shared by the nickel and selenium atoms and that the selenium was in direct coordination with the nickel. An interesting aspect of this study was that the hydrogenase isolated from bacteria grown on excess selenium showed that the presence of selenium was enhanced by a factor of two and that specific activity had increased by a factor of four. It then became necessary to test if this was yet another new hydrogenase, especially because different hydrogenases in a single species of the sulfate-reducing bacteria and multiple forms of hydrogenase had been found in the earlier experiments.[37] Western blot analyses of

[34] S. Yamazaki, *J. Biol. Chem.* **257,** 7926 (1982).

[35] I. Chambers, J. Framton, P. Goldfarb, N. Affara, W. McBain, and P. R. Harrison, *EMBO J.* **5,** 1221 (1986).

[37] T. Lissolo, E. S. Choi, J. Le Gall, and H. D. Peck, Jr., *Biochem. Biophys. Res. Commun.* **139,** 701 (1986).

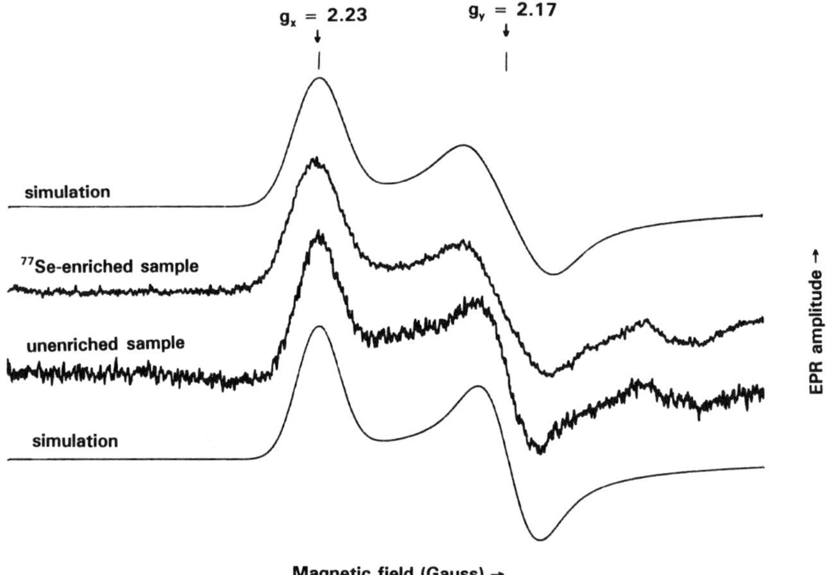

FIG. 1. Effect of ^{77}Se substitution on the nickel signal in reduced *Dsm. baculatum* hydrogenase. (Adapted from Ref. 16.)

the pure hydrogenase using polyclonal antibodies raised against the [Fe] hydrogenase from *D. vulgaris,* the [Ni–Fe] hydrogenase from *D. gigas,* and the [Ni–Fe–Se] hydrogenase of *Dsm. baculatum* (Norway 4) were performed and the hydrogenase did not cross-react with the antibodies of either *D. vulgaris* or *D. gigas* hydrogenase, but it cross-reacted with antibodies to the [Ni–Fe–Se] hydrogenase. This result clearly showed that the hydrogenase purified from the selenium-supplemented medium was indeed the [Ni–Fe–Se] hydrogenase and, also, that the presence of selenium in some manner influenced the increased biosynthesis of this hydrogenase.

An analysis for selenium in the large and small subunits showed that all the selenium was associated with the large subunit. Because the gene encoding the large and small subunits contained only one TGA codon and because the enzyme presumably contained only one selenocysteine residue, the experimentally observed excess selenium was possibly bound loosely to the enzyme molecule or nonspecifically incorporated as selenomethionine. This was resolved by subjecting the purified enzyme to anaerobic dialysis against 0.1 *M* sodium sulfide (pH 9) for various time intervals

(12–108 hr) during which the specific activity, and nickel and iron content, remained unaltered. It was observed that after dialysis for about 12 hr the selenium content decreased from 2.1 to 1.1/mol of hydrogenase, which was clearly consistent with the presence of 1 mol of selenocysteine per mole of hydrogenase.

The magnetic properties of the nickel(II) site in the native active [Ni–Fe–Se] hydrogenase from *Dsm. baculatum* have been studied[38] by multifield saturation magnetization measurements. The methodology for this technique has been described elsewhere.[39] In the native state the enzyme showed full activity in the hydrogen evolution assay and, unlike *D. gigas* [Ni–Fe] hydrogenase, there was no requirement for an activation step. The nickel site, presumably, was in an Ni(II) state because the nickel signal was EPR silent. From the multifield saturation magnetization results the nickel site was found to be diamagnetic Ni(II) and the possiblity of a six-coordinate Ni(II) for the native state was ruled out; because a four-coordinate Ni(II) is always high spin $s = 1$, this also was considered unlikely. Instead a trigonal bipyramidal five-coordinate Ni(II) was favored. This study supported the interpretation of the active Ni-C state as an Ni(II) site in the hydrogenase from *Thiocapsa roseopersicina*.[40] The nickel K edge EXAFS signal in the Ni-C state of *T. rosepersicina* [Ni–Fe] hydrogenase was nearly identical to that of *Dsm. baculatum* [Ni–Fe–Se] hydrogenase in its native state,[36] which was taken as evidence in favor of the argument that the charge density at the metal site did not change significantly with the change in EPR resonance. This raised a question about the electron acceptor site. The electron either could not be accepted by the nickel or the sulfur ligands behaved as electron acceptor. The state of affairs at the active sites appears to be far from clear. *Desulfomicrobium baculatum* [Ni–Fe–Se] hydrogenase was examined by Mössbauer spectroscopy.[41] The reduction potential measurements indicated a value of −315 mV for the two [4Fe-4S] centers, which showed unusual Mössbauer parameters. The reduced [4Fe-4S] centers showed quadruple splittings and isomer shifts typical of [4Fe-4S] centers. However, the hyperfine coupling constants for site 1 in each center were unusually small, as was

[38] C. P. Wang, R. Franco, J. J. G. Moura, I. Moura, and E. P. Day, *J. Biol. Chem.* **267,** 7378 (1992).

[39] E. P. Day, this series, Vol. 227, p. 437.

[40] M. J. Maroney, G. J. Colpas, C. Bagyinka, N. Baidya, and P. K. Mascharak, *J. Am. Chem. Soc.* **113,** 3962 (1991); K. L. Kovács, G. Tigyi, L. T. Thanh, S. Lakatos, Z. Kiss, and C. Bagyinka, *J. Biol. Chem.* **266,** 947 (1991).

[41] M. Teixeira, I. Moura, G. Fauque, D. V. DerVartanian, J. Le Gall, H. D. Peck, Jr., J. J. G. Moura, and B. H. Huynh, *Eur. J. Biochem.* **189,** 381 (1990).

observed earlier for the *D. gigas* hydrogenase [4Fe-4S] centers.[42] This could not be explained from the acquired data and it was suggested that this could be due to a difference in the "arrangement" of cysteine ligation akin to that in nickel hydrogenases and bacterial ferredoxins. The authors of this study suggested that owing to a small internal magnetic field the paramagnetic and diamagnetic [4Fe-4S] clusters are rendered indistinguishable and pointed out to an earlier study on *Dsm. baculatum* (Norway 4) soluble hydrogenase,[30] which had noted that a partial (50%) reduction of these clusters, even under H_2 pressure, did not prove that each enzyme molecule contained one reducible and one nonreducible cluster. Instead, this observation was considered a consequence of the presence of inactive apoprotein molecules without the nickel moiety.

The understanding of the [Fe-S] clusters in *Dsm. baculatum* hydrogenase, in spite of the Mössbauer study,[41] remained ambiguous. It was pointed out that further clarification would be possible only by studying the enzyme samples poised at various redox potentials that might correspond to different oxidation states of the [4Fe-4S] clusters. The need for redox–EPR–Mössbauer sampling experiments was emphasized.

Desulfovibrio vulgaris Hydrogenase

The experimental methods and strategy that led to the siting of [Ni–Fe–Se] hydrogenase in *D. vulgaris*[37] were as follows: the membrane-bound hydrogenases were purified from the crude cell extract solubilized in Triton-X by DEAE-BioGel and hydroxyapatite columns. Three hydrogenases were isolated from the membrane fractions of *D. vulgaris* (Hildenborough). One of the hydrogenases cross-reacted with antibodies raised against *Dsm. baculatum* (Norway 4) and was identified as an [Ni–Fe–Se] hydrogenase.[14] The remaining two hydrogenases cross-reacted with antibodies raised against [Ni–Fe] hydrogenase of *D. gigas*. Polyclonal antibodies raised against a purified sample of *Dsm. baculatum* (Norway 4) cytoplasmic [Ni–Fe–Se] hydrogenase, on a Western blot experiment, showed exclusive cross-reaction with the large subunit of the membrane-bound hydrogenase-3 in *D. vulgaris*.[37] The antibodies raised against the [Ni–Fe] hydrogenase of *D. gigas* cross-reacted with the large subunits of membrane fractions 1 and 2 but not with the periplasmic [Fe] hydrogenase and membrane-bound [Ni–Fe–Se] hydrogenase fractions. The experimental evidence pointed to the presence of two types of nickel-containing hydrogenases, immunologically related to the [Ni–Fe] hydro-

[42] M. Teixeira, I. Moura, A. V. Xavier, J. J. G. Moura, J. Le Gall, D. V. DerVartanian, H. D. Peck, Jr., and B. H. Huynh, *J. Biol. Chem.* **264,** 16435 (1989).

genase of *D. gigas* and to the [Ni–Fe–Se] hydrogenase of *Dsm. baculatum* (Norway 4), respectively.

Redox Potentiometry

Hydrogenases are electron transport enzymes. Their redox characteristics and catalytic role are fundamental properties, the study of which is important and central to our understanding of the metabolism of the sulfate-reducing bacteria; moreover, it may possibly reveal the catalytic mechanism for hydrogen evolution, which is of potential biotechnological value.[43–45] Most oxidations and reductions in biological systems occur through a one-electron transfer reaction. Among the experimental techniques used to obtain information on the redox behavior of metalloenzymes, the one that has proved most useful by far, especially in the swift establishment of the preliminary characteristics of [Fe–S] and other metallocenters, has been potentiometric EPR titration.[46–49]

The midpoint reduction potential, or redox potential, of a biological electron carrier is a measure of its capacity to accept or donate electrons and thus act as an oxidizing or reducing agent. To understand its function we need to know this property. A well-known example is the electron transport chain of mitochondria, which contains a series of carriers of progressively more positive potential, each one reducing the next in succession. Where there is a large enough difference in potential (about 250 mV), the change in free energy is adequate to permit the formation of ATP. In hydrogenase the midpoint potential of the active center metal cluster is expected to match that of the hydrogen substrate in order to be relevant for biological function.

The factors that influence the redox potentials of electron transfer proteins have been discussed previously.[50] A comprehensive treatment of oxidation–reduction reactions is given by Clark.[51]

The theory involved in the oxidation–reduction potentiometry, the requirement for redox mediators, the care of electrodes, the construction

[43] K. K. Rao and D. O. Hall, *Trends Biotechnol.* **2,** 124 (1984).

[43a] K. Sasikala, Ch. V. Ramana, P. R. Rao, and K. L. Kovács, *Adv. Appl. Microbiol.* **38,** 211 (1993).

[44] I. Okura, *Biochimie* **68,** 189 (1986).

[45] P. Cuendet, K. K. Rao, M. Grätzel, and D. O. Hall, *Biochimie* **68,** 217 (1986).

[46] P. L. Dutton, this series, Vol. 54, p. 411.

[47] G. S. Wilson, this series, Vol. 54, p. 396.

[48] R. Cammack and C. E. Cooper, this series, Vol. 227, p. 353.

[49] K. E. Paulsen, M. T. Stankovich, and A. M. Orville, this series, Vol. 227, p. 396.

[50] G. R. Moore, G. W. Pettigrew, and N. K. Rogers, *Proc. Natl. Acad. Sci. U.S.A.* **83,** 4998 (1986).

[51] W. M. Clark, "Oxidation-Reduction Potentials of Organic Systems." Waverly Press, Baltimore, Maryland, 1960.

and interpretation of Nernst curves, and other practical aspects of correlating the redox potential, E_h and redox state have been discussed previously by Dutton[46] and Wilson.[47] Most redox processes in hydrogenases occur through one-electron transfer reaction (i.e., $n = 1$ in the Nernst equation). In some instances a better fit is obtained with $n = 2$. This is thought to be due to spin–spin interaction between the metal clusters and it has been suggested that a strong cooperative effect occurs between the reduction of the clusters.[13c,19] Some examples of redox–EPR titrations and Nernst plots for hydrogenases can be found in Refs. 13c, 19, 41, and 52.

Redox Titration and Electron Paramagnetic Resonance Sampling

The titration vessel equipped with a sampling head is constructed with Pyrex glass (Fig. 2). The dimensions are described in the inset of Fig. 3. The side arm of the titration cell is equipped with a calomel reference electrode. The measuring electrode is a platinum wire wound around the reference electrode. Both electrodes are connected to a millivoltmeter for measurement of the potential. The sampling head is equipped with three or four needles connected to the purified inert gas (N_2/Ar) for constant purging such that there is a positive pressure in the vessel throughout the titration. One purging needle is kept just above the surface of the enzyme solution, and another needle is guided into the EPR tube or above the Mössbauer/MCD/EXAFS/optical absorption sample. The two openings on the sampling head are sealed with serum caps. Two stainless steel tubes (1 mm) serve as vents and microsyringe needle insertion points for sample withdrawal or reductant/oxidant addition.

To commence the titration, the cell assembly is purged with an inert gas for about 15–20 min. An appropriate degassed buffer is introduced into the cell, using a long-needle syringe. The solution is uniformly stirred while the mediator dyes are added. The choice of mediators is such that it covers a range of potential from +55 to −420 mV. It is general practice to use the following mediators for hydrogenase titrations: phenazine ethosulfate (+55 mV), methylene blue (+11 mV), resorufin (−51 mV), indigodisulfonate (−125 mV), 2-hydroxy-1,4-naphthaquinone (−145 mV), anthraquinone 2-sulfonate (−225 mV), phenosafranin (−252 mV), safranin O (−280 mV), neutral red (−340 mV), benzyl viologen (−350 mV), and methyl viologen (−440 mV). All potentials are with reference to a SHE. For an exploratory titration run, mediators covering a wide range of potentials (+150 to −420 mV) are used. Once the knowledge of approximate redox potential for a specific metal center is gained from this, fewer

[52] J. M. C. C. Coremans, J. W. van der Zwaan, and S. P. J. Albracht, *Biochim. Biophys. Acta* **997**, 256 (1989).

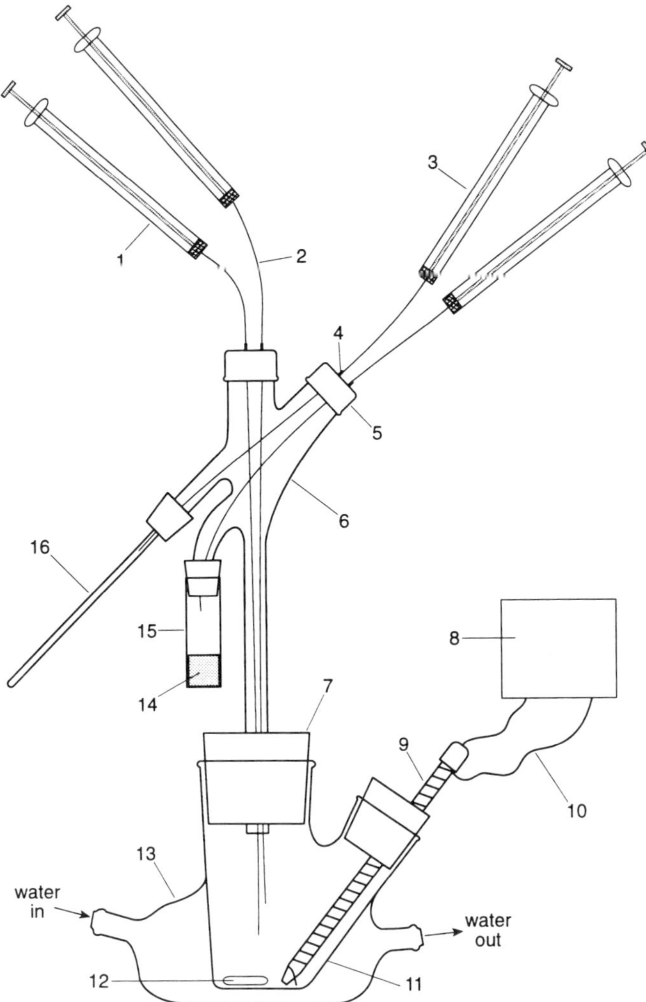

FIG. 2. Redox titration cell assembly (not to scale). (1) Syringe for reductant addition; (2) syringe for oxidant addition; (3) syringe for sample withdrawal for spectroscopy; (4) 1-mm stainless steel tube; (5) serum seal rubber cap; (6) Pyrex glass sampling head (~15-cm length, 0.8-cm o.d. and 0.4-cm i.d.); (7) silicone rubber stopper; (8) millivoltmeter; (9) calomel reference electrode; (10) platinum wire measuring electrode; (11) inner jacket of titration cell (~10-ml capacity; 0.3-ml dead volume below which the solution does not record the accurate potential; diameter at base, ~12 mm); (12) stirring bar; (13) outer jacket of titration cell (height, ~50 mm; base width, ~60 mm); (14) spectroscopic sample container (e.g., Mössbauer sample cup); (15) housing for sample containers (e.g., NMR, EXAFS, magnetic susceptibility, optical absorption); (16) EPR sample tube (3-mm i.d.).

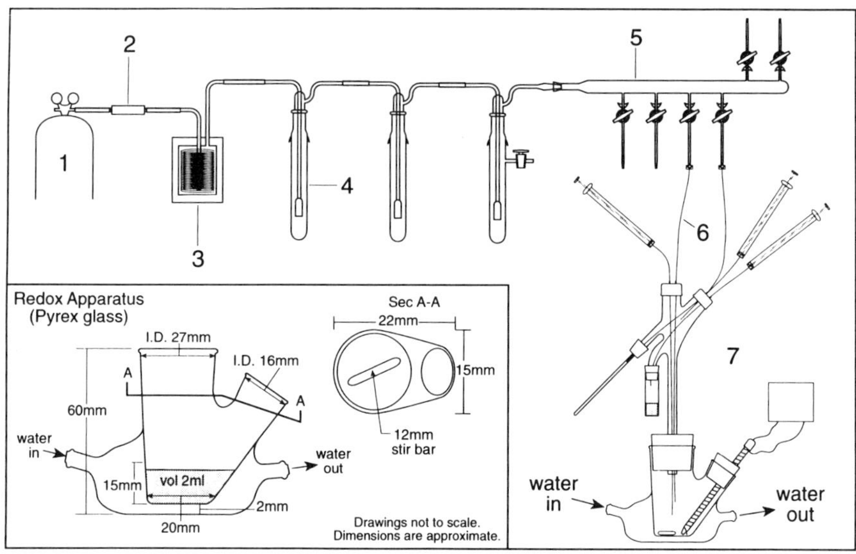

FIG. 3. Anaerobic assembly leading to the titration cell (not to scale). *Inset:* Approximate dimensions of the redox cell. (1) Inert gas cylinder; (2) deoxo catalyst for deoxygenation; (3) heated copper wire; (4) series of bubblers containing 2% alkaline dithionite solution; (5) manifold with taps; (6) needles carrying dry oxygen-free inert gas; (7) titration cell assembly.

mediators covering the specific, narrow range of potential can be employed to determine a refined redox potential for that metal center.

After the addition of mediators to the buffer solution the aliquot is kept stirring under an inert atmosphere for about 15–20 min. The concentrated hydrogenase is then added to the vessel to make up the desired final protein concentration and the enzyme–mediator solution is incubated for further 5–10 min. until a stable potential is read on the voltmeter. The titration can be started by microliter addition of the reductant, sodium dithionite, using a microsyringe with a long (15–20 cm) needle. With each addition of the reductant the decreasing potential should be allowed to equilibrate as indicated by a stable reading. A sample (200 μl) at a stable potential is withdrawn by using a microsyringe and by maneuvering the long needle into the EPR tube. The sample is frozen immediately by plunging it into a liquid nitrogen bath. The frozen EPR tube is replaced immediately with a fresh tube filled with the inert gas. During this operation, because the purging gas exerts a positive pressure in all parts of the cell assembly, there is no risk of an abrupt change of potential. The titration is continued stepwise by further additions of the reductant. Generally, samples are withdrawn at 20- to 30-mV intervals after allowing the potential to stabilize at each potential for about 5 min. Care should be taken to note the changes in potential, if any, before and after a desired sample is taken. After the titration, or if possible after each sample withdrawal, the poised hydrogenase samples are monitored by low-temperature EPR spectroscopy. Most iron-sulfur centers and nickel have well-assigned EPR signatures. The variations in signal intensity with respect to the change in potential are tabulated and plotted to obtain the redox profile of a selected EPR signal corresponding to the center. Using the Nernst equation, the experimental plot is fitted with the calculated Nernst curve with $n = 1$, in most cases, or in some cases with $n = 2$. From this plot the midpoint potential (E_m) for a particular EPR center is calculated. Patil *et al.*,[19] for example, give details of a redox titration for *D. vulgaris* [Fe] hydrogenase EPR centers and Cammack *et al.*,[13c] for example, give details of a redox titration for the nickel signal in *D. gigas* hydrogenase. Normally, preparation of a sample at a given stable potential takes about 3–5 min; and for the whole titration about 12–20 samples are poised at different potentials from oxidized state to the fully reduced state. The entire titration can be performed in about 2–3 hr. A protein concentration of about 50 μM is adequate for EPR titrations. For EXAFS, Mössbauer, and magnetic susceptibility assays, a sample concentration of about 0.5 mM or more is desirable.

Calibration of Redox Cell

It is important that the cell and the reference electrode be calibrated before and after each titration. The calibration is carried out as follows. A saturated solution of quinhydrone in a standard buffer (pH 7) is allowed to equilibrate for about 30 min. The stable potential is recorded. This value is usually +45 to +55 mV. From this value the relative value of the reference electrode (calomel, 244 mV; Ag/AgCl, 200 mV) for that titration can be adjusted, which in turn is used to calculate the redox potentials at which the hydrogenase samples are poised. The redox potential of saturated quinhydrone solution at 25° is 40 and 220 mV in pH 7 and pH 4 standard buffers, respectively. Generally a small shift from these values is observed. This does not affect the measurement as long as the relative difference between the values at pH 7 and pH 4 is kept constant. The redox potential, E_h, is obtained by the relationship

$$E = E_h - E^\circ \text{ (reference)}$$

where E° is the standard redox potential with reference to unit ion activity at 25° and 1-atm gas pressure and E_h is the actual potential measured. E_h is the difference between that of the measuring electrode (platinum) and the reference electrode (calomel). Potentials are referred to an arbitrary standard, the hydrogen electrode, because it is impossible to measure the absolute potential. The oxidation–reduction potentials are expressed relative to the hydrogen electrode reaction, $H_2 = 2H^+ + 2e^-$, and assigned a potential of 0.0000 V.

The measuring electrode (platinum) used in potentiometric titrations often tends to be "poisoned" by proteins, cyanides, sulfides, and oxidant and reductant reagents. Occasionally it may be necessary to restore the electrode by cleaning in chromic acid or aqua regia or polishing the metal surface with fine sandpaper.

Preparation of Reductant/Oxidant Solutions for Titration

Sodium dithionite ($Na_2S_2O_4$) is a widely used reductant in biological chemistry. Commercially available "pure" reagent is only about 80% pure and progressively deteriorates on storage owing to the presence of the oxidation product bisulfite. A method of crystallization of analytically pure reagent has been reported.[53] An incisive study of the effective redox potential of dithionite has been reported.[54] It is important to avoid the direct use of commercially available "pure" reagent because, for example, an equimolar addition of partially oxidized reductant to the oxidized hydrogenase (electron acceptor) does not bring about complete reduction and a further large excess of the reductant addition may be required. Such manipulation may cause complex formation between the by-products of dithionite and hydrogenase. The potential of dithionite in the range of pH 7 is controlled by the p*K* of bisulfite but above pH 8 this effect is diminished. For this reason, in practice, a high-pH buffer (pH 8–9) should be used to dissolve the reductant, which lowers the resultant solution in the region of neutral pH. The concentration of the reductant used in the redox titration is not crucial but addition of a large volume of dilute reagent may influence the final concentration of the enzyme. Generally a 100 m*M* dithionite solution is suitable. The solid reductant should be dissolved in an appropriate buffer (pH 8–9) that has been previously degassed and purged for about 30 min with an oxygen-free inert gas.

Potassium ferricyanide is a widely used oxidant. It is commercially available in the pure crystalline form and can be used directly. However, use of this reagent can cause cluster degradation of [4Fe-4S] centers.[55] Alternative oxidation reagents such as thionin or sodium ascorbate can

[53] C. E. McKenna, W. G. Gutheil, and W. Song, *Biochim. Biophys. Acta* **1075,** 109 (1991).

[54] S. G. Mayhew, *Eur. J. Biochem.* **85,** 535 (1978).

[55] A. J. Thomson, A. E. Robinson, M. K. Johnson, R. Cammack, K. K. Rao, and D. O. Hall, *Biochem. Biophys. Acta* **637,** 423 (1981).

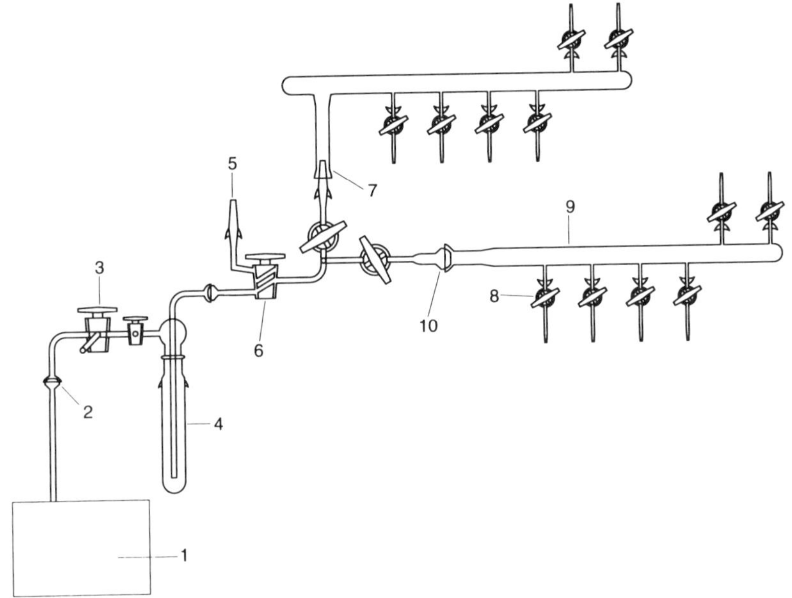

FIG. 4. Anaerobic manifold for sample preparation (not to scale). (1) Vacuum pump; (2) S19 ball-and-socket joint; (3) 8-mm HV stopcock; (4) cold trap, Quickfit 34/35 (Fisher Scientific, Pittsburgh, PA); (5) Quickfit 5-mm joint, for inert gas inlet; (6) 4-mm HV two-way stopcock, for inert gas/vacuum application; (7) Quickfit 14/23 joint; (8) 2-mm HV stopcock; (9) manifold, Pyrex (25-mm i.d.); (10) Quickfit S29 ball-and-socket joint.

be employed for oxidative titrations. A careful choice of an appropriate buffer should be made for use in redox titration and EPR sampling. The phosphate, Tris, and pyrophosphate buffers are prone to an abrupt change in pH on freezing and should be avoided when possible.[56] *N*-2-hydroxyethylpiperazine-*N*′-2-ethanesulfonic acid (HEPES), 3-[cyclohexylamino]-1-propane sulfonic acid (CAPS), Bicine, and Tricine buffers, found to be resistant to pH change on freezing, have been recommended. Methods of preparation of buffers have been described previously.[57]

Sampling for Ligand Reactions Using Anaerobic Manifold and Spin Quantitation of Electron Paramagnetic Resonance Spectrum

Hydrogenase sample (200 μl) is placed in an EPR tube stoppered with a rubber serum cap. It is attached to the manifold via a needle (22 gauge,

[56] D. L. Williams-Smith, R. C. Bray, M. J. Barber, A. D. Tsopanakis, and S. P. Vincent, *Biochem. J.* **167,** 593 (1977).

[57] V. S. Stoll and J. Blanchard, this series, Vol. 182, p. 24.

20 cm). The serum cap is sealed at the point of needle insertion with a touch of grease. The tip of the needle is kept about 2 cm above the enzyme to ensure uniform purging of the inert gas (He/N_2/Ar). The sample is first purged with the inert gas for a few minutes followed by gentle and slow evacuation using the two-way stopcock. During evacuation, air/oxygen is expelled in the form of fine bubbles. Care should be exercised to effect gentle evacuation during this operation to avoid any suction of the protein solution, which can be caused by abrupt application of vacuum. Evacuation is followed by purging of inert gas by using the two-way stopcock. Three to four cycles of the gentle evacuation/purge operation are adequate to render the enzyme oxygen free. Hydrogenases withstand this treatment without any loss of specific activity or degradation. Oxygen-expelled enzyme sample is now ready for further treatment: addition of a reductant, ligand, and reagents such as KCN, $NaNO_2$, CO, and NO. The gas-phase reagents could be added directly using a gas-tight syringe. The reagents in solution are, similarly, rendered oxygen free prior to the treatment with hydrogenase. It is convenient to handle the reagent solutions using a microsyringe equipped with a long (15- to 20-cm) needle. The reagent is injected into the enzyme solution, attached to the manifold, through the serum cap septum. The sealed EPR tube is detached from the manifold and the reaction mixture is allowed to stand for 10–15 min with occasional mixing of the reaction mixture by gentle tapping outside the EPR tube. The sample is rapidly frozen in an isopentane/liquid nitrogen bath and is ready for recording the EPR spectrum at liquid helium temperatures. The sampling procedure is similar for preparation of samples for study by other techniques such as the Mössbauer technique, optical absorption, EXAFS, magnetic susceptibility, and MCD. Readers should refer to special techniques for preparation of EPR samples described previously by Beinert *et al.*[58] and Cammack and Cooper.[48] The technique of expelling molecular oxygen from protein sample for magnetic susceptibility measurements has been described by Day.[39]

Ligand isotopes such as $^{17}O_2$, ^{13}CO, ^{13}CN, ^{77}Se, and $^{2}H_2O$ have been employed as specific probes to study selected metal sites.[59–62] Because of

[58] H. Beinert, W. H. Orme-Johnson, and G. Palmer, this series, Vol. 54, p. 111.

[59] J. W. van der Zwaan, J. M. C. C. Coremans, E. C. M. Bouwens, and S. P. J. Albracht, *Biochim. Biophys. Acta* **1041,** 101 (1990).

[60] Y. H. Huang, J. B. Park, M. W. W. Adams, and M. K. Johnson, *Inorg. Chem.* **32,** 375 (1993).

[61] R. Franko, I. Moura, J. Le Gall, H. D. Peck, Jr., B. H. Huynh, and J. J. G. Moura, *Biochim. Biophys. Acta* **1144,** 302 (1993).

[62] A. Chapman, R. Cammack, E. C. Hatchikian, J. McCracken, and J. Peisach, *FEBS Lett.* **242,** 134 (1988).

the isotope nuclear spin interaction with the electron spin of the metal paramagnetic center, a hyperfine change in line shape of the EPR/ENDOR signal becomes observable; isotopes thus help to extract information about the active metal center and the mode of hydrogen binding in hydrogenase.

The literature on EPR studies of metalloproteins is extensive. Relevant examples are articles by Beinert and Albracht,[63] Ohnishi and King,[64] Fee,[65] Palmer,[66] Hagen,[67] and Aasa and Vänngård[68] and Cammack *et al.*[69] Experimental aspects of biological EPR have been described elsewhere.[48,49,63,64] An account of the considerations and method of spin quantitation of EPR signals of biological material has been discussed previously.[48,49,65,69,69a,70]

A practical consideration, concerning careful control of the cavity/sample temperature in an EPR experimental run, needs further emphasis. In the following discussion, a "run" constitutes the period of instrument operation from the cooldown to the stable constant temperature of the system to the shutdown of the system after several hours of measurement. While recording a spectrum at liquid helium temperatures the operator needs to be constantly vigilant. The major factors/operations that contribute to stabilize a constant temperature are as follows: (1) pressure of the gas phase above the liquid level in the liquid helium tank (normally 3 psi), (2) flow rate setting (normally 50%) of the helium pump, (3) main helium valve opening on the helium transfer line connecting the cryostat, (4) the needle valve on the cryostat, (5) flow rate of the helium gas exiting from the cryostat (normally 40%), and (6) automatic setting of the heating coil of the temperature recording/display device, with respect to helium flow. These operations must be synchronous and finely balanced; otherwise the displayed temperature can be deceptive, and does not accurately represent the cavity temperature. In practice, inaccuracies occur despite a precise calibration of the temperature-recording device with a carbon thermometer against liquid nitrogen and liquid helium temperatures. One occasionally encounters situations in which, for example, a reference standard Cu^{2+} EDTA spectrum (or a protein sample) recorded under a set of nonsaturating instrumental conditions at a constant temperature shows varying signal intensity in comparison with the spectrum of the same sample under

[63] H. Beinert and S. P. J. Albracht, *Biochim. Biophys. Acta* **683,** 245 (1982).

[64] T. Ohnishi and T. E. King, this series, Vol. 53, p. 483.

[65] J. A. Fee, this series, Vol. 49, p. 512.

[66] G. Palmer, *Biochem. Soc. Trans.* **13,** 548 (1985).

[67] W. R. Hagen, *Adv. Inorg. Chem.* **38,** 165 (1992).

[68] R. Aasa and T. Vänngård, *J. Magn. Reson.* **19,** 308 (1975).

[69] R. Cammack, P. K. Knowles, and W. J. Ingledew (eds.), *Biochem. Soc. Trans.* **13**(3), (1985).

[69a] J. R. Pilborw and G. R. Hanson, this series, Vol. 227, p. 331.

identical instrumental conditions and at apparently the same constant temperature in a different EPR run on some other day (see Table I). This leads to serious errors in reporting spin quantitations and it emphasizes the great importance of the careful control of the helium flow in the system, which affects the accuracy of the cavity temperature causing fluctuation of signal intensity. To prove uniform constant cavity temperatures in different EPR runs on different days, of course, one must observe identical signal intensities for a given reference standard under identical EPR conditions. Fluctuating signal intensities measured at the same (apparent) constant stable temperature and under identical instrument conditions make it difficult to develop a benchmark intensity scale for an EPR signal, as a time saving procedure, from which one can estimate the spin concentration of a similar signal measured in a different EPR run or in a different laboratory. However, measurement of several samples, for example, of a redox titration, in one EPR run, the relative intensities of a signal at constant stable temperature and identical instrument conditions remain valid. For this reason, it is good practice to use relative spin intensities measured in a single run to construct the redox profile to fit the Nernst curve. The use of a mix of sample intensities measured in different EPR runs to construct a redox profile and fit a Nernst curve can be misleading and should be avoided. Fluctuation of signal intensity and weak resonance

TABLE I

VARIATION OF ABSOLUTE VALUES OF INTEGRATED AREA WITH RESPECT TO CONSTANT TEMPERATURE READINGS IN DIFFERENT ELECTRON PARAMAGNETIC RESONANCE RUNS[a]

Run 1		Run 2		Run 3	
Temperature (K)	EPR signal integrated area	Temperature (K)	EPR signal integrated area	Temperature (K)	EPR signal integrated area
10.2	6.56	10.5	5.79	10.2	4.15
10.2	6.03	10.5	5.95	10.3	4.14
10.2	5.91	10.5	5.77	10.2	4.11
10.2	5.83	10.5	5.51	10.1	4.05
10.1	5.76	10.5	5.61	10.3	4.03

[a] EPR instrument and cooling system conditions: Cu^{2+} EDTA (1.27 m*M*) reference standard; microwave power, 2 μW; receiver gain, 5×10^4; modulation amplitude, 10 G; phase 90°; liquid helium tank pressure, 3 psi; helium gas flow exit, 40%. Illustrated here is an occasional systematic anomaly associated with the operation of an EPR instrument at liquid helium temperatures. Although the absolute values of the integrated area between the runs, for the same sample, recorded at apparently constant stable temperature (10 K), differ significantly, the relative values within a run show minor variation and hold well. The readings were recorded over a period of 6–8 hr at intervals.

for some metal centers (e.g., nickel signal in hydrogenases often quantitates far less than the amount determined by chemical analysis) make it difficult to judge the significance of a center in metalloenzymes. It is therefore good practice to express the standard deviation in reporting the spin quantitation data of a set of at least five measurements of a signal in one EPR run. The following is a brief outline of the quantitation of spin for an $S = \frac{1}{2}$ EPR species. For detailed information the reader should consult Refs. 48, 63, 65, and 68–70. The area under the absorption envelope of an EPR signal is calculated, using a computer program, by double integration. The spin intensity is worked out by the following relationship:

$$\text{Spin intensity} = AT/\sqrt{P}\,Gg_{\text{av}}$$

where g_{av} is given by the Aasa and Vänngård procedure,[68]

$$g_{\text{av}} = \tfrac{2}{3}[(g_x^2 + g_y^2 + g_z^2)/3]^{1/2} + [\tfrac{1}{3}(g_x + g_y + g_z)/3]$$

where A is the doubly integrated area under the absorbance peak of EPR signal, T is the temperature of the sample (degrees Kelvin), P is the microwave power (in milliwatts), and G is the instrument gain.

From the spin intensities of the sample and the reference, measured under identical conditions, and assuming the reference spin per molecule to be unity, the spin concentration per enzyme molecule is calculated using the following relationship:

$$\text{Spin/molecule} = \frac{S_s P_R D_R^2}{S_R P_S D_S^2}$$

where S_S is the spin intensity of the sample, S_R is the spin intensity of the reference, P_R is the protein concentration of the reference, P_S is the protein concentration of the sample, D_R is the internal diameter of the EPR tube containing reference, and D_S is the internal diameter of the EPR tube containing sample.

From the above relationship it is evident how the experimentally determined protein concentration of the sample, P_S, exerts a significant influence on the calculated spin/molecule result, on which one bases the conclusion about a paramagnetic signal being a minor impurity or a major species.

Concluding Remarks

The discussion presented here provides a few relatively simple guidelines for the growth of *Desulfovibrio* bacteria, hydrogenase purification,

[70] M. L. Randolf, *in* "Biological Applications of Electron Spin Resonance" (H. M. Swartz, J. R. Bolton, and D. C. Borg, eds.), p. 119, Wiley (Interscience), New York, 1972.

enzyme manipulation by redox potentiometry and ligand interaction, which can be monitored by EPR spectroscopy.

Experimental aspects such as the isotope enrichment procedure for the incorporation of an element during bacterial growth and peak growth phase; the choice of chromatography and purification strategy; redox potential poising of intermediate species in a titration, especially at high protein concentrations (>200 μM), for example, for Mössbauer/EXAFS studies; the operational aspects of the EPR instrument and careful control of cavity temperature to record a spectrum for a reliable estimation of spin quantitation; the protein estimation; and sampling protocols for interaction of ligands with metal centers and the choice of spectroscopic (e.g., EPR) conditions to observe isolated metal centers from a mixed species; all these aspects by and large tend to be empirical in nature and a considerable amount of trial and error may be involved. However, execution of a carefully designed experimental strategy promotes a high degree of confidence in data acquisition and helps to arrive at a correct analysis of the results. The empirical aspect of investigative work on hydrogenases requires constant rethinking of conventional procedures, analytical methods, and even interpretation of the results.

Acknowledgments

I wish to thank Angie Stockton for excellent typing of the manuscript. The preparation of this chapter and part of the research work were supported by Department of Energy Grant DE FG0593ER20127, National Institute of Health Grant GM 34903-08, and National Science Foundation Grant MCB 9005734 to Professor H. D. Peck, Jr. I am thankful to numerous colleagues whose names appear in the references and to Professors J. Le Gall and H. D. Peck, Jr., for their encouragement and support.

[7] Pyruvic Acid Phosphoroclastic System

By LARRY L. BARTON

Introduction

Pyruvic acid is an important carbon source for sulfate-reducing bacteria. The transformation of pyruvic acid to acetic acid, CO_2, and H_2 [Eq. (1)] is the product of a complex set of reactions conducted by sulfate-reducing bacteria as well as numerous other anaerobic bacteria[1,2] and by hydrogen-forming protozoans.[3] The bacterial metabolism of pyruvic acid produces either acetyl phosphate, H_2, and CO_2 [Eq. (2)] or acetyl phosphate and formic acid [Eq. (3)]. In general, reactions involving the production of acetyl phosphate from pyruvic acid have been collectively described as phosphoroclastic reactions.[4]

$$\text{Pyruvic acid} + \text{ADP} + P_i \rightleftharpoons \text{acetic acid} + H_2 + CO_2 + \text{ATP} \quad (1)$$
$$\text{Pyruvic acid} + P_i \rightleftharpoons \text{acetyl phosphate} + CO_2 + H_2 \quad (2)$$
$$\text{Pyruvic acid} + P_i \rightleftharpoons \text{acetyl phosphate} + \text{formic acid} \quad (3)$$

Because the use of the term *phosphoroclastic reaction* has evolved over the years, definition is needed to appreciate the nuances of this metabolism. As noted in an earlier review,[5] Utter and Werkman were searching for a reaction in which the condensation of a C_2 and a C_1 compound could be accomplished by extracts of *Escherichia coli.* Their results indicated the presence of a reversible phosphoroclastic split of pyruvic acid into acetyl phosphate and formic acid [Eq. (3)]. This evidence of a hydroclastic reaction with acetyl phosphate supported the hypothesis of Lipmann that energy could be produced from a "phosphoroclastic split" of pyruvic acid. Thus, the phosphoroclastic reaction designated the reversible formation of acetyl phosphate from pyruvic acid. For some time a distinction was made with reference to the reaction products. Phosphoroclastic reactions producing H_2 and CO_2 in addition to acetyl phosphate [Eq. (2)] were designated "clostridial type" whereas those producing formate and acetyl phosphate [Eq. (3)] were called "Enterobacteriaceae type."[4] Equation (2) was catalyzed by pyruvate : ferredoxin 2-oxidoreductase whereas the thiolytic cleavage of pyruvate in Eq. (3) was attributed to pyruvate formate-lyase. The term *phosphoroclastic reaction* became associated only with the reversible transformation of pyruvic acid to acetyl phosphate, H_2, and CO_2.[6]

[1] B. B. Buchanan, *in* "The Enzymes" (P. D. Boyer, ed.), 3rd Ed., Vol. 6, p. 193. Academic Press, New York, 1972.

[2] B. B. Buchanan, *in* "Iron-Sulfur Proteins" (W. Lovenberg, ed.), Vol. 1, p. 129. Academic Press, New York, 1973.

[3] M. Müller, *Annu. Rev. Microbiol.* **29,** 467 (1975).

[4] H. W. Doelle, "Bacterial Metabolism." Academic Press, New York, 1975.

[5] C. H. Werkman, *in* "Bacterial Physiology" (C. H. Werkman and P. W. Wilson, eds.), p. 404. Academic Press, New York, 1951.

[6] G. Gottschalk, "Bacterial Metabolism." Springer-Verlag, New York, 1986.

Reaction Systems

The phosphoroclastic reaction [Eq. (2)] represents the sum of several individual steps [Eqs. (4)–(6)]. Pyruvate:ferredoxin 2-oxidoreductase (PFO) is the CoA-acetylating enzyme that catalyzes Eq. (4) in the presence of thiamine pyrophosphate (TPP) with an unstable intermediate, hydroxyethyl-TPP, found only in association with the enzyme. The metabolic step involving generation of molecular hydrogen [Eq. (5)] is catalyzed by hydrogenase whereas phosphotransacetylase accounts for the formation of acetyl phosphate from acetyl-CoA [Eq. (6)]. Although the phosphoroclastic reaction ends with the formation of acetyl phosphate from pyruvic acid, the continued metabolism of acetyl phosphate may result in the production of acetic acid by acetokinase [Eq. (7)]. Thermodynamic values of these steps in phosphoroclastic reaction are available in an earlier report.[7]

$$\text{Pyruvic acid} + \text{CoA} + \text{ferredoxin}_{\text{oxidized}} \rightleftharpoons \text{acetyl-CoA} + CO_2 + \text{ferredoxin}_{\text{reduced}} + 2H^+ \quad (4)$$

$$\text{Ferredoxin}_{\text{reduced}} + 2H^+ \rightleftharpoons H_2 + \text{ferredoxin}_{\text{oxidized}} \quad (5)$$

$$\text{Acetyl-CoA} + P_i \rightleftharpoons \text{acetyl phosphate} + \text{CoA} \quad (6)$$

$$\text{Acetyl phosphate} + \text{ADP} \rightleftharpoons \text{acetic acid} + \text{ATP} \quad (7)$$

Demonstration of Phosphoroclastic Reaction

Preparation of Cell Extracts

Enzymes of the phosphoroclastic reaction are localized in the cytoplasm[8] and are constitutive in sulfate-reducing bacteria; therefore, special cultivation procedures are not required for enzyme production. After bacteria have been collected from the culture by centrifugation and resuspended in 0.05 *M* Tris-HCl buffer at pH 7.5, bacterial cells can be disrupted by passage through a French pressure cell at 10,000 psi or by sonication treatment for 90 sec. The cell extract is clarified by centrifugation at 10,000 *g* for 20 min followed by centrifugation at 100,000 *g* for 1 hr to resolve the membranes from the soluble protein. The centrifugate contains the soluble protein and is used for enzyme reactions after endogenous compounds are removed by anaerobic dialysis against 0.005 *M* Tris-HCl, pH 7.5. Alternatively the extract can be passed over a Sephadex G-25 column (2.5 × 30 cm) equilibrated with 0.005 *M* Tris-HCl, pH 7.5, operating under

[7] R. K. Thauer, K. Jungermann, and K. Decker, *Bacteriol. Rev.* **41,** 100 (1977).
[8] J. M. Odom and H. D. Peck, Jr., *J. Bacteriol.* **147,** 161 (1981).

an N_2 atmosphere.[9] Cell extracts should be kept under N_2 while being held at 4° or when stored frozen at −20 or −70°.

Removal of Electron Carriers from Extracts

To study electron transport activity [Eqs. (4) and (5)] it is necessary to remove ferredoxin and flavodoxin from the cell extract. This can be accomplished by passing the dialyzed or Sephadex G-25-treated extract over a DEAE-cellulose column (2.5 × 2.5 cm) by the method previously reported.[10] Owing to the extremely high affinity of ferredoxin and flavodoxin for DEAE-cellulose, a protein fraction results that is low in natural electron carriers for this reaction. Therefore, additions of ferredoxin, flavodoxin, or artificial electron acceptors can be made under controlled conditions to the DEAE-treated protein fraction to assess electron transport. Soluble cytochrome can be removed from the protein fraction by procedures described elsewhere in this volume. In a study of the phosphoroclastic reaction in *Desulfovibrio desulfuricans*, cytochrome c_3 was removed from cell extracts by passage over an Amberlite CG-50 type 2 resin.[11]

Enzyme Reactions

Pyruvic : Ferredoxin 2-Oxidoreductase

The enzyme responsible for the oxidation of pyruvic acid with electrons transferred to ferreodoxin is pyruvic : ferredoxin 2-oxidoreductase (PFO) (EC 1.2.7.1).[12] The enzyme catalyzing decarboxylation of pyruvic acid has occasionally been referred to as pyruvic dehydrogenase; however, this designation could lead to confusion and should be avoided. Pyruvic dehydrogenase (EC 1.2.2.2) is the enzyme that uses cytochrome b_1 as an electron acceptor in the formation of acetic acid plus CO_2 and is not associated with the phosphoroclastic system. One of the three enzymes making up the pyruvic dehydrogenase complex of *E. coli* is pyruvic dehydrogenase (EC 1.2.4.0)[13]; however, this enzyme is markedly different from PFO. Additionally, a distinction should be made between PFO and pyruvic

[9] L. L. Barton, J. LeGall, and H. D. Peck, Jr. *Biochem. Biophys. Res. Commun.* **41,** 1036 (1970).

[10] L. L. Barton, J. LeGall, and H. D. Peck, Jr., *in* "Horizons of Bioenergetics" (A. S. Pietro and H. Guest, eds.), p. 33. Academic Press, New York, 1972.

[11] B. Suh and J. M. Akagi, *J. Bacteriol.* **91,** 2281 (1966).

[12] H. D. Peck, Jr., *in* "The Sulfate-Reducing Bacteria: Contemporary Perspectives" (J. M. Odom, Jr., and R. Singleton, Jr., eds.), p. 41. Springer-Verlag, Berlin, 1993.

[13] L. J. Reed, *Acc. Chem. Res.* **7,** 40 (1974).

decarboxylase (EC 4.1.1.1), which catalyzes the formation of acetaldehyde and CO_2.

Assay of Phosphoroclastic Reaction

The phosphoroclastic reaction would usually be conducted in 1- to 3-ml volumes, using stoppered cuvettes or tubes flushed with purified N_2. A typical 1-ml reaction mixture would contain 10 mg of soluble protein, 100 m*M* sodium pyruvate, 0.15 m*M* coenzyme-A, 50 m*M* potassium phosphate at pH 7.0, and 100 m*M* Bis-Tris buffer, pH 7.0. Additions may include 0.1 mg of ferredoxin or desired electron acceptor. Activity of the reaction can be measured by following specific substrates and products of the reaction (Table I).

The author's laboratory has found the following radiometric assay useful to measure the phosphoroclastic reaction in extracts of *Desulfovibrio gigas*. To the main space of Warburg flasks, the following additions are made to give a 2-ml reaction: 0.1 ml of 100 m*M* Bis-Tris buffer at pH 7.0, 0.1 ml of 100 m*M* potassium phosphate adjusted to pH 7.0, 0.1 ml of 1.5 m*M* coenzyme-A, 35–50 mg of cell protein extract, and water to 2

TABLE I
ASSAYS FOR PHOSPHOROCLASTIC REACTION

Compound measured	Method	Relevant equations	Ref.[a]
Pyruvic acid[b]	2,4-Dinitrophenylhydrazine	(1–3)	1
Pyruvic acid[b]	Lactic dehydrogenase	(1–3)	2
Acetyl phosphate	Acetohydroxamic acid	(2), (3), (6)	3
Acetyl-CoA	Spectrophotometric	(7)	4
H_2 evolution/uptake	Standard manometric techniques	(1), (2), (5)	5
CO_2 evolution/uptake	Manometric techniques	(1), (2)	5
$CH_3CHOCOOH$/ CO_2 exchange	2,4-Dinitrophenylhydrazine	(1), (2)	6
Ferredoxin reduction	Spectrophotometric (420 nm)	(4)	7

[a] Key to references: (1) T. E. Friedmann and G. E. Haugen, *J. Biol. Chem.* **147,** 415 (1943); (2) T. Bücher, R. Czok, W. Lambprecht, and E. Latzko, *in* "Methods of Enzymatic Analysis" (H. U. Bergmeyer, ed.), p. 253. Academic Press, New York, 1965; (3) F. Lipmann and L. C. Tuttle, *J. Biol. Chem.* **159,** 21 (1945); (4) D. Decker, *in* "Methods of Enzymatic Analysis" (H. U. Bergmeyer, ed.), p. 419. Academic Press, New York, 1965; (5) W. W. Umbreit, *in* "Manometric Techniques" (W. W. Umbreit, R. H. Burris, and J. F. Stauffer, eds.), Chap. 1. Burgess, Minneapolis, Minnesota, 1949; (6) B. Suh and J. M. Akagi, *J. Bacteriol.* **91,** 2281 (1966); (7) L. E. Mortenson, R. C. Valentine, and J. E. Carnahan, *J. Biol. Chem.* **238,** 794 (1963).

[b] Measurements can also be by HPLC procedures.

ml. The center well contains 0.25 ml of 0.1 *M* ethanolamine to trap CO_2 evolved from the decarboxylation of pyruvate. One side arm contains 6.25 μmol of $K_3Fe(CN)_6$ whereas to the other is added 10 μmol of sodium pyruvate containing $CH_3CO^{14}COOH$ to give an activity of 1×10^6 cpm/μmol of pyruvate. Through the use of syringe needles, the Warburg flasks are flushed with purified N_2 for 15 min. The reaction is initiated by tipping the pyruvate and ferricyanide from the side arms into the main space of the flasks. After incubation for 30 min at 35° in shaking water bath, 0.2 ml of 1 *M* citric acid is injected into the main space to stop the reaction. After another 30 min of additional shaking, 0.1 ml is removed from the center well and transferred to a vial containing liquid scintillation fluid. Radioactivity is then measured in a liquid scintillation counter.

Phosphotransacetylase

The conversion of acetyl-CoA to acetyl phosphate [Eq. (6)] is catalyzed by acetyl-CoA : orthophosphate acetyltransferase (EC 2.3.1.8), commonly called phosphotransacetylase.[14] To date, this enzyme has not been purified from sulfate-reducing bacteria. Information about enzyme purification and reaction parameters is based on the enzyme from *E. coli,*[14] *Clostridium kluyveri,*[15] *Aerobacter aerogenes,*[16] *Lactobacillus fermenti,*[17] *Clostridium acidiurici,*[18] and *Vellonella alcalescens.*[19] A spectrophotometric assay is used to demonstrate the presence of phosphotransferase by following the utilization of acetyl-CoA. The contents in a 3-ml cuvette would be as follows[20]: 85 m*M* Tris-HCl buffer at pH 7.4, 1.7 m*M* reduced glutathione, 0.4 m*M* coenzyme-A, 7.4 m*M* acetyl phosphate, 19 m*M* ammonium sulfate, and cell extract.

Acetokinase

The conversion of acetyl phosphate to acetic acid [Eq. (7)] is by ATP : acetate phosphotransferase (EC 2.7.2.1), commonly called acetylkinase, acetokinase, or acetate kinase. Acetokinase is a highly regulated enzyme that has been examined in numerous bacteria including *Desulfovibrio*

[14] M. Shimizu, T. Suzuki, K. Kameda, and Y. Abiko, *Biochem. Biophys. Acta* **191,** 550 (1969).
[15] H. U. Bergmeyer, G. Holz, H. Klotzsch, and G. Lang, *Biochem. Z.* **338,** 114 (1963).
[16] T. D. K. Brown, C. R. S. Pereira, and F. C. Störmer, *J. Bacteriol.* **112,** 1106 (1972).
[17] T. Nojiri, F. Tanaka, and I. Nakayama, *J. Bacteriol.* **69,** 789 (1971).
[18] J. R. Robinson and R. D. Sagers, *J. Bacteriol.* **112,** 465 (1972).
[19] H. R. Whiteley and R. A. Pelroy, *J. Biol. Chem.* **247,** 1911 (1972).
[20] H. U. Bergmeyer, M. Grabl, and H.-E. Walter, *in* "Methods of Enzymatic Analysis" (H. U. Bergmeyer, ed.), p. 126. VCH Publ., Weinheim, 1988.

vulgaris,[21] *D. desulfuricans*,[22] *A. aerogenes*,[16] *V. alcalescens*,[23], *Clostridium thermoaceticum*,[24] *Streptococcus faecalis*,[25] *E. coli*,[25] and *Archaeoglobus fulgidus*.[26] The regulation of this key enzyme in pyruvate catabolism is important, with ATP stimulating the activity of acetokinase in *D. vulgaris*.[27]

The assay for acetokinase employs a coupled enzyme reaction.[20] ADP [produced via Eq. (8)] is converted to pyruvic acid with pyruvic kinase (EC 2.7.1.40) [Eq. (9)] and finally pyruvic acid is measured by following NADH oxidation with the lactic dehydrogenase (EC 1.1.1.27) reaction [Eq. (10)]. The 3-ml reaction would contain the following: 67 m*M* triethanolamine buffer at pH 7.6, 333 m*M* sodium acetate, 5.4 m*M* ATP, 1.1 m*M* cyclohexylammonium phosphoenolpyruvate, 0.32 m*M* NADH (sodium salt), 1.33 m*M* $MgCl_2$, 2.7 U of pyruvic kinase, 8.3 U of lactic dehydrogenase, 4.8 U of myokinase/adenylate kinase, and enzyme preparation.

A more direct method to measure acetokinase is to use acetate and ATP as initial reactants.[28] The 1-ml reaction mixture would contain 50 m*M* Tris buffer at pH 7.4, 10 m*M* $MgCl_2$, 10 m*M* ATP, 800 m*M* potassium acetate, 700 m*M* hydroxylamine, and the enzyme fraction. The 2 *M* solution of hydroxylamine is prepared by mixing 4 *M* $NH_2OH \cdot HCl$ and 2 *N* KOH. The addition of 1 ml of a 10% trichloroacetic acid (TCA) solution is used to stop the reaction and the deproteinated solution is mixed with an equal volume of ferric chloride solution (5% $FeCl_3$ in 0.1 *N* HCl). The hydroxylamine reacts with acetyl phosphate to give acetohydroxamic acid. After 20 min in the presence of ferric ions, a colored ferric–acetohydroxamic acid complex is produced and read at 540 nm.

$$\text{Acetic acid} + \text{ATP} \rightleftharpoons \text{acetyl phosphate} + \text{ADP} \quad (8)$$

$$\text{ADP} + \text{phosphoenolpyruvate} \rightleftharpoons \text{ATP} + \text{pyruvic acid} \quad (9)$$

$$\text{Pyruvic acid} + \text{NADH} + \text{H}^+ \rightleftharpoons \text{lactate acid} + \text{NAD}^+ \quad (10)$$

Role of Phosphoroclastic Reaction in Sulfate-Reducing Bacteria

Electron and H_2 Metabolism

Reduced ferredoxin generated by the phosphoroclastic reaction may interface with thiosulfate reductase, bisulfite reductase, and other reac-

[21] M. Ogata and T. Yagi, *J. Biochem.* (*Tokyo*) **100,** 311 (1986).

[22] M. S. Brown and J. M. Akagi, *J. Bacteriol.* **92,** 1273 (1966).

[23] R. A. Pelroy and H. R. Whiteley, *J. Bacteriol.* **105,** 259 (1971).

[24] A. Schaupp and L. G. Ljungdahl, *Arch. Mikrobiol.* **100,** 121 (1974).

[25] J. J. I. Thorne and M. E. Jones, *J. Biol. Chem.* **238,** 2992 (1963).

[26] D. Möller-Zinkhan, G. Börner, and R. K. Thauer, *Arch. Microbiol.* **152,** 362 (1989).

[27] M. G. Yates, *Biochem. J.* **103,** 32c (1967).

[28] I. A. Rose, M. Grunberg-Manago, S. R. Korey, and S. Ochoa, *J. Biol. Chem.* **211,** 737 (1954).

tions in which sulfur compounds are used as electron acceptors by the sulfate-reducing bacteria.[29–31] Flavodoxin may substitute for ferredoxin in some of the low-potential reactions;[10,31,32] however, these two electron transport proteins are not entirely interchangeable because flavodoxin cannot transport electrons between hydrogenase and sulfite reductase, nor can it provide an appropriate coupling for oxidative phosphorylation by *D. gigas* in cell-free systems.[10,32,33]

As indicated in Table II, rubredoxin, desulforedoxin, and rubrerythrin would not be expected to interact with the phosphoroclastic reaction because the redox potential is not sufficiently negative. Cytochrome c_3 has been shown to transfer electrons from reduced ferredoxin to the hydrogenase in *D. desulfuricans*.[34] With cell-free extracts, a hydrogen-producing electron transport scheme for sulfate-reducing bacteria has been constructed[12]:

$$\begin{matrix}\text{Phosphoroclastic}\\ \text{reaction}\end{matrix} \longrightarrow \begin{matrix}\text{flavodoxin}\\ \text{or}\\ \text{ferredoxin}\end{matrix} \longrightarrow \text{Cytochrome } c_3 \longrightarrow \text{hydrogenase} \longrightarrow H_2$$

An important product of the phosphoroclastic reaction is the generation of molecular hydrogen because it represents a most efficient means of electron consumption. The metabolism of molecular hydrogen by sulfate-reducing bacteria is important because most can use H_2 as an electron donor to support growth.[30,35] Additionally, a hydrogen cycle has been proposed for the sulfate reducers that would function to energize metabolic processes.[32,36] Discussions of the hydrogen cycle[31,35] activity in bacteria focus, in part, on phosphoroclastic activity.

ATP Utilization

Reviews have discussed the requirement for ATP in the activation of sulfate before electron-accepting activity may proceed.[29–31,35] ATP pro-

[29] H. D. Peck, Jr., and T. Lissolo, *in* "The Nitrogen and Sulphur Cycles" (J. A. Cole and S. J. Ferguson, eds.), p. 99. 42nd Symposium of the Society for General Microbiology, Cambridge Univ. Press, Cambridge, 1988.

[30] J. LeGall and G. Fauque, *in* "Biology of Anaerobic Microorganisms" (A. J. B. Zehnder, ed.), p. 589. Wiley, New York, 1988.

[31] G. Fauque, J. LeGall, and L. L. Barton, *in* "Variations in Autotrophic Life" (J. M. Shively and L. L. Barton, eds.), p. 271. Academic Press, London, 1991.

[32] H. D. Peck, Jr., and J. LeGall, *Philos. Trans. R. Soc. London B* **298,** 443 (1982).

[33] L. L. Barton and H. D. Peck, Jr., *Bact. Proc. 143* American Society for Microbiology, Washington, D.C., 1970.

[34] J. M. Akagi, *J. Biol. Chem.* **242,** 2478 (1967).

[35] F. Widdel and T. A. Hansen, *in* "The Prokaryotes" (A. Balows, H. G. Trüper, M. Dworkin, W. Harder, and K.-H. Schleifer, eds.), 2nd Ed., Vol. 1, p. 583. Springer-Verlag, Berlin, 1992.

[36] J. M. Odom and H. D. Peck, Jr., *Annu. Rev. Microbiol.* **38,** 551 (1984).

TABLE II
INTERFACE OF PHOSPHOROCLASTIC REACTION WITH NITROGEN FIXATION

Redox compound	Potential E_0' (mV)	Catalyst for phosphoroclastic reaction	Electron donor for N_2 fixation	Ref.[a]
Ferredoxin (ox/red)				
[4Fe-4S][b]	−440	Yes	Yes	1
	−398[c]	Yes	Yes	2
[3Fe-4S][d]	−130	Yes	No	1
Flavodoxin (ox/red)				
(FH^*/FH_2)[e]	−440	Yes	Yes	1
(F/FH_2)	−371[f]	Yes	Yes	2
(F/FH^*)[e]	−140	No	No	1
ox/red ($1e^-$)	−115[f]	No	No	2
Rubredoxin (ox/red)	−57[g]	No	No	2
	0 to −50	No	No	3
Desulforedoxin (ox/red)	0 to −50	No	No	1
Rubrerythrin (ox/red)	+230	No	No	1

[a] Key for references: (1) G. Fauque, J. LeGall, and L. L. Barton, *in* "Variations in Autotrophic Life" (J. M. Shively and L. L. Barton, eds.), p. 271. Academic Press, London, 1991; (2) R. K. Thauer, K. Jugernamm, and K. Decker, *Bacteriol. Rev.* **41,** 100 (1977); (3) I. Moura, M. Bruschi, J. LeGall, J. J. G. Moura, and A. V. Xavier, *Res. Biochem. Biophys. Res. Commun.* **75,** 1073 (1977).

[b] Ferredoxin I from *D. gigas*.

[c] Average values for many ferredoxins.

[d] Ferredoxin II from *D. gigas*.

[e] Flavodoxin from *D. gigas*.

[f] Average values for many flavodoxins.

[g] Average values for many rubredoxins.

duced from the phosphoroclastic reaction by substrate-level phosphorylation contributes to the overall energy balance of the sulfate-reduction process when the bacteria are growing on lactic or pyruvic acids.

Fermentation of Pyruvate

Several strains of sulfate-reducing bacteria will grow in sulfate-free media if pyruvic acid is present.[37] The electron acceptor is either acetyl phosphate or acetyl-CoA with the production of ethanol. Pyruvic acid is not reduced to lactic acid in the sulfate-reducing bacteria. With *D. desulfuricans* grown in the presence of pyruvic acid in a sulfate-free medium, the products of fermentation have been reported[38] to include acetic

[37] J. R. Postgate, *Research (London)* **5,** 189 (1952).

[38] J. R. Postgate, *Abstr. 2nd Int. Congr. Biochim., Paris,* p. 92 (1952).

acid, ethanol, and CO_2. The end products of pyruvic acid metabolism with cell-free extracts of *D. vulgaris* Hildenborough have been reported[39] to be acetyl phosphate, ethanol, and CO_2 according to Eq. (11).

$$2\text{Pyruvic acid} + P_i \rightleftharpoons \text{acetyl phosphate} + 2CO_2 + \text{ethanol} \quad (11)$$

The enzymatic formation of ethanol has not been clearly established for sulfate-reducing bacteria but could be explained by a set of reactions involving an aldehyde dehydrogenase and an alcohol dehydrogenase. These reactions would be a reverse of the ethanol acetate fermentation of *C. kluyveri*[6] and could proceed as shown in Eqs. (12) and (13). For both of these reactions, NAD(P)H could provide the necessary reducing equivalents. Even though the system for ethanol production from pyruvic acid is unclear for sulfate reducers, evidence is strongly suggestive that phosphoroclastic activity is required.

$$\text{Acetyl phosphate or acetyl-CoA} \longrightarrow \text{acetaldehyde} \quad (12)$$

$$\text{Acetaldehyde} \longrightarrow \text{ethanol} \quad (13)$$

Nitrogen Fixation

Nitrogen fixation is a metabolic process that requires considerable energy and a low-potential electron donor. As shown for *Clostridium pasteurianum,* electrons from the phosphoroclastic reaction can be carried to the nitrogen-fixation complex by ferredoxin.[36] Presumably ferredoxin can be replaced in the reaction coupling pyruvate oxidation to nitrogen fixation by the fully reduced form of flavodoxin but not by rubredoxin or the 1*e* carrier state of flavodoxin. The levels of suitability of various electron acceptors to interface between the phosphoroclastic system and N_2 fixation system are given in Table II. It appears that electron carriers would have to have a redox potential less than −333 mV to transfer electrons successfully into the nitrogen fixation complex from the phosphoroclastic reaction.

According to studies,[41–43] the sulfate-reducing bacteria capable of nitrogen fixation include the following: *D. gigas* NCIMB 9332; *D. vulgaris* Hildenborough NCIMB 8303, Groningen NCIMB 11779, and Monticello

[39] J. C. Sadana, *J. Bacteriol.* **67,** 547 (1954).

[40] L. E. Mortenson, R. C. Valentine, and J. E. Carnahan, *J. Biol. Chem.* **238,** 794 (1963).

[41] J. R. Postgate, H. M. Kent, S. Hill, and H. Blackburn, *in* "Nitrogen Fixation and CO_2 Metabolism" (P. W. Ludden and J. E. Burris, eds.), p. 225. Elsevier, New York, 1985.

[42] P. A. Lespinat, Y. M. Berlier, G. D. Fauque, R. Toci, G. Denariaz, and J. LeGall, *J. Ind. Microbiol.* **1,** 383 (1987).

[43] I. Moura, G. Fauque, J. LeGall, A. V. Xavier, and J. J. Moura, *Eur. J. Biochem.* **162,** 547 (1987).

2 NCIMB 9442; *D. desulfuricans* Berre Sol NCIMB 8388 and Norway 4 NCIMB 8310, El Agheila Z NCIMB 8310, and Teddington R NCIMB 8312; *Desulfovibrio salexigens* British Guiana NCIMB 8403 and California 43 : 63 NCIMB 8364; and *Desulfovibrio africanus* Benghazi NCIMB 8401 and Wavis Bay NCIMB 8397. Strains of *Desulfobacter*[44] and *Desulfobulbus*[44] also display diazotrophic growth. Although the coupling of the phosphoroclastic reaction to nitrogen fixation has not been demonstrated in sulfate-reducing bacteria, such activity may be expected.

Perspectives and Conclusions

The phosphoroclastic reaction in sulfate-reducing bacteria is an essential component in cell energetics in which metabolism of pyruvate is involved. Knowledge of the basic reactions is available but the intricate mechanisms that are associated with molecular control have not been addressed. Future studies on the phosphoroclastic reaction in sulfate reducers is needed to provide further insight into our understanding of these unique bacteria.

[44] F. Widdel, *Arch. Microbiol.* **148,** 286 (1987).

[8] Monoheme Cytochromes

By TATSUHIKO YAGI

Introduction

Sulfate-reducing bacteria of the genera *Desulfovibrio* (*D.*) and *Desulfomicrobium* (*Dm.*) contain several *c*-type cytochromes, tetraheme cytochrome c_3, polyheme high molecular mass cytochrome *c* (Hmc), and monoheme cytochrome. Monoheme cytochrome c_{553} was first isolated from two strains of *Desulfovibrio vulgaris*, Hildenborough[1] and Miyazaki.[2] In contrast to cytochrome c_3, which is present in all species of these genera, cytochrome c_{553} is found only in a limited number of species; *Desulfovibrio gigas* lacks monoheme cytochrome,[3] for example. Mono-

[1] J. LeGall and M. Bruschi-Heriaud, *in* "Structure and Function of Cytochromes" (K. Okunuki, M. D. Kamen, and I. Sekuzu, eds.), p. 467. Univ. of Tokyo Press, Tokyo, and Univ. Park Press, Baltimore, Maryland, 1968.

[2] T. Yagi, *J. Biochem.* (*Tokyo*) **66,** 473 (1969).

[3] J. Le Gall and N. Forget, this series, Vol. 53, p. 613.

METHODS IN ENZYMOLOGY, VOL. 243

heme cytochrome from *Desulfomicrobium baculatum* Norway 4 (formerly known as *Desulfovibrio baculatus* Norway 4) is known as cytochrome $c_{553(550)}$.[4] These cytochromes are regarded as members of the cytochrome *c* superfamily, because their monoheme iron has histidyl and methionyl side chains as axial ligands.[4–7] This chapter summarizes knowledge and methodology on these monoheme cytochromes in sulfate-reducing bacteria.

Purification of Monoheme Cytochrome

Growth of Organism

Desulfovibrio vulgaris Hildenborough is cultured as described in this series.[3] *Desulfovibrio vulgaris* Miyazaki can be cultured and harvested similarly. This bacterium can also be cultured in a medium containing peptone and yeast extract as nitrogen sources.[8,9] Precautions in handling the harvested cells are given in this series.[3]

Buffer Solution

Tris-HCl buffers used in these experiments are of pH 7.2–7.5 unless otherwise specified.

Chromatographic Supporters

DEAE-cellulose (DE32; Whatman, Clifton, NJ): Treat with 0.1 *M* NaOH, 0.1 *M* HCl, and with 0.1 *M* NaOH again, wash several times with distilled water (simply designated as water hereafter), pack in a column of the desired size, and wash with a continuous flow of 10 m*M* Tris-HCl until the pH of the eluate becomes 7.2–7.5.

CM-cellulose (CM32, Whatman): Treat with 0.1 *M* NaOH and 0.1 *M* HCl, wash several times with water, and make the ammonium form with 0.1 *M* NH_3; pack in a column, and wash with two column volumes of water. It is essential to avoid excessive washing with water, which releases ammonium ion from CM-cellulose to regenerate the acid form.

[4] G. Fauque, M. Bruschi, and J. Le Gall, *Biochem. Biophys. Res. Commun.* **86,** 1020 (1979).

[5] J. Le Gall, M. Bruschi-Heriaud, and D. V. DerVartanian, *Biochim. Biophys. Acta* **234,** 499 (1971).

[6] K. Nakano, Y. Kikumoto, and T. Yagi, *J. Biol. Chem.* **258,** 12409 (1983).

[7] A. Nakagawa, Y. Higuchi, N. Yasuoka, Y. Katsube, and T. Yagi, *J. Biochem.* (*Tokyo*) **108,** 701 (1990).

[8] K. Tsuji and T. Yagi, *Arch. Microbiol.* **125,** 35 (1980).

[9] T. Yagi, *Biochim. Biophys. Acta* **548,** 96 (1979).

Sepharose CL-6B (Pharmacia, Piscataway, NJ): Suspend in ammonium sulfate solution at 80% saturation (3.28 *M*), and pack in a column.

Sephadex G-50 (fine) (Pharmacia): Suspend in 50 m*M* Tris-HCl containing 0.2 *M* NaCl, and wash with the same buffer solution.

Purification of Cytochrome c_{553} from Desulfovibrio vulgaris Miyazaki

Frozen cells (30 g) suspended in 60 ml of deaerated 10 m*M* Tris-HCl containing 1 mg of benzylsulfonyl fluoride and 2 mg of DNase I (Sigma Chemical Co., St. Louis, MO) are disrupted by means of an ultrasonic disintegrator (UR-200P; Tomy Seiko Co., Tokyo) for 9 min (three 3-min operations with sufficient intervals to avoid a temperature rise), and centrifuged at 85,000 *g* for 60 min to separate the cell-free extract from the membrane fraction. Bacterial cells may also be disrupted by means of a French press. The cell-free extract contains various chromophores including heme, flavin, and iron-sulfur cluster, of which cytochrome c_{553} and cytochrome c_3 are cationic and soluble in ammonium sulfate solution at 80% saturation. The purification is based on these properties. The heme concentration is not proportional to be absorbance at the α peak (A_{553}) owing to interference by other chromophores, and is estimated by the α peak height in the spectrum of the sample reduced with crystalline $Na_2S_2O_4$. The purity of the preparation can be conveniently expressed by the purity index:

$$\alpha\text{-peak height} = A_{553} - (6A_{538} + 5A_{571})/11$$
$$\text{Purity index} = (A_{553} - A_{570})_{\text{ferro form}}/A_{280\ \text{ferri form}}$$

For the purification of cytochrome c_{553} from *D. vulgaris* Miyazaki, procedure I is recommended to purify the cytochrome together with other redox proteins such as ferredoxins and rubredoxin, but the much simpler procedure II may be used if other redox components are not required.

Procedure I

1. The cell-free extract (70 ml, A_{260} = 100–130, A_{280} = 70–85, α-peak height = 0.5–1.5, of which cytochrome c_{553} occupies only about 5%[8,9]) is directly passed through short (15 × 30 mm) and long (22 × 300 mm) columns of DEAE-cellulose connected in series, and the columns are washed with deaerated 10 m*M* Tris-HCl. Cytochrome c_{553}, cytochrome c_3, and high molecular mass cytochrome *c* (Hmc) are collected in the eluate/washing (80 ml) from the columns without being adsorbed, whereas ferredoxins I and II[10] and rubredoxin[11] can be recovered and purified

[10] M. Ogata, S. Kondo, N. Okawara, and T. Yagi, *J. Biochem.* (*Tokyo*) **103,** 121 (1988).

[11] F. Shimizu, M. Ogata, T. Yagi, S. Wakabayashi, and H. Matsubara, *Biochimie* **71,** 1171 (1989).

from the short column, and desulfoviridin,[10] soluble hydrogenase,[12] and adenylylsulfate reductase[12] can be recovered from the long column.

2. No precaution is necessary to exclude atmospheric oxygen from this step. Ammonium sulfate is added to 80% saturation (3.28 *M*) to the eluate/washing of the long column of DEAE-cellulose to precipitate Hmc, which may be recovered by centrifugation and purified.[13]

3. The supernatant (73 ml) containing cytochrome c_{553} and cytochrome c_3 in 3.28 *M* ammonium sulfate can be stored at 4°. When 500 ml of the supernatant is accumulated, it is applied to a column (15 × 100 mm) of Sepharose CL-6B. Cytochrome adsorbed on the column is eluted with 150 m*M* Tris-HCl containing 0.2 *M* NaCl, and the eluate directly applied to two successive columns (22 × 1000 mm, each) of Sephadex G-50 (fine) with 150 m*M* Tris-HCl containing 0.2 *M* NaCl as an elution buffer. Cytochrome c_{553} (eluted between 470 and 540 ml), well separated from cytochrome c_3 (eluted between 380 and 450 ml), is dialyzed against water. The dialyzed solution can be stored frozen, or directly applied to a column (15 × 50 mm) of CM-cellulose. The concentration gradient from water (200 ml) to 20 m*M* aqueous ammonia (200 ml) is applied to the column, where cytochrome c_{553} is eluted at a concentration of about 5 m*M*. On evaporation of ammonia *in vacuo*, a concentrated solution of cytochrome c_{553} is obtained. The purity index is 1.24. It can be vacuum dried to give cytochrome c_{553} powder if desired. The yield of cytochrome c_{553} is 0.4–1.2 μmol (3.6–11 mg) from 300 g of frozen cells, the recovery being about 50%. The content of cytochrome c_{553}, as well as other redox components, in the bacterial cells may vary from culture to culture, and cannot be predicted precisely.[3]

Procedure II

The cell-free extract is brought to 45% saturation with ammonium sulfate (1.85 *M*), and the precipitate removed by centrifugation. Then the supernatant is brought to 80% saturation with ammonium sulfate (3.28 *M*), and centrifuged. (Do not try to achieve 80% saturation directly.) The supernatant is treated as described in step 3 of procedure I to obtain cytochrome c_{553} (purity index of 1.24).

Crystallization. Cytochrome c_{553} can be crystallized by a combination of vapor diffusion with ammonia and salting out with ammonium sulfate by a method slightly modified from that reported before.[14] An apparatus for crystallization consists of a screw-capped small jar (50 ml) and an open vial (4 mm i.d. × 45 mm), the inside of which is coated by Sigmacote

[12] M. Ogata and T. Yagi, unpublished observations, 1992.

[13] Y. Higuchi, T. Yagi, and G. Voordouw, this volume [11].

[14] A. Nakagawa, E. Nagashima, Y. Higuchi, M. Kusunoki, Y. Matsuura, N. Yasuoka, Y. Katsube, H. Chihara, and T. Yagi, *J. Biochem.* (*Tokyo*) **99,** 605 (1986).

(Sigma Chemical Co.) to flatten the roughness of the glass wall (Fig. 1). Ammonium sulfate must be finely ground in a mortar made of agate. Cytochrome c_{553} in 50 m*M* Tris-HCl is concentrated to 10 mg/ml by means of a Centricon-3 (Amicon Corp., Danvers, MA). Finely powdered ammonium sulfate (10 mg at a time) is added to 400 μl of ice-cold cytochrome c_{553} solution, and the mixture gently stirred with a thin glass rod to complete dissolution of ammonium sulfate. This is repeated until the concentration of ammonium sulfate reaches 80% saturation (3.28 *M*), that is, 224 mg/400 μl. The protein solution is filtered through a Millipore (Bedford, MA) membrane of pore size 0.45 μm to remove precipitate and bubbles, and is poured into the vial. The vial is placed in a jar containing about 20 ml of 50 m*M* Tris-HCl saturated with ammonium sulfate, the pH of which has been adjusted to 9.0 with ammonia (Fig. 1). The jar is tightly sealed with a screw cap and kept at 10–15°. After vapor equilibrium is achieved in 2 to 3 weeks, the pH of the outer solution of the vial is adjusted to 10.0 by ammonia. Dark red crystals of tetragonal bipyramid appear on the glass wall 2 to 3 weeks later. Typical dimensions of the crystals are approximately 1.0 × 1.0 × 1.2 mm. Some twinned crystals appear that are recognizable under an optical microscope. Single crystals are picked out and sealed quickly in thin glass capillaries to prevent dissolution of the crystals due to evaporation of ammonia from the mother liquid.

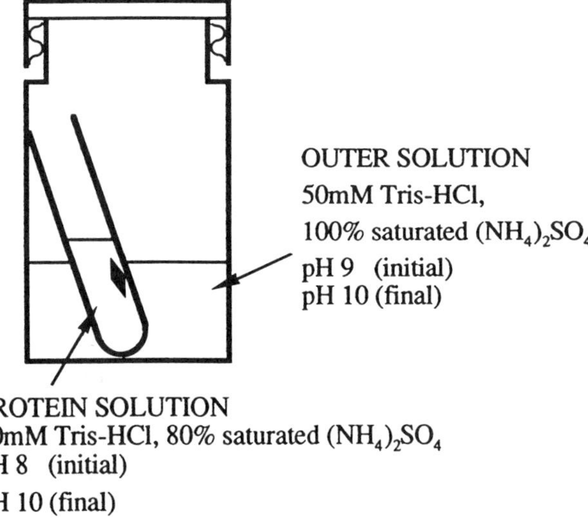

FIG. 1. An apparatus to crystallize cytochrome c_{553}. (Courtesy of Y. Higuchi, Himeji Institute of Technology, Kamigori, Hyogo, Japan.)

Purification of Monoheme Cytochrome from Other Sources

Cytochrome c_{553} from *D. vulgaris* Hildenborough is also cationic. It is purified from the cell-free extract by passing it through a DEAE-BioGel A column to remove acidic proteins; the filtrate is applied to CM-BioGel A, and the adsorbed cytochromes are eluted by a concentration gradient from 10 to 250 m*M* Tris-HCl. Separation from cytochrome c_3 is achieved by chromatography on a hydroxyapatite (BioGel HTP) column (decreasing concentration gradient from 250 to 10 m*M* Tris-HCl). The final step is run on a column of CM-BioGel A to yield a preparation with a purity index of 1.16.[15] Purification of monoheme cytochrome from *Dm. baculatum* Norway 4[4] and *Desulfovibrio desulfuricans* NCIMB 8387[16] can be achieved by a combination of DEAE-cellulose chromatography (not adsorbed), hydroxyapatite, alumina column chromatography, and so on. Cytochrome c_{553} from *D. desulfuricans* NCIMB 8372 is purified from the periplasmic extract by several steps, including hydrophobic chromatography on phenyl-Sepharose.[17]

Structure of Monoheme Cytochromes

Primary Structure

Cytochrome c_{553} is a member of the cytochrome *c* superfamily. The primary structures of cytochromes c_{553} from Miyazaki[6] and Hildenborough[18] strains of *D. vulgaris* are highly homologous, but the cytochrome $c_{553(550)}$ from *Dm. baculatum* Norway 4[19] is only 40% homologous to the former (Fig. 2). In the following discussion, cytochrome $c_{553(550)}$ is included with cytochrome c_{553}.

Tertiary Structure

Cytochrome c_{553} crystals from *D. vulgaris* Miyazaki described above diffract beyond 1.3-Å resolution. A precession photograph as shown in Fig. 3 is obtained by a 24-hr exposure of crystals of a typical size in a monochromated X-ray beam from the Cu rotating anode operated at 40

[15] K. B. Koller, F. M. Hawkridge, G. Fauque, and J. Le Gall, *Biochem. Biophys. Res. Commun.* **145,** 619 (1987).

[16] I. Moura, G. Fauque, J. Le Gall, A. V. Xavier, and J. J. G. Moura, *Eur. J. Biochem.* **162,** 547 (1987).

[17] L. H. Eng and H. Y. Neujahr, *Arch. Microbiol.* **153,** 60 (1989).

[18] G. J. H. van Rooijen, M. Bruschi, and G. Voordouw, *J. Bacteriol.* **171,** 3575 (1989).

[19] M. Bruschi, M. Woudstra, D. Campese, and J. Bonicel, *Biochim. Biophys. Acta* **1162,** 89 (1993).

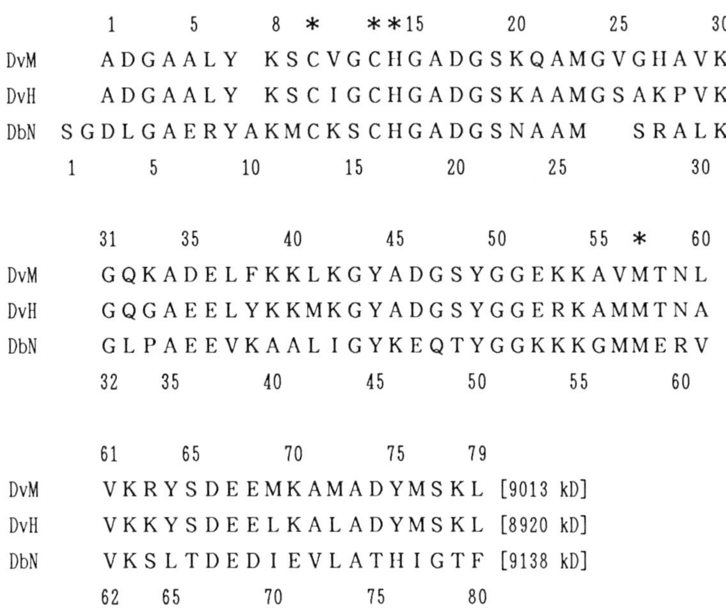

FIG. 2. Comparison of the amino acid sequences of cytochromes c_{553} from *D. vulgaris* Miyazaki (DvM), *D. vulgaris* Hildenborough (DvH), and *Dm. baculatum* Norway 4 (DbN), using single-letter notation. Heme-binding and liganding residues (two cysteines, histidine, and methionine) are asterisked.

kV and 100 mA at precession angles of 18°. The crystals are not damaged for at least 48 hr under these conditions. The diffraction pattern in Fig. 3 shows the ($h0l$) zone of the reciprocal lattice. Crystals are in space group $P4_32_12$, with unit cell dimensions $a = b = 42.7$ Å, $c = 103.4$ Å. It is difficult to prepare heavy atom derivatives of cytochrome c_{553} crystals, necessary for the multiple isomorphous replacement (MIR) method, probably because of the presence of ammonia, which coordinates to heavy atoms. The phase angles can be obtained when three or four X-ray beam wavelengths from synchrotron radiation are available. Phases of intensity data of cytochrome c_{553} are obtained by a multiwavelength anomalous dispersion (MAD) method by use of an anomalous dispersion effect of the iron atom in heme.[7] The small molecular size of cytochrome c_{553} contributed to the phasing power from the iron atom. The iron atom of a heme group was easily located on the Harker section of the anomalous difference Patterson map calculated with the coefficients $|F_{1.743}(+) - F_{1.743}(-)|^2$, where $F_{1.743}(+)$ and $F_{1.743}(-)$ are the Bijoet pairs of the structure factors measured by an X-ray beam of 1.743 Å. Four different X-ray

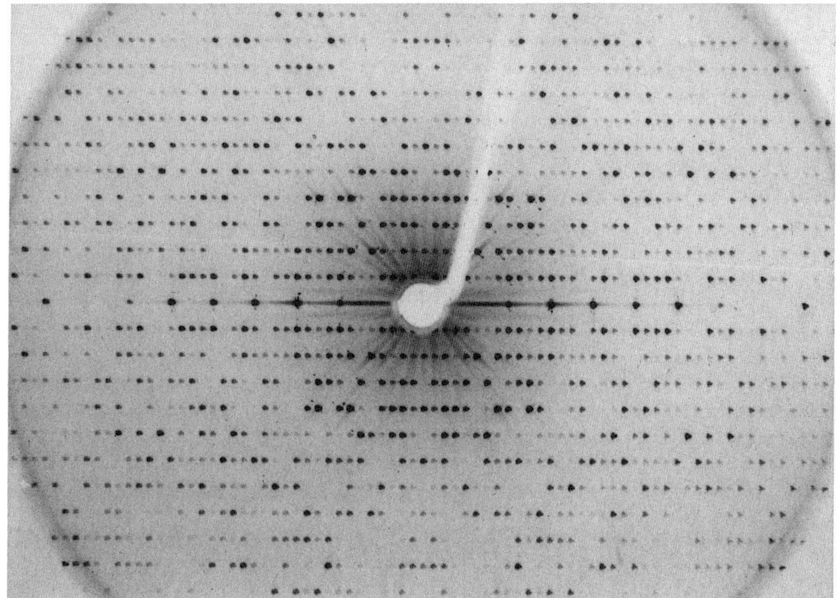

FIG. 3. Precession photograph obtained by a 24-hr exposure of cytochrome c_{553} crystal in a monochromated X-ray beam from a Cu rotating anode (40 kV, 100 mA) at precession angles of 18°, showing the (*h*0*l*) zone of the reciprocal lattice. The crystal is in space group $P4_32_12$, with unit cell dimensions $a = b = 42.7$ Å, $c = 103.4$ Å. (Courtesy of Y. Higuchi, Himeji Institute of Technology, Kamigori, Hyogo, Japan.)

wavelengths (1.746, 1.743, 1.380, and 1.040 Å) were used for the analysis of cytochrome c_{553}, and the structure shown in Fig. 4 was elucidated.[7] The sequences of seven amino acid residues at the single heme-binding site (Lys-8–His-14) of cytochrome c_{553} and the third heme-binding site (Lys-77–His-83) of cytochrome c_3, both from *D. vulgaris* Miyazaki, are identical. The main chain foldings of these peptides are similar, and the α-carbon atoms of both ethylidene side chains of these hemes have *S* chirality, but the relative orientations of the heme planes are different from each other, as shown in Fig. 5.

The structure of ferrocytochrome c_{553} from *D. vulgaris* Hildenborough was elucidated by nuclear magnetic resonance (NMR).[20] The helical fragments in cytochromes c_{553} from the two strains (Miyazaki and Hildenborough) are arranged similarly. The sulfur atom of the heme-liganding methi-

[20] D. Marion and F. Guerlesquin, *Biochemistry* **31,** 8171 (1992).

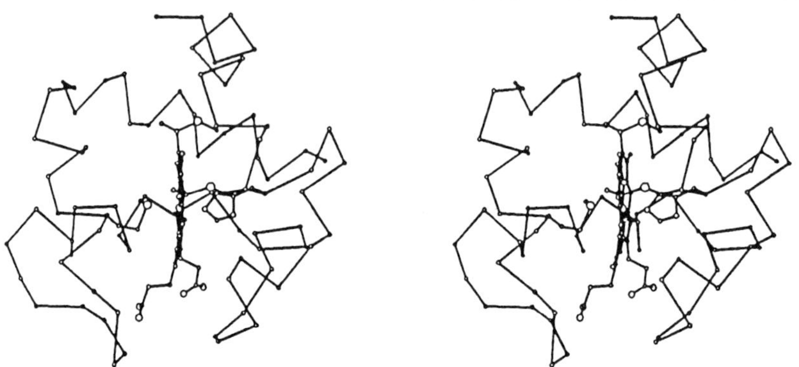

FIG. 4. Stereo drawing of the structure of ferricytochrome c_{553}. Only the α carbons, heme, and heme axial ligands are shown. [From A. Nakagawa, Y. Higuchi, N. Yasuoka, Y. Katsube, and T. Yagi, *J. Biochem. (Tokyo)* **108,** 701 (1990).]

onine has *S* chirality in Miyazaki ferricytochrome, which is in accord with the *S* chirality of ferrocytochromes c_{553} from *D. vulgaris* Hildenborough and *Dm. baculatum* Norway 4,[21] but curiously the latter is reported to have *R* chirality in the ferri forms.[21]

Properties of Monoheme Cytochromes

Spectral Properties

Spectral properties of cytochrome c_{553}, together with other cytochromes of *D. vulgaris* Miyazaki, are given in Table I. These cytochromes are spectrally similar except for the positions of the δ peaks of the ferri forms, and for the presence of an additional peak at 695 nm in cytochrome c_{553} due to the methionyl side chain as an axial ligand to the heme iron.[4–6,20] Spectral properties of cytochrome c_{553} from *D. vulgaris* Hildenborough are similar.[5] The monoheme cytochrome from *Dm. baculatum* Norway 4 has a split α peak (the peak at 553 nm with a shoulder at 550 nm) in the ferro form, from which the name cytochrome $c_{553(550)}$ has been derived.[4]

Standard Midpoint Potential

The standard midpoint potentials of the monoheme cytochromes from various species of *Desulfovibrio* and *Desulfomicrobium* were measured

[21] H. Senn, F. Guerlesquin, M. Bruschi, and K. Wüthrich, *Biochim. Biophys. Acta* **748,** 194 (1983).

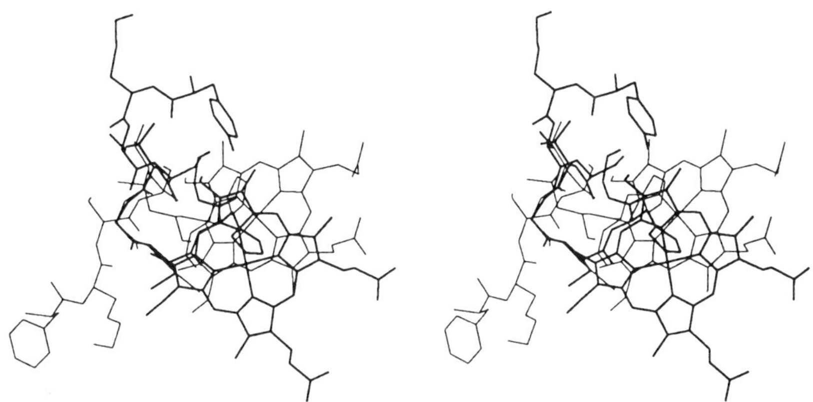

FIG. 5. Stereo drawings of the heme-binding sequence of cytochrome c_{553} (Tyr-7–Gly-15; thick lines) and that of the third heme-binding sequence of cytochrome c_3 (Phe-76–Leu-84; thin lines). (Courtesy of A. Nakagawa, Photon Factory, National Laboratory of High Energy Physics, Tsukuba, Japan.)

by redox titration monitored by electron paramagnetic resonance (EPR)[22] and absorption spectra,[22] or directly by cyclic voltammetry on a gold electrode[23] in the presence of 4,4′-bipyridine,[24,25] or by cyclic voltammetry and derivative cyclic voltabsorptometry on an indium oxide optically transparent electrode without any mediators, promotors, or electrode modifiers.[15] Optically monitored redox titration sometimes gives a midpoint potential significantly more positive than those obtained by other methods.[22] The results are summarized in Table II. As shown, the midpoint potentials are much more negative than those of other members of the cytochrome *c* superfamily ($E°'$, ~250 mV), but are much more positive than those of other *c*-type cytochromes of *Desulfovibrio* and *Desulfomicrobium*.

Chemical Reactivity

Ferricytochrome c_{553} is reduced partially with ascorbate,[1,4,9] and fully with cysteine,[1,9] reflecting its rather positive redox potential (Table II). Ferrocytochrome c_{553} is nonenzymatically oxidized with O_2 and ferricya-

[22] P. Bertrand, M. Bruschi, M. Denis, J. P. Gayda, and F. Manca, *Biochem. Biophys. Res. Commun.* **106,** 756 (1982).

[23] W. R. Heineman, B. J. Norris, and J. F. Goelz, *Anal. Chem.* **47,** 79 (1975).

[24] P. Bianco, J. Haladjian, R. Pilard, and M. Bruschi, *J. Electroanal. Chem.* **136,** 291 (1982).

[25] P. Bianco, J. Haladjian, M. Loutfi, and M. Bruschi, *Biochem. Biophys. Res. Commun.* **113,** 526 (1983).

TABLE I
SPECTRAL PROPERTIES OF *c*-TYPE CYTOCHROMES FROM *D. vulgaris* MIYAZAKI

	Monoheme cytochrome c_{553}		Tetraheme cytochrome c_3		Polyheme Hmc[a]	
	λ (nm)	ε_{mM}	λ (nm)	ε_{mM}	λ (nm)	ε_{mM}
Ferri form						
(Met ligand)	695	0.8	—		—	
α–β peak	526	10.1	532	39.0	529	153
γ peak	410	110	410	464	409	1820
δ peak	360	28.1	350	90.8	355	380
Protein	280	19.3	280	35.6	280	140
Ferro form						
α peak	553	24.7	552	116	553	400
β peak	524	16.1	524	58.5	523	221
γ peak	418	154	419	733	419	2735
δ peak	317	33.1	325	137	325	630
$\Delta\varepsilon^{b}_{mM}$ at α peak	553	18.0	552	86.3	553	286
Purity index	1.24		3.14		2.72	

[a] *D. vulgaris* Miyazaki contains two kinds of polyheme cytochromes, sHmc and mHmc. Both have identical spectra and contain 16 hemes. (Y. Higuchi, T. Yagi, and G. Voordouw, this volume [11]).

[b] Millimolar (reduced minus oxidized) difference absorbance.

nide, but does not complex with CO.[9] Redox equilibrium is attained between the ferri/ferro couple of cytochrome c_{553} and rubredoxin.[11]

The EPR spectra are recorded for cytochrome c_{553} from *D. vulgaris* Hildenborough. The spectrum at 93 K is of the broad low-spin type,[5] similar to that reported for mitochondrial cytochrome *c*, and that recorded at 10 K contains two main paramagnetic components at $g_1 = 3.154$ and $g_2 = 2.065$.[22] Chemical shifts of proton-NMR lines of heme *c* and axial ligands in cytochrome c_{553} from *D. vulgaris* Hildenborough and *Dm. baculatum* Norway 4 were recorded and analyzed.[21]

Function of Cytochrome c_{553}

Cellular Localization of Cytochrome c_{553}

Cytochrome c_{553} from *D. vulgaris* (both Hildenborough and Miyazaki strains) is localized on the periplasm as evidenced by the presence of a signal peptide in the proprotein.[18]

TABLE II
ELECTROCHEMICAL PROPERTIES OF CYTOCHROMES c_{553} FROM VARIOUS SOURCES

Sources	$E^{\circ\prime}$ (mV)	p*I*
D. vulgaris Miyazaki	26[a]	10.2[b]
D. vulgaris Hildenborough	20,[c,d] 18[e]	8.0[c,d]
Dm. baculatum 9974 (DSM 1743)	>−50[f]	
Dm. baculatum Norway 4	50,[g] 40[h]	6.6[g]
D. desulfuricans NCIMB 8372	0[i]	>9[i]
D. desulfuricans NCIMB 8387	>−50[f]	9.2[f]

[a] K. Niki, unpublished observation, 1989. Previously reported value[b] is erroneous due to inadequate standard in the redox titration.
[b] T. Yagi, *Biochim. Biophys. Acta* **548,** 96 (1979).
[c] P. Bertrand, M. Bruschi, M. Denis, J. P. Gayda, and F. Manca, *Biochem. Biophys. Res. Commun.* **106,** 756 (1982).
[d] P. Bianco, J. Haladjian, R. Pilard, and M. Bruschi, *J. Electroanal. Chem.* **136,** 291 (1982).
[e] K. B. Koller, F. M. Hawkridge, G. Fauque, and J. LeGall, *Biochem. Biophys. Res. Commun.* **145,** 619 (1987).
[f] I. Moura, G. Fauque, J. LeGall, A. V. Xavier, and J. J. G. Moura, *Eur. J. Biochem.* **162,** 547 (1987).
[g] G. Fauque, M. Bruschi, and J. LeGall, *Biochem. Biophys. Res. Commun.* **86,** 1020 (1979).
[h] P. Bianco, J. Haladjian, M. Loutfi, and M. Bruschi, *Biochem. Biophys. Res. Commun.* **113,** 526 (1983).
[i] L. H. Eng and H. Y. Neujahr, *Arch. Microbiol.* **153,** 60 (1989).

Biochemical Reactivity of Cytochrome c_{553}

Cytochrome c_{553} accepts electrons from formate and lactate dehydrogenases,[2,9,26] and hydrogenase and is assumed to be an electron acceptor for anaerobic heme synthesis at the step from protoporphyrinogen IX to protoporphyrin IX,[27] but no enzyme system has ever been proved to oxidize its ferro form to regenerate ferricytochrome c_{553}. In this respect, the role of cytochrome c_{553} in the energy metabolism of this bacterium is an enigma.

[26] M. Ogata, K. Arihara, and T. Yagi, *J. Biochem. (Tokyo)* **89,** 1423 (1981).
[27] T. Yagi and M. Ogata, *in* "Microbiology and Biochemistry of Strict Anaerobes Involved in Interspecies Hydrogen Transfer" (J.-P. Bélaich, M. Bruschi, and J.-L. Garcia, eds.), p. 237. Plenum, New York, 1990.

The enzymatic reduction of cytochrome c_{553} can best be monitored spectrophotometrically by an increase in A_{553} (absorbance at 553 nm) in a long-necked optical cell kept anaerobic under a stream of N_2 or H_2 as shown in Fig. 6. This cell is more convenient than the Thunberg-type optical cell, because thorough deaeration with N_2 or saturation with H_2 is possible before the addition of the enzyme. Immediate mixing is achieved by means of a syringe with a multipored needle,[28] as shown in Fig. 6, if the amount of enzyme is more than 0.05 ml. Immediate mixing may be achieved by just tipping in the enzyme to the reaction mixture while bubbling with gas (N_2 or H_2) for only a few seconds if the amount of the enzyme to be added is less than 0.05 ml. Using the millimolar absorbance difference at 553 nm of cytochrome c_{553} (Table I), the reaction rate (micromoles of substrate oxidized per minute, i.e., micromoles of cytochrome c_{553} reduced per minute divided by 2) per milliliter of the reaction mixture is expressed as

$$\text{Reaction rate} = (\Delta A_{553}/\text{min})/(18.0 \times 2)$$

In the following sections the reduction of cytochrome c_{553} by D-lactate dehydrogenase and hydrogenase is given as an example.

Reduction of Cytochrome c_{553} with Lactate Dehydrogenase. D-Lactate dehydrogenase is purified from *D. vulgaris* Miyazaki as follows. The frozen cells (50 g) suspended in 100 ml of deaerated 50 m*M* Tris-HCl containing 0.1 m*M* FAD are disrupted by sonication and centrifuged. The supernatant is brought to 50% saturation with ammonium sulfate (2.05 *M*) and centrifuged to remove the precipitate, and the supernatant is made 70% saturated with ammonium sulfate (2.87 *M*). The precipitate dissolved in 6 ml of deaerated 50 m*M* Tris-HCl containing 0.1 m*M* FAD is centrifuged to remove turbidity, and chromatographed on a column (22 × 1000 mm) of Sephacryl S-200 preequilibrated with deaerated 50 m*M* Tris-HCl containing 0.05 m*M* FAD and 0.2 *M* NaCl. Lactate dehydrogenase eluted between 180 and 205 ml is precipitated with ammonium sulfate at 70% saturation, dissolved in 50 m*M* Tris-HCl, the final concentration of ammonium sulfate being adjusted at 45% saturation (1.85 *M*), and passed through a column (15 × 100 mm) of DE32 preequilibrated with 50 m*M* Tris-HCl containing ammonium sulfate at 45% saturation. The adsorbed lactate dehydrogenase is eluted by a concentration gradient of ammonium sulfate from 45% saturation to 0% in 50 m*M* Tris-HCl, precipitated with ammonium sulfate at 70% saturation, and stored anaerobically under N_2. It is dissolved in deaerated 50 m*M* Tris-HCl containing 0.01 m*M* FAD just before use.[27]

[28] I. Tabushi, T. Nishiya, T. Yagi, and H. Inokuchi, *J. Am. Chem. Soc.* **103,** 6963 (1981).

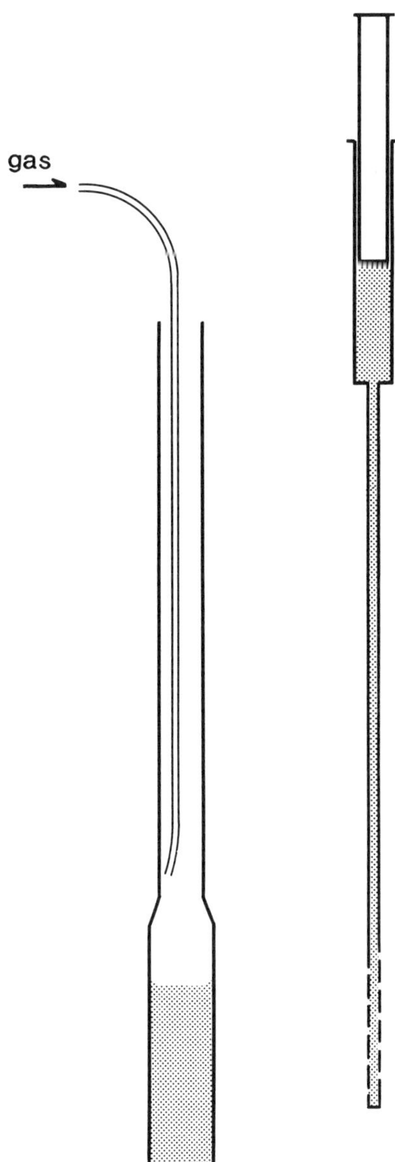

FIG. 6. A long-necked optical cell for monitoring absorbance change under anaerobic conditions (left), and a syringe with a multipored needle for monitoring a rapid reaction (right). The optical path of the cell is 10 mm and the length of the neck is about 100 mm.

A 2.9-ml portion of a reaction mixture containing 0.1 *M* lithium DL-lactate in 0.2 *M* Tris-HCl, 0.01 m*M* FAD, and 0.01 m*M* cytochrome c_{553} is placed in a long-necked optical cell (Fig. 6), and deaerated by bubbling with N_2. The reaction is started by ejecting 0.1 ml of lactate dehydrogenase from a syringe as shown in Fig. 6, and the A_{553} value is monitored. In typical experiments, 0.012 units of lactate dehydrogenase (definition of the enzyme unit is given by Ogata *et al.*[26]) catalyzes the oxidation of 0.014 nmol of D-lactate per minute, but the rate is increased to 1.84 nmol/min if 0.02 m*M* 2-methyl-1,4-naphthoquinone (added as a solution in ethanol) is added to the reaction mixture.[27] The nature of the natural quinone involved in this reaction is unknown.

Reduction of Cytochrome c_{553} with Hydrogenase. In spite of an earlier description that cytochrome c_{553} is not reduced with hydrogen by the action of *Desulfovibrio* hydrogenase,[9] it is reduced by hydrogenase in a mixture that has been thoroughly deaerated and saturated with H_2.

A 3.0-ml portion of a reaction mixture containing 0.033 m*M* cytochrome c_{553} in 50 m*M* Tris-HCl is placed in a long-necked optical cell (Fig. 6), and deaerated by bubbling with H_2. The reaction is started by tipping 3 μl of 0.037 m*M* hydrogenase[29] into the mixture while bubbling with H_2, followed by immediate removal of the gas inlet from the reaction mixture to monitor the change in A_{553}. The reaction rate was 25 nmol of H_2 consumed per minute. Under similar reaction conditions with 0.008 m*M* cytochrome c_3, the reaction rate was 22 nmol of H_2/min. Thus the reduction of cytochrome c_{553} proceeded more rapidly than the reduction of cytochrome c_3 in this case. The reduction of cytochrome c_{553} by hydrogenase is rather non-reproducible, however. Sometimes the reduction of cytochrome c_{553} is interrupted before full reduction is achieved. On the other hand, the reduction of cytochrome c_3 by hydrogenase is consistent and reproducible, and diluted hydrogenase solution is protected by the addition of cytochrome c_3. Cytochrome c_{553} is, therefore, not considered to be a physiological electron carrier of *Desulfovibrio* hydrogenase. Cytochrome c_{553} is ineffective in mediating electron transfer in the H_2 evolution assay of hydrogenase.

[29] Purification of hydrogenase from *D. vulgaris* Miyazaki is given by M. Asso, B. Guigliarelli, T. Yagi, and P. Bertrand, *Biochim. Biophys. Acta* **1122,** 50 (1992). Purification from other sources is given by R. Cammack *et al.,* this volume [5].

[9] Tetraheme Cytochromes

By ISABEL B. COUTINHO and ANTÓNIO V. XAVIER

Introduction

Tetraheme cytochromes c_3 were first isolated in the 1950s from two strains of the species *Desulfovibrio (D.) vulgaris*.[1,2] It was eventually detected in all organisms belonging to the Desulfovibrionaceae family, as well as in several other sulfate-reducing bacteria,[3] in which it is produced in large quantities. It is a small (M_r ~15,000) soluble protein, containing four structurally inequivalent heme *c* moieties, all with bishistidinal axial ligation, which present low, albeit different, redox potentials (between −50 and −400 mV). Although the amino acid composition, the sequence homology, and the formal charges and isoelectric points of these cytochromes may be different, in particular for the proteins isolated from phylogenetically distant organisms,[4] the heme–heme distances and angles and the orientation of the heme substituents established by X-ray crystallography[5,6] and by nuclear magnetic resonance (NMR)[7-9] show that the overall architecture of the heme core is highly conserved in all the proteins studied.[10] The term cytochrome c_3 has been informally used for the nomenclature of several multiheme type-*c* cytochromes isolated from sulfate-reducing bacteria. In this chapter, the term will refer to the tetraheme cytochrome protein only.

[1] J. R. Postgate, *Biochem. J.* **56,** xi (1952).
[2] M. Ishimoto, J. Koyana, and Y. Nagay, *Seikagaku Zasshi* **26,** 303 (1954).
[3] F. Widdel and F. Bak, "The Prokaryotes" (A. Balows, H. G. Trüper, M. Dworkin, W. Harder, and K.-H. Schleifer, eds.), Vol. 4, p. 3353. Springer-Verlag, Berlin, 1992.
[4] R. Devereaux, S.-H. He, C. L. Doyle, S. Oackland, P. A. Stahl, J. LeGall, and W. B. Whitman, *J. Bacteriol.* **172,** 3609 (1990).
[5] R. Haser, M. Pierrot, M. Frey, F. Payan, J. P. Astier, M. Bruschi, and J. LeGall, *Nature (London)* **282,** 806 (1979).
[6] Y. Higushi, Y. S. Bando, M. Kusonoki, Y. Matsuura, N. Yasuoka, M. Kadudo, T. Yamanaka, T. Yagi, and M. Inokushi, *J. Biochem. (Tokyo)* **89,** 1659 (1981).
[7] I. B. Coutinho, D. L. Turner, J. LeGall, and A. V. Xavier, *Eur. J. Biochem.* **209,** 329 (1992).
[8] D. L. Turner, C. A. Salgueiro, J. LeGall, and A. V. Xavier, *Eur. J. Biochem.* **210,** 931 (1992).
[9] A. Piçarra-Pereira, D. L. Turner, J. LeGall, and A. V. Xavier, *Biochem. J.* **294,** 909 (1993).
[10] Y. Higushi, K. Kusonoki, N. Yasuoka, M. Kakudo, and T. Yagi, *J. Biochem. (Tokyo)* **90,** 1715 (1981).

All sequenced cytochromes c_3 have a typical N-terminal signal peptide indicative of their localization in the periplasm,[11] where cytochrome c_3 is proposed to act as a coupling factor to the enzyme hydrogenase because its presence is needed for the reduction *in vitro* of the redox proteins ferredoxin, flavodoxin, and rubredoxin by the system H_2/Hase (molecular hydrogen in the presence of hydrogenase).[12] However, as these proteins are cytoplasmic, this conclusion is open to question, although stable models for the complexes between cytochrome c_3 and all three proteins have been proposed from molecular graphics simulations.[13–15] Cytochromes c_3 isolated from some *Desulfovibrio* organisms, as well as from two strains of *Desulfomicrobium* (*Dsm.*) *baculatum* present an additional sulfur reductase activity.[16,17]

Protein Purification

A common experimental sequence, well established in the literature,[18,19] has been used for the isolation of tetraheme cytochromes c_3; the main steps used for the purification of the protein isolated from *Dsm. baculatum* (DSM 1741, Norway 4) are summarized in Figs. 1 and 2, and experimental details are available in this volume.[17] Most of the cytochrome is obtained at the initial stage of the first ion-exchange chromatographic step applied to the soluble fraction, including the charge and washing procedures. Further purification is achieved by several successively finer steps involving ion-exchange, molecular exclusion, and adsorption chromatography. The sample purification stage is monitored by purity indices obtained from visible spectroscopy data[17]: a rough absorbance ratio $[A_{530}(\text{ox})/A_{280}(\text{ox})]$ used in the initial stages, and a finer ratio $R = \{[A_{\alpha\max}(\text{red}) - A_{560}(\text{red})]/A_{280}(\text{ox})\}$, which attain optimum values of about 1 and 3.4, respectively.[20]

[11] J. LeGall and H. D. Peck, Jr., *FEMS Microbiol. Rev.* **46,** 35 (1987).

[12] J. R. Bell, J. P. Lee, H. D. Peck, Jr., and J. LeGall, *Biochimie* **60,** 315 (1978).

[13] P. E. Stuart, J. LeGall, I. Moura, J. J. G. Moura, H. P. Peck, Jr., A. V. Xavier, P. K. Weiner, and J. E. Wampler, *Biochemistry* **27,** 2444 (1988).

[14] L. Cambillau, M. Frey, J. Mossé, F. Guerlesquin, and M. Bruschi, *Proteins: Struct. Funct. Genet.* **4,** 63 (1988).

[15] D. E. Stewart and J. E. Wampler, *Proteins: Struct. Funct. Genet.* **11,** 142 (1991).

[16] G. Fauque, D. Hervé, and J. LeGall, *Arch. Microbiol.* **121,** 261 (1979).

[17] G. D. Fauque, O. Klimmek, and A. Kröger, this volume [25].

[18] M. Brushi, L. E. Hatchikian, L. A. Golovleva, and J. LeGall, *J. Bacteriol.* **129,** 30 (1977).

[19] M.-C. Liu, C. Costa, I. B. Coutinho, J. J. G. Moura, I. Moura, A. V. Xavier, and J. LeGall, *J. Bacteriol.* **170,** 5545 (1988).

[20] G. Fauque, Thèse de Doctorat d'État, Université de Technologie de Compiègne, 1985.

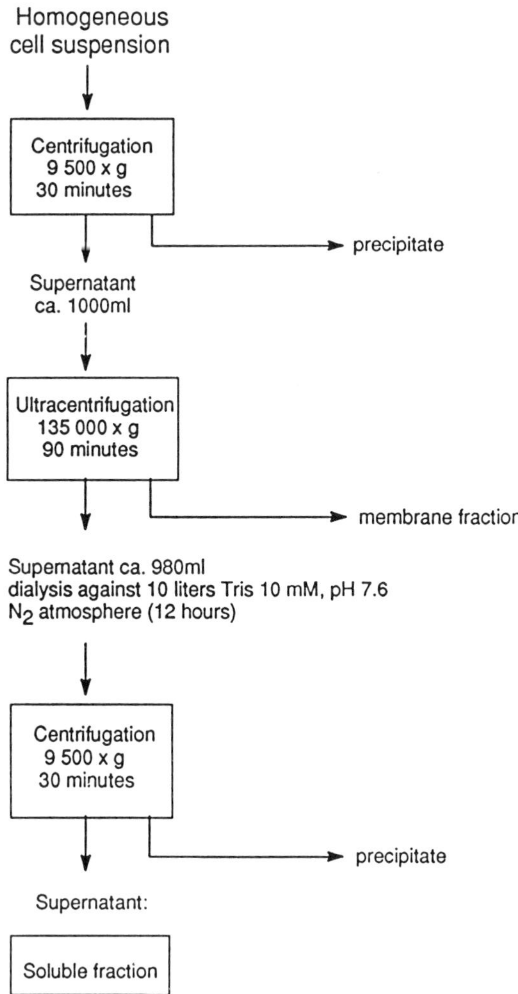

FIG. 1. Preparation of a soluble extract from a cell suspension of *Dsm. baculatum* (Norway 4). The suspension (700 ml, 1 mg of total protein per milliliter) was stirred overnight after dilution 1:1 in Tris (10 m*M*, pH 7.6) in the presence of DNase, homogenized in a French press (9000 psi, 30 min); streptomycin sulfate (0.5 mg/mg protein) was added before the centrifugation sequence.

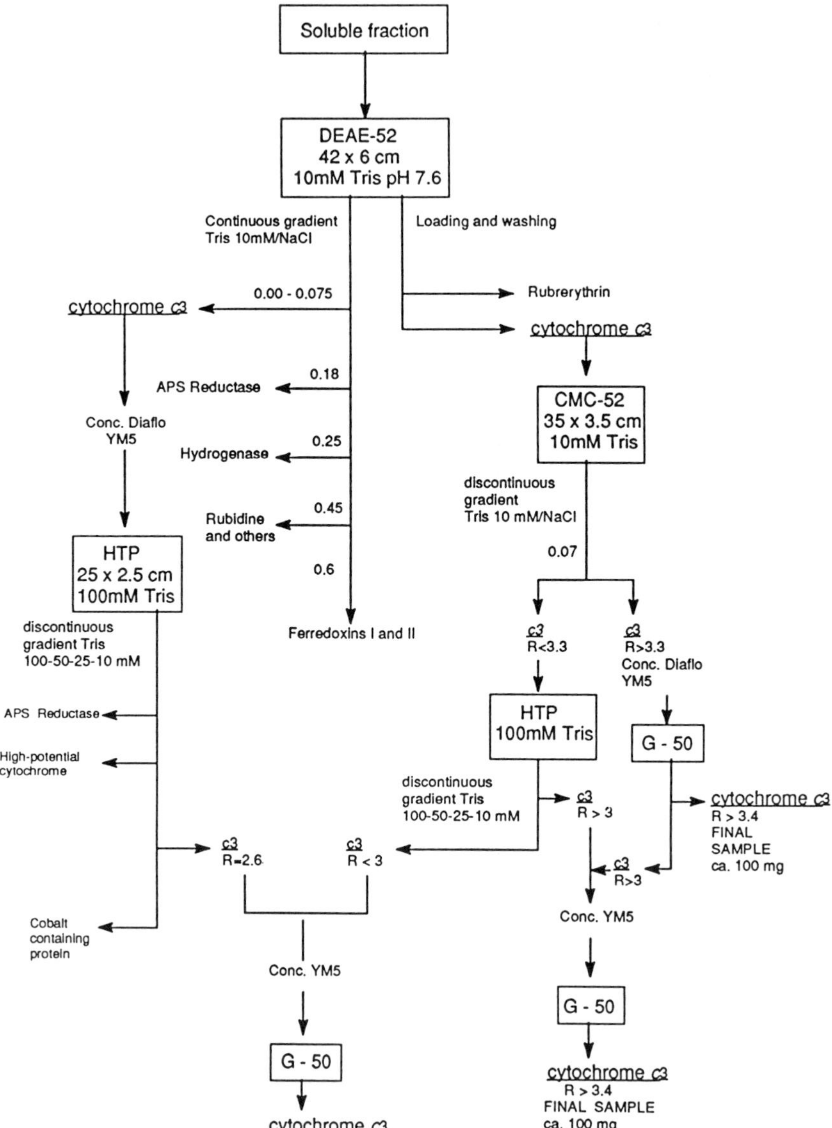

FIG. 2. Separation of the proteins contained in the soluble fraction and purification of cytochrome c_3 from *Dsm. baculatum* (Norway 4). Numbers refer to the approximate eluent concentration (*M*) at which the fraction is eluted. Tris, Trisethylenodiaminomethane; DEAE, diethylaminoethylcellulose; CMC, carboxymethylcellulose; HTP, hydroxyapatite; G-50, Sephadex G50.

Protein Characterization

Structure

The structures of the cytochromes c_3 isolated from *Dsm. baculatum*,[5,21,22] *D. vulgaris* (Miyazaki),[6,23] *D. vulgaris* (Hildenborough)[24,25] (Fig. 3), and *Desulfovibrio gigas*[26] have been determined by X-ray diffraction. It was found in all the structures that the ligation of four hemes adopted a characteristic topology and that the heme core architecture is highly conserved,[10] presenting the same iron-to-iron distances and heme plane orientations; NMR results have shown that the location of the heme substituents is also maintained (see below). The hemes, which shall be labeled HI to HIV according to the numbering of the cysteine residues in the thioether bridges, are ligated to the apoprotein trough sequence segments of the type Cys-X-X-(X-X)-Cys-His, with the sixth axial histidine ligand present in a different region of the polypeptide chain and, in particular, with the sixth histidine ligand to heme HII situated immediately after the fifth histidine ligand to heme HI. Heme-to-heme distances vary between 1.1 and 1.8 nm; the planes defined by hemes HI and HIV are approximately parallel and both are perpendicular to hemes HII and HIII, which are mutually perpendicular. The shortest interporphyrin ring distance is about 0.6 nm and occurs between the edges of hemes HIII and HIV. The heme edges are accessible to the solvent (more than 1 nm^2 compared to about 0.4 nm^2 established for horse cytochrome *c*[27]), following the order HI < HIV < HIII < HII.

Other structural invariants were established for these proteins. They include the strict conservation of specific residues near the heme ligation sites, such as the threonine, glycine, and phenylalanine residues integrated in the segments [*Dsm. baculatum* (Norway 4) sequence numbering] Thr-109–X–Cys-111–(X–X)–Cys-117 (ligation to HIV), Cys-61–X–X–X–Gli-65–Cys-66 (ligation to HII), and Phe-34–X–His-36 (ligation to HI), where, in the last case, the particular orientation of the phenylalanine residue, situated close to heme HI and parallel to one of the axial histidines of

[21] M. Frey, R. Haser, M. Pierrot, M. Bruschi, and J. LeGall, *J. Mol. Biol.* **104,** 741 (1976).

[22] M. Pierrot, P. Haser, M. Frey, F. Payan, and J.-P. Astier, *J. Biol. Chem.* **257,** 14341 (1982).

[23] Y. Higushi, M. Kusonoki, Y. Matsuura, N. Yasuoka, and M. Kakudo, *J. Mol. Biol.* **172,** 109 (1984).

[24] Y. Marimoto, T. Tari, H. Okimura, Y. Higushi, and N. Yasuoka, *J. Biochem.* (*Tokyo*) **110,** 531 (1991).

[25] P. M. Matias, C. Frazäo, J. Morais, M. Coll, and M. A. Carrondo, *J. Mol. Biol.* **234,** 680 (1993).

[26] C. Kissinger, Ph.D. Thesis, Washington University, St. Louis, Missouri (1989).

[27] E. Stellwalgen, *Nature* (*London*) **275,** 73 (1978).

FIG. 3. Stereo view of the three-dimensional structure of cytochrome c_3 from *D. vulgaris* (Hildenborough) [(P. M. Matias, C. Frazão, J. Morais, M. Coll, and M. A. Carrondo, *J. Mol. Biol.* **234,** 680 (1993). The structure was generated using coordinates from the Brookhaven Protein Databank file 1CTH.

heme HIII, has led to the suggestion that it has an important role in the electron-transfer process[28]; other proposed conserved residues, such as Pro-4, Asp-6, Lys-21, and Val-32,[10,26] depend on the sequence alignment strategies. Cytochromes c_3 also present a highly charged, high-lysine content sequence segment situated close to heme HIV in the C-terminal region; the lateral amino groups of spatially close, positively charged lysine side chains form a characteristic patch at the surface in cytochromes c_3 isolated from *D. vulgaris* and *D. gigas*. This patch, which has been proposed as a specific recognition site by the putative redox partner flavodoxin,[13] is absent in the cytochrome from *Dsm. baculatum*, which does not contain flavodoxin.

The heme superstructure from *Dsm. baculatum*,[7] *D. vulgaris* (Hildenborough),[8] and *D. gigas*[9] was also probed using two-dimensional NMR, through the observation of the short-range and longer range interheme nuclear Overhauser effect (NOE) connectivities between those heme substituents oriented toward the interior of the heme core superstructure. These results have shown that, in addition to the heme spatial arrangement conservation described above, the heme equatorial substituents have fixed orientations different from the ones proposed for the cytochrome from *Dsm. baculatum* (on the basis of X-ray crystallography[5,22]) but similar to the coordinates established for the cytochromes from *D. vulgaris*[23–25] and *D. gigas*,[26] and to the heme superstructure independently generated for the cytochrome from *Dm. baculatum* using NMR data only.[29] As the heme geometry is maintained in detail in the cytochromes isolated from such phylogenetically unrelated organisms, it is probable that the same core architecture is conserved in the other, nonstructurally characterized cytochromes c_3.

Some of the common features established for the cytochromes c_3 from *D. baculatum* (Norway 4), *D. vulgaris* (Hildenborough) and (Myazaki), and *D. gigas* are presented in Fig. 4 and in Table I.

Sequence Alignment

Amino acid sequences have been determined for the cytochromes c_3 from *Dsm. baculatum* (Norway 4, DSM 1741; DbN),[30] *Desulfovibrio desulfuricans* (El Agheila Z, NCIB 8380; DdE),[31] *Desulfovibrio salexigens*

[28] R. Haser, M. Fey, and F. Payan, *in* "Crystalography in Molecular Biology" (D. Moras, J. Drenth, B. Strandlberg, and K. Wilson, eds.), p. 425. Plenum, New York, 1987.

[29] I. B. Coutinho, D. L. Turner, J. LeGall, and A. V. Xavier, *Biochem. J.* in press (1993).

[30] M. Bruschi, *Biochim. Biophys. Acta* **671,** 219 (1981).

[31] R. P. Ambler, M. Brushi, and J. LeGall, *FEBS Lett.* **18,** 347 (1971).

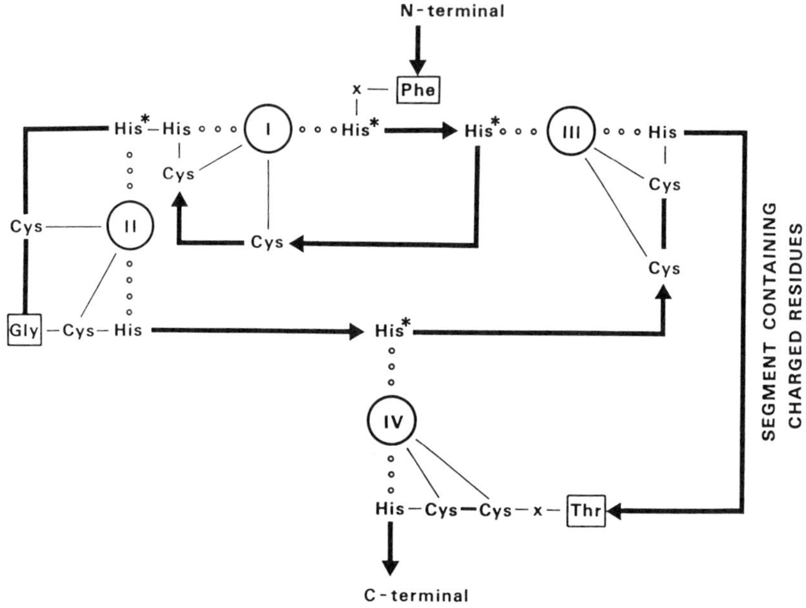

FIG. 4. Diagram of the topological and sequence invariants of the tetraheme cytochromes c_3.

(NCIB 8403; Ds),[32] *D. vulgaris* (Hildenborough, NCIB 8303; DvH),[33] *D. vulgaris* (Miyazaki, IAM 12604; DvM),[34] and *D. gigas* (NCIB 9932; Dg).[35] The sequence corresponding to the cytochrome isolated from *Dsm. baculatum* (DSM 1743) is not fully established yet, but the first 20 N-terminal amino acids are identical to those found in the protein isolated from *Dsm. baculatum* (Norway)[20,36] and, although there is at least 1 amino acid alteration, the substitution of Phe-88 in *Dsm. baculatum* (Norway) by a spatially homologous tyrosine residue,[29] it is assumed that the overall sequence is similar for both *Desulfomicrobium* proteins.

Proposed sequence alignments (Table II) were established considering that all the cytochromes had a heme core arrangement similar to the one determined by X-ray diffraction (see above) and postulating sequence insertions and deletions such as to maximize the homologies between

[32] R. P. Ambler, *Syst. Zool.* **22,** 554 (1973).

[33] R. P. Ambler, *Biochem. J.* **109,** 47 (1968).

[34] W. Shinkai, T. Hase, T. Yagi, and H. Matsubara, *J. Biochem. (Tokyo)* **97,** 1747 (1980).

[35] R. P. Ambler, M. Bruschi, and J. LeGall, *FEBS Lett.* **5,** 115 (1969).

[36] M. Regalla, unpublished results (1993).

TABLE I

CHARACTERISTICS OF CYTOCHROMES c_3 ISOLATED FROM *Desulfomicrobium baculatum* (NORWAY 4) (c_3N),[a] *Desulfovibrio vulgaris* (MIYAZAKI) AND (HILDENBOROUGH) (c_3V),[b,c] AND *Desulfovibrio gigas* (c_3G)[d]

	Ligands				Fe-Fe distance (nm)			Edge-to-edge distance (nm)	Angles between heme planes (degrees)	
Heme	Residue	c_3N	c_3V	c_3G	Heme	c_3N	c_3V	(c_3V)	c_3N	c_3V
I	Cys	44	30	34	I–II	1.28	1.22–1.24	0.64–0.66	80	88.3–89.7
	Cys	47	33	37	I–III	1.09	1.10–1.12	1.10–1.12	85	81.0–81.6
	His	48	34	38	I–IV	1.73	1.77–1.80	1.77–1.80	35	19.2–22.0
	His	36[e]	22[e]	26[e]						
II	Cys	61	46	49	II–III	1.68	1.58–1.61	0.85–0.88		59.7–60.1
	Cys	66	51	54	II–IV	1.63	1.64–1.67	1.02–1.03	60	71.0–74.7
	His	67	52	55						
	His	49[e]	35[e]	39[e]						
III	Cys	92	79	83	III–IV	1.27	1.19–1.20	0.54–0.55	90	80.1–82.0
	Cys	95	82	86						
	His	96	83	87						
	His	39[e]	25[e]	29[e]						
IV	Cys	111	100	104						
	Cys	114	105	109						
	His	115	106	110						
	His	89[e]	70[e]	73[e]						

[a] R. Haser, M. Pierrot, M. Frey, F. Payan, J. P. Astier, M. Bruschi, and J. LeGall, *Nature (London* **282,** 806 (1979).

[b] Y. Higushi, M. Kusonoki, Y. Matsuura, N. Yasuoka, and M. Kakudo, *J. Mol. Biol.* **172,** 109 (1984 .

[c] P. M. Matias, C. Frazão, J. Morais, M. Coll, and M. A. Carrondo, *J. Mol. Biol.* **234,** 680 (1993).

[d] D. Kissinger, Ph.D. Thesis, Washington University, St. Louis, Missouri (1989).

[e] Sixth ligand.

TABLE II

SEQUENCE ALIGNMENTS PROPOSED FOR CYTOCHROMES c_3[a–c]

```
Species   Sequence

                  1               5                2       3       4       5
              5   0               5                0   5   0   5   0   5   0   5
DvM   APKAP ADGL KMD             KTK        QPVV  FNHS THKAVKCGDCHHP VNGKENYQKCAT AGCHDNM
          *  *                      *               *  *  *  *    *  ***    *      *    ***
DvH   APKAP ADGL KME             ATK        QPVV  FNHS THKS VKCGDCHHP VNGKEDYRKCGT AGCHDS M
          *  *                      *               *  *  *  *    *  ***    *      *    ***
Dg    V DVP ADGA KI DFI AG       GEK        NLVV  FNHS THKDVKCBBCHHBP  GBKQYAGCTTDGCHNI L
          *  *                      *               *  *  *  *    *  ***    *      *    ***
Ds    V DAP AD  MVLKAP AGAK  MT K        AP   VDFS HKGHAALDCT KCHHKWDGKAEVKKCS AEGCHVBT
          *  *                      *               *  *  *  *    *  ***    *      *    ***
DdE   V DAP AD  MVI KAP AGAK  VTK        AP   VAFS HKGHAS MDCKTCHHKWDGAGAI QPCGAS GCHANT
          *  *                      *               *  *  *  *    *  ***    *      *    ***
DbN   A DAP GDDY VI SAP EGMKAKP KGDKP GALQKT VP FP HT KHAT VECVQCHHXADG  GAVKKCTTS GCHDS L
                  1        2            3        4        5        6
          5   0    5   0     5      0     5   0    5   0    5   0    5

          6          7         8          9         0
          0    5     0    5    0     5    0    5    0   5
DvM   D KKDK SAKGYY    HAMHD  KGTKFKS CVGCH  LETAGADAAKKKELT GCKGS KCHS
                           *             *  **              *  *     **
DvH   D KKDK SAKGYY    HVMHD  KNTKFKS CVGCH  VEVAGADAAKKKDLT GCKKS KCHE
                           *             *  **              *  *     **
Dg    D KADK SVNSWY    KVVHDAKGGAKPTCI SCHK  DKAGDDKELKKKLT GCKGS ACHP S
                           *             *  **              *  *     **
Ds     SKKGKKST PKFY   SAFHS  KSDI   SCVGCHKALKK       ATGPTKC  G  DCHP KKK
                           *             *  **              *  *     **
DdE   ESKKGDDS    FY   MAFHERKSEK   SCVGCHKS MKK         GPTKC  T  ECHP KN
                           *             *  **              *  *     **
DbN   E FRDKANA KDI KLVESAFHT      QCI DCH  ALKKKD  KKPTGPTAC  G  KCHTTN
        7      8               9       0        1
        0   5  0   5           0   5   0    5   0    5
```

[a] DvM, *D. vulgaris* (Miyazaki); Dg, *D. gigas;* DdE, *D. desulfuricans* (El Agheila); DvH, *D. vulgaris* (Hildenborough); Ds, *D. salexigens;* DbN, *Dsm. baculatum* (Norway 4).

[b] P. M. Matias, C. Frazão, J. Morais, M. Coll, and M. A. Carrondo, *J. Mol. Biol.* **234,** 680 (1993).

[c] C. Kissinger, Ph.D. Thesis, University of Washington, Seattle, Washington (1989).

various sequence segments.[10,26] The lowest homology (about 25%) is found between the cytochromes from *Dsm. baculatum* and *D. vulgaris*, the less phylogenetically related Desulfovibrionaceae organisms,[4] and include just about the heme coordination sites and the sequence invariants pointed out above. It should be mentioned that the amino acid sequence determined for the trihemic cytochrome c_3 (formerly cytochrome c_7) isolated from the sulfur reducer *Desulforomonas acetoxidans*,[37] in which the iron–iron distances proposed from preliminary X-ray studies are similar to those found in cytochromes c_3,[38] present ligating segments Cys-X X (X X)-Cys-His and residues phenylalanine and threonine homologous to Phe-34 and Thr-110 in cytochrome c_3 from *Dsm. baculatum* (Norway 4). These observations have led to the suggestion of a three-heme arrangement similar to the one found in cytochromes c_3, with the deletion of the sequence segment that accommodates heme HII.[39]

Redox Potentials

Cytochromes c_3 are low (less than 0 mV) redox-potential proteins; their full reduction is achieved only by using low-potential chemical reductants such as sodium dithionite (E° of ~ -500 mV at pH 7), or the system H_2/Hase at pH values higher than pH 7. Estimates of the four macroscopic redox potentials for these proteins have been based on electrochemical methods[40–44] and spectroelectrochemical data.[45,46] These studies established values between -50 and -400 mV for the macroscopic redox potentials (Table III).

One-dimensional (1D) NMR redox titration data obtained for cyto-

[37] R. P. Ambler, *FEBS Lett.* **18,** 351 (1971).

[38] R. Haser, F. Payan, R. Bache, M. Bruschi, and J. LeGall, *J. Mol. Biol.* **130,** 97 (1979).

[39] F. Scott Mathews, *Prog. Biophys. Mol. Biol.* **45,** 1 (1985).

[40] M. Bruschi, R. Loufti, P. Bianco, and J. Haladjian, *Biochem. Biophys. Res. Commun.* **120,** 384 (1984).

[41] P. Bianco and J. Haladjian, *Biochim. Biophys. Acta* **26,** 1001 (1981).

[42] K. Niki, K. Kawasaki, N. Nishimura, Y. Higushi, N. Yasuoka, and M. Kakudo, *J. Electroanal. Chem.* **168,** 275 (1984).

[43] I. Mus-Veteau, A. Dolla, F. Guerlesquin, F. Payan, M. Czjek, R. Haser, P. Bianco, J. Haladjian, B. J. Rapp-Giles, J. D. Wall, G. Voordouw, and M. Bruschi, *J. Biol. Chem.* **267,** 16851 (1992).

[44] V. Nivière, E. C. Hatchikian, P. Bianco, and J. Haladjian, *Biochim. Biophys. Acta* **935,** 34 (1988).

[45] K. Fan, H. Akutsu, K. Niki, N. Higuchi, and Y. Kyoguku, *J. Electroanal. Chem.* **278,** 295 (1990).

[46] M. Coletta, T. Catarino, J. LeGall, and A. V. Xavier, *Eur. J. Biochem.* **202,** 1101 (1991).

TABLE III
MACROSCOPIC POTENTIALS DETERMINED FOR CYTOCHROMES c_3

Organism	Potential (mV)			
	E_1	E_2	E_3	E_4
Dsm. baculatum (Norway 4)[a]	−400	−365	−305	−165
Dsm. baculatum (Norway 4)[b,c]	−410	−370	−310	−210
D. vulgaris (Miyazaki)[d]	−360	−322	−300	−242
D. vulgaris (Hildenborough)[e]	−380	−350	−320	−280
D. vulgaris (Hildenborough)[e,f]	−350	−320	−260	−80
D. gigas[g]	−330	−315	−295	−195
D. desulfuricans (Berre Sol)[h]	−375	−335	−305	−225
D. desulfuricans (El Agheila)[h]	−320	−290	−265	−235

[a] P. Bianco and J. Haladjian, *Biochim. Biophys. Acta* **26,** 1001 (1981).

[b] A. Dolla, C. Cambillau, P. Bianco, J. Haladjian, and M. Bruschi, *Biochem. Biophys. Res. Commun.* **147,** 818 (1987).

[c] Cytochrome with the residue Arg-73 chemically blocked.

[d] W. F. Sokol, P. H. Evans, K. Niki, and T. Yagi, *J. Electroanal. Chem.* **108,** 107 (1980).

[e] I. Mus-Veteau, A. Dolla, F. Guerlesquin, F. Payan, M. Czjek, R. Haser, P. Bianco, J. Haladjian, B. J. Rapp-Giles, J. D. Wall, G. Voordouw, and M. Bruschi, *J. Biol. Chem.* **267,** 16851 (1992).

[f] Obtained for a cytochrome subjected to site-directed mutagenesis His-70 → Met-70.

[g] V. Nivière, E. C. Hatchikian, P. Bianco, and J. Haladjian, *Biochim. Biophys. Acta* **935,** 34 (1988).

[h] M. Bruschi, R. Loufti, P. Bianco, and J. Haladjian, *Biochem. Biophys. Res. Commun.* 120, 384 (1984).

chromes c_3 from *D. gigas*,[9,47] *D. vulgaris* (Miyazaki),[48] and *D. vulgaris* (Hildenborough)[49] have shown that the redox potentials of the four hemes (microscopic redox potentials or micropotentials) were similar, that the redox state of one or more of the hemes influenced the micropotential values of the other hemes (measured as heme–heme interaction potentials), and that the individual microscopic midpoint redox potentials are pH dependent; all the individual microscopic midpoint redox potentials and redox-linked pK_a values have been determined for the cytochromes from *D. gigas*[46,47] and *D. vulgaris* (Hildenborough).[50] Spectral interactions

[47] H. Santos, J. J. Moura, I. Moura, J. LeGall, and A. V. Xavier, *Eur. J. Biochem.* **141,** 287 (1984).

[48] K. Fan, H. Akutsu, Y. Kyoguku, and K. Niki, *Biochemistry* **29,** 2257 (1990).

[49] C. Salgueiro, D. L. Turner, H. Santos, J. LeGall, and A. V. Xavier, *FEBS Lett.* **314,** 155 (1992).

[50] D. L. Turner, C. Salgueiro, T. Catarino, J. LeGall, and A. V. Xavier, *Biochim. Biophys. Acta,* in press (1994).

between the hemes have been observed in data obtained by visible spectroscopy,[51] electron paramagnetic resonance (EPR),[52,53] Mössbauer spectroscopy,[54] and resonance Raman spectroscopy.[55] The NMR-determined heme–heme interaction potentials take both positive and negative values, ranging from −40 to 40 mV, and therefore the interactions cannot be explained solely in terms of the electrostatic effects due to the localized charge modification that goes with the redox change of each heme; thus, these results suggest that there is one or several (presently unspecified) structural modifications linked to the redox state of the molecule. On the other hand, the redox titrations obtained for the cytochromes from *Dsm. baculatum* (Norway 4) and *Dsm. baculatum* (DSM 1743) by EPR[56–58] and NMR[59] have yielded three micropotential values similar to the ones determined for *D. vulgaris* and *D. gigas*,[52,53,60,61] but in addition a fourth micropotential considerably higher than the other three, which has no correspondence in the other proteins; no important heme–heme interaction potentials could be detected for these two last proteins.

The redox micropotentials and the heme–heme interaction potentials determined by NMR and EPR for cytochromes c_3 are presented in Tables IV and V, respectively. The four macroscopic redox potentials, which are uniquely determined by the microscopic potentials and by the heme–heme interaction potentials,[62] are comparable with the values estimated independently.

The assignment of each micropotential to the corresponding heme in the three-dimensional structure was established by NMR for the cytochromes from *D. gigas*, *D. vulgaris*, and *Dsm. baculatum* (Table VI). It should be stressed that, although the heme core architecture is maintained, a different order for the redox potentials is observed.

[51] T. Yagi, *Biochim. Biophys. Acta* **767,** 288 (1984).

[52] C. More, J. P. Gayda, and P. Bertrand, *J. Magn. Reson.* **90,** 486 (1990).

[53] M. Teixeira and J. P. Campos, unpublished results (1993).

[54] M. Utuno, K. Ono, K. Kimura, H. Inokuchi, and T. Yagi, *J. Phys. (Paris)* **41,** 957 (1980).

[55] A. L. Verma, K. Kimura, A. Nakamura, T. Yagi, H. Inokuchi, and T. Kitagawa, *J. Am. Chem. Soc.* **110,** 6617 (1988).

[56] J. P. Gayda, P. Bertrand, C. More, F. Guerlesquin, and M. Bruschi, *Biochim. Biophys. Acta* **829,** 262 (1985).

[57] J. P. Gayda, H. Benosman, P. Bertrand, C. More, and M. Asso, *Eur. J. Biochem.* **177,** 199 (1988).

[58] I. Moura, M. Teixeira, B. H. Huynh, J. LeGall, and J. J. G. Moura, *Eur. J. Biochem.* **176,** 365 (1988).

[59] I. B. Coutinho, Ph.D. Thesis, FCT-UNL Lisboa (1993).

[60] A. V. Xavier, J. J. G. Moura, J. LeGall, and D. V. DerVartanian, *Biochimie* **61,** 689 (1979).

[61] H. Berosman, M. Asso, P. Bertrand, T. Yagi, and J.-P. Gayda, *Eur. J. Biochem.* **182,** 51 (1989).

[62] T. Catarino, Undergraduate Research Report, FCT-UNL (1987).

TABLE IV
MICROSCOPIC REDOX POTENTIALS DETERMINED FOR CYTOCHROMES c_3 FROM NUCLEAR MAGNETIC RESONANCE DATA

Organism	Redox potentials (mV)				pH
	e_1	e_2	e_3	e_4	
Dsm. baculatum[a]	−335	−296	−286	−120	7.8
D. vulgaris (Miyazaki)[b]	−340[c]	−328	−302	−270	7.1
	−325[d]	−291	−355	−321	
D. gigas[e,f]	−295[c]	−260	−225	−180	5.5
	−255[d]	−290	−255	−240	
	−360[c]	−305	−280	−205	9.8
	−340[d]	−285	−280	−205	

[a] I. B. Coutinho, Ph.D. Thesis, FCT-UNL Lisboa (1993).
[b] K. Fan, H. Akutsu, Y. Kyoguku, and K. Niki, *Biochemistry* **29,** 2257 (1990).
[c] Redox potentials when the other hemes are oxidized.
[d] Redox potentials when the other hemes are reduced.
[e] H. Santos, J. J. G. Moura, I. Moura, J. LeGall, and A. V. Xavier, *Eur. J. Biochem.* **141,** 287 (1984).
[f] M. Coletta, T. Catarino, J. LeGall, and A. V. Xavier, *Eur. J. Biochem.* **202,** 1101 (1991).

For the cytochromes from *D. gigas* and *D. vulgaris*, which are in slow intermolecular exchange conditions in the NMR time scale, the complete assignment was done by following the two-dimensional (2D) NMR EXSY and ROESY connectivities due to the heme methyl group resonances, previously identified in the fully reduced state, throughout the five individual redox stages up to the fully oxidized state.[9,49,63] The results agree only partially with the ones obtained from the direct interpretation of 2D NMR nuclear Overhauser effect spectroscopy (NOESY) and TOCSY results acquired for the oxidized *D. vulgaris* proteins, which were based mainly on the tentative assignment of specific interheme connectivies.[64–66] An attempt to identify the redox potential of heme HIV of the cytochrome from *D. vulgaris* (Hildenborough) through potentiometric data obtained for the protein subjected to site-directed mutagenesis of His-70 to Met-70[43] is considered ambiguous (see Table III).

[63] A. V. Xavier and D. L. Turner, this series, Vol. **227**.
[64] J.-S. Park, K. Kano, Y. Morimoto, Y. Higushi, N. Yasuoka, M. Ogata, K. Niki, and H. Akutsu, *J. Biomol. NMR* **1,** 271 (1991).
[65] J.-S. Park, K. Kano, K. Niki, and H. Akutsu, *FEBS Lett.* **285,** 149 (1991).
[66] M. Sola and J. A. Cowan, *Inorg. Chem. Acta* **202,** 241 (1992).

TABLE V

MICROSCOPIC REDOX POTENTIALS DETERMINED FOR CYTOCHROMES c_3 FROM ELECTRON PARAMAGNETIC RESONANCE DATA[a]

	Redox potentials (mV)			
Organism	e_1	e_2	e_3	e_4
Dsm. baculatum (DSM 1743)[b]	−355	−300	−280	−70
Dsm. baculatum (Norway 4)[c]	−355	−325	−270	−150
Dsm. baculatum (Norway 4)[d]	−345/−352	−320/−330	−275/−300	150
Dsm. baculatum (Norway 4)[e]	−355	−330	−300	−150
D. vulgaris (Miyazaki)[f]	−355/−360	−335/−330	−325/−320	−250/−220
D. vulgaris (Hildenborough)[g]	−365	−350	−320	−300
D. gigas[h]	−315	−306	−235	−235

[a] The values were obtained for frozen solutions.

[b] I. Moura, M. Teixeira, B. H. Huynh, J. LeGall, and J. J. G. Moura, *Eur. J. Biochem.* **176,** 365 (1988).

[c] P. Gayda, P. Bertrand, C. More, F. Guerlesquin, and M. Bruschi, *Biochim. Biophys. Acta* **829,** 262 (1985).

[d] P. Gayda, H. Benosman, P. Bertrand, C. More, and M. Asso, *Eur. J. Biochem.* **177,** 199 (1988).

[e] Guigliarelli, P. Bertrand, C. More, R. Haser, and J.-P. Gayda, *J. Mol. Biol.* **216,** 161 (1990).

[f] Berosman, M. Asso, P. Bertrand, T. Yagi, and J.-P. Gayda, *Eur. J. Biochem.* **182,** 51 (1989).

[g] M. Teixeira and J. P. Campos (1993).

[h] A. V. Xavier, J. J. G. Moura, J. LeGall, and D. V. DerVartanian, *Biochimie* **61,** 689 (1979).

The intermolecular exchange conditions presented by the cytochromes c_3 from *Dsm. baculatum* are fast in the NMR time scale, and thus no two-dimensional experiments similar to the ones described were performed on these proteins. The assignments of the higher potential heme[29] and of the three lower potential ones[59] were obtained by monitoring the chemical shifts of selected resolved resonances, such as those due to heme *meso* and methine protons as well as to aromatic protons, subjected to small extrinsic pseudocontact effects along redox titration processes followed by 1D NMR. The assignment is identical to one of the solutions (the most favored one) proposed by Guigliarelli and co-workers, based on redox titration results together with single-crystal EPR studies,[67] and agrees with partial attributions based on features such as the exposition of the hemes

[67] B. Guigliarelli, P. Bertrand, C. More, R. Haser, and J.-P. Gayda, *J. Mol. Biol.* **216,** 161 (1990).

TABLE VI
STRUCTURE–REDOX POTENTIAL CORRELATION ESTABLISHED FOR HEMES IN CYTOCHROME c_3

Organism	Heme associated with redox potential			
	e_1	e_2	e_3	e_4
D. gigas[a]	HI	HII	HIII	HIV
D. vulgaris[b]	HIII	HII	HI	HIV
Dsm. baculatum[c,d]	HII	HI	HIV	HIII

[a] A. Piçarra-Pereira, D. L. Turner, J. LeGall, and A. V. Xavier, *Biochem. J.* **294,** 909 (1993).
[b] C. Salgueiro, D. L. Turner, H. Santos, J. LeGall, and A. V. Xavier, *FEBS Lett.* **314,** 155 (1992).
[c] I. B. Coutinho, D. L. Turner, J. LeGall, and A. V. Xavier, *Biochem. J.* **294,** 899 (1993).
[d] I. B. Coutinho, Ph.D. Thesis, FCT-UNL Lisboa (1993).

to the solvent[40] and the inductive effects created by α-helical segments,[28] or the distribution of the charges within the molecule.[29,59] A different assignment for the higher potential heme was proposed from a study of cross-linked fragments of a covalent complex between cytochrome c_3 and ferredoxin from *Dsm. baculatum* (Norway 4),[68,69] considered comparable with the noncovalent model generated by molecular graphics,[70] together with the interpretation of NMR data obtained for a cytochrome c_3 and ferredoxin solution[71] and with the analysis of the redox behavior of the cytochrome subjected to chemical modification in the residue Arg-73.[72] A tentative correlation of the micropotentials with the relative orientation of the axial histidines also yielded a different assignment.[58] As usual with multiredox center proteins, only self-consistent global assignments are reliable (Table VI).

[68] A. Dolla and M. Bruschi, *Biochim. Biophys. Acta* **932,** 26 (1988).
[69] A. Dolla, G. Leroy, F. Guerlesquin, and M. Bruschi, *Biochim. Biophys. Acta* **1058,** 171 (1991).
[70] A. Dolla, F. Guerlesquin, M. Bruschi, and R. Haser, "Microbiology and Biochemistry of Strict Anaerobes Involved in Interspecies Transfer," p. 249. Plenum, New York, 1990.
[71] F. Guerlesquin, M. Noailly, and M. Bruschi, *Biochem. Biophys. Res. Commun.* **103,** 1102 (1985).
[72] A. Dolla, C. Cambillau, P. Bianco, J. Haladjian, and M. Bruschi, *Biochem. Biophys. Res. Commun.* **147,** 818 (1987).

Kinetic Studies

The existence of four hemes in cytochrome c_3 confers to this protein the possibility of undergoing electron transfer (ET) between the hemes within the same molecule (an intramolecular first-order ET reaction) or between hemes belonging to different molecules [an intermolecular (bimolecular) second-order ET reaction].

The intramolecular rate constants for cytochrome c_3 have not been determined. The absence of individualized heme resonances in NMR results obtained in intermediate redox states for samples in slow intermolecular exchange shows that those constants are very high ($>10^5$ sec^{-1}),[73] whereas resonance Raman results on similar samples indicate that they are lower than the spectral time scale.[55] Excluding the rates found for the ET processes within the photosynthetic reaction center, these orders of magnitude are considerably higher than those found for most intramolecular ET processes in proteins containing multiple metallic redox centers.[74] This behavior could be due to the short distances between the iron atoms and the porphyrin edges, as well as to the existence of several shortened pathways involving noncovalent bonds, which connect otherwise separate segments of the sequence between the heme moieties (see Fig. 4), thus facilitating the intramolecular electron flow. The physiological implications of this high intramolecular exchange rate are discussed later.

First-order ET rate constants were calculated for the paired ferredoxin-cytochrome c_3, using stopped-flow (SF) results.[75] In this study, the redox pair was considered to be a dimer, and the rate constants for the direct and backward reaction were, respectively, $k_{dir} = 1.6 \times 10^2$ sec^{-1} and $k_{back} > 10^3$ sec^{-1}. It is important to point out that other intramolecular constants obtained from SF data,[75,76] with values between 1.6 and 23 sec^{-1}, are pseudo-first-order parameters used for the direct adjustment of experimental data and have no readily apparent physical significance.

Early results obtained for the reduction of the cytochrome from *D. vulgaris* with sodium dithionite and carboxyl radicals followed by pulse

[73] H. Santos, Ph.D. thesis, FCT-UNL Lisboa (1984).

[74] G. R. Moore and G. H. Pettigrew, "Cytochromes c: Evolutionary, Structural and Physicochemical Aspects," p. 363. Springer-Verlag, Berlin, 1990.

[75] C. Capeillère-Blandin, F. Guerlesquin, and M. Bruschi, *Biochim. Biophys. Acta* **848,** 279 (1989).

[76] I. Tabushi, T. Nishiya, T. Yagi, and H. Inokuchi, *J. Biochem.* (*Tokyo*) **94,** 1375 (1983).

radiolysis[77,78] (Table VII) provided a first indication that the four hemes in cytochrome c_3 were kinetically nonequivalent with respect to bimolecular reduction with inorganic redox partners. Later, SF data obtained for the reduction of cytochrome c_3 from *Dsm. baculatum* with sodium dithionite[75] established a biphasic kinetic behavior, in which the faster ($k = 6 \times 10^6$ M^{-1} $\sec^{-1}$) and slower ($k = 6 \times 10^5$ M^{-1} $\sec^{-1}$) components correspond, respectively, to the reduction of 25 and 75% of the total protein. Comparable biphasic kinetics and rate constant values have been established for the cytochromes from *D. vulgaris*[75] and *D. gigas*[79]; in the latter case, the conjugation of kinetic data with the independently determined microscopic redox potentials and heme–heme interactions led to the establishment of the intermolecular ET rate constants corresponding to each separate heme. Formally similar results were obtained for the reduction kinetics of cytochromes c_3 from *D. vulgaris* (Miyazaki) and (Hildenborough) with inorganic partners followed by flash photolysis,[80] whereby it was additionally observed that the rate constants were considerably affected by the formal charge of the reductant and by the ionic strength of the solution. The stopped-flow experiments performed using ferredoxin as an electron acceptor or as an electron donor to the cytochrome from *Dsm. baculatum*,[75] the hydrogenase-mediated reduction of cytochrome c_3 from *D. gigas* with molecular hydrogen followed by visible spectroscopy and cyclic voltametry,[44] and the pH- and ionic strength-dependent autoexchange ET process involving the higher potential heme HIII of *Dsm. baculatum* detected by NMR[59] (Table VIII) have yielded rate constant values comparable to the ones determined for the experiments with inorganic redox partners.

It was found that there was a time-dependent reduction of the iron ferricyanide contained within cardiolipin and lecithin liposomes in the presence of excess reductant and cytochrome c_3 from *D. vulgaris*.[81] The behavior was attributed to an electron transfer process through the membrane, eventually mediated by a dimer of cytochrome c_3 stabilized by strong noncovalent forces. Although this proposal is consistent with the

[77] J. W. Van Leuwen, C. Van Djik, H. J. Grande, and C. Veeger, *Eur. J. Biochem.* **127,** 631 (1982).

[78] V. Favaudon, C. Ferradini, J. Pucheault, L. Gilles, and J. LeGall, *Biochem. Biophys. Res. Commun.* **84,** 435 (1978).

[79] T. Catarino, M. Coletta, J. LeGall, and A. V. Xavier, *Eur. J. Biochem.* **202,** 1107 (1991).

[80] H. Akutsu, J. H. Hazzard, R. G. Bartsch, and M. A. Cusanovich, *Biochim. Biophys. Acta* **1140,** 144 (1992).

[81] I. Tabushi, T. Nishiya, T. Yagi, and H. Inokuchi, *J. Am. Chem. Soc.* **103,** 6963 (1981).

TABLE VII
INTERMOLECULAR RATE CONSTANTS FOR ELECTRON TRANSFER PROCESSES BETWEEN CYTOCHROMES c_3 AND INORGANIC REDOX PARTNERS

Reduced species	Reducer	Kinetics	k (mol^{-1} sec^{-1})
Cytochrome c_3 from *D. vulgaris* (Hildenborough)[a]	Methyl viologen	Monophasic	4.5×10^8
Methyl viologen[a]	Cytochrome c_3 from *D. vulgaris* (Hildenborough)	Monophasic	8×10^4
Cytochrome c_3 from *D. vulgaris*[b]	Dithionite	Biphasic (50–50)	6.8×10^6 2.1×10^6
Cytochrome c_3 from *D. vulgaris*[b]	Carboxyl radicals	Monophasic	2.17×10^8
Cytochrome c_3 from *Dsm. baculatus* (Norway 4)[c]	Dithionite	Biphasic (20–80)	$\cong 10^7$ 6×10^5
Cytochrome c_3 from *D. vulgaris*[c]	Dithionite	Biphasic (25–75)	3.2×10^6 6.3×10^5
Cytochrome c_3 from *D. gigas*[d]	Dithionite	Biphasic (25–75)	2.7×10^4–7.5×10^6 (pH 5.5) 2.7×10^4–20×10^6 (pH 9.4)
Cytochrome c_3 from *D. vulgaris* (Hildenborough)[g]	DRFH (neutral)[e]	Monophasic	10^9–4.2×10^8 (low I)[f] 2.6×10^8–8.8×10^8 (high I)
	PDQ (positive)[h]	Monophasic	0.6×10^8–1.6×10^8 (low I) 1.4×10^8–3.3×10^8 (high I)
Cytochrome c_3 from *D. vulgaris* (Miyazaki)[g]	DRFH		5.2×10^8–8.0×10^8 (low I)
	PDQ		0.6×10^8–1.9×10^8 (low I)

[a] J. W. Van Leuwen, C. Van. Djik, H. J. Grande, and C. Veeger, *Eur. J. Biochem.* **127,** 631 (1982).
[b] V. Favaudon, C. Ferradini, J. Pucheault, L. Gilles, and J. LeGall, *Biochem. Biophys. Res. Commun.* **84,** 435 (1978).
[c] C. Capeillére-Blandin, F. Guerlesquin, and M. Bruschi, *Biochim. Biophys. Acta* **848,** 279 (1989).
[d] T. Catarino, M. Coletta, J. LeGall, and A. V. Xavier, *Eur. J. Biochem.* **202,** 1107 (1991).
[e] DRFH, Reduced 5-deazariboflavine.
[f] I, Ionic strength.
[g] H. Akutsu, J. H. Hazzard, R. G. Bartsch, and M. A. Cusanovich, *Biochim. Biophys. Acta* **1140,** 144 (1992).
[h] PDQ, Propylene diquat.

electrical conductivity detected in anhydrous cytochrome c_3 films,[82] the presence of a high amount of charged residues on the protein surface makes it unprobable.[83] The reduction of ferricyanide within micelles was also observed in micelle–cytochrome c_3 systems, using gaseous hydrogen

[82] Y. Nakahava, K. Kimura, H. Inokushi, and T. Yagi, *Chem. Phys. Lett.* **73,** 31 (1980).
[83] G. Fauque, J. LeGall, and L. L. Barton, *in* "Variations in Autotrophic Life" (J. M. Shiveley and L. L. Barton, eds.), p. 271. Academic Press, San Diego, 1991.

TABLE VIII
INTERMOLECULAR RATE CONSTANTS FOR ELECTRON TRANSFER PROCESSES BETWEEN CYTOCHROMES c_3 AND BIOLOGICAL REDOX PARTNERS

Reduced species	Reducer	k (mol^{-1} sec^{-1})
Cytochrome c_3 from *D. gigas*[a]	Hydrogen/hydrogenase	6.6×10^7
Ferredoxin[b]	Cytochrome c_3 from *Dsm. baculatum* (Norway 4)	1.5×10^8
Cytochrome c_3 from *Dsm. baculatum* (Norway 4)[b]	Ferredoxin	6.6×10^7
Cytochrome c_3 from *Dsm. baculatum* high-potential heme HIII[c]	Cytochrome c_3 from *Dsm. baculatum* high-potential heme HIII	4.5×10^5

[a] V. Nivière, E. C. Hatchikian, P. Bianco, and J. Haladjian, *Biochim. Biphys. Acta* **935,** 34 (1988).

[b] C. Capeillère-Blandin, F. Guerlesquin, and M. Bruschi, *Biochim. Biophys. Acta* **848,** 279 (1989).

[c] I. B. Coutinho, Ph.D. Thesis, FCT-UNL Lisboa (1993).

as a reductant in the presence of colloidal platinum.[84,85] In these experiments it was stated that there was a concomitant generation of a through-membrane pH gradient, leading the authors to propose the system as a model for hydrogen-metabolizing bacteria, although no reduction was detected in the presence of cytochrome and hydrogenase. However, it has been proposed that the intramicellar pH could not be measured as indicated.[86]

Genetics

The gene that codes for cytochrome c_3 from *D. vulgaris* has been identified[87,88] and expressed in *Escherichia coli*,[89] *Dsm. baculatum*,[90] and *Rhodobacter sphaeroides*.[91] The genetics of these organisms have been

[84] I. Tabushi, T. Nishya, T. Yagi, and H. Inokuchi, *Tetrahedron Lett.* **22,** 4989 (1981).

[85] I. Tabushi, T. Nishya, M. Shimonura, T. Kunitake, H. Inokushi, and T. Yagi, *J. Am. Chem. Soc.* **106,** 219 (1984).

[86] C. Vandijk, R. Spruijt, I. Laiane, and C. Veeger, *Eur. J. Biochem.* **207,** 587 (1992).

[87] G. Voodrouw and S. Brenner, *Eur. J. Biochem.* **159,** 347 (1986).

[88] G. Voodrouw, H. M. Kent, and J. R. Postgate, *Can. J. Microbiol.* **33,** 1006 (1987).

[89] W. B. R. Pollock, G. Voodrouw, M. E. Forrest, J. T. Beatty, and P. J. Chemerica, *J. Gen. Microbiol.* **135,** 2319 (1989).

[90] G. Voodrouw, W. B. R. Pollock, M. Bruschi, F. Guerlesquin, B. J. Rappgiles, and J. D. Wall, *J. Bacteriol.* **172,** 6122 (1990).

[91] V. Canac, M. S. Caffrey, G. Voodrouw, and M. A. Cusanovich, *Arch. Biochem. Biophys.* **286,** 629 (1991).

reviewed by Wall[92] and Voodrouw[93] and in other chapters of this volume.[94–95]

Concluding Remarks

Being a small soluble protein, cytochrome c_3 is by far the best characterized multiredox center protein. Indeed, an impressive amount of data has been obtained for several cytochromes c_3 from different sulfate-reducing bacteria. However, as mentioned above (see introduction), no unequivocal physiological role has been established for this protein, which is produced in large quantities by all sulfate-reducing bacteria. The fact that it is quickly and fully reduced in the presence of hydrogen and catalytic amounts of hydrogenase suggests that it is a coupling redox partner to this enzyme. Thus, it is interesting to put together all of the data obtained for cytochrome c_3 in the context of its postulated physiological role. Such an approach will consider the data obtained for the *D. gigas* proteins.

Nickel hydrogenase is a periplasmic membrane-bound enzyme that catalyzes the generation of two electrons and two protons from molecular hydrogen. Nuclear magnetic resonance titrations (A. V. Xavier, unpublished work, 1994) have shown that there is a specific interaction between this enzyme and cytochrome c_3 near HIV, the heme closest to the lysine patch on the protein surface. Furthermore, stopped-flow experiments have shown that the higher potential heme HIV is the one that reacts more rapidly with a negatively charged chemical reducer (dithionite) and that electrons are drained (via fast intramolecular ET) at identical rates to the two intermediate-potential hemes HII and HIII.[79] This is due to the interacting redox potential network and, in particular, to the strong positive redox interaction (40 mV) between those two hemes; the resulting cooperativity allows the protein to perform a synchronous two-electron transfer step. On the other hand, the protons generated by the oxidation of molecular hydrogen can have a role in this mechanism. Indeed, the microscopic redox potential of heme HII in the unprotonated species

[92] J. D. Wall, *in* "The Sulfate-Reducing Bacteria: Contemporary Perspectives" (J. M. Odom and R. Singleton, Jr., eds.), p. 77. Brock-Springer Series in Contemporary Biosciences, New York 1993.

[93] G. Voodrouw, *in* "The Sulfate-Reducing Bacteria: Contemporary Perspectives" (J. M. Odom and R. Singleton, Jr., eds.), p. 89. Brock-Springer Series in Contemporary Biosciences, New York 1993.

[94] W. M. A. M. van Dongden, J. P. W. G. Stokkermans, and W. A. M. van den Berg, this volume [22].

[95] C. Dahl, N. Speich, and H. G. Trüper, this volume [23].

when hemes HIV and HIII are reduced (−295 mV) increases by 30 mV on protonation of the protein (pK_a 7.1). This positive e^-/H^+ heterotrophic cooperativity, known as the redox–Bohr[79,96] effect, can be used both for electronic/protonic energy transduction and to facilitate the synchronization of the two-electron transfer step. Subsequently, the specific transport of electrons through membrane carriers can lead to a charge separation necessary for the ATPase-mediated oxidative phosphorylation.[97]

Acknowledgments

We should like to express our sincere admiration to Professor Jean LeGall. Were it not for his constant stimulus, intuition, and helpful discussions the knowledge about cytochrome c_3 would still be quite incipient. We should also like to thank the many collaborators and coauthors cited in the reference list, who contributed enormously to our understanding of this small protein with amazing properties.

[96] A. V. Xavier, *in* "Frontiers in Bioinorganic Chemistry" (A. V. Xavier, ed.), p. 722. VCH Weinheim, 1986.

[97] J. M. Odom and H. D. Peck, Jr., *FEMS Lett.* **12,** 47 (1981).

[10] Cytochrome c_3 (M_r 26,000) Isolated from Sulfate-Reducing Bacteria and Its Relationships to Other Polyhemic Cytochromes from *Desulfovibrio*

By MIREILLE BRUSCHI

Introduction

Sulfate-reducing bacteria of the genus *Desulfovibrio* are able to use oxidized sulfur compounds, such as sulfate, as the terminal electron acceptor of a complex electron transfer chain in which either organic compounds or hydrogen acts as the initial electron donor. This process, whereby the reduced sulfur compounds produced are dissimilated in the environment, is called "dissimilatory sulfate reduction." Various aspects of the *Desulfovibrio* metabolism have been extensively studied[1–3] and a model named "hydrogen cycling" has been proposed to explain the elec-

[1] J. R. Postgate, *in* "The Sulfate Reducing Bacteria," 2nd Ed., Cambridge Univ. Press, Cambridge, 1984.

[2] J. H. Odom and H. D. Peck, *Annu. Rev. Microbiol.* **38,** 551 (1984).

[3] J. LeGall and G. Fauque, *in* "Biology of Anaerobic Microorganisms" (A. J. B. Zehnder, ed.), p. 587. Wiley, New York, 1988.

tron transport and energy conservation in this genus. In the chain that links the perisplasmic oxidation of hydrogen to the cytoplasmic reduction of sulfite, hydrogenases and cytochrome c_3 play a central role together with membrane and cytoplasmic enzymes and other electron carriers.

Although many of the electron carrier proteins have been fully characterized, the data published to date on their functional roles have been contradictory, due to their great diversity, which possibly reflects the existence of multiple terminal enzymes and donor reactions. Moreover, these proteins are not uniformly distributed among *Desulfovibrio* species and the data available have been based on cell-free reconstitution experiments in which the compartmentation of the electron carriers is lost. Because the soluble cytochromes can be easily purified, most studies on *Desulfovibrio* have focused on these cytochromes, although the existence of membrane-associated *c*-type cytochromes has been reported.[4]

Since the initial discovery of cytochrome c_3 in 1954,[4a] several subclasses have by now been described in *Desulfovibrio*, including other types of polyhemic cytochrome c_3, among which cytochrome c_3 (M_r 26,000) has been the least thoroughly studied, probably due to its low solubility. This subclass of cytochrome has been poorly characterized and requires further investigation.

Distribution and Localization of Polyhemic Cytochromes c_3 in *Desulfovibrio*

More is known at present about the biochemical and structural nature of *Desulfovibrio* cytochromes *c* than about their physiological specificity. The published nucleotide sequence of these genes has shown their cellular organization and their diversity. *Desulfovibrio* species contain four different *c*-type cytochromes: the monohemic cytochrome c_{553} (M_r 9000), the tetraheme cytochrome c_3 (M_r 13,000), the octaheme cytochrome c_3 (M_r 26,000), and a high molecular weight cytochrome *c* (Hmc) with 16 hemes and a polypeptide chain with a molecular weight of 65,500. Cytochrome c_{553} is characterized by the presence of a single heme group with a His-Met-coordinated heme iron atom, as in mitochondrial cytochromes *c*, but exhibits a lower molecular weight (M_r 9000) and a lower redox potential. Cytochrome c_3 (M_r 13,000) is a tetrahemic protein in which each heme exhibits a distinct redox potential in the −200 to −400-mV range and all have the same His–His iron atom axial ligands. The occurrence of a soluble high molecular weight cytochrome (Hmc) has been reported only

[4] J. F. Kramer, D. H. Pope, and J. C. Salerno, *FEBS Lett.* **206,** 157 (1986).

[4a] J. R. Postgate, *Biochem. J.* **56,** XI–XII (1954).

in *Desulfovibrio* (*D.*) *vulgaris* strains Miyazaki and Hildenborough.[5] A comparison of the arrangement of heme-binding sites and coordinated histidines between the amino acid sequences of cytochromes c_3 and *D. vulgaris* Hildenborough Hmc has shown that the protein contains 16 hemes distributed among 4 domains, 3 of which are complete cytochrome c_3-like domains, whereas the fourth is an incomplete cytochrome c_3-like domain.[6]

The genes encoding cytochromes c_{553}, c_3, and Hmc from *D. vulgaris* Hildenborough have been cloned and sequenced and, because the amino acid signal sequences of all three cytochromes have the same characteristic positive and hydrophobic regions,[6–8] their location is periplasmic. Purification experiments carried out on these proteins have also indicated that these cytochromes are located in the periplasm of *D. vulgaris* Hildenborough, as well as in the periplasm of *Desulfovibrio desulfuricans* G200, when they are overexpressed in this bacteria.

This finding is in agreement with the rule according to which the maturation of *c*-type cytochromes involves a membrane translocation step during or after which the hemes are inserted covalently by the cytochrome *c* heme-lyase enzyme.

Although information is lacking about the gene sequence of cytochrome c_3 (M_r 26,000), it can be assumed that its location may also be periplasmic. If this hypothesis is true, formate and D. lactate dehydrogenases, which have been thought to be possible electron acceptors for cytoplasmic cytochrome c_{553}, will be physically separate from cytochrome c_{553}. However, cytochromes c_3, which are constitutive periplasmic proteins of all *Desulfovibrio* species, act as electron carriers for the periplasmic [Ni-Fe] and [Fe] hydrogenases. Cytochrome c_3 (M_r 26,000) and Hmc have never been detected together in the various *Desulfovibrio* bacteria studied. Several *Desulfovibrio* genomes have been probed with *D. vulgaris* Hildenborough Hmc gene,[7] and no hybridization bands were observed in the case of *Desulfovibrio gigas*, *Desulfovibrio multispirans*, *Desulfovibrio salexigens* or *D. desulfuricans* El Agueila Z and *Desulfomicrobium* (*Dsm.*) *baculatum* Norway 4.[8a] Single bands were observed, however, in the case of *D. vulgaris* Brockhurst Hill, Hildenborough, Monticello, and Wandle;

[5] T. P. Yagi and T. Ogata, *in* "Microbiology and Biochemistry of Strict Anaerobes Involved in Interspecies Hydrogen Transfer" (J. P. Belaich, M. Bruschi, and J. L. Garcia, eds.), Proceedings of the FEMS Symposium p. 247. Plenum, New York, 1990.

[6] W. B. R. Pollock, M. Loutfi, M. Bruschi, B. J. Rapp-Giles, J. D. Wall, and G. Voordouw, *J. Bacteriol.* **173,** 220 (1991).

[7] G. J. H. Van Rooijen, M. Bruschi, and G. Voordouw, *J. Bacteriol.* **171,** 3375 (1989).

[8] G. Voordouw and S. Brenner, *Eur. J. Biochem.* **159,** 347 (1986).

[8a] We have adopted in this chapter the suggestion that the strain Norway 4 should be now called *Desulfomicrobium baculatum*. However, the reader's attention is brought to the fact that the old name *Desulfovibrio desulfuricans* will be found in several references cited in this chapter such as: 10, 11, 13, 16, 18, and 35.

D. desulfuricans Berre Sol, Canet 41, and Teddington R; and *Desulfovibrio africanus* Benghazi and Walvis Bay. A high molecular weight cytochrome *c* (M_r 67,000) containing 11 hemes from *D. vulgaris* Miyazaki[5] has also been characterized. It can therefore be said that in the strains in which cytochromes c_3 (M_r 26,000) have been described (*D. gigas* and *Dsm. baculatum* Norway 4), the Hmc cytochrome is not present. This fact might have to do with the physiological role of these cytochromes, which needs to be investigated.

Characterization of Cytochrome c$_3$ (M_r 26,000)

Isolation and Purification

Cytochrome c_3 (M_r 26,000) has been purified and characterized only in *D. gigas*[9] and *Dsm. baculatum* Norway 4,[10] but its presence has also been detected in *D. desulfuricans* El Agueila Z and in *D. salexigens* Benghazi. In the past, another cytochrome c_3 (M_r 26,000) was described from *D. vulgaris* Hildenborough as an octaheme homodimer[10,11]: amino acid and nucleotide determination sequence experiments have demonstrated, however, that this cytochrome is a 65.5-kDa protein containing 16 hemes per molecule that cannot be classified in the cytochrome c_3 (M_r 26,000) group.

Cytochrome c_3 (M_r 26,000), which might be a homodimeric cytochrome c_3 isoenzyme (isocytochrome c_3), has been variously named in the literature cytochrome cc_3, cytochrome c_3 (M_r 26,000), and (eight-heme) cytochrome c_3. As the last of these terms implies that the eight hemes are linked on a single polypeptide chain, however, we have decided to adopt here the same name, cytochrome c_3 (M_r 26,000), used in our previous publications. *Desulfomicrobium baculatum* Norway 4 cytochrome c_3 (M_r 26,000) is an acidic cytochrome. After a 2-hr centrifugation of the crude extract (corresponding to 1.7 kg wet weight of bacteria) the supernatant was stirred with 300 ml of silica gel (J. T. Baker, Phillipsburg, NJ) for 3 hr to adsorb cytochromes c_3 and c_{553}. The unadsorbed proteins were stirred overnight with 600 ml of DEAE-cellulose and then eluted with 1 *M* Tris-HCl buffer, pH 7.6. After dialysis, cytochrome c_3 (M_r 26,000) was adsorbed on a DEAE-cellulose column and eluted with a discontinuous gradient from 10 to 500 m*M* Tris-HCl buffer.

[9] M. Bruschi, J. LeGall, E. C. Hatchikian, and M. Dubourdieu, *Bull. Soc. Fr. Physiol. Veg.* **15,** 381 (1969).

[10] F. Guerlesquin, G. Bovier-Lapierre, and M. Bruschi, *Biochem. Biophys. Res. Commun.* **105,** 530 (1982).

[11] M. Loutfi, F. Guerlesquin, P. Bianco, J. Haladjian, and M. Bruschi, *Biochem. Biophys. Res. Commun.* **159,** 670 (1989).

After dialysis, the protein was adsorbed on a calcinated alumina column equilibrated with 10 m*M* Tris-HCl buffer and then eluted after a discontinuous gradient with 300 m*M* phosphate buffer, pH 7.6.

Desulfovibrio gigas cytochrome c_3 (M_r 26,000) is also an acidic protein and the purification procedure involves the same steps as with *Dsm. baculatum* Norway 4 cytochrome.

Spectroscopic Studies and Redox Potentials

The molecular weight of the purified cytochrome was estimated to be about 26,000, based on the results of analytical gel electrophoresis performed on 10% polyacrylamide gels in the presence of sodium dodecyl sulfate. The presence of two 13,500-Da subunits was detected in *Dsm. baculatum* Norway 4 by removing the hemes with mercuric chloride. The subunits were strongly associated and the presence of denaturing agents such as 8 *M* urea or 6 *M* guanidine did not suffice to dissociate the dimer.

The absorption spectrum is similar to those of the tetrahemic cytochromes c_3 previously described and the purity coefficient, defined as $(A_{553}^{red} - A_{570}^{red})/A_{280}^{ox}$, was 3.2. The 695-nm band, which is a marker of the His-Met coordination of the heme iron, was found to be absent. The values of the millimolar extinction coefficients at the reduced α peak were around twice those of cytochrome c_3 (M_r 13,000). [The measured value was 242.28 with *Dsm. baculatum* Norway 4 cytochrome c_3 (M_r 26,000).] Accordingly, this cytochrome has four hemes per subunit and two histidine residues constitute the fifth and sixth heme iron ligands. Although the octaheme nature, molecular weight, and optical spectrum suggest the existence of a simple dimeric relationship between this cytochrome and cytochrome c_3 (M_r 13,000), comparisons on the amino acid compositions, N-terminal analysis, and electron paramagnetic resonance (EPR) spectra between the cytochromes c_3 (M_r 26,000) isolated from *Dsm. baculatum* Norway 4 or from *D. gigas* and cytochromes c_3 (M_r 13,000) isolated from the same organisms showed that the octaheme cytochrome is a different cytochrome.

Cyclic voltammetry has been performed on *Dsm. baculatum* Norway 4 cytochrome c_3 (M_r 26,000) in order to determine the redox potential values.[11] These values (−210, −270, −325, and −365 mV) show unambiguously that the molecule is a dimer in which each subunit has four hemes, which are equivalent and bishistidinyl coordinated, as in cytochrome c_3 (M_r 13,000). Electron paramagnetic resonance studies have made it possible to differentiate between these polyhemic cytochromes.[12,13] The EPR spectrum of *Dsm. baculatum* Norway 4 cytochrome c_3 (M_r 26,000) is clearly different from that of cytochrome c_3 (M_r 13,000) in the low- and high-field regions. The redox titration curve deduced from signal intensity measurements fits the assumption that the spectrum may result from a combination of four pairs of identical hemes. The values obtained (−180, −270, −320, and −390 mV) were slightly different from those obtained by carrying out electrochemical measurements.

The biochemical role of this class of dimeric cytochrome c_3 or isocytochrome c_3 may be different from that of cytochrome c_3 (M_r 13,000) and might be correlated with the high heme content, the strong association of the dimeric molecule, the affinity for different redox partners, and the possible membrane association,[9] which is in agreement with the extremely low solubility of the molecule.

Role of Cytochrome c_3 (M_r 26,000): What Is Known

In *D. gigas*, cytochrome c_3 (eight heme) has been described as being involved in the reduction of thiosulfate in a crude system.[9,14] With purified thiosulfate reductase and pure hydrogenase, the most efficient restoration of activity was obtained in the presence of cytochromes (eight heme) whereas cytochrome c_3 (four heme) is almost inactive.[15] However, although the location of the octaheme cytochrome is not known, it seems likely to be perisplasmic or periplasmic facing, and the reaction with sulfate reductase may not be physiological. In fact, most of the activities reported so far are "stimulating" activities, which do not necessarily prove that the electron transfer protein participates directly in the reaction.

The octaheme cytochrome c_3 from *D. desulfuricans* Norway was studied using cyclic voltammetry with a pyrolytic graphite electrode.[16] The kinetics of the reduction of this cytochrome under an H_2 atmosphere by the [Ni-Fe-Se] hydrogenase purified from the same organism was studied in 100 m*M* Tris-HCl, and the second-order homogeneous rate constant

[12] J. LeGall, M. Bruschi-Heriaud, and D. V. Dervartanian, *Biochim. Biophys. Acta* **234,** 499 (1971).

[13] J. P. Gayda, P. Bertrand, C. More, F. Guerlesquin, and M. Bruschi, *Biochim. Biophys. Acta* **829,** 262 (1985).

[14] E. C. Hatchikian, J. LeGall, M. Bruschi, and M. Dubourdieu, *Biochim. Biophys. Acta* **258,** 701 (1973).

[15] M. Bruschi, E. C. Hatchikian, L. A. Golovleva, and J. LeGall, *J. Bacteriol.* **129,** 30 (1977).

[16] J. Haladjian, P. Bianco, F. Guerlesquin, and M. Bruschi, *Biochem. Biophys. Res. Commun.* **179,** 605 (1991).

of the electron transfer between the two proteins was found to be 8×10^8 M^{-1} sec^{-1}. In similar experiments performed on *Dsm. baculatum* Norway 4 cytochrome c_3 (four heme) and hydrogenase, a value of 6×10^7 M^{-1} sec^{-1} was obtained. These results may indicate that the ionic strength plays a decisive role, especially in the reactivity of the octahemic cytochrome either at the electrode or with hydrogenase. At low buffer concentrations the protein becomes electrochemically inactive and cannot transfer electrons between the hydrogenase and the electrode. Given the marked acidity of the cytochrome and that of the hydrogenase (p*I* 4.8, and 6.0, respectively), a protein–protein electrostatic recognition process would necessitate either very localized charged interacting sites or the strong participation of intermediate ions.

The occurrence of a soluble high molecular weight perisplasmic cytochrome (M_r 70,000) has been reported in *D. vulgaris* strains Miyazaki and Hildenborough.[17] The role of these cytochromes has not been determined, probably due to their instability in atmospheric oxygen. When cytochrome from the Miyazaki strain is purified under anaerobic conditions, however, it is reduced by hydrogenase, in the presence of cytochrome c_3.[8] The amino acid sequence of the high molecular weight cytochrome from the Hildenborough strain has been elucidated and found to contain four cytochrome **c**$_3$-like domains.[6] The role of the cytochrome c_3-like domain structures found to exist in polyhemic cytochrome *c* from *Desulfovibrio* species [Hmc and cytochrome c_3 (M_r 26,000)] therefore needs to be clarified. The first amino acid determination of a cytochrome c_3 (M_r 26,000) would help to elucidate the role of these highly specialized structures, with a view to understanding the interactions between these cytochromes and their specific redox partners and their contribution to the redox channel linking hydrogen oxidation and sulfate reduction.

Amino Acid Sequence of Desulfomicrobium baculatum Norway 4 Cytochrome c_3 (M_r 26,000)

The cytochrome c_3 (M_r 26,000) purified from *Dsm. baculatum* Norway 4 comprises 2 identical subunits of 111 amino acids resulting in a molecular mass of 14,987 Da, including that of the 4 heme groups and 29,974 Da in the case of the dimeric form.[18] In comparison with the amino acid composition of cytochrome c_3 (M_r 13,000) from the same organism, that of the monomer was found to contain eight cysteine and eight histidine residues,

[17] Y. Higuchi, K. Inaka, N. Yasuoka, and T. Yagi, *Biochim. Biophys. Acta* **911,** 341 (1987).

[18] M. Bruschi, G. Leroy, F. Guerlesquin, and J. Bonicel, *Biochim. Biophys. Acta* **1205,** 123–131 (1994).

which is just sufficient to link four hemes, although the number of basic and acidic residues differed, in agreement with the measured isoelectric points (4.8 in the octaheme and 7.0 in the tetrahemic cytochrome).

The alignment of the amino acid sequences of cytochrome c_3 from six different sulfate-reducing bacteria[19] has shown the main structural features of the cytochromes of the c_3 type. Twenty-six amino acid residues are invariant; these are mainly located in the cysteine–histidine clusters that bind each heme. The histidine forming the sixth ligand of each heme is conserved and has been attributed to each heme in the three-dimensional model.[20]

This sequence alignment is similar to that based on a comparison between the X-ray crystallographic structures of *Dsm. baculatum* Norway 4 and *D. vulgaris* Miyazaki.[21] The similarity between the patterns of the heme-binding sites and the histidine-containing regions has led to the suggestion that gene duplication of a two-heme cytochrome may take place[19] at an early stage in the evolution of the molecule. This idea is consistent with the two well-defined domains described in *Dsm. baculatum* Norway 4 cytochrome c_3, which become separated at approximately residue 72.[22]

The amino acid sequence of cytochrome c_3 (M_r 26,000) has similar characteristics (Fig. 1)[18] and, consequently, the histidine residues linked to each heme iron atom are supposedly associated according to the three-dimensional structure of *D. desulfuricans* cytochrome c_3 (M_r 13,000).[20] Heme 1 is covalently linked to Cys-38 and -41 and the fifth and sixth histidine ligands are located in positions 42 and 30. Heme 2 is linked to Cys-54 and -59, which are separated by four amino acids instead of two. The fifth and sixth histidine ligands are in positions 60 and 43. Heme 3 is linked to Cys-86 and -89 and to His-90 and -33, and heme 4 to Cys-105 and -108 and His-109 and -77.

This sequence has been compared to each published cytochrome c_3 (M_r 13,000) sequence, using the computer sequence analysis program described by Devereux *et al.*[23] Our comparison between the amino acid sequences provides definite evidence that cytochrome c_3 (M_r 26,000) and cytochrome c_3 (M_r 13,000) isolated from *Dsm. baculatum* Norway 4 are two different cytochromes with only 34.6% homology.

[19] M. Bruschi, *Biochim. Biophys. Acta* **671,** 219 (1981).

[20] R. Haser, M. Pierrot, M. Frey, F. Payan, J. P. Astier, M. Bruschi, and J. LeGall, *Nature* (*London*) **282,** 806 (1979).

[21] Y. Higuchi, S. Bando, M. Kusunoki, Y. Matsuura, N. Kakudo, T. Yamanaka, T. Yagi, and H. Inokuchi, *J. Biochem.* (*Tokyo*) **89,** 1659 (1981).

[22] M. Pierrot, R. Haser, M. Frey, F. Payan, and J. P. Astier, *J. Biol. Chem.* **257,** 14341 (1982).

[23] J. Devereux, P. Haeberli, and O. Smithies, *Nucleic Acids Res.* **12,** 387 (1984).

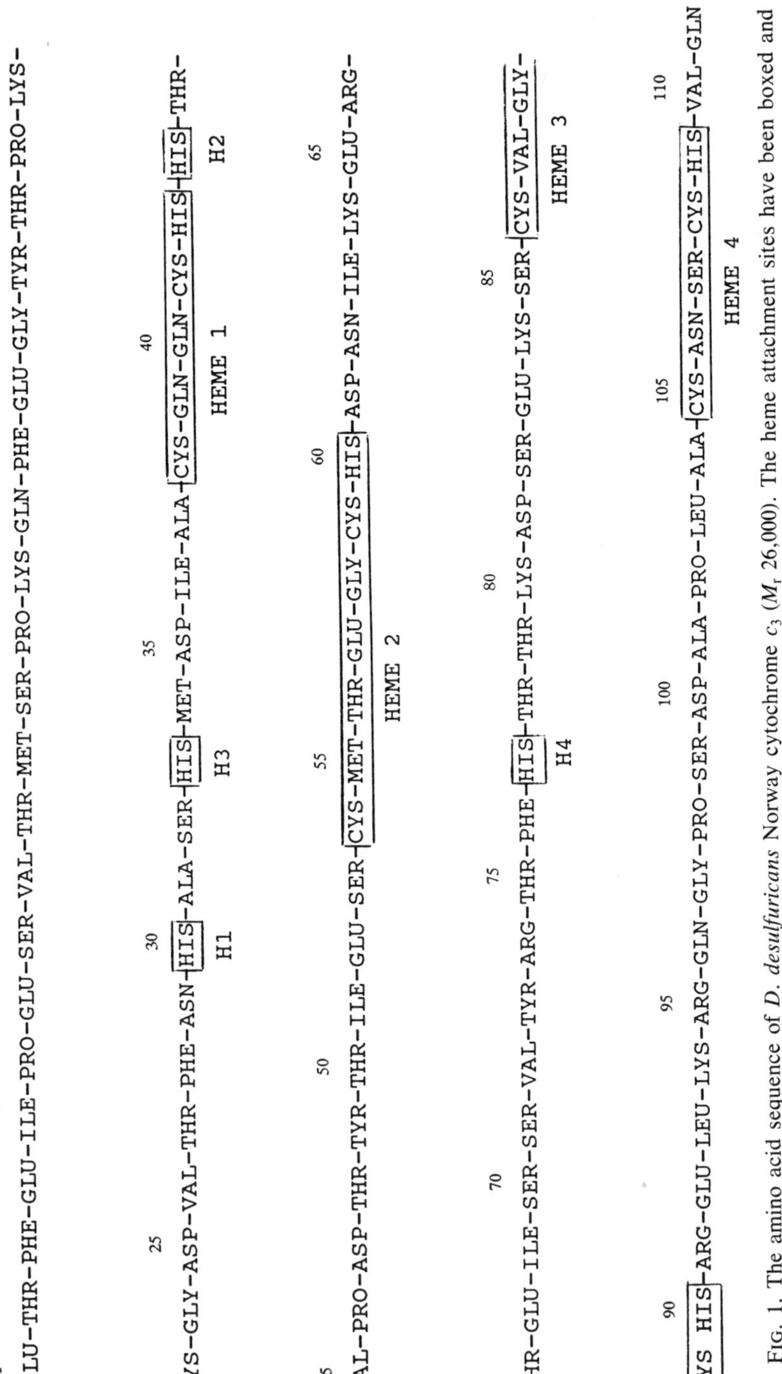

FIG. 1. The amino acid sequence of *D. desulfuricans* Norway cytochrome c_3 (M_r 26,000). The heme attachment sites have been boxed and numbered as they occur in the sequence. The sixth histidine axial ligand is indicated for each heme. The numbers of the histidine residues correspond to the heme-binding site with which they are supposedly associated according to the three-dimensional structure determination of *D. desulfuricans* Norway cytochrome c_3 (M_r 13,000).

In Fig. 2, an alignment of cytochromes c_3 (M_r 13,000) is proposed on the basis of X-ray crystallographic data published by Higuchi *et al.*[21] Thirty-five residues were found to be conserved when the *D. desulfuricans* Norway cytochrome c_3 (M_r 26,000) sequence was aligned with the sequences of cytochromes c_3 (M_r 13,000) from *Dsm. baculatum* Norway 4 and *D. desulfuricans* El Agueila Z. Thirty-three conserved residues were observed between cytochrome c_3 (M_r 26,000) and *D. salexigens* cytochrome c_3, 27 as compared to *D. gigas* cytochrome c_3, 33 as compared to *D. vulgaris* Miyazaki cytochrome c_3, and 32 as compared to *D. vulgaris* Hildenborough cytochrome c_3. From the point of view of the phylogenetic relationships, this comparison is consistent because two different strains from *D. desulfuricans* and *D. vulgaris* species show the same degree of homology.

Among the homologous polyhemic cytochromes c_3, a consensus sequence of 18 amino acids has been found to occur corresponding to the cysteine–histidine clusters that bind each heme and to the sequence Val-X-Phe (positions 33–35). The highly basic sequence observed in all cytochromes c_3 (positions 130–138 in Fig. 2) was not conserved in the octaheme cytochrome. This sequence, together with assymetrically distributed charged residues, has been thought to produce a dipole moment in cytochrome c_3. In the three-dimensional structure of *D. desulfuricans* Norway cytochrome c_3,[22] the latter authors have described the self-association of two molecules in the crystal via the propionyl groups of heme 4 through hydrogen bonds with these conserved lysine residues. The lack of these lysine residues in cytochrome c_3 (M_r 26,000) might possibly play a role in establishing other electrostatic interactions involved in the formation of stable dimers.

Except for the sequence of the high molecular weight cytochrome from *D. vulgaris* Hildenborough, little is known about this class of cytochrome. The partially established sequence of the Hmc cytochrome from *D. vulgaris* Miyazaki shows that the N-terminal sequence is homologous to that of the Hmc from the Hildenborough strain, but the former cytochrome contains 11 covalently linked hemes and the latter contains 16. A comparison between the arrangement of the heme-binding sites and the coordinated residues in the amino acid sequences of cytochrome c_3 and Hmc from *D. vulgaris* Hildenborough has clearly shown the presence of four domains.[6] Three are cytochrome c_3-like domains. The first domain is composed of only three of the four heme-binding sites present in cytochrome c_3 and heme-binding site 12 does not fit into the cytochrome c_3 motif.[6] Two main peptide fragments (M_r 33,100 and 19,100) have been purified from Miyazaki Hmc. The first was red whereas the second was

```
                                  1                 10                  20                  30                    40
                                  |                 |                   |                   |                     |
(1) D. desulfuricans (N)          . A D A P G D D Y V I S A P E G M K A K P K G D K P . . G A L Q K T V P F P H T K H A
(2) D. desulfuricans (EAZ)        . V D A P A D M V I K . A P A G A K V . . . . . . . . . . . T K A P V A F S H K G H A
(3) D. salexigens                 . V D A P A D M V L K . A P A G A K M . . . . . . . . . . . T K A P V D F S H K G H A
(4) D. vulgaris (H)               A P K A P A D . . . . . . . . G L K M E . . . . . . . . . A T K Q P V V F N H S T H K
(5) D. vulgaris (M)               A P K A P A D . . . . . . . . G L K M D . . . . . . . . . K T K Q P V V F N H S T H K
(6) D. gigas                      . V D V P A D . . . . . . . . G A K I D . . . . . F I A G G E K N L V V F N H S T H K
(7) C3 (Mr 26000) D.d. N.         . . . . . . . . E T F E I P E S V T M S P K . Q F E G Y T P K K G D V T F N H A S H M
(8) hmc1 (residues 1-111)         . . . K A L P E G P G E K R A D L I E I G A . M E R F G K L D L P K V A F R H D Q H T
(9) hmc2 (residues 112-220)       . . . . . . . . . . . . . . . . . . . . . . P S A A S S W K E I G F D K S L H Y R H V
(10)hmc3 (residues 229-338)       R E V P R L D R G Q P D A A L I L P V P G K D A P R E M K G T M K P V A F D H K A H E
(11)hmc4 (residues 398-514)       I P E K V V I G S I A K E Y Q P S E F P H R K I V K T L I A G I G E D K L A A T F H I
                                                                                                          H1    H3

                   50                   60                   70                   80                   90
                   |                    |                    |                    |                    |
(1) T V . . . . . . . . . . . E C V Q C H H T L E A D G G A . . . . . . . . V K K C T T S G C H D S L E F R D K A N
(2) S M . . . . . . . . . . . D C K T C H H K W D G A G A . . . . . . . . I Q P C Q A S G C H A N T E . . S K K G
(3) A L . . . . . . . . . . . D C T K C H H K W D G K A E . . . . . . . . V K K C S A E G C H V(B,T,S). . . K(K,G
(4) S V . . . . . . . . . . . K C G D C H H P V N G K E D . . . . . . . . Y R K C G T A G C H D S M . . . D K K D
(5) A V . . . . . . . . . . . K C G D C H H P V N G K E N . . . . . . . . Y Q K C A T A G C H D N M . . . D K K D
(6) D V . . . . . . . . . . . K C B B C H H Z . P G B K Q . . . . . . . . Y A G C T T D G C H N I L . . . D K A D
(7) D I . . . . . . . . . . . A C Q Q C H H T V P D T Y T . . . . . . . . I E S C M T E G C H D N I . . . . . K E
(8) T A . . . . . V T G M G K D C A A C H K S K D G K M S L . . . . . . . . . . . . . . . . . . . . . K F M R L D
(9) A S K A I K P V G D P Q K N C G A C H H V Y D E A S K K L V W G K N K E D S C R A . . C H G E K . P V D K R .
(10)A . . . . . . . . . K A N D C R T C H H V R I D T . . . . . . . . . . . . C T A . . C H T V N G T A D S K F
(11)E . . . . . . . . . K G T L C Q G C H H N S P A S L T . . . . . . . . P P K C A S . . C H G K P F D A D R G D
                                 HEME 1   H2                                     HEME 2
```

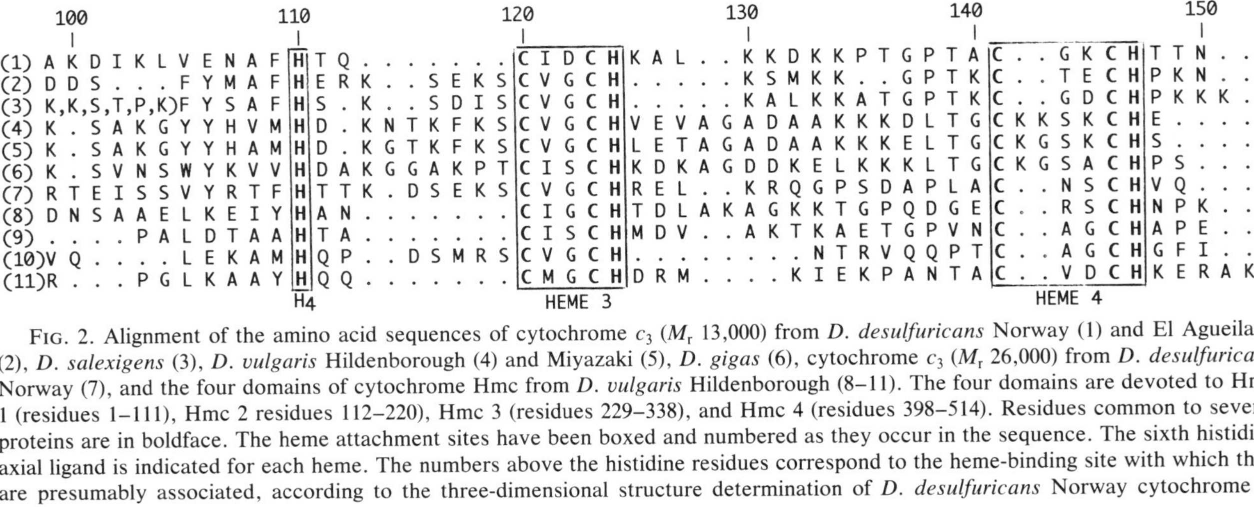

FIG. 2. Alignment of the amino acid sequences of cytochrome c_3 (M_r 13,000) from *D. desulfuricans* Norway (1) and El Agueila Z (2), *D. salexigens* (3), *D. vulgaris* Hildenborough (4) and Miyazaki (5), *D. gigas* (6), cytochrome c_3 (M_r 26,000) from *D. desulfuricans* Norway (7), and the four domains of cytochrome Hmc from *D. vulgaris* Hildenborough (8–11). The four domains are devoted to Hmc 1 (residues 1–111), Hmc 2 residues 112–220), Hmc 3 (residues 229–338), and Hmc 4 (residues 398–514). Residues common to several proteins are in boldface. The heme attachment sites have been boxed and numbered as they occur in the sequence. The sixth histidine axial ligand is indicated for each heme. The numbers above the histidine residues correspond to the heme-binding site with which they are presumably associated, according to the three-dimensional structure determination of *D. desulfuricans* Norway cytochrome c_3 (M_r 13,000).

nearly colorless.[24] On the basis of the sequence homology with Hildenborough Hmc, a complete lack of heme-binding sites in domain 4 or a high degree of variability may be predictable.

If we compare, in Fig. 2, the four domains of Hmc[6] described previously (domain 1, residues 1–111; domain 2, residues 112–220; domain 3, residues 229–338; and domain 4, residues 398–514) with those of the cytochrome c_3 (M_r 13,000 and 26,000) sequences, it can be seen that the heme-binding site sequences (Cys-X-X-Cys-His) and the axial histidines are highly conserved. The cysteines in the heme-binding sites of the four Hmc domains always have a two-amino acid spacer, whereas in cytochromes c_3 there is either a two-amino acid or a four-amino acid spacer. The heme 2-binding site and its ligated histidines are lacking in domain 1 of Hmc, as in the case of the trihemic cytochrome c_3 isolated from *Desulforomonas*.[25] As for the sixth heme ligand, the histidine residues are conserved in all four domains, except that the sixth ligand is missing from heme 1 in domain 4. The structural data on cytochromes c_3 suggest that His-39 (Fig. 2 numbering) may be a heme 1 ligand, His-63 a heme 2 ligand, His-42 a heme 3 ligand, and His-110 a heme 4 ligand. An additional binding site is to be found in residues 339–397 of the Hmc sequence. This sequence has been assigned to the high-spin heme observed in the EPR spectrum,[26] as the sixth axial ligand could not be identified by comparing this part of the sequence with c_3-type cytochrome sequences. In particular, Fig. 2 clearly indicates that domain 3 is the most homologous with the other cytochrome c_3 sequences. Heme 2 is not present in domain 1 and is replaced by heme 12, which differs from cytochrome c_3 heme in its sixth ligand.

Evolutionary and Structural Relationships among the Cytochrome c_3 Superfamily

Subsequent to discovery of an increasing number of cytochromes *c* in sulfate- (and sulfur)-reducing bacteria, Moura *et al.*[27] have proposed a tentative classification based on the number of hemes per monomer, heme axial ligation, heme spin state, and primary structures.

Four families have been defined: the monoheme cytochromes with a low molecular mass (9 kDa) and a covalently attached heme *c* with methio-

[24] C. Tasaka, M. Ogata, T. Yagi, and A. Tsugita, *Protein Sequence Data Anal.* **4,** 25 (1991).

[25] A. Dolla, F. Guerlesquin, M. Bruschi, and R. Haser, *J. Mol. Recognit.* **4,** 27 (1991).

[26] M. Bruschi, P. Bertrand, C. More, G. Leroy, J. Bonicel, J. Haladjian, C. Chottard, W. B. R. Pollock, and G. Voordouw, *Biochemistry* **31,** 3281 (1992).

[27] J. J. G. Moura, C. Costa, M. Y. Liu, I. Moura, and J. LeGall, *Biochim. Biophys. Acta* **1058,** 61 (1991).

nine as sixth ligand, the diheme proteins, the multiheme cytochromes c_3 characterized by bishistidinyl iron coordination and 20 to 40 residues per heme, and the hexaheme proteins such as nitrite reductase. The multiheme cytochromes have been described and classified by Ambler in class III of the *c*-type cytochromes.[28] This group includes triheme, tetraheme, and octaheme proteins from sulfate- and sulfur-reducing bacteria.[29] The amino acid determination of *Dsm. baculatum* Norway 4 cytochrome c_3 (M_r 26,000) indicates that this subclass of dimeric cytochromes is composed of two identical subunits and that the hemes may play a role in the building of the dimer. The amino acid sequence of the monomer is homologous with that of cytochrome c_3 (M_r 13,000) but differs from that of the tetrahemic cytochrome isolated from the same organism. This cytochrome belongs to the cytochrome c_3 superfamily and will be classified among the class III *c*-type cytochromes as described by Ambler.[28] These cytochromes have a low solubility, so that although their localization has not yet been definitely determined, it seems likely that, like Hmc, they are periplasmic cytochromes that may interact with the membrane.

Dickerson has constructed a tentative phylogenetic tree relating various prokaryotic families with eukaryotic cytochromes *c* on the basis of molecular size and sequence similarity.[30] Although the mode of attachment of the heme groups is the same, the multiheme cytochromes c_3 cannot be related to this family. Gene duplication accompanied by gene fusion is more likely to have occurred after divergence of a mono- or diheme cytochrome. This mechanism has been lined up with the nucleotide sequence of genes coding for different redox proteins isolated from sulfate-reducing bacteria.[31] The *D. vulgaris rbo* gene product, which has a 4-kDa desulforedoxin domain as the NH_2 terminus, may therefore have arisen as the result of gene fusion.

The NH_2-terminal sequence of the [Fe] hydrogenase α subunit was found to be homologous to that of bacterial ferredoxin and suggested the hypothesis that gene fusion may have occurred.[32] In the case of rubrerythrin,[33] a 4-kDa desulforedoxin domain acting as the NH_2 terminal may have been generated by gene fusion. In the same way, the emergence of

[28] R. P. Ambler, *in* "From Cyclotrons to Cytochromes" (A. B. Robinson and N. O. Kaplan, eds.), p. 263. Academic Press, London, 1980.

[29] G. R. Moore and G. W. Pettigrew, "Cytochromes *c*: Evolutionary, Structural and Physicochemical Aspects," p. 161. Springer-Verlag, Berlin, 1990.

[30] R. E. Dickerson, "Evolution of Protein Structure and Function," p. 173. Academic Press, New York, 1980.

[31] M. J. Brumlik, G. Leroy, M. Bruschi, and G. Voordouw, *J. Bacteriol.* **172,** 7289 (1990).

[32] G. Voordouw and S. Brenner, *Eur. J. Biochem.* **148,** 515 (1987).

[33] M. J. Brumlik and G. Voordouw, *J. Bacteriol.* **171,** 4996 (1989).

c-type cytochromes with up to 16 covalently bound hemes may have resulted from shuffling and fusion of genes among redox domains. The stereochemical constraints involved in binding 8 heme groups in cytochrome c_3 (M_r 26,000) or 16 hemes in the Hmc may, however, have prevented a repeated folding pattern from being conserved.

Crystallographic Data

Preliminary X-ray studies have been carried out on the octaheme cytochromes c_3 from *D. gigas*[34] and *Dsm. baculatum* Norway 4.[35]

The crystals of the cytochrome from *D. gigas* belonging to the trigonal system have either a space group $P3_1$ or the enantiomorph $P3_2$ and the cell parameters are $a = b = 57.4$ Å, c = 97.3 Å, c = 120°. In the case of two molecules with a molecular weight of 26,000 per asymmetric unit, the volume per dalton (Vm) is 1.77; for one molecule per asymmetric unit, $V_m = 3.54$. Although this protein is not very soluble after crystallization, it becomes stable and can be recycled numerous times for crystal growth experiments.

X-Ray structural investigations on the octaheme cytochrome from *D. desulfuricans* Norway are now in progress. The crystals are trigonal, space group $P3_121$ (or its enantiomorph $P3_221$) with cell dimensions $a = b = 72.9$ Å and $c = 62.7$ Å. The asymmetric unit probably contains one monomer and has a solvent content of about 60%. With one monomer per asymmetric unit, the crystallographic two-fold symmetry axis must link the two monomer subunits of the dimer. The diffraction was up to 2.0 Å.

Attempts to solve the structure of the octaheme cytochromes and that of the *D. vulgaris* Hildenborough Hmc cytochrome, which have been reported to be in progress,[36] should provide us with some essential information about the roles of these cytochromes in the electron transfer system of *Desulfovibrio*. These polyhemic systems provide a structural basis for understanding the factors that control the redox potentials and therefore the intra- and intermolecular electron processes, and for elucidating the possible processes whereby the domains are recognized by specific redox partners. Determining these structures could help us to understand the mechanisms involved in the folding and evolution of the polyhemic cytochromes.

[34] L. C. Sieker, L. H. Jensen, and J. LeGall, *FEBS Lett.* **209,** 261 (1986).

[35] M. Czjzek, F. Guerlesquin, V. Roig, F. Payan, M. Bruschi, and R. Haser, *J. Mol. Biol.* **228,** 995 (1992).

[36] Y. Higuchi, K. Inaka, N. Yasuoka, and T. Yagi, *Biochim. Biophys. Acta* **911,** 341 (1987).

The roles of these various cytochromes and the compartmentation of the redox process in which they are involved between the periplasm and the cytoplasm, as well as the nature of the membrane-bound redox channel, are all questions on which future research will no doubt throw interesting light.

[11] Hexadecaheme Cytochrome c

By YOSHIKI HIGUCHI, TATSUHIKO YAGI, and GERRIT VOORDOUW

Introduction

High molecular weight cytochrome *c* (HMC) was first found in *Desulfovibrio vulgaris* Miyazaki (DvM) in 1969.[1] Its characteristic was a relatively high molecular weight (about 70,000), and an absorption spectrum with an α band at 553 nm. Its instability in aqueous solution has made the purification of this protein difficult. Considerable confusion existed with regard to its heme content and molecular weight. It was reported as cytochrome cc_3[2] with two subunits, as octaheme cytochrome c_3,[3,4] and as cytochrome c_3 with a molecular weight of 26,000.[5] Finally, high molecular weight cytochrome *c* in *D. vulgaris* Hildenborough (DvH) was purified and crystallized, and characterized to be a 75-kDa protein with 16 *c*-type hemes.[6] The sequence of the gene for this protein indicated that the molecule had a molecular weight of 65,500, and consisted of three tetraheme cytochrome c_3-like domains with C-X_1-X_2-C-H heme-binding sites, and a single three heme-binding domain.[7] The sixteenth heme-binding site was found to be located outside these four domains. Although the function of this protein is not yet completely clear, work has indicated that the gene for HMC is the first in an operon that encodes five other proteins, which together form a transmembrane redox protein complex that allows

[1] T. Yagi, *J. Biochem. (Tokyo)* **66,** 473 (1969).

[2] M. Loutfi, F. Guerlesquin, P. Bianco, J. Haladjian, and M. Bruschi, *Biochem. Biophys. Res. Commun.* **159,** 670 (1989).

[3] I. Moura, G. Fauque, J. LeGall, A. V. Xavier, and J. J. G. Moura, *Eur. J. Biochem.* **265,** 547 (1987).

[4] M.-C. Liu, C. Costa, I. B. Coutinho, J. J. G. Moura, I. Moura, A. V. Xavier, and J. LeGall, *J. Bacteriol.* **170,** 5545 (1988).

[5] J. M. Odom and H. D. Peck, Jr., *Annu. Rev. Microbiol.* **149,** 447 (1984).

[6] Y. Higuchi, K. Inaka, N. Yasuoka, and T. Yagi, *Biochim. Biophys. Acta* **911,** 341 (1987).

[7] W. B. R. Pollock, M. Loutfi, M. Bruschi, B. J. Rapp-Giles, J. D. Wall, and G. Voordouw, *J. Bacteriol.* **173,** 220 (1991).

electrons to flow from the periplasm to the cytoplasm or vice versa.[8] This chapter presents the methodology for purification and current physicochemical data for hexadecaheme cytochrome *c* from DvH and DvM. These cytochromes were originally named high molecular weight cytochrome *c* and will therefore be abbreviated as HMC in this chapter.

Purification of Hexadecaheme Cytochrome c

Chemicals

All chemicals used for the purification and crystallization procedure should be biochemical grade, whereas those for the cultivation of bacteria can be of a less pure grade. The buffer solution used for the purification and crystallization procedure is Tris-HCl (pH 7.4) unless otherwise specified.

DEAE Toyopearl 650S (Toso Co., Tokyo, Japan) Toyopearl HW-55S (Toso Co.), Sephacryl S-200 HR (Pharmacia Fine Chemicals, Piscataway, NJ), CM Toyopearl 650S (Toso Co.), and CM- and DEAE-cellulose (Whatman, Clifton, NJ) are equilibrated with at least two bed volumes of Tris-HCl buffer, pH 7.4. Hydroxyapatite (BioGel HTP; Bio-Rad, Richmond, CA) is equilibrated with 25 m*M* phosphate buffer containing 20 m*M* NaCl.

Bacterial cells of DvH are obtained following growth on medium C according to Postgate,[9] and those of DvM can be cultured similarly.

Purification of Hexadecaheme Cytochrome c from Desulfovibrio vulgaris Hildenborough

Wet cells (1 kg) are suspended in 2–2.5 vol of the 25 m*M* buffer solution, and disrupted with an ultrasonic disintegrator for 10 min. The cells should be stirred and cooled by ethanol–dry ice during sonication. The sonicate is centrifuged at 100,000 *g*. The cell-free supernatant is loaded onto a column (5 × 40 cm) of DEAE Toyopearl 650S and the column is washed with 25 m*M* buffer. The basic proteins including various cytochromes are passed through the column and separated from the acidic proteins such as ferredoxin, rubredoxin, and hydrogenase adsorbed on the column. Solid ammonium sulfate is added to the fraction of basic proteins to 80% saturation. The HMC is precipitated and separated from small cytochromes such as cytochrome *c*-553 and cytochrome c_3 by centrifugation at 40,000 *g* for 20 min. The pellet is suspended in the 25 m*M* buffer

[8] M. Rossi, W. B. R. Pollock, M. W. Reij, R. G. Keon, R. Fu, and G. Voordouw, *J. Bacteriol.* **175,** 4699 (1993).

[9] J. R. Postgate, *in* "The Sulphate-Reducing Bacteria," 2nd Ed., p. 32. Cambridge Univ. Press, Cambridge, 1984.

and dialyzed against the same buffer for 24 hr. The dialyzed solution is concentrated to 20–30 ml on an Amicon (Danvers, MA) YM30 membrane, using a Diaflo cell. The remaining cytochromes c_3 and *c*-553 are removed thoroughly by passing them through a column (5 × 40 cm) of Toyopearl HW-55S. The elution volume of HMC on this column is about 1000 ml. The red fractions are combined and applied to a column (5 × 40 cm) of CM-Toyopearl 650S, and washed with two bed volumes of 25 m*M* buffer. The adsorbed substances, which show as a sharp red band on the column, are eluted with the buffer by increasing the concentration of NaCl from 0 to 200 m*M* with a total volume of four times the column bed volume. The eluted cytochrome fractions with a purity index $\{[A_{553}(\text{red}) - A_{570}(\text{red})]/A_{280}(\text{ox})\}$ larger than 2.5 are combined and concentrated to 20 ml and dialyzed against 25 m*M* potassium phosphate buffer containing 20 m*M* NaCl. The dialyzed protein solution is adsorbed onto a column (3 × 20 cm) of hydroxyapatite. The cytochrome is eluted by a linear gradient of phosphate from 25 to 300 m*M* with a total volume of 1000 ml. The HMC is eluted at about 100 m*M* phosphate.

Purification of Soluble and Membrane-Bound Hexadecaheme Cytochrome c from Desulfovibrio vulgaris Miyazaki

HMC from Soluble Fraction. Frozen cells are disrupted and treated as described for the purification of cytochrome *c*-553 to step 2.[10] The precipitate of HMC from the soluble fraction (sHMC) at step 2 is dissolved in 8 ml of 10 m*M* buffer solution, centrifuged to remove turbidity, and applied immediately to a column (2.2 × 100 cm) of Sephacryl S-200 with 50 m*M* buffer containing 200 m*M* NaCl as the elution solvent. Delay of the chromatographic separation will result in proteolytic degradation of sHMC. Elution of sHMC is monitored by its δ peak at 355 nm (A_{355}), and the fractions of $A_{355}/A_{280} > 0.6$ eluted between 185 and 220 ml are collected. The proteolytic activity is eluted immediately following sHMC. The sHMC is dialyzed against 10 m*M* buffer, and is applied to a CM-cellulose column (1.5 × 8 cm) preequilibrated with 10 m*M* buffer solution. The sHMC is eluted by a linear gradient from 10.8 m*M* buffer containing 4 m*M* NaCl to 18 m*M* buffer containing 40 m*M* NaCl (200 ml each). Fractions of purity index 2.72 eluted at about 90 ml of the effluent can be used for experiments.

HMC from Membrane Fraction. High molecular weight cytochrome *c* in the membrane fraction (mHMC) can be solubilized by alkali treatment at pH 11 for 30 min and separated from insoluble materials by centrifugation at 85,000 *g* for 60 min. The solubilized mHMC is purified by a proce-

[10] T. Yagi, this volume [8].

dure similar to that described above, that is, removal of anionic proteins by chromatography on DEAE-cellulose, precipitation of mHMC by salting out with 80% saturated ammonium sulfate, gel-filtration chromatography on a Sephacryl S-200 column, and ion-exchange chromatography on CM-cellulose, as described above for sHMC. The mHMC behaves in a manner identical to sHMC except for the last step: mHMC is eluted at about 130 ml from the CM-cellulose.

Both sHMC and mHMC are unstable, and readily undergo proteolytic degradation during purification. Purified HMC is denatured by bubbling with gas such as H_2 or O_2, and irreversibly precipitated on thorough dialysis against water.

Properties of Hexadecaheme Cytochrome *c*

Spectral Properties

The absorption spectra of HMC from DvH and DvM are compared in Table I. Both cytochromes are easily reduced by sodium dithionite. The

TABLE I
COMPARISON OF SPECTRAL PROPERTIES OF MULTIHEME CYTOCHROMES FROM *Desulfovibrio vulgaris*

	Hexadecaheme cytochrome *c* from DvH[a]		Hexadecaheme cytochrome c from DvM[b]		Cytochrome c_3 from DvM[c]	
	λ (nm)	ε (m*M*)	λ (nm)	ε (m*M*)	λ (nm)	ε (m*M*)
Oxidized form						
γ peak	410	1573	409	1820	410	464
δ peak	355	332	355	380	350	90.8
Protein	280	110	280	140	280	35.6
Reduced form						
α peak	553	374	553	400	552	116
β peak	523	223	523	221	524	58.5
γ peak	419	2471	419	2735	419	733
Purity index $[A_{\alpha(red)} - A_{570(red)}]/A_{280(ox)}$	3.0		2.72		3.14	

[a] Y. Higuchi, K. Inaka, N. Yasuoka, and T. Yagi, *Biochim. Biophys. Acta* **911,** 341 (1987). Note that the millimolar absorption coefficients of hexadecaheme cytochrome *c* from DvH here are recalculated from the original data, using a molecular weight of 65,500 obtained by amino acid sequence analysis data by W. B. R. Pollock, M. Loutfi, M. Bruschi, B. J. Rapp-Giles, J. D. Wall, and G. Voordouw, *J. Bacteriol.* **173,** 220 (1991).

[b] M. Ogata, N. Kiuchi, and T. Yagi, *Biochimie* **75,** 977 (1993).

[c] T. Yagi and K. Maruyama, *Biochim. Biophys. Acta* **243,** 214 (1971).

absorption spectra are typical for *c*-type cytochromes, and the position of each absorption peak is almost identical to that of tetraheme cytochrome c_3 except for the δ peak. Both sHMC and mHMC from DvM have identical spectral properties.[11] Neither HMC from DvM and DvH have an absorption peak near 695 nm. Purity indices of HMC (3.0 for DvH,[6] 2.72 for DvM[11]) are similar to those of tetraheme cytochrome c_3 (around 3.0), and higher than those of monoheme cytochrome *c* or *c*-553 (around 1.0).[10]

Molecular Weight and Heme Content

Both HMCs consist of a single polypeptide chain with molecular weights of about 65,500 (DvH)[7] and 65,000 (DvM).[11] SHMC and mHMC from DvM have identical molecular weight. The molecular weight of HMC from DvH was determined by amino acid sequence analysis, whereas that of HMC from DvM was estimated by sodium dodecyl sulfate-polyacrylamide gel electrophoresis (SDS–PAGE). The molecular weight of the DvH protein calculated from the elution volume on gel filtration was slightly larger.[6]

The total iron content of these cytochromes *c* is analyzed by atomic absorption spectroscopy using a standard curve obtained from properly diluted iron ammonium sulfate solutions. The heme content can be determined from the pyridine ferrohemochrome spectrum, using the millimolar absorbance coefficient at 550 nm of 29.1 for heme *c*. The heme content can also be estimated from the number of -C-X_1-X_2-C-H- segments in the amino acid sequence of cytochromes, because this segment is the typical heme attachment site of *c*-type cytochromes. The heme content of HMC from DvH was determined as 16 from the amino acid sequence[7] (see Cloning and Sequencing of the Gene for Hexadecaheme Cytochrome *c*, below), 15 by the pyridine hemochrome spectrum, and 14 by atomic absorption spectroscopy.[6] The HMC from DvM was estimated to have the same number of heme groups as that from DvH.[11]

Amino Acid Composition

The amino acid composition of HMC and cytochrome c_3 from DvH, obtained from the gene or protein sequence, is compared with that directly determined for HMC from DvM in Table II. Amino acid composition for the latter was obtained after hydrolysis of purified protein in 6 *M* HCl at 110° for 24 hr in an evacuated glass tube. The amino acid compositions of HMC from DvH and DvM are similar, but definitely different from that of cytochrome c_3[12] in spite of the spectral similarity.

[11] M. Ogata, N. Kiuchi, and T. Yagi, *Biochimie* **75,** 977 (1993).

[12] R. P. Ambler, *Biochem. J.* **109,** 47 (1968).

TABLE II
AMINO ACID COMPOSITION OF MULTIHEME CYTOCHROMES FROM *Desulfovibrio vulgaris*

	Hexadecaheme cytochrome *c* from DvH[a]	Hexadecaheme cytochrome *c* from DvM (sHMC)[b]	Cytochrome c_3 from DvH[c]
Ala	63	71	10
Asx	47	52	12
Arg	23	17	1
Cys	32	—	8
Gly	37	45	9
Glx	44	44	5
His	31	29	9
Ile	18	13	0
Leu	25	26	2
Lys	54	54	20
Met	12	15	3
Phe	13	12	2
Pro	33	35	4
Ser	24	28	6
Thr	24	27	5
Trp	2	—	0
Tyr	5	8	3
Val	27	34	8
Total:	514	510	107

[a] From the gene sequence by W. B. R. Pollock, M. Loutfi, M. Bruschi, B. J. Rapp-Giles, J. D. Wall, and G. Voordouw, *J. Bacteriol.* **173,** 220 (1991).

[b] The amino acid composition of sHMC was shown to be similar to that of mHMC [M. Ogata, N. Kiuchi, and T. Yagi, *Biochimie* **75,** 977 (1993)]. Cysteine and tryptophan were not determined.

[c] R. P. Ambler, *Biochem. J.* **109,** 47 (1968).

Redox Properties

The redox properties of HMC were studied by electrochemical or redox titration methods. The redox potentials of HMC from DvH in 50 m*M* sodium phosphate buffer (pH 7.0) solution at 25° were measured with classic electrode cell methods.[13] The redox potentials of hemes in HMC from DvH fall in three different classes with 0, −100, and −250 mV [vs normal hydrogen electrode (NHE)].[13] The low-potential values are similar to those observed for the four hemes of cytochromes c_3 from *Desulfovibrio*

[13] M. Bruschi, P. Bertrand, C. More, G. Leroy, J. Bonicel, J. Haladjian, G. Chottard, W. B. R. Pollock, and G. Voordouw, *Biochemistry* **31,** 3281 (1992).

species. This could mean that these values are contributed by histidine–histidine-liganded hemes like the hemes in cytochromes c_3. In contrast, the value of 0 mV, which is comparable to those found in monoheme cytochromes *c*-553,[10] could be due to a heme that lacks bishistidinyl ligands. The redox potentials of the hemes of sHMC of DvM were evaluated spectroscopically by redox titration in the presence of standard redox dyes.[11] The electrochemical method cannot be used for this HMC because of denaturation by stirring and/or gas bubbling during the measurement. A mixture of sHMC and a redox dye in 100 m*M* Tris-HCl buffer was gradually reduced by adding increasing amounts of a sodium dithionite solution under a stream of N_2 in an optical cell. The percentage reduction of sHMC was monitored by the net height of the α peak, and that of dye was determined by the specific absorbance of each dye. The redox potentials of 16 hemes were estimated by a simulation of calculated potentials against the plots of observed reduction percentages between sHMC and each dye. They were estimated to be 60, 15, $-120 \sim -135$ (seven hemes), $-190 \sim -205$ (five hemes), and -260 (two hemes) mV.[11]

Spin State

A spin state analysis of the hemes in DvH HMC has been performed by electron paramagnetic resonance (EPR) spectroscopy. The EPR spectrum of 0.18 m*M* DvH HMC in 300 m*M* phosphate buffer (pH 7.0) has distinctive signals at g = 1.52, 2.26, 2.65, 2.94, 3.08, 3.66, and 5.57.[13] The ratio of the number of low-spin and high-spin hemes was determined by numerical simulation of the high-spin component of the EPR spectrum according to a g-strain statistical procedure.[14] The HMC from DvH was estimated to have 1 or 2 high-spin hemes and 14 or 15 low-spin hemes per molecule.[13]

Crystallization

High molecular weight cytochrome *c* from DvH can be crystallized by the microdialysis method using 2-methyl-2,4-pentanediol (MPD) as precipitating agent. The apparatus used for crystallization consists of microdialysis cells (50 μl) made of acrylic resin, and a small vial with a screw cap (20 ml). The cell interiors are treated with ethyl acetate to reduce wall roughness. The initial protein solution for crystallization is 0.6 ml of a 10-mg/ml HMC solution. The MPD (0.4 ml) is gradually added to the above protein solution in ice, and the mixture is gently stirred with

[14] W. R. Hagen, D. O. Hearshen, R. H. Sands, and W. R. Dunham, *J. Magn. Reson.* **61,** 220 (1985).

a thin glass rod to keep the solution homogeneous. The protein solution, containing 40% MPD, is filtered through Millipore (Bedford, MA) membrane of pore size 0.45 μm to remove insoluble precipitates. The filtered solution is injected into the microdialysis cells and covered with dialysis membrane, and sealed with O rings. Dialysis membranes are boiled in 10 m*M* ethylenediaminetetraacetic acid (EDTA) solution and rinsed with deionized water thoroughly before use. The microdialysis cells are submerged in a small vial containing 50% MPD buffer solution, and tightly sealed with a screw cap. Dark red crystals in the shape of a hexagonal pyramid or bipyramid appear in 2–3 weeks at 10–15°. Typical dimensions of the crystals are approximately $0.7 \times 0.7 \times 0.3$ mm.

The crystals diffract beyond 3.0-Å resolution and are usually stable against X-ray beam damage at 5°. They tolerate radiation damage by an X-ray beam from a Cu rotating anode operated at 40 kV and 100 mA for at least 24 hr. Crystals are in space group of $P6_4$ or $P6_2$ with unit cell dimensions of $a = b = 227.8$ Å, $c = 105.7$ Å, and $\gamma = 120°$.[6] The unit cell volume, 4.75×10^6 Å^3, shows that there are four or five molecules per asymmetric unit.

Cloning and Sequencing of Gene for Hexadecaheme Cytochrome c

Cloning of hmc Gene

The gene (*hmc*) for HMC from DvH was cloned by Pollock *et al.*,[7] using a synthetic deoxyoligonucleotide probe. The probe, a 47-mer, 5′ AAGATCGA(AG)AAGCC(CG)GC(CG)AACAC(CG)GC(CG)TGCGT (CG)GACTGCCACAAGGA, was designed on the basis of partial amino acid information on HMC and recognizes DNA encoding the COOH-terminal region of the protein (Fig. 1). The *hmc* gene was cloned as a 3.7-kb *Xho*I fragment in the *Sal*I site of vector pUC8 in plasmid pP6A.[15] The complete nucleotide sequence of the insert of plasmid pP6A was determined and the amino acid sequence deduced from the sequence of the gene is shown in Fig. 1. This sequence was in part (85%) confirmed by direct protein sequencing of HMC isolated from DvH.[2,7] The nucleic acid-derived sequence confirmed that HMC is synthesized with an N-terminal signal sequence (Fig. 1, residues −31 to −1), required for export of HMC to the periplasm. It is of interest in this regard that two smaller *c*-type cytochromes, the tetrahemic cytochrome c_3 and the mono-

[15] J. Vieira and J. Messing, *Gene* **19,** 259 (1982).

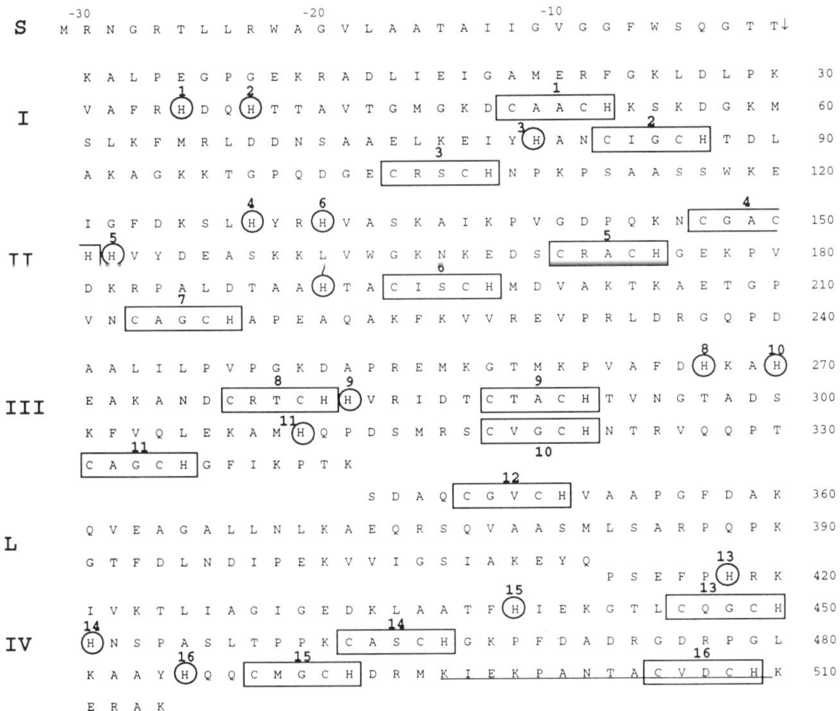

FIG. 1. Amino acid sequence of HMC (514 residues) from *D. vulgaris* Hildenborough deduced from the sequence of the gene.[7] Heme-binding sites are boxed and histidine residues are circled. The signal peptidase cleavage site in the signal peptide S is indicated (↓), Domains I, II, III, and IV and the connecting peptide L are explained in text. A deoxyoligonucleotide probe corresponding to the underlined protein sequence near the C terminus was used as the probe to clone the gene.

hemic cytochrome *c*-553 which are also periplasmic proteins, are also synthesized with N-terminal signal peptides.[16,17]

Amino Acid Sequence

The amino acid sequence of mature HMC comprises 514 residues. The calculated molecular weight of the apoprotein is 55,700. Covalent binding of heme to each of the 16 C-X_1-X_2-C-H sites, where X_1 and X_2 are variable amino acid residues, gives a holoprotein of 65,500. These heme-binding sites have been boxed in Fig. 1. Their sequence-derived number corresponds to that determined directly for the purified protein.[6,13]

[16] G. Voordouw and S. Brenner, *Eur. J. Biochem.* **159,** 347 (1986).

[17] G. J. H. van Rooijen, M. Bruschi, and G. Voordouw, *J. Bacteriol.* **171,** 3575 (1989).

The arrangement of the heme-binding sites and histidine residues, which serve as ligands to the sixth coordination position of heme iron, in the HMC sequence can be understood by considering the positions of these elements in the sequence of the tetrahemic cytochrome c_3, the structure of which is known.[18,19] In the sequence of cytochrome c_3 from DvH these elements are placed as follows: (1) at position 22, sequence HSTH, H-22 coordinates to heme 1, H-25 to heme 3; (2) at position 30, sequence CGDCHH, CGDCH is heme-binding site 1, H-35 coordinates to heme 2; (3) at position 46, sequence CGTAGCH, binding heme 3; (4) at position 70, H-70, coordinating heme 4; (5) at position 79, sequence CVGCH, binding heme 3; (6) at position 100, sequence CKKSKCH, binding heme 4. An unusual feature of the cytochrome c_3 sequence is the separation of the cysteine residues in heme-binding sites 2 and 4 by four rather than two amino acid residues. In HMC all heme-binding sites are of the conventional type CX_1X_2CH. Inspection of the HMC sequence indicates that the cytochrome c_3 sequential pattern of heme-coordinating and heme-binding elements (1) to (6) has been conserved in three domains indicated as II, III, and IV in Fig. 1. This suggests that HMC comprises three complete cytochrome c_3 domains in which the four hemes all have histidine as the sixth coordination position ligand. The heme-binding sites have been numbered sequentially as 1 to 16 in Fig. 1 and the histidines have been numbered to indicate the heme to which they may be expected to coordinate on the basis of the cytochrome c_3 sequential pattern. Domain I lacks two of the elements defined above. It has only three hemes and three histidine residues. A fourth heme (heme 12) is present in the sequence (L) that links domains III and IV. It can be readily confirmed by spectroscopic methods that most of the hemes in HMC are bishistidine coordinated.[6,13] However, assuming the presence of 16 *c*-type hemes in HMC, it appears that at least 1 of these cannot have bishistidinyl coordination, because HMC contains only 15 histidine residues, in addition to the 16 residues present in the 16 heme-binding sites. Heme 12 is a candidate for having a sixth coordination position ligand other than histidine.

The domain structure of HMC explains some of the confusion with respect to the structure of this protein in the earlier literature. Loutfi *et al.*[2] isolated cytochrome cc_3 from DvH, for which they reported a molecular weight of 43,300, eight hemes, and two 20-kDa subunits. Partial sequencing has shown that this protein is a proteolytic degradation product of HMC. We do not want to suggest that HMCs with eight hemes always

[18] M. Pierrot, R. Haser, M. Frey, F. Payan, and J. Astier, *J. Biol. Chem.* **257,** 14341 (1982).

[19] Y. Higuchi, M. Kusunoki, Y. Matsuura, N. Yasuoka, and M. Kakudo, *J. Mol. Biol.* **172,** 109 (1984).

arise as a result of proteolytic digestion.[20] However, in DvH it appears that no multiheme cytochrome other than cytochrome c_3 and HMC are present and that additional octa- and tetrahemic cytochromes[21] arise by proteolysis of HMC.

The domain structure and suggested heme ligands of HMC must be confirmed by the determination of the structure of this interesting cytochrome. X-Ray crystallographic and biophysical studies on this protein are greatly helped by developments in molecular genetics of *Desulfovibrio*. Recloning of the *hmc* gene in broad host range vector pJRD215[22] and conjugation of the resulting plasmid, pBPHMC-1, into *Desulfovibrio desulfuricans* G200 led to threefold overexpression of HMC. Expression can probably be further boosted by exchanging the *hmc* promoter for a more powerful one. This work is currently in progress.

[20] M. Bruschi, this volume [10].

[21] J. A. Tan and A. Cowan, *Biochemistry* **29,** 4886 (1990).

[22] J. Davison, M. Heusterpreute, N. Chevalier, H. T. Vinh, and F. Brunel, *Gene* **51,** 275 (1987).

[12] Ferredoxins

By José J. G. Moura, Anjos L. Macedo, and P. Nuno Palma

Introduction

This chapter[1] focuses on ferredoxins isolated from sulfate-reducing bacteria (SRB) and presents an update review on the following topics: types and distribution of ferredoxins; purification methods; cluster-binding motifs; electronic and magnetic properties of the iron-sulfur clusters involved; cluster interconversions, including heterometal cluster formation; and the physiological role. The literature survey includes mainly the published work referring to ferredoxins isolated from SRB. Review articles on related topics have been published.[1a–6]

[1] Professor Jean LeGall described a new sulfate-reducing bacteria, *Desulfovibrio gigas* NCIB 9332 [J. LeGall, *J. Bacteriol.* **86,** 1120 (1963)] and purified the first ferredoxin from these bacterial group [J. LeGall and N. Dragoni, *Biochem. Biophys. Res. Commun.* **23,** 145 (1966)]. We would like to dedicate to him this chapter not only for his enormous contributions to the field, but also for the discovery of this organism and this protein (and so many others) which has given us years of stimulating and exciting research.

[1a] G. Fauque and J. LeGall, *in* "Biology of Anaerobic Microorganisms" (A. J. Zehnder, ed.), p. 587. Wiley, New York, 1988.

TABLE I
COMPARISON OF PROPERTIES OF FERREDOXINS ISOLATED FROM SULFATE-REDUCING BACTERIA

SRB	Fd	Number of amino acids	Cys	Structure	Cluster type	Redox potential (mV)	Ratio (UV–vis)	Extinction coefficient (mM^{-1} cm^{-1}) (400 nm)
Dg[e]	I	58	6	Tetramer	[3Fe]	−130	0.77	15.7
	II	58	6	Dimer	[4Fe]	−455[a]	0.68	16.0
DvM	I	61	7	Dimer	[3Fe],[4Fe]	−430[b]	0.69	35.0
	II	63	7	Dimer	[4Fe]	−405	0.52	17.0
Dmb	I	59	6	Dimer	[4Fe]	−374	0.80	17.5
	II	59	8	Dimer	2 × [4Fe]	−500[c,d]	0.72	31.6
Da	I	60	4	Dimer	[4Fe]	nd	0.56	28.5
	III	61	7	Dimer	[3Fe],[4Fe]	−140, −410	0.78	28.6

[a] *Desulfovibrio gigas* FdI can accommodate a [3Fe-4S] center as a minor species and in variable amounts (around 10%), E_0 = −50 mV.

[b] The redox potential of the [3Fe-4S] center is determined in the range −10 to −140 mV (T. Yagi, personal communication, 1993).

[c] A minor amount (5%) of a [3Fe-4S] is also found in DmbFdII, E_0 = −115 mV.

[d] After exposure of the protein to a low redox potential, the cluster–cluster interaction is no longer observed and the modified clusters have E_0 values around −440 mV.[20]

[e] *Desulforismo gigas* Fds have the same amino acid sequence.

Ferredoxins (Fds) are simple iron-sulfur proteins that contain prosthetic groups composed of iron and sulfur atoms and play a functional role in electron transfer processes relevant for SRB metabolism (phosphoroclastic reaction and sulfite reduction). They are characterized by the presence of iron and sulfide (not necessarily in equimolar amounts), low molecular masses (~6 kDa), a preponderance of acidic amino acid residues and a low content in aromatic amino acid residues, low midpoint redox potential, characteristic electronic spectra with Fe-S charge-transfer bands contributing at around 400 and 300 nm, and typical electron paramagnetic resonance (EPR) signals observed in oxidized and/or reduced states.

Four distinct types of ferredoxins are found in SRB-containing [3Fe-4S], [4Fe-4S], [3Fe-4S] plus [4Fe-4S], and 2 × [4Fe-4S] clusters (see Table I). A common structural feature shared by these centers is that each iron atom is tetrahedrally coordinated and contains bridging inorganic sulfur atoms. The terminal ligands for the clusters are in general cysteinyl sulfur

[2] J. J. G. Moura, *in* "Frontiers in Iron-Sulfur Proteins" (H. Matsubara, Y. Katsube, and K. Wado, eds.), p. 149. Osaka Japan Sci. Soc. Publ. and Springer-Verlag, Berlin, 1986.

[3] H. Matsubara and S. Kazuhiko, *Adv. Inorg. Chem.* **38,** 223 (1992).

[4] R. Cammack, *Adv. Inorg. Chem.* **38,** 281 (1992).

[5] M. Bruschi and F. Guerlesquin, *FEMS Microbiol. Rev.* **54,** 155 (1988).

[6] H. Beinert and A. J. Thomson, *Arch. Biochem. Biophys.* **222,** 333 (1983).

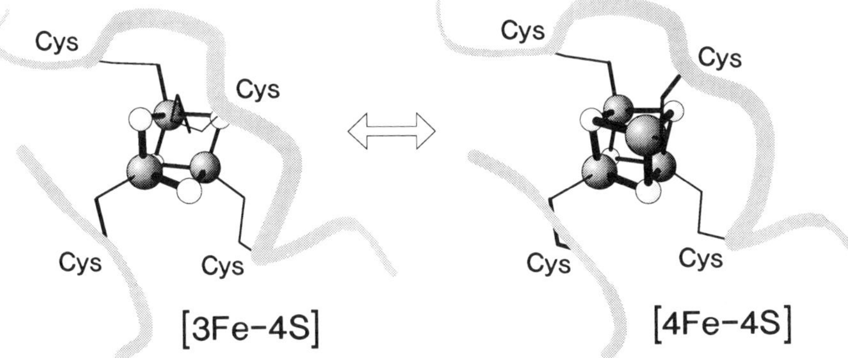

FIG. 1. Structural relationship between [3Fe-4S] and [4Fe-4S] clusters. The portions of the polypeptide chain shown were extracted from [3Fe-4S] *D. gigas* FdII and [4Fe-4S] *B. thermoproteolyticus* Fd structures, respectively.

atoms,[7-11] but other ligands containing O (or N) may be involved (i.e., aspartic acid).

A close structural relationship is apparent between the [3Fe-4S] and [4Fe-4S] clusters and will be relevant for the following considerations (Fig. 1). In particular, the study of SRB ferredoxins enables the survey of the different properties of simple iron-sulfur proteins including electron transfer, flexibility in coordination chemistry, and ability to undergo cluster interconversions. Most of the observations can be extrapolated to more complex situations.

General Survey of Sulfate-Reducing Bacteria Ferredoxins

Distribution and Ferredoxin Types

Seven ferredoxins are well-characterized in SRB and are the focus of this chapter. Different SRB species have now been found to contain more than one type of ferredoxin.[3] *Desulfovibrio gigas* (Dg) is the only exception, reported so far, in which a single polypeptide chain is used to build

[7] K. Fukuyama, H. Matsubara, T. Tsukihara, and Y. Katsube, *J. Mol. Biol.* **210,** 383 (1989).

[8] C. D. Stout, *J. Mol. Biol.* **205,** 545 (1989).

[9] G. H. Stout, S. Turley, L. C. Sieker, and L. H. Jensen, *Proc. Natl. Acad. Sci. U.S.A.* **85,** 1020 (1988).

[10] C. R. Kissinger, E. T. Adman, L. C. Sieker, L. H. Jensen, and J. LeGall, *FEBS Lett.* **244,** 447 (1989).

[11] K. Fukuyama, T. Hase, S. Matsumoto, T. Tsukihara, Y. Katsube, N. Tanaka, M. Kakudo, K. Wada, and H. Matsubara, *Nature (London)* **286,** 522 (1980).

up two ferredoxins with distinct iron-sulfur centers. Only [3Fe-4S] (1+, 0 oxidation states) and [4Fe-4S] (2+, 1+ oxidation states, bacterial ferredoxin type) centers are found in the active sites of ferredoxins isolated from SRB.

Desulfovibrio gigas Fds have been extensively studied by different experimental approaches and spectroscopic techniques and is used here as a reference system.

Desulfovibrio gigas NCIB 9332

Ferredoxin I (DgFdI) and ferredoxin II (DgFdII) are two distinct iron-sulfur proteins isolated from Dg.[12–14] Both ferredoxins are composed of the same polypeptide chain (58 amino, 6 cysteines),[15] but they are reported to have significantly different molecular masses due to different states of oligomerization. *Desulfovibrio gigas* FdI is a dimer[15a] and contains mainly a single $[4Fe\text{-}4S]^{2+,1+}$, whereas the same monomeric unit of the tetrameric DgFdII contains a single $[3Fe\text{-}4S]^{1+,0}$ cluster. The $[4Fe\text{-}4S]^{2+,1+}$ center in DgFdI exhibits a midpoint redox potential of −450 mV but the $[3Fe\text{-}4S]^{1+,0}$ center has a midpoint redox potential of −130 mV.[14,16] Other minor forms are present.[14]

Desulfomicrobium baculatum Norway 4

Two ferredoxins have been isolated from *Desulfomicrobium baculatum* (Dmb) Norway 4.[16a,17,18] Ferredoxin I (DmbFdI) (59 amino acids, 6 cysteines) has 1 conventional $[4Fe\text{-}4S]^{2+,1+}$ center with a midpoint redox potential of −374 mV. *Desulfomicrobium* FdII the most acidic, is unstable toward oxygen exposure, and contains 2 × $[4Fe\text{-}4S]^{2+,1+}$ cores in a poly-

[12] E. C. Hatchikian and J. LeGall, *Ann. Inst. Pasteur (Paris)* **118,** 288 (1970).

[13] J. LeGall and N. Dragoni, *Biochem. Biophys. Res. Commun.* **23,** 145 (1966).

[14] M. Bruschi, E. C. Hatchikian, J. LeGall, J. J. G. Moura, and A. V. Xavier, *Biochim. Biophys. Acta* **449,** 275 (1976).

[15] M. Bruschi, *Biochim. Biophys. Res. Commun.* **91,** 623 (1979).

[15a] *Desulfovibrio gigas* FdI was previously described as a trimer.[14] Measurements show that this is a dimeric structure of the same basic subunit that builds up the tetrameric FdII (J. J. G. Moura, A. L. Macedo, and P. Nuno Palma, unpublished results, 1994).

[16] R. Cammack, K. K. Rao, D. O. Hall, J. J. G. Moura, A. V. Xavier, M. Bruschi, J. LeGall, A. Deville, and J. P. Gayda, *Biochim. Biophys. Acta* **490,** 311 (1977).

[16a] *Desulfovibrio (D.) desulfuricans* Norway 4 was renamed *D. baculatus* Norway 4. The organism was reclassified as *Desulfomicrobium baculatum* (Dmb) Norway 4 [R. Devereux, S. H. He, C. L. Doyle, S. Orkland, D. A. Atahl, J. LeGall, and W. B. Whitman, *J. Bacteriol.* **172,** 3609 (1990)].

[17] M. Bruschi, E. C. Hatchikian, L. A. Golovleva, and J. LeGall, *J. Bacteriol.* **129,** 30 (1977).

[18] F. Guerlesquin, M. Bruschi, G. Bovier-Lapierre, and G. Fauque, *Biochim. Biophys. Acta* **626,** 127 (1980).

peptide chain of 59 amino acids and 8 cysteines (average $E_0 = -500$ mV).[19,20]

Desulfovibrio africanus Benghazi

Three ferredoxins were isolated and characterized from *Desulfovibrio africanus* (Da). The proteins are dimers of subunits with a molecular mass of ~6 kDa. *Desulfovibrio africanus* FdI contains a single $[4Fe\text{-}4S]^{2+,1+}$ center bound to a polypeptide structure of 60 amino acids with only 4 cysteines. This is the minimal requirement for the [4Fe-4S] cluster binding. *Desulfovibrio africanus* FdII is a minor component not so well characterized and seems to contain a [4Fe-4S] center also.[21,22] *Desulfovibrio africanus* FdIII is a dimer containing 61 amino acids and 7 cysteine residues per subunit. The protein contains a $[3Fe\text{-}4S]^{1+,0}$ cluster ($E_0 = -140$ mV) and a $[4Fe\text{-}4S]^{2+,1+}$ cluster ($E_0 = -410$ mV).[23–25]

Desulfovibrio vulgaris Miyazaki

Two ferredoxins were isolated and purified from *Desulfovibrio vulgaris* Miyazaki, (DvM). The major form, FdI, contains two redox centers with distinct behavior[26] and a high sequence homology to DaFdIII. The protein is a dimer of a polypeptide chain of 61 amino acids with 7 cysteines. The clusters have midpoint redox potentials in the range −140 to −10 mV (assumed $[3Fe\text{-}4S]^{1+,0}$ center) and −430 mV [assumed $[4Fe\text{-}4S]^{2+,1+}$ center (T. Yagi, personal communication)]. *Desulfovibrio vulgaris* Miyazaki FdII is a dimer of 63 amino acids; it contains 7 cysteines but only one $[4Fe\text{-}4S]^{2+,1+}$ cluster with a midpoint redox potential of −405 mV.[26–28]

From a related organism, *Desulforomonas acetoxidans*, a 2 × $[4Fe\text{-}4S]^{2+,1+}$ ferredoxin was isolated[29]; ferredoxins are present, but not yet

[19] M. Bruschi, F. Guerlesquin, G. E. Bovier-Lapierre, J. J. Bonicel, and P. M. Couchoud, *J. Biol. Chem.* **260,** 8292 (1985).

[20] F. Guerlesquin, J. J. G. Moura, and R. Cammack, *Biophys. Biochim. Acta* **679,** 422 (1982)..

[21] E. C. Hatchikian, H. E. Jones, and M. Bruschi, *Biochim. Biophys. Acta* **548,** 471 (1979).

[22] M. Bruschi and E. C. Hatchikian, *Biochimie* **64,** 503 (1982).

[23] E. C. Hatchikian and M. Bruschi, *Biochim. Biophys. Acta* **634,** 41 (1981).

[24] G. Bovier-Lapierre, M. Bruschi, J. Bonicel, and E. C. Hatchikian, *Biochim. Biophys. Acta* **913,** 20 (1987).

[25] F. A. Armstrong, S. J. George, R. Cammack, E. C. Hatchikian, and A. J. Thomson, *Biochem. J.* **264,** 265 (1989).

[26] M. Ogata, S. Kondo, N. Okawara, and T. Yagi, *J. Biochem. (Tokyo)* **103,** 121 (1988).

[27] N. Okawara, M. Ogata, T. Yagi, S. Wakabayashi, and M. Matsubara, *J. Biochem. (Tokyo)* **104,** 196 (1988).

[28] N. Okawara, M. Ogata, T. Yagi, S. Wakabayashi, and M. Matsubara, *Biochimie (Tokyo)* **70,** 1815 (1988).

[29] I. Probst, J. J. G. Moura, I. Moura, M. Bruschi, and J. LeGall, *Biochim. Biophys. Acta* **502,** 38 (1978).

fully characterized, in the following SRB: *D. vulgaris* Hildenborough, *Desulfovibrio salexigens, Desulfovibrio desulfuricans* ATCC 27774 and *Dsm. baculatum* 9974, and *Desulfotomaculum*[30,31] (J. J. G. Moura, A. L. Macedo, and P. Nuno Palma, unpublished results, 1994).

Purification Methods

The isolation of ferredoxins is largely based on the acidic characteristic of these proteins. In general they are purified by adsorption on ion exchangers such as DEAE 52-cellulose (Whatman, Clifton, NJ) and desadsorbed using gradients (i.e., Tris-HCl or NaCl). Because of their low molecular mass, they can be easily separated from other, higher molecular mass components by gel filtration [Sephadex G-75 (Pharmacia, Piscataway, NJ) or BioGel (Bio-Rad, Richmond, CA)]. Other purification steps may include calcium phosphate gel chromatography. Purification is, in general, carried out at 4° and the buffers [Tris-HCl and phosphate buffers are used (pH 7.6) at appropriate molarities] should be flushed with an inert gas, because some of these proteins are oxygen unstable. Purity indices are determined by the ratio between Fe-S charge-transfer bands centered around 300 nm (or the protein absorption at 280 nm) and 400 nm. The low content in aromatic residues allows in most cases the definition of the second charge-transfer band in the ultraviolet (UV) spectral region.

Desulfovibrio gigas ferredoxins are used here as an example (adapted and updated from Refs. 13 and 14) in order to illustrate the purification of a ferredoxin and the separation of oligomeric forms.

The starting material is a frozen cell paste. All operations are carried out at 4° unless otherwise stated. The outline procedure is scaled up to 1 kg of cells (wet weight). The acidic protein extract, obtained as described in Ref. 14 and centrifuged at 13,000 *g* (1 hr) and at 140,000 *g* (2 hr), is adsorbed, after overnight dialysis, on a large DEAE-cellulose column (4 × 35 cm) equilibrated with 10 m*M* Tris-HCl buffer. The proteins are eluted with a continuous gradient of Tris-HCl (0.01 to 0.5 *M*, 3 liters each). This procedure allows separation of a band mainly constituted by ferredoxin (well separated from flavodoxin, the next most acidic component). Ferredoxins, recognized by their brown color, are located at the top of the column and elute at about 0.4 *M* Tris. Dialysis of this fraction against 10 m*M* buffer [or an intermediate concentration step in a Diaflo (membrane YM10; Amicon, Danvers, MA) while simultaneously lowering

[30] J. R. Postgate, "Sulfate Reducing Bacteria," 2nd Ed., Cambridge Univ. Press, London, 1984.

[31] J. LeGall and J. R. Postgate, *Adv. Microb. Physiol.* **20,** 81 (1973).

the ionic strength] and application to a second DEAE-52 or BioGel column is a useful purification step. Gel filtration (Sephadex G-50) has been used before.[14] However, the separation is better accomplished with DEAE-Sephadex A-50 equilibrated at 0.45 *M* Tris-HCl. The fraction from the BioGel is brought about the same buffer molarity. Elution is performed with a narrow gradient (0.45 to 0.55 *M*). Ferredoxin separates into four bands. Ferredoxin I and FdII are the major components. Other forms, FdI′ and FdII′, are minor bands that are less well characterized. Calcinated alumina was previously used in the purification of ferredoxins, particularly for the removal of "260-nm" contaminants.[13,11] High-pressure liquid chromatography can be used to advantage.

Cluster-Binding Motifs

The only high-resolution crystallographic study performed on ferredoxins isolated from SRB has involved the [3Fe-4S] DgFdII structure.[10,32,33] There are no three-dimensional (3D) data on [4Fe-4S] Fds isolated from this bacterial group.

The definition of the cluster ligands of the [3Fe-4S] cluster and of the related structural features are useful in predicting cluster types in other Fds of known sequence, as well as in determining the nature of the cluster coordinating atoms (and variability) and their control of the type and performances of the metal sites, in particular in terms of cluster stability, cluster interconversion capability, and acceptance of other metals at the cluster.

The three-dimensional structure of DgFdII revealed that the main chain fold is similar to that of 2 × [4Fe-4S] *Peptococcus aerogenes* (Pa) Fd,[34] [3Fe-4S] + [4Fe-4S] *Azotobacter vinelandii* (Av) FdI,[35,36] and [4Fe-4S] *Bacillus thermoproteolyticus* (Bt) Fd.[7] On the basis of sequence homology and secondary structure predictions, the chain fold in other [4Fe-4S] proteins was predicted, including DgFdII.[37] An extended α helix, found in BtFd, was proved to be present in DgFdII, and experimental evidence was given for its presence in DmbFdI by 2D nuclear magnetic resonance (NMR).[38] Preliminary crystallographic data were obtained for [4Fe-4S]

[32] L. C. Sieker, E. T. Adman, L. H. Jensen, and J. LeGall, *J. Mol. Biol.* **134,** 375 (1984).

[33] E. T. Adman, L. C. Sieker, and L. H. Jensen, *J. Biol. Chem.* **248,** 3987 (1973).

[34] J. N. Tsunoda, K. Yasunobu, and H. R. Whiteley, *J. Biol. Chem.* **243,** 6262 (1968).

[35] G. H. Stout, S. Turley, L. C. Sieker, and L. H. Jensen, *Proc. Natl. Acad. Sci. U.S.A.* **85,** 1020 (1988).

[36] C. D. Stout, *J. Biol. Chem.* **263,** 9256 (1988).

[37] K. Fukuyama, Y. Nagahara, T. Tsukihara, Y. Katsube, T. Hase, and H. Matsubara, *J. Mol. Biol.* **199,** 183 (1988).

[38] D. Marion and F. Guerlesquin, *Biochem. Biophys. Res. Commun.* **159,** 592 (1989).

DmbFdI (the crystal had an octahedral form with space group $P4_23_2$),[39] and a structural prediction was shown based on molecular homology modeling.[40]

The [3Fe-4S] core is now considered a unique basic iron-sulfur core whose structure was determined in three different proteins: aconitase,[41–43] AvFd,[3,35,36] and DgFdII,[10,32,33] and it was indicated by spectroscopic methods [for the [3Fe-4S] center, in particular, by extended X-ray absorption fine structure (EXAFS)[44]] that a strong homology exists between them, reminiscent of the [4Fe-4S] cluster in PaFd.[34] The clusters in these proteins have Fe-Fe and Fe-S distances around 2.8 and 2.2 Å. The [3Fe-4S] cluster has a cuboidal geometry (in which a corner of the cube is missing) with an Fe : labile S stoichiometry of 3 : 4.[41,43]

The X-ray structure analysis performed for DgFdII[10] indicates that the cluster is bound to the polypeptide chain by three cysteinyl residues: Cys-8, Cys-14, and Cys-50 (see Fig. 2). The residue Cys-11 (a potential ligand for a fourth site, if the cluster were a [4Fe-4S] cluster) is not bound and is tilted toward the solvent, away from the cluster. An unexpected observation from X-ray data analysis is that Cys-11 appears to be chemically modified, possibly by a methanethiol group. It is currently undetermined if this chemical modification is present in the native protein; it most probably is an artifact of the crystallization procedure. A previously undetected feature in this structure is the presence of a disulfide bridge between Cys-18 and Cys-42 (connecting the remaining two cysteines of the four that bind a second iron sulfur core in two clusters containing Fd, i.e., PaFd). The loss of a second cluster is associated with the disappearance of Cys-41 and -47. Because BtFd does not have a disulfide bridge and has folding similar to that of DgFdII, the disulfide bridge does not seem to be a full requirement to maintain the overall structure. Additional information was the detection of an extra amino acid after position 55 (proposed to be a valine residue), making the total amino acids in the subunit of the DgFdII tetramer equal to 58.[10,15]

Amino acid sequences are available for one-cluster DgFdI and FdII

[39] F. Guerlesquin, M. Bruschi, J. P. Astier, and M. Frey, *J. Mol. Biol.* **168,** 203 (1983).

[40] C. Cambillau, M. Frey, J. Mosse, F. Guerlesquin, and M. Bruschi, *Proteins: Strut. Funct. Genet.* **4,** 63 (1988).

[41] H. Beinert and M. C. Kennedy, *Eur. J. Biochem.* **186,** 5 (1989).

[42] A. H. Robbins and C. D. Stout, *Proc. Natl. Acad. Sci. U.S.A.* **86,** 3639 (1989).

[43] M. C. Kennedy and C. D. Stout, *Adv. Inorg. Chem.* **38,** 323 (1992).

[44] M. R. Antonio, B. A. Averill, I. Moura, J. J. G. Moura, W. H. Orme-Johnson, B. K. Teo, and A. V. Xavier, *J. Biol. Chem.* **257,** 6646 (1982).

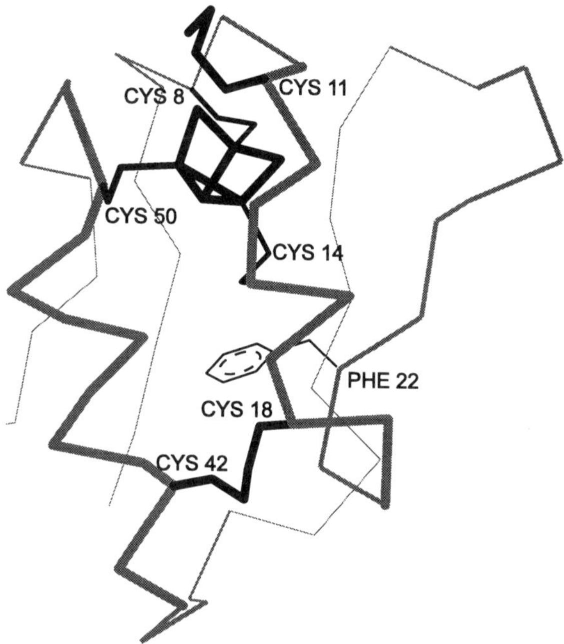

FIG. 2. General outline of [3Fe-4S] *D. gigas* FdII 3D structure. Highlighted are the cysteines that coordinate to the cluster, the cysteines that participate in the formation of the disulfide bridge, the relative position of the aromatic residue (Phe-22) in relation to the cluster, and the strategic positioning of Cys-11, not bound to the cluster, providing the fourth coordinating site when cluster conversion takes place. The α-helix structure motif is also shown. (Coordinates of DgFdII kindly provided by L. C. Sieker, Seattle, Washington.)

oligomeric forms,[15] DmbFdI,[19] DvMFdII,[28] and DaFdI,[22] and two-cluster DvMFdI,[27] DaFdIII,[24] and DmbFdII[5] (Fig. 3).

Comparison of DgFd amino acid sequence with those of DmbFdI and DvMFdII clearly indicates that Cys-8, Cys-11, Cys-14, and Cys-50 are involved in cluster binding and that two extra cysteines (Cys-18 and Cys-42) may be involved in the formation of an intramolecular disulfide bridge, although this is not proved. (Sequence numbering used refers to the DgFd sequence.) The DaFdI sequence displays the minimal cysteine content for the binding of a [4Fe-4S] core (Cys-8, Cys-11, Cys-14, and Cys-50). The DaFdIII sequence is a particularly interesting one: the sequence shows the presence of seven cysteines. Thus the [4Fe-4S] center can be chelated to Cys-18, Cys-40, Cys-43, and Cys-46, as usual, and the [3Fe-4S] center to Cys-8, Cys-14, and Cys-50. However, it was clearly demonstrated that cluster interconversion may take place in the presence of

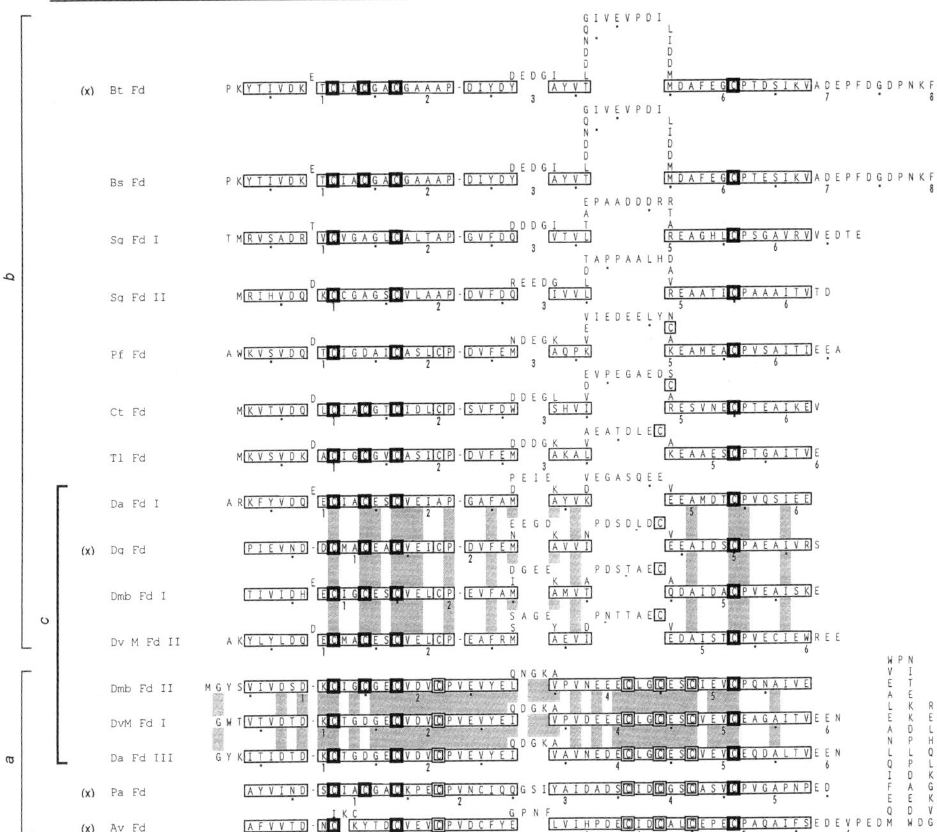

FIG. 3. Amino acid sequences of bacterial ferredoxins containing cubane-type Fe-S clusters: (*a*) ferredoxins containing two [3,4Fe-4S] clusters, (*b*) one [3,4Fe-4S] cluster, and (*c*) *Desulfovibrio* ferredoxins. The amino acid alignments were done on the basis of sequence and structural homology, as explained in text. Horizontal portions of the sequences, enclosed in boxes, represent (predicted) structurally conserved regions, whereas segments with variable tertiary conformation are represented as outstanding loops. Shading emphasizes amino acid homologies between the *Desulfovibrio* ferredoxins. Cysteine residues responsible for binding the cluster(s) are marked with heavy and double-line boxes. Potential disulfide-forming cysteines are also indicated in some ferredoxins containing a single cluster. Bt, *Bacillus thermoproteolyticus*; Bs, *Bacillus stearothermophilus*; Sg, *Streptomyces griseolus*; Pf, *Pyrococcus furiosus*; Ct, *Clostridium thermoaceticum*; Tl, *Thermus thermophilus*; Da, *D. africanus*; Dg, *D. gigas*; Dmb, *Desulfomicrobium baculatum* Norway 4; DvM, *D. vulgaris* Miyazaki; Pa, *Peptococcus aerogenes*; Av, *Azotobacter vinelandii*. More details on other ferredoxins can be found, for example, in Ref. 3.

ferrous ions (see Cluster Interconversions, below), and this polypeptide chain can accommodate two [4Fe-4S] cores.[45] It was then proposed that an aspartic residue in position 11, replacing one of the potential cysteine cluster ligands, is the fourth ligand of the cluster. It should be noted that proline in position 51 is changed to a glutamic acid. A highly homologous situation is found in the DvMFdII amino acid sequence. *Desulfomicrobium baculatum* Norway 4 FdII contains 2 × [4Fe-4S] and the eight cysteinyl residues required to bind these cores (cluster I, Cys-8, Cys-14, and Cys-50; cluster II, Cys-18, Cys-40, Cys-43, and Cys-46). *Desulfovibrio vulgaris* Miyazaki FdI shows another special situation. It contains six cysteines in homologous positions and a seventh cysteine quite close the C terminal in position 54, not involved in cluster binding. Conservative isoleucine residues are observed in the C terminus.[3]

The amino acid sequence alignment of ferredoxins from *Desulfovibrio* spp. shown in Fig. 3 is a proposal based on sequence and structural homology, done by superimposing the known three-dimensional structures of four ferredoxins (Pa, Av, Bt, and Dg) along with sequences of other *Desulfovibrio* ferredoxins. Gaps were inserted where loops or nonorganized secondary structure force the spatial positions of amino acids to diverge from each other.

General trends are observed: ferredoxins containing two Fe-S clusters per monomeric unit have higher sequence homology; among all ferredoxins containing cubane-type clusters, whenever a [3Fe-4S] center is present, the missing corner (Fe) of the cube is associated with substitution or displacement of the cysteine residue corresponding to position 11.

A consensus sequence for the binding of two cubane-type clusters in *Desulfovibrio* ferredoxins is proposed:

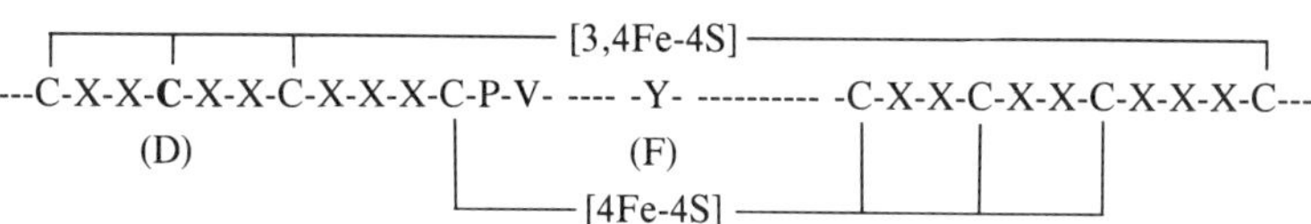

1. The second cysteine residue can be replaced by an aspartic acid (as in DvMFdI and DaFdIII), which can under certain conditions coordinate a fourth iron atom to build a [4Fe-4S] center.

2. A valine residue is always conserved after the first -C-P- group, in sequences of bacterial ferredoxins containing two cubane-type clusters, not present in one cluster containing ferredoxins.

[45] S. J. George, F. A. Armstrong, E. C. Hatchikian, and A. J. Thomson, *Biochem. J.* **264,** 275 (1989).

3. An aromatic residue is highly conserved (except for PaFd), positioned between the two clusters, which may be involved in intramolecular electron transfer.

4. The proline residue after the second group of cysteines is not always conserved.

A consensus sequence for the binding of one cubane-type cluster in *Desulfovibrio* ferredoxins is

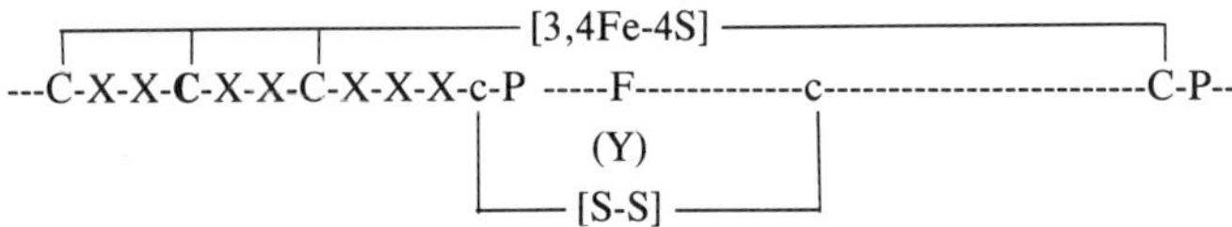

1. The second cysteine can be replaced by aspartate or alanine.

2. The two cysteine residuess shown in the lower case are not necessarily conserved, but their presence seems to be interdependent, that is, when one of them appears, the other is also present. They may be involved in the formation of an intramolecular disulfide bridge, as in the case of DgFdII. This observation stands for all the nine sequences of bacterial ferredoxins containing one cubane-type cluster (not shown[3]).

3. The proline residue after the fourth cluster binding cysteine seems to be conserved, not only within all *Desulfovibrio* ferredoxins, but also within all bacterial ferredoxins containing one cubane cluster.

These observations are put together in Fig. 4, which shows schematically a three-dimensional consensus structure common to bacterial ferredoxins containing one or two clusters.

Electronic and Magnetic Properties of [3Fe-4S] and [4Fe-4S]

The identification of center type is crucial when analyzing complex enzymatic systems. The study of ferredoxins, as simple cases, enable one to obtain information that can be used as a tool for cluster type identification. In particular, the [3Fe-4S] center in DgFdII has been a case study for this type of center and the magnetic and electronic properties explored by different spectroscopic techniques: EPR,[16] Mössbauer spectroscopy (MB),[46,47] resonance Raman,[48] magnetic circular dichroism (MCD),[49]

[46] B. H. Huynh, J. J. G. Moura, I. Moura, I., T. A. Kent, J. LeGall, A. V. Xavier, and E. Münck, *J. Biol. Chem.* **255,** 3242 (1980).

[47] V. Papaefthymiou, J.-J. Girerd, I. Moura, J. J. G. Moura, and E. Münck, *J. Am. Chem. Soc.* **109,** 4703 (1987).

[48] M. K. Johnson, J. W. Hare, T. G. Spiro, J. J. G. Moura, A. V. Xavier, and J. LeGall, *J. Biol. Chem.* **256,** 9806 (1981).

[49] A. J. Thomson, A. E. Robinson, M. K. Johnson, I. Moura, J. J. G. Moura, A. V. Xavier, and J. LeGall, *Biochim. Biophys. Acta* **670,** 93 (1981).

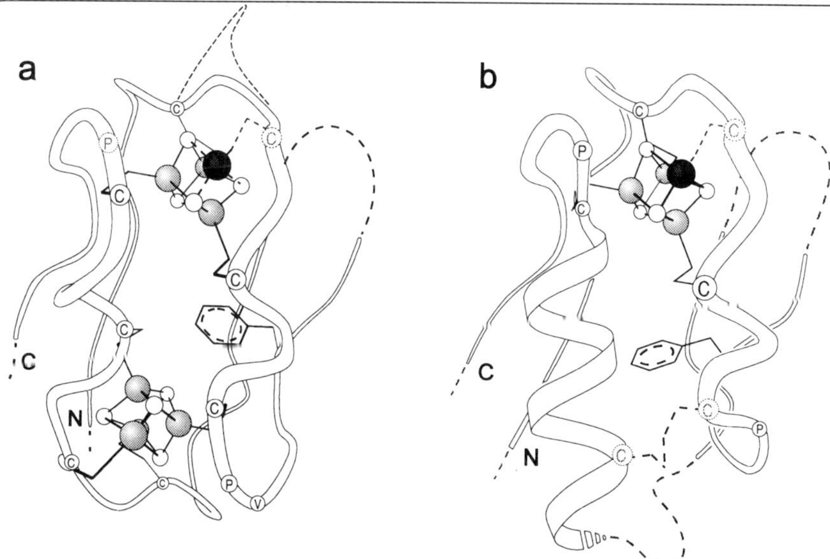

FIG. 4. Tentative schematic 3D consensus structure, common to bacterial ferredoxins containing two (a) and one (b) cubane-type Fe-S cluster. Dashed portions of the backbone chain represent regions of variable length and conformation. An aromatic residue (phenylalanine or tyrosine) is shown in a position that is highly conserved between the two clusters or between one cluster and a hypothetical disulfide bridge, Darker spheres in the upper cluster represent the iron atom that is invariably missing in the [3Fe-4S] clusters. A possible disulfide bridge is shown (dashed line) between two closely positioned cysteinyl residues, as in the case of DgFdII. In *A. vinelandii* Fd, Cys-11 is displaced by an extra loop, shown at the top in (a).

EXAFS,[44] saturation magnetization,[50] electrochemistry,[51] and NMR.[52,53] Some of these results can be used as fingerprints of the cluster. The [4Fe-4S] center is also well characterized and surprising information has been obtained in relation to cluster interconversions and noncysteinyl coordination, as illustrated for DgFdI and DaFdIII.

The [3Fe-4S] cluster can be stabilized in two oxidation states, 1+ and 0. In the oxidized state the $[3Fe\text{-}4S]^{1+}$ cluster contains three high-

[50] E. P. Day, J. Peterson, J. J. Bonvoisin, I. Moura, and J. J. G. Moura, *J. Biol. Chem.* **263,** 3684 (1988).

[51] C. Moreno, A. L. Macedo, I. Moura, J. LeGall and J. J. G. Moura, *J. Inorg. Biochem.* **53,** 219 (1994).

[52] A. L. Macedo, I. Moura, J. J. G. Moura, J. LeGall and B. H. Huynh, *Inorg. Chem.* **32,** 1101 (1993).

[53] A. L. Macedo, P. N. Palma, I. Moura, J. LeGall, V. Wray, and J. J. G. Moura, *Magn. Reson. Chem.* **31,** 559 (1993).

spin ferric ions spin-coupled to form an $S = 1/2$ state[46,54,55] and exhibits an isotropic EPR signal centered around $g = 2.02$.[46] Mössbauer spectra taken at $T > 40$ K exhibit one sharp quadrupole doublet with quadrupole splitting ($\Delta E_Q = 0.54$ mm/sec and isomer shift $\Delta E_Q = 0.27$ mm/sec) typical of Fe^{3+} ions in a tetrahedral environment of thiolate ligands. The effective $S = 1/2$ results from antiferromagnetic coupling of three high-spin ferric rubredoxin-type ($S_1 = S_2 = S_3 = 5/2$) ions.[47] Reduction by one electron yields a $[3Fe\text{-}4S]^0$ cluster [$E_m = -130$ mV, vs normal hydrogen electrode (NHE)] with integer spin ($S = 2$)[21] proved by MCD studies[49] and in agreement with the Mössbauer predictions leading to the detection of a $\Delta m_s = 4$ EPR transition, at around $g = 12$.[47] At 4.2 K (zero-field) the Mössbauer spectrum of the reduced cluster exhibits two quadrupole doublets with intensity ratios of 2:1. The more intense doublet, representing two iron sites (site I), has $\Delta E_Q(\mathrm{I}) = 1.47$ mm/sec and $\delta(\mathrm{I}) = 0.46$ mm/sec. The second doublet (site II) has $\Delta E_Q(\mathrm{II}) = 0.52$ mm/sec and $\delta(\mathrm{II}) = 0.32$ mm/sec. These values suggest that the iron atom associated with doublet (II) is high-spin ferric in character, similar to the iron sites in oxidized FdII. The parameters of doublet I indicate a formal oxidation state between 2+ and 3+. The two iron atoms are in the oxidation level $Fe^{2.5+}$. The reduced core is an example of a mixed valence compound with one localized site Fe^{3+} in site II and a delocalized iron pair. Spectroscopic evidence is consistent with the delocalized pair having a spin of 9/2 coupled to the high-spin ferric ion, forming a cluster spin of 2.[47] This is the simplest iron cluster to exhibit this equal sharing of an electron by more than one iron site, a feature in common with the [4Fe-4S] cluster.

A further two-electron reduction was suggested to occur in DgFdII originating an all-ferrous trinuclear core,[51,56] and has been studied by Armstrong and co-workers in DaFdIII.[57] Electrochemical studies of DaFdIII adsorbed with neomycin as an electroactive film on a pyrolytic graphite edge shows that the $[3Fe\text{-}4S]^0$ can undergo a two-electron, further reduced step at -720 mV, in a pH-dependent process. The observation of similar voltammetric waves in other [3Fe]-containing proteins suggests that stabilization of an all-ferrous cluster may be a common feature of this cluster.

[54] T. A. Kent, B. H. Huynh, and E. Münck, *Proc. Natl. Acad. Sci. U.S.A.* **77,** 6574 (1980).

[55] M. H. Emptage, T. A. Kent, B. H. Huynh, J. Rawlings, W. H. Orme-Johnson, and E. Münck, *J. Biol. Chem.* **255,** 1793 (1980).

[56] F. A. Armstrong, *Adv. Inorg. Chem.* **38,** 117 (1992).

[57] F. A. Armstrong, J. N. Butt, S. J. George, E. C. Hatchikian, and A. J. Thomson, *FEBS Lett.* **259,** 15 (1989).

The [4Fe-4S] center in DgFdI (or reconstituted FdII or Fe^{2+}-activated FdII; see below) is indistinguishable.[58] The center is diamagnetic in the oxidized state.[58a] On one-electron reduction, an EPR signal develops at $g = 1.91$, 1.94, and 2.07, typical of a reduced $[4Fe\text{-}4S]^{1+}$ cluster. The MB spectrum of oxidized DgFdI (we refer to material labeled with ^{57}Fe) consists of two sharp quadrupole doublets (labeled I and II) with intensity ratio 1 : 3, characteristics of a diamagnetic state ($S = 0$). The quadrupole splitting and isomer shift of doublet II is typical of a subsite of a [4Fe-4S] cluster in the 2+ state ($\Delta E_Q = 1.32$ mm/sec, $\delta = 0.45$ mm/sec). The doublets reflect inequivalent subsites of the $[4Fe\text{-}4S]^{2+}$ cluster. The reduced spectrum exhibits paramagnetic hyperfine structure (in agreement with the EPR active species detected, $S = 1/2$). The Mössbauer data are almost identical to the one reported for [4Fe-4S] *B. polymyxa* Fd and could be fitted with the same set of parameters.[60]

One-dimensional nuclear Overhauser effect and relaxation measurements in native DgFdII and analysis of temperature-dependence studies of the NMR signals due to one pair of β-CH_2 protons of cluster-bound cysteine were used to calculate the coupling constant between the iron sites.[52] Moreover, the complementary use of the available X-ray crystallographic coordinates enabled specific assignment of Cys-50. A full structural assignment of the three cysteines ligands of the [3Fe-4S] core in DgFdII was possible by 2D NMR methodology.[53] In the absence of a crystallographic structure for DgFdI, multidimensional NMR was used to identify the four cysteines that are cluster ligands. The presence of four pairs of geminal β-CH_2 protons for DgFdI unambiguously proves the occupancy of the fourth site of the trinuclear complex and implied the coordination of Cys-11 at the cluster.[53]

The two ferredoxins isolated from Dmb were investigated by EPR spectroscopy.[20] The $[4Fe\text{-}4S]^{2+,1+}$ center of FdI shows a well-defined rhombic EPR signal in the reduced form at low temperature with g values of 1.902, 1.937, and 2.068. *Desulfomicrobium baculatum* Norway 4 FdII,

[58] J. J. G. Moura, I. Moura, T. A. Kent, J. D. Lipscomb, B. H. Huynh, J. LeGall, A. V. Xavier, and E. Münck, *J. Biol. Chem.* **257,** 6259 (1982).

[58a] Three stable oxidation states, 3+, 2+, and 1+, have been observed for the [4Fe-4S] cluster. The 3+ and 2+ states are stable in high-potential iron-sulfur proteins, whereas the 2+ and 1+ states are detected for bacterial ferredoxins and other [4Fe] cluster-containing proteins.[59] The clusters discovered in ferredoxins from SRB belong to this last category. Both the 3+ and 1+ states are paramagnetic with $S = 1/2$. The 2+ state is diamagnetic.

[59] C. W. Carter, Jr., *in* "Iron-Sulfur Proteins" (W. Lovenberg, ed.), Vol. 3, p. 157. Academic Press, New York, 1977.

[60] P. Middleton, D. P. E. Dickson, C. E. Johnson, and J. D. Rush, *Eur. J. Biochem.* **88,** 135 (1978).

as prepared, shows a weak contribution of a $[3Fe\text{-}4S]^{1+,0}$ center (5%, E_0 = 115 mV) and on reduction a complex EPR spectrum indicating an interaction between two $[4Fe\text{-}4S]^{2+,1+}$ clusters located in the same protein subunit. On extensive exposure of the protein to reductant the intercluster interaction was not observed accompanying an irreversible change in line shape. The midpoint redox potentials reported for the native $[4Fe\text{-}4S]^{2+,1+}$ centers were about −500 mV.

The EPR spectrum of native DaFdIII shows an almost isotropic signal centered around $g = 2.01$, similar to the one observed in proteins containing a $[3Fe\text{-}4S]^{1+}$ cluster.[25] The temperature dependence of this signal and low-temperature MCD spectra and magnetization properties are identical to the ones reported for DgFdII.[49] On one-electron reduction a $g = 12$ signal develops, characteristic of the $[3Fe\text{-}4S]^0$ state. Two-electron reduction originates an EPR active species with an axial EPR spectrum with g values at 1.93 and 2.05, consistent with the presence of a $[4Fe\text{-}4S]^{1+}$ center.

Cluster Interconversions

The [3Fe-4S] core present in DgFdII can be interconverted into a cubane structure ([4Fe-4S]). This facile conversion occurs on incubation of the protein with an excess of Fe^{2+} in the presence of a reducing agent (e.g., dithiothreitol). A summary of the interconversion pathways as well as the potential of the method for specific labeling of an iron-sulfur core were previously discussed[58,61] (see Scheme I). Combination of ^{57}Fe isotopic enrichment and specific introduction of a fourth iron atom into the [3Fe-4S] core produces different isotopic labeled clusters.

Electrochemical studies performed in the 7-Cys–Asp-14 DaFdIII indicate that the reduced [3Fe-4S] center can react rapidly with Fe^{2+} to form a [4Fe-4S] core that must include noncysteinyl coordination.[57] The carboxylate side chain of Asp-14 was proposed as the most likely candidate, because this amino acid occupies the cysteine position in the typical sequence of an eight-iron protein, as indicated before. The magnetic properties of the interconverted protein indicate that the [4Fe-4S] is $S = 1/2$ in the reduced state, and the novel cluster is diamagnetic in the oxidized state and $S = 3/2$ in the one-electron reduced state. The novel [4Fe-4S] with mixed S and O coordination has a midpoint redox potential of −400 mV.[45] This novel coordinated state with oxygen ligation to the iron-sulfur core is a plausible model for a [4Fe-4S] core showing unusual spin states

[61] T. A. Kent, J. J. G. Moura, I. Moura, J. D. Lipscomb, B. H. Huynh, J. Le Gall, A. V. Xavier, and E. Münck, *FEBS Lett.* **138,** 55 (1982).

Cluster reconstitution in apoprotein

$$[3^{56}Fe\text{-}4S] \rightarrow [4^{56}Fe\text{-}4S] \text{ or } [4^{57}Fe\text{-}4S]$$

Cluster subsite isotopic labeling

$$[3^{56}Fe\text{-}4S] \rightarrow [3^{56,57}Fe\text{-}4S]$$

$$[3^{57}Fe\text{-}4S] \rightarrow [3^{57}Fe, {}^{56}Fe\text{-}4S]$$

Cluster oxidative conversion

$$[4Fe\text{-}4S] \rightarrow [3Fe\text{-}4S] \qquad ({}^{56}Fe \text{ or } {}^{57}Fe)$$

Heterometal cluster formation

$$[3Fe\text{-}4S] \rightarrow [M,3Fe\text{-}4S] \qquad ({}^{56}Fe \text{ or } {}^{57}Fe;\ M = \text{Co, Ni, Zn, Cd, Ga, Tl})$$

SCHEME I. [3Fe-4S] and [4Fe-4S] interconversion possibilities and their potential use for cluster labeling and formation of heterometal-containing clusters.

present in complex proteins.[62,63] The [4Fe-4S] cluster in aconitase was indicated to use a water molecule as a fourth ligand.[43]

Experimental Procedures

Cluster Reconstitution from Apoprotein. The procedure to remove the [3Fe-4S] center from DgFdII and to rebuild an iron-sulfur center in the resulting monomeric apoprotein was adapted from Hong and Rabinowitz.[64] The protein is precipitated at 4° with trichloroacetic acid (5% final concentration) in the presence of 0.5 *M* mercaptoethanol and under argon. After 2 hr the precipitate is collected by centrifugation and then dissolved in 0.5 *M* Tris base containing 60 m*M* mercaptoethanol to a concentration of 5 mg of protein/ml. A second precipitation may be desirable, in order to deplete the protein of the cluster. The resulting apoprotein solution shows no optical spectrum and iron quantitation indicates that essentially all iron is removed. The apoprotein solution is kept under argon at 4° for 1 hr. Then iron (^{57}Fe enriched to 99.5%, if the sample is used for MB studies) and sulfide are added simultaneously in equimolar concentrations. Reconstitutions are performed with Fe^{2+}/protein ratios of 3–5 to 1. The mixture is allowed to react under argon for 30 min. After exposure to air the solution

[62] I. Moura, A. L. Macedo, and J. J. G. Moura, *in* "Advanced EPR and Applications in Biology and Biochemistry" (A. J. Hoff, ed.), p. 813, Elsevier, Amsterdam, 1989.

[63] W. R. Hagen, *Adv. Inorg. Chem.* **38,** 164 (1992).

[64] J. S. Hong and J. C. Rabinowitz, *Biochem. Biophys. Res. Commun.* **29,** 246 (1967).

is passed through a Sephadex G-25 column (1 × 15 cm) equilibrated with 10 m*M* Tris buffer. The resulting protein is loaded at 25° onto a DEAE-cellulose column (0.5 × 10 cm) and eluted with 0.4 *M* Tris-HCl, pH 7.6. The protein is then desalted on another Sephadex column and finally concentrated (by evaporation under argon or using Centricons) to the desired concentration.[58]

Oxidation with Ferricyanide. A sample of reconstituted DgFdII or DgFdI (350 μ*M*) in 500 μl in 100 m*M* Tris-HCl buffer (pH 7.6) is incubated with a fivefold excess of $K_3Fe(CN)_6$ at 4° for 3 hr. No protein precipitation occurs.[58,65] The solution is then passed through a Sephadex G-25 column (1 × 15 cm) equilibrated with 10 m*M* buffer and loaded onto a DEAE-52 column (0.5 × 10 cm) equilibrated with 10 m*M* buffer. The column is developed with a linear gradient (0.1–0.4 *M*). The protein is eluted at 0.4 *M* and readily separates from the band of Prussian blue formed during the reaction and retained in the column. The protein is then desalted and concentrated.

Incubation of [3Fe-4S] Ferredoxin with Iron. A sample can be prepared by incubating native FdII with 5 m*M* dithiothreitol and $^{57}Fe^{2+}$ and S^{2-} (four- to fivefold excess) for 6 hr under anaerobic conditions.[58] The sample is allowed to reoxidize in air and purified on Sephadex G-25 and DEAE-52 columns as previously indicated. The EPR samples can be prepared without further purification of reagents, an advantage for a rapid screening of the metal incorporation procedure.

Synthesis of Heterometal Clusters

The facility of the [3Fe-4S]/[4Fe-4S] conversions has suggested that incorporation of other metals in the vacant site of [3Fe-4S] core may take place, thus generating a series of novel clusters containing heterometals. A number of proteins containing [3Fe-4S] centers form cubane-like clusters of the type [M,3Fe-4S].[66–71] The [3Fe-4S] core present in DgFdII was

[65] A. J. Thomson, A. E. Robison, M. K. Johnson, R. Cammack, K. K. Rao, and D. O. Hall, *Biochim. Biophys. Acta* **637,** 423 (1981).

[66] I. Moura, J. J. G. Moura, E. Münck, V. Papaefthymiou, and J. LeGall, *J. Am. Chem. Soc.* **108,** 349 (1986).

[67] K. K. Surerus, E. Münck, I. Moura, J. J. G. Moura, and J. LeGall, *J. Am. Chem. Soc.* **109,** 3805 (1987).

[68] J. N. Butt, J. N., F. A. Armstrong, J. Breton, S. J. George, A. J. Thomson, and E. C. Hatchikian, *J. Am. Chem. Soc.* **113,** 6663 (1991).

[69] J. N. Butt, A. Sucheta, F. A. Armstrong, J. Breton, A. J. Thomson, and E. C. Hatchikian, *J. Am. Chem. Soc.* **113,** 8948 (1991).

[70] R. C. Conover, J.-B. Park, M. W. W. Adams, and M. K. Johnson, *J. Am. Chem. Soc.* **112,** 4562 (1990).

[71] R. C. Conover, A. T. Kowal, W. Fu, J.-B. Park, S. Aono, M. W. W. Adams, and M. K. Johnson, *J. Biol. Chem.* **265,** 8563 (1990).

the first used for the synthesis of heterometal cores inside a protein matrix and the first derivative synthesized was the [Co,3Fe-4S] core.[65] Similar synthetic products have also been extended to DaFdIII[68,69] and *Pyrococcus (P.) furiosus* Fd.[70,71]

Preparation of Heterometal Clusters

The procedure for converting a [3Fe-4S] core into an [M,3Fe-4S] was described.[66,67] Purified DgFdII is incubated with an excess of the metal to be incorporated, as for the cluster interconversion procedures. Typically 0.5 ml of dithionite-reduced protein, 0.5 m*M* in [3Fe-4S] core, is anaerobically incubated for 6–12 hr with 15 m*M* $M(NO_3)_2$ and 5 m*M* dithiothreitol and then repurified as previously described.

Spectroscopic Properties of Heterometal Clusters Formed in Desulfovibrio gigas Ferredoxin II and Desulfovibrio africanus Ferredoxin III

Cobalt(II) ion was introduced into a heterometal cluster, assuming a paramagnetic configuration ($S = 3/2$) at the fourth site of a cubane structure.[66–69] The newly formed cluster was studied in the oxidized and reduced states. In the oxidized state the cluster exhibits an $S = 1/2$ EPR signal with g values of 1.82, 1.92 and 1.98 (g values obtained after spectral simulation). The well-resolved ^{59}Co hyperfine at the g_z line ($A_z \sim 4.4$ mT) is also broadened by ^{57}Fe isotopic substitution, indicating that iron and cobalt share a common unpaired electron. On one-electron reduction ($E_m = -220$ mV, vs NHE) the cluster has an integer spin ($S = 1$) as indicated by Mössbauer and MCD measurements. The Mössbauer spectrum of the oxidized cluster exhibits at 4.2 K two distinct spectral subsites with an intensity ratio of 2 : 1, indicating that the three iron atoms reside in the same cluster, suggesting the presence of an Fe^{3+} site and again a delocalized pair.[65] On one-electron reduction the third iron is formally Fe^{2+}.

The introduction of Cu^{2+}, Zn^{2+}, Cd^{2+}, and Ga^{3+} was anticipated with great interest owing to the possibility of introducing a diamagnetic site at the cubane.[71] The zinc-incubated product was shown to be EPR active (spin system, $S = 5/2$) with $E/D \sim 0.25$ and $D = -2.7$ cm^{-1} and shows ^{57}Fe hyperfine broadening.[66] The spectrum produced by Mössbauer studies performed at 4.2 K consists of two spectral components with an intensity ratio of 2 : 1, and the MB parameters suggests again a delocalized Fe^{2+}/Fe^{3+} pair, as observed for the reduced [3Fe-4S] core. The formal oxidation states involved were assigned: two Fe^{2+}, one Fe^{3+}, and, being a Kramer system ($S = 5/2$), the zinc site must be divalent. Incorporation of the zinc atom at the vacant site of the [3Fe-4S] core implies a previous reduction step of the cluster plus an extra electron for metal incorporation. There

TABLE II
SPIN AND OXIDATION STATES OF HETEROMETAL CLUSTERS IN PROTEINS

Cluster	Oxidation states	Spin state	Ref.
[Co, 3Fe-4S]	+2, +1	1/2, 1	65, 74
[Zn, 3Fe-4S]	+2, +1	5/2, 2	68, 72, 74
[Ni, 3Fe-4S]	+2	3/2	51, 73, 74
[Cd, 3Fe-4S]	+2	5/2	68, 72, 74
[Tl, 3Fe-4S]	+2	1/2	69

is also preliminary evidence for the incorporation of Cd^{2+}, which yields a similar $S = 5/2$ system under reducing conditions.[72] A similar $[Ni,3Fe\text{-}4S]^{1+}$ was produced in *P. furiosus* Fd[70] and DgFdII.[73] The dithionite-reduced Fd after anaerobic incubation with excess Ni(II) in the presence of dithiothreitol shows EPR spectral features dominated by the presence of an $S = 3/2$ species. Zinc and nickel heterometal-containing clusters were studied in detail in *P. furiosus* Fd.[74]

The redox properties of a series of heterometal clusters were accessed by electrochemical measurements and EPR. The derivatives formed in DgFdII were measured by direct square-wave voltammetry promoted by Mg(II) at a vitreous carbon electrode and the following redox transitions were detected: Cd (−495 mV), Fe (−420 mV), Ni (−360 mV), and Co (−245 mV) (vs NHE). The values agree well with independent measurements.[51,73] Similar derivatives generated in DaFdIII measured by cyclic voltammetry in film and bulk solution yield the following: Cd (−590 mV), Zn (−490 mV), and Fe (−400 mV).[68] A novel addition to these series was obtained by the introduction of a monovalent ion.[69] The $[Tl,3Fe\text{-}4S]^{2+}$ core shows a redox transition at 80 mV.

Table II presents the variable spin states found in the heterometal cubane-type cores so far formed in simple iron-sulfur proteins, indicating the effect of different *d*-level occupancy of the fourth site. A series of model compounds have been synthesized[75] and their properties compared with those of the clusters formed within the polypeptide chains.

[72] E. Münck, V. Papaefthymiou, K.. K. Surerus, and J.-J. Girerd, *in* "Metal Ions in Proteins," (L. Que, ed.), ACS Symposium Series 372, Chap. 15. Washington, D.C., 1988.

[73] C. Moreno, A. L. Macedo, K. K. Surerus, E. Münck, J. LeGall, and J. J. G. Moura, Abstract of the Third International Conference on Molecular Biology of Hydrogenases, Tróla, Portugal, 1991.

[74] K. K. P. Srivastava, K. K. Surerus, R. C. Conover, M. K. Johnson, J.-B. Park, M. W. W. Adams, and E. Münck, *Inorg. Chem.* **32,** 927 (1993).

[75] R. H. Holm, S. Ciurli, and J. A. Weigel, *Prog. Inorg. Chem.* **38,** 1 (1990).

Physiological Role

Ferredoxins function primarily as electron transfer proteins. The need for ferredoxin in a biological reaction is demonstrated by preparing crude cell extracts lacking these electron carriers and passing the extracts through a DEAE-cellulose column at low ionic strength under anaerobic conditions. Because of its acidic properties the ferredoxin (and flavodoxin) remains adsorbed on the column and the unadsorbed fraction is ferredoxin depleted. The activity is analyzed by reconstitution of the electron transfer chain by adding purified ferredoxin. Therefore the major role of ferredoxins is inferred from the stimulation of an activity and comparing it with the complete and the depleted systems. Transfer of electrons in the phosphoroclastic reaction (from pyruvate via pyruvate dehydrogenase to the cytochrome c_3/hydrogenase complex with consequent hydrogen evolution), as well as in the transfer of electrons from molecular hydrogen via the cytochrome c_3/hydrogenase complex to the reduction of sulfite (sulfite reductase), are relevant metabolic reactions in which ferredoxins were shown to be involved.[12,13,76]

Activity Measurements

Phosphoroclastic Reaction. The pyruvate dehydrogenase activity and the coupling effect of the ferredoxin are determined by measuring the hydrogen evolution by manometric assay under the following conditions: the main compartment contains 150 μmol of phosphate buffer (pH 7.0), 100 μmol of electron carrier (Fd), pure hydrogenase, 4 μmol of (CoA), 5 μmol of thymine PP_i (TPP), 20 μmol of $MgCl_2$, 10 mmol of mercaptoethanol, and pyruvate dehydrogenase-containing extract in a final volume of 3 ml. The side arm contains 30 μmol of sodium pyruvate and the center well contains 0.05 ml of 10 *M* NaOH. The flasks are incubated for 30 min at 37°.

The pyruvate dehydrogenase-containing extract devoid of the acidic electron carriers (mainly Fd and flavodoxin) is obtained by passing the *D. gigas* acidic extract[12] (10 ml containing 30 mg of protein/ml) on a small DEAE column (10 × 10 mm) equilibrated with 10 m*M* Tris-HCl buffer at pH 7.6. Of this extract about 30 mg is used in each assay. Pure *D. gigas* hydrogenase (400 μg), prepared as indicated in Ref. 77, with a specific activity of 440 μmol of H_2 produced/min/mg protein, is added to ensure an excess of this activity.

Sulfite Reduction. A manometric assay is utilized to determine the participation of ferredoxin in the reduction of sulfite by hydrogen. The

[76] J. J. G. Moura, A. V. Xavier, E. C. Hatchikian, and J. LeGall, *FEBS Lett.* **89,** 177 (1978).

[77] J. LeGall, P. O. Ljungdahl, I. Moura, H. D. Peck, Jr., A. V. Xavier, J. J. G. Moura, M. Teixeira, B. H. Huynh, and D. V. DerVartanian, *Biochem. Biophys. Res. Commun.* **106,** 610 (1982).

main compartment of each manometric vessel contains 150 μmol of phosphate buffer (pH 7.0), pure hydrogenase (400 μg), and the reductase preparation (17 mg of protein). Freshly prepared sodium sulfite (4 μmol) is added from a side arm after incubation of the flasks for 30 min under hydrogen at 37°. The center well contains 0.1 ml of 10 *M* NaOH. Enzyme activity is calculated from the initial rates of hydrogen utilization. The sulfite reductase-containing extract devoid of the acidic electron carriers is prepared as indicated above (or by ammonium sulfate precipitation).[12] Ferredoxin (10 to 100 μg) is added to detect the stimulation of hydrogen consumption.

Coupling Activity of Ferredoxin in Relevant Sulfate-Reducing Bacteria Metabolic Pathways. It has been shown that DgFdII is capable of mediating electron transfer between cytochrome c_3 and the sulfite reductase system.[14] The other oligomeric form, FdI, serves as an electron carrier in the phosphoroclastic reaction.[76] Because the two different oligomers can accommodate both types of clusters the observations implied that both forms may be biologically active and may interconvert *in vivo*. It was discussed that the different cluster types stabilized by the different oligomers, controlling the redox potentials involved, would be an important factor in determining their differentiated biological activity.[76] Particularly intriguing is the observation that FdII stimulates the phosphoroclastic reaction after a long lag phase when added to crude cell extracts depleted of Fd. It was reported that this stimulation is accompanied by the concomitant appearance of a $g = 1.94$ type EPR signal, suggesting that cluster conversions take place in crude cell extracts.[78]

Both DmbFds were found to be active in the pyruvate phosphoroclastic reaction as well as electron donors to sulfite reductase.[78]

The evolution of hydrogen from pyruvate was observed in a reconstructed system containing hydrogenase, cytochrome c_3, DvMFdI, partially purified pyruvate dehydrogenase, and CoA.[26] The hydrogen sulfite-reducing system can be reconstructed from purified hydrogenase, cytochrome c_3, DvMFdI, and sulfite reductase (desulfoviridin), but the reaction is slow compared with that observed for the crude extract. Ferredoxin II is 40% as effective as FdI as an electron carrier for pyruvate dehydrogenase coupled with hydrogenase and cytochrome c_3. *Desulfovibrio vulgaris* Miyazaki FdI is part of the electron transfer chain of the phosphoroclastic reaction composed of pyruvate dehydrogenase and the hydrogenase-cytochrome c_3 system.[26]

The data on the biological activity of DaFds indicate that the three proteins function with similar effectiveness as electron carriers in the phosphoroclastic reaction and in the hydrogen sulfite reductase system.[23]

[78] J. J. G. Moura, J. Le Gall, and A. V. Xavier, *Eur. J. Biochem.* **141,** 319 (1984).

Final Remarks

The simple iron-sulfur proteins described in this chapter are typical bacterial ferredoxins containing [3Fe-4S], [4Fe-4S], [3Fe-4S] plus [4Fe-4S], and 2 × [4Fe-4S] clusters. They are recognized to participate in the two main pathways of SRB, transferring electrons between carriers and enzymes. Most of their biological significance is extracted by activity stimulation studies of depleted systems. Other electron carriers can replace ferredoxin. Flavodoxin (see [13] in this volume) is a well-known case. Remarkably, the two redox transitions observed in flavodoxin (about −150 and −450 mV, both pH dependent) are in the range of the redox events observed for the [3Fe-4S] and [4Fe-4S] centers. Also, the iron concentration in the culture medium regulates the relative amounts of these two electron carriers, further suggesting their interplay in the physiological pathways. *Desulfovibrio gigas* Fds are unique: two oligomeric forms, having different roles, seem to indicate that the type of structure involved may be a control device.[76] Stimulated by these observations and the elegant work developed in the aconitase system,[41,43] the interconversion pathways between [3Fe-4S] and [4Fe-4S] centers have gained in importance, in addition to the possibility of novel heterometallic cluster synthesis. The observation that nonredox roles could be carried out by these structures, in particular the evidence provided by Thomson[79] that some ferredoxins may function as DNA-binding proteins involved in gene expression, advanced the hypothesis (using AvFdI as a model system) that this process could be regulated by Fe(II) concentration and the redox status of the cell. A protein that interacts with the iron-responsive element of mRNA (IRE-binding protein), which encodes ferritin, is prevented from binding at high Fe(II) concentrations.[80–85] This protein is homologous in some ways to aconitase, and the proposal that cluster formation takes place and regulates the binding process is exciting. The presence of a surface-exposed α helix (a motif present in some ferredoxins) seems important

[79] A. J. Thomson, *FEBS Lett.* **285,** 230 (1991).

[80] T. A. Roualt, C. D. Stout, S. Kaptain, J. B. Harford, and R. D. Kluaiusner, *Cell (Cambridge, Mass.)* **64,** 881 (1991).

[81] M. W. Hentze and P. Argos, *Nucleic Acids Res.* **19,** 1739 (1991).

[82] T. Saito and R. J. P. Williams, *Eur. J. Biochem.* **197,** 43 (1991).

[83] D. J. Haile, T. A. Rouault, J. B. Harford, M. C. Kennedy, G. A. Blondin, H. Beinert, and R. D. Klausner, *Proc. Natl. Acad. Sci. USA* **89,** 11735 (1992).

[84] J. C. Drapier, H. Hirling, J. Wietzerbin, P. Kaldy, and L. C. Kühn, *EMBO J.* **12,** 3643 (1993).

[85] G. Weiss, B. Goosen, W. Doppler, D. Fuchs, K. Pantopolous, G. Werner-Felmayer, H. Wachter, and M. Hentze, *EMBO J.* **12,** 3651 (1993).

for recognition. As a final remark, it is remarkable to notice that the interconversion process is a sequence of redox and coordination events:

$$[3Fe\text{-}4S]^{1+} \xrightarrow[E_0\ (3Fe)]{e} [3Fe\text{-}4S]^{0} \xrightarrow{Fe(2+)} [4Fe\text{-}4S]^{2+} \xrightarrow[E_0\ (4Fe)]{e} [4Fe\text{-}4S]^{1+}$$

The fact that an increase in affinity for iron is detected when [3Fe-4S] is reduced,[69,79] and the order of redox potential observed [E_0(3Fe) > E_0(4Fe)], led to the conclusion that uptake or loss of an iron site from the cluster is related to iron concentration and redox potential. Also, the observation that an additional ligand is required (in general an extra cysteine) when the fourth site of the cube is filled makes the iron atom trigger a redox-linked conformational change. The inorganic element, iron, is a sensor, linked to the redox state of the cell.[79,82]

These considerations motivate the search for the physiological role of the [3Fe-4S] centers and further studies on the significance of the cluster interconversions observed. The SRB ferredoxins and the metabolic pathways involved seem to be an adequate system for gathering answers to these questions.

[13] Flavodoxins

By JACQUES VERVOORT, DIRK HEERING, SJAAK PEELEN, and WILLEM VAN BERKEL

Introduction

Flavodoxins are a group of relatively small monomeric flavoproteins (M_r 15,000–22,000) containing a single molecule of noncovalently bound riboflavin 5′-phosphate (FMN) (Fig. 1). This cofactor functions as a redox center and is involved in electron transfer, which is the main biological function of flavodoxins. Flavodoxins can be found in many prokaryotes and also in some eukaryotic algae. In *Desulfovibrio* spp. the flavodoxins have been suggested to be involved in (1) the electron transport from hydrogenase to the sulfite-reducing system, using molecular hydrogen as the electron donor, or (2) as electron carrier from the pyruvate phosphoroclastic system toward the sulfite-reducing system, using pyruvate as electron donor.[1]

[1] J.-H. Kim and J. M. Akagi, *J. Bacteriol.* **163,** 472 (1985).

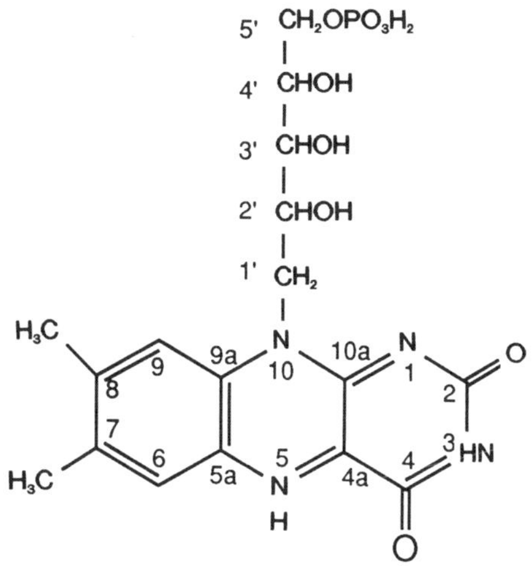

FIG. 1. Structure of riboflavin 5′-phosphate (FMN).

In the presence of iron in the culture medium of sulfate-reducing bacteria, however, the biosynthesis of flavodoxins is suppressed at the expense of ferredoxin.[2,3] Flavodoxins seem to be able to take over the role of ferredoxins as electron donor in the sulfite-reducing system.

Purification

The purification of *Desulfovibrio* flavodoxins can be done according to the procedures of Le Gall and Hatchikian,[4] Moura *et al.*,[5] and Irie *et al.*[6]

Reagents and Materials

DEAE-52 cellulose
Sephadex G-50
Tris-HCl buffer (50 m*M*), pH 7.6

All steps are carried out at 4° unless specified otherwise. The *Desulfovibrio* cells are suspended in 50 m*M* Tris-HCl buffer (about 1 g of cells/ml buffer). The cells are subsequently disrupted by a French press and centrifuged at 30,000 *g* for 30 min. The supernatant is passed over a DEAE-52 cellulose

[2] H. D. Peck, Jr., and J. LeGall, *Philos. Trans. R. Soc. London B.* **298,** 443 (1982).
[3] S. G. Mayhew and M. L. Ludwig, *Enzyme* **12,** 57 (1975).
[4] J. LeGall and E. C. Hatchikian, *C.R. Acad. Sci. Paris* **264,** 2580 (1967).
[5] I. Moura, J. J. G. Moura, M. Brusch, and J. LeGall, *Biochim. Biophys. Acta* **591,** 1 (1980).
[6] K. Irie, K. Kobayashi, M. Kobayashi, and M. Ishimoto, *J. Biochem.* (*Tokyo*) **73,** 353 (1973).

column, equilibrated with 50 m*M* Tris-HCl buffer. The bound flavodoxin is eluted from the column with 0.15 *M* Tris-HCl buffer, 0.4 *M* KCl, pH 7.6. This fraction is desalted over a Sephadex G-50 column (in 50 m*M* Tris-HCl, pH 7.6) or alternatively dialyzed against 50 m*M* Tris-HCl buffer, pH 7.6. The protein is then brought onto a DEAE-52 cellulose column (volume > 150 ml) equilibrated with Tris-HCl buffer (50 m*M*, pH 7.6) and the flavodoxin is eluted from the column by a linear gradient of Tris-HCl (50 m*M* to 0.5 *M*). The flavodoxin fraction elutes at 0.30–0.35 *M* Tris-HCl. As the flavodoxin fraction can still contain other protein components (mainly rubredoxin) this fraction is dialyzed against 50 m*M* Tris-HCl buffer (pH 7.6) and brought onto a second DEAE-52 cellulose column, which is then eluted with a linear gradient of Tris-HCl (0.25 to 0.40 *M*, pH 7.6).

Redox Potentials

Flavodoxins can occur in three redox states (Fig. 2). Of particular physiological importance[3] is the transition between the one electron-re-

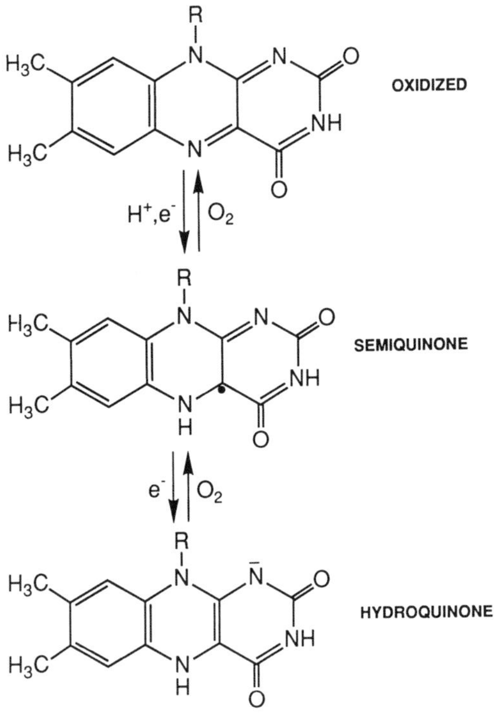

Fig. 2. The structures of oxidized (quinone), one electron-reduced (semiquinone), and two electron-reduced (hydroquinone) flavin.

TABLE I
REDOX POTENTIALS OF FLAVODOXINS (IN MILLIVOLTS)[a]

	E_2	E_1	Ref.
Desulfovibrio vulgaris	−143	−435	7
Clostridium beijerinckii MP	−92	−399	8
Megasphaera elsdenii	−115	−372	9
Anabaena variabilis	−195	−390	10
Anacystis nidulans	−221	−447	11
Azotobacter vinelandii	−165	−458	12
Azotobacter chroococcum	−115	−520	13
Escherichia coli	−240	−410	14
Klebsiella pneumoniae	−158	−412	13
Riboflavin 5′-phosphate (FMN)	−314	−124	15

[a] E_1, the transition from semiquinone to the hydroquinone form; E_2, the transition from the oxidized to the semiquinone form. All values indicated are at pH 7.

duced state and the two electron-reduced state. The redox potentials of this transition are among the lowest observed in nature, ranging from −305 to −520 mV (Table I).[7–15]

Reduction of oxidized flavodoxin to the semiquinone form can easily be achieved by the addition of (chemical) reducing agents. However, owing to the low redox potential of the semiquinone–hydroquinone transition (historically called E_1) complete reduction of *Desulfovibrio vulgaris* flavodoxin is difficult to achieve and depends on the relative redox potentials of the electron acceptor (flavodoxin) and electron donor.

The most widely used electron donor is dithionite. At low concentrations of flavodoxin (<0.5 m*M*) reduction can be accomplished rather easily by the near stoichiometric addition of a freshly prepared dithionite solution in phosphate buffer, pH 7 (100–200 mg/ml). It is advisable to flush the buffer with (oxygen-free) argon before addition of dithionite. However,

[7] G. P. Curley, M. C. Carr, P. A. O'Farell, S. G. Mayhew, and G. Voordouw, *in* "Flavins and Flavoproteins 1990" (B. Curti, S. Ronchi, and G. Zanetti, eds.), p. 429. de Gruyter, Berlin, 1991.

[8] S. G. Mayhew, *Biochim. Biophys. Acta* **235,** 276 (1971).

[9] S. G. Mayhew, G. P. Foust, and V. Massey, *J. Biol. Chem.* **244,** 803 (1969).

[10] M. F. Fillat, G. Sandmann, and C. Gomez-Moreno, *Biochem. Biophys. Res. Commun.* **1040,** 301 (1990).

[11] B. Entsch and R. M. Smillie, *Arch. Biochem. Biophys.* **151,** 378 (1972).

[12] M. F. Taylor, W. H. Boylan, and D. E. Edmondson, *Biochemistry* **29,** 6911 (1990).

[13] J. Deistung and R. N. F. Thorneley, *Biochem. J.* **239,** 69 (1986).

[14] H. Vetter, Jr., and J. Knappe, *Hoppe-Seyer's Z. Physiol. Chem.* **352,** 433 (1971).

[15] R. F. Anderson, *Biochim. Biophys. Acta* **772,** 158 (1983).

at high flavodoxin concentrations (>0.5 m*M*) reduction to a 100% hydroquinone form is virtually impossible to achieve at pH values below 7.5 and at ionic strengths below 100 m*M*.[16] At high concentrations of flavodoxins (>1 m*M*) complete reduction to the hydroquinone form by dithionite (at 200-mg/ml stock solution) can be achieved when keeping the pH of the sample above pH 8. (The flavodoxin sample should be oxygen free, as reduction of the residual oxygen by dithionite gives rise to pH changes that make reduction more difficult.) The contamination of sodium dithionite with small amounts of (bi)sulfite may cause oxidation of flavodoxin.[16] Therefore it is essential to use the highest purity of dithionite available.

Reduction of flavodoxins can also be accomplished by illumination (50- to 100-W tungsten lamp) of an anaerobic solution of flavodoxin in the presence of ethylenediaminetetraacetic acid (EDTA) (50 m*M*) and of a catalytic amount of 5-deazariboflavin. The reduction to the semiquinone form is completed typically within 1 hr. Continued illumination (several hours) yields the hydroquinone. A major advantage over reduction by dithionite is that by using EDTA and light no spectral disturbance occurs in the region of 280–380 nm.[17]

The flavodoxin redox potentials can be determined by controlled titration with dithionite and subsequent measurement of the concentration of the semiquinone radical by electron paramagnetic resonance (EPR).[18,19] Typically a deaerated and argon-flushed buffered solution of 20 to 50 μM flavodoxin and a 10 μM series of mediator dyes is titrated with dithionite in buffer under a constant flow of argon. The potential of the solution is measured between a platinum electrode and a calomel or Ag/AgCl reference electrode. After each addition of dithionite, and after allowing the solution to equilibrate, a sample is anaerobically injected into an EPR tube and frozen in liquid nitrogen. The potential range of the mediators must be sufficient to buffer the potential at any point of the titration. The use of mediators accepting two electrons is recommended to avoid interference of mediator radical signals with the protein semiquinone signal. However, for the titration of E_1, no two-electron mediators are available. Mediators accepting one electron can be used if the concentrations of these mediators are low with respect to the protein and the sharp mediator radical signal can be subtracted from the broader protein signal. The relative concentration of semiquinone as determined from the EPR spectra (corrected for dilution) is plotted versus the potential and fitted to the Nernst equation to determine the redox potentials.

[16] S. G. Mayhew, *Eur. J. Biochem.* **85,** 535 (1978).
[17] V. Massey and P. Hemmerich, *Biochemistry* **17,** 9 (1978).
[18] G. S. Wilson, this series, Vol. 54, p. 396.
[19] P. L. Dutton, this series, Vol. 54, p. 411.

Alternatively, a potentiometric titration can be performed by measuring the concentrations of quinone, semiquinone, and hydroquinone spectrophotometrically. The reduction is performed by adding dithionite (E_2), by hydrogen in the presence of hydrogenase (E_1), or by photoreduction (E_2 and E_1). The potential of the solution can be calculated from the amount of reductant added (or partial hydrogen pressure) or from the redox state of a reference mediator dye, or may be directly measured as described.[17-21]

The quinone–semiquinone potential can also be determined by the spectrophotometric xanthine/xanthine oxidase titration described by Massey.[22] The method uses the low potential of the xanthine/ureate couple (−350 mV at pH 7) together with the ability of xanthine oxidase and flavoproteins to exchange electrons with redox dyes. Neither xanthine nor ureate interfere in the near ultraviolet (UV)–visible range. Typically, a buffered solution of 10 to 20 μM flavodoxin, 200–300 μM xanthine, a redox dye with a potential within 30 mV from the unknown, and 2 μM methyl or benzyl viologen is placed in an anaerobic cuvette, deaerated, and flushed with argon. It is preferable to use a redox dye with a measurable difference in absorption between the oxidized and reduced form at an isosbestic point of the flavodoxin. At $t = 0$ an anaerobic solution of xanthine oxidase is added and spectra are recorded. The amount of xanthine oxidase must be low (10–50 nM) to ensure equilibrium conditions. The complete reduction is typically completed within 1 to 2 hr and about 50–100 spectra are recorded. The potential of the flavodoxin can be determined by plotting its log(ox/red) value versus the log(ox/red) value of the redox dye. The slope of the plot is unity when the two couples involve the same number of electrons. The difference between the potentials can be calculated (using the Nernst equation) from the log(ox/red) value of the flavodoxin at the point where the log(ox/red) value of the dye is zero.

As an alternative for the titrative determination of redox potentials, direct electrochemistry of redox proteins has become feasible over the last 10 years.[23-31] Direct electrochemistry of flavodoxins has been reported

[20] M. Dubourdieu, J. LeGall, and V. Favaudon, *Biochim. Biophys. Acta* **376,** 519 (1975).

[21] S. G. Mayhew and V. Massey, *Biochim. Biophys. Acta* **315,** 181 (1973).

[22] V. Massey, "Proceedings of the Tenth International Symposium on Flavins and Flavoproteins," p. 59. de Gruyter, Berlin, 1991.

[23] C. Van Dijk, J. W. Van Leeuwen, and C. Veeger, *Bioelectrochem. Bioenerg.* **9,** 743 (1982).

[24] W. R. Hagen, *Eur. J. Biochem.* **182,** 523 (1989).

[25] L. H. Guo, H. A. O. Hill, G. A. Lawrence, G. S. Sanghera, and D. J. Hopper, *J. Electroanal. Chem.* **266,** 379 (1989).

[26] P. Bianco, J. Haladjian, A. Manjaoui, and M. Bruschi, *Electrochim. Acta* **33,** 745 (1988).

[27] F. A. Armstrong, J. N. Butt, and A. Sucheta, this series, Vol. 227, p. 479.

[28] F. A. Armstrong, H. A. O. Hill, B. N. Oliver, and N. J. Walton, *J. Am. Chem. Soc.* **106,** 921 (1984).

by Van Dijk,[23] Armstrong, Hill, and co-workers.[27,28] Van Dijk,[23] using differential and normal pulse polarography, determined the quinone/semiquinone (−114 mV, pH 7.4) and semiquinone/hydroquinone (−392 mV, pH 7.4) couples of *M. elsdenii* flavodoxin adsorped at the mercury electrode. At the pyrolytic graphite electrode, Armstrong *et al.*[28] only found the semiquinone/hydroquinone couple of *M. elsdenii* flavodoxin (−318 mV, pH 5.0) using square wave voltammetry and in the presence of $MgCl_2$ and $Cr(NH_3)_6{}^{3+}$. Bianco *et al.*[26] determined the semiquinone/hydroquinone couple of *D. vulgaris* (Hildenborough) flavodoxin (−430 mV, pH 7.6) at the pyrolytic graphite electrode using differential pulse voltammetry and cyclic voltammetry. Cyclic voltammetry of the flavodoxins isolated from *Azotobacter choococcum* at a polished edge-plane graphite electrode has been performed[29] in the presence of the cationic aminoglycoside neomycin. The semiquinone–hydroquinone potentials of two distinct flavodoxins were found (−305 and −520 mV at pH 7.4). Our group has investigated the electrochemistry of *D. vulgaris* Hildenborough.[30,31] In the classic three-electrode setup and semiinfinite diffusion, a bulk solution of about 1 ml and a protein concentration of 100 μM is required. In the setup developed in our laboratory the volume of the bulk is reduced to less than 20 μl with full retention of semiinfinite diffusion down to a scan rate of 1 mV/sec.[24]

The working electrode is a dismountable inverted disk of glassy carbon (15-mm diameter and 2-mm height, type V25; obtained from Le Carbon Loraine, Paris). Prior to each electrochemical measurement the disk is polished firmly with polishing cloth with 6-μm diamond lapping compound (Engis, Ltd., Kent, England), rinsed thoroughly with water, and dried. To activate the electrode it is exposed for 30 sec to a methane flame from a Bunsen burner. The polished working surface is never in direct contact with the flame, and if any inhomogeneity is detected on the surface the polishing and glowing are repeated. The counterelectrode is a microplatinum electrode (P-1312; Radiometer) and the reference electrode is a saturated calomel electrode (K-401; Radiometer). After mounting the electrodes the cell is flushed with wet argon. A deaerated, argon-flushed sample is then transferred with a gas-tight Hamilton syringe to the tip of the reference electrode and the working electrode holder is pushed up until the shape of the droplet is approximately cylindrical. The electrodes are connected to an Autolab 10 potentiostat (Eco Chemie) controlled by GPES software (version 2.0) (Eco Chemie) on a PC.

[29] S. Bagby, P. D. Barker, H. A. O. Hill, G. S. Sanghera, B. Dunbar, G. A. Ashby, R. R. Eady, and R. N. F. Thorneley, *Biochem. J.* **277,** 313 (1991).

[30] H. A. Heering and W. R. Hagen, *J. Inorg. Biochem.* **51,** 25 (1993).

[31] H. A. Heering and W. R. Hagen, unpublished results (1994).

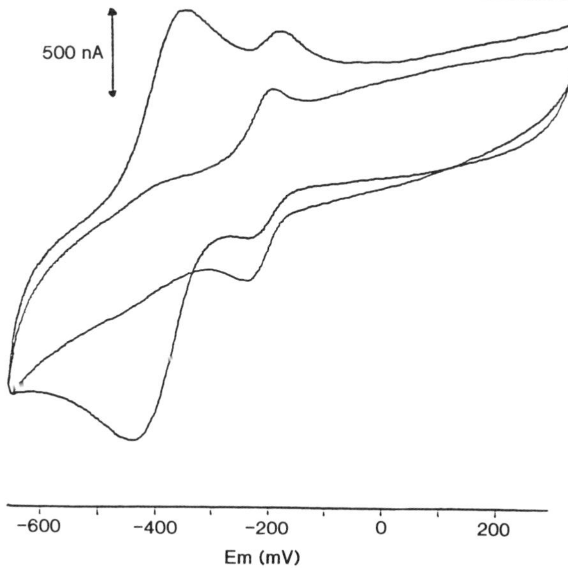

FIG. 3. Staircase cyclic voltammogram (20 mV/sec) of 6 μl of 0.14 m*M D. vulgaris* flavodoxin in 20 m*M* potassium phosphate, pH 7.0: before and after addition of 0.5 μl of 50 m*M* neomycin.

Staircase cyclic voltammetry (SCV) was done with steps of 2.44 mV. Differential pulse polarography (DPP) was done in both directions (after a 60-sec equilibration at the starting potential) with steps of 2.44 mV and 250 msec, modulated by pulses of 20 mV and 100 msec (Fig. 3). The experiments were performed at a temperature of 22 ± 1° and the potentials have been recalculated with respect to the normal hydrogen electrode (NHE), using the potential of −246 mV (NHE) of the saturated calomel electrode (SCE). The promotor neomycin B (Sigma, St. Louis, MO) is added from a 50 m*M* solution, titrated to pH 7.0 with NaOH. The pH dependence is measured with SCV and DPP. The SCV measurements are performed on droplets of 5 μl of 0.11 m*M* flavodoxin and 3.3 m*M* neomycin in 20 m*M* citrate, 20 m*M* bistrispropane, and 20 m*M* CAPS, titrated with HCl or NaOH and adjusted to ionic strength $\mu = 0.2$ with NaCl. The DPP measurements were performed on droplets of 5 μl 0.13 m*M* flavodoxin and 3 m*M* neomycin in the same buffer. In all experiments the fully oxidized flavodoxin (as determined spectroscopically) was used.

As in the experiments of Armstrong *et al.*[27,28] and of Bagby *et al.*,[29] the first reduction (quinone to semiquinone) is not detectable at the electrode and the second reduction step is detected only after addition of a cationic promotor such as neomycin. The cations are believed to form a

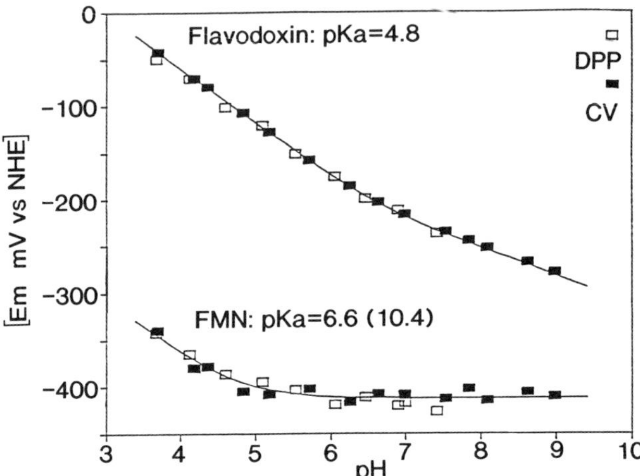

FIG. 4. pH dependence of the midpoint potentials of FMN (adsorbed at the electrode) and of *D. vulgaris* flavodoxin E_1 in the presence of neomycin and at constant ionic strength (μ = 0.2). (□) DPP; (■) CV.

bridge between the negative charges of both the electrode and flavodoxin (pI = 3.6). Without neomycin (at pH 7.0) only one redox couple is found at a potential of −217 mV (NHE) with a peak current proportional to the scan rate. This response is caused by FMN, dissociated from the protein and adsorbed onto the electrode. The potential and the pK_a of 6.6 are equal to those measured for free FMN in solution. After addition of neomycin one additional response is observed with a peak current proportional to the square root of the scan rate and the characteristics of a one-electron transition. The potential of −409 mV (NHE) is near the chemically determined potential of the semiquinone-to-hydroquinone reduction.[20,32] The observed pK_a of 4.8 is, however, much lower than the reported value of 6.6[20] to 6.8.[32] This difference between the pK_a at the electrode and the one reported for the chemically reduced flavodoxin might be caused by a conformational change near the electrode. The measured pK_a of 4.8 can be due to the altered environment of the amino acid residue responsible for the redox-linked pK_a. Alternatively, the residue might be blocked or no longer in the vicinity of the isoalloxazine. The measured pK_a of 4.8 can be due to protonation of the isoalloxazine at N-1. See Fig. 4.

The midpoint potential of the quinone to semiquinone reduction (as determined by EPR titration with dithionite) is −113 mV (NHE) at pH 7.0. This value is in agreement with reported potentials.[20,32] The absence of the quinone to semiquinone couple in the voltammograms can be ex-

[32] G. P. Curley, M. C. Carr, S. G. Mayhew, and G. Voordouw, *Eur. J. Biochem.* **202,** 1091 (1991).

plained[30,31] by the known fast comproportionation mechanism of one fully reduced and one fully oxidized flavodoxin to two semiquinone flavodoxin molecules[20] in combination with a very slow (virtually not occurring) reduction of the oxidized to the semiquinone form.[21] By consequence, only a few molecules (catalytic amount) of semiquinone are sufficient to generate equilibrium concentrations of semiquinone and hydroquinone near the electrode surface and the development of the voltammogram without a visible first redox couple. A catalytic amount of free FMN may also act as a mediator to generate the first molecules of semiquinone.

Amino Acid Sequences and Structure

The amino acid sequences of several flavodoxins are known (Table II). The homology between these sequences ranges from 20 to 40%. As alignment solely on the basis of sequences is known to be difficult, structural alignment was investigated using the structures of the flavodoxins from *Chondrus crispus*, *D. vulgaris*, and *Clostridium beijerinckii*. The structurally conserved regions were then determined. On the basis of these conserved regions, the alignment of the other flavodoxin sequences was performed using the Clustal V alignment program.[33] Conserved residues are marked in bold in Table II.[34–46] As is evident from this alignment only five residues are totally conserved. The first four of these residues are embedded close to each other in a hydrophobic pocket in the protein. These four residues may either be of importance for electron transfer processes or may be important in the folding process of flavodoxins. As can be inferred from the amino acid sequences, flavodoxins are acidic proteins. Their net negative charges ranges from −10 to −20, and their isoelectric points are situated around pH 4. The low isoelectric points, high

[33] D. G. Higgins and P. M. Sharp, *Gene* **73,** 237 (1988).

[34] M. Dubourdieu and J. L. Fox, *J. Biol. Chem.* **252,** 1453 (1977).

[35] L. R. Helms and R. P. Swenson, *Biochim. Biophys. Acta* **1089,** 417 (1991).

[36] L. R. Helms, G. D. Krey, and R. P. Swenson, *Biochem. Biophys. Res. Commun.* **168,** 809 (1990).

[37] M. Tanaka, M. Haniu, K. T. Yasunobu, and S. G. Mayhew, *J. Biol. Chem.* **249,** 4393 (1974).

[38] D. Santangelo, D. T. Jones, and D. R. Woods, *J. Bacteriol.* **173,** 1088 (1991).

[39] M. Tanaka, M. Haniu, K. T. Yasunobu, S. G. Mayhew, and V. Massey, *J. Biol. Chem.* **249,** 4397 (1974).

[40] S. Wakabayashi, T. Kimura, K. Fukuyama, H. Matsubara, and L. J. Rogers, *Biochem. J.* **263,** 981 (1989).

[41] K. G. Leonhardt and N. A. Straus, *Nucleic Acids Res.* **17,** 4384 (1989).

[42] D. E. Laudebach, M. E. Reith, and N. A. Straus, *J. Bacteriol.* **170,** 258 (1988).

[43] L. T. Bennett, M. R. Jacobson, and D. R. Dean, *J. Biol. Chem.* **263,** 1364 (1988).

[44] C. Osborne, L. M. Chen, and R. G. Matthews, *J. Bacteriol.* **173,** 1729 (1991).

[45] W. Arnold, A. Rump, W. Klipp, U. B. Priefer, and A. J. Puehler, *J. Mol. Biol.* **203,** 715 (1988).

[46] Y. Jouanneau, P. Richard, and C. Grabau, *Nucleic Acids Res.* **18,** 5284 (1990).

TABLE II
SEQUENCE ALIGNMENT OF FLAVODOXINS[a]

Species	Sequence	
D. vulgaris	MPK-ALIVYGSTTGN**T**EYT**A**ETIARELADAG-YEVDSRDAASVEAG---GLFEGFDLVLL	(55)
D. desulfuricans	MSK-VLIVFGSSTGN**T**ESI**A**QKLEELIAAGGH-EVTLLNAADASAE---NLADGYDAVLF	(55)
D. salexigens	MSKSL-IVYGSTTGN**T**ETA**A**EYVAEAFENK-EIDVELKNVTDVSVA---DLGNGYDIVLF	(55)
C. beijerinckii	----MKIVYWSGTGN**T**EKM**A**ELIAKGIIESG-KDVNTINVSDVNID---E-LLNEDILIL	(51)
C. acetobutylicum	M--KISILYSSKTGK**T**ERV**A**KLIEEGVKRSGNIEVKTMNLDAV-DK---KFLQESEGIIF	(54)
M. elsdenii	M---VEIVYWSGTGN**T**EAM**A**NEIEAAVKAAGA-DVESVRFEDTNVD---DVASK-DVILL	(52)
C. crispus	--K-IGIFFSTSTGN**T**TEV**A**DFIGKTLG--A-KADAPIDVDDVTDP---QALKDYDLLFL	(51)
A. variabilis	MSKKIGLFYGTQTGK**T**ESV**A**EIIRDEFGNDV-VTLHDVSQAEVTD------LNDYQYLII	(53)
A. nidulans	MAK-IGLFYGTQTGV**T**QTI**A**ESIQQEFGGES-IV-DLNDIANADAS---D-LNAYDYLII	(53)
A. vinelandii	MAK-IGLFFGSNTGK**T**RKV**A**KSIKKRFDDET-MS-DALNVNRVSAE---D-FAQYQFLIL	(53)
E. coli	MAIT-GIFFGSDTGN**T**ENI**A**KMIQKQLGK----DVADVHDIAKSSK---EDLEAYDILLL	(52)
K. pneumoniae	MAN-IGIFFGTDTGK**T**RKI**A**KMIHKQL---GELADAPVNINRTTLD---DFM-AYPVLLL	(52)
R. capsulatus	L----LIFYVSAYAA**T**AHV**A**QAIHDGAAESPDVRVSLFDLEGGEITPFLDLIEEADGIAL	(56)
D. vulgaris	**G**CSTWGD------DSIELQDDFIPLFDSL-EETGAQGRKVAC**F**GCGDS-SYE-YFCGA-V	(105)
D. desulfuricans	**G**CSAW------GMEDLEMQDDFLSLFEE-FNRIGLAGRKVAA**F**ASGDQ-EY-EHFCGA-V	(105)
D. salexigens	**G**CSTWGEEEI------ELQDDFIPLYDS-LENADLKGKKVSV**F**GCGDS-DYT-YFCGA-V	(105)
C. beijerinckii	**G**CSAMGD------EVLEES-EFEPFIEEIS--TKISGKKVAL**F**GSYG-W--G--DGKW-M	(96)
C. acetobutylicum	**G**TPTYYA------NISWEMKKWI----DESSEFNLEGKLGAA**F**STANSIAGGSDI--A-L	(101)
M. elsdenii	**G**CPAMGSEEL-------EDSVVEPFFTDLAPK--LKGKKVGL**F**GSYG-W--G--SGEW-M	(97)
C. crispus	**G**APTWNTGA----DTERSGTSWDEFLYDKLPEVDMKDLPVAI**F**GLGDAEGYPDNFCDA-I	(106)
A. variabilis	**G**CPTWNIGEL--------QSDWEGLYSE-LDDVDFNGKLVAY**F**GTGDQIGYADNFQDA-I	(103)
A. nidulans	**G**CPTWNVGEL--------QSDWEGIYDD-LDSVNFQGKKVAY**F**GAGDQVGYSDNFQDA-M	(103)
A. vinelandii	**G**TPTLGEGELPGLSSDCENESWEEFL-PKIEGLDFSGKTVAL**F**GLGDQVGYPENYLDA-L	(111)
E. coli	**G**IPTWYYGE--------AQCDWDDFF-PTLEEIDFNGKLVAL**F**GCGDQEDYAEYFCDA-L	(102)
K. pneumoniae	**G**TPTLGDGQLPGLEAGCESESWSEFISG-LDDASLKGKTVAL**F**GLGDQRGYPDNFVSG-M	(110)
R. capsulatus	**G**TPTINGDAVRTI--------WEML--AALVDIETRGKLGAA**F**GSYGWSGEAVRLVETRL	(106)

Species	Sequence	
D. vulgaris	DAIEEKLKNL**G**AEIVQD---------------------GLRIDGDP--RAARDDIVGWAH	(142)
D. desulfuricans	PAIEERAKEL**G**ATIIAE---------------------GLKMEGDASNDPEA--VASFAE	(142)
D. salexigens	DAIEEKLEKM**G**AVVIGD---------------------SLKIDGDPER----DEIVSWGS	(140)
C. beijerinckii	RDFEERMNGY**G**CVVVET---------------------PLIVQMEP--DEAEQDCIEFGK	(133)
C. acetobutylicum	LTILNHLMVK**G**MLVYS----GGVAFGKPKTHLG---YV--HINEIQENEDENARI--FGE	(150)
M. elsdenii	DAWKQRTEDT**G**ATVIG----------------------TAIVNEMPDNAPECKELGEAAA	(135)
C. crispus	EEIHDCFAKQ**G**AKPVGFSNPDDYDYEESKSVRDGK-FLGLPLDMVNDQIPMEKRVAGWVE	(165)
A. variabilis	GILEEKISQR**G**GKTVGYWSTDGYDFNDSKALRNGK-FVGLALDEDNQSDLTDDRIKSWVA	(162)
A. nidulans	GILEEKISSL**G**SQTVGYWPIEGYDFNESKAVRNNQ-FVGLAIDEDNQPDLTKNRIKTWVS	(162)
A. vinelandii	GELYSFFKDR**G**AKIVGSWSTDGYEFESSEAVVDGK-FVGLALDLDNQSGKTDERVAAWLA	(170)
E. coli	GTIRDIIEPR**G**ATIVGHWPTAGYHFEASKGLADDDHFVGLAIDEDRQPELTAERVEKWVK	(162)
K. pneumoniae	RPLFDALSAR**G**AQMIGSWPNEGYEFSASSAL-EGDRFVGLVLDQDNQFDQTEARLASWLE	(169)
R. capsulatus	QGLKMRLPEP**G**LRV--KLHPSAAELEEGRAF--GRRLA--------------DHLTGRAA	(148)

Species	Sequence	
D. vulgaris	DVRGAI--------	(148)
D. desulfuricans	DVLKQ--------L	(148)
D. salexigens	GIADKI--------	(146)
C. beijerinckii	KIANI---------	(138)
C. acetobutylicum	RIANKVKQ----IF	(160)
M. elsdenii	KA------------	(137)
C. crispus	AVVSETGV------	(173)
A. variabilis	QLKSEFG------L	(170)
A. nidulans	QLKSEFG------L	(170)
A. vinelandii	QIAPEFGL----SL	(180)
E. coli	QISEELHLDEILNA	(176)
K. pneumoniae	EIKRTV-------L	(176)
R. capsulatus	P--REVDFAEIAAR	(160)

[a] References: *Desulfovibrio vulgaris*,[34] *Desulfovibrio desulfuricans*,[35] *Desulfovibrio salexigens*,[36] *Clostridium beijerinckii*,[37] *Clostridium acetobutylicum*,[38] *Megasphaera elsdenii*,[39] *Chondrus crispus*,[40] *Anabaena variabilis*,[41] *Anacystis nidulans*,[42] *Azotobacter vinelandii*,[43] *Escherichia coli*,[44] *Klebsiella pneumoniae*,[45] *Rhodobacter capsulatus*.[46]

stability, and high charge density make it possible to prepare flavodoxin solutions of up to 400 mg/ml. At these high concentrations no aggregation phenomena can be observed as is evident from ^{1}H nuclear magnetic resonance (NMR) studies at these concentrations.[47,48] This property, in combination with their low molecular weight, makes flavodoxins good candidates for structure determination using multidimensional NMR techniques. From such multidimensional NMR studies it became clear that the solution and crystal structures of *D. vulgaris* flavodoxin are virtually identical.[48]

The tertiary structures of all flavodoxins share a common polypeptide fold consisting of a central parallel β sheet with five strands surrounded on both sites by helices. The FMN is strongly bound via H bonds on one side of the protein with the isoalloxazine moiety exposed to the solvent. The loop regions that surround the isoalloxazine ring (residues 59–64 and 94–104 in *D. vulgaris* flavodoxin) have hardly any conserved sequences. Nevertheless, in general the isoalloxazine ring seems to be sandwiched between two aromatic residues (W60 and Y98 in *D. vulgaris* flavodoxin).

In some flavodoxins one of the two aromatic residues is replaced by a hydrophobic residue (M, L, or I). It appears, on comparison of the known structures and amino acid sequences (Table II), that the isoalloxazine ring needs to be shielded from the solvent by these residues for the establishment of the low redox potentials.[49] The reader is referred to detailed reviews on the three-dimensional structure characteristics of flavodoxins.[50]

Apoflavodoxin Preparation

Apoflavodoxin can be prepared according to the following procedure (this procedure contains slight modifications of the original procedure of Wassink and Mayhew[51]).

Procedure 1. Cold trichloroacetic acid (TCA)($T = 4°$) at a concentration of 40% (w/v) is slowly added to a cold flavodoxin solution (concentration 0.2–0.4 m*M*, 4°) in 0.1 *M* phosphate buffer (pH 7–8)–0.3 m*M* EDTA, to give a final concentration of 5% TCA. The flavodoxin solution should preferably be shielded from light. After the addition of TCA the solution

[47] S. S. Wijmenga and C. P. M. van Mierlo, *Eur. J. Biochem.* **195,** 807 (1991).

[48] S. Peelen, J. Vervoort, and J. LeGall, unpublished results (1994).

[49] J. Vervoort, *Curr. Opin. Struct. Biol.* **1,** 889 (1991).

[50] M. L. Ludwig and C. L. Luschinsky, *in* "Chemistry and Biochemistry of Flavoenzymes" (F. Müller, ed.), Vol. 3, p. 427. CRC Press, Boca Raton, Florida, 1991.

[51] J. H. Wassink and S. G. Mayhew, *Anal. Biochem.* **68,** 609 (1975).

is centrifuged at 10,000 *g* for 10 min. The white precipitate can be dissolved in 0.1 *M* potassium phosphate–0.3 m*M* EDTA (pH 7–8). As the procedure does not give 100% apoprotein in one run, it is advisable, when necessary, to repeat the procedure.

The apoprotein prepared in this way is stable for a long period: more than 1 month at 4° up to 1 week at room temperature.

Procedure 2. A milder procedure for apoflavodoxin preparation is as follows: Dialyze 0.2–0.4 m*M* flavodoxin in 0.1 *M* potassium phosphate–0.3 m*M* EDTA (pH 7–8) four times against 250 ml of 2 *M* KBr in 0.1 *M* sodium acetate buffer, pH 3.9, with 0.3 m*M* EDTA (total time, 48 hr). The precipitate can be redissolved in 0.1 *M* potassium phosphate (pH 7.0)–0.3 m*M* EDTA. The disadvantage of this procedure is the long time span needed to prepare the apoprotein.

The apopreparation procedure does not lead to irreversible modifications of the apoprotein. ^{31}P and ^{1}H NMR studies of flavodoxins from *D. vulgaris* and also from *M. elsdenii*, before (native) and after apopreparation followed by reconstitution with FMN, give identical spectra. From this the firm conclusion can be drawn that the recombined protein (apoprotein recombined with FMN) and the native protein have the same conformation.

Reconstitution of Holoflavodoxin from Apoprotein and Riboflavin 5′-Phosphate

A wealth of information is available about the reconstitution of holoflavodoxin from its constituents, the apoprotein and prosthetic group. For details about the thermodynamics and kinetics of flavin binding the reader is referred to the review of Mayhew and Ludwig.[3] Apoflavodoxins from *Clostridium* species and *M. elsdenii* are specific for binding of flavins at the FMN level.[52,53] The apoflavodoxins from *Azotobacter vinelandii* and *D. vulgaris* form tight complexes not only with FMN but also with riboflavin and lumiflavin analogs.[54,55] Apoflavodoxins tightly bind FMN at neutral pH. For both *M. elsdenii* and *D. vulgaris* flavodoxin, the dissociation constant (K_d) of the complex is in the range of 10^{-10} *M*.[52,56] Stabilization of the interaction between apoprotein and prosthetic group is de-

[52] S. G. Mayhew, *Biochim. Biophys. Acta* **235,** 289 (1971).

[53] J. A. D'Anna and G. Tollin, *Biochemistry* **11,** 1073 (1972).

[54] D. E. Edmondson and G. Tollin, *Biochemistry* **10,** 113 (1971).

[55] D. E. Edmondson and G. Tollin, *Biochemistry* **10,** 133 (1971).

[56] J. Vervoort, W. J. H. van Berkel, S. G. Mayhew, F. Müller, A. Bacher, P. Nielsen, and J. Legall, *Eur. J. Biochem.* **161,** 749 (1986).

creased at pH values below 5 and is also dependent on the ionic strength and type of anions present in solution.[57] Binding of FMN at neutral pH is rapid and results in almost complete quenching of flavin fluorescence.[52] This property has been used with the apoflavodoxin from *M. elsdenii* to assay for FMN and FAD in mixtures and to analyze commercial FMN preparations for their content.[51]

Commercial FMN preparations contain 25–30% of fluorescent impurities.[51,58] Synthesis of FMN by chemical phosphorylation of riboflavin yields 4′-FMN as a main contaminant.[59] By ^{31}P NMR it was shown that commercial FMN also contains considerable amounts of the 3′ and 2′ isomers and other phosphorus-containing compounds.[60] The impurities (including riboflavin) and 5′-FMN are conveniently separated by reversed-phase high-performance liquid chromatography (HPLC)[61] and preparative amounts of each compound can be obtained in pure form.[62,63] On the basis of various chemical properties and enzymatic hydrolysis, the structures of the phosphorylated impurities have been assigned to[61,62] riboflavin 4′-phosphate (4′-FMN), riboflavin 3′-phosphate (3′-FMN), riboflavin 2′-phosphate (2′-FMN), riboflavin 4′,5′-bisphosphate (4′,5′-FBP), riboflavin 3′,5′-bisphosphate (3′,5′-FBP), riboflavin 3′,4′-bisphosphate (3′,4′-FBP), riboflavin 2′,5′-bisphosphate (2′,5′-FBP), riboflavin 2′,4′-bisphosphate (2′,4′-FBP) and riboflavin 4′,5′-cyclophosphate (4′,5′-FCP), respectively. The occurrence of these flavin isomers is explained by the acid-catalyzed migration of the phosphoric acid groups in both the riboflavin mono- and bisphosphates.[59,62] At neutral pH, the isomerization reactions are slow as compared to hydrolysis of the phosphoester bond.[62]

For a long time it was thought that apoflavodoxin is highly specific for 5′-FMN.[51] Separation of the isomeric riboflavin mono- and bisphosphates, however, revealed that apoflavodoxin from *M. elsdenii* also binds 3′,5′-FBP tightly.[61] This compound makes up 2% of the total amount of commercial FMN.[62] *Megasphaera elsdenii* apoflavodoxin reconstituted with 3′,5′-FBP is fully active as an electron carrier by transferring reduction equivalents from H_2 via hydrogenase to metronidazole.[61] This suggested that the additional phosphate group does not influence the properties of the complex. A more thorough kinetic, thermodynamic, and spectral analysis of the complexes of 3′,5′-FBP and the apoflavodoxins from *M. elsdenii*

[57] R. Gast, B. E. Valk, F. Müller, S. G. Mayhew, and C. Veeger, *Biochim. Biophys. Acta* **446,** 463 (1976).

[58] V. Massey and B. E. P. Swoboda, *Biochem. Z.* **339,** 474 (1963).

[59] G. Scola-Nagelschneider and P. Hemmerich, *Eur. J. Biochem.* **66,** 567 (1976).

[60] C. T. W. Moonen and F. Müller, *Biochemistry* **21,** 408 (1982).

[61] P. Nielsen, P. Rauschenbach, and A. Bacher, *Anal. Biochem.* **130,** 359 (1983).

[62] P. Nielsen, J. Harksen, and A. Bacher, *Eur. J. Biochem.* **152,** 465 (1985).

[63] P. Nielsen, P. Rauschenbacher, and A. Bacher, this series, Vol. **122,** p. 209.

and *D. vulgaris* confirmed this idea.[56] Both apoflavodoxins bind 3′,5′-FBP somewhat more weakly than 5′-FMN owing to slower association rate constants. The redox potentials of the artificial complex, however, are similar to the values reported for the native protein. ^{31}P NMR experiments revealed that the 3′-phosphate group is accessible for Mn^{2+} and indicates that this group is located close to the protein surface.[56] This is in sharp contrast with the 5′-phosphate group, which is dianionic[60] and buried in both proteins. The 5′-phosphate group therefore determines the specificity of flavin binding, whereas introduction of the 3′-phosphate group only slightly influences the protein conformation. The latter conclusion is in full accordance with results obtained from the crystal structure of *D. vulgaris* flavodoxin as well as ^{1}H NMR results, which show that the 3′-OH of the ribityl side chain is oriented toward bulk solvent.[48,64]

[64] W. Watt, A. Tulinsky, R. P. Swenson, and K. D. Wautenpaugh, *J. Mol. Biol.* **218,** 195 (1991).

[14] Rubredoxin in Crystalline State

By LARRY C. SIEKER, RONALD E. STENKAMP, and JEAN LEGALL

General Remarks

Rubredoxin (Rd) is one of the simplest of iron proteins and has been found, thus far, only in certain microorganisms.[1] The initial report and characterization of a rubredoxin was done by Lovenberg and Sobel in 1965.[2] Rubredoxins are composed of 45 to 54 amino acid residues with molecular weights ranging from 5000 to 6000 and contain 1 iron atom liganded by 4 cysteine residues. The iron center can be reversibly reduced at a redox potential near 0 mV.[1]

Although many rubredoxins have been detected and isolated from a variety of bacteria, only 13 of the rubredoxins have had amino acid sequences determined. Figure 1 shows the amino acid sequence alignment of the 13 rubredoxins. Because this chapter is primarily directed to the sulfate-reducing bacteria we have chosen to divide these rubredoxins into three categories. Figure 1a lists the Rds from the sulfate-reducing *Desulfovibrio* species, Fig. 1b shows the Rds from a mixed assortment of bacteria, and Fig. 1c contains the thermophilic Rds.

[1] T. G. Spiro, "Iron-Sulfur Proteins." Wiley (Interscience), New York, 1982.

[2] W. Lovenberg and B. E. Sobel, *Proc. Natl. Acad. Sci. U.S.A.* **54,** 193 (1965).

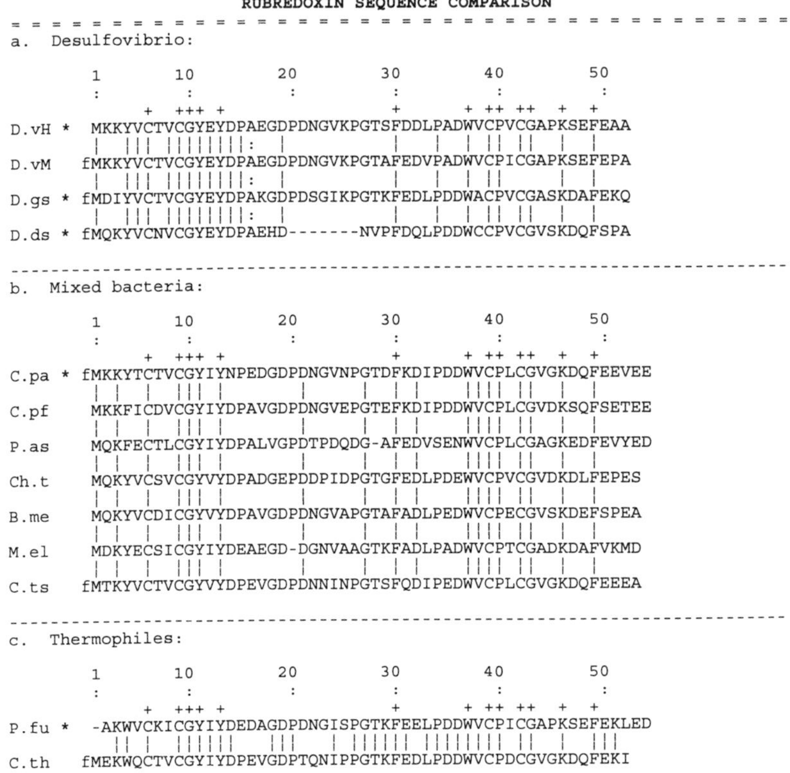

FIG. 1. *, Crystal structures; |, conserved residues among the various types of rubredoxins; +, residues that are strictly conserved throughout the 13 rubredoxins. (a) *Desulfovibrio*: D.vH, *Desulfovibrio vulgaris* Hildenborough [M. Bruschi, *Biochem. Biophys. Acta* **434,** 4 (1976); G. Voordouw, *Gene* **67,** 75 (1988)]; D.vM, *Desulfovibrio vulgaris* Miyazaki [F. Shimizu, M. Ogata, T. Yagi, S. Wakabayashi, and H. Matsubara, *Biochemie* **71,** 1171 (1989)]; D.gs, *Desulfovibrio gigas* [M. Bruschi, *Biochem. Biophys. Res. Commun.* **70,** 615 (1976)]; D.ds, *Desulfovibrio desulfuricans* [S. Hormel, K. A. Walsh, B. C. Prickril, K. Titani, and J. Le Gall, *FEBS Lett.* **201,** 147 (1986)]. (b) Mixed bacteria: C.pa, *Clostridium pasteurianum* [K. T. Yasunobu and M. Tanaka, *in* "Iron-Sulfur Proteins" (W. Lovenberg, ed.), p. 27. Academic Press, New York, 1973; I. Mathieu, J. Meyer, and J.-M. Moulis, *Biochem. J.* **285,** 255 (1992)]; C.pf, *Clostridium perfringens* [Y. Seki, S. Seki, M. Satoh, A. Ikeda, and M. Ishimoto, *J. Biochem.* (*Tokyo*) **106,** 336 (1989)]; P.as, *Peptostreptococcus asaccharolyticus* (formerly *Peptococcus aerogenes*) [H. Bachmeyer, A. M. Bensen, K. T. Yasunobu, W. T. Garrard, and H. R. Whitely, *Biochemistry* **7,** 986 (1967)]; Ch.t, *Chlorobium thiosulfatophilum* [K. J. Woolley and T. E. Meyer, *Eur. J. Biochem.* **163,** 161 (1987)]; B.me, *Butyribacterium methylotrophicum* [K. Saeki, Y. Yao, S. Wakabayashi, G. J. Shen, J. G. Zeikus, and H. Matsubara, *J. Biochem.* (*Tokyo*) **106,** 656 (1989)]; M.el, *Megasphaera elsdenii* [H. Bachmeyer, K. T. Yasunobu, J. L. Peel, and S. Mayhew, *J. Biol. Chem.* **243,** 1022 (1968)]; C.st, *Clostridium sticklandii* [sequence reported by I. Mathieu, J. Meyer, and J.-M. Moulis, *Biochem. J.* **285,** 285 (1992)]. (c) Thermophiles: P.fu, *Pyrococcus furiosus* [P. R. Blake,

Physiological Role

The redox potentials of Rds isolated from sulfate-reducing bacteria are relatively high (−5 to 0 mV)[3] whereas dissimulatory sulfate reduction requires electrons from −400 to −200 mV.[4,5] Consequently, the search for the electron transfer reaction(s) catalyzed by Rds is a difficult task. Rubredoxins have been proposed to accept electrons from carbon monoxide dehydrogenase (EC 1.2.99.2) in other anaerobes such as *Clostridium thermoaceticum* or *Acetobacter woodii*.[6] For the sulfate-reducing bacteria, such an activity had not been detected, nor had electron donors been clearly linked to sulfate respiration. It was shown several years ago that the Rd from *Desulfovibrio gigas* is reduced by the tetraheme cytochrome c_3 from the same organism[7] in the presence of hydrogenase. A hypothetical model of the complex formed between these two proteins from *Desulfovibrio vulgaris* Hildenborough has been proposed utilizing computer graphic modeling and nuclear magnetic resonance (NMR) spectroscopy.[8] According to this model, the iron atom of Rd is in close proximity to heme 1 (the most positive heme) of the cytochrome. More importantly, *D. gigas* cells that have been grown on a lactate-sulfate medium contain an NADH-rubredoxin oxidoreductase (NRO).[9] This enzyme is composed of two subunits of 27 and 32 kDa, respectively[10] and contains both FAD and riboflavin 5′-phosphate (FMN). It induces the specific reduction of *D. gigas* Rd. Rubredoxins from other *Desulfovibrio* species show low reaction rates with this enzyme. Such a specificity can be explained by the differences that exist between Rds in terms of the external residues.

[3] I. Moura, J. J. G. Moura, H. M. Santos, A. V. Xavier, and J. Le Gall, *FEBS Lett.* **107,** 419 (1979).

[4] J. Le Gall, J. J. G. Moura, H. D. Peck, Jr., and A. V. Xavier, *in* "Iron-Sulfur Proteins" (T. G. Spiro, ed.), p. 177. Wiley, New York, 1982.

[5] J. Le Gall, D. V. DerVartanian, and H. D. Peck, Jr., *Curr. Top. Bioenerg.* **9,** 237 (1979).

[6] S. W. Ragsdale, L. G. Ljungdahl, and D. V. DerVartanian, *J. Bacteriol.* **155,** 1224 (1983).

[7] G. R. Bell, J.-P. Lee, H. D. Peck, Jr., and J. Le Gall, *Biochimie* **60,** 315 (1978).

[8] D. E. Stewart, J. Le Gall, I. Moura, J. J. G. Moura, H. D. Peck, Jr., A. V. Xavier, P. K. Weiner, and J. E. Wampler, *Eur. J. Biochem.* **185,** 695 (1989).

[9] J. Le Gall, *Anal. Inst. Pasteur (Paris)* **114,** 109 (1968).

[10] L. Chen, M.-Y. Liu, J. Le Gall, P. Fareleira, H. Santos, and A. V. Xavier, *Eur. J. Biochem.* **216,** 443 (1993).

J.-B. Park, F. O. Bryant, A. Shigetoshi, J. K. Magnuson, E. Eculston, J. B. Howard, M. F. Summers, and M. W. W. Adams, *Biochemistry* **30,** 10885 (1991); P. R. Blake, J.-B. Park, F. O. Bryant, S. Aono, J. K. Magnuson, E. Eccleston, J. B. Howard, M. F. Summers, and M. W. W. Adams, *Biochemistry* **30,** 10885 (1991)]; C.th, *Clostridium thermosaccharolyticum* [J. Meyer, J. Gagnon, L. C. Sieker, A. Van Dorsselaer, and J.-M. Moulis, *Biochem. J.* **271,** 839 (1990)].

The first report of a physiological electron acceptor for *D. gigas* Rd has been published.[11] It is a rubredoxin-oxygen oxidoreductase (ROO), a homodimer of 43 kDa per monomer. The protein is a flavohemoprotein because it contains both FAD and a new type of heme group. Because the product of oxygen reduction by this protein is water, the following electron chain scheme has been proposed:

$$
\begin{array}{ccccl}
\text{NADH} & \longrightarrow & \text{NRO} & \xrightarrow{2e\ +\ O_2} & H_2O_2\ \text{(slow)} \\
 & & \downarrow & & \\
 & & \text{Rd} \longrightarrow \text{ROO} & \xrightarrow{4e\ +\ O_2} & H_2O\ \text{(fast)}
\end{array}
$$

It is proposed that this electron transfer pathway is sufficient to explain the observation that ATP is formed from the degradation of polyglucose in the presence of oxygen.[12]

The similarity between the utilization of reduced pyridine nucleotides by *D. gigas* and the hydroxylation of hydrocarbons by *Pseudomonas oleovorans* is striking.[13] This could indicate that ROO may also play a role in a mixed oxygenase reaction that is still to be discovered. Another protein has been proposed to have an oxidoreductase activity toward Rd. This is the product of the *rbo* gene, called a rubredoxin oxidoreductase.[14,15] This gene has been found in *D. vulgaris* Hildenborough. The protein, named desulfoferrodoxin,[16] has no NADH oxidoreductase activity.

Because Rd is found in *D. vulgaris* Hildenborough cells, which have no NRO activity,[9] more physiological roles for Rds remain to be discovered. Rubredoxin acts as an intermediate electron carrier in the reduction of nitrates by *Clostridium perfringens*.[17] Such a function in strains of *Desulfovibrio* that are capable of the dissimilatory reduction of nitrates should

[11] L. Chen, M.-Y. Liu, J. Le Gall, P. Fareleira, H. Santos, and A. V. Xavier, *Biochem. Biophys. Res. Commun.* **193,** 100 (1993).

[12] H. Santos, P. Fareleira, A. V. Xavier, L. Chen, M.-Y. Liu, and J. Le Gall, *Biochem. Biophys. Res. Commun.* **195,** 551 (1993).

[13] T. Ueda, E. T. Lode, and M. J. Coon, *J. Biol. Chem.* **247,** 2109 (1972).

[14] M. J. Brumlik and G. Voordouw, *J. Bacteriol.* **171,** 4996 (1989).

[15] G. Voordouw, *in* "The Sulfate-Reducing Bacteria: Contemporary Perspectives" (J. M. Odom and R. Singleton, eds.), p. 88. Brock/Springer Series in Contemporary Bioscience, New York, (1993).

[16] I. Moura, P. Tavares, J. J. G. Moura, N. Ravi, B.-H. Huynh, M.-Y. Liu, and J. Le Gall, *J. Biol. Chem.* **263,** 21596 (1990).

[17] S. Seki, A. Ikeda, and M. Ishimoto, *J. Biochem. (Tokyo)* **103,** 583 (1988).

be possible. Indeed, *Desulfovibrio desulfuricans* 27774, which performs such a reaction, produces much more Rd when growing at the expense of nitrate that when it reduces sulfate (J. Le Gall, M.-Y. Liu, J. J. G. Moura, and I. Moura, unpublished observations, 1994).

Crystal Structures

High-resolution crystal structure studies of three rubredoxins from sulfate-reducing *Desulfovibrio* species,[18–20] one clostridium,[21] and one hyperthermophilic pyrococcus[22] have been reported. The crystal structure of another thermophilic rubredoxin is under investigation.[23]

The first crystal structure determination of a rubredoxin was reported in 1970 on the rubredoxin from *Clostridium pasteurianum*.[24] This analysis was one of the first to make effective use of the anomalous scattering of the heavy atom derivatives coupled with the isomorphous differences to solve the structure at 2.5-Å resolution and ultimately to 1.2 Å.[21] The 1.5-Å crystal structure of this rubredoxin was the vehicle used to demonstrate that heavy atom-derived phases could be dispensed with, once an effective model was established and that refinement of the model of a protein could be performed providing some kind of restraints kept the model and its calculated structure factors close to the observed values.[25,26]

Structure Comparisons

The initial part of this section discusses the crystal structure of the rubredoxin from *D. vulgaris* because this structure has been taken to 1.0-Å resolution. This crystallographic model was refined using a re-

[18] M. Frey, L. Sieker, F. Payan, R. Haser, M. Bruschi, G. Pepe, and J. Le Gall, *J. Mol. Biol.* **197,** 525 (1987).

[19] R. E. Stenkamp, L. C. Sieker, and L. H. Jensen, *Proteins: Struct. Funct. Genet.* **8,** 352 (1990).

[20] E. T. Adman, L. C. Sieker, and L. H. Jensen, *J. Mol. Biol.* **217,** 337 (1991).

[21] K. D. Watenpaugh, L. C. Sieker, and L. H. Jensen, *J. Mol. Biol.* **131,** 509 (1979).

[22] M. W. Day, B. T. Hsu, L. Joshua-Tor, J.-B. Park, Z. H. Zhou, M. W. W. Adams, and D. C. Rees, *Protein Sci.* **1,** 1494 (1992).

[23] J. Meyer, J. Gagnon, L. C. Sieker, A. van Dorsselaer, and J.-M. Moulis, *Biochem. J.* **271,** 839 (1990).

[24] J. R. Herriott, L. C. Sieker, L. H. Jensen, and W. Lovenberg, *J. Mol. Biol.* **50,** 391 (1970).

[25] K. D. Watenpaugh, L. C. Sieker, J. R. Herriott, and L. H. Jensen, *Cold Spring Symp. Quant. Biol.* **34,** 359 (1972).

[26] K. D. Watenpaugh, L. C. Sieker, J. R. Herriott, and L. H. Jensen, *Acta Crystallogr. Sect. B: Struct. Crystallogr. Cryst. Chem.* **B29,** 943 (1973).

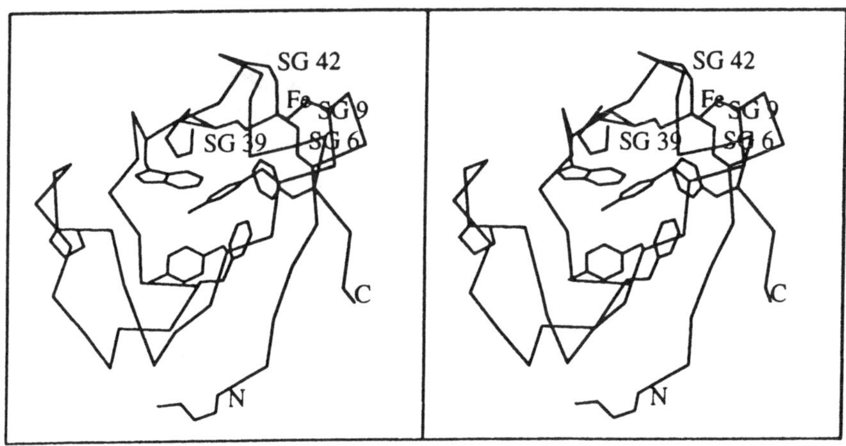

FIG. 2. Stereo view showing the overall polypeptide fold of *Desulfovibrio vulgaris* rubredoxin, the Fe–Cys-4 complex, and the invariant residues listed in Fig. 1.

strained refinement procedure.[27] This allows a direct comparison with all the other rubredoxin crystal structures that were refined by similar techniques. Diffraction data of *D. vulgaris* Rd have subsequently been extended to 0.9 Å where the structural model has been refined by normal crystallographic (free atom) techniques (G. Sheldrick, private communication, 1994). This analysis is still in progress and is not considered here. We then compare the crystal structures of the different rubredoxins, taking into consideration structural features that appear to be important for maintaining the integrity of the rubredoxin molecule.

The comparison of these heterologous proteins with their respective residue changes provides the possibility of determining some general rules concerning the structure of other rubredoxins for which the amino acid sequence has been determined and in mapping the recognition sites of NRO, ROO, and other redox partners.

Figure 2 shows a stereo plot of the α-C chain of the structure of the 1-Å model of *D. vulgaris* Rd, including the iron atom coordinated to the four-cysteine side chain groups and the invariant aromatic groups in the core of the molecule. Briefly described, the molecule contains β-sheet structure, several 3_{10} helical turns, and some glycine-type turns encompassing the aromatic dominated core.[20–23] This specific core arrangement appears to be essential for the integrity of the metalloprotein structure and probably plays a role in controlling or stabilizing the redox states of

[27] Z. Dauter, L. C. Sieker, and K. S. Wilson, *Acta Crystallogr. Sect. B: Struct. Sci.* **B48,** 42 (1992).

this type of metal center. The high variability in the sequence region 16 to 29 (flap of chain on the left in Fig. 2) is consistent with some of this polypeptide region missing in the "unique" *D. desulfuricans* Rd molecule.

The crystal structures of all the Rds reported to date are similar, the only major changes in their tertiary structure being the deleted flap in *D. desulfuricans* Rd and one or two deletions in this region in other molecules and either a deletion at the N terminus (*Pyrococcus furiosus* Rd) or at the C terminus of some Rds. The structures can all be superimposed using the α-C atoms and the root mean square (rms) distance, for each residue from the reference molecule *D. vulgaris* Rd, can be calculated (see Fig. 3). An indication that the structures are similar is the low overall rms deviation between any two of them, on the order of 0.75 Å. A measure of the precision of the structure determinations is indicated by the values between the independently determined structural models of *D. vulgaris* Rd done at 1.0 and 1.5 Å. It is encouraging to see that independent studies agree so well. At this point, it is difficult to judge the significance of the differences of 0.1–0.2 Å in comparing these different structures. This becomes important in understanding the structural basis for the differences in redox potential of electron transfer proteins.

C-*x*-*y*-C-G-*z* Chain Segments

The amino acid sequences of all rubredoxins show two sets of the -C-*x*-*y*-C-G-*z*- sequence around the iron center, where each cysteine is a ligand to the iron atom. The first set has significant changes at *x* and *y*, but *z* is maintained as a tyrosine residue. This aromatic residue appears to protect/maintain the Fe–4Cys center relative to the interior of the molecule. The second set has the invariant Pro-40 in the *x* position. This strictly invariant proline appears to be in this location for a purpose. We suggest that it may be involved in an association with some part of the rubredoxin redox partner facilitating the exchange of the electron between the redox centers. Figure 1 shows a significant number of changes at *y* and a few changes at *z*.

In addition, there is always an aromatic residue two residues before the first cysteine of either -C-*x*-*y*-C-G-*z*- chain segment. This aromatic residue can be a tyrosine, phenylalanine, and tryptophan at position 4, which is somewhat near the surface of the molecule. The equivalent residue, in the second chain segment, is an invariant tryptophan at position 37. This tryptophan is located toward the interior of the molecule and is close to the Fe–4Cys center and is in association with the aforementioned tyrosine at position 11 in the polypeptide chain. It appears to protect or poise the iron center for its activity. The aromatic residue at position 4

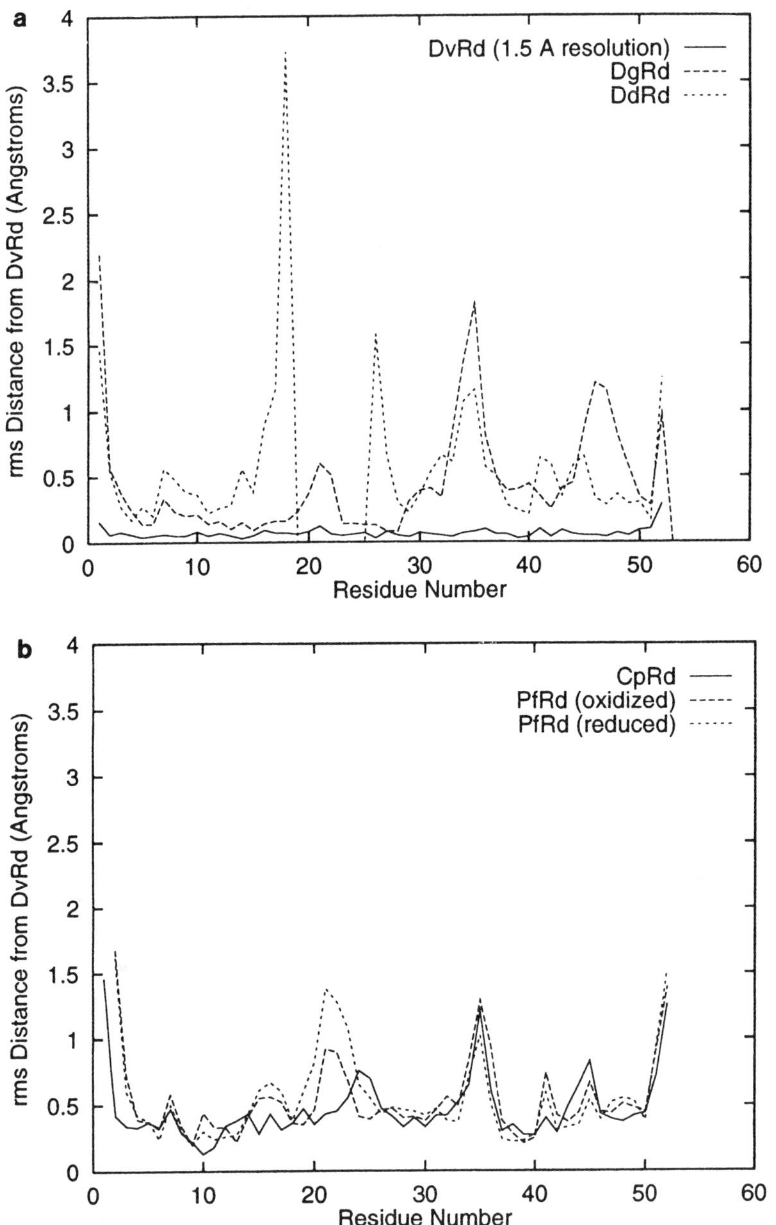

FIG. 3. Root mean square differences between backbone atoms of the rubredoxin crystal structures after superposition of the α-carbon atoms on *D. vulgaris* Rd. (a) *Desulfovibrio* rubredoxins. (Note the large differences associated with the residues bridging the deleted segment in *D. desulfuricans* Rd.) (b) Mixed bacterial rubredoxins.

can be variable, in accordance with its location near the surface of the molecule, but the tryptophan is strictly invariant, being located in the interior of the molecule and somewhat closely associated with the iron center. Of the 13 residues that are strictly invariant, the two -C-*x*-*y*-C-G- segments account for 6 of these residues. In addition, there are five aromatic residues, one proline, and one lysine that are conserved.

Invariant Lys-46

Lys-46 presents an interesting situation for discussion. From the sequence comparisons it is shown to be the only strictly invariant hydrophilic residue in all the Rds. Except for the *D. vulgaris* Rd structure, the crystal structures of the other four Rds show that Lys-46 extends across to the neighboring chain, making an H bond to the carbonyl oxygens of residue 30 and residue 33, presumably contributing to the stability of the molecule. This is not the situation in the crystal structure of *D. vulgaris* Rd, in which the side chain of Lys-46 bends around to make an H bond with N46 and with an oxygen atom of a sulfate ion, which in turn makes an H bond to N47 of the main chain. It seems clear that this invariant residue does more than stabilize the molecule.

Structure and Redox Function

There are several structural features that have been suggested as playing a role in modulating the redox properties of the metal center. These include the metal complex itself, the NH--S hydrogen bonds of the protein residues with the metal complex, and some of the aromatic side chains in the core of the molecule.[18,28] No major differences have been found in any of these features. This is not surprising because the differences in redox potential among the different Rds are rather modest.

On the other hand, these same features indicate what is necessary to maintain the rubredoxin structure and the general range of redox potentials. The structure of *D. desulfuricans* Rd indicates that residues 20–26 of the chain are not necessary to maintain a stable and functional Rd. *Pyrococcus furiosus* Rd shows a tryptophan in place of tyrosine or phenylalanine at position 4, but this is at the opposite end of the molecule where somewhat larger structural adjustments can be accommodated without

[28] E. T. Adman, K. D. Watenpaugh, and L. H. Jensen, *Proc. Natl. Acad. Sci. U.S.A.* **72,** 4854 (1975).

disrupting the integrity of the molecule. Figure 4 shows the superimposed structures for the aromatic side chains in this region. Day *et al.*[22] point out that the additional H bond supplied by this tryptophan residue probably contributes to the thermal stability of *P. furiosus* Rd.

Fe–4Cys Center

Small differences between the molecules are also seen when comparing the bond lengths and angles of the metal center (see Table I). The differences seen in the Fe—S bond distances show a trend but, owing to the target values and the restraints used in the refinement of these crystal structures, the accuracy of these individual values should not be overemphasized. On the other hand, subject to the decisions made on target values and restraints, the average of these distances may have some significance relative to the individual values. Although the spread of the Fe—S bond distances is relatively large among the oxidized structures, it is interesting to note that the average distance of the symmetry-related pair Fe-SG6 and Fe-SG39 is about 2.30 Å compared to an average distance of about 2.26 Å for the pair Fe-SG9 and Fe-SG42. Although the accuracy of these values can be questioned they do show that the twofold symmetry is maintained around the iron atom and that the Fe—S bond distances of the bonds closer to the center of the molecule are longer than those to the exterior. Adman *et al.* pointed out that there are two NH--S hydrogen bonds to SG6 and SG39 but only one to SG9 and SG42.[28] The NH--S

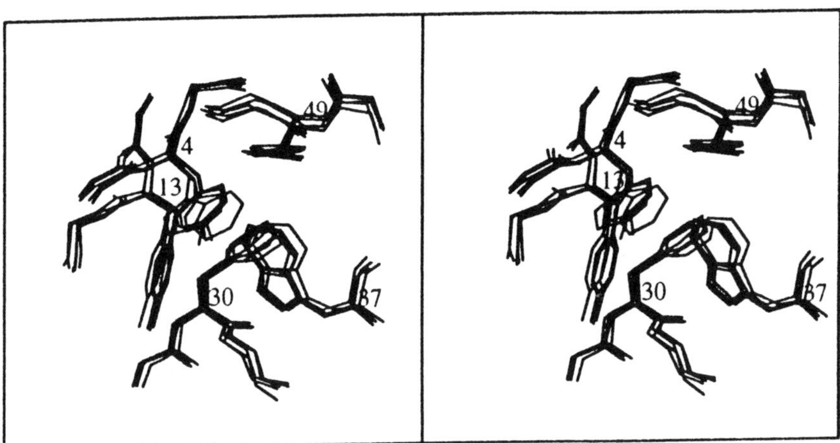

FIG. 4. Stereo view of the aromatic side chains at position 4 and the immediate core environment, showing the positional variability of the superimposed structures.

TABLE I
BOND LENGTHS AND ANGLES OF Fe–4Cys CENTER

Center	8RXN	7RXN	1RDG	6RXN	4RXN	5RXN	1CAA	1CAD	Ave. oxidized[a]
Bond distances (Å)									
Fe-SG6	2.29	2.33	2.32	2.28	2.34	2.32	2.32	2.34	2.31
Fe-SG9	2.26	2.29	2.29	2.26	2.29	2.29	2.25	2.29	2.27
Fe-SG39	2.29	2.29	2.28	2.30	2.31	2.30	2.33	2.35	2.30
Fe-SG42	2.26	2.27	2.27	2.25	2.23	2.25	2.25	2.29	2.26
Bond angles (degrees)									
SG6-Fe-SG9	114.8	113.8	114.5	111.4	113.8	114.3	113.0	112.6	113.6
SG6-Fe-SG39	110.9	110.2	111.3	112.8	108.6	109.3	111.8	113.4	110.9
SG6-Fe-SG42	105.8	106.2	106.1	104.7	104.1	104.0	102.6	104.6	104.9
SG9-Fe-SG39	104.5	104.0	103.4	100.8	103.7	103.9	102.4	102.7	103.2
SG9-Fe-SG42	109.2	109.6	109.8	114.1	114.4	113.6	115.2	111.2	111.9
SG39-Fe-SG42	111.8	113.2	111.9	113.4	112.4	111.9	112.3	112.7	112.4
NH--S hydrogen bonds distances (Å)									
SG6--N8	3.54	3.56	3.53	3.49	3.63	3.67	3.51	3.34	3.55
SG6--N9	3.55	3.53	3.54	3.61	3.67	3.67	3.58	3.43	3.58
SG9--N11	3.42	3.47	3.53	3.54	3.44	3.46	3.47	3.41	3.48
SG39--N41	3.58	3.57	3.59	3.51	3.55	3.58	3.42	3.36	3.54
SG39--N42	3.62	3.64	3.64	3.56	3.72	3.61	3.52	3.47	3.60
SG42--N44	3.50	3.41	3.49	3.86	3.90	3.84	3.49	3.50	3.60

[a] The unconstrained 4RXN is not included in the averages.

[b] *C. pasteurianum* Rd, 4RXN, 1.2 Å [Watenpaugh *et al.*, 1979 (unconstrained model)[21]; *C. pasteurianum* Rd, 5RXN, 1.2 Å (K. D. Watenpaugh (unpublished work; 1984); *D. gigas* Rd, 1RDG, 1.4 Å (Frey *et al.*)[18]; *D. desulfuricans* Rd, 6RXN, 1.5 Å (Stenkamp *et al.*)[19]; *D. vulgaris* Rd, 7RXN, 1.5 Å (Adman *et al.*)[20]; *D. vulgaris* Rd, 8RXN, 1.0 Å (Dauter *et al.*)[27]; *P. furiosus* Rd, 1CAA, 1.8 Å (Day *et al.*)[22]; *P. furiosus* Rd, 1CAD, 1.8 Å (Day *et al.*[22])]

bonds of SG6 to backbone amides of 8 and 9 and the one from SG9 to N11 are on one side of the twofold axis of symmetry whereas the NH--S bonds of SG9 to the amides at positions 41 and 42 and of SG42 to position 44 are on the other side. The SGs with the two NH--S bonds have the longer Fe—S bond length whereas the SGs with the single NH--S bond have the shorter Fe—S bond length. Confirmation of this interesting pattern of NH--S bonds above and below the twofold axis of symmetry requires additional higher resolution structure analyses.

Considerable variation is seen in the NH--S hydrogen bond distances among the individual oxidized Rds. Relative to the local twofold axis of symmetry, the average values of these distances (see Table I) in the oxidized molecules show no obvious symmetry. The average distance is about 3.56 Å. Figure 5 shows a stereo view of the twofold configuration around the iron center in *D. vulgaris*.

Reduced Rubredoxin

Because the reduced Pf structure has essentially the same metal site geometry as the oxidized forms, it appears that only small structural changes occur with the change in oxidation state. In these comparisons of the oxidized versus reduced models, the lower resolution of the *P. furiosus* Rd structures is the limiting factor. The Fe—S bonds of the reduced molecule appear to be about 0.03 Å longer than the averages of the oxidized Fe—S bond distances. The average of NH--S bond distances of the reduced *P. furiosus* Rd is 3.42 Å and is therefore 0.14 Å less than the average of the oxidized forms. Within the precision of the current values of the bond lengths for the different Rds, this decrease in the NH--S distance on reduction of *P. furiosus* Rd is probably significant.

Specificity

Figure 6 shows the charged side chains on the surface of the Rd molecules. Most of the variability in residue distribution occurs at the opposite end of the protein from the iron center (bottom of this stereo

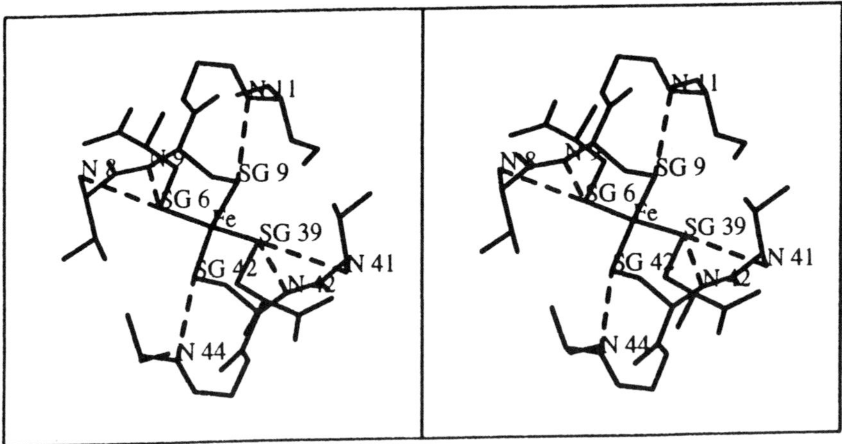

FIG. 5. Stereo view of the Fe–4Cys center, showing the pseudotwofold symmetry and the NH--S hydrogen bonds.

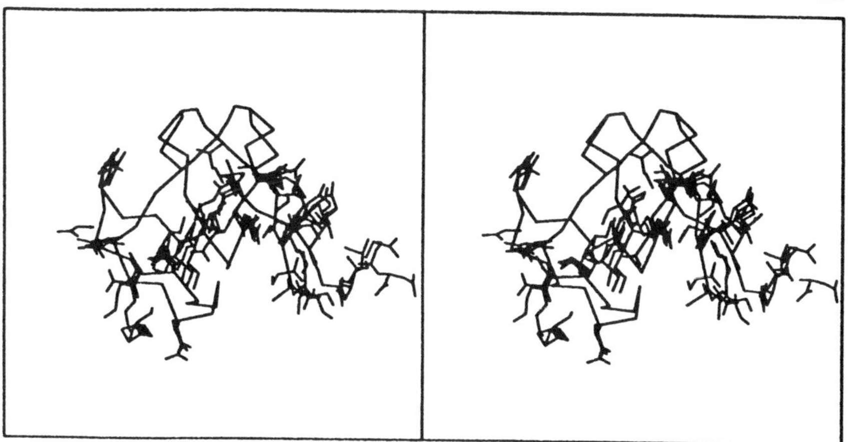

FIG. 6. Stereo view of the charged side chains relative to the hydrophobic Fe–4Cys center.

view). The iron center is located in the molecule in a region surrounded by hydrophobic residues. The external residues are the most likely candidates for specificity interactions with its redox partners. As mentioned above in the discussion of the physiological role of the Rds the *D. gigas* Rd exchanges electrons very well with the NRO from *D. gigas* but *D. vulgaris* Rd does not.[9,10] Obviously some differences in the molecules are responsible for this change in electron transfer. The iron center is in a rather invariant hydrophobic environment at the opposite end of the molecule from the variable charged residues. A reasonable proposal is that the invariant hydrophobic region provides the common docking and electron exchange region whereas the more variable region of Rd provides the specificity to interact with its redox partner. The Rd molecule is sufficiently small that this should not require an extensive docking region of its much larger redox partner. The variable residue at position 7 could play a special role in docking or the electron exchange. *Desulfovibrio gigas* Rd is distinct from *D. vulgaris* Rd in several ways, as shown in Fig. 1a. It is interesting to note that, among the 13 Rds, only *D. gigas* Rd has residue 3 as isoleucine in place of lysine. This hydrophobic surface residue is probably contributing to some of the specificity of this particular protein.

Conclusion

The high-resolution crystal structures of five different rubredoxins show that these iron proteins have a strong conservation of the stereochemical geometry around the Fe–4Cys center. Although more flexible, the arrangement of the aromatic core is conserved. The polypeptide chain folds into a rather rigid configuration contributing to the stability of the core residues and the iron center. In spite of the small number of amino acids, the molecule shows a large diversity in the surface residues. These surface residues likely provide the specificity for docking the most invariant hydrophobic portion of the molecule, which contains the Fe–4Cys redox center.

At the current state of analysis of several structural models (1.0 to 1.8 Å) two new features seem to be apparent. The cysteinyl sulfurs with two NH--S bonds appear to have longer Fe—S bonds than the cysteinyl sulfur with one NH--S bond. An average decrease of about 0.17 Å in the NH--S hydrogen bond distance occurs when the Fe–4Cys center goes from the oxidized state to the reduced state.

Acknowledgments

We acknowledge the use of Molscript[29] for Figs. 2, 4, 5, and 6. X-Ray structures are from the Brookhaven Protein Data Bank.[30,31]

[29] P. J. Kraulis, *J. Appl. Crystallogr.* **24,** 956 (1991).

[30] F. C. Berstein, T. F. Koetzle, G. J. B. Williams, E. F. Meyer, Jr., M. D. Brice, J. R. Rodgers, O. Kennard, T. Shimanouchi, and M. Tasumi, *J. Mol. Biol.* **112,** 535 (1977).

[31] E. E. Abola, F. C. Bernstein, S. H. Bryant, T. F. Koetzle, and J. Weng, *in* "Crystallographic Databases—Information Content, Software Systems, Scientific Applications" (F. H. Allen, G. Bergerhoff, and R. Sievers, eds.), p. 107. International Union of Crystallography, Bonn/Cambridge/Chester, 1987.

[15] Characterization of Three Proteins Containing Multiple Iron Sites: Rubrerythrin, Desulfoferrodoxin, and a Protein Containing a Six-Iron Cluster

By ISABEL MOURA, PEDRO TAVARES, and NATARAJAN RAVI

Introduction

It has been demonstrated that a number of proteins isolated from anaerobic sulfate reducers contain many novel combinations of iron centers. The unusual and diversified metal centers found in these proteins are more than a surprise, as no one could have envisaged the possibility of the existence of such metal centers, and needless to mention they offer a plethora of research pursuits. In this chapter we focus on three proteins, with relevance to biochemical, electron paramagnetic resonance (EPR), and Mössbauer spectroscopic characteristics in order to unravel the structural and redox properties of the active sites. The following three proteins display the presence of redox metal centers with metal compositions spanning from mono- through bi- to hexanuclearity: (1) a protein designated as rubrerythrin (Rr), a contraction of rubredoxin (Rd) and hemerythrin, is a trivial name given to this nonheme iron protein, which has a rubredoxin-type and hemerythrin-type center; (2) desulfoferrodoxin (Dfx) contains a desulforedoxin (Dx) center (a distorted rubredoxin-type center found in the protein isolated from *Desulfovibrio gigas*) and a new monoiron center in which the iron is bound mainly to nitrogen and/or oxygen ligands; (3) an [Fe—S] protein in which the metal centers manifest themselves in the form of a [6Fe] cluster of unknown structure.

Rubrerythrin

Rubrerythrin (Rr) was first characterized by LeGall *et al.*[1] from *Desulfovibrio (D.) vulgaris* (Hildenborough). Rubrerythrin has also been found in *D. desulfuricans* ATCC 27774.[2]

Physicochemical Characterization

Purification

All purification procedures are performed at 4° and pH 7.6. The presence of Rr is judged by a change in absorbance at 490 nm after ascorbate reduction. Purity is determined during the later steps by using the ratio A_{280}/A_{490}. The preparation of the crude extract and the growth of *D. vulgaris* (Hildenborough) are performed as previously described.[1] The crude extract (700 ml) is loaded onto a diethylaminoethyl (DEAE)-BioGel column (4.5 × 35 cm) equilibrated with 10 m*M* Tris-HCl buffer and washed with 2 liters of the same buffer. A fraction containing Rr and cytochromes is collected during this washing and is loaded onto a hydroxylapatite column (6 × 24 cm) equilibrated with 0.01 *M* Tris-HCl buffer. The column is washed successively with 200 ml each of 0.01 and 0.001 *M* Tris-HCl buffer and then with 0.001 *M* potassium phosphate buffer (KPB). The two linear KPB gradients are applied (0.001–0.2 *M* with a total volume of 3 liters and 0.2–0.4 *M* with a total volume of 2 liters), and a fraction contain-

[1] J. LeGall, B. Prickril, I. Moura, A. V. Xavier, J. J. G. Moura, and B. H. Huynh, *Biochemistry* **27,** 1636 (1988).

[2] I. Moura, P. Tavares, and N. Ravi, unpublished results.

ing Rr and cytochrome is eluted at about 0.03 *M*. This fraction is concentrated to a volume of 10 ml, degassed, and loaded onto a Sephadex G-75 column (5.4 × 85 cm) equilibrated with 0.05 *M* Tris-HCl buffer. At a flow rate of 25 ml/hr the Rr fraction, which has an A_{280}/A_{490} ratio of 15, is twice concentrated to 5 ml and diluted with water to 50 ml in order to lower the ionic strength. The Rr fraction is then loaded onto a DEAE-52 column (4.5 × 29 cm), and a linear Tris-HCl gradient is applied (0.01–0.1 *M*, with a total volume of 1.5 liters). The Rr, which elutes at about 0.03 *M*, is then applied to a second hydroxylapatite column (4 × 23 cm) equilibrated with 0.001 *M* KPB. A linear gradient is applied (0.001–0.1 *M* with a total volume of 1.5 liters), and the purified Rr, collected at about 0.03 *M*, has an A_{280}/A_{490} ratio of 7.0.

General Properties

Rubrerythrin is composed of two identical subunits of molecular mass (MM) 21.9 kDa [determined by sodium dodecyl sulfate–polyacrylamide gel electrophoresis (SDS–PAGE)]. The molecular mass of the purified protein was determined to be 45.2 kDa by equilibrium ultracentrifugation. The first reported iron analysis on this protein performed by plasma emission showed that it contains four iron atoms per homodimer.[1] The optical, EPR, and Mössbauer properties showed that two of the iron atoms belong to FeS_4 centers similar to Rd-type centers[3] and the other two belong to an exchange-coupled binuclear center. Because only one binuclear cluster was found per dimer it was first thought that it could have a structural role in binding the two subunits. Subsequent Mössbauer measurements of samples from different preparations indicated that the ratio of the binuclear center/rubredoxin center was found to vary between 0.5 and 0.7. More recently Rr has also been found in *D. desulfuricans*, and from the preliminary Mössbauer data it appears that it may contain two diiron clusters per homodimer.[2] It has been reported that Rr from *D. vulgaris*[4] contains an approximately 1 : 1 ratio of diiron cluster to mononuclear FeS_4, indicating three iron atoms per subunit despite the fact that metal analysis shows five iron atoms per homodimer.

Iron analysis of ^{57}Fe-reconstituted Rr gives a value of 7.5 ± 0.4 mol of iron per mole of Rr homodimer. This value is higher than the value of 6.0 expected for fully occupied iron sites, suggesting that approximately

[3] I. Moura, M. Bruschi, J. LeGall, J. J. G. Moura, and A. V. Xavier, *Biochem. Biophys. Res. Commun.* **75,** 1037 (1977).

[4] A. J. Pierik, R. B. G. Wolbert, G. L. Portier, M. F. J. M. Verhagen, and W. R. Hagen, *Eur. J. Biochem.* **212,** 237 (1993).

20% of the iron is adventitiously bound to the reconstituted Rr. Mössbauer data convincingly prove this suggestion.[5]

Amino Acid Sequence Studies

After the initial characterization, the *D. vulgaris* Rr has been sequenced both by classic amino acid sequencing methods[6] and also by genetic methods.[7] Both methods reveal an identical sequence of a polypeptide chain of 191 amino acid residues. The C-terminal part of the protein (positions 153–191) shows the typical sequence features of Rd, a protein with a nonheme iron center that is also present in other *Desulfovibrio* species. The Cys-x-x-Cys spacing characteristic of Rds is preserved in Rr but the 12-residue spacing between the cysteine pairs in Rr is less than half that seen in Rds. The N-terminal portion, residues 1 to 152, contains two highly homologous regions containing a conserved Glu-x-x-His sequence. As similar sequences provide the histidine and glutamate ligands to the binuclear cluster in *Escherichia coli* ribonucleotide reductases (RNR-R2),[8] it was conceived that in Rr, the N-terminal region residues Glu-53, His-56, Glu-128, and His-131 might provide ligands to this center.[7] The gene encodes a polypeptide of 191 amino acids, and a normal ribosome-binding site is located upstream (nucleotides −6 to −11) from the translational start of the gene, which implies a cytoplasmic location of the protein.

Reconstitution of Active Centers in Desulfovibrio vulgaris Rubrerythrin

Reconstitution was carried out[5] using a method similar to that described by Zhang *et al.*[9] for reconstitution of the diiron cluster of hemerythrin. All the operations were carried out at room temperature under strict anaerobic conditions.

Procedure for Preparing Apoprotein. As-isolated Rr [0.6 ml (7 mg)] in 0.1 *M* Tris-HCl (pH 7.6), 0.1 ml of 0.1 *M* 2,2′-dipyridyl solution, and 0.73 g of guanidine-HCl (GuHCl) are introduced through a septum into an anaerobic glass vial. Solid sodium dithionite (10 mg) is added directly

[5] N. Ravi, B. C. Prickril, D. M. Kurtz, Jr., and B. H. Huynh, *Biochemistry* in press (1993).

[6] J. J. Van Beeumen, G. Van Driessche, M.-Y. Liu, and J. LeGall, *J. Biol. Chem.* **266,** 20645 (1991).

[7] B. C. Prickril, D. M. Kurtz, Jr., J. LeGall, and G. Voordouw, *Biochemistry* **30,** 11118 (1991).

[8] P. Nordlund, B.-M. Sjöberg, and H. Eklund, *Nature (London)* **3,** 593 (1990).

[9] J.-H. Zhang, D. M. Kurtz, Jr., Y.-M. Xia, and P. G. Debrunner, *Biochemistry* **30,** 583 (1991).

to the vial anaerobically. Formation of the iron complex is indicated when the solution changes in color to bright red. The solution is kept anaerobic for 30 min and is exposed to air subsequently. The volume is adjusted to 5 ml with 6 *M* GuHCl transferred to a 10-ml Diaflo apparatus (Amicon, Danvers, MA) and concentrated to less than 1 ml. This process is repeated three times to ensure the removal of excess reagent. The apoprotein solution is then transferred to a Centricon-10 microconcentrator, concentrated to 0.1 ml, and stored at 4°. One-half milliliter of an ^{57}Fe solution (0.7 mg of iron/ml) containing 85 m*M* 2-mercaptoethanol is added dropwise to a 10-ml vial containing 0.5 ml of the anaerobic apo-Rr solution over a period of 5 min. The amount of ^{57}Fe added represents a 10-fold molar excess of iron over that in as-isolated Rr. The solution is allowed to equilibrate for 30 min and anaerobic 0.1 *M* Tris-HCl buffer, pH 7.6, is added to make a final volume of 5 ml. This solution is then concentrated in a Diaflo apparatus until the volume decreases to less than 1 ml. This concentrated solution is then diluted with the same buffer to 5 ml to remove GuHCl, excess Fe(II), and 2-mercaptoethanol and reconcentrated. The final concentration is then done in a Centricon-10 microconcentrator with 2 additional 30-fold concentration/dilution cycles.

Spectroscopic Studies

Ultraviolet–Visible Spectroscopy

Figure 1 shows an ultraviolet (UV)–visible absorption spectrum of the as-isolated Rr from *D. vulgaris* (spectrum A). The spectrum has maxima at 492, 365, and 280 nm with shoulders at 570 and 350 nm. Apart from the higher A_{365}/A_{492} ratio, the Rr spectrum appears similar to that of Rd.[3] The molar extinction coefficient at 492 nm is 10.4 $mM^{-1}\ cm^{-1}$. To examine the contribution of the binuclear center we subtracted from the Rr visible spectrum the contribution of the rubredoxin center (spectrum B) and the spectrum obtained (spectrum C) has an absorption maximum at 365 nm and a shoulder at 460 nm. The estimated molar extinction coefficient at 365 nm is 5.3 $mM^{-1}\ cm^{-1}$. This spectrum bears some resemblance to that of met-Hr, which was reported to show an absorption maximum at 355 nm ($\varepsilon_{355} = 6.4\ mM^{-1}\ cm^{-1}$) and shoulders at 480 and 580 nm.[10] The visible spectrum of the reconstituted Rr is similar to that of as-isolated Rr with maxima at 492 and 365 nm and shoulders at 570 and 350 nm.[5] However, the reconstituted Rr shows higher absorption below 450 nm. The increase

[10] K. Garbett, C. E. Johnson, I. M. Kloz, M. Y. Okamura, and R. J. P. Williams, *Arch. Biochem. Biophys.* **142,** 574 (1971).

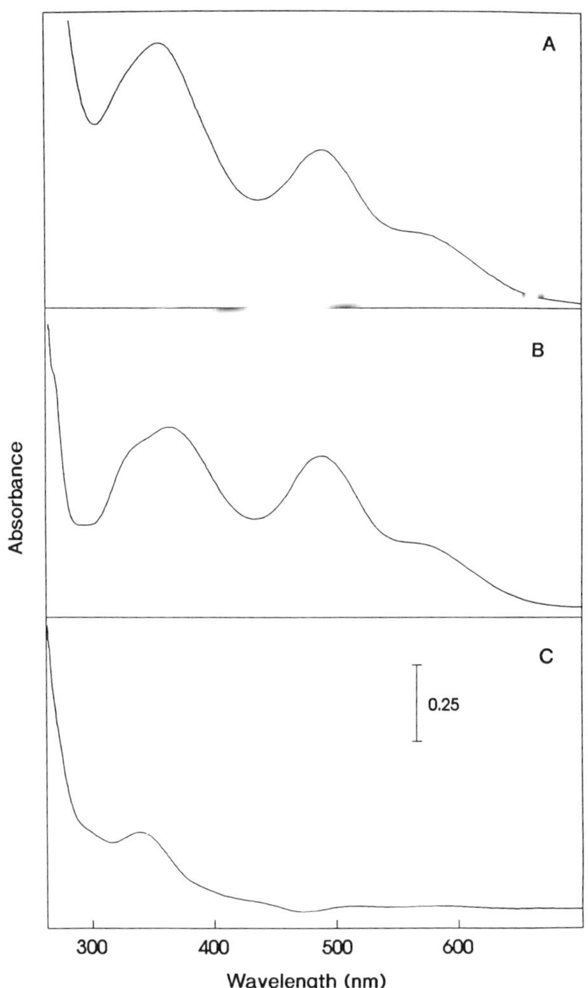

FIG. 1. Ultraviolet–visible absorption spectrum of the as-isolated rubrerythrin from *D. vulgaris* (A) and from *D. gigas* rubredoxin (B). Trace C represents the difference spectrum (trace A − trace B) obtained after normalization of spectrum B in order to match the intensities at 490 nm.

in absorption in this region is therefore an indication of an increase in the amount of the diiron cluster relative to that of the FeS_4 center.

Electron Paramagnetic Resonance Spectroscopy

A low-temperature (8 K) EPR spectrum of the as-isolated Rr from *D. vulgaris* is shown in Fig. 2. Two dominant EPR signals are observed with

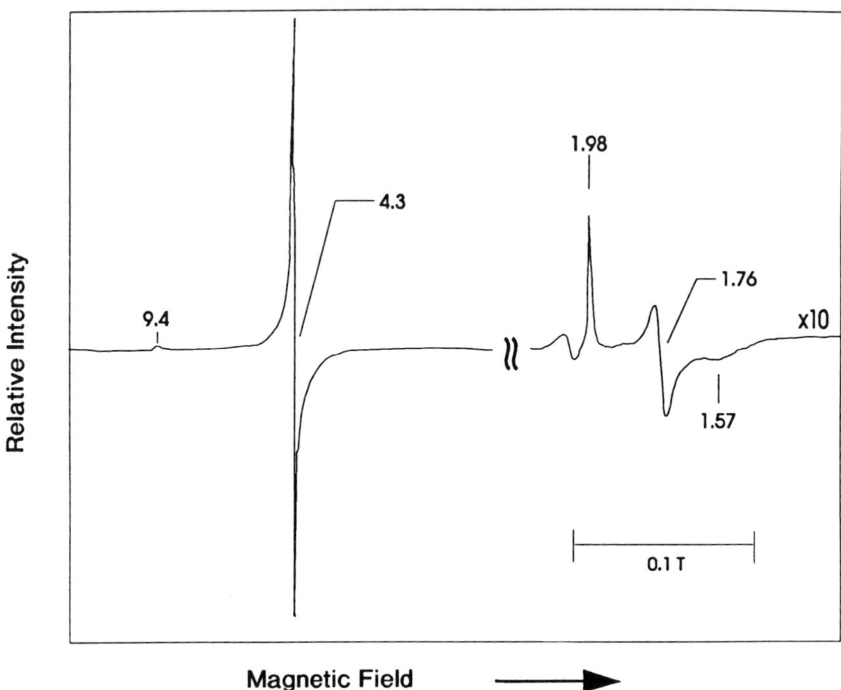

FIG. 2. The EPR spectrum of *D. vulgaris* rubrerythrin. The spectrum was measured at 8 K. Other experimental conditions are as follows: microwave frequency, 9.435 GHz; microwave power, 2 mW; modulation amplitude, 1 mT; receiver gain, 5×10^4.

g values at 9.4 and 4.3 arising from the ground—and the first excited state—of the Kramers doublets of the Rd-type center. Another rhombic signal with *g* values at 1.98, 1.76, and 1.57 is detected and is characteristic of a mixed valence diiron cluster. This signal is weak in the as-isolated protein but it accounts for ~0.5 spin per homodimer in the reconstituted protein. The diiron cluster in the diferric state is EPR silent and is the major species in the as-isolated form of the protein.

Mössbauer Spectroscopy

The Mössbauer spectrum of the as-isolated Rr from *D. vulgaris* at 4.2 K with an external magnetic field of 50 mT applied parallel to the γ beam shows two major spectral components. One of the components is a magnetic spectrum with six lines extending from −5.5 to 6.4 mm/sec and is similar to that observed for *Clostridium pasteurianum* Rd[11] and is

[11] C. Schulz and P. G. Debrunner, *J. Phys.* (*Paris*) **37**(66), 163 (1976).

presumably due to the rubredoxin center, an observation in conformity with the EPR signal observed for this center discussed above. The other component is a quadrupole doublet with $\Delta E_Q = 1.47 \pm 0.05$ mm/sec and $\delta = 0.52 \pm 0.03$ mm/sec, values typical of high-spin ferric ions with nitrogenous and/or oxygenous ligands. This central doublet is attributed to the binuclear cluster in a diamagnetic state as expected for an antiferromagnetically coupled diferric cluster. Spectra measured with strong applied magnetic fields indeed show the diamagnetic nature of this center.

The reduced form was also analyzed by Mössbauer spectroscopy. The Mössbauer spectrum of the reduced form measured at 4.2 K in the absence of a magnetic field shows two quadrupole doublets labeled doublet I and doublet II. Parameters of doublet I are $\Delta E_Q = 3.15 \pm 0.04$ mm/sec, $\delta = 0.70 \pm 0.03$ mm/sec, and for doublet II are $\Delta E_Q = 3.14 \pm 0.04$ mm/sec, $\delta = 1.30 \pm 0.03$ mm/sec. Both sets of parameters are typical of high-spin ferrous ion ($S = 2$). The smaller isomer shift, 0.7 mm/sec for doublet I, is characteristic of a high-spin ferrous ion with tetrahedral coordination of sulfur ligands and is attributed to the rubredoxin center. The larger isomeric shift, 1.30 mm/sec for doublet II, is indicative of an octahedral coordination with oxygen/nitrogen ligands. The ratio of the percent absorptions of doublet I to doublet II is 1.07. Consequently the Mössbauer data on the reduced protein show that the stoichiometry is two rubredoxin centers and one binuclear cluster.

The reconstituted protein was also studied by Mössbauer spectroscopy.[5] The 4.2 K Mössbauer spectrum in a parallel field of 50 mT is similar to that of the as-isolated Rr except for the increased intensity of the central quadrupole doublet and the presence of an additional broad magnetic spectrum due to adventitiously bound iron. A careful analysis was made in order to estimate the contribution of each type of center. The amount of rubredoxin and binuclear cluster was estimated to be one rubredoxin per binuclear cluster. The stoichiometry of one diiron cluster per subunit in the reconstituted protein is in accordance with the prediction based on amino acid sequence analysis.

It may also be worth pointing out that a least-squares fit of the central quadrupole doublet reveals that the ΔE_Q values for the two iron sites are 1.74 and 1.38 mm/sec, suggesting a possibility of a minor structural or environmental change involving the two iron sites. However, despite the striking homology of diiron-binding sites Glu-x-x-His seen in Rr and RNR-R2, the Mössbauer data of the latter show that the iron sites are distinct as seen in the quadrupole splitting of 2.44 and 1.62 mm/sec with an isomer shift of 0.45 and 0.55 mm/sec, respectively, for the two iron sites.[12] Taking

[12] J. B. Lynch, C. J.-Garcia, E. Münck, and L. Que, Jr., *J. Biol. Chem.* **264,** 8091 (1989).

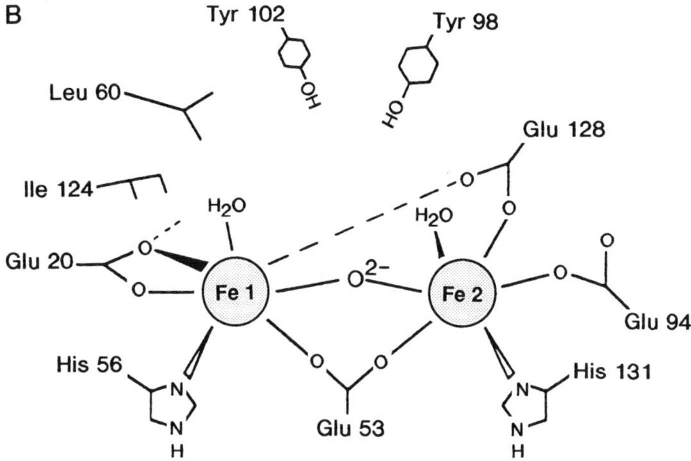

FIG. 3. Schematic representation of (A) the diiron site in *E. coli* ribonucleotide reductase (RNR-R2) and (B) the proposed structure of the diiron site in *D. vulgaris* rubrerythrin.

all these points together, a structure has been proposed for the binuclear center in Rr[5] and is depicted in Fig. 3. A comparison has also been made with RNR-R2 subunit structure.

At this juncture, a comparison between the binuclear sites of RNR-R2 and Rr merits attention. The key amino acid residues proposed to be at or near the diiron site are strikingly similar to the spacing of analogous residues in RNR-R2. Thus Glu-53–x–x–His-56 and Glu-128–x–x–His-131 sequences are proposed to be analogous to Glu-115–x–x–His-118 and Glu-238–x–x–His-241 sequences, respectively, in RNR-R2. Furthermore, the spacings of nearest neighbors carboxylate ligands in RNR-R2 (Glu-238–Glu-204, 34 residues; Glu-115–Asp-84, 31 residues) are either the same or similar in Rr (Glu-128–Glu-94, 34 residues; Glu-53–Glu-20, 33 residues). Glu-94 and Glu-20 are proposed to provide terminal ligands in Rr. Relative to the more G-terminal Glu-x-x-His sequences, residues Tyr-98, Tyr-102, and Ile-124 in Rr are sequentially homologous to Phe-208, Phe-212, and Ile-234 in RNR-R2.

Desulfoferrodoxin

Desulfoferrodoxin was first isolated by Moura *et al.*[13] from *D. desulfuricans* ATCC 27774 and from *D. vulgaris* (Hildenborough). This protein has an unusual composition of prosthetic groups: in the as-isolated form it contains a desulforedoxin (Dx)-like FeS_4 (center I),[14] and a mononuclear ferrous site (center II). Because of this unusual combination of metal centers, this novel protein was named desulfoferrodoxin (Dfx).

Physicochemical Characterization

Purification

All purification procedures are performed at 4° and pH 7.6. The purity of Dfx is determined during the later steps by the A_{279}/A_{495} ratio. We describe here the purification of this protein from cells of *D. desulfuricans*. The preparation of the crude extract and the growth of *D. desulfuricans* are performed as previously described.[15] In a typical preparation 800 g of cells is suspended in 10 m*M* Tris-HCl, pH 7.6, and ruptured in a French press at 9000 psi. The extract is then centrifuged at 19,000 *g* for 30 min and then at 180,000 *g* for 75 min. The crude extract (1500 ml) is loaded onto a DEAE-52 column (6 × 32.5 cm) equilibrated with 10 m*M* Tris-HCl buffer. A linear gradient in Tris-HCl (0.01–0.50 *M*) is applied with a total volume of 6000 ml. The fraction containing mainly cytochromes is eluted between 0.10 and 0.20 *M* and is collected in a final volume of 1000 ml. This fraction is dialyzed overnight against distilled water and loaded onto another DEAE-52 column (4 × 30 cm) equilibrated with 10 m*M* Tris-HCl buffer. A linear Tris-HCl gradient (0.01–0.20 *M*) is applied with a total

[13] I. Moura, P. Tavares, J. J. G. Moura, N. Ravi, B. H. Huynh, M.-Y. Liu, and J. LeGall, *J. Biol. Chem.* **265,** 21596 (1990).

[14] I. Mours, B. H. Huynh, R. P. Hausinger, J. LeGall, A. V. Xavier, and E. Münck, *J. Biol. Chem.* **255,** 2493 (1980).

[15] M. C. Liu and H. D. Peck, Jr., *J. Biol. Chem.* **256,** 13159 (1981).

volume of 3000 ml. A fraction containing mainly cytochrome and a brown protein is collected at an ionic strength of 0.15 *M* Tris.

This fraction is concentrated in a Diaflo apparatus (Amicon) with a YM5 membrane and loaded onto a hydroxylapatite column (2.5 × 15 cm) equilibrated at the same ionic strength. The column is washed with 250 ml of 0.15 *M* Tris-HCl buffer and a descending linear gradient (0.15–0.01 *M*) of Tris-HCl with a total volume of 500 ml is applied, followed by an ascending linear gradient (0.001–0.200 *M*) of potassium phosphate buffer with a total volume of 1000 ml. A pink fraction descending at 50 m*M* potassium phosphate is collected in a total volume of 50 ml. At this stage this protein already has a distinct visible spectrum that resembled that of Dx.[3] The ratio of A_{279}/A_{495} is 16.6. This fraction is concentrated again on a Diaflo apparatus with a YM5 membrane and the ionic strength is decreased by adding successively several volumes of distilled water. The last step of purification is performed on a high-performance liquid chromatograph from Beckman (Fullerton, CA) on a Protein-Pack Column-DEAE SPW from Waters Associates (Milford, MA). A linear gradient is applied (0.01–0.02 *M* NaCl) in 0.01 *M* potassium phosphate, pH 7, with a flow of 2 ml/min for 3 hr. After this step Dfx has an A_{279}/A_{495} ratio of 7.8.

General Properties

Desulfoferrodoxin is a monomer of molecular mass 14 kDa (determined by SDS–PAGE and gel filtration). The iron determination shows that this protein contains about two iron atoms per molecule. No labile sulfide was found in this protein. The optical, EPR, and Mössbauer properties showed that these two iron sites are inequivalent: one is an FeS_4 site similar to that found in Dx from *D. gigas*[14] and the other is an octahedrally coordinated high-spin ferrous site, presumably with nitrogen/oxygen ligands.

Genetic Studies

An analysis of the transcriptional unit encoding the gene for rubredoxin from *D. vulgaris* (Hildenborough) showed an additional 378-bp open reading frame that terminates 16 nucleotides from the translational start of the rubredoxin gene and encodes a polypeptide of 14 kDa.[16] The NH_2 terminus of this polypeptide was found to be homologous to the *D. gigas* Dx, indicating that it is also a redox protein. On the basis of these observations, this redox protein was suggested to be a redox partner of rubredoxin, and was tentatively termed rubredoxin oxidoreductase (Rbo). The N-terminal sequence of 41 amino acids of Dfx from *D. desulfuricans* was determined

[16] M. J. Brumlik and G. Voordouw, *J. Bacteriol.* **171,** 4996 (1989).

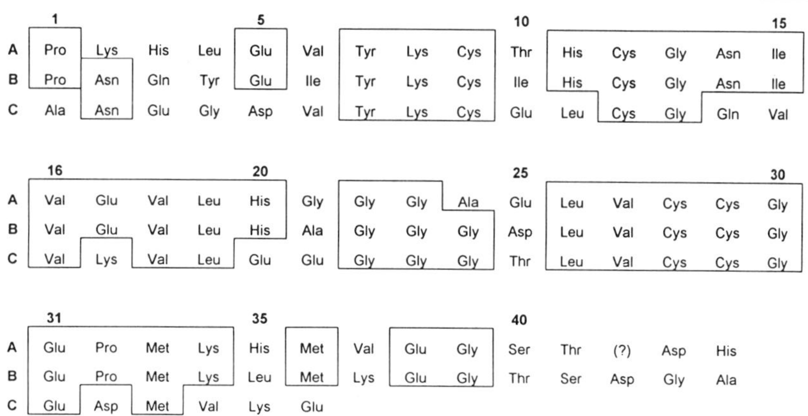

FIG. 4. Comparison of the N-terminal amino acid sequence of desulfoferrodoxin from *D. desulfuricans* ATCC 27774 (A) with the *rbo* gene product[15] of *D. vulgaris* Hildenborough (B) and *D. gigas* desulforedoxin (C).

(see Fig. 4). For comparison, the NH_2-terminal sequence of *D. vulgaris* rubredoxin oxodireductase gene product and *D. gigas* Dx is shown. Twenty-eight of 41 residues are similar between Dfx from *D. desulfuricans* (ATCC 27774) and the rubredoxin oxidoreductase gene product. Eighteen of 36 residues of Dx from *D. gigas* were found to be identical to Dfx from *D. desulfuricans*. The amino acid composition of Dfx shows the presence of six or seven cysteine residues and the sequence analysis shows that four of them (Cys-9, Cys-12, Cys-28, and Cys-29) are in the same position as those found in Dx from *D. gigas* and should be responsible for the binding of the desulforedoxin-like center.

The Dx from *D. gigas* is a dimer and two possible models of coordination of the two iron atoms by the eight cysteine residues have been proposed[3]: (1) the four cysteine residues of a single subunit could coordinate to the same iron, in which case dimer formation is achieved by noncovalent interaction of the two subunits, or (2) three cysteine residues of one subunit (e.g., Cys-9, Cys-12, and Cys-28 from one and Cys-29 from the other) could coordinate to one iron atom, causing a covalent connection of the two subunits via two iron sites. Because Dfx is a monomer, the binding of the desulforedoxin center must be due to the four cysteines of the same subunit.

Cloning and sequencing of the gene encoding Dx from *D. gigas* showed that it is formed by expression of an autonomous gene of 111 bp, not by processing of a 14-kDa protein. These results led Brumlik *et al.*[17] to con-

[17] M. J. Brumlik, G. Leroy, M. Bruschi, and G. Voordouw, *J. Bacteriol.* **172,** 7289 (1990).

clude that the *rbo* gene product, which has a 4-kDa Dfx domain at the NH_2 terminus, may have arisen by gene fusion.

Shuffling and fusion of genes for redox protein domains could then explain the large variety of redox proteins found in sulfate-reducing bacteria.[17] We found that Dfx has a strong sequence homology to the sequence of the *rbo* gene product. However, Dfx cannot accept electrons from the reduced pyridine nucleotides and is therefore probably not a pyridine-linked rubredoxin oxidoreductase.

Spectroscopic Studies

Ultraviolet–Visible Spectroscopy

As-isolated Dfx exhibits an optical spectrum with maxima at 495, 368, and 279 nm (Fig. 5). This spectrum is similar to that of *D. gigas* Dx, indicating the presence of a common chromophore for both proteins. On the basis of protein determination, the estimated molar extinction coefficient at 495 nm (4980 M^{-1} cm^{-1}) is almost identical to the value of 4580 M^{-1} cm^{-1} found for the subunit of Dx. These observations suggest that each molecule of Dfx contains one Dx-like FeS_4 center. Because the

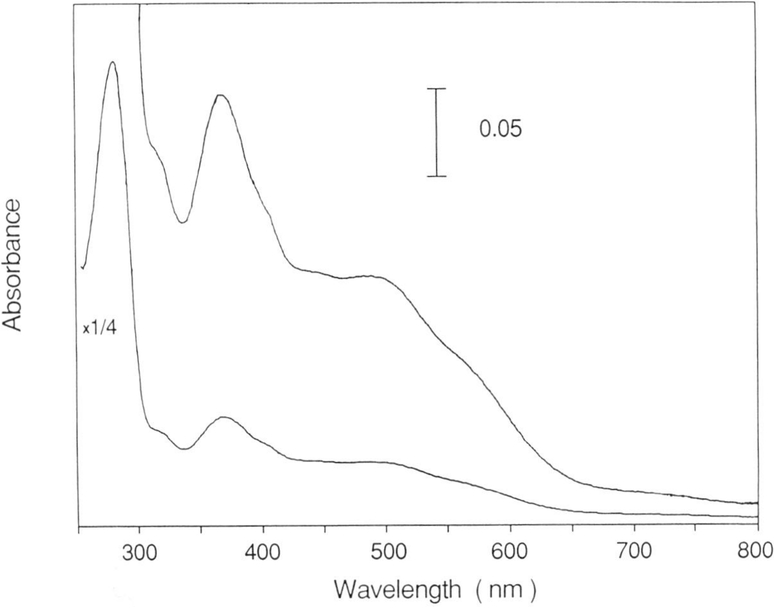

FIG. 5. Optical spectra of desulfoferrodoxin from *D. desulfuricans* (ATCC 27774) obtained for the as-isolated half-reduced (pink) form.

iron determination yields two iron atoms per molecule of Dfx, a second iron site with little or no visible absorption must be present. It is shown in the section Mössbauer Spectroscopy (below) that the as-isolated Dfx does contain two distinct iron sites: a Dx-like $Fe^{III}S_4$ and a ferrous site. This form corresponds to the semireduced form. This is unusual because, under aerobic conditions, mononuclear iron centers in proteins generally exist in the ferric state. This protein was also isolated in the totally oxidized form.[18] In this form both metal centers are seen in the ferric state. The visible spectrum presents extra contributions at 335 and 640 nm imparting a gray color to the protein. The semireduced protein is pink. From this point on the terms *pink* and *gray* mean semireduced and completely oxidized protein, respectively. This gray form can be separated in the last step of purification in HPLC on an anionic exchange column. The gray form is less acidic than the pink form and elutes first from the column. The pink form can be converted to the gray form by oxidation with potassium ferricyanide and the gray form can be converted to the pink form by reduction with sodium ascorbate. Both forms can be completely reduced by sodium dithionite.

Electron Paramagnetic Resonance Spectroscopy

Figure 6 shows the 4.3 K EPR spectrum of the pink form of Dfx. The EPR spectrum of the pink form has features at $g = 7.7, 5.7, 4.1$, and 1.8 that can be attributed to center I. These signals are due to a high-spin ferric ion ($S = 5/2$) with positive D and $E/D = 0.08$. The small signal at $g = 4.3$ is attributed to a minor fraction of center II in the oxidized form and is typical of a high-spin ferric ion ($S = 5/2$). The dithionite-reduced sample shows a broad derivative signal at $g = 10.9$ that is typical of an $S = 2$ system. It is inferred that this signal must be due to the reduced form of center I, as center II shows no such EPR signal in the pink form.[13]

Mössbauer Spectroscopy

Mössbauer studies were performed[13] on both pink and reduced forms. The Mössbauer spectrum of the pink form recorded at 4.2 K with a magnetic field of 50 mT applied parallel to the γ beam, shown in Fig. 7A, has two major components: a dominant central quadrupole doublet (center II) and a major magnetic spectral component extending from -4 to 5 mm/sec (center I). The Mössbauer absorption intensities for the quadrupole doublet and the major magnetic component were estimated to be $44 \pm 4\%$ and $40 \pm 4\%$, respectively, of the total absorption. The Mössbauer

[18] P. Tavares, N. Ravi, M. Y. Liu, J. LeGall, B. H. Huynh, J. J. G. Moura, and I. Moura, *J. Inorg. Biochem.* **43,** 264 (1991).

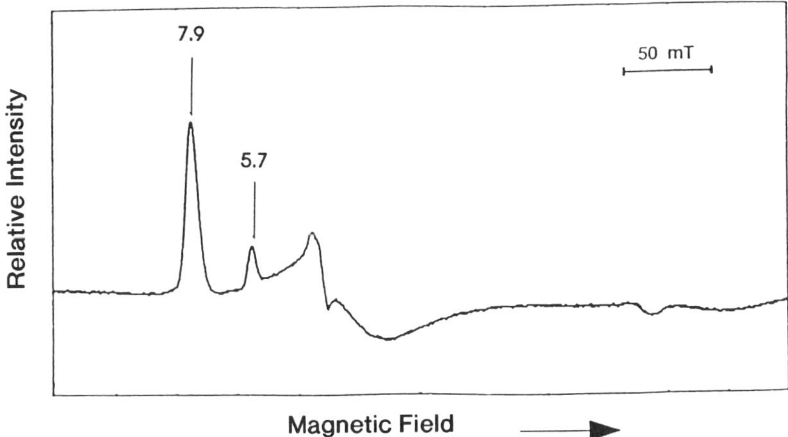

FIG. 6. EPR spectra of desulfoferrodoxin from *D. desulfuricans* (ATCC 27774) obtained for the half-reduced (pink) form. The experimental conditions are as follows: temperature, 4.3 K; microwave frequency, 9.45 GHz; microwave power, 2.37 mW; modulation amplitude, 0.5 mT; receiver gain, 2.0×10^4.

parameters obtained for the quadrupole doublet ($\Delta E_Q = 2.8$ mm/sec and $\delta = 1.04$ mm/sec) are characteristic of high-spin ferrous ion ($S = 2$) with octahedral nitrogenous/oxygenous coordination. In strong applied fields (>0.5 T) this ferrous ion exhibits paramagnetic properties indicative of mononuclear character. The spectral shape of the magnetic component is similar to that of the ferric Dx from *D. gigas*, and we have used the hyperfine parameters reported previously for that protein and the resulting simulation shown in Fig. 7A agrees well with the experimental spectrum.

There are two other magnetic components in the Mössbauer spectrum of the pink form, each one accounting for ~8% of the total absorption. The component with the largest magnetic splitting can be attributed to a small amount of center II existing in the oxidized form as described in EPR Spectroscopy (above). The hyperfine parameters used to simulate this magnetic component[13] are typical of a ferric site with nitrogenous and/or oxygenous ligands. The other component with smaller magnetic splitting is consistent with a ferric site with $D = 1.5$ cm^{-1}, $E/D = 0.14$, and $A/g_n\beta_n = 17.0$ T. This value of $A/g_n\beta_n$ is typical for iron with tetrahedral sulfur coordination, suggesting that a slightly different iron environment exists for the Dx-like FeS_4 center. The reduced form shows a Mössbauer spectrum, at 4.2 K in the absence of a magnetic field, composed of two sharp quadrupole doublets (Fig. 7B). The two quadrupole doublets

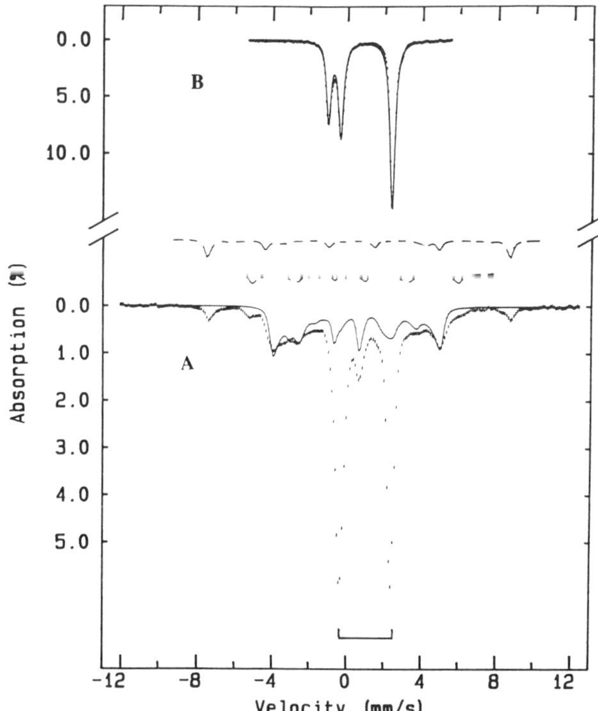

FIG. 7. Mössbauer spectra of the ^{57}Fe *D. desulfuricans* (ATCC 27774) desulfoferrodoxin: (A) pink form recorded at 4.2 K with 50-mT magnetic field applied parallel to the γ radiation. The solid line on the spectrum is a simulation using the parameters of the desulforedoxin from *D. gigas* (14). Simulated spectra of the two minor components are plotted as dashed lines at the top. (B) Reduced desulfoferrodoxin recorded at 4.2 K. The solid line is a least-squares fit of the experimental data assuming two quadrupole doublets.

have the following parameters: ΔE_Q = 3.51 ± 0.03 mm/sec and δ = 0.69 ± 0.02 mm/sec (center I) and ΔE_Q = 2.87 ± 0.03 mm/sec and δ = 1.04 ± 0.02 mm/sec (center II). The parameters of the first doublet compare very well with those of the reduced *D. gigas* Dx, confirming that Dfx contains a desulforedoxin-like FeS_4 site. The parameters of the other doublet are identical to the parameters of high-spin ferrous doublet observed in the pink form.

Redox Properties

The redox potentials of center I and center II were evaluated by a visible redox titration followed at two different wavelengths.[18] At pH 7.6

the values were found to be $E_0 = 4 \pm 10$ mV and $E_0 = 240 \pm 10$ mV for centers I and II, respectively. These values are in accordance with the fact that during purification we obtain the protein in two different oxidation states: totally oxidized (gray form) or semireduced (pink form).

A Protein Containing a Six-Iron Cluster

A novel iron sulfur protein was purified from the extract of *D. desulfuricans* (ATCC 27774)[19] and also from *D. vulgaris* (Hildenborough) by Hagen and co-workers.[20,21] Here we describe the properties of the protein isolated from *D. desulfuricans* and some comparisons are made to the protein isolated from *D. vulgaris*.

Physicochemical Characterization

Purification

The purification procedure is the same as performed for Dfx. In the second step of the purification scheme of Dfx a brown fraction is eluted from the hydroxylapatite column at an ionic strength of 80 m*M* phosphate buffer. This fraction is collected in a total volume of 100 ml. At this stage the protein already shows a distinct visible spectrum resembling that of an iron-sulfur protein and the ratio of A_{400}/A_{280} is determined to be 0.24. The fraction is concentrated on a Diaflo apparatus with a YM5 membrane and its ionic strength is decreased by adding distilled water during concentration. The last step of purification is performed by high-performance liquid chromatography on a Protein-Pack Column-DEAE SPW (Waters). A linear gradient of 0.01–0.2 *M* NaCl in 0.01 *M* potassium phosphate, pH 7, is applied with a flow of 2 ml/min for 3 hr. After this step the protein is found to have an A_{400}/A_{280} ratio of 0.33.

General Properties

The purified protein is a monomer of 57-kDa molecular mass. It contains comparable amounts of iron and inorganic labile sulfur atoms. The number of iron atoms per protein molecule of 57 kDa varies from 4.8 to 8.3, showing no particular dependence on the protein preparations or on the methods used for protein determination. The average value is 6.5 ±

[19] I. Moura, P. Tavares, J. J. G. Moura, N. Ravi, B. H. Huynh, M.-Y. Liu, and J. LeGall, *J. Biol. Chem.* **267,** 4489 (1992).

[20] W. R. Hagen, A. J. Pierik, and C. Veeger, *J. Chem. Soc. Faraday Trans. I* **85,** 4083 (1989).

[21] A. J. Pierik, R. B. G. Wolberg, P. H. A. Mutsaers, W. R. Hagen, and C. Veeger, *Eur. J. Biochem.* **206,** 697 (1992).

0.9 iron atoms per molecule. The visible spectrum of the as-isolated protein, which exhibits maxima at 400, 305, and 280 nm, is typical of an iron-sulfur protein. The molar extinction coefficient at 400 nm is determined to be within the range of 25–34 mM^{-1} cm^{-1} using protein determination based on the Lowry method.

Genetic Studies

The gene encoding this protein has been cloned from *D. desulfuricans* ATCC 27774 and sequenced by Stokkermans *et al.*[22] An open reading frame was found encoding a 545-amino acid protein (*MM* 58,496). The amino acid sequence is highly homologous with that of the corresponding protein from *D. vulgaris* (Hildenborough).[23] The two proteins have 66% of the residues being identical. The homology is highest in the C-terminal region (residues 140–540) and in the first 45 residues. There are 11 cysteines in the sequence of the protein isolated from *D. desulfuricans* whereas in the *D. vulgaris* protein the number of cysteines is 9. There is a cysteine-motif on the conserved N-terminal domain of both proteins (Cys-x_2-Cys-x_{12}-Gly-x-Cys-Gly) that is also found in the α subunit of the CO dehydrogenase from *Methanothrix soehngenii*.[22] There are also six histidines in both proteins that are conserved in the same positions. Some of them may be involved in iron coordination. The gene encoding the [6Fe] protein from *D. vulgaris* was inserted into broad host range vector pSUP104. The recombinant plasmid, pJSP104, was transferred to *D. vulgaris* by conjugal plasmid transfer. In the transconjugant *D. vulgaris* cells the six-iron protein was 25-fold overproduced.[24]

Spectroscopic Studies

Mössbauer Spectroscopy

Figure 8 shows the Mössbauer spectra of the as-isolated protein measured at 140 K (spectrum A) in the absence of a magnetic field, and in 50 mT and 3 T fields applied parallel to the γ radiation at 1.5 K and 4.2 K (spectra B and C, respectively). The 1.5 K spectrum (Fig. 8, spectrum B) shows two spectral components: a paramagnetic and a quadrupole doublet of equal intensity. Mössbauer spectra measured with an applied field from 1 to 8 T unambiguously prove that the central quadrupole doublet is a spin-

[22] J. P. W. G. Stokkermans, W. A. M. van den Berg, W. M. A. M. van Dongen, and C. Veeger, *Biochim. Biophys. Acta* **1132,** 83 (1992).

[23] J. P. W. G. Stokkermans, A. J. Pierik, R. B. G. Wolbert, W. R. Hagen, W. M. A. M. van Dongen, and C. Veeger, *Eur. J. Biochem.* **208,** 435 (1992).

[24] J. P. W. G. Stokermans, P. H. J. Houba, A. J. Pierik, W. R. Hagen, W. M. A. M. van Dongen, and C. Veeger, *Eur. J. Biochem.* **210,** 983 (1993).

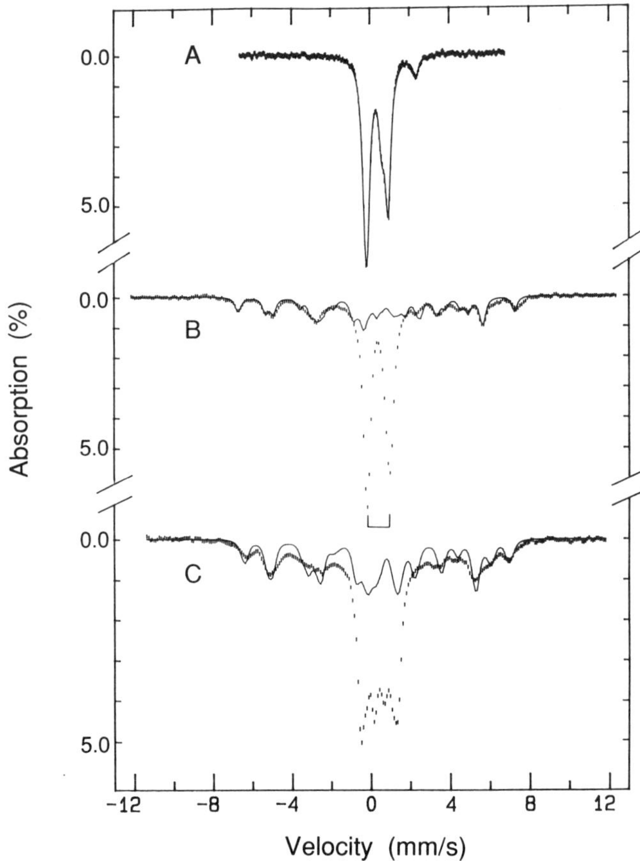

FIG. 8. Mössbauer spectra of the as-isolated new iron-sulfur protein from *D. desulfuricans* (ATCC 27774). The data were recorded at 140 K in the absence of a magnetic field (spectrum A), at 1.5 K in a parallel field of 50 mT (spectrum B), and at 4.2 K in a parallel field of 3 T (spectrum C). The solid line in (A) is a least-squares fit to the data, assuming three quadrupole doublets; the solid lines plotted in (B) and (C) are spectral simulations of the paramagnetic [6Fe] cluster, using the parameters listed in Table I. The simulated spectra in (B) and (C) are normalized to 48% of the total absorption.

coupled cluster with diamagnetic properties ($S = 0$), and the paramagnetic component reveals six distinct antiferromagnetically coupled iron sites as judged by the inward and outward movements of the absorption lines. The lack of direction of the field dependence of the Mössbauer spectrum, indicating a uniaxial property with either $D < 0$ or $E/D = 1/3$ of this paramagnetic center, has been exploited in order to analyze this complex spectral pattern within the spin Hamiltonian framework. Analysis of these strong-field spectra and the weak-field spectrum has been performed by using the following spin Hamiltonian:

$$\hat{H} = D\left[S_z^2 - \frac{S(S+1)}{3} + \frac{E}{D}(S_x^2 - S_y^2)\right] + \beta \mathbf{S}\tilde{g}\mathbf{H} + \mathbf{S}\tilde{A}\mathbf{I} + \frac{eQV_{zz}}{4}\left[I_z^2 - I(I+1)/3 + \frac{\eta}{3}(I_x^2 - I_y^2)\right] - g_n\beta_n\mathbf{HI} \quad (1)$$

The magnetic component is treated as a superposition of six spectral components, each corresponding to an iron site. All six sites share the same electronic state (i.e., the same D, E/D, and S values) but have different magnetic hyperfine coupling $\tilde{A}$ tensor and electric field gradient $\tilde{V}$ tensor. On the basis of EPR data presented below, we have assigned a spin $S = 9/2$ and an $E/D = 0.062$ for this magnetic cluster.[19]

In Fig. 8 (spectra B and C, respectively), 50 mT and 3T spectra are shown along with the simulation. The parameters used for the simulation are given in Table I. This analysis provides direct spectroscopic evidence for the presence of a [6Fe] cluster in this newly purified protein (cluster

TABLE I

HYPERFINE PARAMETERS DETERMINED AT 1.5 K FOR SIX-IRON SITES OF PARAMAGNETIC CLUSTER IN AS-ISOLATED IRON-SULFUR PROTEIN FROM *Desulfovibrio desulfuricans* (ATCC 27774)[a]

Iron site	ΔE_Q (mm/sec)	δ (mm/sec)	η	$A/g_n\beta_n(T)$
1	−1.0	0.40	8.5	−10.0
2	−1.0	0.40	3.0	−7.8
3	−1.0	0.27	3.0	7.0
4	2.67	1.15	5.0	5.5[b]
5	−1.0	0.27	0	−4.0
6	1.0	0.27	0	−1.5

[a] Other parameters used in the analysis are $S = 9/2$, $D = -1.3\ \text{cm}^{-1}$, $E/D = 0.062$, and full-width at half-maximum = 0.35 mm/sec.

[b] The magnetic hyperfine A tensor for a high-spin ferrous ion is generally anisotropic. In this analysis, however, an isotropic A tensor is assumed for the ferrous site because the system is uniaxial, and the spectrum is sensitive only to the A value along the uniaxis.

I). In Fig. 9, we also show the same analysis for the six iron sites after subtracting 50% of the diamagnetic spectra from the raw experimental data with the corresponding individual sites plotted above the 50 mT spectrum.

One of the iron sites exhibits parameters (ΔE_Q = 2.67 mm/sec and δ = 1.09 mm/sec at 140 K; Fig. 8, spectrum A) typical for high-spin ferrous

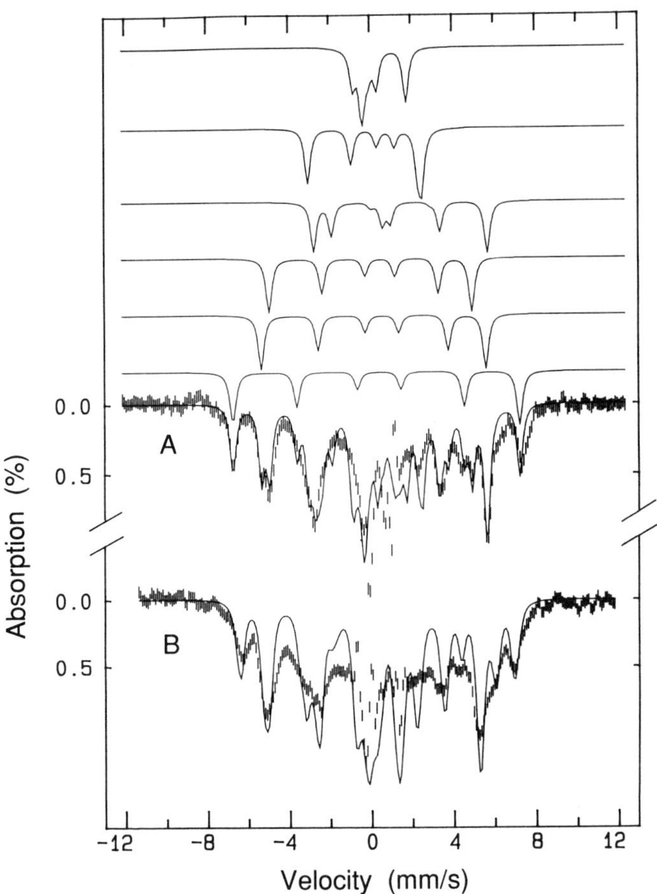

FIG. 9. Mössbauer spectra of the paramagnetic [6Fe] cluster. The spectra are prepared from the raw data shown in Fig. 8. The experimental conditions are (A) 1.5 K with a parallel field of 50 mT and (B) 4.2 K with a parallel field of 3 T. For comparison, spectral simulations for the [6Fe] cluster are plotted as solid lines over the experimental data. Simulated 1.5 K spectra for the six individual iron sites at an applied field of 50 mT are shown on top of spectrum A. These spectra are plotted in a descending order with respect to increasing absolute value of the magnetic hyperfine coupling constant A. The iron sites are labeled in an order corresponding to decreasing absolute value of A (see Table I).

ion accounting for 8% of the total absorption. The observed large isomer-shift indicates an iron environment that is distinct from the tetrahedral sulfur coordination commonly observed for the iron atoms in iron-sulfur clusters and is consistent with a penta- or hexacoordination containing N and/or O ligands. The other five sites are presumably in the high-spin ferric state. Three of them show parameters characteristic for tetrahedral sulfur coordination while the parameters of the other two are suggestive of N/O coordination in coordination sphere higher than a tetrahedron.

Electron Paramagnetic Resonance Spectroscopy

The EPR spectrum (Fig. 10) of this iron-sulfur protein in its as-isolated form shows (1) a resonance peak at $g = 15.3$, (2) a group of resonances

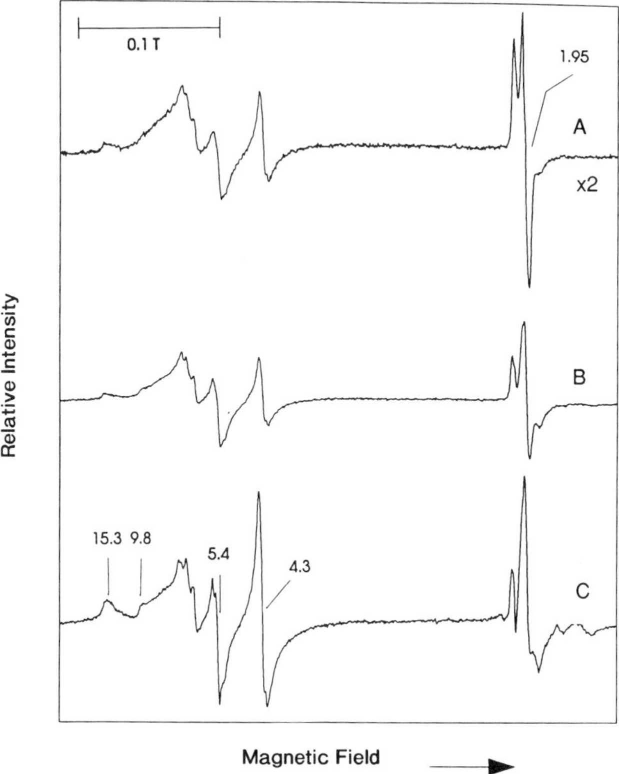

FIG. 10. EPR spectra of the as-isolated *D. desulfuricans* (ATCC 27774) [6Fe]-containing protein. (A) 25 K; (B) 16 K; (C) 4.4 K. Other experimental conditions are as follows: microwave frequency, 9.46 GHz; microwave power, 2 mW; modulation amplitude, 1 mT; receiver gain, 1.6×10^5 for spectrum (C), and 1.0×10^5 for spectra (A) and (B).

in the region between g = 9.8 and g = 5.4, (3) a signal at g = 4.3, and (4) several signals in the g = 2 region. The signal at g = 4.3 is typical for adventitiously bound ferric ions and may represent a minor impurity. The signals at the g = 2 region are most probably originating from S = 1/2 species. At least two sets of g values can be identified in this region because their temperature-dependent relaxation behaviors are different. At temperatures above 25 K, one set of g values at 2.02, 1.98, and 1.95 is detected, and at lower temperatures an additional set of signals at g = 1.97, 1.94 and 1.90 is observed. Spin quantitation of these signals amounts to not more than 0.07 spin/molecule.

We can associate the magnetic component seen in the Mössbauer spectrum with the other EPR resonances. For a half-integer spin system, the smallest spin value that could yield a g value of 15.3 is S = 9/2. This spin system forms five Kramers doublets, and their principal g values can be calculated using the electronic part of the Hamiltonian described in Eq. (1), assuming an isotropic g tensor of 2.0. Table II lists the calculated principal g values and the energies for all five Kramers doublets using zero-field parameters $D = -1.3\ \text{cm}^{-1}$ obtained from the Mössbauer analysis and E/D = 0.062. According to this analysis, the g = 15.3 signal could be arising from the highest doublet ($S_z = \pm 1/2$) whereas the resonances between g = 9.8 and 5.4 could be originating from the other excited doublets. The g = 18 value for the ground doublet is not observable because its transition probability is almost nil. The temperature dependence of the intensity of the g = 15.3 signal indicates that it arises from an excited state. We believe by examining redox titrations followed by EPR and Mössbauer spectroscopies that cluster I can be stabilized in five oxidation states whereas cluster II can exist in two. The redox data are extremely complicated and have not yet been completely understood.

TABLE II

Principal g Values and Energies for Five Kramers Doublets of Spin 9/2 System with $D = -1.3\ \text{cm}^{-1}$ and $E/D = 0.062$

Doublet	Energy (cm^{-1})	g_x	g_y	g_z
1	0.0	0.0	0.0	18.0
2	10.4	0.01	0.01	13.97
3	18.1	0.54	0.51	9.89
4	23.1	5.73	5.80	5.22
5	26.4	3.72	15.3	1.31

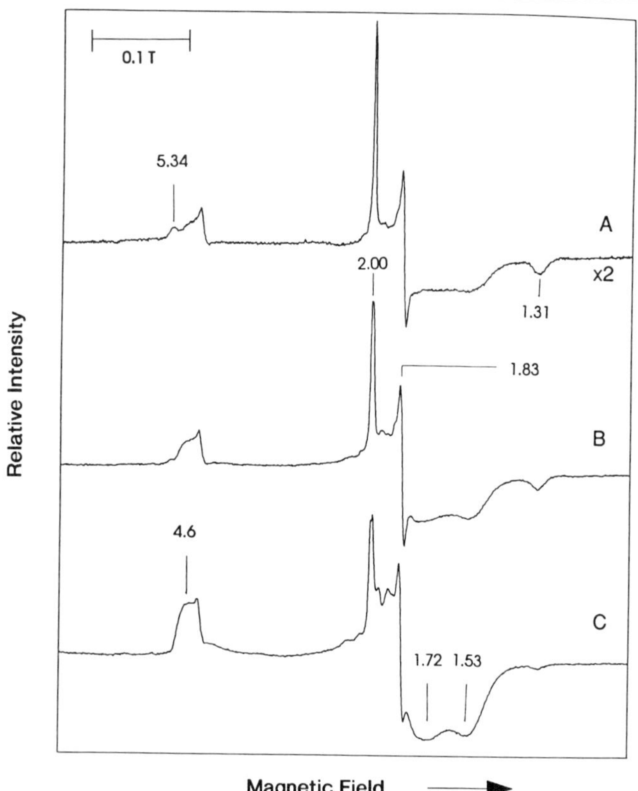

FIG. 11. EPR spectra of the dithionite-reduced *D. desulfuricans* (ATCC 27774) [6Fe]-containing protein. (A) 16 K; (B) 7.5 K; (C) 4.3 K. Other experimental conditions are as follows: microwave frequency, 9.46 GHz; microwave power, 2 mW; modulation amplitude, 1 mT; receiver gain, 4×10^4.

The EPR spectrum of a dithionite-reduced sample is presented in Fig. 11. In the high-field region, an intense $S = 1/2$ rhombic signal with g values at 2.00, 1.83, and 1.31 is observed. A similar signal has been reported for the analogous protein from *D. vulgaris* (Hildenborough) and this signal has been suggested to represent a prismane [6Fe-6S] cluster based on the similar g values observed for the synthetic model compounds.[25] Spin quantitation of this signal yields near stoichiometric concentration, approximately 0.8 spin/molecule. At lower temperature additional features are observed at $g = 1.72$ and 1.53. At 4.2 K, the signal

[25] M. G. Kanatzidis, W. R. Hagen, W. R. Dunham, R. K. Lester, and D. Couconvanis, *J. Am. Chem. Soc.* **107,** 953 (1985).

becomes complex and the peak at $g = 2.0$ splits into two peaks with a peak-to-peak separation of 2.4 mT. These observations suggest the presence of magnetic interaction between this $S = 1/2$ species and another nearby paramagnetic center.

In addition to the above-mentioned high-field features in the EPR spectrum, a broad signal at around $g = 4.7$ is observed at low temperature (below 7 K). With increasing temperature an additional peak at $g = 5.34$ appears. These g values and temperature behavior are consistent with an electronic system of spin $S = 3/2$ with positive D and $E/D = 0.24$, and an intrinsic isotropic g_0 of 1.87. For such a system the theoretical g values for the ground and excited Kramers doublets are (2.35, 4.84, and 1.58) and (1.39, 1.10, and 5.32), respectively. Whether these signals arise from two different clusters or the same cluster is not yet fully understood. Mössbauer investigations are in progress to clarify this situation. Mössbauer and EPR studies of the analogous protein from *D. vulgaris* were also performed.[26] In spite of the similarity between the properties of the two proteins, the data were analyzed in a different manner and it was concluded by the authors that this protein contains only one [6Fe] cluster.

Final Remarks

Sulfate reducers, as we have demonstrated in this chapter, contain a variety of proteins with novel iron centers. The role of these proteins is not known but the study of their spectroscopic properties and comparison with other known enzymes will contribute to the search for their biological functions. These three new proteins offer the unique opportunity of exploring the meaning of the association between simple metallic centers (i.e., rubredoxin and desulforedoxin) forming more complex situations as well as opening new insights in the research domain of iron-sulfur clusters, by revealing the chemical properties of the new six iron-containing protein.

Acknowledgments

We would like to acknowledge Dr. Jean LeGall who, in the years of collaboration, has been able to transmit to us the knowledge and the spirit of searching new proteins. It has been a rewarding experience to discover, for the first time, some of the proteins described here. We are also deeply indebted to Dr. B. H. Huynh for his active participation, encouragement, and illuminating discussions during the course of this study. One of us (N.R.) wishes to acknowledge the financial support by National Institutes of Health Grant GM 47295 to Dr. B. H. Huynh.

[26] A. J. Pierik, W. R. Hagen, W. R. Dunham, and R. H. Sands, *Eur. J. Biochem.* **206,** 705 (1992).

[16] Adenylylsulfate Reductases from Sulfate-Reducing Bacteria

By JORGE LAMPREIA, ALICE S. PEREIRA, and JOSÉ J. G. MOURA

Introduction

Inorganic sulfur usually enters biosynthetic pathways as a series of oxidative reactions producing sulfate from reduced forms of sulfur such as $S_2O_3^{2-}$, S^0, and S^{2-}. The sulfur cycle is completed by a series of reductive reactions involving the generation of sulfide from sulfate. Multiple enzymatic systems are required, reflecting the specific use of sulfate in each case. Sulfate can be considered both as a terminal electron acceptor in anaerobic respiration (*dissimilatory sulfate reduction*) or as raw material for the biosynthesis of cysteine (*assimilatory sulfate reduction*).[1]

The sulfate molecule is rather inert chemically and must be activated in order to enter any of the above pathways. All the organisms that utilize sulfate contain the enzyme ATP-sulfurylase, which catalyzes the formation of adenylylsulfate (APS) (Fig. 1) and inorganic pyrophosphate (PP_i) from ATP and sulfate. The APS molecule has twice the energy of the comparable ADP molecule, and the equilibrium of the reaction lies in the direction of ATP and sulfate. It is assumed that an inorganic pyrophosphatase shifts the equilibrium toward APS formation by hydrolyzing the PP_i molecules.

In most organisms, APS is phosphorylated in the 3′ position to ATP by APS kinase to yield 3′-phosphoadenylyl sulfate [3′-phosphoadenosine 5′-phosphosulfate (PAPS)]. This sulfur-containing molecule is the sulfate donor for the sulfotransferase reactions leading to the biosynthesis of sulfate esters. Adenylylsulfate kinase has not been found in sulfate-reducing bacteria that directly reduce APS to sulfite and AMP. Sulfite is then further reduced to sulfide by bisulfite reductase.

Assimilatory Sulfate Reduction Pathways

The key reaction in assimilatory sulfate reduction that differentiates it from dissimilatory sulfate reduction is the transferring of the sulfonate group of APS or PAPS to a thiol group producing the corresponding nucleotide (AMP or PAP) plus a thiosulfonate:

$$\text{APS (PAPS)} + \text{RSH} \rightleftharpoons \text{AMP (PAP)} + \text{RSSO}_3^{1-} + \text{H}^+ \qquad (1)$$

[1] J. LeGall and G. Fauque *in* "Biology of Anaerobic Microorganisms," (Alexander J. B. Zehnder, Ed.), p. 587, John Wiley and Sons, New York, 1988.

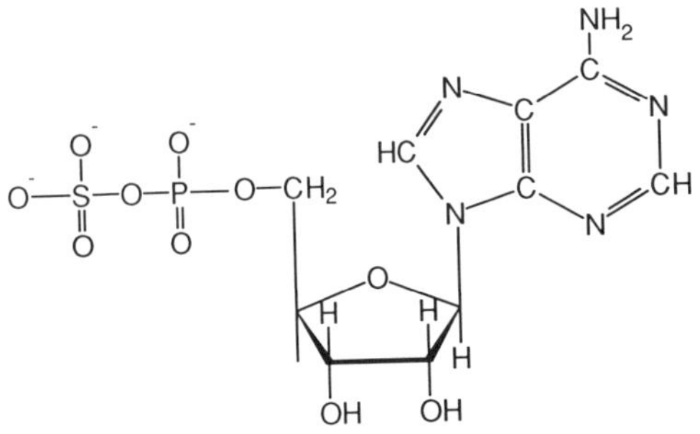

FIG. 1. The adenylylsulfate (APS) molecule.

This thiol is either a low molecular weight compound such as glutathione,[2] or a protein such as thioredoxin[3,4] or glutaredoxin.[5]

The pathways of assimilatory sulfate reduction differ in terms of their specificity for APS or PAPS and the nature of the acceptor thiol group. The PAPS sulfotransferase pathway is found in most eubacteria, yeast, and fungi and utilizes thioredoxin or glutaredoxin as its thiol acceptor.[6] The PAPS thioredoxin sulfotransferase belongs to the cysteine regulon in enterobacteria and is encoded in the *cys* operon. The APS sulfotransferase pathway is found in green plants,[7] algae,[8,9] and some photosynthetic bacteria.[10,11] The APS sulfotransferase is highly specific for APS; in the few cases it has been investigated in green plants, the APS sulfotransferase is regulated by the end products of the reductive process, H_2S and/or cysteine.[12,13]

On the basis of their mechanism, function, and regulation, one would expect that APS and PAPS thiol sulfotransferases would prove to be a group of homologous proteins and completely unrelated to the APS reductases, which would form a homologous group on their own.

[2] M. L.-S. Tsang and J. A. Sciff, *Plant Sci. Lett.* **11,** 177 (1978).
[3] L. G. Wilson, T. Asahi, and R. S. Bandvurski, *J. Biol. Chem.* **236,** 1822 (1961).
[4] P. G. Porque, A. Baldestein, and P. Reichard, *J. Biol. Chem.* **245B,** 2731 (1961).
[5] M. Rassel, P. Model, and A. Holmgren, *J. Bacteriol.* **172,** 1923 (1990).
[6] H. D. Peck, Jr., and T. Lissolo, *Symp. Gen. Microbiol.* **42,** 99 (1988).
[7] A. Schmidt, *Plant Sci. Lett.* **5,** 407 (1975).
[8] M. L.-S. Tsang and J. A. Shiff, *Plant Sci. Lett.* **4,** 301 (1975).
[9] A. Schmidt, *FEMS Microbiol. Lett.* **1,** 137 (1977).
[10] H. G. Trüper and V. Fischer, *Philos. Trans. R. Soc. London B* **298,** 99 (1982).
[11] A. Schmidt, *Arch. Microbiol.* **112,** 264 (1977).
[12] C. Brunold and A. Schmidt, *Planta* **133,** 85 (1976).
[13] B. E. Jenni, C. Bruunold, J.-P. Zrÿd, and P. Lavanchy, *Planta* **150,** 140 (1980).

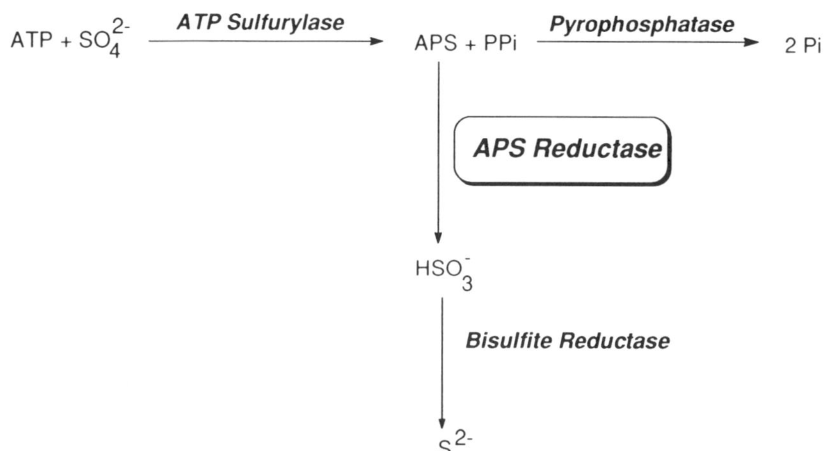

FIG. 2. Pathway of dissimilatory sulfate reduction.

Dissimilatory Sulfate Reduction Pathway

The possibility of utilizing sulfate as a terminal electron acceptor in a respiratory-like process is restricted and defines a large group of prokaryotes known as sulfate-reducing bacteria (SRB). The dissimilatory sulfate reduction implies the formation of large amounts of H_2S and the process has never been shown to be regulated in any way. In all SRB studied to date, APS reductase is present at 1 to 3% of the total soluble protein and appears to be the key enzyme in respiratory sulfate reduction. The pathway of dissimilatory sulfate reduction involves four enzymes: ATP-sulfurylase, inorganic pyrophosphatase (PP_i), APS reductase, and bisulfite reductase as shown in Fig. 2. The specific electron donors required for the reduction of APS and bisulfite are yet unknown.

The bisulfite reductases, treated in detail elsewhere in this volume can produce thiosulfate and trithionate during the reduction of bisulfite.[14] On the basis of their visible spectra, four different types are normally characterized: desulfoviridin,[15] desulfofuscidin,[16] P-590[17] desulforubidin.[18]

Adenylylsulfate reductases are also found in some thiobacilli and anoxygenic phototrophs, which oxidize sulfite to sulfate by a reverse pathway

[14] K. Kobavashi, Y. Seki, and M. Ishimoto, *J. Biochem.* (*Tokyo*) **75,** 519 (1973).

[15] J.-P. Lee and H. D. Peck, Jr., *Biochem. Biophys. Res. Commun.* **45,** 583 (1973).

[16] E. C. Hatchikian and J. G. Zeikus, *J. Bacteriol.* **153,** 1211 (1983).

[17] P. A. Trudinger, *J. Bacteriol.* **104,** 158 (1970).

[18] J.-P. Lee, C. S. Yi, J. LeGall, and H. D. Peck, Jr., *J. Bacteriol.* **115,** 453 (1973).

similar to the one presented in Fig. 2, and they are the subject of two chapters in this volume.

Adenylylsulfate Reductases from Sulfate-Reducing Bacteria

We focus our attention on APS reductases from SRB belonging mainly to the genus *Desulfovibrio*, although it has been found in the genus *Desulfotomaculum*[19] and in *Desulfobacter postgatei*, *Desulfococcus multivorans*, *Desulfobulbus propionicus* and *Desulfosarcina variabilis*.[20] An extensive characterization of the reductase follows, with emphasis on its occurrence, active site composition, and interaction with natural substrates in order to enumerate the known details contributing to a general understanding of its mechanism.

Physicochemical Characterization

Purification. The purification of APS reductase must be carried out in such a way that it maximizes its final purity and yet minimizes the time required to accomplish it. This reductase has poor stability and its activity will begin to decrease as soon as it is purified from the crude extract. The purity of the reductase is checked during the procedure by activity measures, but mainly by visible spectral characteristics, the value of the ratio (the lower this value, the purer the enzyme), and, in the final steps of the purification, by sodium dodecyl sulfate–polyacrylamide gel electrophoresis (SDS–PAGE).

The bacterial crude extract is applied to a DEAE-cellulose column equilibrated with 10 m*M* Tris-HCl, pH 7.6. The column is washed with 200 ml of 20–30 m*M* Tris-HCl to wash out all the nonadsorbed proteins, and a linear salt gradient (30–300 m*M* Tris-HCl, pH 7.6) is then applied. Adenylylsulfate reductase is eluted around 0.17 m*M*. The enzyme solution is dialyzed overnight against 10 m*M* Tris buffer. This fraction is applied to a BioGel-DEAE agarose column and the previous steps repeated. A further BioGel column is used depending on the initial amount of the crude extract. One final purification step is carried out by high-performance liquid chromatography (HPLC) with a 2.1 × 20 cm DEAE TSK 5PW [Tsoba (Tokyo, Japan) or Waters (Milford, NJ)] or Mono Q (Pharmacia, Piscataway, NJ) column and, if necessary, an extra step using a 60 × 2.1 cm TSK G-3000 gel-permeation column. The reductase is then

[19] W. G. Skyring, *Can. J. Microbiol.* **19,** 375 (1973).

[20] W. Stille and H. G. Trüper, *Arch. Microbiol.* **137,** 145 (1984).

$$APS + 2\ \text{methyl viologen}^{-} \longrightarrow AMP + SO_3^{2-} + 2\ \text{methyl viologen}$$

$$SO_3^{2-} + AMP + 2\ Fe(CN)_6^{3-} \longrightarrow APS + 2\ Fe(CN)_6^{2-}$$

$$SO_3^{2-} + AMP + 2\ \text{cytochrome}\ c_{ox} \longrightarrow APS + 2\ \text{cytochrome}\ c_{red}$$

FIG. 3. Adenylylsulfate reductase activity assays.

concentrated and frozen in small beads in liquid nitrogen, owing to its poor stability.[21] All the purification steps are carried out at 4–6°.

This general procedure is slightly modified with some extracts of *Desulfovibrio gigas* and *Desulfovibrio desulfuricans* ATCC 27774, in which an HTP column is utilized, as it appears to be the only way to remove the catalase enzyme. Normally APS reductase is eluted in this column with the descendent gradient of Tris-HCl buffer, pH 7.6 (0.2 to 10 m*M*).

Stability. The purified enzyme is unstable during storage and loses much of its activity after 30 days at −20°. Even in liquid nitrogen, it aggregates and precipitates in less than 6 months. Activity is not recovered on solubilization of the precipitate. Thus, and as a general rule, APS reductase is to be purified as needed (the activity is more stable in crude extract fractions).

Activity Studies. As previously stated, APS reductase catalyzes the reaction of adenylylsulfate to sulfite and AMP in the presence of reduced methyl viologen, but the reverse reaction can be accomplished in the presence of ferricyanide or O_2/cytochrome *c*[22] (Fig. 3).

In general the activity studies of APS reductase are carried out using the reverse reaction and ferricyanide as electron acceptor. This method is used to allow some consistency in data between different groups and also because this activity is known to be related to the presence of iron-sulfur centers.[23]

The reaction mixture contains 1 m*M* ferricyanide, 3 m*M* sodium sulfite in 5 m*M* ethylenediaminetetraacetic acid (EDTA), 3.3 m*M* AMP, and 0.5 m*M* mercaptoethanol in 100 m*M* Tris-HCl, pH 7.6, and the enzyme in a final volume of 1 ml. The enzyme is incubated in this medium minus sulfite for 10 min and sulfite is then added to initiate the reaction. The reduction of ferricyanide is followed at 420 nm ($\varepsilon = 1.02\ mM^{-1}\ cm^{-1}$), using the

[21] J. Lampreia, I. Moura, M. Teixeira, H. D. Peck, Jr., J. LeGall, B. H. Huynh, and J. J. G. Moura, *Eur. J. Biochem.* **188,** 653 (1990).

[22] R. N. Bramlett and H. D. Peck, Jr., *J. Biol. Chem.* **250,** 250 (1975).

[23] R. N. Bramlett, Ph.D. Thesis, University of Georgia, Athens (1975).

mixture without enzyme as the blank. The APS reductase specific activity unit is defined as 1 μmol of APS reacted min^{-1} mg^{-1}.

General Characterization. Adenylylsulfate reductase is a soluble cytoplasmic component. However, it was shown that in *Desulfovibrio thermophilus* the reductase is clearly membrane associated.[24] The reductases from the studied SRB are [Fe-S] flavin (FAD)-containing proteins with eight iron atoms, determined by plasma emission and the TPTZ colorimetric method,[25] and eight sulfur atoms per molecule arranged in two [4Fe-4S] clusters as is shown below, and 1 FAD/mol determined after extraction (protein precipitation with trichloroacetic acid) using $\varepsilon = 11.3\ mM^{-1}\ cm^{-1}$. Their molecular mass ranges between 150 and 180 kDa as determined by SDS–PAGE and HPLC gel-filtration chromatography, using Pharmacia molecular mass standards in both cases. They possess two different subunits with molecular masses of around 70 and 20 kDa and seem to be arranged in an $\alpha_2\beta$ pattern, with the exception of the reductase from *Archaeglobus fulgidus*, an extremely thermophilic archaebacterium,[26] which is described as being a dimer, each unit of which has a molecular mass of 80 kDa.

Spectroscopic Studies

Spectroscopic studies have been carried out extensively in a large number of different APS reductases in the native form, in the presence of natural substrates (AMP and sulfite) and in the reduced form. The APS reductase from *D. gigas* will be used as an example.

Ultraviolet–Visible Spectroscopy. The ultraviolet (UV)–visible spectrum of the native APS reductase shows a broad maximum around 392 nm ($\varepsilon = 50.0\ mM^{-1}\ cm^{-1}$) with shoulders at 445 and 475 nm and a protein absorption peak at 278 nm. The overall spectrum indicates the presence of [Fe-S] centers, and is consistent with the presence of a flavin group (Fig. 4).

The addition of sulfite to the reductase causes an absorption decrease between 500 and 340 nm and a slight increase in absorbance around 320 nm, indicating the formation of an adduct between the FAD molecule and sulfite[27] [Fig. 5 (left), spectrum B]. Further addition of AMP leads to a minor decrease in the absorbance of the overall spectrum [Fig. 5 (left),

[24] D. R. Kremer, M. Veenhuis, H. D. Peck, Jr., J. LeGall, J. Lampreia, J. J. G. Moura, and T. A. Hansen, *Arch. Microbiol.* **150,** 296 (1988).

[25] D. S. Fisher and D. C. Price, *Clin. Chem.* **10,** 10 (1969).

[26] N. Speich and H. G. Trüper, *J. Gen. Microbiol.* **134,** 1419 (1988).

[27] G. B. Michaels, J. T. Davidson, and H. D. Peck, Jr., *Biochem. Biophys. Res. Commun.* **39,** 321 (1970).

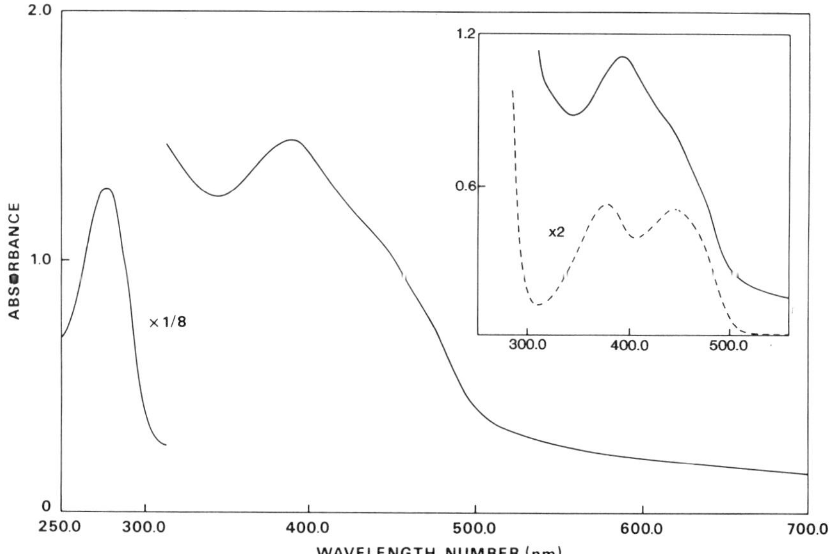

FIG. 4. Absorbance spectrum of *D. gigas* APS reductase. *Inset:* Spectrum of flavin extracted as described in text.

spectrum C]; however, the presence of AMP alone does not alter in any way the reductase visible spectrum. Addition of dithionite to the enzyme causes further bleaching of the spectrum [Fig. 5 (left), spectrum D].

Difference spectra were obtained in order to understand the role of the chromophore reactivity toward sulfite, AMP, and dithionite (Fig. 5, right). A flavin-like spectrum is obtained when the native spectrum is run against the sulfite-reacted reductase. The negative absorption value in the 320-nm region indicates the formation of the described flavin-sulfite adduct [Fig. 5 (right), spectrum A].[28] A broad peak in the 400- to 450-nm region is detected in the difference spectrum between AMP plus sulfite-reacted enzyme and the sulfite-reacted enzyme [Fig. 5 (left), spectrum B minus spectrum A], indicating that AMP, in the presence of sulfite, may be involved in the reduction of the [Fe-S] centers present [Fig. 5 (right), spectrum C].

Electron Paramagnetic Resonance Spectroscopy. Electron paramagnetic resonance spectroscopic studies were also conducted with the addition of the same substrates and reductants.

The native enzyme shows an almost isotropic signal with a *g* value centered around 2.02 and detectable at temperatures below 18 K (Fig.

[28] F. Müller and V. Massey, *J. Biol. Chem.* **244,** 4007 (1969).

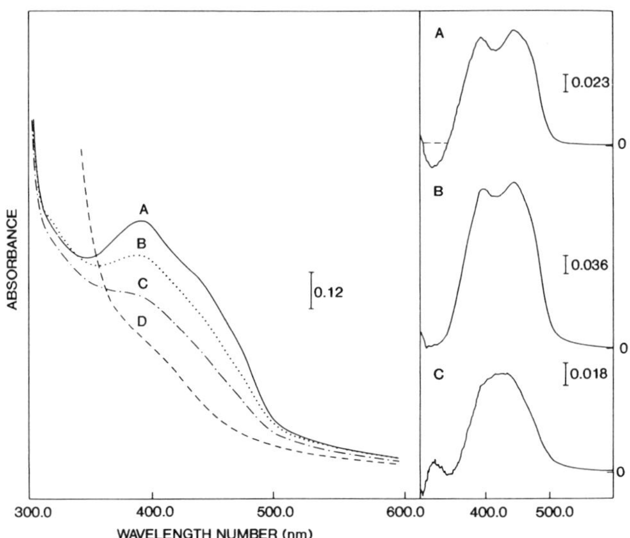

FIG. 5. Effects of substrates and reductants on the visible spectrum of *D. gigas* APS reductase. *Left:* A, native enzyme; B, after sulfite addition; C, spectrum B after AMP addition; D, spectrum C after dithionite addition. *Right* (difference spectra): A, native minus sulfite reacted; B, native minus (sulfite + AMP reacted); C, sulfite reacted minus (sulfite + AMP reacted) (spectrum B − spectrum A).

6A). The integration value for this signal varies (0.10–0.25 spins/mol). This signal is undoubtedly due to iron as confirmed by the observation of a 1.5 mT line broadening in an EPR spectrum run on APS reductase purified from an ^{57}Fe-enriched batch.[20] On the other hand, no other metal was detected besides iron. The spectral shape, g value, and temperature dependence indicate that this signal can be attributed to a [3Fe-4S] center; however Mössbauer studies were not able to detect the presence of this center, probably due to its low concentration.[20]

The addition of AMP and APS (Fig. 6B and C) causes no change in the native APS reductase spectrum. The addition of sulfite alone to the native reductase causes a very weak "$g = 1.94$"-type signal (Fig. 6D), depending on the samples, and is generally attributed to the existence of residual and endogenous AMP.[22] The fact that the addition of sulfite alone causes almost no change in the EPR spectrum is in great contrast with what is observable in visible spectroscopy, in which it is seen that sulfite produces a strong decrease in absorbance. However, the addition of AMP to the sulfite-reacted enzyme causes a major change in the EPR spectrum in which the isotropic signal starts to decrease and a rhombic signal rises with g values of 2.096, 1.940, and 1.890 characteristic of a reduced [Fe-S] center (center I), with an intensity around 0.35–0.50 spin/mol (Fig. 6E).[20]

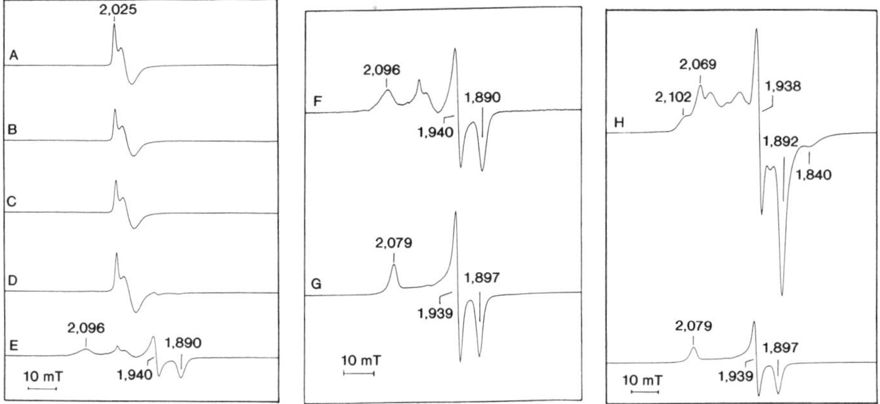

FIG. 6. EPR spectra of *D. gigas* APS reductase. A, Native enzyme; B, native enzyme plus AMP; C, native enzyme plus APS; D, native enzyme plus sulfite; E and F, native enzyme plus sulfite plus AMP; G, native enzyme plus dithionite (15 sec) (center I); H, native enzyme plus dithionite (>30 min) ("fully reduced"). Modulation, 1 mT; microwave power, 2 mW; temperature, 8 K.

The addition of chemical reductants such as dithionite (1 g of sodium dithionite to 50 ml of degassed 0.2 *M* Tris-HCl, pH 9.5) for a short reduction time (15 sec) or reduced methyl viologen to the native enzyme causes the total disappearance of the isotropic signal and the evolution of the rhombic signal designated as center I to its full intensity (0.78–1.0 spin/mol) with *g* values of 2.079, 1.939, and 1.897 (Fig. 6F). Mössbauer spectroscopy shows that this signal is due to a reduced [4Fe-4S] center and that the sample is in a half-reduced state, indicating the existence of yet another [Fe-S] center in the oxidized form.[20]

The full reduction of the reductase is accomplished only with the addition of dithionite for at least 30 min, and it presents a complex EPR spectrum similar to those describing interacting [Fe-S] centers,[29] and confirms the presence of another [4Fe-4S] center (center II) (Fig. 6G). The total signal accounts for 1.5 to 1.96 spin/mol, depending on the samples. However, center II is never fully reduced under the usual experimental conditions.

Temperature dependence studies indicate that the two centers are different with respect to their relaxation properties. Whereas center I still exists at temperatures around 45 K, center II is visible only at temperatures below 25 K.[20]

[29] R. Cammack, D. P. E. Dickinson, and C. E. Johnson, *in* "Iron-Sulfur Proteins" (W. Lovenberg, ed.), Vol. 3, p. 283. Academic Press, New York, 1977.

Mössbauer Spectroscopy. A detailed Mössbauer study of *D. gigas* APS reductase was published previously,[20] and thus only a summary of the Mössbauer data pertinent to the comprehension of the visible and EPR results is discussed here.

The Mössbauer spectrum of the native protein confirms that the majority of the iron atoms are in a diamagnetic environment (Fig. 7). Its spectroscopic characteristics [ΔE_Q = 1.3 mm/sec and isomer shift (δ) = 0.45 mm/sec], the reduction of ΔE_Q with increasing temperature, the diamagnetism, and the overall pattern of the spectra are all characteristic of [4Fe-4S] clusters, showing that the iron atoms present in the reductase are organized

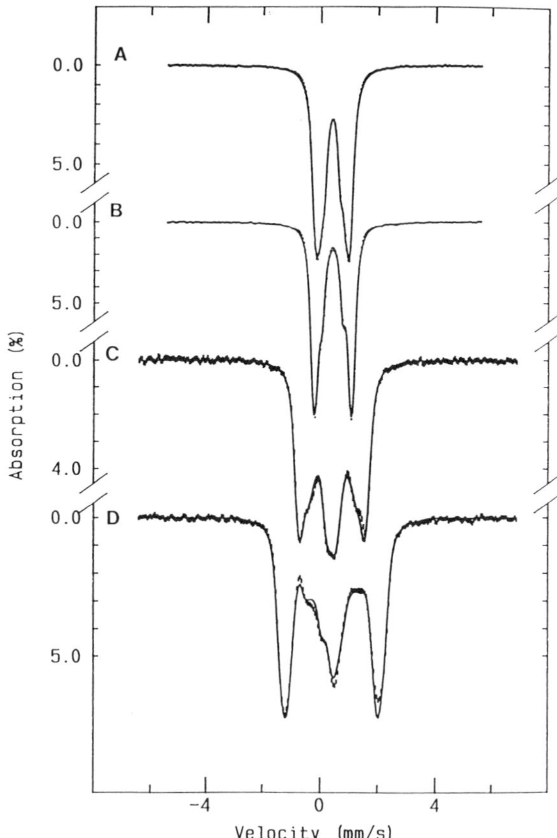

FIG. 7. Mössbauer spectra of native *D. gigas* APS reductase. Data were recorded at 100 K (A) and 4.2 K (B–D); in zero magnetic field (A) and with parallel applied field of 50 mT (B), 4.0 T (C), and 8.0 T (D). The solid lines in spectra A and B are the result of a least-squares fitting with four doublets of equal intensity; the solid lines in spectra C and D are the result of a simulation for diamagnetic species.

in the form of two [4Fe-4S] centers. The detectable isotropic EPR signal in the native form must be caused by an interconversion of one of the [4Fe-4S] centers to a [3Fe-4S] cluster undetectable by Mössbauer spectroscopy.

The spectra of the partially methyl viologen- or dithionite (15 sec)-reduced sample show the existence of one oxidized diamagnetic cluster and one reduced magnetic cluster, which in association with the EPR data proves to be center I.

Mössbauer spectroscopy also shows that center I is spectroscopically different when obtained with chemical reductants or AMP plus sulfite, which is in accordance with the EPR data previously presented. Again, it indicates that there is some kind of direct interaction between AMP and center I when sulfite is present.

The Mössbauer spectrum of the "fully reduced" (dithionite, 30 min) reductase shows that 85–92% of the total iron absorption originated from reduced [4Fe-4S] centers, indicating that center II is now reduced. It also shows that the two centers are indeed interacting (Fig. 8C), producing the complex EPR spectrum shown before (Fig. 6G). However, this spin–spin interaction is relatively weak, as it is easily broken in the presence of an applied field of 1.0 T (Fig. 8D). It also shows that the two reduced centers are spectroscopically distinguishable (Fig. 8G).

Redox Properties

Extensive redox titrations followed by EPR spectroscopy were carried out in a number of APS reductases from several *Desulfovibrio* strains.

These titrations were conducted in an anaerobic cell slightly modified from the one designed by Dutton.[30] The samples are kept at 25° in a positively pressurized argon atmosphere purified by flushing it through two vessels containing an alkaline solution of sodium dithionite. The system is calibrated with quinhydrone and the pH is buffered using 0.1 *M* Tris-HCl, pH 9.5. The redox potentials are varied by adding aliquots of 0.1 *M* sodium dithionite with a gas-tight Hamilton syringe. The redox potentials of the solution are stabilized using dye mediators as described.[31] The samples are transferred to the EPR tubes and immediately frozen in liquid nitrogen.

From the obtained data it is possible to ascertain that center I has an atypical redox potential (0 to −50 mV) (reminiscent of one of the [Fe-S] clusters [H-cluster] detected in *Desulfovibrio vulgaris* hydrogenase[32]), which is in agreement with the observed EPR data, namely the coexistence

[30] P. L. Dutton, *Biochim. Biophys. Acta* **226,** 63 (1971).

[31] I. Moura, A. V. Xavier, R. Cammack, M. Bruschi, and J. LeGall, *Biochim. Biophys. Acta* **533,** 156 (1978).

[32] D. Patil, J. J. Moura, H. S. He, M. Teixeira, B. C. Pickrill, D. V. Dervartanian, H. D. Peck, Jr., J. Le Gall, and B. H. Huynh, *J. Biol. Chem.* **263,** 18732 (1988).

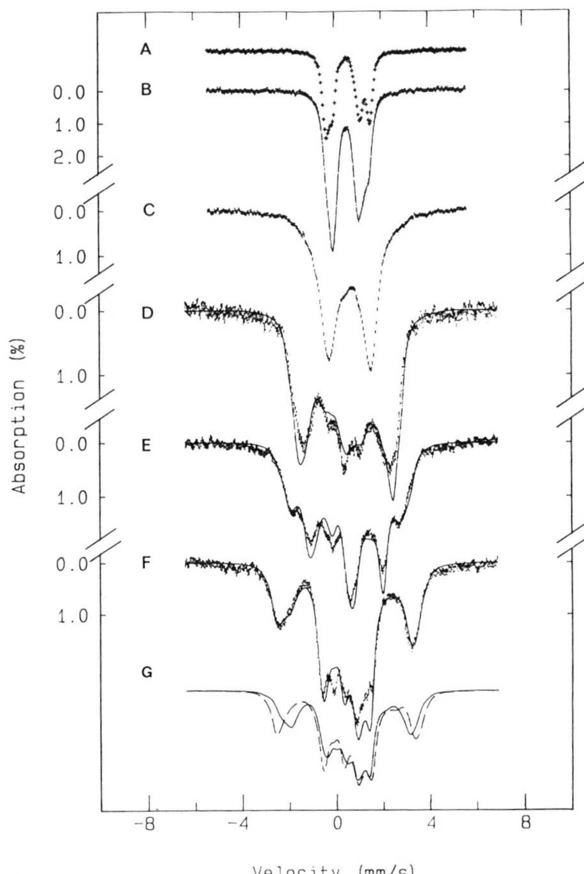

FIG. 8. Mössbauer spectra of "fully reduced" *D. gigas* APS reductase (B–F). A-Prepared spectrum of center I in a methyl-viologen reduced sample. Temperature: 100 K (A, B) and 4.2 K (C–F). A–C were recorded in the absence of an external magnetic field. D–F were recorded with parallel applied magnetic fields of 1.0 T (D), 4.0 T (E), and 8.0 T (F). The simulated 8.0-T spectra are shown in (G); center I, dashed line; center II, solid line.

of the isotropic signal and center I in the sulfite plus AMP-reacted reductase.

The midpoint redox potential of center II can only be estimated to be lower than −400 mV. However, in the cases of *D. salexigens* and *D. desulfuricans* Berre-eau, an estimation indicates that it is lower than −450 mV[33] (Fig. 9).

[33] J. Lampreia, I. Moura, A. V. Xavier, J. Le Gall, H. D. Peck, Jr., and J. J. G. Moura, *in* "Chemistry and Biochemistry of Flavoenzymes" (F. Müller, ed.), Vol. 3, 333, CRC Press, Boca Raton, Florida, 1992.

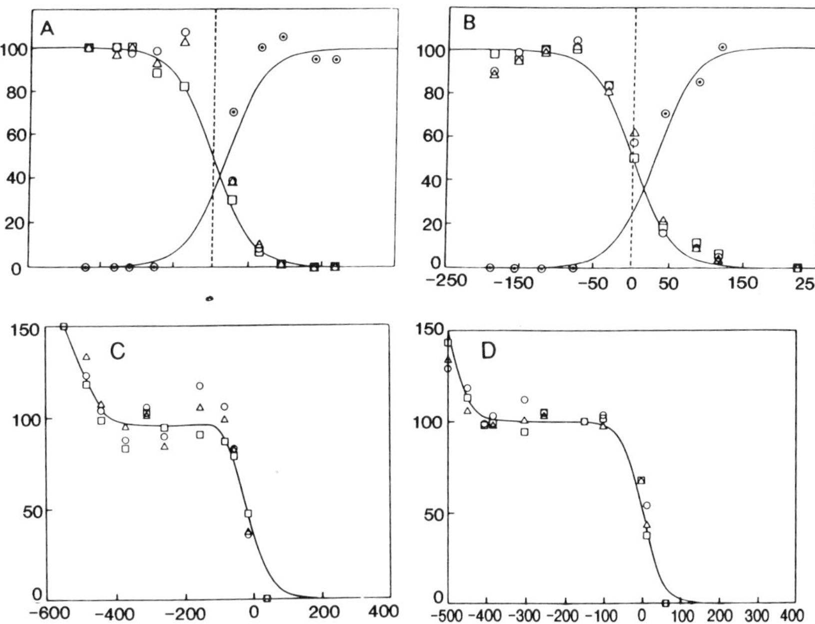

FIG. 9. EPR redox titrations using the procedure described in text. (A) *D. gigas* APS reductase; (B) *D. desulfuricans* (ATCC 27774) APS reductase; (C) *D. desulfuricans* (Berre-eau) APS reductase; (D) *D. salexigens* APS reductase. X-axis, redox potential (mV); y-axis, relative intensity.

Comparison of Several Adenylylsulfate Reductases

Adenylylsulfate reductases from seven other species and strains of *Desulfovibrio* (*D. vulgaris* Hildenborough, *D. desulfuricans* (Berre-eau), *D. desulfuricans* (ATCC 27774), *Desulfomicrobium baculatum* (NCIB 9774) (formerly *Desulfovibrio baculatus*), *Desulfovibrio salexigens* (British Guiana), *Desulfovibrio multispirans*, and *D. thermophilus*, and one from *A. fulgidus*[34] have been purified to homogeneity and their properties and spectroscopic characteristics determined. As can be seen in Tables I and II and Figs. 10–12, they have a high degree of homology not only regarding their physicochemical properties but also in their visible and EPR spectra.

The composition of their active centers is conserved (two [4Fe-4S] centes per flavin), as are their EPR and visible spectra. Also, the unusual midpoint redox potential of center I is characteristic of all the APS reductases studied so far.

[34] J. Lampreia, G. Fauque, N. Speich, C. Dahl, I. Moura, H. G. Truper, and J. J. G. Moura, *Biochem. Biophys. Res. Commun.* **181,** 342 (1991).

TABLE I
PHYSICOCHEMICAL PROPERTIES OF ADENYLYLSULFATE REDUCTASE[a]

	Molecular mass (KDa)	Iron (Fe/mol)	FAD (FAD/mol)	Specific activity (μmol APS min^{-1} mg^{-1})	pH optimum	K_m sulfite (m*M*)	K_m AMP (m*M*)	Redox potential		
								g = 202	Center I	Center II
D. gigas	160	8–9	1	6	7.4	0.34	0.16	≈60	0	<−400
D. vulgaris (Hildenborough)	180, 220[21]	9–10, 12[21]	1[21]	1.25	7.4	0.6[22]	0.3[22]	nd	nd	nd
D. desulfuricans (ATCC 27774)	170	8–9	nd	1.1	7.5	0.78	0.16	≈60	−30	<−400
D. desulfuricans (Berre-eau)	nd	7–8	nd	0.75	7.4	nd	nd	≈60	−25	<−450
D. baculatus	170	7–8	1	0.6	nd	nd	nd	nd	nd	nd
D. salexigens	180	nd	1	2.5	7.5	0.76	0.31	≈60	0	<−450
D. multispirans	170	7–8	nd	nd	nd	nd	nd	nd	nd	nd
D. thermophilus	175	8–9	nd	nd	7.6	nd	nd	nd	nd	nd
A. fulgidus[25]	160	8	1	1.65	8.0	1.3	nd	nd	nd	nd

[a] nd, Not determined.

TABLE II

VISIBLE AND ELECTRON PARAMAGNETIC RESONANCE SPECTROSCOPIC PROPERTIES OF ADENYLYLSULFATE REDUCTASE[a]

	Visible spectroscopy		EPR spectroscopy						
			g values			Spin quantitation			
	Maximum (nm)	Adduct	Native ($g = 2$)	AMP + sulfite (center I)	Semireduced (center I)	Native ($g = 2$)	AMP + sulfite (center I)	Semireduced (center I)	Fully reduced (center I + center II)
D. gigas	392	+	2.025 2.002 2.002	2.096 1.940 1.890	2.079 1.939 1.897	0.1–0.2	0.3–0.52	0.75–1.13	1.50–1.79
D. vulgaris (Hildenborough)	375	+	2.031 2.002 2.002	2.092 1.945 1.899	2.070 1.942 1.891	0.21	nd	0.70–0.95	1.80
D. desulfuricans (ATCC 27774)	388	+	2.023 2.002 2.002	2.086 1.934 1.899	2.077 1.935 1.894	0.14–0.025	0.16–0.28	0.98–1.1	1.45–1.85
D. desulfuricans (Berre-eau)	380	+	2.021 2.002 2.002	2.073 1.936 1.887	2.072 1.935 1.886	0.1–0.22	0.15–0.30	0.85–1.10	1.60–1.90
D. baculatus	385	+	2.022 2.002 2.002	2.085 1.939 1.895	2.085 1.939 1.895	0.15	nd	0.85–1.15	1.65–1.85
D. salexigens	392	+	2.027 2.008 2.008	2.093 1.944 1.896	2.083 1.939 1.899	0.038–0.1	0.15–0.25	0.90–1.00	1.80–1.96
D. multispirans	393	nd	2.026 2.007 2.007	2.096 1.946 1.897	2.079 1.941 1.897	nd	nd	nd	nd
D. thermophilus	394	+	2.025 2.004 2.004	2.097 1.938 1.888	2.088 1.939 1.897	nd	nd	nd	nd
A. fulgidus	400	nd	2.040 2.020 2.020	2.095 1.947 1.909	2.095 1.947 1.909	0.0035	0.17	1.00	1.75

[a] nd, Not determined.

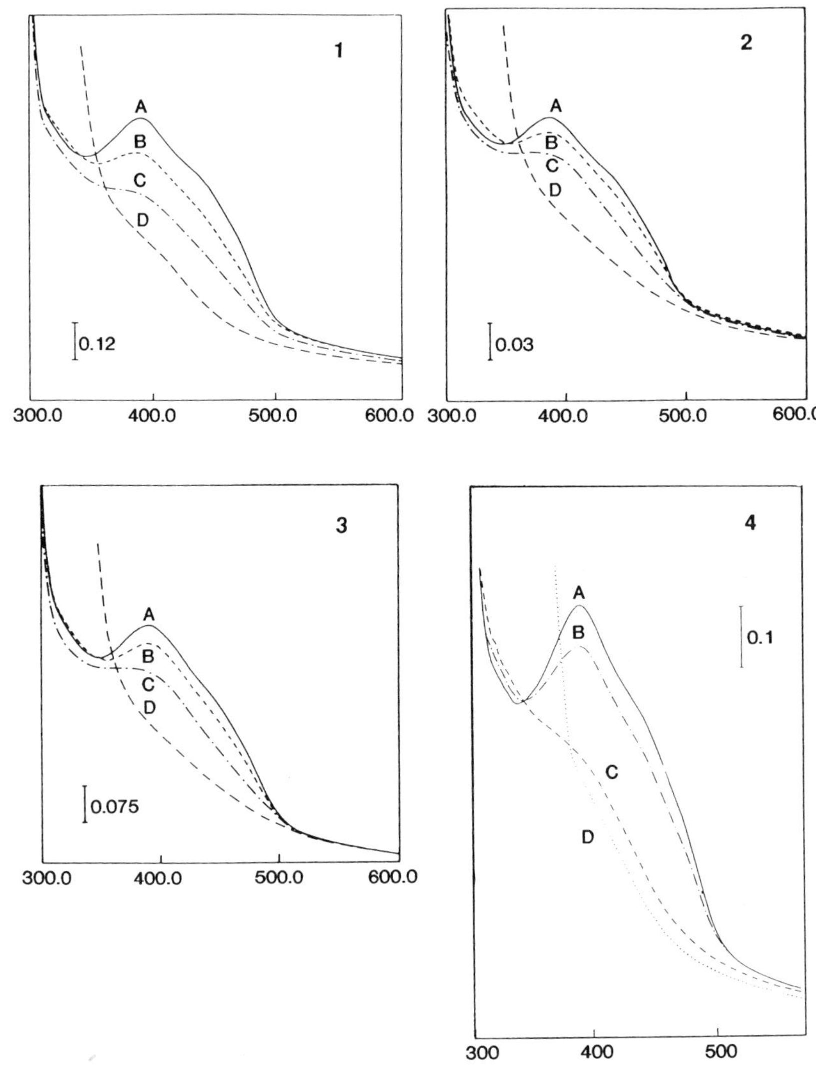

FIG. 10. Visible spectra of several APS reductases. (1) *D. gigas*; (2) *Dm. baculatum*; (3) *D. salexigens*; (4) *A. fulgidus*. The spectra are labeled as in Fig. 5 (left). X-axis, wavelength number (nm); y-axis, absorbance.

It is evident that APS is a well-conserved enzyme that represents an optimized solution for the crucial enzymatic reaction it must accomplish in SRB. It is probably due to this degree of homology that Odom devised a technique for the rapid detection of SRB in the environment, based on polyclonal antibodies to APS reductase (paper [41] in this volume).

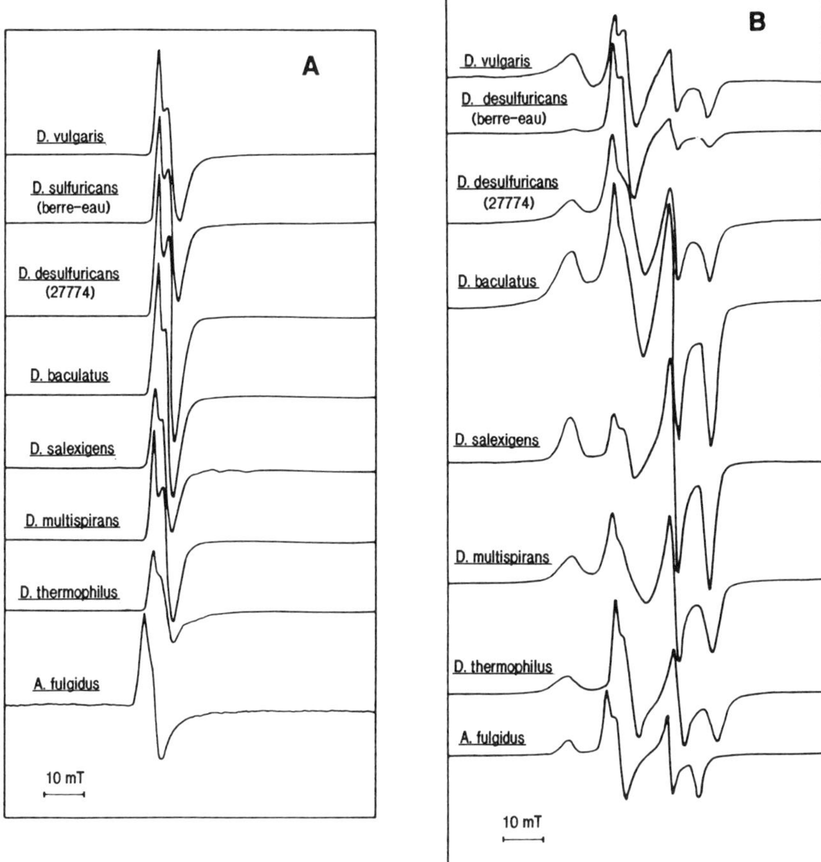

FIG. 11. EPR spectra of several APS reductases. (A) Native enzyme; (B) native enzyme plus sulfite and AMP.

On the Mechanism of Adenylylsulfate Reductase: Final Remarks

The general aspects of the mechanism of APS reductase in SRB have been described since 1970,[26] especially concerning the formation of the flavin–sulfite adduct. However, it was only in 1982 that the possible involvement of an iron-sulfur center was put forward.[35] Here we summarize the current understanding of the active sites of APS reductase in SRB.

[35] H. D. Peck, Jr., and R. N. Bramlett, *in* "Flavins and Flavoproteins" (V. Massey and C. H. Williams, eds.), p. 851, Elsevier, Amsterdam, 1982.

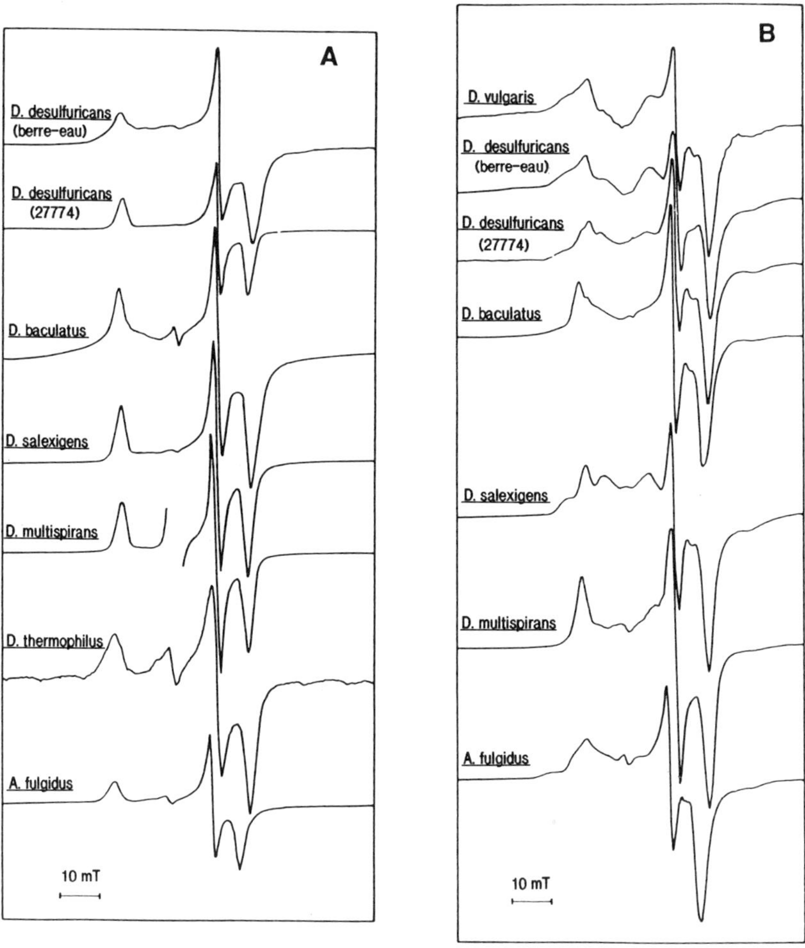

FIG. 12. EPR spectra of several APS reductases. (A) Native enzyme plus dithionite (15 sec); (B) native enzyme plus dithionite (>30 min).

1. Adenylylsulfate reductase is a protein containing one FAD per molecule and eight iron atoms arranged as two [4Fe-4S] clusters.
2. Sulfite forms an adduct with the flavin moiety present in the reductase, detectable by an increase in absorbance in the 320-nm region of the visible spectrum.
3. The addition of AMP alone to the native reductase produces no change in either the visible or EPR spectrum.

4. The addition of AMP to a sulfite-reacted sample causes a further decrease in absorbance in the visible spectrum.
5. The addition of sulfite to the enzyme induces no changes in the EPR spectrum but further addition of AMP produces the clear formation of center I and a strong decrease in the intensity of the isotropic native signal.
6. On addition of sulfite to the reductase, the formation of a semiquinone is never observed, in EPR or in the difference-visible spectrum.
7. The addition of p-hidroxymercurybenzoate (pHMB) to the enzyme destroys the reductase activity with ferricyanide.
8. *g* values of center I when obtained on addition of sulfite plus AMP are different than when the sample is reduced chemically.
9. Temperature dependence of center I when obtained on addition of sulfite plus AMP is different than when the sample is reduced chemically.
10. Further addition of dithionite to the sulfite plus AMP-reacted sample causes an increase of intensity in the EPR spectrum without changing the *g* values.
11. Mössbauer parameters for center I when obtained on addition of sulfite plus AMP are different than when center I parameters are obtained on chemical reduction.
12. The redox potential of center I is positive (0 to -50 mV) and atypical of [4Fe-4S] centers.
13. Center II is a [4Fe-4S] cluster with a redox potential lower than -400 mV.

In spite of some yet unresolved aspects of the mechanism of the reductase, it seems to be definitely proved that there is an interaction between center I and AMP once the FAD moiety of the enzyme has been reacted with sulfite enzyme. Exactly how this interaction takes place remains unsolved. On the other hand, the role of center II is yet unknown. One could imagine that this center might be involved in accepting electrons from the oxidation of $FADH_2$. If that is the case, because a [4Fe-4S] center is a one-electron acceptor, a flavin semiquinone should be easily detectable by EPR or visible spectroscopy. This is still a question under consideration.

A common feature of all the APS reductases from SRB is the observed perturbation of the EPR spectral features of center I after its reaction with sulfite and AMP as well as its atypical midpoint redox potential when compared with other [4Fe-4S] centers. The existence of coordinating amino acids other than cysteine at this center has been presented as an explanation for this unusual redox potential.[36]

The role of center II is yet unsolved. It is known that sulfite never accumulates in the growth medium of SRB; therefore it is elegant to speculate that a sulfite-bound APS reductase complex can be the substrate for bisulfite reductase *in vivo*.[37] Center II of APS reductase can then be involved in the interaction between this formed complex and bisulfite reductase. However, this hypothesis is far from being proved.

A further understanding of the APS reductase mechanism also requires a more extensive structural characterization of APS reductases from organisms that oxidize reduced sulfur compounds, such as the phototrophic bacteria and thiobacilli (Section IV in this volume) in order to detect similarities and differences among the enzymes characterized in SRB.

[36] A. J. Thomson, S. J. George, A. Armstrong, and C. Hatchikian, *in* "Proceedings of the XIII International Conference on Magnetic Resonance in Biological Systems." Madison, Wisconsin, 1988.

[37] H. D. Peck, Jr., and J. LeGall, *Philos. Trans. R. Soc. London B* **298,** 298 (1982).

[17] Thiosulfate and Trithionate Reductases

By J. M. AKAGI, H. L. DRAKE, JAE-HO KIM, and DIANE GEVERTZ

Trithionate and thiosulfate have both been implicated as intermediates during the respiratory sulfate reduction process occurring in dissimilatory sulfate-reducing bacteria. According to this hypothesis, bisulfite, originating from the reduction of adenylyl sulfate (APS), is reduced to sulfide sequentially through trithionate and thiosulfate as intermediates.[1] This pathway would require the participation of the three enzymes bisulfite reductase, trithionate reductase, and thiosulfate reductase. Bisulfite reductase reduces bisulfite to trithionate,[2] which in turn is reduced to thiosulfate. Two enzyme systems have been described that reduce trithionate to thiosulfate. These were designated the thiosulfate-forming enzyme[3] and the trithionate reductase system[4]; the latter activity has also been referred to as bisulfite reductase-dependent trithionate reductase.[5] Thiosulfate reductase

[1] K. Kobayashi, S. Tachibana, and M. Ishimoto, *J. Biochem.* (*Tokyo*) **65**, 155 (1969).

[2] J. P. Lee and H. D. Peck, *Biochem. Biophys. Res. Commun.* **45,** 583 (1971).

[3] H. L. Drake and J. M. Akagi, *J. Bacteriol.* **132,** 132 (1977).

[4] J. H. Kim and J. M. Akagi, *J. Bacteriol.* **163,** 472 (1985).

[5] J. M. Akagi, *in* "Biology of Inorganic Nitrogen and Sulfur" (H. Bothe and A. Trebst, eds.), p. 178. Springer-Verlag, Berlin, Heidelberg, and New York, 1981.

has been isolated from aerobic and anaerobic bacteria and its activity in microorganisms has been known for many years. This chapter describes how to purify and assay for trithionate and thiosulfate reductases from *Desulfovibrio vulgaris* (Hildenborough) NCIMB 8303.

The cultivation, harvesting, and preparation of cell extracts were described previously.[6]

Thiosulfate-Forming Enzyme

Assay Method

Principle. The most convenient method for assaying thiosulfate-forming enzyme (TF) under anaerobic conditions is by using manometry. Because the electrons for reducing bisulfite or trithionate to thiosulfate are derived from molecular hydrogen, the stoichiometry of the reaction may be determined. Thiosulfate and trithionate are measured according to Kelly *et al.*[7]

Reagents

$KHSO_3$, 0.1 *M*
$K_2S_3O_6$, 0.1 *M*
Potassium phosphate buffer (pH 6.0), 1 *M*
Methyl viologen, 0.01 *M*
KCN, 0.25 *M*
$CuSO_4$, 0.1 *M*
$Fe(NO_3)_3$, 1.5 *M* in 4 *N* $HClO_4$
$K_2S_3O_6$, prepared according to Roy and Trudinger[8]
Particulate hydrogenase prepared according to Suh and Akagi,[9] 10 mg/ml
Bisulfite reductase prepared as described previously[9]

Procedure. To determine the reduction of bisulfite to thiosulfate, both bisulfite reductase and TF are required. In the main compartment of a single-arm Warburg flask of 8-ml capacity is added 0.1 ml each of buffer, methyl viologen, hydrogenase, bisulfite reductase (0.5 mg), and TF. To the side arm is added 0.1 ml of $KHSO_3$. The total volume of the reaction mixture is 1.2 ml adjusted with water. The flask is gassed with H_2 for 10 min and allowed to equilibrate for 5 min at 37°. At zero time the $KHSO_3$

[6] J. M. Akagi and L. L. Campbell, *J. Bacteriol.* **84,** 1194 (1962).
[7] D. P. Kelly, L. A. Chambers, and P. A. Trudinger, *Anal. Chem.* **41,** 898 (1969).
[8] A. B. Roy and P. A. Trudinger, "The Biochemistry of Inorganic Compounds of Sulphur." Cambridge Univ. Press, Cambridge, 1970.
[9] B. Suh and J. M. Akagi, *J. Bacteriol.* **99** 210 (1969).

is tipped into the main compartment and the reaction is allowed to proceed for 30–60 min; the course of the reaction may be followed by observing the utilization of H_2. At the end of the experiment, a final reading is taken and the flasks are disconnected from their manometers. The contents of the flasks are transferred to a graduated test tube and the flasks are washed with approximately 1 ml of water and pooled with the reaction mixture. The volume of the mixture is made to 2.5 ml with water. One milliliter of this is removed and analyzed for thiosulfate according to Kelly *et al.*[7] If desired, another milliliter may be removed and analyzed for trithionate.[7] The formation of thiosulfate from bisulfite is shown in Table I. To assay for trithionate reduction to thiosulfate by TF, bisulfite reductase is not required but, in addition to $KHSO_3$, $K_2S_3O_6$ is also added to the side arm. Table II shows the requirement for bisulfite by TF in reducing trithionate to thiosulfate.

Purification of Thiosulfate-Forming Enzyme

The formation of thiosulfate from bisulfite was observed by Suh and Akagi[9] during their studies on the bisulfite-reducing system of *Desulfovibrio vulgaris*. They found that bisulfite was reduced to thiosulfate by a two-enzyme system consisting of desulfoviridin (bisulfite reductase) and another protein designated FII (fraction II). The FII was subsequently purified to homogeneity by Drake and Akagi[3] and the active fraction was

TABLE I
PRODUCTS FORMED BY THIOSULFATE-FORMING SYSTEM[a]

System	Products formed (μmol)		
	$S_3O_6^{2-}$	$S_2O_3^{2-}$	S^{2-}
BR[a] + HSO_3	2.37	0	0.23
BR + TF[b] + HSO_3^-	0.27	3.68	0.20
BR + $S_3O_6^{2-}$	—	0	0.10
BR + TF + $S_3O_6^{2-}$	—	2.56	0
TF + $S_3O_6^{2-}$	—	0	0
TF + HSO_3^-	0	0	0
TF + HSO_3^- + $S_3O_6^{2-}$	—	5.33	0
Boiled TF[c] + HSO_3^- + $S_3O_6^{2-}$	—	0	0

[a] BR, Bisulfite reductase.
[b] TF, Thiosulfate-forming enzyme. Reaction mixture as described in Procedure. BR, 1.2 mg; TF, 0.39 mg; $NaHSO_3$, 10 μmol; $K_2S_3O_6$, 11 μmol. Time, 60 min.
[c] TF placed in a boiling water bath for 10 min.

TABLE II
EFFECT OF SUBSTRATE CONCENTRATION ON THIOSULFATE FORMATION BY THIOSULFATE-FORMING ENZYME[a]

Substrate (μmol)		Thiosulfate formed (μmol)
$S_3O_6^{2-}$	HSO_3^-	
11.0	0.5	4.1
11.0	1	5.0
11.0	2	5.7
11.0	4	6.3
11.0	10	7.6
1.2	10	1.4
2.3	10	2.4
3.4	10	3.2
11.0	0	0
0	10	0

[a] TF concentration, 0.41 mg; time, 100 min. Assay conditions as described in Procedure.

given the trivial name thiosulfate-forming enzyme, or TF. The role of desulfoviridin in thiosulfate formation from bisulfite was elucidated when FII alone was shown to form thiosulfate from bisulfite plus trithionate. In the overall reaction of bisulfite reduction to thiosulfate, the role of desulfoviridin was to reduce bisulfite to trithionate, which was subsequently reduced to thiosulfate by FII. By using ^{35}S-labeled substrates, the mechanism of the FII-catalyzed reaction was elucidated:

$$HS'O_3^- + O_3S{-}S^*{-}SO_3^{2-} + H^+ \xrightarrow{2e^-} S^*{-}S'O_3^{2-} + 2HSO_3^-$$

Confirmation for the existence of TF was made by Peck and LeGall; however, because they did not detect this enzyme in extracts from *Desulfovibrio gigas*,[10] they questioned the validity of the existence of a trithionate pathway for bisulfite reduction in sulfate-reducing bacteria.

Preparation of Cell Extract. All purification steps are carried out at 0–4°. When crude extracts of *D. vulgaris* are fractionated to purify TF, the fractions containing other activities, that is, bisulfite reductase and cytochrome c_3, are separated and saved for later fractionations. When frozen wet cells of *D. vulgaris* are thawed, the cell suspensions are usually viscous and difficult to manipulate. Add a small amount of DNase to the cell suspension and hold in an ice bath until the viscosity decreases. Cells are then disrupted with a French pressure cell as described previously.[9]

[10] H. D. Peck and J. LeGall, *Philos. Trans. R. Soc. London B* **298,** 443 (1982).

Ammonium Sulfate Precipitation. To 600 ml of crude extract solid ammonium sulfate is added to 0.3 saturation and the mixture is centrifuged at 27,000 *g* for 20 min. The precipitate is discarded and the supernatant fraction is made to 0.5 saturation with ammonium sulfate and centrifuged. The precipitate is dissolved in a minimum amount of distilled water and dialyzed against 30 vol of 1 m*M* buffer, pH 7.0 (potassium phosphate buffer is used throughout this purification).

Ion-Exchange Chromatography. The dialyzed fraction is applied to a diethylaminoethyl (DEAE)-cellulose column (5 × 30 cm, chloride form) and the column is washed with 1 liter of 10 m*M* buffer, pH 7.5. A fraction collector with a 280-nm detector should be employed to monitor proteins eluting from the column. The TF activity is eluted with 50 m*M* buffer, pH 7.5, followed by dialysis against 1 m*M* buffer until free of salts, and lyophilized.

Sephadex Chromatography. One-half of the lyophilized material is dissolved in a minimum amount of 20 m*M* buffer, pH 7.5, and applied to a Sephadex G-100 column (4 × 40 cm) equilibrated against the same buffer. Three distinct yellow bands will migrate down the column; the middle band, containing all of the TF activity, is collected, dialyzed as described above, and lyophilized.

DEAE-BioGel A Chromatography. The dried material is dissolved in a minimum amount of 20 m*M* buffer, pH 7.5, and applied to a DEAE-BioGel A column (2.5 × 30 cm) equilibrated against the same buffer. A linear buffer gradient (10 to 150 m*M*) is passed through the column at a rate of 25 ml/hr for 16 hr. The TF fraction, eluting as a shoulder just behind the second major protein peak, is collected, dialyzed against 1 m*M* buffer, pH 7.5, and lyophilized.

Second Sephadex Chromatography. The pooled lyophilized material from two DEAE-BioGel A chromatography is dissolved in a small amount of 20 m*M* buffer, pH 7.5, and applied to a Sephadex G-100 column (2.5 × 30 cm) equilibrated against the same buffer.

Second DEAE-BioGel A Chromatography. The TF fraction eluting from this column is dialyzed as above, concentrated by lyophilization, and applied to a second DEAE-BioGel A column (2.5 × 30 cm). A linear gradient, as described previously, is used to elute the TF activity, which is collected, dialyzed against 1 m*M* buffer, pH 7.5, and lyophilized.

Properties

The enzyme is stable to repeated lyophilization or freezing or thawing. The enzyme solution is colorless, and a single protein band is observed after polyacrylamide gel electrophoresis. The pH optimum for TF activity is 6.0. The estimated molecular weight by the Sephadex gel-filtration method[11] is 43,000.

In Table I the formation of $S_2O_3^{2-}$ (2.56 μmol) from $S_3O_6^{2-}$ by bisulfite reductase plus TF can be explained by the fact that, under the conditions of the assay, bisulfite reductase can form HSO_3^- from S_3O_6.[2,12] This indicates that the reduction of bisulfite to trithionate, by bisulfite reductase, is a reversible reaction.

Electron Carrier. Cytochrome c_3 can participate in the TF reaction, receiving electrons from hydrogenase and transferring them to TF.

Bisulfite Reductase-Dependent Trithionate Reductase

Assay Method

Principle. The assay for trithionate reduction to thiosulfate by the bisulfite reductase-dependent trithionate reductase (TR-1) system depends on the presence of both bisulfite reductase and TR-1. The reduction of trithionate by this system does not depend on the presence of bisulfite, as does TF, and the mechanism of this reaction is not known.

Reagents. All of the reagents for this assay are the same as for the TF assay except that TR-1 replaces TF.

Procedure. The conditions for determining trithionate reductase activity are essentially the same as described for TF. To the main compartment of a Warburg flask is added 0.1 ml each of methyl viologen, buffer, hydrogenase, bisulfite reductase, and TR-1. After gassing the system with hydrogen, the trithionate solution in the side arm is tipped into the main compartment. After an appropriate incubation time, the reaction is terminated by opening the flasks and analyzing the contents for thiosulfate as described for TF activity.

Purification of Trithionate Reductase

All purification steps are carried out at 0–4° and, unless otherwise indicated, potassium phosphate buffer, pH 7.0, is used throughout the purification.[4] *Desulfovibrio vulgaris* crude extract (210 ml; 61 mg/ml protein) is centrifuged at 55,000 *g* for 2 hr to remove particulate matter.

Heat Treatment of Cell Extract. The soluble fraction is quickly heated to 60° in a boiling water bath and held at that temperature for 60 sec. It is cooled rapidly in an ice bath and centrifuged at 27,000 *g* for 20 min to remove denatured proteins.

[11] J. R. Whitaker, *Anal. Chem.* **35,** 1950 (1963).

[12] H. L. Drake, Ph.D. Thesis, University of Kansas, Lawrence (1978).

Ion-Exchange Chromatography. The supernatant fraction is applied to a DEAE-cellulose column (2.6 × 15 cm) equilibrated against 1 m*M* buffer and the column is washed successively with 400 ml of 10 m*M* buffer, 200 ml of 100 m*M* buffer, and 50 ml of 300 m*M* buffer. The brown fraction that elutes with the 100 m*M* buffer is concentrated with an Amicon EC-20 unit (Amicon Corp., Lexington, MA) with a PM-30 filter.

Desalting with Sephadex G-25. The concentrated material is passed through a Sephadex G-25 gel-filtration column (4 × 25 cm) equilibrated against water and then passed through an Amberlite CG-50 column (2.5 × 3 cm) to remove residual cytochrome c_3.

Second DEAE Chromatography. The unabsorbed fraction is applied to a second DEAE column (2.6 × 7 cm) equilibrated against 1 m*M* buffer. The column is washed with 100 ml of 40 m*M* buffer and the amber fraction containing TR-1 activity is eluted with 70 m*M* buffer.

Sephadex Chromatography. It is concentrated by filtration with a PM-30 filter and applied to a Sephadex G-100 column (1.6 × 100 cm) equilibrated against 10 m*M* buffer. When the same buffer is passed through this column, two major yellow bands and a faint yellow band between the two major bands should be observed.

Third DEAE Chromatography. The faint yellow band, which contains trithionate-reducing activity, is collected, concentrated by PM-30 filtration, and applied to a third DEAE-cellulose column (1.6 × 3 cm). This column is washed with 30 ml of 20 m*M* buffer and the enzyme is removed with 50 m*M* buffer. It is concentrated by filtration through PM-30 filter and stored at −20°. From 100 g (wet weight) of cells, the yield of TR-1 is approximately 4–5 mg. By polyacrylamide gel electrophoresis a single band should be observed indicating the purity of this preparation.

Properties

Solutions of TR-1 are stable during storage at −20° for at least 2 months. Repeated freezing and thawing or storage at 4° results in progressive loss of activity during this time. The purified enzyme is stable to lyophilization, but dialysis for 24 hr at 4° under any condition will result in considerable loss of activity. Purified TR-1 solutions exhibit no unusual absorption spectrum in the visible and ultraviolet (UV) regions. The molecular weight of TR-1 is 30,000. By using the value of 30,000 for the molecular weight of TR-1 and 220,000 for bisulfite reductase, the optimum ratio of TR-1 to bisulfite reductase for maximum thiosulfate formation is 1 : 1.

The TR-1 isolated from *Desulfotomaculum* (*Dt.*) *nigrificans*-catalyzed trithionate reduction when P-582 was the other component. It also worked well with desulfoviridin from *D. vulgaris*, indicating that the bisulfite reductase-dependent trithionate reductase can use bisulfite reductase isolated from other sulfate-reducing bacteria.[13]

Inhibitors. The trithionate-reducing activity by the two-enzyme system is inhibited by several sulfhydryl agents. At a concentration of 1 mM, N-ethylmaleimide, iodoacetate, p-hydroxymercuribenzoate, silver nitrate, and mercurous chloride all inhibit thiosulfate formation ranging from 17% (N-ethylmaleimide) to 94% (iodoacetate) inhibition. These reagents do not inhibit bisulfite reductase activity, indicating that the inhibitors affect the TR-1 moiety of the dual-enzyme catalysis.

Electron Carriers. The electron carrier that participates in the reduction of trithionate to thiosulfate with the two-enzyme system is flavodoxin. Cytochrome c_3 does not participate in this reaction.

Thiosulfate Reductase

Assay Method

Principle. The assay for thiosulfate reduction can be accomplished by spectrophotometric methods or by manometry. The former method depends on the oxidation of a reduced electron carrier, such as methyl viologen, in the presence of thiosulfate and thiosulfate reductase.[14] The assay described below measures the production of sulfide from thiosulfate as described by the authors during their purification of thiosulfate reductase from *D. vulgaris*.[15]

Reagents

$FeCl_3$, 0.023 M in 1.2 M HCl
H_3PO_4, 20 N
N,N-Dimethyl-p-phenylenediamine sulfate, 0.5 g in 500 ml of 5.5 N HCl
Potassium phosphate buffer (pH 7.0), 1 M
Methyl viologen, 0.01 M
Cadmium chloride solution (20%, v/v)
Sodium thiosulfate, 0.1 M
Particulate hydrogenase, 10 mg/ml

Procedure. Double-arm Warburg flasks are used for this assay (8-ml capacity). Into the center well is placed a fluted filter paper and 0.1 ml of $CdCl_2$ solution is added to the center well, saturating the filter paper. Into the main compartment is added 0.1 ml each of the buffer, methyl viologen,

[13] D. Gevertz, Ph.D. Thesis, University of Kansas, Lawrence (1983).
[14] W. Badziong and R. K. Thauer, *Arch. Microbiol.* **125,** 167 (1980).
[15] R. H. Haschke and L. L. Campbell, *J. Bacteriol.* **106,** 603 (1971).

hydrogenase, thiosulfate reductase, and water to a final volume of 0.9 ml. One side arm contains 0.1 ml of the thiosulfate solution while the other arm contains 0.1 ml of 20 *M* H_3PO_4. After gassing with hydrogen for 10 min, the flasks are closed and allowed to equilibrate at 37° for 5 min. At zero time, the thiosulfate is tipped into the main compartment and readings are taken. After sufficient hydrogen is utilized, the reaction is terminated by tipping in the H_3PO_4. The flasks are shaken for an additional 10 min and the filter papers are analyzed for H_2S by the method of Fogo and Popowski.[16]

Purification Procedure

The procedure described here is taken from the report by Haschke and Campbell.[15] Other workers who purified thiosulfate reductase from sulfate-reducing bacteria are Hatchikian,[17] Nakatsukasa and Akagi,[18] and Aketagawa *et al.*[19]

Preparation of Cell Extracts. A cell paste of *D. vulgaris* is suspended in 1.5 vol (w/v) of 0.10 *M* tris(hydroxymethyl)aminomethane, pH 7.5. The suspension is passed through a continuous-flow attachment (maintained at 0°) of a Branson sonifier at maximum power at a flow rate of 60 to 80 ml/hr. The cell extract is centrifuged at 34,000 *g* for 45 min and the pellet is discarded.

Ammonium Sulfate Precipitation. Solid ammonium sulfate (special enzyme grade; Mann, New York, NY) is added to the supernatant fraction to 0.5 saturation and centrifuged at 19,000 *g* for 20 min. The supernatant fraction is thoroughly dialyzed against three to four changes of demineralized water for 20 hr.

DEAE Chromatography. The dialyzed extract is applied on a DEAE-cellulose column (5 × 20 cm) at a flow rate of approximately 120 ml/hr. All unadsorbed material is washed off with distilled water and the adsorbed protein is eluted from the column with 0.8 *M* NaCl in 0.05 *M* Tris buffer, pH 7.5. This fraction is dialyzed against distilled water for several hours and concentrated with a Diaflo ultrafilter with a UM-1 membrane (cutoff, M_r 10,000, Amicon Corporation, Lexington, MA) under an N_2 pressure of 85 psi.

Preparatory Gel Electrophoresis. The concentrated fraction is subjected to preparatory polyacrylamide electrophoresis.[15] The gel column should be approximately 4.7 × 8 cm and should contain 7.5% (w/v) acryl-

[16] J. K. Fogo and M. Popowski, *Anal. Chem.* **21,** 732 (1949).

[17] E. C. Hatchikian, *Arch. Microbiol.* **105,** 249 (1975).

[18] W. Nakatsukasa and J. M. Akagi, *J. Bacteriol.* **98,** 429 (1969).

[19] J. Aketagawa, K. Kobayashi, and M. Ishimoto, *J. Biochem.* (*Tokyo*) **97,** 1025 (1985).

amide at pH 8.9. The concentrated NaCl fraction (100–300 mg) is layered on each gel and electrophoresed at 60–70 mA for 7 hr at 4°. The thiosulfate reductase should be observed as a yellow band. The band is sliced from the gel and collected by electrophoresis into a dialysis bag.[15]

Guanidine Hydrochloride Treatment. The enzyme from several preparative electrophoresis runs is pooled and rerun under the same conditions. Elution is performed as previously described and guanidine hydrochloride (6 *M* final concentration) is added to the enzyme. After incubation for 20 hr at 4°, a 2.5-ml sample is applied to a Sephadex G-25 column (2.5 × 40 cm) equilibrated in distilled water. The enzyme is eluted with water, using a 280-nm monitor to locate the protein. The contents of the peak tubes are pooled, lyophilized, and stored at −20°.

Properties

The enzyme catalyzes the reduction of thiosulfate to sulfide and sulfite. The optimum pH for its activity is 8.0–9.0 and the molecular weight, by sedimentation equilibrium studies, is 16,300, and by amino acid analysis is 15,500. These values are different from those of thiosulfate reductase isolated from *D. vulgaris* Miyazaki F (M_r 85,000–89,000)[19] and *Desulfovibrio gigas* (M_r 220,000).[17]

Drake[12] observed that the thiosulfate reductase from *D. vulgaris* was extremely sensitive to oxygen, and partially purified this enzyme from cell extracts by performing all manipulations in the absence of air. The *D. vulgaris*[12] and *Dt. nigrificans*[20] thiosulfate reductases were more stable in phosphate buffer than in Tris, borate, or triethanolamine buffer.

Inhibitors. Thiosulfate reductase from *D. vulgaris* Miyazaki F,[19] *D. gigas*,[17] and *Dt. nigrificans*[18] is inhibited by sulfhydryl agents, indicating that at least one sulfhydryl group is involved in the catalysis. In addition, the enzyme from *D. vulgaris* Miyazaki F is stimulated by ferrous ions and inhibited by the iron-chelating agents *o*-phenanthroline and 2,2′-bipyridine.[19] Sulfite inhibits thiosulfate reductase of *D. gigas*[17] and *Dt. nigrificans*.[18] A potent inhibitor of thiosulfate reductase activity is trithionate.[21,22]

Nakatsukasa[20] found that the thiosulfate reductase isolated from *Dt. nigrificans* contained FAD as a prosthetic group. Using the pyruvic phosphoroclastic system and ferredoxin from *Dt. nigrificans*, he observed that the apoenzyme was not active in reducing thiosulfate unless FAD was added to the reaction mixture. If an artificial dye, such as methyl viologen, was substituted for ferredoxin, the requirement for FAD disappeared.

[20] W. Nakatsukasa, Ph.D. Thesis, University of Kansas, Lawrence (1969).

[21] R. Fitz and H. Cypionka, *Arch. Microbiol.* **154,** 400 (1990).

[22] J. M. Akagi, *Biochem. Biophys. Res. Commun.* **117,** 530 (1983).

Concluding Remarks

If dissimilatory sulfate-reducing bacteria reduce bisulfite to sulfide by the trithionate pathway the presence of trithionate and thiosulfate reductases in these organisms is not surprising. However, the presence of enzyme systems that reductively form thiosulfate from trithionate may not necessarily mean that bisulfite is reduced to sulfide by the trithionate pathway. Although trithionate and thiosulfate are products of bisulfite reduction by bisulfite reductase, it is possible that the true mechanism for bisulfite reduction to sulfide is a direct, six-electron reduction catalyzed by an assimilatory (bi)sulfite reductase. The reaction catalyzed by bisulfite reductase, forming trithionate and thiosulfate, may be the result of alternate (side) reactions occurring in sulfate-reducing bacteria. If these occur, then the roles of enzymes described in this chapter may be to metabolize trithionate and thiosulfate to prevent any intracellular accumulation of these compounds.

The elucidation of the pathway for dissimilatory bisulfite reduction should clarify this problem.

[18] Desulforubidin: Dissimilatory, High-Spin Sulfite Reductase of *Desulfomicrobium* Species

By DANIEL V. DERVARTANIAN

Introduction

Desulforubidin is a novel sulfite reductase that catalyzes the six-electron reduction of sulfite to sulfide, a key reaction in the dissimilatory sulfate reduction pathway. Other dissimilatory sulfite reductases with similar physical properties are desulfoviridin from *Desulfovibrio* (*D.*) *gigas*, *Desulfovibrio vulgaris*, and *Desulfovibrio salexigens*,[1–3] P-582 from *Desulfotomaculum* (*Dt.*) *nigrificans* and *Desulfotomaculum ruminis*,[4–5] and desul-

[1] J. P. Lee and H. D. Peck, Jr., *Biochem. Biophys. Res. Commun.* **45,** 583 (1971).

[2] J. P. Lee, J. LeGall, and H. D. Peck, Jr., *J. Bacteriol.* **115,** 529 (1973).

[3] M. Czechowski, A. V. Xavier, B. A. S. Barata, A. R. Lino, and J. LeGall, *J. Ind. Microbiol.* **1,** 139 (1986).

[4] P. A. Trudinger, *J. Bacteriol.* **104,** 158 (1970).

[5] J. M. Akagi and V. Adams, *J. Bacteriol.* **116,** 392 (1973).

TABLE I
PROPERTIES OF DESULFORUBIDIN

Property	Value/characteristic
Subunit molecular weight[10]	45,000
	50,000
Partial specific volume[10]	0.730 ml/g
Molecular weight[8]	225,000 (based on $s_{20,w} = 9.8$)
Type of subunit structure[10]	$\alpha_2\beta_2$
Light absorption spectra[8]	Peaks at 392, 545, 580 nm
Plus CO[11]	Peaks shifted to 550 and 593 nm
Plus pyridine[9]	Characteristic sirohydrochlorin maxima at 378, 510, 545, 588, and 630 nm
Plus $S_2O_4^{2-}$ [10]	Reduction causes shift from 545 to 610 nm
Nonheme ion content	14.8 mol/mol protein; determination by Biuret method (cf. Ref. 10)
	21 ± 2 mol/mol protein; determination by Lowry procedure (cf. Ref. 9)
Labile sulfide content[10]	14.7 mol/mol desulforubidin
Siroheme	2.0 mol/mol desulforubidin[10]
	2.2 ± 0.3 mol/mol desulforubidin[9]

fofuscidin from *Desulfovibrio thermophilus*[6] and *Thermodesulfobacterium commune*[7] (see [19] in this volume).

Desulforubidin has been purified to homogeneity from *Desulfomicrobium baculatum* Norway 4[8] and DSM 1743C.[9] The various dissimilatory sulfite reductases share many similarities. They are high molecular weight proteins (>200,000) and consist of two different subunits in an $\alpha_2\beta_2$ configuration. Each enzyme contains multiple [Fe-S] clusters (14–21 nonheme irons, plus an equivalent sulfide content). In addition, two sirohemes have been detected in these enzymes (see Table I[10,11]).

At highly reduced methyl viologen concentrations desulforubidin reduces sulfite to sulfide. At less reduced methyl viologen concentrations, the product was found to be trithionate.[8] The electron donor to methyl

[6] G. Fauque, A. R. Lino, M. Czechowski, L. Kang, D. DerVartanian, J. J. G. Moura, J. LeGall, and I. Moura, *Biochim. Biophys. Acta* **1040,** 112 (1990).

[7] E. C. Hatchikian and J. G. Zeikus, *J. Bacteriol.* **153,** 1211 (1983).

[8] J. P. Lee, Ching-Sui Yi, J. LeGall, and H. D. Peck, Jr., *J. Bacteriol.* **115,** 453 (1971).

[9] I. Moura, J. LeGall, A. R. Lino, H. D. Peck, Jr., G. Fauque, A. V. Xavier, D. DerVartanian, J. J. G. Moura, and B. H. Huynh, *J. Am. Chem. Soc.* **110,** 1075 (1988).

[10] Chi-Li Liu, Ph.D. Dissertation, University of Georgia, Athens (1981).

[11] M. J. Murphy, L. M. Siegel, H. Kamin, D. V. DerVartanian, J.-P. Lee, J. LeGall, and H. D. Peck, Jr., *Biochem. Biophys. Res. Commun.* **54,** 82 (1973).

viologen may be sodium dithionite or H_2 plus hydrogenase (e.g., from *D. gigas*).

Purification of Desulforubidin from *Desulfovibrio desulfuricans* Norway

Frozen or fresh cells are suspended in a 0.1 *M* phosphate buffer, pH 7.6, at a ratio of 1 g of cells to 2 ml of buffer.[10] The cell suspension is passed through a Manton–Gaulin homogenizer (Gaulin Co., Wilmington, MA) or an Aminco French press (Silver Springs, MD) at 8000 psi. The preparation is centrifuged for 30 min in a Sorvall (Newtown, CT) centrifuge at 25,000 *g*. The combined supernatant is centrifuged in an ultracentrifuge at 100,000 *g* for 1.5 hr. To this supernatant 0.5 mg of streptomycin sulfate per milligram total protein is added to remove nucleic acid. The nucleic acid-free supernatant is dialyzed overnight against 0.1 *M* phosphate buffer, pH 7.6, and then fractionated with solid ammonium sulfate (enzyme grade) from 0–50% and 50–80% saturation. The resulting precipitates are dissolved in a minimum amount of 0.01 *M* phosphate buffer (pH 7.6) and dialyzed overnight against several changes of the same buffer. These preparations are then absorbed on an anion-exchange gel column (DEAE-BioGel A) preequilibrated with 0.01 *M* phosphate buffer, pH 7.6. The enzyme is eluted by means of a linear phosphate gradient (pH 7.6) from 0.01 to 0.4 *M*. The desulforubidin fractions are pooled and concentrated with a Diaflo PM30 membrane (Amicon, Danvers, MA) to less than 10 ml and are then passed through an Ultrogel AcA-34 column that is preequilibrated with 0.01 *M* phosphate buffer. Desulforubidin is eluted with the same buffer from this gel-filtration column and is applied directly to a BioGel HTP column. The final purification step for desulforubidin is on a second DEAE-BioGel A column.

Purification of Desulforubidin from *Desulfovibrio baculatus* DSM 1743

The bacterium is grown in lactate-sulfate medium as described by Huynh and co-workers[9] and stored at −80°. Cells (300 g wet weight) are suspended in 300 ml of 10 m*M* (pH 7.6) Tris-HCl buffer. After addition of DNase, the cells are subjected to a French pressure cell at 62 MPa. Centrifugation of the extract is carried out for 30 min at 12,000 rpm and 4°. This yields 500 ml of crude cell extract. This cell extract is passed through a DEAE-BioGel A column (5 × 50 cm). Elution is achieved with 2 liters of a linear gradient of Tris-HCl buffer (10–400 mM). Desulforubidin is obtained as a red protein solution of 500 ml between 250 and 300 m*M* Tris-HCl. Dialysis of this red fraction against distilled water is conducted overnight. The dialyzed red fraction is applied to a second DEAE-BioGel A column (5 × 50 cm). Application of the same linear gradient of Tris-

HCl results in a similar red desulforubidin fraction but with an absorbance ratio A_{280}/A_{543} of 9.6. Concentration of this red fraction is achieved with an Amicon Diaflo ultrafilter (YM30 membrane). After concentration the red fraction is applied to a high-performance liquid chromatography (HPLC) apparatus with a Spherogel TSK-3000 preparative column. The purified desulforubidin, which was eluted with 0.1 *M* phosphate buffer, pH 7.0 (containing 0.1 *M* NaCl), exhibits an absorbance ratio A_{280}/A_{543} of 7.1.

Enzymatic Assay

Sulfite reductase activity, involving the six-electron reduction of sulfite to sulfide, is measured at 37° by a manometric procedure according to Schedel *et al.*[12] It utilizes hydrogenase (in the presence of H_2) to reduce methyl viologen, which then provides low-potential electrons for the reduction of sulfite to sulfide. The main compartment of the Warburg manometric vessel contains, at final concentration, 0.05 *M* potassium phosphate buffer (pH 6.5), 0.075% (w/v) methyl viologen, 20–40 μM pure *D. gigas* hydrogenase, 2 mg of bovine albumin, and 0.5–2 mg of desulforubidin. The center well is filled with 50 μl of 10 *M* NaOH and 0.05 ml of 10% (w/v) calcium acetate. The side arm contains 0.1 ml of a 0.1 *M* fresh solution of Na_2SO_3.

The monometric vessel is placed under a hydrogen atmosphere and shaken at 37° for 30 min. The reaction is started by inversion of the vessel in order to mix the sulfite from the side arm with the main reaction mixture vessel compartment. A specific activity of 0.41 μmol of H_2 consumed per minute per milligram is obtained.

Light Absorption Spectra

Lee *et al.*[8] reported that desulforubidin exhibited absorption maxima at 392, 545, and 580 nm with a weak absorption maximum at 720 nm. Desulforubidin was slowly reduced by sodium dithionite, which caused a shift in maximum from 345 to 610 nm. Reduction in the presence of methyl viologen and an electron donor (sodium dithionite or H_2 plus hydrogenase) caused a more complete reduction.

Dithionite reduction of desulforubidin in the presence of CO resulted in shifts in maxima to 550 and 593 nm.[13]

[12] M. Schedel, J. LeGall, and J. Baldensperger, *J. Arch. Microbiol.* **105,** 339 (1975).

[13] Chi Li Liu, D. DerVartanian, and H. D. Peck, Jr., *Biochem. Biophys. Res. Commun.* **91,** 962 (1979).

The siroheme nature of desulforubidin was established by Murphy *et al.*[11] by extraction of the enzyme chromophore and its transfer to pyridine. Moura *et al.*[9] determined for their desulforubidin from *Desulfomicrobium baculatum* (DSM 1743) the characteristic absorption maximum of the extracted heme in pyridine and identified the heme as sirohydrochlorine (maxima at 378, 510, 545, 588, and 638 nm). They determined a value of 2.2 plus 0.3 mol of sirohemes per mole desulforubidin. The [Fe-S] clusters did not contribute extensively to the light absorption spectrum (see Table I).

Spectroscopic Studies

Electron Paramagnetic Resonance Studies

Desulforubidin in the as-isolated state exhibited complex high-spin ferric siroheme resonances with major g values at 6.33 and 5.29[13] or at 6.43, 5.34, and 1.97,[9] and a minor species of high-spin ferric siroheme with resonances at $g = 6.92$ and 5.72[13] or at 6.94, ~5.0, and ~1.9.[9] The latter species has been estimated to represent less than 10% of the major component. Quantitation of the ferric siroheme resonances accounts for two sirohemes per mole desulforubidin (Table II).

Reduction of desulforubidin with sodium dithionite or borohydride is slow, requiring 30 min to achieve slight effects on the light absorption or electron paramagnetic resonance (EPR) spectra. However, in the presence of 1 μM methyl viologen plus dithionite, all high-spin ferric siroheme resonances disappeared within 1 min and a small reduced [4Fe-4S] signal appeared with g values at 2.03, 1.93, and 1.89. This latter signal accounted for only a 3% recovery of EPR-detectable iron when compared to the nonheme iron content determined chemically.[13]

When desulforubidin was reduced by H_2 plus hydrogenase in the presence of 1 μM methyl viologen, a complete loss of the initial ferric heme resonances occurred. However, no reduced [4Fe-4S] cluster signals could be detected. An EPR signal at $g = 2.00$ was attributed to the methyl viologen radical.[13]

Mössbauer Spectroscopy

Mössbauer spectral analysis[9] shows the presence of four [4Fe-4S] clusters that are in the diamagnetic 2+ oxidation state. The two sirohemes are in the high-spin ferric state, in agreement with the EPR studies. Mössbauer

TABLE II
MAGNETIC SPECTROSCOPIC PROPERTIES OF DESULFORUBIDIN

Spectroscopy	Properties: As isolated	Properties: Reduced[a]
EPR spectroscopy	Major species with *g* values at 6.33, 5.29[13] Minor species at g = 6.92, 5.72 or Major species, *g* values at 6.47, 5.34, 1.97[9] Minor species, *g* values 6.94, ~5.0, ~1.9 These signals are characteristic for high-spin ferric siroheme Quantitation: Accounts for two sirohemes[13]	All heme resonance signals disappear none or negligible reduced "g = 1.94" type iron-sulfur signals
Mössbauer spectroscopy[9]	1. Two [4Fe-4S] clusters are exchange coupled to the two paramagnetic sirohemes. Two uncoupled $[4Fe\text{-}4S]^{2+}$ clusters are also present 2. Sirohemes are in the high-spin ferric state 3. The coupled [4Fe-4S] cluster is similar to the uncoupled cluster in a typical [4Fe-4S] cluster and resembles the ferredoxin-type cluster 4. One rhombic high-spin Fe(III) species is detected	1. Sirohemes are reduced to the high-spin ferrous state 2. Oxidation state for the [4Fe-4S] clusters remains at 2+

[a] In the presence of $S_2O_4^{2-}$ plus methyl viologen or H_2 plus hydrogenase.

studies have shown that each high-spin ferric siroheme is exchange coupled to a $[4Fe\text{-}4S]^{2+}$ cluster.

Reduction of desulforubidin in the presence of small amounts of hydrogen and hydrogenase resulted in the sirohemes exclusively acquiring donated electrons. The sirohemes were reduced to a high-spin ferrous state. In addition, the electronic states of the exchange-coupled siroheme–[Fe-S] clusters and the uncoupled [4Fe-4S] clusters did not change on reduction. This observation indicates that the coupling between the siroheme and [Fe-S] cluster is maintained in all oxidation–reduction states examined.

A solitary iron center not associated with either the sirohemes or [4Fe-4S] clusters was detected. In the as-isolated state this solitary iron was high-spin ferric whereas in the reduced state it was high-spin ferrous.

Acknowledgment

This work was supported by NIH Grant GM34903.

[19] Desulfofuscidin: Dissimilatory, High-Spin Sulfite Reductase of Thermophilic, Sulfate-Reducing Bacteria

By E. CLAUDE HATCHIKIAN

Four different types of dissimilatory sulfite reductase, including desulfoviridin,[1] desulforubidin,[2] P-582,[3] and desulfofuscidin,[4,5] have been found in various genera of sulfate-reducing bacteria.[6] Their substrate is actually the bisulfite ion as deduced from the optimum pH of activity of these enzymes.[4,5,7] Desulfofuscidin, which is the subject of this chapter is present only in the extreme thermophilic sulfate reducers of the genus *Thermodesulfobacterium*.[8] As with the other dissimilatory sulfite reductases,[1,2,9] desulfofuscidin catalyzes the reduction of sulfite mainly to trithionate with concomitant formation of thiosulfate and sulfide under assay conditions.[4]

[1] J.-P. Lee and H. D. Peck, Jr., *Biochem. Biophys. Res. Commun.* **45,** 83 (1971).
[2] J.-P. Lee, C.-S. Yi, J. LeGall, and H. D. Peck, Jr., *J. Bacteriol.* **115,** 453 (1973).
[3] P. A. Trudinger, *J. Bacteriol.* **104,** 158 (1970).
[4] E. C. Hatchikian and J. G. Zeikus, *J. Bacteriol.* **153,** 1211 (1983).
[5] G. Fauque, A. R. Lino, M. Czechowski, L. Kang, D. V. DerVartanian, J. J. G. Moura, J. LeGall, and I. Moura, *Biochim. Biophys. Acta* **1040,** 112 (1990).
[6] J. LeGall and G. Fauque, *in* "Biology of Anaerobic Microorganisms" (A. J. B. Zehnder, ed.), p. 587. Wiley, New York, 1988.
[7] B. Suh and J. M. Akagi, *J. Bacteriol.* **99,** 210 (1969).
[8] F. Widdel, *in* "The Prokaryotes" (A. Balows, H. G. Trüper, M. Dworkin, W. Harder, and K.-H. Schleicher, eds.), 2nd Ed., Vol. 4, p. 3390. Springer-Verlag, New York, 1992.
[9] J. M. Akagi and V. Adams, *J. Bacteriol.* **116,** 392 (1973).

Characteristics of Genus *Thermodesulfobacterium*

The genus *Thermodesulfobacterium* includes only two species, namely *Thermodesulfobacterium commune* (DSM 2178, ATCC 33708)[10] and *Thermodesulfobacterium mobile* (DSM 1276), formerly called *Desulfovibrio thermophilus*.[11] Both strains have been isolated from extreme environments and are only distantly related to other sulfate-reducing eubacteria.[8] They are identified by their high-temperature optimum for growth (65 to 70°), rod-shaped cells that do not form spores, the presence of unusual nonisoprenoid ether-linked lipids, and the utilization of H_2, formate, lactate, and pyruvate as electron donors for sulfate reduction. They contain desulfofuscidin as specific dissimilatory sulfite reductase[8] and a tetrahemic cytochrome c_3, which appears to be homologous to the *Desulfovibrio* species tetrahemic cytochrome c_3.[12,13]

Assay Method

Principle. Bisulfite reductase activity can be determined manometrically as hydrogen uptake in a system containing hydrogenase, bisulfite reductase, methyl viologen (MV), and sulfite.[1] This assay requires saturating concentration of reduced methyl viologen, which is achieved under the assay conditions by using a large excess of hydrogenase as compared to bisulfite reductase. The reduced dye then serves as the electron donor for the bisulfite reductase as well as other reductases present in the extracts. The rate of hydrogen consumption is proportional to the amount of bisulfite reductase added to the reaction mixture when enzyme activity is calculated from the initial rate of hydrogen utilization.

Except for the concentrations of substrates, similar reaction conditions are used to determine other reductase activities present in the extracts, such as thiosulfate reductase and trithionate reductase.[4] These will interfere with the activity measurements in cell extracts (Table I).

Reagents

Potassium phosphate buffer (0.5 *M*), pH 6.0
Desulfovibrio gigas pure hydrogenase, 1 mg/ml (specific activity, 180 units/mg)

[10] J. G. Zeikus, M. A. Dawson, T. E. Thompson, K. Ingvorsen, and E. C. Hatchikian, *J. Gen. Microbiol.* **129,** 1159 (1983).

[11] E. P. Rozanova and T. A. Pivovarova, *Mikrobiologiya* **57,** 102 (1988).

[12] E. C. Hatchikian, P. Papavassiliou, P. Bianco, and J. Haladjian, *J. Bacteriol.* **159,** 1040 (1984).

[13] G. Fauque, J. LeGall, and L. L. Barton, *in* "Variations in Autotrophic Life" (J. M. Shively and L. L. Barton, eds.), p. 271. Academic Press, London, 1991.

Methyl viologen, 80 m*M*
Bovine serum albumin, 10 mg/ml
Sodium sulfite, 110 m*M*, prepared just before use
NaOH, 10 *N*

Procedure. The main compartment of the manometric vessel contains 150 μmol of phosphate buffer (pH 6.0), 2.1 μmol of MV, 3 mg of bovine serum albumin, 50 μg of pure *D. gigas* hydrogenase, and a variable amount of enzyme. The side arm contains 22 μmol of sodium sulfite (freshly prepared), and the center well receives 0.1 ml of 10 *N* NaOH. The enzymatic activity was routinely measured at 45° and the final volume is 3 ml. After preincubation under hydrogen for 30 min, the reaction is started by tipping in 0.2 ml of 0.11 *M* sulfite from the side arm.

Hydrogen sulfide produced during the reaction is scavenged by 10 *N* NaOH in the center well. It can be determined using the method of Fogo and Popowsky.[14] Trithionate and thiosulfate, which accumulate in the reaction mixture, can be estimated according to the procedure of Kelly *et al.*[15]

Definition of Unit and Specific Activity. One unit of enzyme is the enzyme activity catalyzing the consumption of 1 μmol of hydrogen per minute under the assay conditions. Specific activity is expressed as units per milligram of protein. Protein is determined by the method of Lowry *et al.*[16] with bovine serum albumin as standard.

Procedure for Purification of *Thermodesulfobacterium commune* Desulfofuscidin

Growth Conditions and Preparation of Cell Extract. The stock cultures are kept in the low phosphate-buffered basal medium with 0.1% (w/v) yeast extract, 0.4% (w/v) Na_2SO_4, and 0.4% (w/v) sodium lactate in Hungate tubes under an atmosphere of nitrogen.[10] For large-scale cultures in a 300-liter fermentor (model 674 − 300 − 72; Chemap) *Thermodesulfobacterium commune* is grown in a basic lactate-sulfate medium containing sodium lactate [60% solution (v/v), 7.5 ml], NH_4Cl (1 g), $MgSO_4 \cdot 7H_2O$ (2 g), $Na_2SO_4 \cdot 10\ H_2O$ (4 g), K_2HPO_4 (0.5 g), trace elements solution (1 ml),[17] distilled water (1000 ml). The pH is adjusted to 7.2. After sterilization, the medium is supplemented with 2 ml/liter of a 10% (w/v) sodium sulfide solution kept in tubes under vacuum and the bacteria are grown

[14] J. K. Fogo and M. Popowsky, *Anal. Chem.* **21,** 732 (1949).
[15] D. P. Kelly, L. A. Chambers, and P. A. Trudinger, *Anal. Chem.* **41,** 898 (1969).
[16] O. H. Lowry, N. J. Rosebrough, A. L. Farr, and R. J. Randall, *J. Biol. Chem.* **193,** 265 (1951).
[17] T. Bauchop and S. R. Elsden, *J. Gen. Microbiol.* **23,** 457 (1960).

at 65° under an atmosphere of nitrogen. The cells are harvested at the stationary phase after 48 hr of growth, by centrifugation in a Sharples centrifuge (model AS 16). The growth medium is cooled at 10° in a refrigerated container prior to centrifugation.

Freshly thawed cells of *T. commune* (430 g wet weight) are suspended in 250 ml of 20 m*M* tris(hydroxymethyl)aminomethane-hydrochloride (Tris-HCl) buffer, pH 7.6, containing 1 μM deoxyribonuclease and are passed twice through a French pressure cell at 15,000 lb/in^2. The extract is centrifuged at 30,000 *g* for 30 min, and the supernatant fluid is then centrifuged for 2 hr at 198,000 *g*. The pellet is washed once with 100 ml of 20 m*M* Tris-HCl buffer, pH 7.6, and centrifuged under the same conditions. The soluble extract constituted by the mixture of the two 198,000 *g* supernatants is used as the source of the enzyme.

All subsequent steps are carried out at 4° and all buffers are at pH 7.6. Bisulfite reductase is monitored in the extracts both by its enzymatic activity and its typical absorption peak at 576 nm.

Silica Gel Fractionation. A settled volume of silica gel [Merck (Rahway, NJ) type 60] equal to 70 ml, equilibrated with 20 m*M* Tris-HCl (pH 7.6), is added to the soluble extract and the mixture is stirred for 3 hr. This suspension is subsequently poured into a column, and the unadsorbed proteins are collected. After washing the gel with 100 ml of the same buffer, the adsorbed protein fraction, which contains the tetrahemic cytochrome c_3 (M_r 13,000), is eluted with 1 *M* K_2HPO_4 containing 1 *M* NaCl and stored at −80°.[12]

DEAE-Cellulose Fractionation. The unadsorbed proteins are subsequently treated with DEAE-cellulose (DE-52; Whatman, Clifton, NJ) by a batchwise technique. About 100 ml of DEAE-cellulose equilibrated with 20 m*M* Tris-HCl buffer (pH 7.6) and previously dried in a Büchner funnel *in vacuo* is added to the extract, and the mixture is stirred for 5 hr.[18] The suspension is then poured into a column and, after washing the resin with the equilibration buffer, the adsorbed acidic proteins are eluted with 1 *M* Tris-HCl buffer. The acidic extract is then dialyzed overnight against 10 liters of distilled water.

First DEAE-Cellulose Column Chromatography. The dialyzed fraction from the previous step is applied to a column (4.2 × 16 cm) of DEAE-cellulose equilibrated with 20 m*M* Tris-HCl buffer. After washing, the proteins are eluted with a discontinuous gradient (1000 ml) with 100-ml aliquots of Tris-HCl buffer (in 50 m*M* steps from 50 to 500 m*M*). The bisulfite reductase exhibiting a brown color is eluted with 300 m*M* Tris-HCl buffer, and the acidic electron carriers (rubredoxin and ferredoxin)

[18] J. Le Gall and N. Forget, this series, Vol. 53, p. 613.

are eluted with 350 and 450 m*M* Tris-HCl buffers, respectively.[19] At this stage, the extract containing the enzyme (21 mg/ml) is divided into three parts of equal volume, and each fraction is purified separately.

First Ultrogel AcA 34 Gel Filtration. The previous bisulfite reductase-containing fraction is concentrated to 45 ml by ultrafiltration using a PM-30 membrane (Amicon, Danvers, MA) and filtered through an Ultrogel AcA 34 (IBF) column (5 × 100 cm) equilibrated with 20 m*M* Tris-HCl. The protein is collected in 8-ml fractions and the tubes that have a purity index ($A_{279\ nm}/A_{389\ nm}$) lower than 2.7 are combined.

Second DEAE-Cellulose Column Chromatography. The bisulfite reductase is subsequently adsorbed on a second DEAE-cellulose column (3 × 12 cm) equilibrated with 20 m*M* Tris-HCl. The column is eluted with a discontinuous gradient (500 ml), with 70 ml aliquots of Tris-HCl buffer (in 50 m*M* steps from 50 to 350 m*M*). The enzyme elutes at about 300 m*M* with a purity index of 2.46.

Second Ultrogel AcA 34 Gel Filtration. The active fraction is concentrated to 30 ml by ultrafiltration as indicated previously and filtered once again through an Ultrogel AcA 34 column (5 × 100 cm) equilibrated with 20 m*M* Tris-HCl.

Hydroxylapatite Column Chromatography. Finally, the protein is adsorbed onto a hydroxylapatite (BioGel HTP; Bio-Rad, Richmond, CA) column (2.8 × 8 cm) equilibrated with 20 m*M* Tris-HCl buffer and eluted with a discontinuous gradient (660 ml) of potassium phosphate buffer (pH 7.6) with 60 ml of 2, 10, 20 . . . up to 100 m*M*.

This procedure yields a dark brown-colored bisulfite reductase preparation exhibiting a purity index of 2.14. The protein fractions are pooled and stored at −80°. The purified enzyme has a specific activity of 0.20 μmol of H_2 utilized per minute per milligram protein at 45° and the yield is 117 mg of pure bisulfite reductase. Typical data obtained at each purification step are summarized in Table I. The fivefold purification of desulfofuscidin as compared to the specific activity of acidic protein fraction (Table I) is indicative of the abundance of this enzyme in the crude extract of *T. commune*.

The apparently higher specific activity observed with fractions from steps 1 and 2 as compared to the acidic protein extract from step 3 (Table I) is due to the presence in these fractions of trithionate and thiosulfate reductase activities that subsequently reduce the products of bisulfite reduction (trithionate and thiosulfate) to sulfide. After the batch treatment with DEAE-cellulose, trithionate and thiosulfate reductase activities are

[19] P. Papavassiliou and E. C. Hatchikian, *Biochim. Biophys. Acta* **810,** 1 (1985).

TABLE I
PURIFICATION OF BISULFITE REDUCTASE FROM *Thermodesulfobacterium commune*

Step	Volume (ml)	Total protein (mg)	Total activity (units)	Specific activity (units/mg)	Recovery (%)
1. Soluble extract[a]	380	17,100	1,075	0.063	
2. Silica gel fractionation	385	15,732	1,070	0.068	
3. DEAE-cellulose fractionation	543	8,688	363.8	0.042	100[b]
4. First DEAE-cellulose column	180	3,780	215	0.057	59.1
5. First Ultrogel AcA34	218	810.9	81	0.100	22.1
6. Second DEAE-cellulose chromatography	51	525.3	64	0.122	17.6
7. Second Ultrogel AcA34	240	216	37.6	0.174	10.3
8. Hydroxylapatite chromatography	195	117	23.4	0.2	6.4

[a] From 430 g of wet cell paste of *T. commune*. In practice, steps 5–8 of the purification procedure were performed from one-third of the original volume of the eluate of step 4.

[b] The activity of the eluate of step 3 is considered as 100% because it is devoid of trithionate and thiosulfate reductase activities, in contrast to the fractions from steps 1 and 2.

separated from the bisulfite reductase as shown by the presence of these activities in the unadsorbed protein fraction.[4]

Purification of *Thermodesulfobacterium mobile* Desulfofuscidin

This procedure is taken from that described by Fauque *et al.*[5] *Thermodesulfobacterium mobile* is grown at 65° on a standard lactate-sulfate medium.[20] The cells (450 g of wet cell paste) are suspended in 10 m*M* Tris-HCl buffer (pH 7.6) and disrupted by passing twice through a French pressure cell at 7000 lb/in^2. The extract is centrifuged at 35,000 *g* for 2 hr. The crude extract is obtained after centrifugation of the supernatant twice at 40,000 *g* for 2 hr. All the purification procedures are carried out at 0–4° and Tris-HCl or phosphate buffers (pH 7.6) are used.

The enzyme has been purified in four chromatographic steps. The crude extract is applied on a DEAE-BioGel A (Bio-Rad) column (7 × 40 cm) equilibrated with 10 m*M* Tris-HCl. The proteins are eluted using a continuous Tris-HCl gradient (2 liters of 10 m*M* Tris-HCl and 2 liters of

[20] R. L. Starkey, *Arch. Mikrobiol.* **8**, 268 (1938).

500 mM Tris-HCl). The fraction containing mainly desulfofuscidin (400 ml) is concentrated to 16 ml on an ultrafilter (Diaflo; Amicon) using a YM-30 membrane and passed twice through a Sephacryl S-200 column (5 × 77 cm) equilibrated with 50 mM Tris-HCl. At this stage, the active fraction exhibits a purity index ($A_{280\ nm}/A_{392\ nm}$) of 4.78. It is dialyzed overnight and adsorbed onto a second DEAE-BioGel A column (5 × 21 cm) equilibrated with 10 mM Tris-HCl. The bisulfite reductase is eluted with a linear gradient of Tris-HCl (600 ml of 10 mM and 600 ml of 350 mM) and concentrated to 12 ml by ultrafiltration using a YM-30 membrane. The previous fraction is then adsorbed onto a hydroxylapatite (BioGel HTP; Bio-Rad) column (4.5 × 15 cm) equilibrated with 300 mM Tris-HCl. The Tris concentration is decreased to 5 mM after washing the column with a descending gradient. A continuous potassium phosphate gradient (400 ml of 5 mM and 400 ml of 200 mM) is subsequently applied for elution. The desulfofuscidin elutes at about 75 mM with a purity coefficient of 2.31. The protein is finally submitted to preparative electrophoresis using a 6% (w/v) polyacrylamide gel.[21] The brown-colored protein band that is collected exhibits a purity coefficient equal to 2.18. This procedure yields 51 mg of pure bisulfite reductase from 450 g of cells.

Properties

Purity. After polyacrylamide disk gel electrophoresis of the purified *T. commune* bisulfite reductase, one brownish band is observed in the gels, and staining with Coomassie blue on a duplicate run gives a single band with the same R_f as the brownish pigment. The protein band in the gels directly catalyzes the sulfite-dependent oxidation of reduced methyl viologen.[4] The purified bisulfite reductase at a protein concentration of 1 mg/ml in 20 mM Tris-HCl (pH 7.6) containing 0.1 M NaCl sedimented in the ultracentrifuge as a single symmetrical boundary and the brown color appeared to sediment with the single peak.

Molecular Weight and Subunit Structure. The molecular weight of desulfofuscidin from *T. commune*, determined by analytical ultracentrifugation at equilibrium sedimentation,[22] is found to be 167,000, using a partial specific volume (0.736) derived from amino acid analysis. The molecular weight of the subunits of bisulfite reductase is estimated to be 47,000 by sedimentation equilibrium centrifugation of the enzyme extensively dialyzed against 4 M guanidine-hydrochloride containing 0.1 M 2-mercaptoethanol, whereas from polyacrylamide gel electrophoresis in sodium

[21] J. M. Brewer and R. B. Ashworth, *J. Chem. Educ.* **46,** 41 (1969).

[22] D. A. Yphantis, *Biochemistry* **3,** 297 (1964).

dodecyl sulfate a molecular weight of 48,000 is obtained. The enzyme appears to be composed of four similar but not identical subunits as shown by the N-terminal sequence of the protein, which is indicative of the presence of two different polypeptidic chains in the molecule (see Amino Acid Composition and N-Terminal Amino Acid Sequence, below). The native protein therefore is a tetramer exhibiting a quaternary structure of an $\alpha_2\beta_2$ type. Desulfofuscidin from *T. mobile* exhibits similar molecular weight and subunits structure.[5] Its molecular mass was found to be 175 kDa by gel filtration and 190 kDa by sedimentation equilibrium whereas on sodium dodecyl sulfate (SDS)-gel electrophoresis the enzyme presents a band of 44 to 48 kDa indicative of the presence of four subunits with very close molecular masses per molecule of native protein.

Absorption Spectra and Coefficients. The absorption spectra of oxidized and dithionite-reduced desulfofuscidin from *T. commune* are shown in Fig. 1. The spectrum of the oxidized form is typical of a siroheme-containing protein and exhibits maxima at 576, 389, and 279 nm, with a weak absorption band at about 693 nm and a shoulder at 532 nm. After addition of dithionite, the peaks at 576 and 693 nm decrease, and a new absorption band appears around 605 nm. Furthermore, the Soret peak

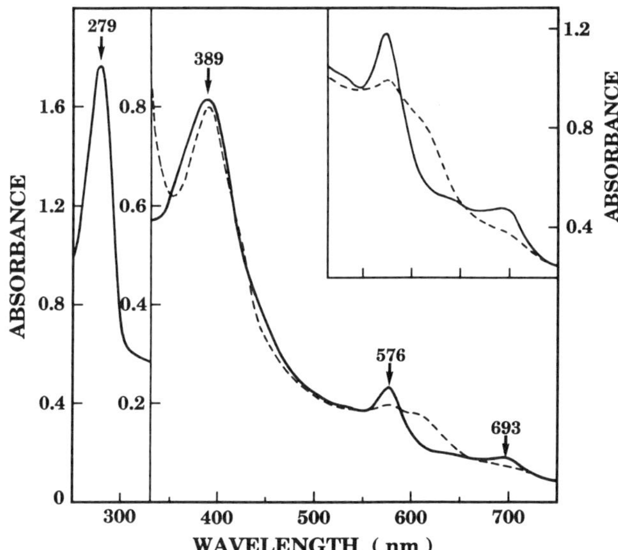

FIG. 1. Absorption spectra of *T. commune* desulfofuscidin [from E. C. Hatchikian and J. G. Zeikus, *J. Bacteriol.* **153,** 12211 (1983)].(—) Oxidized desulfofuscidin, 50 mM in Tris-HCl (pH 7.6), 2.6 μM; (---) reduced with dithionite. *Inset*: oxidized (—) and reduced (---) desulfofuscidin, 13.4 μM.

shifts from 389 to 391 nm, and a shoulder appears in the 420-nm region. At room temperature, the spectrum reaches stability 30 min after the addition of dithionite. The oxidized spectrum is regenerated by gently shaking the enzyme preparation with air. The molar extinction coefficients of the purified enzyme at 576, 389, and 279 nm are determined to be 89,000, 310,000, and 663,000 M^{-1} cm^{-1}, respectively. The reduced desulfofuscidin reacts with various ligands to give complexes that induce modifications of the optical spectrum of the protein. In the presence of dithionite, CO reacts with the enzyme to give a green complex exhibiting a typical spectrum with maxima at 593 and 548 nm, with a Soret peak shifted to 395 nm (Fig. 2).

Thermodesulfobacterium mobile desulfofuscidin exhibits absorption maxima at 578.5, 392.5, and 281 nm, with a small band around 700 nm. The molar extinction coefficients at 578.5 and 392.5 nm are 94,200 and 323,600 M^{-1} cm^{-1}, respectively.[5] After photoreduction of the oxidized enzyme, a decrease in absorption at 578.5 nm and a concomitant increase in absorption at 607 nm are observed.[5] In addition to CO complex, which shows absorption maxima at 592.5, 550, and 395 nm (Fig. 3D), CN^-, S^{2-}, and SO_3^{2-} react with the photoreduced bisulfite reductase. CN^- and S^{2-}

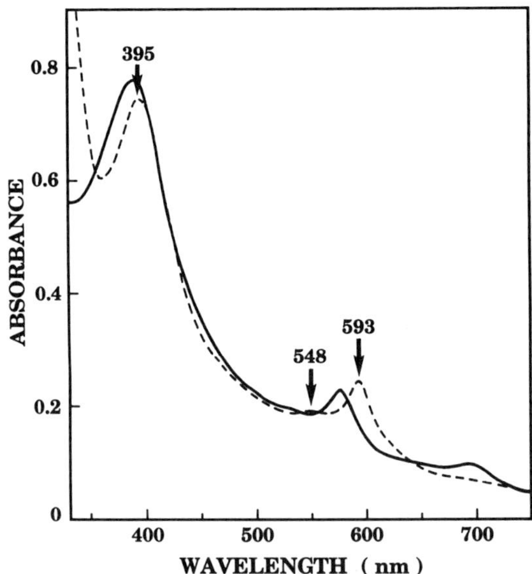

FIG. 2. Absorption spectrum of the CO complex of *T. commune* desulfofuscidin [from E. C. Hatchikian and J. G. Zeikus, *J. Bacteriol.* **153,** 1211 (1983)]. (—) Oxidized desulfofuscidin, 2.5 μM in 50 mM potassium phosphate (pH 7.7); (---) complex between reduced desulfofuscidin and carbon monoxide.

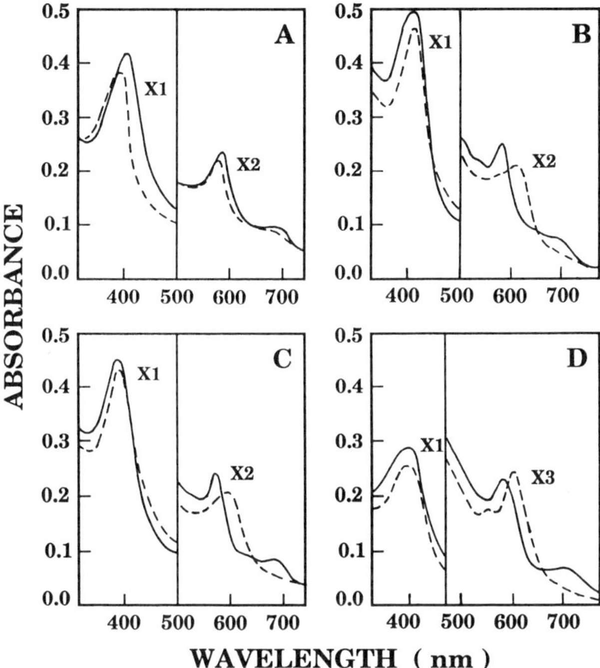

FIG. 3. Absorption spectra of *T. mobile* desulfofuscidin. Complex between photoreduced desulfofuscidin (7 μM) in the presence of deazaflavin and sulfite (A), cyanide (B), sulfide (C), and carbon monoxide (D). [From G. Fauque, A. R. Lino, M. Czechowski, L. Kang, D. V. DerVartanian, J. J. G. Moura, J. LeGall, and I. Moura, *Biochim. Biophys. Acta* **1040,** 112 (1990).]

give complexes with absorption maxima at 605 and 603 nm, respectively (Fig. 3B and C), whereas the complex between the reduced enzyme and SO_3^{2-} presents absorption maxima at 567 and 375 nm (Fig. 3A).

Identification of Heme Prosthetic Group as Siroheme. The spectrum of the supernatant obtained after centrifugation of the extracted chromophore from *T. commune* desulfofuscidin with acetone hydrochloride (0.015 *M*)[23] exhibits typical absorption peaks at 371 and 594 nm (Fig. 4A). The absorption spectrum of the extracted heme on transfer to pyridine exhibits the characteristic siroheme absorption spectrum, with wavelength maxima at 401 and 557 nm and a shoulder at about 520 nm (Fig. 4B). Desulfofuscidin from both species of *Thermodesulfobacterium*[4,5] contains 4 moles of siro-

[23] M. J. Murphy, L. M. Siegel, and H. Kamin, *J. Biol. Chem.* **248,** 2801 (1973).

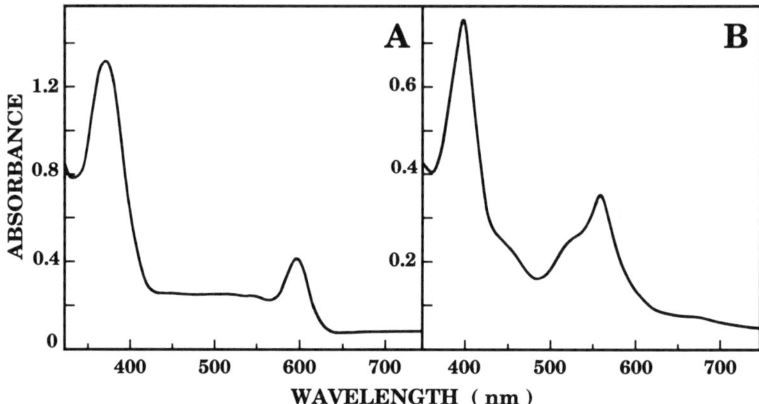

FIG. 4. Absorption spectrum of the extracted heme chromophore of *T. commune* desulfofuscidin in acetone-hydrochloride (A) and pyridine (B). [From E. C. Hatchikian and J. G. Zeikus, *J. Bacteriol.* **153,** 1211 (1983).]

heme per mole of enzyme, based on the α peak absorption of acetone-hydrochloride-extracted heme.[24]

Composition. Desulfofuscidin, like the other sulfite reductases, is an iron-containing enzyme.[4,5] Murphy *et al.*[24] have identified the heme prosthetic group in both assimilatory and dissimilatory sulfite reductases as siroheme, an iron tetrahydroporphyrin of the isobacteriochlorin type with eight carboxylic acid-containing side chains. This heme is common to ferredoxin-nitrite reductases.[25] The average values of iron and acid-labile sulfur content of *T. commune* desulfofuscidin are found to be 20 and 16 atoms per molecule of enzyme, respectively, based on the concentration of the protein estimated from amino acid analysis. On the other hand, desulfofuscidin from *T. mobile* contains 32 ± 2 iron atoms per molecule.[5] Taking into account the content of siroheme, the composition of active sites of *T. commune* enzyme is indicated to be four sirohemes and four [4Fe-4S] centers per molecule.[4] In the case of *T. mobile* desulfofuscidin, which also contains four sirohemes per molecule, the number of [4Fe-4S] clusters per molecule has been reported to be twice that of *T. commune* desulfofuscidin.[5] The nature of the iron core is substantiated by the EPR characteristics of the iron-sulfur signals detected.

[24] M. J. Murphy, L. M. Siegel, H. Kamin, D. V. DerVartanian, J.-P. Lee, J. LeGall, and H. D. Peck, Jr., *Biochem. Biophys. Res. Commun.* **54,** 82 (1973).

[25] M. J. Murphy, L. M. Siegel, S. R. Tove, and H. Kamin, *Proc. Natl. Acad. Sci. U.S.A.* **71,** 612 (1974).

Electron Paramagnetic Resonance Spectra of Desulfofuscidin. The electron paramagnetic resonance (EPR) spectrum of *T. commune* desulfofuscidin as isolated (Fig. 5) exhibits resonances typical of a rhombically distorted high-spin ferric heme with $g_x = 7.02$, $g_y = 4.81$, and $g_z = 1.93$. In addition, it shows minor EPR signals at g values of 6, 4.3, and 2.02. The signal at $g = 6$ is probably due to contaminant free heme, and the resonances at $g = 4.3$ and $g = 2.02$ are assigned to "adventitious" ferric iron and oxidized $[3Fe\text{-}4S]^{1+}$ clusters, respectively. Only minor changes in the EPR spectra are observed after an 8-min reaction with dithionite. On addition of dithionite plus methyl viologen to desulfofuscidin, high-spin ferric heme resonances disappear within 1 min, owing to reduction to the diamagnetic ferrous state (not shown). In the $g = 2$ region, a small, complex "$g = 1.94$" signal appears that can be assigned to $[4Fe\text{-}4S]^{1+}$ clusters.

Oxidized *T. mobile* desulfofuscidin shows an EPR spectrum with resonance absorption at g values 7.26, 4.78, and 1.92 (Fig. 6A).[5] The heme EPR signals, as well as the $g = 4.3$ line, disappear when the sample is anaerobically photoreduced in the presence of deazaflavin with the concomitant appearance of a "$g = 1.94$"-type signal characteristic of a [4Fe-4S] cluster (Fig. 6B). The EPR spectrum of the desulfofuscidin CO complex shows minor modifications in the $g = 2$ region (Fig. 6C).

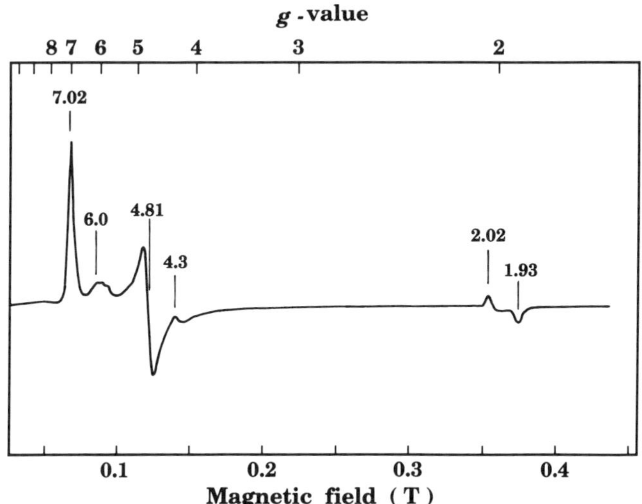

FIG. 5. EPR spectrum of *T. commune* desulfofuscidin as isolated [by permission of R. Cammack, unpublished data (1982)]. Experimental conditions: temperature, 10 K; modulation amplitude, 10 G; microwave frequency, 9.175 GHz; microwave power, 20 mW; protein concentration, 59.6 μM in 250 mM Tris-HCl (pH 8).

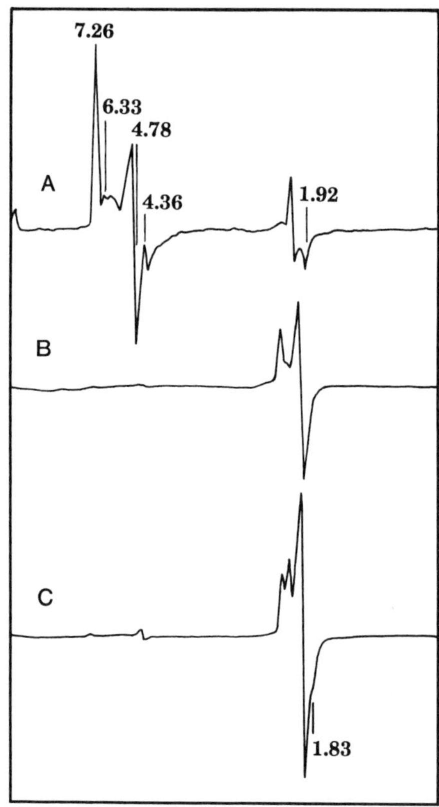

FIG. 6. EPR spectra of *T. mobile* desulfofuscidin [from G. Fauque, A. R. Lino, M. Czechowski, L. Kang, D. V. DerVartanian, J. J. G. Moura, J. LeGall, and I. Moura, *Biochim. Biophys. Acta* **1040,** 112 (1990)]. Experimental conditions: temperature, 10 K; modulation amplitude, 1 mT; microwave frequency, 9.239 GHz; microwave power, 1 mW. Spectrum A, as isolated enzyme; spectrum B, photoreduced enzyme; spectrum C, enzyme photoreduced as in spectrum B and anaerobically reacted with CO.

Amino Acid Composition and N-Terminal Amino Acid Sequence. The amino acid composition of desulfofuscidin from *T. commune* and *T. mobile* are presented in Table II. The two proteins contain a preponderance of acidic amino acids. The ratio (Asx + Glx)/(Lys + Arg) is equal to 1.7 for *T. commune* enzyme. The number of cysteine residues of each protein, 36 and 52 residues, is higher than the number required to link four [4Fe-4S] or eight [4Fe-4S] clusters reported to be present in *T. com-*

TABLE II
AMINO ACID COMPOSITION OF DESULFOFUSCIDIN FROM *Thermodesulfobacterium commune* AND *Thermodesulfobacterium mobile*

	Desulfofuscidin	
Amino acid[a]	*T. commune*[b]	*T. mobile*[c]
Aspartic acid	160	136
Threonine	92	83
Serine	60	68
Glutamic acid	200	163
Proline	120	98
Cysteine	36	52
Glycine	144	127
Alanine	100	93
Valine	108	112
Methionine	36	23
Isoleucine	136	101
Leucine	104	109
Tyrosine	68	60
Phenylalanine	72	65
Histidine	60	40
Lysine	128	109
Arginine	84	81
Tryptophan	ND[d]	34

[a] Moles of amino acid per mole of enzyme.
[b] From E. C. Hatchikian and J. G. Zeikus, *J. Bacteriol.* **153,** 1211 (1983).
[c] From G. Fauque, A. R. Lino, M. Czechowski, L. Kang, D. V. DerVartanian, J. J. G. Moura, J. Le Gall, and I. Moura, *Biochim. Biophys. Acta* **1040,** 112 (1990).
[d] ND, Not determined.

mune and *T. mobile* bisulfite reductase, respectively.[4,5] The N-terminal sequence of *T. commune* bisulfite reductase indicates the presence of two distinct polypeptide chains in the molecule, with threonine and serine as N-terminal residues, respectively (Table III). This N-terminal sequence shows strong homologies with that of desulfofuscidin from *T. mobile* (Table III).

Catalytic Properties. Thermodesulfobacterium commune desulfofuscidin reduces sulfite but not thiosulfate, trithionate, or tetrathionate. It also catalyzes the reduction of nitrite or hydroxylamine to NH_3 as reported

TABLE III
N-Terminal Sequencing Data of Desulfofuscidin from *Thermodesulfobacterium commune* and *Thermodesulfobacterium mobile*

	Residues identified	
Step	*T. commune*[a]	*T. mobile*[b]
1	Thr, Ser	Gly, Pro
2	Glu, Ile	Glu, Ile
3	Val, Glu	Val, Glu
4	Lys, Lys	Lys, Lys
5	Phe, Lys	Phe, Tyr
6	Lys, Lys	Lys, Lys
7	Glu, Asp	
8	Leu, Thr	
9	Asp, Asp	
10	Pro, Lys	

[a] From E. C. Hatchikian and J. G. Zeikus, *J. Bacteriol.* **153,** 1211 (1983).
[b] From G. Fauque, A. R. Lino, M. Czechowski, L. Kang, D. V. DerVartanian, J. J. G. Moura, J. Le Gall, and I. Moura, *Biochim. Biophys. Acta* **1040,** 112 (1990). Residues 7 to 10 have not been determined.

with other sulfite reductases.[26] This is related to the activity of siroheme, which is the active site common to sulfite and nitrite reductases.[25] Enzymatic activity is obtained with methyl or benzyl viologen. Desulfofuscidin shows a sharp optimum of activity around pH 5.8–6.0,[4,5] indicating that bisulfite is probably the active species in sulfite reduction. The specific activity of *T. commune* bisulfite reductase is 2 μmol of H_2 consumed per minute per milligram of protein at 65° as compared to 1.48 units/mg at 60° for *T. mobile* enzyme.[4,5]

Effect of Temperature on Bisulfite Reductase Activity. The effect of temperature on *T. commune* bisulfite reductase activity is shown in Fig. 7. The enzymatic activity increased from 35° to nearly 70°, and no deflection point is observed in the slope of the curve. The maximum activity occurs between 65 and 70°. The effect of temperature on bisulfite reductase activity is associated with unusually high Q_{10} values of 3.6 and 3.2 in the

[26] J.-P. Lee, J. LeGall, and H. D. Peck, Jr., *J. Bacteriol.* **115,** 529 (1973).

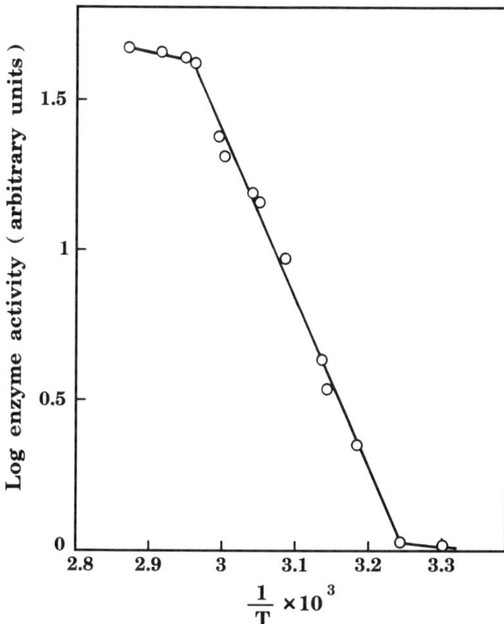

FIG. 7. Arrhenius plot of *T. commune* desulfofuscidin. [From E. C. Hatchikian and J. G. Zeikus, *J. Bacteriol.* **153,** 1211 (1983).]

35–50 and 50–65° temperature ranges, respectively. The value of the energy of activation, calculated from the slope of the data for the V_{max} plots, between 35 and 65°, was 99.58 J/mol. The enzyme exhibits very low activity below 35°, whereas the activity becomes less dependent on temperature above 65°.

Thermostability. Thermodesulfobacterium commune bisulfite reductase shows higher stability than desulforubidin, the homologous protein from the mesophilic sulfate-reducing bacterium *Desulfomicrobium baculatum* Norway 4 strain.[2] With the *T. commune* enzyme, no absorbance change of the chromophore occurs until 70° whereas such a change occurs above 50° with desulforubidin. The greater thermostability of desulfofuscidin may be related to its higher content of siroheme (four sirohemes per molecule) as compared to desulforubidin (two sirohemes per molecule).[27] The discrepancy existing between the unstability temperature of the pure bisulfite reductase (70°) and the maximum growth temperature of *T. com-*

[27] C. L. Liu, D. V. DerVartanian, and H. D. Peck, Jr., *Biochem. Biophys. Res. Commun.* **91,** 962 (1979).

TABLE IV
PRODUCTS OF BISULFITE REDUCTION CATALYZED BY BISULFITE REDUCTASE OF *Thermodesulfobacterium commune*[a,b]

SO_3^{2-} added (μmol)	H_2 consumed (μmol)	Amount of product (μmol)			S (%)
		$S_3O_6^{2-}$	$S_2O_3^{2-}$	H_2S	
22	8.1	7.0	0.30	0.10	98
10	3.4	2.8	0.20	0.11	92
	3.8	2.9	0.21	0.14	93
4	1.8	1.1	0.15	0.09	92

[a] From E. C. Hatchikian and J. G. Zeikus, *J. Bacteriol.* **153,** 1211 (1983).
[b] Bisulfite reductase activity is measured by the manometric assay after hydrogen uptake. The reaction employs 2.2 mg of pure enzyme at 45° and the incubation time is 45 min. When no more hydrogen is consumed by the reaction, the products are determined colorimetrically.

mune (80°)[10] may be attributed to the higher stabilization of this protein within the cell.

Intracellular Location. Desulfofuscidin from *T. commune* is recovered mainly in the soluble protein extract after cell disruption and centrifugation.[4] Immunocytochemical localization experiments indicate that desulfofuscidin from *T. mobile*, like the homologous bisulfite reductases from *Desulfovibrio* mesophilic species, is located in the cytoplasm.[28]

Product Ambiguity. At its optimum pH, *T. commune* desulfofuscidin catalyzes the reduction of sulfite mainly to trithionate.[4] An average value of 88% of sulfite sulfur appears as trithionate whereas trace amounts of thiosulfate and sulfide are produced (Table IV). Sulfide and thiosulfate slightly increase when low sulfite concentrations are used. No change in the products of the reaction is observed between 35 and 70°. The homologous dissimilatory bisulfite reductases including desulfoviridin,[1] desulforubidin,[2] and P-582,[9] either produce trithionate as the sole product[1] or trithionate as the major product with thiosulfate and sulfide as minor products.[2,9] It was shown that the proportions of the products of sulfite reduction are dependent on the pH and the electron donor and electron acceptor concentration.[29–31] The mechanism of dissimilatory sulfite reduction is still

[28] D. R. Kremer, M. Veenhuis, G. Fauque, H. D. Peck, Jr., J. LeGall, J. Lampreia, J. J. G. Moura, and T. A. Hansen, *Arch. Microbiol.* **150,** 296 (1988).
[29] K. Kobayashi, Y. Seki, and M. Ishimoto, *J. Biochem.* (*Tokyo*) **75,** 519 (1974).
[30] H. E. Jones and G. W. Skyring, *Biochim. Biophys. Acta* **377,** 52 (1975).
[31] H. L. Drake and J. M. Akagi, *J. Bacteriol.* **126,** 733 (1976).

a matter of debate.[6,32] Two mechanisms of dissimilatory bisulfite reduction to sulfide have been proposed:

$$SO_3^{2-} \xrightarrow{6e^-} S^{2-} \quad (1)$$

$$3SO_3^{2-} \xrightarrow{2e^-} S_3O_6^{2-} \xrightarrow{2e^-} S_2O_3^{2-} \xrightarrow{2e^-} S^{2-} \quad (2)$$

$$\text{(with } SO_3^{2-} \text{ released at the second and third steps and returned to } 3SO_3^{2-})$$

The first mechanism is the six-electron reduction of sulfite to H_2S in one step [Eq. (1)],[33–35] catalyzed by bisulfite reductase without any free intermediates, as was demonstrated with the assimilatory-type sulfite reductases.[36,37] In the second postulated mechanism [Eq. (2)], sulfite is reduced to sulfide in three steps via the free intermediates trithionate and thiosulfate.[38] The trithionate pathway requires three separate enzymes, namely sulfite reductase, trithionate reductase, and thiosulfate reductase.[7,29,30,39] A specific thiosulfate reductase has been purified from various sulfate-reducing bacteria,[40–42] and a trithionate reductase system has been isolated in *Desulfovibrio vulgaris* Hildenborough.[43] Evidence for the trithionate pathway as opposed to a direct six-electron reduction of sulfite to sulfide has been obtained by Fitz and Cypionka,[44] using deenergized cells of *Desulfovibrio desulfuricans* Essex 6. Whole cells of this sulfate-reducing bacterium produce thiosulfate and trithionate during sulfite reduction coupled to oxidation of physiological electron donors. Most of their observations can be explained on the basis of the redox potentials of the intermediates. Further evidence for intermediate formation during sulfite reduction was obtained from proton translocation coupled to the reduction of trithionate, inhibition of thiosulfate reductase by trithionate, and reverse trithionate reductase activity.[44]

The question of whether additional free intermediates occur between

[32] H. D. Peck, Jr., and J. LeGall, *Philos. Trans. R. Soc. London B* **298,** 443 (1982).
[33] P. A. Trudinger and L. A. Chambers, *Biochim. Biophys. Acta* **293,** 26 (1973).
[34] L. M. Siegel, *in* "Metabolic Pathways. Metabolism of Sulfur Compounds" (D. M. Greenberg, ed.), Vol. 7, p. 217. Academic Press, New York, 1975.
[35] L. A. Chambers and P. A. Trudinger, *J. Bacteriol.* **123,** 36 (1975).
[36] K. Asada, G. Tamura, and R. S. Bandurski, *J. Biol. Chem.* **244,** 4904 (1969).
[37] L. M. Siegel, P. S. Davis, and H. Kamin, *J. Biol. Chem.* **249,** 1572 (1974).
[38] K. Kobayashi, S. Tachibana, and M. Ishimoto, *J. Biochem.* (*Tokyo*) **65,** 155 (1969).
[39] J. M. Akagi, M. Chan, and V. Adams, *J. Bacteriol.* **120,** 240 (1974).
[40] W. Nakatsukasa and J. M. Akagi, *J. Bacteriol.* **98,** 429 (1969).
[41] R. H. Haschke and L. L. Campbell, *J. Bacteriol.* **106,** 603 (1971).
[42] E. C. Hatchikian, *Arch. Microbiol.* **105,** 249 (1975).
[43] J.-H. Kim and J. M. Akagi, *J. Bacteriol.* **163,** 472 (1985).
[44] R. M. Fitz and H. Cypionka, *Arch. Microbiol.* **154,** 400 (1990).

TABLE V
PHYSICOCHEMICAL AND CATALYTIC PROPERTIES OF DISSIMILATORY SULFITE REDUCTASES

Property	Desulfofuscidin		Desulfoviridin[c]	Desulforubidin[d]	P-582[e]
	T. commune[a]	*T. mobile*[b]	(*D. gigas*)	(*Dsm. baculatum* DSM 1741)	(*Dm. nigrificans*)
Molecular mass (kDa)	167	190	200	225	194
Subunit structure	$\alpha_2\beta_2$	$\alpha_2\beta_2$	$\alpha_2\beta_2$	$\alpha_2\beta_2$	$\alpha_2\beta_2$
Absorption maxima (nm)	389, 576, 693	392, 578, 700	390, 408, 580, 628	392, 545, 580	392, 582, 700
Extinction coefficient (mM^{-1} cm^{-1}) (Soret peak)	310	323	200	200	ND[f]
Iron content[g]	20–21	32	16.5	16.6	16
Labile sulfide[g]	16	ND[f]	14	14.7	14
Siroheme[h]	4	4		2	1.3
Siroporphyrin[h]			2		
[4Fe-4S] cluster	4	8	4	4	4
Cysteine residues	36	52	ND[f]	ND[f]	ND[f]
Reaction with CO	+	+	−	+	+
Specific activity[i]	2	1.48	0.63	0.41	ND[f]
Major product	$S_3O_6^{2-}$	ND[f]	$S_3O_6^{2-}$	$S_3O_6^{2-}$	S^{2-}
Unstability temperature	70°	ND[f]	ND[f]	50°[j]	ND[f]

[a] From E. C. Hatchikian and J. G. Zeikus, *J. Bacteriol.* **153,** 1211 (1983).

[b] From G. Fauque, A. R. Lino, M. Czechowski, L. Kang, D. V. DerVartanian, J. J. G. Moura, J. Le Gall, and I. Moura, *Biochim. Biophys. Acta* **1040,** 112 (1990).

[c] From J.-P. Lee and H. D. Peck, Jr., *Biochem. Biophys. Res. Commun.* **45,** 583 (1971); H. D. Peck, Jr., and J. Le Gall, *Philos. Trans. R. Soc. London B* **298,** 443 (1982); I. Moura, J. Le Gall, A. R. Lino, H. D. Peck, Jr., G. Fauque, A. V. Xavier, D. V. DerVartanian, J. J. G. Moura, and B. H. Huynh, *J. Am. Chem. Soc.* **110,** 1075 (1988).

[d] From J.-P. Lee, C. Yi, J. Le Gall and H. D. Peck, Jr., *J. Bacteriol.* **115,** 453 (1973); C. L. Liu, D. V. DerVartanian and H. D. Peck, Jr., *Biochem. Biophys. Res. Commun.* **91,** 962 (1979); H. D. Peck, Jr., and J. Le Gall, *Philos. Trans. R. Soc. London B* **298,** 443 (1982).

[e] From P. A. Trudinger, *J. Bacteriol.* **104,** 158 (1970); J. Le Gall, J. J. G. Moura, H. D. Peck, Jr., and A. V. Xavier, *in* "Iron-Sulfur Proteins" (T. G. Spiro, ed.), Vol. 4, p. 177. Wiley, New York, 1982.

[f] ND, Not determined.

[g] Atoms per enzyme molecule.

[h] Chromophore extracted with acetone-hydrochloride.

[i] Expressed as micromoles of hydrogen consumed per minute per milligram protein.

[j] Unpublished results.

sulfite and sulfide remains open. Several modes of dissimilatory reduction of bisulfite may be possible in the sulfate-reducing bacteria. One pathway would be the trithionate pathway.[38] An alternate pathway functioning in different environmental conditions may involve another mechanism such as the six-electron reduction of bisulfite catalyzed either by the well-known bisulfite reductase or by a distinct enzyme. Isolation of mutants that will either be altered with respect to both reductase activities, or to only one of them, would afford definitive information on the bisulfite reduction pathway.

Thiosulfate and trithionate reductase activities appear to be present in the crude extract of *T. commune*.[4] These enzymatic activities are separated from bisulfite reductase during the DEAE-cellulose fractionation step of the purification procedure (Table I). These observations, together with the formation of trithionate as the main product of sulfite reduction by desulfofuscidin, are indicative of the presence of the trithionate pathway of bisulfite reduction in this microorganism.

Conclusion

Desulfofuscidin is the specific dissimilatory sulfite reductase of extreme thermophilic sulfate-reducing eubacteria (maximum temperature for growth, 85°) as compared to desulfoviridin and desulforubidin isolated from mesophilic species of the genus *Desulfovibrio* and P-582 isolated from the moderate thermophile spore-forming species *Desulfotomaculum nigrificans*. The main physicochemical properties of the four types of dissimilatory sulfite reductase are reported in Table V. They exhibit similar molecular weights and quaternary structure but they differ by their absorption spectrum. Electron paramagnetic resonance spectroscopy indicates that all these enzymes contain high-spin siroheme and [4Fe-4S] clusters as prosthetic groups. In contrast to desulfoviridin, which yields a siroporphyrin, the other bisulfite reductases yield a siroheme on acetone-acid extraction.[4–6] The content of siroheme in desulfofuscidin is twice that of the other bisulfite reductases. Desulfofuscidin shows higher thermostability than desulforubidin and exhibits its maximum activity between 65 and 70°.

Acknowledgments

The author wishes to thank Professor J. G. Zeikus, who initiated the study on desulfofuscidin from *Thermodesulfobacterium commune*, and Professor R. Cammack for his critical review of the manuscript and valuable comments. Thanks are also due to N. Forget for skillful technical assistance.

[20] Low-Spin Sulfite Reductases

By ISABEL MOURA and ANA ROSA LINO

Introduction

Sulfite reductase catalyzes the six-electron reduction of SO_3^{2-} to S^{2-}. This enzyme contains an iron tetrahydroporphyrin prosthetic group, termed *siroheme*, in addition to nonheme iron. On the basis of physiological function, two types of sulfite reductases can be defined: (1) the assimilatory type, which is involved in the synthesis of sulfur-containing compounds, and (2) the dissimilatory one, which participates in the respiratory pathway for sulfate-reducing bacteria. One characteristic of the dissimilatory sulfite reductases is the fact that under certain assay conditions, in addition to sulfide, trithionate and thiosulfate are irreversibly produced.[1,2] The assimilatory type can catalyze the six-electron reduction without the formation of free intermediates.[3,4] The dissimilatory sulfite reductases are described in previous chapters (see [18–37] in this volume). Here, we describe the properties of the assimilatory-type sulfite reductases that were isolated from *Desulfovibrio* (*D.*) *vulgaris* (Hildenborough), *Methanosarcina* (*Ms.*) *barkeri*, and *Desulforomonas* (*Drm.*) *acetoxidans*.[5–7] These three enzymes have low molecularmass and the siroheme is in a low-spin state, representing a new class of nonheme iron-siroheme proteins common to anaerobic bacteria. The general properties of the enzyme isolated from *D. vulgaris* (Hildenborough) are described and compared with the properties of the other two enzymes.

[1] J. M. Akagi, *in* "Biology of Inorganic Nitrogen and Sulfur" (H. Bothe and A. Trebst, eds.), p. 169. Springer-Verlag, Berlin, 1981.

[2] J. P. Lee and H. D. Peck, Jr., *Biochem. Biophys. Res. Commun.* **45,** 583 (1971).

[3] L. M. Siegel and H. Kamin, *in* "Flavins and Flavoproteins" (K. Yagi, ed.), p. 15. Univ. Park Press, Baltimore, Maryland, 1968.

[4] K. Prabhakararau and D. J. D. Nicholas, *Biochim. Biophys. Acta* **180,** 253 (1969).

[5] B. H. Huynh, L. Kang, D. V. DerVartanian, H. D. Peck, Jr., and J. LeGall, *J. Biol. Chem.* **259,** 15373 (1984).

[6] J. J. G. Moura, I. Moura, H. Santos, A. V. Xavier, M. Scandellari, and J. LeGall, *Biochem. Biophys. Res. Commun.* **108,** 1002 (1982).

[7] I. Moura, A. R. Lino, J. J. G. Moura, A. V. Xavier, G. Fauque, H. D. Peck, Jr., and J. LeGall, *Biochem. Biophys. Res. Commun.* **141,** 1032 (1986).

Growth of Organism and Preparation of Crude Extract

Desulfovibrio vulgaris (Hildenborough, NCIB 8303) is grown in lactate-sulfate medium as previously described,[8] containing 1 mg of iron per liter. Cells cultured for 38 hr are used for these experiments. The wet cells are washed with 0.01 *M* Tris-HCl buffer (pH 7.6, 5°) to remove the periplasmic proteins. The washed cells are then mixed with 0.05 *M* Tris-HCl buffer (pH 7.6, 5°) to give a 1 : 1 (w/v) ratio cell suspension and are broken by passing them through a Manton-Gaulin homogenizer three times at 9000 psi. A few milligrams of DNase I and DNase II are added to decrease the viscosity of the extract. The preparation is treated with neutralized streptomycin sulfate (0.5 mg/mg protein), stirred at 4° for 15 min, and centrifuged at 13,200 *g* in a Sorvall (Newtown, CT) RC-5B refrigerated centrifuge for 1 hr. The supernatant is centrifuged again at 144,000 *g* for 90 min in a Sorvall-Beckman (Fullerton, CA) L5-75B ultracentrifuge. The resulting supernatant is the crude extract used for the purification of the sulfite reductase.

Purification of Sulfite Reductase

All purification procedures are carried out at 4° and all buffers are adjusted to pH 7.6. The crude extract is applied to a hydroxylapatite (Bio-Rad, Richmond, CA) column (6 × 24 cm) equilibrated with 0.1 *M* Tris-HCl buffer. The column is washed with a descending gradient to reduce the Tris concentration to 0.02 *M*. A continuous potassium phosphate gradient (0.01–0.5 *M* with a total volume of 4 liters) is then applied for elution. Sulfite reductase is eluted at about 0.3 *M* phosphate buffer. The eluted enzyme is concentrated on a YM-10 membrane using a Diaflo apparatus (Amicon, Danvers, MA) and dialyzed against 0.02 *M* Tris-HCl buffer. Cytochromes present in the sulfite reductase-containing fraction are removed by loading the dialyzed preparation onto a DEAE-BioGel column (4.5 × 20 cm) equilibrated with 0.02 *M* Tris-HCl buffer. Two bed volumes are used to wash out the nonabsorbed cytochromes. A linear gradient of Tris-HCl buffer (0.02–0.2 *M* with a total volume of 2 liters) is applied for elution and sulfite reductase is eluted at a concentration of about 0.1 *M*. The eluted fraction is concentrated using a Diaflo apparatus with YM-30 membrane. A minor flavoprotein contained in the sulfite reductase preparation is removed by molecular sieving through a Sephadex G-75 column. Purity of the sulfite reductase is established by normal

[8] B. H. Huynh, M. H. Czechowski, H.-J. Krüger, D. V. DerVartanian, H. D. Peck, Jr., and J. LeGall, *Proc. Natl. Acad. Sci. U.S.A.* **81,** 3728 (1984).

TABLE I
COMPARISON OF PROPERTIES OF LOW-SPIN SULFITE REDUCTASES FROM *Desulfovibrio vulgaris*, *Desulforomonas acetoxidans*, AND *Methanosarcina barkeri* (DSM 800)

Property	Reductase from:		
	D. vulgaris	*Drm. acetoxidans*	*Ms. barkeri*
Molecular mass (kDa)	27.2	23.5	23.0
Absorption maxima (in the visible range)	590, 545, 400	587, 540, 400	590, 545, 395
Specific activity (mU/mg; pH = 6.0)	900	906	2790
Total Fe	4.7	5.3	5.2
Siroheme (Spins/mol)	0.80	0.96	0.80
Moles of siroheme per mole of protein	0.91	0.94	0.85
EPR *g* values (native form)	2.44, 2.36, 1.77	2.44, 2.33, 1.81	2.40, 2.30, 1.88

and by sodium dodecyl sulfate-polyacrylamide gel electrophoresis. The purified enzyme is stable for six months in 0.05 *M* Tris-HCl buffer at −80°.

Extraction and Quantitation of Siroheme

Siroheme is extracted from the associated protein by addition of 9 vol of ice-cold acetone-HCl, with stirring, to 1 vol of protein solution.[9] After vigorous mixing leave the mixture to stand in the dark, for 5 min at 0°, and then centrifuge it to remove the precipitated protein at 4,000 g for fifteen minutes at 4°C.

The extracted heme can be stabilized by addition of 0.5 vol of pyridine to each volume of extracted heme solution. The concentration of siroheme is determined by optical spectroscopy, using an extinction coeficient of 15.4 mM^{-1} cm^{-1} at 557 nm (Table I). The obtained visible spectrum is similar to that of siroheme extracted from *Escherichia coli* sulfite reductase.[10]

General Properties

The purified sulfite reductase has a molecular mass of 27.2 kDa, determined by amino acid composition. A concentration of 4.6 ± 0.7 mol of acid-labile sulfur atoms determined by Siegel's method[11] was found and 4.7 ± 0.7 iron atoms per mole of purified enzyme was determined by

[9] L. M. Siegel, M. J. Murphy, and H. Kamin, this series, Vol. 52, p. 436.
[10] L. M. Siegel, M. J. Murphy, and H. Kamin, *J. Biol. Chem.* **248,** 251 (1973).
[11] L. M. Siegel, *Anal. Biochem.* **11,** 126 (1965).

plasma emission. The iron content is consistent with the existence of a single [4Fe-4S] cluster and a siroheme. The sulfur content suggests, because the [4Fe-4S] cluster contains only four labile sulfurs, that an extra sulfur atom is present. This sulfur could be the bridging ligand between siroheme and the [4Fe-4S] cluster. The optical spectrum of the purified sulfite reductase exhibits maxima at 280, 400, 545, and 590 nm. The ratio of absorbance at 280 nm to that at 590 nm is 3.4. The molar extinction coefficient at 590 nm was determined to be 22 mM^{-1} cm^{-1}, based on a molecular mass of 27.2. The purified *D. vulgaris* sulfite reductase does not show absorption maxima beyond 700 nm. A band in this region[12] is an indication of the high-spin ferric ($S = 5/2$) complex of isobacteriochlorins through model compound studies.[13] The absence of such an absorption maximum for *D. vulgaris* sulfite reductase indicates that its siroheme is in a different spin state.

In Table I some of the described properties for *D. vulgaris* enzyme are compared with those of the enzymes isolated from *Drm. acetoxidans* and *Ms. barkeri*.

Spectroscopic Properties of Extracted Siroheme

The properties of siroheme extracted from sulfite reductases have been investigated.[14] Optical absorption electron paramagnetic resonance (EPR) and magnetic circular dichroism (MCD) spectroscopy studies were performed on the siroheme isolated from the low-spin sulfite reductase of *D. vulgaris* (Hildenborough). Siroheme exhibits a marked pH dependence, and two pK_a values of 4.2 and 9.0 were observed.

The binding of strong-field ligands CO, NO, and cyanide were investigated by ultraviolet (UV)–visible absorption, and for cyanide complex by EPR and MCD spectroscopies. CO and NO were able to reduce and bind siroheme without additional reducing agent. The isolated siroheme exhibits an EPR axial signal, typical of high-spin ferric heme ($S = 5/2$), whereas the cyanide complex shows a low-spin ferric heme ($S = 1/2$) signal. The MCD data are consistent with the results obtained by EPR.

The results for ferrosiroheme indicate that siroheme remains high spin ($S = 2$) and low spin ($S = 0$) on reduction of the as-isolated and cyanide complexes, respectively.

[12] L. M. Siegel, D. C. Rueger, M. J. Barber, R. J. Krueger, N. R. Orme-Johnson, and W. H. Orme-Johnson, *J. Biol. Chem.* **257,** 6343 (1982).

[13] A. M. Stolzenbach, S. H. Strauss, and R. H. Holm, *J. Am. Chem. Soc.* **103,** 4763 (1981).

[14] L. Kang, J. LeGall, A. T. Kowal, and M. K. Johnson, *J. Inorg. Biochem.* **30,** 273 (1987).

Enzymatic Activity

The specific activity was determined by the manometric method described by Schedel *et al.*[15] It requires the generation of reduced methyl viologen by an excess of hydrogenase activity under hydrogen atmosphere. The reduced dye then serves as electron donor to the reductase.

Freshly prepared sodium sulfite (0.1 ml of a 0.01 *M* solution) is added from the side arm to the main compartment of each manometric vessel containing 0.1 ml of 1 *M* phosphate buffer (pH as required); 0.1 ml methyl viologen, 2% (w/v); hydrogenase (approximately 40 μg of protein) and 0.75 ml of the enzyme solution; water is added to a final volume of 2.8 ml. The center well contains 0.05 ml of 10 *N* NaOH and 0.05 ml of 10% (w/v) $Cd(CH_3COO)_2$. Hydrogenase from *D. gigas*, purified as described by LeGall *et al.*,[16] is added to the system to ensure an excess of this activity. Before addition of sodium sulfite the flask is incubated under hydrogen atmosphere for 30 min at 37°. In typical experiments, the blue color of reduced methyl viologen appears 3–5 min after incubation under H_2.

Table I presents the specific activities of the three low-spin sulfite reductases (1 mU represents the enzyme activity catalyzing the consumption of 1 μmol of H_2 per minute at 30°).

Genetic Studies

The nucleotide sequence encoding the structural gene (651 bp) and flanking regions for the assimilatory-type sulfite reductase from the sulfate-reducing bacterium *D. vulgaris* (Hildenborough) was determined after cloning a 1.4-kb *Hin*d III/*Sal*I genomic fragment possessing the gene into Bluescript pBS(+) KS (Stratogene). [17] The primary structure of the protein was deduced, and the molecular mass of the apoprotein was estimated as 24 kDa. The amino acid sequence of the polypeptide shows some similarities at putative [4Fe-4S] cluster-binding sites in comparison with the heme protein subunit of the larger *E. coli* and *Salmonella typhimurium* sulfite reductases and spinach nitrite reductase.

EPR and NMR Studies

The EPR spectra of the three sulfite reductases from *D. vulgaris*, *Drm. acetoxidans*, and *Ms. barkeri* show a rhombic signal with the *g* values,

[15] M. Schedel, J. LeGall, and J. Baldensperger, *Arch. Microbiol.* **105,** 339 (1975).

[16] J. LeGall, P. O. Ljungdahl, I. Moura, H. D. Peck, Jr., A. V. Xavier, J. J. G. Moura, M. Teixeira, B. H. Huynh, and D. V. DerVartanian, *Biochem. Biophys. Res. Commun.* **106,** 6100 (1982).

[17] J. Tan, L. R. Helms, R. P. Swenson, and J. A. Cowan, *J. Biol. Chem.* **30,** 9900 (1991).

presented in Table I, that are characteristic of low-spin ferriheme.[5,7] The fact that the siroheme is in the low-spin ferric state is a unique feature. The siroheme of other sulfite reductases[6] is high-spin ferric. Electron paramagnetic resonance studies on model complexes have shown that ferric isobacteriochlorins with a single axial ligand are always high spin whereas ferric isobacteriochlorins with two axial ligands are low spin. Because in the three sulfite reductases the siroheme is low-spin ferric, it could be six coordinated.

Preliminary nuclear magnetic resonance (NMR) studies were published by Cowan and Sola[18] on the low-spin sulfite reductase from *D. vulgaris*. They presented a series of hyperfine shifted resonances that they attributed to methylene groups on the acetate and propionate functions around the siroheme ring and from the CH_2 groups of the cysteines binding the [4Fe-4S] cluster. They also suggested that histidine could be the sixth ligand of the siroheme. An unusual resonance was also observed at ~92.5 ppm, which they tentatively interpret as a result of the magnetic coupling between the paramagnetic low-spin Fe(III) of the siroheme and the [4Fe-4S] cluster through the bridge.

Mössbauer Studies

At 150 K the Mössbauer spectrum of the low-spin sulfite reductase from *D. vulgaris* shows two quadrupole doublets.[5] The Mössbauer parameters of the siroheme, $\Delta E_Q = 2.49\ 0.02$ mm/sec and $\delta = 0.31 \pm 0.02$ mm/sec, are typical of low-spin ferric complexes. The [4Fe-4S] cluster is in the 2+ state. The Mössbauer parameters, $\Delta E_Q = 0.95 \pm 0.02$ mm/sec and $\delta = 0.38 \pm 0.02$ mm/sec, are almost identical to those observed for the $[4Fe\text{-}4S]^{2+}$ cluster in the hemoprotein subunit of the sulfite reductase from *E. coli*. Similar to the hemoprotein subunit of *E. coli* sulfite reductase, low-temperature Mössbauer spectra of *D. vulgaris* sulfite reductase recorded with weak and strong applied fields also show evidence of an exchange-coupled siroheme–[4Fe-4S] unit.

Mechanistic Studies

A detailed reaction pathway for the six-electron reduction of SO_3^{2-} to S^{2-} by the low-spin sulfite reductase from *D. vulgaris* was deduced from experiments with ^{35}S-labeled enzyme and the relative reaction rates of

[18] J. A. Cowan and M. Sola, *Inorg. Chem.* **29,** 2176 (1990).

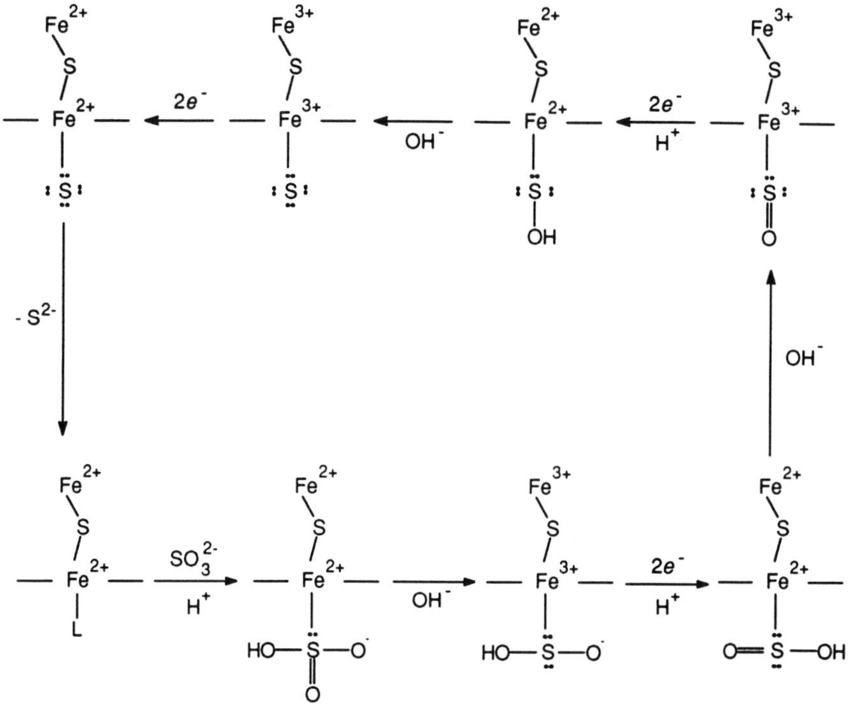

FIG. 1. Proposed pathway for the reduction of SO_3^{2-} to S^{2-}. L, Protein ligand that is displaced by the substrate. Only one iron atom from the [4Fe-4S] cluster is represented and the oxidation and reduction of the siroheme is represented by the charge of the iron atom.

nitrogenous substrates by Tan and Cowan.[19] The proposed pathway is shown in Fig. 1. The ligand bridging the prosthetic [4Fe-4S]–siroheme center is exchanged by $^{35}S^{2-}$ in both oxidized and reduced enzyme, confirming what was suggested previously.[5] This bridging ligand is retained in the course of SO_3^{2-} reduction and the substrate binds the nonbridging axial site of the siroheme. The mechanism suggests that SO_3^{2-} binds to the Fe^{2+} of the siroheme through a sulfur atom followed by a series of two-electron reductive cleavages of S—O bonds (Fig. 1). The protonation of oxygen facilitates the bond cleavage originating hydroxide as the leaving group. The bridging between the siroheme and the iron-sulfur cluster remains intact during the course of the reaction, providing an efficient coupling pathway for electron transfer between the cluster and the siroheme.

[19] J. Tan and J. A. Cowan, *Biochemistry* **30,** 8910 (1991).

Final Comments

The low-spin sulfite reductases from *D. vulgaris*, *Ms. barkeri*, and *Drm. acetoxidans* belong to a new class of sulfite reductases that may be common to anaerobic microorganisms.

They are all low molecular weight proteins with one siroheme and one [4Fe-4S] center per polypeptide chain and in the native state their siroheme is low-spin ferric. They also differ from the *Desulfovibrio* dissimilatory type of sulfite reductase, such as desulfoviridin, desulforubidin, or desulfofuscidin, by their much simpler oligomeric structure. In contrast to the *E. coli* type of assimilatory sulfite reductase, they do not appear to be part of a multiprotein complex.

These low molecular mass sulfite reductase enzymes, being simpler enzymes, are potential models for the understanding of the properties of the more complex sulfite reductases.

[21] Hexaheme Nitrite Reductase from *Desulfovibrio desulfuricans* (ATCC 27774)

By MING-CHEH LIU, CRISTINA COSTA, and ISABEL MOURA

$$NO_2^- + 6e \rightarrow NH_4^+ + 2H_2O$$

Hexaheme nitrite reductase is the second enzyme involved in dissimilatory nitrate reduction performed by a number of anaerobically grown bacteria. It was first discovered in nitrate-respiring *Desulfovibrio desulfuricans* (ATCC 27774).[1] Similar hexaheme nitrite reductases were subsequently identified in *Escherichia coli* K-12,[2,3] *Wolinella succinogenes*,[4,5] and *Vibrio fischeri*.[6] This chapter focuses on the description of the purification and various properties of the enzyme from *D. desulfuricans*.

[1] M.-C. Liu and H. D. Peck, Jr., *J. Biol. Chem.* **256,** 13159 (1981).

[2] M.-C. Liu, H. D. Peck, Jr., A. Abou-Jaoude, M. Chippaux, and J. LeGall, *FEMS Lett.* **10,** 333 (1981).

[3] S. Kajie and Y. Anraku, *Eur. J. Biochem.* **154,** 457 (1986).

[4] M.-C. Liu, M.-Y. Liu, W. J. Payne, H. D. Peck, Jr., and J. LeGall, *FEMS Lett.* **19,** 201 (1983).

[5] R. S. Blackmore, A. M. Roberton, and T. Brittain, *Biochem. J.* **233,** 547 (1986).

[6] M.-C. Liu, B. W. Bakel, M.-Y. Liu, and T. N. Dao, *Arch. Biochem. Biophys.* **262,** 259 (1988).

Growth of Cells

Desulfovibrio desulfuricans (ATCC 27774), isolated originally by M. P. Bryant, is routinely grown at 37° under pure argon and the stock cultures are transferred on nitrate slants twice a week using the Hungate technique.[7] The medium contains the following components per liter: sodium lactate [60% (w/v), 12.5 ml], $Mg(NO_3)_2 \cdot 6H_2O$ (0.73 g); $NaNO_3$ (2.465 g), $CaCl_2$ (0.2 g), $FeCl_2 \cdot 4H_2O$ (3.55 mg), K_2HPO_4 (0.5 g), resazurin [0.1% (w/v) solution; 1.0 ml], yeast extract (1 g; Difco, Detroit, MI), cysteine-HCl (0.5 g), $Na_2S \cdot 9H_2O$ (0.5 g), and agar (15 g). For the growth of large amounts of bacteria, *D. desulfuricans* is transferred using the Hungate technique from a slant to a tube containing 10 ml of the above medium minus agar and incubated for 5–7 days. After reaching the log phase of growth, the bacteria are transferred to a 500-ml flask containing 200 ml of the same medium. After a 2 to 3-day incubation, the bacteria are transferred to a 2-liter flask containing 1.5 liters of the above medium modified per liter as follows: $Na_2S \cdot 9H_2O$ (0.26 g), cysteine-HCl (0.26 g), and resazurin (0.26 ml of a 0.1% solution). After 24 hr of incubation, the bacteria are transferred to a 20-liter carboy containing 18 liters of medium with cysteine-HCl and $Na_2S \cdot 9H_2O$ reduced to 0.19 g each per liter. After 24 hr, the content of the carboy is used to inoculate 400 liters of medium in a New Brunswick fermentor in which the cells are grown under dinitrogen with slow stirring for 24 hr at 37°. The cells are harvested at room temperature in a chilled Sharples centrifuge and stored at −20°. Usually 750 g to 1 kg of wet cells is obtained from a 400-liter culture.

To establish anaerobiosis, the flasks and carboys are flushed with argon during the preparation of the medium. The medium, in a 500-ml flask, is prepared by adding all the constituents of the medium except for the reducing agents, and heating to the boiling point. The reducing agents are then added and the pH adjusted to 6.8–7.2 using 4 *N* HCl or 2.5 *M* NaOH solution. More stringent precautions are required to establish and preserve anaerobic conditions when these organisms are grown on nitrate than are necessary for growth on sulfate, which is reduced to H_2S and aids in maintaining anaerobiosis. The smaller amounts of medium are autoclaved for 20 min at 20 psi, whereas the 20-liter carboys are autoclaved for 1 hr at 20 psi.

Assay Method

The assay procedure for the nitrite reductase involves two separate steps: first, the conversion of nitrite to ammonia by the nitrite reductase,

[7] R. E. Hungate, *in* "Methods in Microbiology" (J. R. Norris and D. W. Ribbons, eds.), Vol. 3B, p. 117. Academic Press, New York, 1969.

and second, the quantitative determination of ammonia by the indophenol reaction.[8] For the enzymatic step, a 1.0-ml reaction mixture is employed that contains phosphate buffer (pH 7.6, 0.2 *M*), $NaNO_2$ (10 m*M*), methyl viologen (0.5 m*M*), $Na_2S_2O_4$ (14.5 m*M*), $NaHCO_3$ (58 m*M*), and the appropriate diluted enzyme. The reaction is initiated by the addition of the reducing agent in $NaHCO_3$ and, after 15 min at 37°, terminated by the addition of 1 ml of 1 *N* H_2SO_4. For the determination of ammonia, 0.1 ml of the reaction mixture is added to 5.0 ml of phenol reagent, which is prepared by adding 5 g of phenol and 25 mg of sodium nitroprusside to 500 ml of distilled H_2O and then sequentially adding 5 ml of 0.5 *M* potassium phosphate buffer (pH 12.1) and 42 μl of commercial bleach. The mixture is immediately mixed by means of a Vortex-Genie (Scientific Industries, Inc., Bohemia, N.Y.) and placed in a 37° water bath. After 20 min, the absorbance at 630 nm is determined. For each assay, a zero time control is prepared to which the 1 *N* H_2SO_4 is added before the enzyme. One unit of enzyme activity is defined as the amount of enzyme that catalyzes the reduction of 1 μmol of nitrite to ammonia per minute at 37°.

Purification

Unless otherwise indicated, the operations described below are all carried out at 4° and the buffers used are all at pH 7.6.

Membrane Isolation

Frozen cells (250 g wet weight) are mixed with 0.1 *M* potassium phosphate buffer to give a 1 : 4 (w/v) suspension and passed three times through a Manton–Gaulin homogenizer at 9000 psi. A few crystals of DNase are added to lessen the viscosity of the homogenate and, after 5–10 min, the preparation is centrifuged at 13,200 *g* for 30 min. The supernatant collected is treated with neutralized streptomycin sulfate (0.5 mg/mg protein). After stirring for 15 min, the preparation is centrifuged again at 13,200 *g* for 30 min. The supernatant collected is further centrifuged at 100,000 *g* for 2 hr. The membrane pellet obtained is washed five or six times with a volume of 0.1 *M* phosphate buffer equivalent to the volume of the crude homogenate. Each time the membranes are recovered by centrifugation at 100,000 *g* for 1 hr. The washed membranes are suspended in a volume of 0.1 *M* phosphate buffer equal to one-fifth the volume of the crude homogenate.

[8] R. L. Searcy, N. M. Simms, J. A. Foreman, and L. M. Bergquist, *Clin. Chim. Acta*. **12**, 170 (1965).

Detergent Extraction

Sodium choleate is first added to the washed membrane suspension at a concentration of 3 mg/ml. The preparation is incubated for 24 hr and centrifuged at 100,000 *g* for 1 hr. The membrane pellet obtained is again washed five or six times and resuspended in 0.1 *M* potassium phosphate buffer as described above. The hexaheme nitrite reductase is then extracted by sodium choleate at a concentration of 4 mg/ml. The suspension is incubated for 24 hr and centrifuged at 100,000 *g* for 2 hr. The supernatant collected exhibits an intense red color and routinely contains over 95% of the nitrite reductase activity originally associated with the membranes.

Purification Procedure

The red sodium choleate extract is first fractionated with ammonium sulfate at 35, 47, and 52% saturation. The precipitates are recovered by centrifugation at 13,200 *g* for 30 min and dissolved separately in minimum amounts of 50 m*M* potassium phosphate buffer. The fractions are dialyzed for 18 hr against 4 liters of 50 m*M* potassium phosphate buffer with three changes of buffer, and tested for nitrite reductase activity. Usually, the 47–52% fraction contains most of the activity and the red color. This fraction is applied to a column of Ultrogel AcA-34 (2.5 × 100 cm) prepared for ascending flow and equilibrated with 50 m*M* potassium phosphate buffer. The enzyme is eluted with the same buffer and the red fractions containing the hexaheme nitrite reductase are pooled and concentrated using an Amicon (Danvers, MA) ultrafiltration cell. The enzyme, stored at −20°, is stable for at least 6 months.

The purification of the hexaheme nitrite reductase from 250 g of wet cells of *D. desulfuricans* is shown in Table I. Over 54% of the nitrite-reducing activity of the crude homogenate was found in the membrane fraction. However, reduced methyl viologen used in the nitrite reductase assay can serve as electron donor for the reduction of nitrite to ammonia by bisulfite reductase (desulfoviridin), which constitutes between 5 and 10% of the soluble protein[9] and probably accounts for most of the soluble nitrite reductase activity. Because of this uncertainty, yield calculations are not included in Table I. Nevertheless, the final recovery of the membrane-associated nitrite reductase activity was over 86%. On the basis of the data shown in Table I, the hexaheme nitrite reductase was purified approximately 55-fold. The purified nitrite reductase, on sodium dodecyl sulfate-polyacrylamide gel electrophoresis (SDS–PAGE), showed a single band after staining with Coomassie Brilliant Blue. When the purified nitrite

[9] J. P. Lee, J. LeGall, and H. D. Peck, Jr., *J. Bacteriol.* **115,** 529 (1973).

TABLE I
PURIFICATION OF NITRITE REDUCTASE FROM NITRATE-GROWN *Desulfovibrio desulfuricans* (ATCC 27774)[a]

Step	Volume (ml)	Protein (mg)	Total activity (units[b])	Specific activity (units/mg)
Crude extract (250 g wet cells)	1,220	50,820	314,600[c]	6.2
Washed membrane suspension	250	15,000	170,000	11.3
Sodium choleate solubilization	212.5	2,009	191,780	95.3
Ammonium sulfate precipitation (47–52%)	13.1	628.8	176,980	282.0
Ultrogel AcA-34	16.4	428.1	147,600	344.6

[a] Reprinted with permission from ASBMB (Liu and Peck, *J. Biol. Chem.* **256,** 13159 (1981)).
[b] One unit is defined as the amount of nitrite reductase needed to convert 1 μmol of NO_2^- to 1 μmol of NH_4^+ in 1 min at 37°.
[c] No correction was made for the NO_2^--reducing activity of desulfoviridin.

reductase was subjected to disc gel electrophoresis (without SDS), a single dark blue region on the top of the gel with nothing beneath the top portion of the gel was found after staining with Amido black. This indicated that the nitrite reductase has a tendency to form aggregates, which prevents the enzyme from penetrating the gel during electrophoresis.

Properties

Absorption Spectra

As shown in Fig. 1, the purified nitrite reductase exhibited a typical *c*-type cytochrome absorption spectrum. The oxidized enzyme showed, in addition to the protein peak at 279 nm, absorption maxima at 409 nm (Soret) and 531 nm. On reduction with dithionite, the Soret band intensified and shifted to 420 nm and at the same time α and β bands appeared at 552.5 and 523 nm, respectively. The millimolar extinction coefficient of nitrite reductase at 409 nm (oxidized Soret band) was 649.9 mM^{-1} cm^{-1}. This value is 6.07 times that of horse heart cytochrome *c*.[10]

Table II shows the absorption maxima of the oxidized and reduced forms (i.e., pyridine hemochromogen and cyanide hemochromogen) of the nitrite reductase as compared with those of horse heart cytochrome

[10] D. L. Drabkin, *J. Biol. Chem.* **146,** 605 (1942).

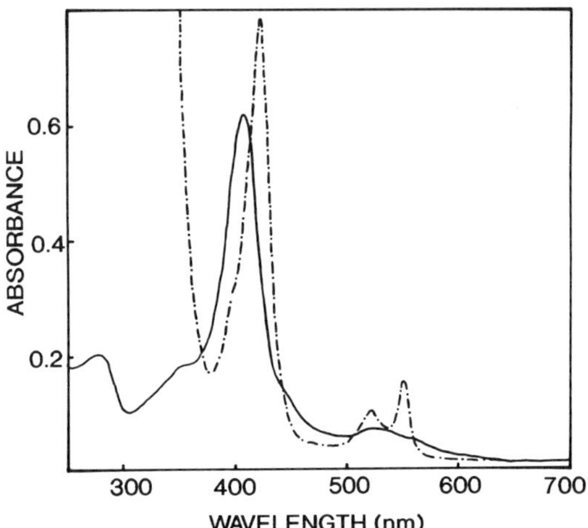

FIG. 1. Absorption spectra of *D. desulfuricans* nitrite reductase. Spectra were measured in 0.1 *M* potassium phosphate buffer, pH 7.6. Enzyme protein concentration, 0.037 mg/ml; (—) oxidized; (-.-.-), reduced with dithionate. (Reprinted with permission from ASBMB, Liu and Peck, *J. Biol. Chem.* **256,** 13159 (1981).)

c. As with horse heart cytochrome *c*, the spectrum of the pyridine hemochromogen formed directly from nitrite reductase differed from that obtained from the separated heme group. The absorption maxima of pyridine hemochromogen after reduction with dithionite were at 416, 522, and 550 nm (as compared to 416, 521, and 550 nm for horse heart cytochrome *c*). The cyanide hemochromogen also shows the same similarity with absorption maxima (after dithionite reduction) at 556, 527, and 422 nm for the nitrite reductase and 555, 527, and 421 nm for horse heart cytochrome *c*.

Heme Prosthetic Group

The heme of the purified nitrite reductase could not be dissociated from the polypeptide backbone by extraction with HCl-acetone,[11] but a hemin was released from the protein on treatment with Ag_2SO_4.[12] The separated heme showed a Soret band at 389 nm (in 0.1 *M* potassium phosphate buffer, pH 7.0), as compared with 390 nm for horse heart cytochrome *c*.[13] These results, together with the spectral properties of

[11] H. Shichi and D. P. Hackett, *J. Biol. Chem.* **237,** 2959 (1962).

[12] K. Paul, *Acta Chem. Scand.* **4,** 239 (1950).

[13] J. E. Falk, *in* "Porphyrins and Metalloporphyrins," p. 240. Elsevier, Amsterdam, 1964.

TABLE II
SPECTRAL PROPERTIES OF *Desulfovibrio desulfuricans* NITRITE REDUCTASE AND HORSE HEART CYTOCHROME *c*[a]

Spectra	Nitrite reductase absorption bands (nm)	Cytochrome *c* absorption bands (nm)
Oxidized (pH 7.0)		
Soret	410	(410)[b]
	279	(280)
	531	(528)
Reduced (pH 7.0)		
α	553	(550.2)
β	524	(520.5)
Soret	521	(416)
Minimum	539	(535.2)
Pyridine hemochromogen		
Oxidized	413	409
Reduced		
α	552	(51)
β	522	(522)
Soret	416	416
Minimum	537	(539)
Cyanide hemochromogen		
Oxidized	416	416
Reduced		
α	556	(555)
β	527	(527)
Soret	422	421
Minimum	543	(545)

[a] Reprinted with permission from ASBMB, Liu and Peck, *J. Biol. Chem.* **256,** 13159 (1981).

[b] Figures in parentheses are obtained from D. L. Drabkin, *J. Biol. Chem.* **146,** 605 (1942).

the nitrite reductase described above, indicated the presence of protoporphyrin IX as the prosthetic group. The number of heme groups per molecule of nitrite reductase was determined on the basis of alkaline pyridine hemochrome and iron content. The extinction coefficient of the dithionite-reduced nitrite reductase pyridine hemochromogen at 550 nm was 5.96 times that of horse heart cytochrome *c*. The heme iron content (no nonheme iron was detected) of nitrite reductase was 0.47%, which is equivalent to 5.5 iron atoms per molecule of nitrite reductase. It is thus clear that the nitrite reductase is a multiheme *c*-type cytochrome and the number of heme prosthetic groups is six. Table III summarizes the heme and iron content of the nitrite reductase.

TABLE III
HEME AND IRON CONTENT OF *Desulfovibrio desulfuricans* NITRITE REDUCTASE[a]

Heme content of nitrite reductase[b]		
Hemochrome	$\varepsilon_{550\ \mathrm{nm}}^{\mathrm{reduced}}$	Heme content[c] (mol/mol of protein)
Nitrite reductase	173.6	5.96
Iron content of nitrite reductase		
Protein	Iron content	Atoms of iron per molecule[d]
Nitrite reductase	0.47%	5.49

[a] Reprinted with permission from ASBMB, Liu and Peck, *J. Biol. Chem.* **256,** 13159 (1981).

[b] On the basis of pyridine hemochrome data.

[c] Heme content was estimated based on the $\varepsilon_{550\ \mathrm{nm}}^{\mathrm{reduced}}$ (29.1 $\mathrm{cm}^{-1}/\mu\mathrm{mol}$ per milliliter) of heme *c*.[13]

[d] Estimated on the basis of the minimal molecular weight of 66,000.

Molecular Weight

Because of the tendency of the purified nitrite reductase to form aggregates, the attempt to determine the molecular weight by the sedimentation equilibrium method was not successful. With SDS-gel electrophoresis, however, the polypeptide molecules could be separated and migrated independently. The minimal molecular weight (SDS disintegrated the possible polymeric structure) was determined to be 66,000.

Amino Acid Composition

Amino acid composition of the purified nitrite reductase, together with those of three other hexaheme nitrite reductases, are listed in Table IV. The values for serine and methionine were corrected for destruction during hydrolysis by extrapolation to zero time. In all 4 cases, there are 12 cysteine residues per monomeric enzyme molecule, just enough to account for 6 covalently bonded heme *c* groups. Although the ligand arrangement of the six-heme irons has not been extensively studied, the number of histidine and methionine residues, 19 and 25, respectively, is more than enough to account for 6 histidine-histidine or 6 histidine-methionine ligand arrangements per molecule. And, except for the tryptophan residues, for which data are not available for the other three nitrite reductases, the four nitrite reductases exhibit surprisingly close similarities in all major groups of amino acids when the data are expressed in moles per 100 mol.

TABLE IV
AMINO ACID COMPOSITIONS OF HEXAHEME NITRITE REDUCTASES[a]

Amino acid	*Desulfovibrio desulfuricans*[b]	*Escherichia coli*[c]	*Wolinella succinogenes*[d]	*Vibrio fischeri*[e]
Asp + Asn	12.4	10.9	9.8	11.8
Thr	4.3	4.3	4.3	5.5
Ser	3.0	2.2	3.3	3.1
Glu + Gln	11.7	15.7	14.7	15.8
Pro	5.8	4.7	5.1	3.2
Gly	6.8	4.0	3.1	3.9
Ala	10.2	10.3	5.8	8.5
Val	5.7	4.2	3.9	5.9
Met	4.2	2.5	2.7	3.0
Ile	2.8	4.8	3.8	3.0
Leu	7.3	6.0	7.0	4.5
Tyr	4.2	3.3	6.2	4.2
Phe	3.3	4.3	4.4	3.3
His	3.2	5.7	4.5	4.9
Lys	7.9	9.1	14.4	10.5
Arg	4.3	5.7	4.7	6.3
Cys	2.0	2.3	2.1	2.3
Trp	0.8	ND[f]	ND	ND

[a] Expressed as mol/100 mol.
[b] From Liu and Peck.[1]
[c] From Liu *et al.*[2]
[d] From Liu *et al.*[4]
[e] From Liu *et al.*[6]
[f] ND, Not determined.

These results clearly indicate that the four hexaheme nitrite reductases belong to the same homologous group of proteins that display the same physiological functions.

Topography

Experiments using high-pH, high ionic strength, or detergent-containing buffer solutions for the extraction of the nitrite reductase clearly indicated the integral membrane protein nature of the enzyme.[1] In a whole-cell study, the active site of the nitrite reductase was shown to be located on the periplasmic side of the cytoplasmic membrane. The fact that the nitrite reductase accepts electrons from reduced viologens preferentially from the cytoplasmic side of the cytoplasmic membrane further indicates its identity as a transmembrane protein.[14]

[14] D. J. Steenkamp and H. D. Peck, Jr., *J. Biol. Chem.* **256,** 5450 (1981).

Substrate Specificity and Electron Donors

The purified nitrite reductase could catalyze the reduction of nitrite or hydroxylamine to form ammonia. The Michaelis constants (K_m) determined for nitrite and hydroxylamine are 1.14 m*M* and 0.11 *M*, respectively. Because the K_m for hydroxylamine is approximately two orders of magnitude greater than that for nitrite, it seems unlikely that hydroxylamine is a free intermediate during the reduction of nitrite to ammonia by nitrite reductase. The purified enzyme did not accept electrons from NADH or NADPH (5 m*M* in reaction mixture) for the reduction of nitrite. No absolute requirement for added thiol compounds was found. Purified hydrogenase from *Desulfovibrio vulgaris* could catalyze the transfer of electrons from H_2 to the nitrite reductase with FAD as electron transfer mediator. The change of color from red (oxidized) to pink (reduced) could be observed when the nitrite reductase was reduced in this way. Nitrite reductase could also be reduced by dithionite, but only slightly by ascorbate and not at all by cysteine. The ascorbate-reduced enzyme was not autooxidizable, thus suggesting that some of the heme *c* groups have higher redox potentials than others. The oxidized spectrum returned when air was bubbled through a solution of dithionite-reduced enzyme even in the presence of 10 m*M* potassium cyanide. This indicated that the nitrite reductase was autooxidizable and potassium cyanide had no effect on the autooxidizability of the enzyme. On addition of nitrite, the reduced enzyme appeared to be rapidly oxidized. However, when hydroxylamine was added, the reduced enzyme was only partially reoxidized. Nitrate, selenite, and sulfite did not oxidize the reduced enzyme.

The ability of various electron transfer proteins and cofactors to mediate the electron transfer from H_2 gas to nitrite through *D. vulgaris* hydrogenase and the nitrite reductase was studied. It was found that *Desulfovibrio gigas* ferredoxin I, *D. gigas* flavodoxin, *D. vulgaris* cytochrome c_3, *D. gigas* cytochrome c_3, and flavin mononucleotide (FMN) could not function in the electron transfer process, whereas FAD could mediate the electron transfer from *D. vulgaris* hydrogenase to the nitrite reductase.

pH Optimum and Effect of Temperature

The pH dependence of nitrite reductase activity was determined in potassium phosphate buffer in the pH range 6.0–8.0 and 9.5–11.5, and in barbital buffer in the range 7.5–9.5. No significant difference in activity was observed whether phosphate or barbital was used as the buffer at overlapping pH values. Maximum activity was found in the pH range 8.0–9.5. Below pH 7.5 it was found that the reduction of methyl viologen by sodium dithionite in the assay system was not as efficient as that at

TABLE V
SPECIFIC ACTIVITY[a] OF *D. desulfuricans* NITRITE REDUCTASE UTILIZING DIFFERENT ELECTRON DONORS AND NITRITE OR NITRIC OXIDE[b] AS ELECTRON ACCEPTOR[c]

Source of electrons	Nitrite	Nitric oxide
Dithionite	778	860
Ascorbate + PMS	28	35
NADPH + crude extract	17	—

[a] Specific activity is defined as the number of enzymatic units per milligram of nitrite reductase.
[b] One enzymatic unit equals the amount of nitrite reductase needed to convert 1 μmol of Nitric oxide to 1/2 μmol of N_2O per minute.
[c] Reprinted with permission from Elsevier, Costa *et al., FEBS Lett.*, **276,** 67–70 (1990).

higher pH. The effect of temperature on the activity of nitrite reductase was studied between 20 and 70°. Maximum activity was obtained at 57° and there was a 1.09-fold increase in activity for a 10° rise in temperature between 20 and 30°, and a 0.78-fold increase between 30 and 40°. The energy of activation calculated from the Arrhenius plot was 11,053 cal/mol.

New Catalytic Activities of the Hexaheme Nitrite Reductase

In addition to the enzymatic activity of the hexaheme nitrite reductase to reduce nitrite to ammonia in a six-electron reaction, a new catalytic activity has been detected.[15] Using dithionite as the electron donor, *D. desulfuricans* hexaheme enzyme yielded comparable rates of ammonia production when nitrite or nitric oxide (NO) served as the electron acceptor. With ascorbate-phenazine methosulfate (PMS) as the electron donor, only small amounts of ammonia were produced from either of the two acceptors. No precedents have been reported that an enzyme can act on both nitrite and nitric oxide. And, the enzymatic activity catalyzing the production of ammonia from nitric oxide has never before been detected for other enzymes (Table V). By employing mass spectrometry,[15] it was shown that the hexaheme nitrite reductases from *D. desulfuricans*, *W. succinogenes*,[4] and *E. coli*,[2] when reduced with ascorbate-PMS, reduced NO (mass peak 46) to nitrous oxide (N_2O) (mass peak 31). These enzymes could also reduce NO generated by *Achromobacter cycloclastes*

[15] C. Costa, A. Macedo, I. Moura, J. Le Gall, Y. Berlier, M.-Y. Liu, and W. J. Payne, *FEBS Lett.* **276,** 67 (1990).

TABLE VI
MASS SPECTROMETRIC MEASUREMENT OF SPECIFIC ACTIVITY[a] OF *Desulfovibrio desulfuricans* NITRITE REDUCTASE USING NITRIC OXIDE AS SUBSTRATE AND ASCORBATE PLUS PMS AS ELECTRON DONOR[b]

Source of hexaheme nitrite reductase	NO (gas)	NO[c]
D. desulfuricans	535	360
W. succinogenes	—	133
E. coli	—	135

[a] Specific activity is defined as the number of enzymatic units per milligram of nitrite reductase. One enzymatic unit equals the amount of nitrite reductase needed to convert 1 μmol of NO to 1/2 μmol of N_2O per minute.
[b] Reprinted with permission from Elsevier, Costa *et al.*, *FEBS Lett.*, **276**, 67–70 (1990).
[c] Generated from the reaction catalyzed by *A. cycloclastes* nitrite reductase using nitrite as the substrate.

copper-containing nitrite reductase, with the production of N_2O. The latter results have been corrected for the low background rate of N_2O production resulting from the prolonged action of the copper-containing nitrite reductase. The specific activities of the three hexaheme enzymes in catalyzing the reduction of NO to N_2O were found to be unexpectedly higher than that reported for the nitric oxide reductase purified from *Pseudomonas stutzeri*[16] (Table VI). The possible pathways for the reduction of various nitrogen compounds catalyzed by the hexaheme nitrite reductases are shown in Fig. 2.

Electron Paramagnetic Resonance Spectroscopic Properties

The electron paramagnetic resonance (EPR) spectrum of purified *D. desulfuricans* nitrite reductase is complex (Fig. 3A). The only signals that can be identified as typical of the heme spectrum are the resonances detectd at g = 2.96, 2.28, and 1.50. These g values are characteristic for low-spin ferric hemes. The other prominent features include a broad absorption-type signal at g = 3.9 and a derivative-type signal with zero crossing at g = 4.8. Other resonances are observed throughout the spectrum from g = 9.36 to g = 1.50. Similar complex EPR spectra have also

[16] B. Heiss, K. Frunzke, and W. G. Zumft, *J. Bacteriol.* **171,** 3288 (1989).

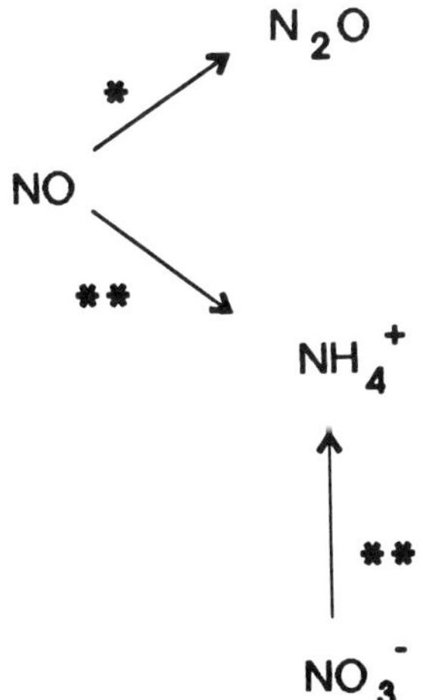

FIG. 2. Proposed pathways for the reduction of various nitrogen compounds catalyzed by the hexaheme nitrite reductases. *, High-potential electron donors; **, low-potential electron donors. (Reprinted with permission from ASBMB, Costa *et al.*, *J. Biol. Chem.*, **265,** 14382 (1990).)

been reported for other hexaheme nitrite reductases[17,18] and a tetraheme cytochrome c_{554} from *Nitrosomona europaea*.[19] The complexity of the spectrum is attributed to spin–spin interaction between the heme groups, which is also indicated by the Mössbauer data presented below.

Figure 3 also reveals the EPR spectra of *D. desulfuricans* nitrite reductase as a function of the redox potential. Below 170 mV and with decreasing redox potential, the interacting signal at $g = 3.9$ decreased in intensity first, followed by the decline of the low-spin ferric heme signal ($g = 2.96$, 2.28, and 1.50) and the signal with zero crossing at $g = 4.8$. A new low-spin ferric heme signal at $g = 3.2$ and 2.14 and a high-spin ferric signal at $g = 6.30$, 5.36, and 1.99 appeared. At lower redox potentials, these two signals increased, passed through a maximum in intensity, and then decreased. The transient nature of these signals can be easily explained by heme–heme interaction. In the isolated state of the enzyme, the heme

[17] R. S. Blackmore, T. Brittain, P. M. A. Gadsby, C. Greenwood, and A. J. Thomson, *FEBS Lett.* **219,** 244 (1987).

[18] M.-C. Liu, M.-Y. Liu, W. J. Payne, H. D. Peck, Jr., J. LeGall, and D. V. DerVartanian, *FEBS Lett.* **218,** 227 (1987).

[19] K. K. Anderson, J. D. Lipscomb, M. Valentine, E. Munck, and A. B. Hooper, *J. Biol. Chem.* **261,** 1126 (1986).

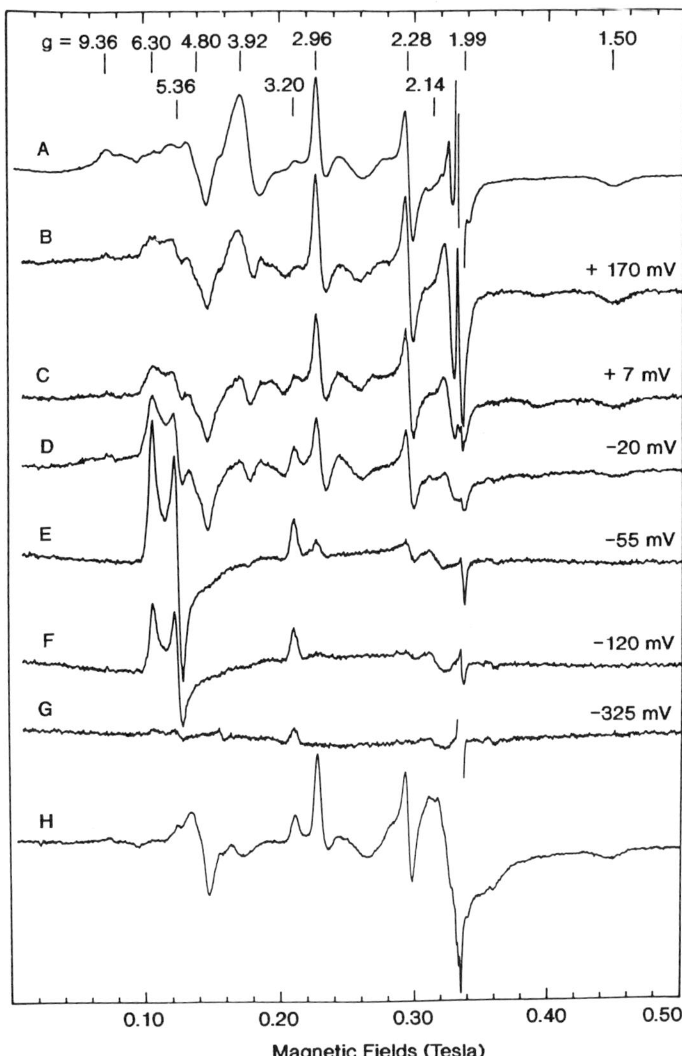

FIG. 3. EPR spectra of *D. desulfuricans* nitrite reductase. (A) Nitrite reductase as isolated; (B–G) nitrite reductase poised at redox potentials as indicated; (H) nitrite-reacted enzyme. Experimental conditions: microwave frequency, 9.43 GHz; microwave power, 2 mW; temperature, 9.5 K; modulation amplitude, 1 mT; receiver gain, 1.0×10^5 for spectra A and H, 2.5×10^5 for spectra B–G; protein concentration, 250 μM for spectra A and H, 130 μM for spectra B–G. (Reprinted with permission from ASBMB, Costa *et al.*, *J. Biol. Chem.*, **265,** 14382 (1990).)

groups responsible for these two signals were spin coupled to other hemes, resulting in the unusually complex EPR signals (Fig. 3). When the potential was lowered, some of the coupled heme groups were reduced to low-spin ferrous, which is diamagnetic, resulting in the decoupling of the hemes and the appearance of typical ferric heme signals that were not detected in the isolated state of the enzyme. When the redox potential was lowered further, more hemes were reduced and the new signals disappeared.

The hexaheme nitrite reductase could be reduced by chemical reductants, such as dithionite or reduced methyl viologen, or by hydrogenase under H_2 atmosphere. The EPR spectrum of the fully reduced enzyme interacting with nitrite is shown in Fig. 3H. The low-spin ferric heme signal (at $g = 2.96$, 2.28, and 1.50), the derivative-type signal at $g = 4.8$, and all the lesser prominent features, which were observed when the enzyme was in its isolated state, reappeared in the case of the nitrite-reacted sample, indicating that most of the hemes were reoxidized. The broad resonance at $g = 3.9$, however, was absent and an intense signal with hyperfine structure was observed at $g = 2$. Using ^{15}N-labeled nitrite, it was confirmed that this signal is due to the nitric oxide-bound ferrous heme. As discussed below, Mössbauer data indicate that a high-spin heme is involved in the nitric oxide binding. The broad signal at $g = 3.9$ was absent for the nitrite-reacted enzyme. On the basis of this, it is obvious that one of the hemes (the high-spin one) must be involved in the heme–heme interaction that produces the 3.9 signal.

Mössbauer Spectroscopic Properties

The native hexaheme nitrite reductase contains six ferric heme groups, with one in a high-spin state ($S = 5/2$) and the other five in low-spin states ($S = 1/2$). The signals of two low-spin ferric hemes ($g = 2.96$, 2.28, and 1.50; $g = 3.20$ and 2.14) and the high-spin ferric heme ($g = 6.30$, 5.36, and 1.99) were observed in EPR measurements. The g values of the other three low-spin ferric hemes were revealed by analyzing the strong-field Mössbauer spectra. Among the three, two low-spin hemes were found to have unusually large g_{max} values of, respectively, 3.5 and 3.6. The third low-spin ferric heme was assigned a g_{max} value of 3.0. The crystal field and hyperfine parameters of the ferric heme groups that were used to simulate the high-field Mössbauer spectra[21] are shown in Table VII.

The Mössbauer spectra of the native enzyme[20] recorded at 4.2 K in zero field clearly demonstrate that four of the six heme groups including the high-spin ferric heme are magnetically coupled, suggesting a close proximity among these heme groups. The coupled heme groups could be

[20] C. Costa, J. J. G. Moura, I. Moura, M.-Y. Liu, H. D. Peck, Jr., J. LeGall, Y. Wang, and B. H. Huynh, *J. Biol. Chem.* **265,** 14382 (1990).

TABLE VII
CRYSTAL FIELD AND HYPERFINE PARAMETERS OF FERRIC HEME GROUPS IN *Desulfovibrio desulfuricans* NITRITE REDUCTASE[a]

Parameter	Low-spin hemes: Heme 1	Heme 2	Heme 3	Heme 4	Heme 5	High-spin heme
D (cm^{-1})						5.5 (5)[b]
E/D						0.015
g_x	0.40	0.70	1.10	1.50	1.50 (3)	2.0
g_y	1.70	1.80	2.14 (3)	2.18	2.28 (3)	2.0
g_z	3.60	3.50	3.20 (3)	3.00	2.96 (3)	2.0
$A_x/g_n\beta_n$ (T)	−33.5	−34.0	−35.5	−40.0	−40.8	−20.9 (5)
$A_y/g_n\beta_n$ (T)	26.5	22.5	22.0	14.4	18.1	−20.9 (5)
$A_z/g_n\beta_n$ (T)	91.0	86.0	66.0	54.0	50.5	−20.9 (5)
ΔE_Q(mm/sec)	2.0 (2)	2.0 (2)	2.0 (2)	2.0 (2)	2.0 (2)	0.8 (2)
δ	0.22 (4)	0.22 (4)	0.22 (4)	0.22 (4)	0.22 (4)	0.4 (1)

[a] Reprinted with permission from ASBMB, Costa *et al.*, *J. Biol. Chem.*, **265**, 14382 (1990).
[b] The numbers in parentheses represent the estimated uncertainties in the last significant digits.

completely decoupled at fields of about 4 T. The Mössbauer spectrum of a nitrite-reacted sample recorded at 4.2 K in a parallel applied field of 8 T shows that the central portion of the spectrum is different from that of the native enzyme, whereas the outer regions appear to be similar. By making a difference spectrum between the nitrite-reacted sample and the native enzyme, the only lines that were modified belong to the hemes that have reacted with nitrite. Analysis of this difference spectrum revealed that only the high-spin component was affected and an NO-bound ferrous heme was observed. These results confirm the high-spin heme being the substrate-binding site.

Electrochemical Properties

The kinetics of electron transfer between three different mediators and the *D. desulfuricans* hexaheme nitrite reductase were investigated by cyclic voltammetry and by chronoamperometry.[21] The three mediators methyl viologen, *D. vulgaris* (Hildenborough) cytochrome c_3, and *D. desulfuricans* (ATCC 27774) cytochrome c_3 differ in structure, redox potential, and charge. The reduced form of each mediator could exchange electrons with the nitrite reductase. Second-order rate constants, k, were calculated on the basis of the results obtained by cyclic voltammetry in

[21] C. Moreno, C. Costa, I. Moura, J. LeGall, M.-Y. Liu, W. J. Payne, C. van Dijk, and J. J. G. Moura, *Eur. J. Biochem.* **212,** 79 (1993).

comparison with those obtained by chronoamperometry. Calculated k values are in the range 10^6–10^8 M^{-1} sec^{-1}, and are, in increasing order: *D. desulfuricans* cytochrome $c_3 \rightarrow$ *D. vulgaris* cytochrome $c_3 \rightarrow$ methyl viologen.

Direct, unmediated electrochemical response of this enzyme was observed only with the voltammetric technique of differential pulse and square wave. The experimental data were simulated using a deconvolution method tested on multiheme cytochromes.[22] The redox potentials used to simulate the experimental curve were −50, −120, −255, −300, and −350 ± 10 mV versus normal hydrogen electrode.

Independent visible and Mössbauer static redox titrations (Moura *et al.*, unpublished results, 1994) indicate that one of the hemes has a considerably high reduction potential (150 ± 15 mV versus normal hydrogen electrode). The contribution of this heme with a positive redox potential was absent in the voltammogram. This could have been due to certain structural characteristics of the heme that made it inaccessible to the electrode surface.

[22] C. Moreno, A. Campos, M. Teixeira, J. LeGall, M. I. Montenegro, I. Moura, C. van Dijk, and J. J. G. Moura, *Eur. J. Biochem.* **202,** 385 (1991).

[22] Genetic Manipulation of *Desulfovibrio*

By WALTER M. A. M. VAN DONGEN, JACK P. W. G. STOKKERMANS, and WILLY A. M. VAN DEN BERG

Introduction

Anaerobic sulfate reducers are a widely divergent group of microorganisms with representatives in both the archaeal and bacterial kingdoms. Genetic manipulation of sulfate reducers has been successful until now only with species of the genus *Desulfovibrio*. The first systems for conjugal transfer of broad host range plasmids belonging to incompatibility group Q from *Escherichia coli* to *Desulfovibrio* were described in 1989.[1,2] Since then, cloning vectors derived from IncQ plasmids have found application

[1] B. Powell, M. Mergeay, and N. Christofi, *FEMS Microbiol. Lett.* **59,** 269 (1989).

[2] W. A. M. van den Berg, J. P. W. G. Stokkermans, and W. M. A. M. van Dongen, *J. Biotechnol.* **12,** 173 (1989).

for the expression of (mutated) genes in *Desulfovibrio*.[3-6] This has been proved especially useful for study of the biochemistry of metal proteins. In several cases it has been observed that expression in *E. coli* of genes for *Desulfovibrio* metal proteins results in the accumulation of apoproteins, sometimes as insoluble complexes, without the metal (clusters) incorporated, due to lack of proper processing of such proteins.[6-9] Introduction of such genes on expression vectors in *Desulfovibrio*, on the other hand, has resulted in a considerable (over)production of the proteins in their native form, allowing purification in high amounts and facilitating biochemical analysis. Up to 25-fold overproduction of a protein with a complex FeS cluster has been obtained with one of these vectors.[6] Furthermore, production of proteins in which deliberate changes in the primary structure have been made by site-directed mutagenesis is possible now, allowing study of the implications of structural changes on functional aspects of these proteins.[5]

Plasmid transfer by conjugation has now been described for four *Desulfovibrio* species.[1,2,10] The basic principles (conjugal mating procedures, IncQ plasmids for stable maintenance) are quite similar for different desulfovibrios. Nevertheless, especially because of its anaerobic nature, genetic manipulation of *Desulfovibrio* remains experimentally complicated. Furthermore, selection procedures (in order to discriminate *Desulfovibrio* cells that have received plasmids from those that have not and from the *E. coli* donor) present a major difficulty: many of the common antibiotic resistance genes do not perform well in *Desulfovibrio*.[2,10,11] This is probably the main reason why attempts to apply more sophisticated genetic techniques (e.g., transposon mutagenesis with suicide plasmids or disrup-

[3] G. Voordouw, W. B. R. Pollock, M. Bruschi, F. Guerlesquin, B. J. Rapp-Giles, and J. D. Wall, *J. Bacteriol.* **172,** 6122 (1990).

[4] M. Bruschi, P. Bertrand, C. More, G. Leroy, J. Bonicel, J. Haladjian, G. Chottard, W. B. R. Pollock, and G. Voordouw, *Biochemistry* **31,** 3281 (1992).

[5] I. Mus-Veteau, A. Dolla, F. Guerlesquin, F. Payan, M. Czjzek, R. Haser, P. Bianco, J. Haladjian, B. J. Rapp-Giles, J. D. Wall, G. Voordouw, and M. Bruschi, *J. Biol. Chem.* **267,** 16851 (1992).

[6] J. P. W. G. Stokkermans, P. H. J. Houba, A. J. Pierik, W. R. Hagen, W. M. A. M. van Dongen, and C. Veeger, *Eur. J. Biochem.* **210,** 983 (1992).

[7] G. Voordouw, W. R. Hagen, K. M. Krüse-Wolters, A. van Berkel-Arts, and C. Veeger, *Eur. J. Biochem.* **162,** 31 (1987).

[8] W. van Dongen, W. Hagen, W. van den Berg, and C. Veeger, *FEMS Microbiol. Lett.* **50,** 5 (1988).

[9] W. B. R. Pollock, P. J. Chemerika, M. E. Forrest, J. T. Beatty, and G. Voordouw, *J. Gen. Microbiol.* **135,** 2319 (1989).

[10] J. L. Argyle, B. J. Rapp-Giles, and J. D. Wall, *FEMS Microbiol. Lett.* **94,** 255 (1992).

[11] B. J. Powell, Academic Thesis, Dept. of Biological Sciences, Napier Polytechnic, Edinburgh, Scotland and Dept. of Biology, SCK/CEN, Belgium (1989).

tion of specific genes by "marker exchange") have not yet been successful (Ref. 10 and W. M. A. M. van Dongen, J. P. W. G. Stokkermans, and W. A. M. van den Berg, unpublished observations, 1994), with one notable exception.[12] However, implementation of these techniques is expected to be a matter of time.

Therefore, the main differences between protocols for plasmid transfer to the various *Desulfovibrio* species are in the strategies for selection of transconjugant cells. A detailed protocol for plasmid transfer to *D. vulgaris* Hildenborough is described in the following sections, with brief reference to variations in procedures for other species.

A seemingly trivial prerequisite for genetic experiments is that bacteria can be grown on plates as colonies originating from single cells. Cultivation of anaerobes such as *Desulfovibrio* as clones on plate is not common practice (it is not even possible for all species) and requires special precautions. Therefore, a separate section is dedicated to this subject.

Cultivation of *Desulfovibrio* as Clones Originating from Single Cells on Solid Media

General Considerations

Growth Medium. The rich tryptic soy agar of Iverson[13] is a suitable medium for plating *Desulfovibrio* (see the next section for the composition). It contains lactate as the carbon source and sulfate as the terminal electron acceptor. Plating efficiencies of *Desulfovibrio vulgaris* on this medium are high (0.7 ± 0.27),[2] and colonies are at least 2 mm in size after 4–5 days of growth at 37°. Satisfactory results have also been obtained with other media (the LS medium of Wall and co-workers[10] and a synthetic medium[1]). To prevent accumulation of the inhibitory sulfide, sulfate might be replaced by nitrate for strains that are able to use nitrate as a terminal electron acceptor (many *Desulfovibrio desulfuricans* strains).

Besides anaerobicity (see below), *Desulfovibrio* also requires a low redox potential for growth (less than −100 mV).[14] Therefore, reducing agents such as $Na_2S_2O_4$, Na_2S, or a mixture of sodium thioglycolate and ascorbic acid are added to media in concentrations of 0.2–0.5 μg/ml. The redox potential can be monitored by including resazurin (0.5 μg/ml), which reduces to the colorless resorufin below approximately 110 mV. Resazurin should be reduced within 1 day after having started a culture on plate.

[12] M. Rousset, Z. Dermoun, M. Chippaux, and J. P. Bélaich, *Mol. Microbiol.* **5,** 1735 (1991).
[13] W. P. Iverson, *Appl. Microbiol.* **14,** 529 (1966).
[14] E. S. Pankhurst, *in* "Isolation of Anaerobes" (D. A. Shapton and R. G. Board, eds.), p. 223. Academic Press, London and New York, 1971.

Plating of *Desulfovibrio* cells in a soft-agar overlay is recommended to prevent swarming of the cells.

Anaerobiosis. Although many *Desulfovibrio* strains can tolerate exposure to air when resting, traces of oxygen in the medium may delay or prevent growth. An anaerobic glove box may be used for preparation and storage of plates and for inoculation, but is not strictly required and often inconvenient. Precautions to exclude oxygen, as described below, are required when experiments are done without a glove box.

Agar plates are either prepared directly before use with freshly autoclaved medium (in which the oxygen level is low) or are kept in an anaerobic jar. Vented petri dishes are recommended for faster gas exchange.

For growth of colonies, plates are incubated in anaerobic jars. Commercially available jars in which anaerobiosis is maintained with H_2 and a palladium catalyst are less suitable, owing to rapid poisoning of the catalyst by H_2S. We use home-made containers, made from 0.6-cm-thick, 32-cm-long Perspex pipe and a Perspex bottom, having an inner diameter of 12 cm and provided with a tightly fitting lid (i.e., sealed by a rubber O ring). The lid contains a rubber septum through which hypodermic needles can be pierced for gas exchange.

The jar should be flushed for at least 2 hr with N_2 to obtain a sufficiently anoxic atmosphere. N_2 can be scrubbed of O_2 by passing it over a column with BASF catalyst R3-11 at 120° and through a wash bottle with 500 ml of a solution of 100 m*M* Tris-HCl, 100 m*M* ethylenediaminetetraacetic acid (EDTA), pH 10, containing 1 m*M* methyl viologen and approximately 5 mg of proflavine, which is illuminated to keep the methyl viologen reduced.[15] The last traces of oxygen are removed from the jar by including a plate with full-grown *Desulfovibrio* colonies[13] and an open petri dish with 10 ml of 0.1 *M* $Na_2S_2O_4$. A vial with methylene blue in the jar [Tris base (200 mg/ml), glucose (13.3 mg/ml), and methylene blue (67 μg/ml)] may serve as an indicator for anaerobicity; this solution should be colorless within 1 day.

Protocol for Growing Desulfovibrio on Solid Medium

Preparation of plates with TSAS medium[2] (which is essentially the medium described by Iverson[13] with some extra additions) is described in detail. Alternative media can be found in Refs. 1 and 10.

1. Bottom agar: Combine 40 g of tryptic soy agar (dehydrated; Difco, Detroit, MI), 2.5 g of sodium lactate, 2.0 g of $MgSO_4 \cdot 7H_2O$, 0.5 g of $(NH_4)_2Fe(SO_4)_2$, and add distilled water up to 1 liter and autoclave. The

[15] V. Massey and P. Hemmerich, *Biochemistry* **17,** 9 (1978).

pH should be 7.2–7.5. After cooling to approximately 50°, add the following filter-sterilized solutions: 10 ml of resazurin (50 mg/ml), 10 ml of a freshly prepared sodium thioglycolate/ascorbic acid stock (both 20 mg/ml), and antibiotics, when required. Pour the agar into vented petri dishes; store the plates under N_2 atmosphere until use.

2. Soft top agar: Replace the tryptic soy agar with 3 g of tryptic soy broth (dehydrated; Difco) and 0.75 g of agar per 100 ml. Pipette 3 ml of molten soft agar into sterile glass tubes and keep at 45° in a heating block or water bath.

3. Make serial dilutions of *Desulfovibrio* cells (from a liquid culture, from a conjugation) in phosphate-buffered saline (PBS; 10 m*M* sodium phosphate, pH 7.2, 150 m*M* NaCl), add 100-μl portions to the soft agar tubes, and mix and pour onto the bottom agar. This may be done aerobically, but work quickly as the viability of *Desulfovibrio* cells in diluted suspensions decreases under aerobic conditions.

4. Place the plates upside down in a jar immediately after the soft agar has solidified, with an open petri dish on top of the pile. If available, also add a plate with already full-grown *Desulfovibrio* colonies.[13] Also include a tube with a few milliliters of methylene blue solution as an oxygen indicator.

5. Close the lid and flush the jar for at least 2 hr with oxygen-free N_2. Inject 10 ml of a freshly prepared 0.1 *M* $Na_2S_2O_4$ solution through the septum into the open petri dish on top of the pile. Replace the gas with N_2/2.5% CO_2 and grow colonies at 30 or 37°.

6. Check now and then for reduction of methylene blue and resazurin. When colonies develop, black spots due to FeS precipitation become visible after approximately 2–3 days. Full development of the colonies requires 4–5 days; the number of colonies on a plate strongly influences the size, probably due to the inhibitory effect of sulfide.

Plasmids and Selection

Transfer of broad host range plasmids belonging to different incompatibility groups (IncP, IncQ, and IncW) to *Desulfovibrio* species has been tested, but stable transconjugants, from which the plasmid can be isolated, have been found only with IncQ plasmids.[1,2,10] However, not all *Desulfovibrio* species appear able to receive and/or maintain IncQ plasmids: conjugal transfer has been found to *Desulfovibrio* sp. Holland SH-1 (NCIMB 8301),[1] *D. desulfuricans* Norway 4 (NCIMB 8312),[1] *D. vulgaris* Hildenborough (NCIMB 8303),[2] and *D. desulfuricans* G200,[10] but attempts to introduce IncQ plasmids into *D. desulfuricans* (ATCC 27774),[10] *Desulfovibrio gigas* (NCIMB 9332), or *D. vulgaris* subsp. *oxamicus* Monticello (NCIMB

TABLE I
MINIMAL INHIBITORY CONCENTRATION (μg/ml) OF DIFFERENT ANTIBIOTICS FOR FOUR *Desulfovibrio* SPECIES[a]

Antibiotic	*Desulfovibrio* sp. Holland SH-1[11]	*D. vulgaris* Hildenborough[2]	*D. desulfuricans* G200[10]	*D. desulfuricans* Norway 4[11]
Ampicillin	V	V	ND	V
Chloramphenicol	5	3	6	ND
Kanamycin	200	>100	100	200
Gentamycin	ND	ND	20	ND
Tetracycline	10	10	2	ND
Streptomycin	50	>100	ND	20
Sulfonamide	100	100	ND	100

[a] V, Variable resistance, dependent on inoculum size; ND, not determined.

9442) (W. M. A. M. van Dongen, J. P. W. G. Stokkermans, and W. A. M. van den Berg, unpublished observations, 1994) have not yet been successful. An IncQ plasmid introduced in *Desulfovibrio fructosovorans*, on the other hand, behaved as a suicide plasmid and has been used for marker exchange.[12]

A serious problem for the introduction of plasmids into *Desulfovibrio* by conjugal transfer from *Escherichia coli* is the limited availability of suitable markers for selection of *Desulfovibrio* transconjugants and for counterselection against the *E. coli* donor. Table I shows the minimal inhibitory concentrations of several common antibiotics for four *Desulfovibrio* species, derived from data in Refs. 2, 10, and 11. Ampicillin resistance is variable and therefore not a useful marker. Introduction of IncQ plasmids containing the tetracycline resistance gene of pSC101[16] into *D. vulgaris* results in only a marginal, unselectable increase of the background resistance (W. M. A. M. van Dongen, J. P. W. G. Stokkermans, and W. A. M. van den Berg, unpublished observations, 1994), although the *tet* promoter appears to be active in *Desulfovibrio*.[17] Chloramphenicol resistance caused by expression of the chloramphenicol acetyltransferase gene of Tn*9* is used as a reliable selective marker for plasmid transfer to *D. vulgaris* Hildenborough. However, this marker should be applied with caution, as the resistance of *D. vulgaris* transconjugants that have acquired plasmids with this gene is only slightly higher than background level (5–10 vs 3 μg/ml),[2] and nontransconjugant *Desulfovibrio* colonies develop with

[16] A. C. Y. C. Chang and S. N. Cohen, *J. Bacteriol.* **134,** 1141 (1978).

[17] W. A. M. van den Berg, W. M. A. M. van Dongen, and C. Veeger, *J. Bacteriol.* **173,** 3688 (1991).

a delay of only a few days, possibly by a slow, acetyltransferase-independent detoxification of chloramphenicol through reduction of the nitro group.[10,18] Kanamycin resistance is used as a marker for selection of *D. desulfuricans* G200 transconjugants having acquired plasmids encoding aminoglycoside 3′-phosphotransferase from transposon Tn*5*, which increases the kanamycin resistance level from 100 μg/ml (endogenous level) to 175 μg/ml.[10] Streptomycin has been used to select for plasmid transfer to *D. desulfuricans* Norway 4 and to *Desulfovibrio* sp. Holland SH-1, but again, the resistance of the plasmid-containing strains was only slightly higher than the endogenous level.[1]

Transfer of several plasmids with the replicon and mobilization functions derived from the naturally occurring IncQ plasmid RSF1010 (also known as R300B)[19] has been tested; among these are pJRD21,[20] pSUP104,[21] pGSS33K,[10,22] pDSK519,[23] and pKT230.[24] Two cloning vectors appeared especially useful for (over)production of proteins from cloned (mutated) genes in *Desulfovibrio* (Fig. 1). One is pJRD215,[20] which has a kanamycin resistance gene for selection and has been used in *D. desulfuricans* G200. pJRD215 was constructed as a broad host range cosmid vector and has a polylinker containing several unique restriction sites. This vector has been used for overproduction of (mutated) *D. vulgaris* cytochromes in *D. desulfuricans*.[3–5] *Desulfovibrio desulfuricans* clones containing *D. vulgaris* cytochrome genes in pJRD215 produced three- to fivefold higher levels of these cytochromes than *D. vulgaris* does from the chromosomal copies of the genes.

The second vector is pSUP104,[21] which has a chloramphenicol resistance gene for selection and has been used in *D. vulgaris* Hildenborough. The promoter of the *tet* gene in pSUP104 appears to be active in this host. Unique restriction sites in the *tet* gene (*Bam*HI, *Sph*I, *Sal*I, and *Nru*I) have been used to insert genes for metal proteins, resulting in an up to 25-fold overproduction of these proteins in *D. vulgaris*.[6] Insertion of genes in reversed orientation toward the *tet* promoter may result in the production of antisense RNA; *D. vulgaris* clones producing hydrogenase anti-

[18] E. Soutschek-Bauer, L. Hartl, and W. L. Staudenbauer, *Biotechnol. Lett.* **7,** 705 (1985).

[19] P. Guerry, J. van Embden, and S. Falkow, *J. Bacteriol.* **117,** 619 (1974).

[20] J. Davison, M. Heusterspreute, N. Chevalier, V. Ha-Thi, and F. Brunel, *Gene* **51,** 275 (1987).

[21] U. Priefer, R. Simon, and A. Pühler, *in* "Proceedings of the Third European Congress on Biotechnology," p. III-207, Verlag Chemie, Weinheim, 1984.

[22] G. S. Sharpe, *Gene* **29,** 93 (1984).

[23] N. T. Keen, S. Tamaki, D. Kobayashi, and D. Trollinger, *Gene* **70,** 191 (1988).

[24] Bagdasarian, M. R. Lurz, B. Ruckert, F. C. H. Franklin, M. M. Bagdasarian, J. Frey, and K. N. Timmis, *Gene* **16,** 237 (1981).

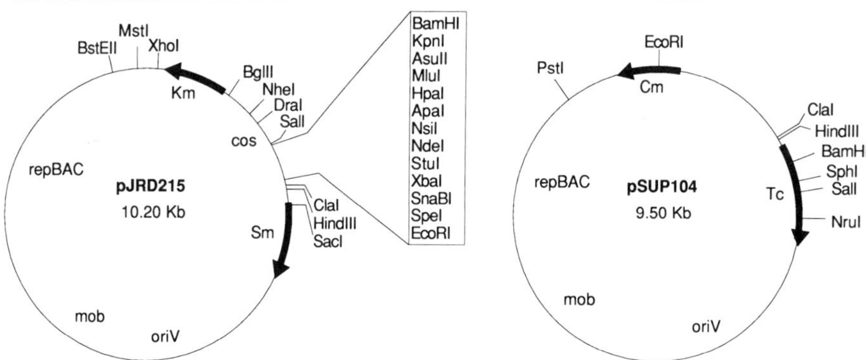

FIG. 1. Physical maps of IncQ cloning vectors pJRD215[20] and pSUP104.[21] All restriction sites shown are unique in each vector. *Mob*, *repB,A,C*, and *oriV* represent mobilization and replication genes and the replication origin of IncQ plasmid RSF1010. *Cos* is the *cos* site of bacteriophage λ. Kanamycin, streptomycin, chloramphenicol, and tetracycline resistance genes are indicated by Km, Sm, Cm, and Tc, respectively.

sense RNA have been used to suppress synthesis of the iron hydrogenase and contributed to the analysis of the physiological role of this enzyme.[17]

There is evidence that IncP plasmids can also be transferred by conjugation to *Desulfovibrio*, although they apparently do not replicate.[2,11] However, attempts to use IncP plasmids or mobilizable, narrow host range *E. coli* plasmids for exchange of *Desulfovibrio* genes with cloned, plasmid-localized genes that were interrupted with an antibiotic resistance gene ("marker exchange")[2] or for transposon mutagenesis with Tn*5*[10] have not yet been successful. The only reported case of marker exchange was with *D. fructosovorans*, in which genes for the nickel hydrogenase have been inactivated by exchange with the homologous genes carrying an insertion of the kanamycin resistance gene on an IncQ plasmid.[12]

Transfer of Plasmids by Conjugation

IncQ plasmids are not self-transmissable and require *trans*-acting functions for mobilization. These *trans*-acting functions (*tra* genes) are normally derived from the IncP plasmid RP4[25] and are either incorporated into the genome of the *E. coli* donor (e.g., in *E. coli* S17-1)[26] or provided on helper plasmids such as pRK2013[27] or pRK2073.[28] In the case of *E.*

[25] C. J. Thomas, *Plasmid* **5,** 10 (1981).
[26] R. Simon, U. Priefer, and A. Pühler, *Bio/Technology* **1,** 784 (1983).
[27] D. H. Figurski and D. R. Helinski, *Proc. Natl. Acad. Sci. U.S.A.* **76,** 1648 (1979).
[28] C. H. Kim, D. R. Helinski, and G. Ditta, *Gene* **50,** 141 (1986).

coli S17-1 as donor, the plasmid to be transferred is introduced into this strain by transformation, followed by conjugation of the donor with the *Desulfovibrio* recipient (double mating). When a helper plasmid is used, the conjugative plasmid is either introduced into the *E. coli* strain that contains the helper plasmid (provided that selection for the presence of both plasmids is possible) or into a different *E. coli* strain, followed by mating of the (two) *E. coli* strain(s) with the receptor (double or triple mating). The frequency of plasmid transfer to *D. desulfuricans* G200 was found to be 10-fold higher in triple matings with pRK2013 as a helper plasmid compared to double matings with *E. coli* S17-1 as a donor.[10]

Matings between *E. coli* and *Desulfovibrio* are done on nitrocellulose membranes or on the surface of agar plates, for 4–16 hr. Remarkably, at least for *D. vulgaris* and *D. desulfuricans* G200, it does not appear necessary to perform the mating under anaerobic conditions: transfer frequencies after anaerobic or aerobic overnight conjugation were similar, with only a small decrease in viability of *Desulfovibrio* after aerobic mating.[2,10]

After mating, cells are plated on selective agar to select for plasmid containing *Desulfovibrio* (see the previous section) and against the *E. coli* donor. Counterselection against *E. coli* is sometimes difficult. *Escherichia coli* cells may survive and develop colonies in coculture with *Desulfovibrio* in the presence of concentrations of antibiotics that are fully inhibitory for pure cultures of *E. coli*, both on media with sulfate and with nitrate.[2,10,11] Various strategies for selection against the donor have been developed, such as the use of an eightfold auxotrophic *E. coli* strain and selection for *Desulfovibrio* transconjugants on a defined minimal medium,[1] or conjugation for short times with low ratios of donor to acceptor cells, followed by counterselection with antibiotics.[10] For conjugal transfer of Cm^R plasmids from *E. coli* S17-1 to *D. vulgaris* Hildenborough, we use kanamycin and streptomycin to select against the donor.[2] The occasional *E. coli* or mixed *E. coli*/*Desulfovibrio* colony can be easily distinguished from *Desulfovibrio* colonies on sulfate-containing selection medium by its translucent appearance due to the absence of FeS precipitates. In any case, *Desulfovibrio* colonies developing on selective medium should be checked for the absence of donor cells and should be restreaked at least once on fresh selection plates.

Frequencies for transfer of different IncQ plasmids to *Desulfovibrio* strains, expressed as the number of transconjugants per recipient cell, vary between 10^{-3} and 1 and are generally around 10^{-2}.[1,2]

Anaerobic electroporation may become an alternative approach for the introduction of plasmids into *Desulfovibrio*, by which some of the difficulties with selection might be avoided. Unfortunately, electroporation has to date been described only once for *Desulfovibrio* (for *D. fructo-*

sovorans).[12] General applicability of this technique still must be demonstrated.

Contrary to the efforts required to select *Desulfovibrio* cells that have acquired IncQ plasmids, transconjugants are easy to maintain. It has been shown that the IncQ cloning vector pSUP104 is maintained in approximately 10–12 copies per *D. vulgaris* cell.[2] Plasmid loss was not observed after cultivation of *D. vulgaris* (pSUP104) for more than 30 generations without selection for the plasmid-encoded antibiotic resistance. Plasmids appeared stable after storage of the recombinant desulfovibrios for several months in soft-agar tubes at 4°.[2]

Protocol for Conjugation and Recovery of Desulfovibrio vulgaris Transconjugants

We use the following protocol for transfer of pSUP104-derived plasmids to *D. vulgaris* Hildenborough. Owing to the difficult selection procedure, we advise that proper controls be done as indicated in the protocol. Protocols for transfer of IncQ plasmids to other desulfovibrios might be slightly different with respect to the media used, selection of transconjugants and counterselection against *E. coli*, and the optional use of the triple mating technique with helper plasmids.[1,10]

1. Donor strain: *E. coli* S17-1[26] (*thi pro hsdR recA* RP4-2[Tc::Mu, Km::Tn*7*]). This strain has the *trans*-acting mobilization functions and tetracycline resistance gene of IncP plasmid RP4-2 integrated in the genome. The tetracycline resistance gene contains a Mu-integrate with a kanamycin resistance gene interrupted by Tn*7*. As the strain is slightly unstable and might lose its mobilization functions, always start with a colony picked from a fresh plate containing trimethoprim (25 μg/ml), which selects for the presence of Tn*7*. pSUP104 or recombinant plasmids derived from pSUP104 are introduced into *E. coli* S17-1 by standard transformation procedures. (*Note*: Transformation frequencies of S17-1 made competent with $CaCl_2$ are only approximately 100–1000 transformants per microgram of plasmid DNA.) Pick a colony appearing on plates with chloramphenicol (20 μg/ml) and check for the presence and integrity of the plasmid. Grow overnight at 37° in a shaker (200 rpm) in 10 ml of TY broth (10 g of tryptone, 5 g of yeast extract, 5 g of NaCl per liter, pH 7.4 with NaOH) with chloramphenicol.
2. Acceptor strain: *D. vulgaris* Hildenborough (NCIMB 8303). *Desulfovibrio vulgaris* stocks are maintained in soft agar tubes according to Postgate.[29] Liquid cultures are grown at 37° in stoppered serum flasks

[29] J. R. Postgate, *J. Gen. Microbiol.* **5,** 714 (1951).

with Saunders medium[30]: 0.06 g of $MgCl_2 \cdot 6H_2O$, 1.0 g of sodium citrate $\cdot$ $2H_2O$, 0.06 g of $CaCl_2 \cdot 2H_2O$, 0.5 g of KH_2PO_4, 1.0 g of NaCl, 7.0 g of $(NH_4)_2SO_4$, 1.0 g of yeast extract, 6 g of sodium lactate, and 0.01 g of $(NH_4)_2Fe(SO_4)_2 \cdot 6H_2O$ per liter. Adjust the pH to 7.2 with 5 *M* KOH after autoclaving. Replace the air by alternate evacuation and flushing with N_2. Inoculate with a thick *D. vulgaris* suspension (e.g., 1–5 ml from a culture maintained in a soft-agar tube per 100 ml of medium); add $Na_2S_2O_4$ (100 mg/liter). Repeat the evacuation/N_2 flushing cycles and grow the culture at 37° with gentle stirring on a magnetic stirrer.

3. Inoculate 10 ml of fresh TY broth with antibiotics with 100 μl of an overnight culture of plasmid containing *E. coli* S17-1; also grow *E. coli* S17-1 without plasmid for a mock mating as control. Grow at 37° in a shaker (200 rpm) to midexponential phase ($OD_{600} \sim 1$ on most spectrophotometers). Inoculate 10 ml of fresh Saunders medium with 0.5 ml of a full-grown culture of *D. vulgaris*. Grow at 37° to end-exponential phase ($OD_{600} \sim 0.8$).

4. Collect cells from 2 ml of the exponential *E. coli* and *D. vulgaris* cultures by centrifugation (5 min, 6500 rpm); wash with 1 ml of PBS. Resuspend the cells in 100 μl of PBS. Anoxic working conditions are not required at this stage.

5. Mix equal volumes of both suspensions and spot 50 μl of the mixture on a well-dried plate with TSAS bottom agar (without antibiotics, resazurin, thioglycolate, and ascorbic acid). Incubate overnight at 30° in either an anaerobic jar (anaerobic mating) or air (aerobic mating). Also perform a mock mating between *D. vulgaris* and *E. coli* S17-1 without plasmid, grown under similar conditions.

6. After mating, collect the cells from the plate in 1 ml of PBS and make serial 10-fold dilutions in PBS (10^{-2}–10^{-6}). Mix 100 μl of the diluted suspensions from mating and mock mating with 3 ml of TSAS soft agar at 45° (prepared as described in a previous section), containing (a) kanamycin (50 μg/ml) and streptomycin (50 μg/ml) and (b) kanamycin (50 μg/ml), streptomycin (50 μg/ml), and chloramphenicol (10 μg/ml) (kanamycin and streptomycin select against *E. coli*, chloramphenicol selects for plasmid containing *D. vulgaris*). Pour immediately onto TSAS bottom agar plates containing the same antibiotics. Grow at 30 or 37° in an anaerobic jar.

7. Open the jar when colonies are visible on the plates (after approximately 5 days). Black *Desulfovibrio* colonies are easily distinguished from the occasional translucent *E. coli* colony that has escaped selection. Transconjugant desulfovibrios should appear at a frequency of 10^{-2}–10^{-3}, as estimated from the relative amounts of black colonies on plates with and

[30] G. F. Saunders, L. L. Campbell, and J. R. Postgate, *J. Bacteriol.* **87,** 1073 (1964).

without chloramphenicol; the mock mating should not give *Desulfovibrio* colonies on chloramphenicol plates.

8. Cut a few well-isolated chloramphenicol-resistant *Desulfovibrio* colonies from the overlay of a relatively uncrowded plate, put each colony in a separate Eppendorf tube with 1 ml of PBS, and squeeze the agar to elute the cells.

9. Make serial 10-fold dilutions of the eluted cells in PBS and plate 100-μl portions of 10^{-3}–10^{-6} dilutions on TSAS plates with kanamycin, streptomycin, and chloramphenicol as described under step 6. Grow again for approximately 5 days.

10. Elute cells from a few well-isolated colonies as under step 8. To check for the absence of *E. coli*, inoculate a TY agar plate with approximately 50 μl of this suspension. No colonies should develop overnight after aerobic incubation at 37°.

11. Use the rest of the suspension to inoculate a tube of Postgate soft-agar medium[29] with chloramphenicol (10 μg/ml) to prepare a stock of the transconjugant *D. vulgaris* (which can be kept at least 6 months at 4°) and/or a 10-ml liquid culture in Saunders medium with chloramphenicol (10 μg/ml) to check for the presence of the plasmid.

Isolation of IncQ Plasmids from Desulfovibrio (Minipreparation)

Given the marginal differences in antibiotic resistance between wild-type and transconjugant desulfovibrios, the presence (and integrity) of the conjugative plasmids in cells that have been selected as transconjugants should be verified by isolation of the plasmids and checking the restriction pattern on agarose gel. IncQ plasmids can be isolated from *Desulfovibrio* with the standard alkaline lysis method of Birnboim and Doly.[31] The best plasmid preparations are obtained from cultures that are not full grown. Plasmid yields are 5–10 μg/10-ml culture (i.e., approximately 8 mg of cells).

[31] H. C. Birnboim and J. Doly, *Nucleic Acids Res.* **7,** 1513 (1979).

[23] Enzymology and Molecular Biology of Sulfate Reduction in Extremely Thermophilic Archaeon *Archaeoglobus fulgidus*

By CHRISTIANE DAHL, NORBERT SPEICH, and HANS G. TRÜPER

The ability to use sulfate as an external electron acceptor is a common feature of dissimilatory sulfate-reducing bacteria. All sulfate-reducing prokaryotes have been assumed to belong to the eubacteria; however, in 1987 an archaeal member of this physiological group was isolated and described by Stetter *et al.*[1] This organism, *Archaeoglobus fulgidus*, is an extreme thermophile and grows at temperatures between 64 and 92°, with an optimum at 83°. Phylogenetically, *A. fulgidus* belongs to the kingdom of the Euryarchaeota[2] and is grouped with the Methanomicrobiales/extreme-halophile cluster.[3] The organism carries out sulfate reduction via the pathway originally proposed for bacterial species.[4–6] Owing to its chemical inertia, sulfate needs first to be activated to adenylylsulfate [adenosine-5′-phosphosulfate (APS)] by ATP-sulfurylase [sulfate adenylyltransferase, EC 2.7.7.4; reaction (1)], with formation of pyrophosphate. The equilibrium of the ATP-sulfurylase reaction lies far to the left and it is proposed that the activation of sulfate *in vivo* is promoted by the hydrolysis of the inorganic pyrophosphate[7] [reaction (2)] and the favorable adenylylsulfate reductase [reaction (3)]. The enzyme adenylylsulfate (APS) reductase (EC 1.8.99.2) catalyzes the reduction of APS to sulfite and AMP. Sulfite is finally reduced to sulfide by sulfite reductase (EC 1.8.99.1) [reaction (4)]. Alternative routes, such as a direct six-electron reduction to sulfide[8] or a pathway involving trithionate and thiosulfate as intermediates,[9,10] are discussed for sulfite reduction in sulfate-reducing bacteria.

[1] K. O. Stetter, G. Lauerer, M. Thomm, and A. Neuner, *Science* **236,** 822 (1987).

[2] C. R. Woese, O. Kandler, and M. L. Wheelis, *Proc. Natl. Acad. Sci. U.S.A.* **87,** 4576 (1990).

[3] C. R. Woese, L. Achenbach, P. Rouviere, and L. Mandelco, *Syst. Appl. Microbiol.* **14,** 364 (1991).

[4] H. D. Peck, Jr., *Bacteriol. Rev.* **26,** 67 (1962).

[5] N. Speich and H. G. Trüper, *J. Gen. Microbiol.* **134,** 1419 (1988).

[6] C. Dahl, H.-G. Koch, O. Keuken, and H. G. Trüper, *FEMS Microbiol. Lett.* **67,** 27 (1990).

[7] G. Fauque, J. LeGall, and L. L. Barton, *in* "Variations in Autotrophic Life" (J. M. Shively and L. L. Barton, eds.), p. 271. Academic Press, New York, 1991.

[8] L. A. Chambers and P. A. Trudinger, *J. Bacteriol.* **123,** 35 (1975).

[9] H. Drake and J. M. Akagi, *J. Bacteriol.* **132,** 139 (1977).

[10] R. M. Fitz and H. Cypionka, *Arch. Microbiol.* **154,** 400 (1990).

$$ATP + SO_4^{2-} \leftrightharpoons APS + PP_i \qquad \Delta G^{\circ\prime} = 46.4 \text{ kJ/mol} \quad (1)$$
$$PP_i + H_2O \leftrightharpoons 2P_i \qquad \Delta G^{\circ\prime} = -21.7 \text{ kJ/mol} \quad (2)$$
$$APS + 2[H] \leftrightharpoons SO_3^{2-} + AMP \qquad \Delta G^{\circ\prime} = -68.6 \text{ kJ/mol} \quad (3)$$
$$SO_3^{2-} + 6[H] \leftrightharpoons H_2S \qquad \Delta G^{\circ\prime} = -171.5 \text{ kJ/mol} \quad (4)$$

Enzyme Assays

Auxiliary enzymes are widely applied in the determination of ATP-sulfurylase and sulfite reductase activities from mesophilic organisms. They cannot, however, be used in continuous assays of the respective enzymes from the extreme thermophile *A. fulgidus*, because most commercially available auxiliary enzymes are inactive at the high temperatures required for optimum activity of *A. fulgidus* enzymes.

ATP-Sulfurylase

Principle. ATP-sulfurylase from *A. fulgidus* is measured in the thermodynamically favored direction of ATP generation from the reaction of APS with pyrophosphate. The ATP formed is then spectrophotometrically determined. It is difficult to assay ATP-sulfurylase accurately in cell-free extracts and early purification fractions. The problems are caused by the competing reaction of pyrophosphatase, which catalyzes the rapid depletion of millimolar levels of PP_i. The easiest solution to overcome this complication is to assay ATP-sulfurylase at the suboptimum temperature of 22°. In contrast to pyrophosphatase, which is inactive at this temperature, ATP-sulfurylase retains 7.5% of its maximum activity at 22° and can readily be detected.[6] After pyrophosphatase has been removed or reduced to manageable levels, the further purification of ATP-sulfurylase can be followed and the characterization of the kinetics of the pure enzyme can be performed at 85°.

Procedure. The reaction mixture, a 1.0-ml total volume, contains 100 m*M* Tris-HCl, pH 8.0, 1 m*M* APS[11] (or varied for kinetic studies), 1 m*M* $MgCl_2$, 1 m*M* PP_i (or varied for kinetics), ATP-sulfurylase (~7 ng of pure enzyme), and distilled water. The reaction mixture is preincubated in small centrifuge tubes at the temperature of choice for 5 min. To minimize evaporation the tubes should be sealed with marbles. The reaction is started by addition of appropriately diluted extract. After the desired incubation period, the reaction is terminated with 0.2 ml of 2 *N* NaOH (ATP-sulfurylase shows no activity at pH 12 and ATP is stable under the conditions used) and afterwards neutralized with 0.2 ml of 1 *M* sodium acetate buffer, pH 4.8. Denatured protein is removed by centrifugation

[11] C. Dahl and H. G. Trüper, this volume [28].

at 25,000 *g* for 10 min. An aliquot of supernatant is used for quantitative determination of generated ATP by a standard hexokinase and glucose-6-phosphate dehydrogenase-coupled spectrophotometric test system.[12]

Adenylylsulfate Reductase

Principle. Adenylylsulfate reductase activity is measured in a continuous spectrophotometric assay in the direction of APS formation from sulfite and AMP with ferricyanide as electron acceptor. Each molecule of SO_3^{2-} that reacts with AMP to form APS leads to the reduction of two molecules of ferricyanide. Cytochrome *c*, which has been used successfully for assaying APS-reductases from thiobacilli[13] and phototrophic bacteria,[14] cannot serve as an electron acceptor for the archaeal enzyme because it is denatured at the reaction temperature of 85°. The assay described here requires a recording spectrophotometer that can be thermostatted to 85° and cuvettes that can be sealed in order to minimize evaporation.

Procedure. The sulfite- and AMP-dependent reduction of ferricyanide is monitored at 85° in a reaction mixture (1.0-ml total volume) containing the following final concentrations of reagents: Tris-HCl, pH 8.0 (50 m*M*), AMP, pH 7.0 (2 m*M*, or varied as necessary for kinetic studies), $K_3Fe(CN)_6$ (5 m*M*, or varied as necessary), Na_2SO_3 [30 m*M*, or varied as necessary; the sulfite solution must be freshly prepared in 50 m*M* Tris-HCl (pH 8.0) plus 5 m*M* ethylenediaminetetraacetic acid (EDTA)], and adenylylsulfate reductase. The reaction is started by adding AMP after the background rate of ferricyanide reduction has been established. The reduction of ferricyanide is monitored by following the decrease in absorbance at 420 nm ($\varepsilon_{420} = 1.09$ cm^2 μmol^{-1}).

Thin-layer chromatography can be used to prove formation of APS, by cochromatography of reference nucleotides.[11]

During purification procedures APS reductase can also be monitored by its characteristic ultraviolet (UV)–visible spectrum. The enrichment is followed by the A_{280}/A_{390} ratio.

Sulfite Reductase

Principle. Sulfite reductase activity is measured in a continuous spectrophotometric assay in the direction of sulfite reduction with electrochemically reduced methyl viologen as electron donor.

[12] W. Lamprecht and I. Trautschold, *in* "Methoden der enzymatischen Analyse" (H. U. Bergmeyer, ed.), p. 2034. Verlag Chemie, Weinheim, 1970.

[13] M. R. Lyric and J. Suzuki, *Can. J. Biochem.* **48,** 344 (1970).

[14] H. G. Trüper and L. A. Rogers, *J. Bacteriol.* **108,** 1112 (1971).

Procedure. Sulfite reductase activity in crude extracts is assayed in cuvettes sealed with rubber stoppers at 85° with N_2 as a gas phase. Additions are made with microliter syringes. In a volume of 0.9 ml the assay contains the following: extract in 50 m*M* potassium phosphate buffer, pH 7.0, 0.5 m*M* dithioerythritol (DTE), 50 m*M* potassium phosphate buffer, pH 7.0, and 2 m*M* electrochemically reduced methyl viologen. The reactions are started by addition of 100 μl of oxygen-free sulfite solution (100 m*M* Na_2SO_3 in 50 m*M* potassium phosphate buffer, pH 7.0, 5 m*M* EDTA). The disappearance of methyl viologen is monitored by following the decrease in absorbance at 600 nm ($\varepsilon = 13.0\ cm^2\ \mu mol^{-1}$). During purification procedures sulfite reductase can most conveniently be followed by measuring the absorbances at 394, 545, and 593 nm. The enrichment is monitored by the A_{280}/A_{593} ratio. Reduction of sulfite reductase is performed under anaerobic conditions, using cuvettes with rubber stoppers. Concentrations of sulfite reductase are based on a molecular weight of 178,200, which is calculated from the deduced amino acid sequence of the α and β subunits and the $\alpha_2\beta_2$ structure of the enzyme (see p. 339).

Purification of ATP-Sulfurylase, Adenylylsulfate Reductase, and Sulfite Reductase from *Archaeoglobus fulgidus*

The following procedures consistently provide homogeneous enzymes from cells of *A. fulgidus* (DSM 4304^T). With minor modifications we have used the procedures to obtain the APS-reductase and sulfite reductase from *Thermodesulfobacterium commune*.

Growth Conditions. The medium[1] contains, per liter, 0.34 g of KCl, 2.75 g of $MgCl_2 \cdot 2H_2O$, 3.45 g of $MgSO_4 \cdot 7H_2O$, 0.25 g of NH_4Cl, 0.14 g of $CaCl_2 \cdot 2H_2O$, 0.14 g of K_2HPO_4, 18 g of NaCl, 5 g of $NaHCO_3$, 2 mg of $Fe(NH_4)_2(SO_4)_2$, 1 mg of resazurin, 10 ml of trace mineral solution,[15] 0.5 g of yeast extract (Difco, Detroit, MI), 1 g of L(+)-lactate, and 0.5 g of Na_2S. The pH is adjusted with H_2SO_4 to 6.9. The cells are grown anaerobically with a gas phase of N_2/CO_2 (80 : 20; 3 atm) at 85°. Mass cultures for enzyme production are grown in 300-liter enamel-protected fermentors that are gassed with N_2/CO_2 under stirring. Cells are allowed to grow for 40 hr, harvested, and stored as packed deep-frozen cells. The yield is around 0.4 g (wet mass) per liter. We obtained deep-frozen cell mass from K. O. Stetter (Universität Regensburg, Germany).

Preparation of Crude Extracts. Packed cells are taken up in about twice their volume of 50 m*M* Tris-HCl, pH 8.0, and disrupted by ultrasonic

[15] W. E. Balch, G. E. Fox, L. J. Magrum, C. R. Woese, and R. S. Wolfe, *Microbiol. Rev.* **43,** 260 (1979).

treatment. The broken cell mass is centrifuged at 17,000 *g* for 20 min and the supernatant is subjected to ultracentrifugation (140,000 *g*, 2 hr). Addition of 1 m*M* 2-mercaptoethanol and 0.5 m*M* ethylenediaminetetraacetic acid (EDTA) to the standard buffer, 50 m*M* Tris-HCl, pH 8.0, used throughout the purification procedures slightly increases enzyme yields.

Purification of Adenylylsulfate Reductase. The purification procedure originally described by Speich and Trüper[5] yielded an enzyme preparation that on sodium dodecyl sulfate-polyacrylamide gel electrophoresis (SDS–PAGE) analysis showed a single band with an apparent molecular weight of 80,000. We developed a modified purification procedure omitting isoelectric focusing as the final step and reevaluated the structure of the enzyme. We found a further subunit with an apparent molecular weight of 18,500 that was probably lost during isoelectric focusing. Subunits of comparable size have also been described for the adenylylsulfate reductases from *Desulfovibrio vulgaris*, *Desulfovibrio gigas*, and *Thermodesulfobacterium mobilis*.[16–19] Furthermore, cloning and sequencing revealed that the two genes (*aprA* and *aprB*) encoding the two different subunits of the archaeal APS reductase are separated by only 17 bp and are probably localized in the same operon (see p. 347), supporting that the small subunit is an authentic part of *A. fulgidus* APS reductase. We here describe two methods yielding homogeneous APS reductase still consisting of its two different subunits.

All purification steps are performed aerobically at 4°. Ammonium sulfate is added to the supernatant of ultracentrifugation to 50% saturation at 0°. The precipitated protein is separated from the supernatant by centrifugation (17,000 *g*, 20 min) and discarded. The supernatant is applied to a column of Phenyl-Sepharose CL-4B (1.5 × 20 cm; flow rate, 20 ml/hr) equilibrated with 50 m*M* Tris-HCl, pH 8.0, containing ammonium sulfate at 50% saturation. The enzyme is eluted with a linear gradient of ammonium sulfate between 50 and 0% saturation (460 ml). The active enzyme fractions are pooled and desalted by Sephadex G-25 column. The eluate is applied to a column of DEAE-cellulose (5 × 10 cm; flow rate, 15 ml/hr) equilibrated with 50 m*M* Tris-HCl, pH 8.0, and eluted with a gradient of 0–300 m*M* NaCl in the same buffer. Active APS reductase fractions are pooled,

[16] R. N. Bramlett and H. D. Peck, Jr., *J. Biol. Chem.* **250,** 2979 (1975).

[17] J. Lampreia, I. Moura, M. Teixeira, H. D. Peck, Jr., J. LeGall, B. Huynh, and J. J. G. Moura, *Eur. J. Biochem.* **188,** 653 (1990).

[18] G. Fauque, M. H. Czechowski, L. Kang-Lissolo, D. V. DerVartanian, J. J. G. Moura, I. Moura, J. Lampreia, A. V. Xavier, and J. LeGall, *Abstr. Ann. Meet. Soc. Ind. Microbiol. San Francisco*, 92 (1986).

[19] D. R. Kremer, M. Venhuis, G. Fauque, H. D. Peck, Jr., J. LeGall, J. Lampreia, J. J. G. Moura, and T. A. Hansen, *Arch. Microbiol.* **150,** 296 (1988).

desalted, and subjected to ion-exchange chromatography on Mono Q HR 5/5. The column is equilibrated with 50 m*M* Tris-HCl, pH 8.0, and after application of the protein washed with the same buffer containing 100 m*M* NaCl. Via a linear gradient from 100 to 300 m*M* NaCl (20 ml) APS reductase is recovered at 190–230 m*M* NaCl with a purity coefficient of 5.0 (A_{280}/A_{390}). Table I summarizes the purification of adenylylsulfate reductase.

Alternatively the supernatant of the 50% ammonium precipitation can be loaded onto a Phenyl-Sepharose CL-4B FastFlow (low substituted) column (2.6 × 17 cm; flow rate, 3 ml/min) equilibrated with 50% ammonium sulfate in 50 m*M* Tris-HCl, pH 8.0. After washing the column with 40% ammonium sulfate in the same buffer a linear gradient from 40 to 0% ammonium sulfate (800 ml) is applied. Adenylylsulfate reductase elutes between 16 and 9% $(NH_4)_2SO_4$. Active fractions are pooled and desalted by dialysis against 50 m*M* Tris-HCl, pH 8.0, followed by chromatography on Q-Sepharose FastFlow (1.6 × 10 cm; flow rate, 2 ml/min) equilibrated in standard buffer. With a gradient from 0 to 400 m*M* NaCl (600 ml) APS reductase is retrieved at 240–280 m*M* NaCl. Fractions containing the enzyme are combined, desalted by dialysis, and further purified by chromatography on Mono Q as described above.

Properties of Adenylylsulfate Reductase. The enzyme constitutes 1.5% of the total cellular soluble protein. The temperature optimum for activity is 85° under the assay conditions used. Higher temperatures are inhibitory and no activity is found below 55°. The pH optimum is 8.0. At pH 8.0 and 85° the specific activity (V_{max}; units per milligram protein) is 2.3. The K_m values for AMP and ferricyanide are 1 and 0.4 m*M*, respectively. Nucleotide specificity is determined by replacing AMP with other nucleotides. The enzyme shows the lowest activities with pyrimidine nucleotides (0% with UMP, 24% with CMP). DeoxyAMP and GMP replace AMP with 74 and 68% efficiency, whereas the enzyme shows 41% activity with IMP.

TABLE I
PURIFICATION OF ADENYLYLSULFATE REDUCTASE FROM *Archaeoglobus fulgidus*

Step	Volume (ml)	Protein (mg)	Activity (units)	Specific activity (units/mg)
1. Crude extract	60.0	2364.0	217.5	0.092
2. Supernatant (140,000 *g*)	53.0	1654.8	213.5	0.129
3. 50% $(NH_4)_2SO_4$	44	822.8	159.0	0.193
4. Phenyl-Sepharose	52	115.0	143.7	1.25
5. DE-52	48	71.0	134.9	1.9
6. Mono Q	2	28.3	65.1	2.3

The enzyme shows 50 and 24% activity with ADP and ATP, respectively. A probable reason for this is the hydrolysis of these nucleotides at the reaction temperature of 85°.

The APS reductase from *A. fulgidus* is a slightly acidic protein (p*I* 4.8); analytical gel filtration gives an apparent molecular weight of 160,000. The native enzyme consists of two different subunits with molecular weights of 73,300 and 17,100 (deduced from the nucleotide sequence, see p. 346). Densitometric analyses of SDS-polyacrylamide gels give a ratio of 2.5 : 1 for the 73.3- to 17.1-kDa subunits. The calculated molecular mass of an $\alpha_2\beta$-structured APS reductase is 163.7 kDa, which is in agreement with the value obtained by analytical gel filtration. The oxidized enzyme exhibits an absorption maximum at 390 nm. On reduction the absorption between 400 and 500 nm decreases significantly. Analysis of flavin content[20,21] yields 1 mol of noncovalently bound FAD per mole of APS reductase. Additionally, each mole of the enzyme contains 8 mol of nonheme iron[22] and 6 mol of labile sulfide.[23] The iron and sulfur atoms are arranged as two distinct [4Fe-4S] clusters (centers I and II).[24] Center I, which has a high redox potential, is reduced by AMP and sulfite, and center II has a negative redox potential.

Separation of ATP-Sulfurylase and Sulfite Reductase. With the exception of the initial cell disrupture and centrifugations, all purification steps are carried out aerobically at room temperature. The supernatant of the ultracentrifugation of the crude extract is desalted on Sephadex G-25 and made 40% saturated with solid ammonium sulfate at 0°. After equilibration overnight at 0° the precipitated protein is separated from the supernatant by ultracentrifugation (140,000 *g*, 4°, 120 min). The supernatant is applied to a Pharmacia-FPLC (fast protein liquid chromatography) system and aliquots of 1000 mg total protein are separated on a Phenyl-Sepharose CL-4B FastFlow column (1.5 × 18 cm; flow rate, 3 ml/min) equilibrated with 40% ammonium sulfate in 50 m*M* Tris-HCl, pH 8.0. Using a linear decreasing gradient from 40 to 0% ammonium sulfate (370 ml) ATP-sulfurylase elutes between 27 and 21%, and sulfite reductase elutes between 20 and 14% ammonium sulfate.

Further Purification of ATP-Sulfurylase. The active enzyme fractions from phenyl-Sepharose CL-4B are pooled and desalted by a Sephadex G-25 column. Portions of less than 20 mg total protein are applied to a

[20] G. L. Kilgour, S. P. Felton, and F. M. Huennekens, *J. Am. Chem. Soc.* **79,** 2254 (1957).

[21] N. A. Rao, S. P. Felton, and F. M. Huennekens, this series, Vol. 10, p. 494.

[22] Ferrozine method, Sigma.

[23] T. E. King and R. O. Morris, this series, Vol. 10, p. 634.

[24] J. Lampreia, G. Fauque, N. Speich, C. Dahl, I. Moura, H. G. Trüper, and J. J. G. Moura, *Biochem. Biophys. Res. Commun.* **181,** 342 (1991).

column of Mono Q HR 5/5, equilibrated with 50 mM Tris-HCl buffer, pH 8.0, at a flow rate of 1.0 ml/min. The column is washed with 3 vol of the above-mentioned buffer containing 150 mM NaCl, and a linear gradient of 150–300 mM NaCl in the same buffer is applied. Active ATP-sulfurylase fractions are pooled and concentrated by ultrafiltration (XM50 membranes). The concentrated protein solution is transferred into 50 mM Tris-HCl, pH 8.0, containing 150 mM NaCl via a Pharmacia (Piscataway, NJ) PD-10 column and subjected to gel filtration on Superose 6, equilibrated with the same buffer, at a flow rate of 0.4 ml/min. Table II summarizes the purification of ATP-sulfurylase.

Properties of ATP-Sulfurylase. On examination by analytical gel filtration the enzyme elutes as a major peak at a molecular weight of 460,000 and a second peak at a molecular weight of 150,000, suggesting that the monomeric enzyme has a strong tendency to aggregate and form catalytically active trimers. The enzyme consists of two different subunits of 50 and 53 kDa in an $\alpha_2\beta$ structure. The pH optimum for activity is 8.0, and the optimum reaction temperature is 90°. At pH 8.0 and 85° (the growth temperature of *A. fulgidus*) the specific activity of ATP-sulfurylase (V_{max}; units per milligram protein) is 480. The apparent K_m values for adenylylsulfate and pyrophosphate are 0.17 and 0.13 mM, respectively. The isoelectric point of the archaeal ATP-sulfurylase is 4.3.

Further Purification of Sulfite Reductase. Fractions from Phenyl-Sepharose CL-4B FastFlow containing sulfite reductase with a purity index of less than 16 (A_{280}/A_{593}) are combined and dialyzed against 50 mM Tris-HCl, pH 8.0. Portions of less than 20 mg total protein are passed through a Mono Q HR 5/5 column equilibrated with the same buffer. The column

TABLE II
PURIFICATION OF ATP-SULFURYLASE FROM *Archaeoglobus fulgidus*

Step	Volume (ml)	Protein (mg)	Activity (units)[a]	Specific activity (units/mg)[a]
1. Crude extract	59.0	3,787.8	10,151.0	2.7
2. Supernatant (140,000 *g*)	49.0	3,248.0	10,071.0	3.3
3. 40% $(NH_4)_2SO_4$	24.5	742.4	3,711.0	5.0
4. Phenyl-Sepharose	10.0	85.4	685.0	7.6
5. Mono Q	4.0	7.7	192.0	25.0
6. Superose 6	5.0	1.6	54.3	33.9[b]

[a] For comparative reasons enzyme activity was always measured at 22°.
[b] Determined to be 480 units/mg at 85°.

is washed with 50 m*M* Tris-HCl, pH 8.0, containing 280 m*M* NaCl. Via a linear gradient from 280 to 380 m*M* NaCl (20 ml) sulfite reductase is recovered at 310–330 m*M* NaCl with a purity coefficient of 6.2. The fractions containing the enzyme are pooled and desalted. The ion-exchange chromatography is repeated, resulting in an electrophoretically homogeneous protein (A_{280}/A_{593}, 5.4; A_{280}/A_{394}, 2.18). Sulfite reductase has been applied to Superose 6 gel-filtration columns in 50 m*M* Tris-HCl, pH 8.0, containing 150 m*M* NaCl. The enzyme yields a single peak with an apparent molecular weight of 215,000. As no increase in the purification index is obtained this step is usually omitted. On the average ~10 mg of pure sulfite reductase is obtained per 10 g of cells. Table III summarizes the purification of sulfite reductase.

Properties of Sulfite Reductase. Cell extracts of *A. fulgidus* catalyze sulfite-dependent oxidation of reduced methyl viologen with a specific activity of 70 mU/mg protein at 85°. This activity decreases to 10% when the reaction temperature is lowered to 65° and no activity is found below 55°.

Sulfite reductase from *A. fulgidus* consists of two different subunits with molecular masses of 47.4 and 41.7 kDa (deduced from their amino acid sequence, see p. 345), arranged in a quaternary structure of $\alpha_2\beta_2$. The calculated molecular mass of the native enzyme is 178,200 Da. Electrofocusing shows an isoelectric point of 4.2.

The oxidized enzyme exhibits absorption maxima at 281, 394, 545, and 593 nm with shoulders at 430 and 625 nm and a weak band around 715 nm. On reduction with dithionite the α band is shifted to 598 nm while its absorption decreases, the Soret peak shifts from 394 to 390 nm, and the band at 715 nm disappears. The spectra of oxidized and reduced sulfite

TABLE III
PURIFICATION OF SULFITE REDUCTASE FROM *Archaeoglobus fulgidus*

Step	Volume (ml)	Protein (mg)	Total A_{593}	A_{280}/A_{593}[a]
1. Extract	62.0	3992.8	146.1	73
2. Supernatant (140,000 *g*)	58.5	3287.7	119.0	70
3. 40% $(NH_4)_2SO_4$	25.6	750.0	40.9	48.5
4. Phenyl-Sepharose	26.3	260.9	16.3	16.0
5. Second Mono Q eluate	2.9	20.4	8.3	5.4

[a] Besides sulfite reductase cell extracts of *A. fulgidus* contain high concentrations of various colored compounds and proteins (e.g., APS reductase). These substances increase the absorption of extracts at 593 nm. Therefore, the purification factor achieved for sulfite reductase is probably substantially higher than can be deduced here.

reductase indicate that the archaeal enzyme contains siroheme, which is mainly in the high-spin state.[25] The presence of high-spin ferric heme in the protein is supported by preliminary electron paramagnetic resonance (EPR) spectroscopy. The molar extinction coefficients of *A. fulgidus* sulfite reductase at 281, 394, and 593 nm are estimated to be 395,000, 184,000, and 60,000, respectively.

Analysis of siroheme content[26] yields the presence of 2 mol of siroheme per mole of the archaeal sulfite reductase. An analysis for acid labile sulfide[23] gives a value of 20 mol of sulfide per mole sulfite reductase. The complete release of iron from the enzyme requires precipitation of the protein with trichloroacetic acid prior to analyses.[27] Quantitation of the data results in a value of 22–24 nonheme iron atoms per enzyme molecule. The content of iron and acid-labile sulfide indicates that the native sulfite reductase contains six [4Fe-4S] clusters per $\alpha_2\beta_2$ molecule. The nature of the iron cores is suggested by the occurrence of amino acid sequences characteristic of [4Fe-4S] ferredoxins (see below) and by preliminary EPR spectroscopic evidence.[28]

Antiserum prepared against the purified sulfite reductase from *A. fulgidus* cross-reacts with desulfofuscidin from *T. commune*[29] and with sulfite reductase from the extremely thermophilic sulfite reducer *Pyrobaculum islandicum*.[30]

Cloning of Genes for Sulfite Reductase and Adenylylsulfate Reductase

Studies on the molecular biology of sulfur metabolism have almost exclusively been confined to the pathway of assimilatory sulfate reduction.[31-36] We have extended this work to dissimilatory sulfate reduction

[25] A. M. Stolzenberg, S. H. Strauss, and R. H. Holm, *J. Am. Chem. Soc.* **103,** 4763 (1981).

[26] L. M. Siegel, M. J. Murphy, and H. Kamin, this series, Vol. 52, p. 436.

[27] V. Massey, *J. Biol. Chem.* **229,** 763 (1957).

[28] J. Lampreia, unpublished results (1992).

[29] C. Dahl, N. M. Kredich, R. Deutzmann, and H. G. Trüper, *J. Gen. Microbiol.* **139,** 1817 (1993).

[30] M. Molitor, personal communication (1993).

[31] N. M. Kredich, *in* "*Escherichia coli* and *Salmonella typhimurium*: Cellular and Molecular Biology" (F. C. Neidhardt, J. L. Ingraham, K. B. Low, B. Magasanik, M. Schaechter, and H. E. Umbarger, eds.), p. 419. American Society for Microbiology, Washington, D.C., 1987.

[32] J. Ostrowski, M. J. Barber, D. C. Rueger, B. E. Miller, L. M. Siegel, and N. M. Kredich, *J. Biol. Chem.* **264,** 15796 (1989).

[33] J. Ostrowski, J.-Y. Wu, D. C. Rueger, B. E. Miller, L. M. Siegel, and N. M. Kredich, *J. Biol. Chem.* **264,** 15726 (1989).

[34] J. Tan, L. R. Helms, R. P. Swenson, and J. A. Cowan, *Biochemistry* **30,** 9900 (1991).

[35] T. S. Leyh, T. F. Vogt, and Y. Suo, *J. Biol. Chem.* **267,** 10405 (1992).

[36] R. R. Karkhoff-Schweizer, M. Bruschi, and G. Voordouw, *Eur. J. Biochem.* **211,** 501 (1993).

and cloned and sequenced the genes encoding APS reductase[37,38] and sulfite reductase[29,39] from the dissimilatory sulfate reducer *Archaeoglobus fulgidus*.

The basic techniques of gene isolation from *A. fulgidus* are similar if not identical to the processes applied to other prokaryotes and eukaryotes. The following discussion therefore emphasizes DNA isolation from *A. fulgidus*, construction of phage libraries, and preparation of DNA probes from *A. fulgidus* suitable to detect related genes in other organisms. For information on general recombinant DNA methods the reader is referred to a number of books and laboratory manuals,[40,41] including several volumes in this series.[42–44]

Cloning Strategy

For identification and isolation of the genes for APS reductase and sulfite reductase we relied on the use of synthetic oligonucleotide probes whose sequences were derived from chemically determined amino acid sequences of the respective proteins. We excluded heterologous hybridization with related genes from other organisms as a tool for detection of the siroheme-sulfite reductase genes because all siroheme-containing proteins sequenced so far were isolated from bacteria or eukaryotes with very distant phylogenetic relationship to the archaeon *A. fulgidus*. The method was also inapplicable for detection of the APS reductase genes because no such genes have been cloned or sequenced so far.

The cloning strategy we applied can be summarized as follows: (1) radioactively labeled oligonucleotide probe is hybridized to blots of electrophoretically separated genomic DNA fragments in order to determine size and number of hybridizing fragments and the appropriate hybridizing conditions; (2) probes are hybridized to a library of genomic DNA fragments; (3) positive clones are identified and purified; (4) DNA is extracted from positive clones and characterized by restriction mapping; (5) cloned genes are further characterized by nucleotide sequence analysis; (6) if the

[37] N. Speich, C. Dahl, P. Heisig, A. Klein, F. Lottspeich, K. O. Steller, and H. G. Trüper, *Microbiology* (*UK*) **140,** in press (1994).

[38] The sequence has been deposited in the EMBL data library under accession number X63435.

[39] The sequence has been deposited in the GenBank database under accession number M95624.

[40] J. Sambrook, E. F. Fritsch, and T. Maniatis, "Molecular Cloning: A Laboratory Manual." Cold Spring Harbor Laboratory, Cold Spring Harbor, New York, 1989.

[41] L. G. Davis, M. D. Dibner, and J. F. Battey, "Basic Methods in Molecular Biology." Elsevier, New York, 1986.

[42] This series, Vol. 68.

[43] This series, Vols. 100 and 101.

[44] This series, Vols. 152–155.

clone isolated does not contain the complete gene, libraries are rescreened for positive clones containing DNA fragments with additional sequences. This is repeated until the gene is completely sequenced.

Cloning of Genes for Sulfite Reductase

Step 1. Determination of Amino-Terminal and Internal Amino Acid Sequences. For protein sequencing the subunits of highly purified sulfite reductase (50–100 μg) are separated by SDS–PAGE. For N-terminal amino acid sequencing, subunits are electrophoretically transferred onto siliconized glass fiber membranes (Glassybond; Biometra, Göttingen, Germany), visualized by Coomassie Brilliant Blue staining, excised, and subjected to automated Edman degradation with an Applied Biosystems (Foster City, CA) 477A gas-phase sequencer. For tryptic digests, bands are excised from the gel, washed with 90% ethanol, lyophilized, soaked in 0.2 *M* NH_4HCO_3, and digested with1 mg of trypsin/100 μl overnight at 37°. Tryptic peptides are extracted from the gel matrix with additional 0.2 *M* NH_4HCO_3, and this solution is lyophilized. The residue is dissolved in 6 *M* guanidine-HCl, and peptides are fractionated on a Vydac (The Separations Group, Hesperia, CA) C_{18} reversed-phase high-performance liquid chromatography (HPLC) column and sequenced. Analyses of the α subunit of sulfite reductase gave an amino-terminal sequence of 35 residues and sequences for 6 tryptic peptides for a total of 115 residues. Amino-terminal sequence could not be obtained from the β subunit, but 2 tryptic peptides were sequenced for a total of 28 residues.

Step 2. Synthesis of Degenerate Oligonucleotide Probe. One of the tryptic peptides of the α subunit contained the amino acid sequence FMFFEKD, from which a mixture of 32 different 20-mer oligonucleotides was derived. Using a model 380A automated DNA synthesizer from Applied Biosystems all potential oligodeoxynucleotide sequences were synthesized as a mixture and afterward radiolabeled with [γ-^{32}P]ATP and T_4-polynucleotide kinase.

Step 3. Purification of DNA from Archaeoglobus fulgidus. To obtain unsheared high molecular weight genomic DNA of *A. fulgidus*, 1 g of frozen cell paste is suspended in 5 ml of 10 m*M* Tris-HCl, pH 8.0, 10 m*M* EDTA. Proteinase K at 20 μg/ml and 0.5% (w/v) sodium dodecyl sulfate (SDS) is added, and cells are lysed by incubation at room temperature for 120 min with occasional swirling. It is important to perform the proteinase K incubation below 37° because otherwise the DNA is unspecifically digested, probably due to thermophilic *A. fulgidus* DNases, which become active at elevated temperatures. The mixture is then extracted with phenol, phenol–chloroform–isoamyl alcohol (25 : 24 : 1), and chloroform–isoamyl alcohol (24 : 1), and DNA is purified by CsCl density gradient ultracentrifugation.[40] The DNA yield is about 1–1.2 mg/g of cells.

Step 4. Southern Hybridization of Archaeoglobus fulgidus DNA. The DNA is digested to completion with a two- to fivefold excess of restriction endonuclease and the resulting fragments are fractionated on a 0.8% (w/v) agarose gel (5–10 μg of DNA per lane). The DNA is transferred to nitrocellulose (Hybond-C; Amersham, Arlington Heights, IL) or nylon filters (Hybond-N; Amersham) by the method of Southern.[45] After transfer, the filters are prehybridized for 5 hr at 38° in 6× NET (1× NET is 0.15 *M* NaCl, 1 m*M* EDTA, 15 m*M* Tris-HCl, pH 8.3), 6× Denhardt's solution[46] [50×: 1% (w/v) bovine serum albumin, 1% (w/v) polyvinylpyrrolidone, 2% (w/v) Ficoll 400], 0.1% (w/v) SDS, and nicked and denatured salmon sperm DNA (200 μg/ml).[47] Hybridization is carried out for 16–20 hr at 38° in the same solution lacking salmon sperm DNA but containing sodium pyrophosphate (0.5 mg/ml) and labeled oligodeoxynucleotide (0.02 μg/ml; 5×10^8 cpm/μg). Filters are washed three or four times for 30 min each at 23° in 6× NET, 0.1% SDS, and then once for 7 min at 40° in 2× NET, 0.1% SDS. Autoradiographs are exposed at −70° with intensifying screens for 36–48 hr. Under the conditions described, *A. fulgidus* DNA showed reactivity of a 3.2-kb *Hin*dIII fragment with the probe to the FMFFEKD sequence.

Step 5. Preparation of Library of Archaeoglobus fulgidus DNA in the λ FixII Vector. High molecular weight DNA from *A. fulgidus* is partially digested with restriction endonuclease *Sau*IIIA1 and restriction sites are partially filled in with dGTP, dATP, and the Klenow fragment of *Escherichia coli* DNA polymerase. The DNA is fractionated by electrophoresis on a 0.8% (w/v) agarose gel and fragments with sizes in the range 9–20 kb are recovered by electroelution,[40] followed by phenol extraction and ethanol precipitation. These fragments are then ligated into λ FixII,[48] which has previously been linearized with *Xho*I and partially filled in with dCTP and dTTP. Ligated DNA is packaged *in vitro* with a phage λ packaging extract (Gigapack Gold; Stratagene, San Diego, CA). λ FixII phage are plated in top agar consisting of NZCYM[40] containing 0.7% (w/v) agarose. Recombinant phage show a Spi$^-$ phenotype and can therefore be selected by their ability to grow on *E. coli* harboring a phage P2 lysogen. We use *E. coli* P2PLK17[49] as a host and found 70% recombinant phage

[45] E. Southern, *J. Mol. Biol.* **98,** 503 (1975).

[46] D. T. Denhardt, *Biochem. Biophys. Res. Commun.* **23,** 641 (1966).

[47] T. D. Sargent, this series, Vol. 152, p. 432.

[48] E. Elgin, C. Mackman, M. Pabst, P. A. Kretz, and A. Greener, *Stratagene, Strategies* **4,** 8 (1991).

[49] E. Raleigh and G. Wilson, *Proc. Natl. Acad. Sci. U.S.A.* **83,** 9070 (1986).

in the *A. fulgidus* genomic DNA library. For amplification the phage are plated on *E. coli* P2PLK17 at a density of approximately 2×10^3 PFU/150-cm^2 petri dish. Phage are collected by washing the plates and stored over chloroform at 4° in SM buffer.[40]

Step 6. Screening of Library of Genomic DNA. For screening the amplified λ FixII library is replated at 1 to 10×10^3 PFU/150-cm^2 petri dish, and after 8 to 16 hr at 37° plaques are transferred and bound to nylon membranes (Duralon; Stratagene) by ultraviolet (UV) irradiation.[40] The probe and the hybridization conditions are as described above. Positive plaques are identified by radioautography and purified by replating. Recombinant phage are grown in *E. coli* P2PLK17 and DNA prepared from them according to Chisholm.[50]

Step 7. Sequencing of Positive DNA Fragment. Ten positive clones, each containing an 18.4-kb insert, were isolated from the λ FixII library and a simple restriction map was constructed by digestion of the phage DNA with combinations of restriction enzymes, electrophoretic separation of the fragments, transfer to filters, and hybridization with the oligonucleotide probe. For further analysis DNA fragments were purified from an agarose gel, subcloned into the polylinker region of the pT7T3 phagemid[51] (Pharmacia-LKB, Bromma, Sweden), and sequenced by the dideoxy chain termination method[52] from either single-stranded or double-stranded DNA templates using T7 polymerase and [α-^{32}P]dATP. *Escherichia coli* NM522[53] was the host for pT7T3 phagemid derivatives. The strain was grown on double-strength YT medium supplemented with 100 μg of ampicillin per milliliter for plasmid-containing derivatives and 70 μg of kanamycin per milliliter for the production of single-stranded phagemid DNA. Each of the positive clones isolated contained the same insert, which was found to be a cloning artifact consisting of an 18-kb fragment of λ DNA and a 470-bp portion of *A. fulgidus* DNA encoding the FMFFEKD peptide.

Step 8. Synthesis of DNA Probe. Our failure to isolate more than a single clone from the λ FixII library prompted us to screen a second library. A polymerase chain reaction (PCR)-generated 364-bp *Taq*I fragment from the DNA cloned, which contained only *A. fulgidus* sequence, was used as a hybridization probe. For PCR we use a reagent kit from Perkin-Elmer Cetus (Norwalk, CT). Reaction mixtures contain 5 ng of template DNA, 100 pmol of each oligodeoxynucleotide primer, and 2.5 U of *Taq* polymerase in 100 μl of 10 m*M* Tris-HCl, pH 8.3, 50 m*M* KCl, 1.5 m*M* $MgCl_2$,

[50] D. Chisholm, *Biofeedback* **7,** 21 (1989).

[51] D. A. Mead, E. Szszesna-Skorupa, and B. Kemper, *Protein Eng.* **1,** 67 (1986).

[52] F. Sanger, S. Nicklen, and A. R. Coulson, *Proc. Natl. Acad. Sci. U.S.A.* **74,** 5463 (1977).

[53] D. Hanahan, *J. Mol. Biol.* **166,** 1 (1983).

0.2 m*M* of each deoxynucleotide triphosphate, 0.01% gelatin and are incubated for 25 cycles at 93° for 0.5 min, 55° for 1.0 min, and 72° for 1.5 min. After a final cycle at 93° for 0.5 min, 55° for 1 min, and 72° for 3 min, mixtures are extracted with phenol–chloroform–isoamyl alcohol (25 : 24 : 1) and precipitated with ethanol.

Step 9. Preparation of Library of 3- to 4-kb HindIII Fragments in λ ZAPII Vector. High molecular weight DNA from *A. fulgidus* is completely digested with *Hin*dIII and the resulting DNA fragments are filled in with Klenow fragment and blunt-end ligated to an *Not*I/*Eco*RI adaptor. Size-fractionated fragments of 3 to 4 kb are ligated into λ ZAPII[54] (Stratagene), which has been digested with *Eco*RI and dephosphorylated. After *in vitro* packaging the library is plated on *E. coli* XL1-Blue[55] at a density of 1 × 10^3 PFU/150-cm^2 petri dish. 5-Bromo-5-chloro-3-indolyl-β-D-galactopyranoside (X-Gal) and isopropyl-β-D-thiogalactopyranoside (IPTG) are included in solid media to identify recombinant phage containing inserts in the α portion of *lacZ*. We obtained approximately 5 × 10^5 plaques with an insert frequency of 90%.

Step 10. Screening of λ ZAPII Library. The λ ZAPII library was screened directly without amplification. For screening with homologous DNA probes as in the case here, filters are prehybridized for 5 hr at 42° in 5× SSC (1× SSC is 0.15 *M* NaCl, 15 m*M* sodium citrate, pH 8.0), 1× Denhardt's solution, 0.1% SDS, containing nicked and denatured salmon sperm DNA (200 μg/ml) and 50% formamide. Hybridization is performed for 12–16 hr at 42° in the same solution lacking the salmon sperm DNA but containing sodium pyrophosphate (0.5 mg/ml) and random-primed [^{32}P]DNA probe[56] at 5 × 10^6 cpm/ml. Filters are washed three times for 30 min each at 23° in 2× NET, 0.1% SDS, and then twice for 45 min at 60° in 0.2× NET, 0.1% SDS. Ten positive clones were isolated from the λ ZAPII library and found to contain the same 3.17-kb *Hin*dIII insert.

Step 11. Sequencing of Complete dsrAB Operon. The *Hin*dIII fragment was cloned into pT7T3 and portions were subcloned and sequenced until the nucleotide sequence was completely determined on both strands. The 3175-bp fragment contains two open reading frames (ORFs) of 1257 bp (*dsrA*) and 1101 bp (*dsrB*), encoding peptides of 47.4 and 41.7 kDa, respectively. Comparison of the deduced amino acid sequence with those of the chemically determined peptides showed identity of these peptides with the α and β subunits of sulfite reductase from *A. fulgidus*. The two genes are contiguous in the order *dsrA dsrB* and probably comprise an operon,

[54] J. M. Short, J. M. Fernandez, J. A. Sorge, and W. Huse, *Nucleic Acids Res.* **16,** 7583 (1988).
[55] W. O. Bullock, J. M. Fernandez, and J. M. Short, *BioTechniques* **5,** 376 (1987).
[56] A. P. Feinberg and B. Vogelstein, *Anal. Biochem.* **132,** 6 (1983).

because *dsrA* is preceded by sequences characteristic of promotors in methanogenic archaea, and *dsrB* is followed by a sequence resembling termination signals in extremely thermophilic sulfur-dependent archaea.

Step 12. Analysis of Gene Products. Analyses with the programs BestFit, Compare, DotPlot, and Gap from the Sequence Analysis Software package of the Genetics Computer Group (University of Wisconsin, Madison, WI)[57] showed that *dsrA* and *dsrB* encode peptides that are homologous (25.6% amino acid sequence identity), indicating that the two genes may have arisen by duplication of an ancestral gene. Comparison of the protein sequences with those in the Genetic Sequence Data Bank (GenBank) revealed that each deduced peptide contains two cysteine clusters resembling those postulated to bind siroheme–[4Fe-4S] complexes in sulfite reductases and nitrite reductases from other species.[33] The *dsrB*-encoded peptide lacks a single cysteine residue in one of the two clusters, suggesting that only the α subunit binds a siroheme–[4Fe-4S] complex. Both deduced peptides also contain an arrangement of cysteine residues characteristic of [4Fe-4S] ferredoxins. Four of the six [4Fe-4S] clusters present per $\alpha_2\beta_2$ enzyme probably bind to these ferredoxin-like sites whereas the other two are associated with siroheme.

Cloning of Genes for Adenylylsulfate Reductase

The genes encoding APS reductase from *A. fulgidus* were cloned using virtually the same methods as described for the sulfite reductase genes. N-Terminal sequencing of the purified protein gave 2 different sequences of 24 and 5 residues. These two sequences could unambiguously be distinguished on the basis of their relative yields, which were close to 2 : 1. A mixture of 128 different 20-mer oligonucleotides, which was designed from residues 3 to 9 (YYPKKYE) of the major amino acid sequence, was used to analyze Southern blots of genomic DNA. Hybridizations in 5× SSC, 1× Denhardt's solution, 0.1% SDS, containing nicked and denatured calf thymus DNA (100 μg/ml) at 42° and subsequent washing at 42° in 5× SSC, 0.1% SDS (four times for 15 min each) showed reactivity of a 1.3-kb *Hin*dIII fragment. Therefore a nonamplified λ ZAPII library of 1- to 1.5-kb *Hin*dIII fragments was screened with the same probe. A positive clone was isolated and found to contain a 1.3-kb insert. A portion (697 bp) of the 3′-terminal half of this fragment was sequenced on both strands and found to carry an ORF of 453 bp encoding a 150-residue, 17.1-kDa peptide. Amino acids 2–6 of this peptide match the minor sequence obtained by Edman degradation of native APS reductase, indicating that the

[57] J. Devereux, P. Haeberli, and O. Smithies, *Nucleic Acids Res.* **12,** 387 (1989).

corresponding ORF is the gene for the β subunit (*aprB*). Seventeen base pairs downstream of *aprB* 138 bp of the 5′ terminus of a second ORF were detected encoding the complete chemically determined major N-terminal amino acid sequence of APS reductase, indicating that this truncated ORF was part of the gene for the α subunit (*aprA*). To clone the complete gene the amplified λ Fix II library of *A. fulgidus* (see above) was screened with the radioactively labeled 1.3-kb *Hin*dIII fragment as a probe. A positive clone was isolated that contained a 14-kb insert harboring both complete APS reductase genes.

The nucleotide sequence of both strands of a 2720-bp portion of the phage insert was determined. In addition to *aprB* (residues 73–525) an ORF of 1935 bp (*aprA*) spanning residues 542–2476 was identified that encodes a 644-residue, 73.3-kDa peptide. Both *aprB* and *aprA* are immediately preceded by putative ribosome-binding sites. Although probable promotor sequences could not be found in the 72 bp of 5′-untranslated sequence preceding *aprB*, a sequence identical to a putative transcription termination signal is located 11 nucleotides downstream of the *aprA* termination codon.

The amino acid sequence of the *aprA*-encoded polypeptide shows significant overall similarities with the flavoprotein subunits of the succinate dehydrogenases from *E. coli* and *Bacillus subtilis* and the corresponding flavoprotein of *E. coli* fumarate reductase. Part of the homologous peptide stretches could be assigned to domains that are involved in binding of the FAD prosthetic group: an amino acid sequence that is probably involved in binding the ADP portion of FAD is located at the amino terminus of the *aprA*-encoded peptide, whereas a sequence putatively binding the ribityl chain of the cofactor is found around residues 431–441. Because chemical determinations yielded only 1 mol of FAD per mole holoenzyme, the FAD-binding site in each subunit may represent a form of "half of sites reactivity" in the $\alpha_2\beta$-structured APS reductase. The polypeptide encoded by *aprB* represents an iron-sulfur protein, seven cysteine residues of which are arranged in two clusters typical of ligands of the iron-sulfur centers in ([3Fe-4S][4Fe-4S]) seven-iron ferredoxins.[58]

Detection of Genes Homologous to *aprA* and *dsrAB* in Other Organisms

The procedure of heterologous DNA hybridization is a versatile method in the detection of related genes from different species. Its application to the detection and identification of structural genes encoding enzymes of sulfur metabolism shows a considerable degree of homology

[58] G. H. Stout, *J. Biol. Chem.* **263,** 9256 (1988).

among the respective DNA of *A. fulgidus*, the archaeal sulfite reducer *P. islandicum*, and the bacterial sulfate reducer *T. commune*, all of which are extremely thermophilic, and the phototrophic sulfur-oxidizing bacterium *Thiocapsa* (*Tc.*) *roseopersicina*. The conditions described below have been developed empirically in our laboratory.[59]

Probe Preparation

The best results with heterologous hybridizations are obtained when gene internal fragments corresponding to highly conserved regions of the protein are employed as probes. For detection of sulfite reductase genes we therefore use a 384-bp DNA probe derived from *dsrA* spanning a region that encodes both highly conserved siroheme-binding cysteine motifs. Because an *aprB*-derived probe may detect ferredoxin genes unrelated to those encoding iron protein subunits of APS reductases, we currently employ only *aprA* as a probe for APS reductase genes. We use the whole gene because it has not been possible to identify stretches of bases specific for APS reductase genes up to now.

Preparation and Labeling of Probes with Digoxigenin by Polymerase Chain Reaction. As a template for PCR we use 400 ng of *A. fulgidus* DNA that is denatured at 94° for 5 min before the first PCR cycle. Oligonucleotide primers are derived from the nucleotide sequences of *dsrA* and *apr*. The PCR is performed as described above except for replacement of one-tenth of the dTTP by digoxigenin (DIG)-11-dUTP. The DIG-labeled DNA fragment synthesized is purified by agarose gel electrophoresis, elution from the gel, and phenol extraction.

Heterologous Hybridizations

Preparation of Southern Blots. DNA from *P. islandicum* is isolated following the procedure for *A. fulgidus* DNA. DNA from *T. commune* and *Tc. roseopersicina* is prepared by sarcosyl lysis.[60] In all cases the DNA yield is 0.13–0.15 mg/100 mg of cells. After digestion with restriction endonucleases and agarose gel electrophoresis (10 μg of DNA per lane) the DNA is blotted onto nylon membranes and UV irradiated.

Hybridization Conditions. Membranes are prehybridized for 4 hr in 5× SSC, 2% (v/v) blocking reagent (Boehringer GmbH, Mannheim, Ger-

[59] M. Walbröl, M. Molitor, I. Faath, B. Kallinger, C. Dahl, and H. G. Trüper. *Abstr. VAAM-DGHM-Frühjahrstagung Hannover, Bioengineering* **2194,** 80 (1994).

[60] M. Bazaral, D. R. Helinski, *J. Mol. Biol.* **36,** 185 (1968).

many), 0.1% (v/v) *N*-lauroylsarcosine, and 0.2% (w/v) SDS at 55°. The heat-denatured probe DNA is added directly to the prehybridization solution. We use 200–250 ng of DIG-labeled DNA per milliliter of hybridization solution. Hybridizations are performed at 55° for 14–18 hr.

Wash Conditions. Membranes are washed in 2× SSC, 0.1% (w/v) SDS twice for 5 min each at room temperature and twice for 15 min each in 0.1× SSC, 0.1% (w/v) SDS at 55°. Chemiluminescent detection[61] is done with CSPD[62] (Tropix, Inc., Bedford, MA) as a substrate following the protocol provided by the manufacturer.

Results. With the archaeal sulfite reductase probe a single hybridizing band could be detected at 1.5 kb for *Hin*dIII-digested *P. islandicum* DNA. *Thermodesulfobacterium commune* and *Tc. roseopersicina* DNA showed reactivity with the APS reductase probe at 4.0 and 2.6 kb after restriction with *Hin*cII and *Cla*I, respectively. Because of potential artifacts in heterologous hybridization experiments, the identities of detected genes must be independently established. Available approaches include nucleotide sequence analysis and comparison with known sequences or previously chemically determined amino acid sequences of the protein encoded by the gene under investigation.

Acknowledgments

We wish to thank Karl O. Stetter for kindly supplying us with frozen cell material. Support of this work by the Deutsche Forschungsgemeinschaft and by the Fonds der Chemischen Industrie is gratefully acknowledged. C. Dahl wishes to thank Nicholas M. Kredich for help and hospitality extended to her. We thank Rainer Deutzmann and Friedrich Lottspeich for help with peptide sequencing and are grateful to Peter Heisig for helpful discussions.

[61] I. Bronstein, J. C. Voyta, K. G. Lazzari, O. Murphy, B. Edwards, and L. J. Kricka, *BioTechniques* **8,** 310 (1990).

[62] Disodium 3-(4-methoxyspirol{1,2-dioxetane-3,2′-(5′-chloro)tricyclo[3.3.1.$1^{3,7}$]decan}-4-yl) phenyl phosphate.

Section III

Dissimilatory Sulfur Reduction

[24] Sulfur Reductase from Thiophilic Sulfate-Reducing Bacteria

By GUY D. FAUQUE

Introduction

An important contribution to our knowledge of the biological sulfur cycle has been the discovery that some microorganisms are able to use elemental sulfur as terminal electron acceptor.[1,2] Indeed, several genera of eubacteria and archaebacteria are able to gain energy for growth by dissimilatory reduction of elemental sulfur to hydrogen sulfide.[3] The dissimilatory eubacterial sulfur reducers comprise both true (or strict) and facultative microorganisms.[4] The facultative sulfur-reducing bacteria are able to utilize elemental sulfur as a respiratory substrate in the absence of other possible electron acceptors, such as nitrate, sulfite, or sulfate. Two groups of facultative mesophilic sulfur reducers are known and comprise some spirilloid microorganisms and some sulfate-reducing bacteria.[4] The mesophilic spirilloid sulfur-reducing bacteria belong mainly to the genera *Wolinella*, *Campylobacter*, and *Sulfurospirillum* and the properties of their sulfur reductases are treated in [25] of this volume. Even if the majority of sulfate-reducing bacteria cannot grow by sulfur reduction, some thiophilic sulfate reducers are able to use elemental sulfur as an alternative electron acceptor.[4] They belong mainly to the genera *Desulfomicrobium* and *Desulfovibrio*, including *Desulfomicrobium* (*Dsm.*) *baculatum* Norway 4 and DSM 1743,[2] *Desulfovibrio* (*D.*) *gigas*,[2] *D. sapovorans*,[5] *D. fructosovorans*,[6] and "*D. multispirans*."[7]

[1] N. Pfennig and H. Biebl, *Arch. Microbiol.* **110,** 3 (1976).

[2] H. Biebl and N. Pfennig, *Arch. Microbiol.* **112,** 115 (1977).

[3] F. Widdel, *in* "Biology of Anaerobic Microorganisms" (A. J. B. Zehnder, ed.), p. 469. Wiley, New York, 1988.

[4] F. Widdel and N. Pfennig, *in* "The Prokaryotes. A Handbook on the Biology of Bacteria: Ecophysiology, Isolation, Identification, Applications" (A. Balows, H. G. Trüper, M. Dworkin, W. Harder, and K.-H. Schleifer, eds.), 2nd ed., Vol. 4, p. 3379. Springer-Verlag, New York, 1992.

[5] H. J. Nanninga and J. C. Gottschal, *Appl. Environ. Microbiol.* **53,** 802 (1987).

[6] B. Ollivier, R. Cord-Ruwisch, E. C. Hatchikian, and J. L. Garcia, *Arch. Microbiol.* **149,** 447 (1988).

[7] S.-H. He, Ph.D. Thesis, University of Georgia, Athens (1987).

We report here on the purification and characterization of the sulfur reductase from *Dsm. baculatum* Norway 4 and on the oxidative phosphorylation coupled to the dissimilatory reduction of colloidal sulfur by *D. gigas*.

Assay Methods

Preparation of Colloidal Sulfur

Sulfur is the element with the largest number of allotropes, most of which consist of cyclic nonplanar molecules.[8,9] The stable form of elemental sulfur is the S_8 ring, which constitutes orthorhombic crystals and which is nearly insoluble in water (5 μg of S_8 per liter at 25°[10]). Sulfur flower or precipitated sulfur may be utilized for studies of the growth of sulfur-reducing microorganisms but for metabolic or biochemical experiments hydrophilic suspensions of sulfur are necessary. Colloidal sulfur, the "soluble form" of elemental sulfur, is stable at low salt concentrations (less than 100 m*M*) and exists in hydrophilic and hydrophobic forms according to the method of preparation.[11] Two methods have been utilized to prepare hydrophilic colloidal sulfur: (1) the first one follows the method described by Roy and Trudinger.[12] The colloidal suspension of hydrophilic sulfur is obtained from thiosulfate acidification with concentrated H_2SO_4. Ten milliliters of concentrated H_2SO_4 is added dropwise, with constant stirring, to 30 ml of 3 *M* $Na_2S_2O_3$ immersed in an ice bath. The milky suspension of sulfur is then precipitated by the addition of 40 ml of saturated NaCl and the product is collected by centrifugation for 5 min at 10,000 *g*. The precipitate is washed twice by resuspending it in 50 ml of distilled water and reprecipitating with 50 ml of saturated NaCl. The sulfur is finally resuspended in 50 ml of distilled water and dialyzed overnight against 2 liters of distilled water to remove NaCl. This suspension consists of very finely divided or colloidal sulfur particles that floculate on the addition of salts but nevertheless remains highly reactive for 2 months in biological systems. This suspension contains 110 μmol of active elemental sulfur

[8] R. Steudel, *in* "Elemental Sulfur and Related Homocyclic Compounds and Ions" (A. Müller and B. Krebs, eds.), p. 3. Elsevier Science Publ., Amsterdam, 1984.

[9] R. Steudel, *in* "Autotrophic Bacteria" (H. G. Schlegel and B. Bowien, eds.), p. 289. Springer-Verlag, Berlin, 1989.

[10] J. Boulegue, *Phosphorus Sulfur* **5,** 127 (1978).

[11] A. LeFaou, B. S. Rajagopal, L. Daniels, and G. Fauque, *FEMS Microbiol. Rev.* **75,** 351 (1990).

[12] A. B. Roy and P. A. Trudinger, "The Biochemistry of Inorganic Compounds of Sulphur." Cambridge Univ. Press, Cambridge, 1970.

per milliliter; (2) the second method is that described by Fehér,[13] with a slight modification.[14] A stable, monodispersed colloidal sulfur solution is obtained by mixing acidified Na_2S and Na_2SO_3 solutions of the proper concentration. Solutions of 6.4 g of $Na_2S \cdot 9H_2O$ and 7.2 g of $Na_2SO_3 \cdot 7H_2O$, each in 50 ml of distilled water, are prepared separately. Then 1.5 ml of the Na_2SO_3 solution is added slowly to the Na_2S solution. A mixture of 2.7 g of concentrated H_2SO_4 and 10 ml of distilled water is subsequently added in drops with constant stirring up to the point of incipient turbidity (a total of 8 ml of the mixture is needed). Then 5.5 g of concentrated H_2SO_4 is added to the remaining Na_2SO_3 solution and the Na_2S solution is finally poured in with constant stirring. The mixture is allowed to stand for 1 hr and then is centrifuged for 15 min at 22,000 *g*. The precipitate is washed thrice with 200 m*M* NaCl and resuspended in 30 ml of distilled water. The biggest sulfur particles are removed by centrifuging for 5 min at 4000 *g*, and the supernatant may be utilized for biological activity for 4 to 6 weeks.

Assay of Sulfur Reduction Activity

Principle. Sulfur reductase activity is determined manometrically by the Warburg respirometric technique as hydrogen consumption in a system containing hydrogenase, sulfur reductase, and colloidal sulfur according to reactions (1) and (2):

$$H_2 \rightarrow 2H^+ + 2e^- \tag{1}$$

$$2H^+ + 2e^- + S^\circ \rightarrow H_2S \tag{2}$$

This assay requires a large excess of hydrogenase as compared to sulfur reductase. Sulfur reductase activity is calculated from the initial rate of hydrogen utilization. The conditions of the assay are as follows.

Reagents

Potassium phosphate buffer (1 *M*), pH 7.0
NaOH, 10 *N*
Pure periplasmic [Ni-Fe] hydrogenase from *D. gigas*,[15] 1.4 mg/ml (specific activity, 90 μmol of hydrogen evolved per minute per milligram of protein)
Colloidal sulfur (11 μmol)

[13] F. Fehér, *in* "Handbook of Preparative Inorganic Chemistry" (G. Brauer, ed.), p. 341. Academic Press, New York, 1963.

[14] G. Fauque, Doctoral Thesis, University of Aix-Marseille I, France (1979).

[15] E. C. Hatchikian, M. Bruschi, and J. LeGall, *Biochem. Biophys. Res. Commun.* **82**, 451 (1978).

Procedure. The reaction is performed in 15-ml Warburg flasks under hydrogen at 37°. The 3 ml of reaction mixture in the manometric vessel contains 100 μmol of potassium phosphate buffer (pH 7.0), 40 μg of pure periplasmic *D. gigas* hydrogenase, and 14 nmol of sulfur reductase in the main compartment. The center well receives 0.1 ml of 10 *N* NaOH and the side arm contains 0.1 ml of colloidal sulfur (11 μmol). After preincubation under a hydrogen atmosphere for 30 min, the reaction is initiated by the addition of colloidal sulfur in the main compartment. The hydrogen uptake is measured every minute.

Hydrogen sulfide produced during the reaction is scavenged by 10 *N* NaOH in the center well. To determine the stoichiometry of the reaction, the NaOH solution is removed from the center well and its sulfide content determined using the colorimetric method of Fogo and Popowsky[16] as modified by Lovenberg *et al.*[17]

Definition of Unit and Specific Activity. One unit of sulfur reductase activity is defined as the amount of enzyme that consumes 1 μmol of hydrogen per minute under the assay conditions. Specific sulfur reductase activity is expressed in terms of micromoles of hydrogen consumed per minute per milligram of protein. The protein content of all cell fractions is determined by the method of Lowry *et al.*,[18] using bovine serum albumin as the standard.

Measurement of Oxidative Phosphorylation. Cultures of *Dsm. baculatum* Norway 4 and *D. gigas* are grown in a standard lactate-sulfate medium (see Growth Conditions, below). Membranes are isolated from the crude extracts using differential sucrose gradient centrifugation according to the method described by Barton *et al.*[19] Anaerobic energy coupling is assessed by determining the amount of ATP synthesized for each micromole of molecular hydrogen oxidized using standard manometric procedures. Esterification of P_i is measured by using the hexokinase system as described by Barton *et al.*[19]

Purification Procedure of Sulfur Reductase from *Desulfomicrobium baculatum* Norway 4

Because four times as much hydrogen sulfide is produced when elemental sulfur replaces sulfate as terminal electron acceptor, according to reactions (3) and (4),

[16] J. K. Fogo and M. Popowsky, *Anal. Chem.* **21,** 732 (1949).

[17] W. Lovenberg, B. B. Buchanan, and J. C. Rabinowitz, *J. Biol. Chem.* **238,** 3899 (1963).

[18] O. H. Lowry, N. J. Rosebrough, A. L. Farr, and B. J. Randall, *J. Biol. Chem.* **193,** 265 (1951).

[19] L. L. Barton, J. LeGall, and H. D. Peck, Jr., *Biochem. Biophys. Res. Commun.* **41,** 1036 (1970).

$$8[H] + 4S^\circ \rightarrow 4H_2S \quad (3)$$
$$8[H] + H_2SO_4 \rightarrow H_2S + 4H_2O \quad (4)$$

the maximal sulfide concentration tolerated by the cells is reached faster and at a lower cell density with elemental sulfur than with sulfate-grown cultures.[2] The sulfur reductase being a constitutive enzyme in *Dsm. baculatum* Norway 4 (see Catalytic Properties below), we decided to utilize a lactate-sulfate medium for its growth rather than a lactate (or ethanol)-elemental sulfur medium.

Growth Conditions

Desulfomicrobium baculatum Norway 4 (NCIB 8310, DSM 1741) is grown under anaerobic conditions at 35° in a standard lactate-sulfate medium containing the following (per liter of distilled water): 6.2 ml of sodium lactate [60% (w/v) solution], 2 g of $MgSO_4 \cdot 7H_2O$, 1 g of NH_4Cl, 4 g of $Na_2SO_4 \cdot 10H_2O$, 0.5 g of K_2HPO_4, 1 g of yeast extract (Difco, Detroit, MI), and 1 ml of a trace mineral solution. The trace element solution is modified from that described by Bauchop and Elsden[20] and contains (per liter of distilled water) 10.75 g of MgO, 2 g of $CaCO_3$, 6 g of $FeC_6H_5O_7$, 1.44 g of $ZnSO_4 \cdot 7H_2O$, 1.12 g of $MnSO_4 \cdot 4H_2O$, 0.25 g of $CuSO_4 \cdot 5H_2O$, 0.28 g of $CoCl_2 \cdot 6H_2O$, 0.05 g of $Na_2SeO_3 \cdot 5H_2O$, 0.05 g of $NiCl_2 \cdot 6H_2O$, 0.10 g of $(NH_4)_6Mo_7O_{24} \cdot 4H_2O$, 0.06 g of H_3BO_3, and 51.3 ml of concentrated HCl. The pH is adjusted to 7.2 and the medium is autoclaved for 30 min at 120°. The medium is then supplemented with 2.5 ml/liter of a 5% (w/v) sodium sulfide solution as the reducing agent. Bacterial growth is monitored by measuring the increase in absorbance at 450 nm. The purity of the strain is checked by microscopic examination of the cultures. For short-term preservation, exponentially growing cultures are cooled to 4° in ice water and then stored at 2–4°. Maintenance transfer of cultures should be done every 6 to 8 weeks.

Large-scale cultures of *Dsm. baculatum* Norway 4 are obtained by growing the bacteria under an atmosphere of nitrogen in a 200-liter Biogen fermentor (American Sterilizer Company) inoculated with a 20-liter flask of growing cells. The cells are harvested at the end of the exponential growth phase (after 44 hr) by centrifugation in a Sharples (Rueil Malmaison, France) centrifuge (model AS 16) and stored as a wet paste at −60° until use. Approximately 0.25 g of cell protein (about 1.2 g wet weight) is obtained per liter of growth medium.

The same lactate-sulfate medium has been utilized for the growth of the other strains of sulfate-reducing bacteria mentioned in this chapter: *D. gigas* (NCIB 9332, DSM 1382), *D. vulgaris* Hildenborough (NCIB

[20] T. Bauchop and S. R. Elsden, *J. Gen. Microbiol.* **23,** 457 (1960).

8303, DSM 644), *D. desulfuricans* Berre-Eau (NCIB 8387), *D. salexigens* British Guiana (NCIB 8403, ATCC 14822, DSM 2638), *D. africanus* Benghazi 1 (NCIB 8401, DSM 2603), "*D. multispirans*" (NCIB 12078), and *Dsm. baculatum* DSM 1743 [formerly called *D. baculatus* 9974 (an isolate from the mixed culture "*Chloropseudomonas ethylica* N_2")]. *Desulfuromonas* (*Drm.*) *acetoxidans* 5071 (DSM 1675) is cultured on the basal salt medium described by Pfennig and Biebl,[1] containing 0.05% (v/v) ethanol as carbon and energy source and 0.2% (v/v) DL-sodium malate as terminal electron acceptor.

Preparation of Crude Extract

Six hundred grams (wet weight) of frozen packed *Dsm. baculatum* Norway 4 cells are softened overnight in the cold room and homogenized with 2 vol (w/v) of 10 m*M* Tris(hydroxymethyl)aminomethane hydrochloride (Tris-HCl) buffer, pH 7.6, to which 10 mg of deoxyribonuclease (type I from bovine pancreas; Sigma, St. Louis, MO) is added. The cells are disrupted by two passages through a continuous French pressure cell at about 15,000 psi. The extract is centrifuged at 4° for 1 hr at 25,000 *g* and the tightly packet pellet is resuspended in about 150 ml of 10 m*M* Tris-HCl buffer, pH 7.6, and centrifuged under the same conditions. The crude extract is obtained after centrifugation of the combined supernatants at 40,000 *g* for 1.5 hr at 4°.

Purification Procedure

All operations of the purification procedure are carried out aerobically at 4° and Tris-HCl or potassium phosphate buffers, pH 7.6, at appropriate molarities are utilized. Sulfur reductase is monitored in the extracts both by its enzymatic activity and its purity coefficient, which is defined as $[A_{553\ nm(reduced)} - A_{570\ nm(reduced)}]/A_{280\ nm(oxidized)}$. The absorbances at 553 and 570 nm in the reduced state are measured in the presence of a slight excess of sodium dithionite, obtained by adding a few crystals of the solid. The purification of sulfur reductase from *Dsm. baculatum* Norway 4 is performed essentially as described by Fauque.[14] The sulfur reductase has been purified in four chromatographic steps.

Step 1: Silica Gel Fractionation. The crude extract is stirred overnight with 150 ml of silica gel [Merck (Rahway, NJ) type 60 GF_{254}] equilibrated with 10 m*M* Tris-HCl buffer and the gel with the cytochromes adsorbed is separated by decantation. The unadsorbed protein fraction shows the two absorption peaks at 545 and 580 nm characteristic of desulforubidin, the dissimilatory bisulfite reductase present in *Dsm. baculatum* Norway 4 (see [18], this volume). The silica gel is washed several times with 5 vol of 10 m*M* Tris-HCl buffer until the supernatant is free of proteins. The cytochrome fraction is then eluted from the gel with 1 *M* potassium monohydrogen phosphate containing 1 *M* NaCl. The eluate is dialyzed for 12 hr against 10 liters of distilled water with two changes of dialysate (this volume is sufficient to bring the concentration of phosphate ions to about 10 m*M*).

Step 2: DEAE-Cellulose Column Chromatography. A glass column (diameter, 4.5 cm; height, 30 cm) is packed with diethylaminoethyl (DEAE)-cellulose (DE-52; Whatman, Clifton, NJ) and equilibrated with 10 m*M* Tris-HCl buffer. The dialyzed solution containing the cytochromes from step 1 (1120 ml with a purity coefficient of 0.84 and a specific activity of 0.49 units/mg protein) is applied to the column, which is then washed with 500 ml of 10 m*M* Tris-HCl buffer until the effluent is nearly colorless. During the washing a cytochrome fraction containing mainly the monohemic split α cytochrome $c_{553(550)}$ is obtained.[21] The column is then eluted with a discontinuous gradient (900 ml) consisting of 150-ml aliquots of 50, 100, 150, 200, 250, and 300 m*M* Tris-HCl buffer. The cytochrome fraction containing the sulfur reductase is eluted at about 150 m*M* Tris-HCl buffer with a purity index of 1.80. Its specific activity, equal to 2.11 units/mg protein, is now about 30 times that of the crude extract (Table I).

Step 3: Alumina Column Chromatography. The previous sulfur reductase-containing fraction is dialyzed for 8 hr against 10 liters of 10 m*M* Tris-HCl buffer with one change of dialysate and subsequently adsorbed onto a calcined alumina column (Merck type 60 GF_{254}; diameter, 5 cm; height, 11 cm) equilibrated with 10 m*M* Tris-HCl buffer. After washing with 200 ml of 10 m*M* Tris-HCl buffer, the column is eluted with a discontinuous gradient (975 ml) of potassium phosphate buffer with 75-ml aliquots in 20 m*M* increments from 10 to 250 m*M*. The sulfur reductase elutes at about 170 m*M* potassium phosphate buffer with a purity coefficient of 2.45 and a specific activity of 3.63 units/mg protein.

Step 4: Sephadex G-50 Gel Filtration. In the last step of purification, the sulfur reductase is filtered in two passages through a Sephadex G-50 column (diameter, 4.5 cm; height, 100 cm) equilibrated with 10 m*M* Tris-HCl buffer. This procedure yields a red-colored sulfur reductase preparation that exhibits a purity index of 3.20. The yield is 162 mg of pure sulfur reductase from 600 g of cells with a specific activity of 10.56 μmol of hydrogen consumed per minute per milligram of protein. Its specific activity is 167.6 times that of the crude extract (Table I). The sulfur reductase activity is purified together with the tetraheme cytochrome c_3 with a yield of 59.9%. Main data obtained at each step in the purification of *Dsm.*

[21] G. Fauque, M. Bruschi, and J. LeGall, *Biochem. Biophys. Res. Commun.* **86,** 1020 (1979).

TABLE I
PURIFICATION OF SULFUR REDUCTASE FROM *Desulfomicrobium baculatum* NORWAY 4

Purification step	Total protein (mg)	Total activity (units)	Specific activity (units/mg protein)	Recovery (%)	Purification (fold)	Purity index[a]
Crude extract	45,333	2,856	0.063	100	1.0	ND[b]
Silica gel fractionation	4,052	1,985	0.49	69.5	7.8	0.84
DEAE-cellulose column	882	1,861	2.11	65.2	33.5	1.80
Alumina column	492	1,786	3.63	62.5	57.6	2.45
Sephadex G-50 column	162	1,710	10.56	59.9	167.6	3.20

[a] The purity index is defined as $[A_{553\ \mathrm{nm(reduced)}} - A_{570\ \mathrm{nm(reduced)}}]/A_{280\ \mathrm{nm(oxidized)}}$.
[b] ND, Not determined.

baculatum Norway 4 sulfur reductase are summarized in Table I. Similar results have been obtained with the sulfur reductase from *Dsm. baculatum* DSM 1743, which has been purified to 156 times its original level, with a purity coefficient of 3.10 and a specific activity of 10.80 μmol of hydrogen utilized per minute per milligram of protein.[14]

Properties of *Desulfomicrobium baculatum* Norway 4 Sulfur Reductase (Tetraheme Cytochrome c_3)

Purity and Stability. The procedure for purification of *Dsm. baculatum* Norway 4 sulfur reductase described above yields a protein that is homogeneous by the following criteria[22]: (1) electrophoresis in a sodium dodecyl sulfate-10% (w/v) polyacrylamide gel at pH 8.8, (2) the presence of a single symmetrical peak on elution from a Sephadex G-50 column, (3) the presence of a single band on 7% (w/v) polyacrylamide gel electrophoresis at pH 8.9, and (4) the presence of a single protein species migrating with an apparent isoelectric point of 7.0 on isoelectric focusing. Purified sulfur reductase can be stored for a few months at −20° or longer at −80° without any effect on its catalytic or spectral properties.

Physicochemical Properties. The molecular weight of the native tetraheme cytochrome c_3 was estimated to be ~16,000 by gel filtration on a Sephadex G-50 column.[22] The minimum molecular weight calculated from the amino acid composition is 15,066, including a content of 4 hemes per molecule. The average value of iron content determined by plasma emission spectroscopy is 4.2 iron atoms per protein molecule. This tetraheme cytochrome c_3 contains 118 amino acid residues and possesses the required number of cysteines (eight residues) to bind four hemes per molecule as well as the required number of histidines (eight residues), because each iron coordination complex is of the bishistidinyl type.[23] The amino acid sequence[24] and the three-dimensional structure[25–27] with the relative orientation of the four hemes of the sulfur reductase are known (see [9], this volume). The four hemes of *Dsm. baculatum* Norway 4 tetraheme cytochrome c_3 have four different midpoint redox potentials

[22] G. Fauque, Doctorat d'Etat Thesis, University of Technology of Compiègne, France (1985).

[23] G. Fauque, J. LeGall, and L. L. Barton, *in* "Variations in Autotrophic Life" (J. M. Shively and L. L. Barton, eds.), p. 271. Academic Press, London, 1991.

[24] M. Bruschi, *Biochim. Biophys. Acta* **671,** 219 (1981).

[25] R. Haser, M. Pierrot, M. Frey, F. Payan, J. P. Astier, M. Bruschi, and J. LeGall, *Nature (London)* **282,** 806 (1979).

[26] M. Pierrot, R. Haser, M. Frey, F. Payan, and J. P. Astier, *J. Biol. Chem.* **257,** 14341 (1982).

[27] I. B. Coutinho, D. L. Turner, J. LeGall, and A. V. Xavier, *Eur. J. Biochem.* **209,** 329 (1992).

(−150, −300, −330, and −355 mV[28]). the highest value being well separated from the three others.

Effect of Temperature and pH. The optimal temperature of sulfur reductase activity is 37° and the pH optimum is between 7.0 and 7.6 with 75% loss of enzyme activity below pH 6.5 or above pH 8.5.

Spectral Properties. The visible spectrum of the oxidized tetraheme cytochrome c_3 from *Dsm. baculatum* Norway 4 exhibits a broad absorption band centered around 531 nm (β band), a Soret peak with a maximum at 408 nm (γ band), and another broad peak at 351 nm (δ band). The sulfur reductase is not reduced by sodium ascorbate but is fully reduced by sodium dithionite, showing absorption maxima at 552.5 nm (α band), 523 nm (β band), and a Soret peak at 418.5 nm (γ band). The millimolar extinction coefficient values at the maxima absorption of the oxidized and reduced forms of tetraheme cytochrome c_3 are indicated in Table II. The sulfur reductase from *Dsm. baculatum* DSM 1743 presents properties similar to those of the homologous protein from *Dsm. baculatum* Norway 4.[22,29] The main physicochemical and spectral properties of the sulfur reductase from these two strains of *Dsm. baculatum* are listed in Table II.

Catalytic Properties. The purified sulfur reductase is able to utilize colloidal sulfur as terminal electron acceptor but not sulfur flower or precipitated sulfur. The colloidal sulfur prepared according to Roy and Trudinger[12] is 2.5 times more active than sulfur obtained by the method described by Fehér.[14] The theoretical stoichiometry of 1 mol of H_2S produced per mole of H_2 oxidized was obtained using 3.6 μmol of colloidal sulfur, corresponding to reaction (2). The purified enzyme has neither thiosulfate nor tetrathionate reductase activities but is able to reduce a preparation of polysulfides.[22] The sulfur reductase is a constitutive enzyme in *Dsm. baculatum* Norway 4 and DSM 1743,[14] as is also the case for the sulfur reductase from the sulfur-reducing eubacteria *Drm. acetoxidans* DSM 1675, *Wolinella succinogenes* DSM 1740,[30] and *Sulfurospirillum deleyianum* DSM 6946 (formerly called "Spirillum" strain 5175[31,32]).

Mechanism of Attack of Colloidal Sulfur. An important turbidity occurs in the manometric vessels during the lag phase preceding the hydrogen oxidation and the appearance of this precipitate is essential for the starting

[28] B. Guigliarelli, P. Bertrand, C. More, R. Haser, and J. P. Gayda, *J. Mol. Biol.* **216,** 161 (1990).

[29] I. Moura, M. Teixeira, B. H. Huynh, J. LeGall, and J. J. G. Moura, *Eur. J. Biochem.* **176,** 365 (1988).

[30] G. D. Fauque, O. Klimmek, and A. Kröger, this volume [25].

[31] A. Zöphel, M. C. Kennedy, H. Beinert, and P. M. H. Kroneck, *Arch. Microbiol.* **150,** 72 (1988).

[32] W. Schumacher, P. M. H. Kroneck, and N. Pfennig, *Arch. Microbiol.* **158,** 287 (1992).

TABLE II
PHYSICOCHEMICAL AND SPECTRAL PROPERTIES OF *Desulfomicrobium baculatum* SULFUR REDUCTASE

Property	*Desulfomicrobium baculatum*	
	Norway 4[a]	DSM 1743[a]
Isoelectric point	7.0	7.0
Molecular weight	15,066	15,500
Redox potential (mV)	−150, −300, −330, −355[b]	−70, −280, −300, −355[c]
Extinction coefficient (mM^{-1} cm^{-1})		
Oxidized form		
β band	49.8	50.4
γ band	568.9	617.3
δ band	116.7	123.9
Reduced form		
α band	130.2	144.4
β band	66.7	72.4
γ band	911.2	975.9

[a] G. Fauque, Doctorat d'Etat Thesis, University of Technology of Compiègne, France (1985).
[b] B. Guigliarelli, P. Bertrand, C. More, R. Haser, and J. P. Gayda, *J. Mol. Biol.* **216,** 161 (1990).
[c] I. Moura, M. Teixeira, B. H. Huynh, J. LeGall, and J. J. G. Moura, *Eur. J. Biochem.* **176,** 365 (1988).

of the reduction of colloidal sulfur by *Dsm. baculatum* Norway 4 tetraheme cytochrome c_3.[33] To determine if this turbidity is associated with the formation of a complex between colloidal sulfur and tetraheme cytochrome c_3 an electron paramagnetic resonance (EPR) spectroscopic study was carried out but failed to provide any clue to the identity of the sulfur-binding site.[34] During this reaction there is also a radical, with a g value of 2.02, which is sometimes observed in the EPR spectrum and could be a transient sulfur radical species. The following mechanism of attack of colloidal sulfur has been proposed[34]: the polysulfide chains of elemental sulfur are rapidly attacked by reduced tetraheme cytochrome c_3, leading to a collapse of the micelles with the precipitation of S_8 molecules; this would explain the appearance of the abundant turbidity at the beginning of the reaction. The sulfide that has been produced by reduction of the polysulfides opens up the S_8 rings by a nucleophilic attack, leading again to the production of new molecules of polysulfides, which are themselves quickly reduced to sulfide by tetraheme cytochrome c_3.

[33] G. Fauque, D. Hervé, and J. Le Gall, *Arch. Microbiol.* **121,** 261 (1979).
[34] R. Cammack, G. Fauque, J. J. G. Moura, and J. LeGall, *Biochim. Biophys. Acta* **784,** 68 (1984).

Reduction of Colloidal Sulfur by Different Tetraheme Cytochromes c_3 from *Desulfovibrio* Species

Different tetraheme cytochromes c_3 from *Desulfovibrio* and *Desulfomicrobium* species have been tested in the reduction of colloidal sulfur. The tetraheme cytochrome c_3 from *Dsm. baculatum* Norway 4 and DSM 1743 is the most active in the colloidal sulfur reduction (see above). The tetraheme cytochrome c_3 is also the sulfur reductase of several strains of the genus *Desulfovibrio*,[11] such as *D. gigas*, "*D. multispirans*," *D. africanus* Benghazi 1, *D. desulfuricans* Berre-Eau, and *D. salexigens* British Guiana. Some properties of tetraheme cytochrome c_3 acting as sulfur reductase in *Desulfovibrio* and *Desulfomicrobium* species are presented in Table III. In contrast, a strong inhibition by the product of the reaction, namely hydrogen sulfide, was observed with the tetraheme cytochrome c_3 from *D. vulgaris* Hildenborough.[33] This fact is sufficient to explain why this strain is unable to grow using elemental sulfur instead of sulfate as a respiratory substrate. This shows that a drastic change in the function of the tetraheme cytochrome c_3 took place in *D. vulgaris* Hildenborough where it has only the role of an electron carrier in the electron transfer

TABLE III

PROPERTIES OF SULFUR REDUCTASE (TETRAHEME CYTOCHROME c_3) FROM *Desulfovibrio* AND *Desulfomicrobium* SPECIES

Species	Purity index	Isoelectric point	Number of amino acids	Specific activity[a]
Dsm. baculatum DSM 1743[b]	3.10	7.0	120	10.80
Dsm. baculatum Norway 4[b]	3.20	7.0	118	10.56
D. gigas[c]	2.94	5.2	111	9.10
"*D. multispirans*"[d]	3.35	ND[e]	103	8.25
D. africanus Benghazi[f]	2.90	8.5	109	6.70
D. desulfuricans Berre-Eau[g]	3.22	8.6	ND	5.95
D. salexigens British Guiana[h]	2.73	10.8	106	4.70

[a] Specific activity is expressed in micromoles of hydrogen consumed per minute per milligram of protein.

[b] G. Fauque, Doctorat d'Etat Thesis, University of Technology of Compiègne, France (1985).

[c] J. LeGall, G. Mazza, and N. Dragoni, *Biochim. Biophys. Acta* **99,** 385 (1965).

[d] S.-H. He, Ph.D. Thesis, University of Georgia, Athens (1987).

[e] ND, Not determined.

[f] R. Singleton, L. L. Campbell, and F. M. Hawkridge, *J. Bacteriol.* **140,** 893 (1979).

[g] I. Moura, G. Fauque, J. LeGall, A. V. Xavier, and J. J. G. Moura, *Eur. J. Biochem.* **162,** 547 (1987).

[h] H. Drucker, E. B. Trousil, and L. L. Campbell, *Biochemistry* **9,** 3395 (1970).

chain instead of being a terminal reductase. The sensibility of tetraheme cytochrome c_3 from *D. vulgaris* Hildenborough to sulfide has been utilized as a possible explanation for the fact that this strain diverts an important amount of its reducing power to the reduction of protons instead of using it exclusively for the dissimilatory reduction of sulfate.[35]

Other *c*-type cytochromes from *Desulfovibrio* and *Desulfomicrobium* species are not active in the reduction of colloidal sulfur; such is the case of the monohemic cytochrome c_{553} from *D. vulgaris* Hildenborough and cytochrome $c_{553(550)}$ from *Dsm. baculatum* Norway 4 and DSM 1743 and also of the octaheme cytochrome c_3 from *D. gigas* and *D. vulgaris* Hildenborough.[22] The triheme cytochrome c_3 (previously named cytochrome c_7 or cytochrome $c_{551.5}$) isolated from the obligate anaerobic sulfur-reducing eubacterium *Drm. acetoxidans* DSM 1675[36] presents obvious structural similarity to *Desulfovibrio* and *Desulfomicrobium* tetraheme cytochrome c_3.[37] This triheme cytochrome c_3 is unable to reduce colloidal sulfur itself; the complete reaction requires another cytochrome fraction from *Drm. acetoxidans*. The loss of sulfur reduction activity could be due to the absence of very low redox potential hemes in *Drm. acetoxidans* triheme cytochrome c_3.[38]

Oxidative Phosphorylation Linked to Dissimilatory Reduction of Colloidal Sulfur by *Desulfovibrio gigas*

The sulfate-reducing bacteria of the genus *Desulfovibrio* are the first nonphotosynthetic obligate anaerobic microorganisms in which phosphorylation coupled to electron transfer has been clearly demonstrated.[39] With the membrane fraction of *D. gigas*, oxidative phosphorylation was shown to be associated with the oxidation of molecular hydrogen and the concomitant reduction of sulfite,[40] thiosulfate,[40] fumarate,[41] nitrite,[42] and hydroxylamine.[42] The different electron donor–acceptor systems that can drive ATP synthesis in *D. gigas* and the $P/2e^-$ values obtained with the different

[35] A. S. Traoré, C. E. Hatchikian, J. P. Belaich, and J. Le Gall, *J. Bacteriol.* **145,** 191 (1981).

[36] I. Probst, M. Bruschi, N. Pfennig, and J. LeGall, *Biochim. Biophys. Acta* **460,** 58 (1977).

[37] J. LeGall and G. Fauque, *in* "Biology of Anaerobic Microorganisms" (A. J. B. Zehnder, ed.), p. 587. Wiley, New York, 1988.

[38] M. D. Fiechtner and R. J. Kassner, *Biochim. Biophys. Acta* **579,** 269 (1979).

[39] H. D. Peck, Jr., *J. Biol. Chem.* **235,** 2734 (1960).

[40] H. D. Peck, Jr., *Biochem. Biophys. Res. Commun.* **22,** 112 (1966).

[41] L. L. Barton, J. Le Gall, and H. D. Peck, Jr., *in* "Horizons of Bioenergetics" (A. San Pietro and H. Gest, eds.), p. 33. Academic Press, New York and London, 1972.

[42] L. L. Barton, J. LeGall, J. M. Odom, and H. D. Peck, Jr., *J. Bacteriol.* **153,** 867 (1983).

TABLE IV
COMPARISON OF PHOSPHORYLATION ACTIVITY COUPLED TO FIVE RESPIRATORY SYSTEMS OF *Desulfovibrio gigas*

Reaction	$\Delta G^{\circ\prime}$ [a] (kcal/mol)	H_2 oxidized (μmol)	P_i esterified (μmol)	$P/2e^-$	Ref.
$\text{Fumarate}^{2-} + 2H_2 \rightarrow \text{succinate}^{2-}$	−20.6	14.3	4.89	0.34	*b*
$NO_2^- + 2H^+ + 3H_2 \rightarrow NH_4^+ + 2H_2O$	−104.3	4.0	1.27	0.32	*c*
$HSO_3^- + 3H_2 \rightarrow HS^- + 3H_2O$	−41.0	13.9	1.82	0.13	*d*
$S^\circ + H_2 \rightarrow HS^- + H^+$	−6.7	2.7	0.26	0.10	*e*
$NH_2OH + H^+ + H_2 \rightarrow NH_4^+ + H_2O$	−59.4	14.8	1.12	0.08	*c*

[a] R. K. Thauer, K. Jungermann, and K. Decker, *Bacteriol. Rev.* **41,** 100 (1977).
[b] L. L. Barton, J. LeGall, and H. D. Peck, Jr., *Biochem. Biophys. Res. Commun.* **41,** 1036 (1970).
[c] L. L. Barton, J. LeGall, J. M. Odom, and H. D. Peck, Jr., *J. Bacteriol.* **153,** 867 (1983).
[d] H. D. Peck, Jr., *Biochem. Biophys. Res. Commun.* **22,** 112 (1966).
[e] G. D. Fauque, L. L. Barton, and J. LeGall, *in* "Sulphur in Biology," Ciba Foundation Symposium 72, p. 71. Excerpta Medica, Amsterdam, 1980.

substrates are listed in Table IV. The system giving the greatest values for $P/2e^-$ is the H_2–fumarate couple, and the H_2–NH_2OH system gives the lowest values.

Hydrogenase and tetraheme cytochrome c_3 from *Dsm. baculatum* Norway 4 and *D. gigas* form a soluble complex that is capable of transferring electrons from molecular hydrogen to colloidal sulfur. Because both strains are able to grow using elemental sulfur as terminal electron acceptor, it was of interest to check for oxidative phosphorylation in this sulfur reduction system. Membranes isolated from *Dsm. baculatum* Norway 4 and *D. gigas* contain some tetraheme cytochrome c_3 and hydrogenase.[43] The addition of these membranes to colloidal sulfur results in hydrogen oxidation. If the $\Delta G^{\circ\prime}$ value of −6.7 kcal/mol[44] is correct for reaction (5)

$$H_2 + S^\circ \rightarrow HS^- + H^+ \tag{5}$$

sufficient energy is released to account for the production of one ATP for two electrons. The particulate fraction from *D. gigas* was observed to couple esterification of P_i to electron flow from molecular hydrogen to colloidal sulfur.[43] A $P/2e^-$ ratio of 0.10 was obtained (Table IV), a value similar to the one reported with the hydrogen sulfite system. It is to be noted that the electron transfer chain from hydrogen to colloidal sulfur can be reconstituted using two soluble components, namely tetraheme

[43] G. D. Fauque, L. L. Barton, and J. LeGall, *in* "Sulphur in Biology," Ciba Foundation Symposium 72, p. 71. Excerpta Medica, Amsterdam, 1980.
[44] R. K. Thauer, K. Jungermann, and K. Decker, *Bacteriol. Rev.* **41,** 100 (1977).

cytochrome c_3 and hydrogenase, whereas similar attempts to reconstitute either fumarate or sulfite reduction have failed so far.

Uncoupling of phosphorylation was observed with pentachlorophenol and by the addition of *D. gigas* tetraheme cytochrome c_3 or methyl viologen to the membrane fraction.[43] The effect of adding these two soluble electron carriers may be explained by a short-circuiting of the membraneous electron transfer system with a concomitant loss of coupling activity.

Conclusion

Tetraheme cytochrome c_3 is the sulfur reductase of some strains of thiophilic sulfate-reducing bacteria from which the sulfur reductase activity can be purified together with the tetraheme cytochrome c_3. This hemoprotein acts as a terminal reductase for the dissimilatory reduction of elemental sulfur in some *Desulfovibrio* and *Desulfomicrobium* species. Membrane preparations of *D. gigas* are able to couple esterification of P_i to electron flow from molecular hydrogen to elemental sulfur. Sufficient tetraheme cytochrome c_3 and hydrogenase must be associated with the *D. gigas* cytoplasmic membrane in the correct conformation to generate proton translocation sufficient for chemiosmotic synthesis of ATP. The thiophilic sulfate-reducing bacteria have a broader ecological valence and, therefore, represent a different physiological–ecological group. These species have a selective advantage in ecological niches in which elemental sulfur is available either directly or as a metabolic product of phototrophic green sulfur bacteria.

[25] Sulfur Reductases from Spirilloid Mesophilic Sulfur-Reducing Eubacteria

By Guy D. Fauque, Oliver Klimmek, and Achim Kröger

Introduction

Several genera of archaebacteria and eubacteria have been reported to be able to gain energy for their growth by a dissimilatory reduction of elemental sulfur to hydrogen sulfide in a respiratory type of metabolism.[1–3]

[1] F. Widdel, *in* "Biology of Anaerobic Microorganisms" (A. J. B. Zehnder, ed.), p. 469. Wiley, New York, 1988.

This chapter focuses only on the dissimilatory sulfur-reducing eubacteria, which are physiological assemblages of mostly anaerobic microorganisms comprising both true (or obligate) and facultative bacteria. The true or strict dissimilatory sulfur reducers utilize elemental sulfur as the only inorganic electron acceptor and belong to the mesophilic genus *Desulfuromonas*[4] (type species *Desulfuromonas acetoxidans*) and the moderately thermophilic genus *Desulfurella*[5]; this category of microorganisms is not discussed in this chapter. The facultative dissimilatory sulfur-reducing bacteria utilize elemental sulfur in the absence of other potential electron acceptors such as sulfite, sulfate, or nitrate. The facultative mesophilic sulfur reducers are represented by two groups of bacteria: some spirilloid non-sulfate-reducing microorganisms and some sulfate-reducing bacteria.[6,7] The properties of sulfur reductase from thiophilic sulfate reducers are described in [24] of this volume. The facultatively microaerobic, spirilloid sulfur reducers comprise the "free-living *Campylobacter* species" of Laanbroek *et al.*,[8] *Sulfurospirillum deleyianum* (formerly called "Spirillum" strain 5175[9]) and *Wolinella succinogenes*.[10]

We report in this chapter on the composition of the electron transport chain catalyzing sulfur reduction in *W. succinogenes* and on the purification and preliminary characterization of the *S. deleyianum* sulfur oxidoreductase.

Electron Transport Chain Catalyzing Sulfur Reduction in *Wolinella succinogenes*

Wolinella succinogenes (DSM 1740) can grow at the expense of elemental sulfur reduction by molecular hydrogen [reaction (1)].

---

[2] F. Widdel and N. Pfennig, *in* "The Prokaryotes. A Handbook on the Biology of Bacteria: Ecophysiology, Isolation, Identification, Applications" (A. Balows, H. G. Trüper, M. Dworkin, W. Harder, K.-H. Schleifer, eds.), 2nd Ed., Vol. 4, p. 3379. Springer-Verlag, New York, 1992.

[3] A. LeFaou, B. S. Rajagopal, L. Daniels, and G. Fauque, *FEMS Microbiol Rev.* **75,** 351 (1990).

[4] N. Pfennig and H. Biebl, *Arch. Microbiol.* **110,** 3 (1976).

[5] E. A. Bonch-Osmolovskaya, T. G. Sokolova, N. A. Kostrikina, and G. A. Zavarzin, *Arch. Microbiol.* **153,** 151 (1990).

[6] F. Widdel and T. A. Hansen, *in* "The Prokaryotes. A Handbook on the Biology of Bacteria: Ecophysiology, Isolation, Identification, Applications" (A. Balows, H. G. Trüper, M. Dworkin, W. Harder, K.-H. Schleifer, eds.), 2nd Ed. Vol. 1, p. 583. Springer-Verlag, New York, 1992.

[7] H. Biebl and N. Pfennig, *Arch. Microbiol.* **112,** 115 (1977).

[8] H. J. Laanbroek, L. J. Stal, and H. Veldkamp, *Arch. Microbiol.* **119,** 99 (1978).

[9] R. S. Wolfe and N. Pfennig, *Appl. Environ. Microbiol.* **33,** 427 (1977).

[10] J. M. Macy, I. Schröder, R. K. Thauer, and A. Kröger, *Arch. Microbiol.* **144,** 147 (1986).

$$H_2 + S^\circ \rightarrow HS^- + H^+ \quad (1)$$

Hydrogen can be replaced by formate.[10] Growth of the bacteria demonstrates that reaction (1) must be coupled to ATP synthesis. *Wolinella succinogenes* also grows with polysulfide as electron acceptor[11] [reaction (2)].

$$nH_2 + S_{n+1}{}^{2-} \rightarrow (n + 1)HS^- + (n - 1)H^+ \quad (2)$$

Polysulfide is formed from elemental sulfur and sulfide [reaction (3)],

$$nS^\circ + HS^- \rightarrow S_{n+1}{}^{2-} + H^+ \quad (3)$$

and appears to be the soluble intermediate in the sulfur reduction catalyzed by *W. succinogenes*[11,12] [reaction (1)]. The electron transport chain catalyzing reaction (2) is made up of hydrogenase[13] and polysulfide reductase,[14,15] which are integrated in the bacterial membrane (Fig. 1). A quinone is not involved. Hydrogenase and polysulfide reductase are constitutive enzymes[16] and can also be isolated from *W. succinogenes* grown with formate and fumarate.

Using antisera raised against the major subunits of polysulfide reductase (PsrA) and hydrogenase (HydB), the corresponding genes were cloned from a genomic library of *W. succcinogenes.*[13,15] The gene *psrA* forms an operon together with two additional genes (*psrB* and *psrC*) and the promotor in front of *psrA*. Polysulfide reductase A is homologous to six bacterial oxidoreductases that contain molybdenum bound to a pterin dinucleotide.[17] Polysulfide reductase B is predicted to carry [Fe-S] centers and PsrC should be extremely hydrophobic (Table I). The hydrogenase genes form a large operon containing at least eight open reading frames.[13] The three structural genes are located next to the promotor in the order *hydA,B,C*. The derived amino acid sequences are homologous to those predicted from the genes of several other membraneous [Ni-Fe] hydrogenases. Hydrogenases A and B are also homologous to the subunits of the periplasmic [Ni-Fe] hydrogenases of sulfate-reducing bacteria of the genus *Desulfovibrio*.

[11] O. Klimmek, A. Kröger, R. Steudel, and G. Holdt, *Arch. Microbiol.* **155,** 177 (1991).

[12] R. Schauder and A. Kröger, *Arch. Microbiol.* **159,** 491 (1993).

[13] F. Dross, V. Geisler, R. Lenger, F. Theis, T. Krafft, F. Fahrenholz, E. Kojro, A. Duchêne, D. Tripier, K. Juvenal, and A. Kröger, *Eur. J. Biochem.* **206,** 93 (1992).

[14] I. Schröder, A. Kröger, and J. M. Macy, *Arch. Microbiol.* **149,** 572 (1988).

[15] T. Krafft, M. Bokranz, O. Klimmek, I. Schröder, F. Fahrenholz, E. Kojro, and A. Kröger, *Eur. J. Biochem.* **206,** 503 (1992).

[16] J. P. Lorenzen, A. Kröger, and G. Unden, *Arch. Microbiol.* **159,** 477 (1993).

[17] J. L. Johnson, N. R. Bastian, and K. V. Rajagopalan, *Proc. Natl. Acad. Sci. U.S.A.* **87,** 3190 (1990).

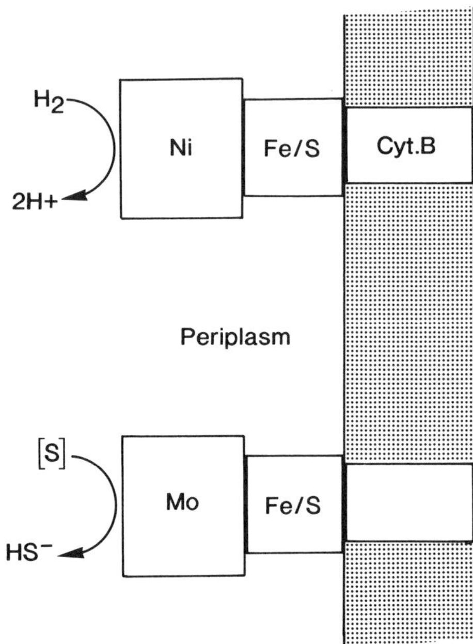

FIG. 1. Proposed composition of the *Wolinella succinogenes* electron transport chain catalyzing reaction.[2] The largest subunits of the enzymes are exposed toward the periplasm and carry the substrate sites as well as nickel (Ni, hydrogenase) and molybdenum (Mo, polysulfide reductase). Fe/S, iron-sulfur centers; Cyt.b, cytochrome b; [S], polysulfide.

TABLE I
PROPERTIES OF POLYSULFIDE REDUCTASE AND HYDROGENASE PROTEINS FROM *Wolinella succinogenes* AS PREDICTED BY GENES

Protein	Number of amino acids	M_r from: Gene	M_r from: SDS–PAGE	Hydropathy index[a]
PsrA[b]	729	81,263	85,000	−0.4
PsrB	191	20,942	23,000	−0.3
PsrC	317	34,131	—	1.0
HydA[b]	318	34,295	30,000	−0.13
HydB	576	63,974	60,000	−0.20
HydC	230	26,828	23,000	0.4

[a] According to J. Kyte and R. F. Doolittle, *J. Mol. Biol.* **157,** 105 (1982).
[b] The size refers to the mature protein.

Here we report on the isolation of polysulfide reductase and of hydrogenase from *W. succinogenes* as well as on the incorporation of these enzymes into liposomes. The incorporation resulted in partial restoration of electron transport activity.

Assay of Enzymatic Activities

Electron Transport. The activity of polysulfide reduction with H_2 [reaction (2)] is measured by recording the absorbance of polysulfide at 360 nm.[11] The extinction coefficient based on polysulfide sulfur (equivalent to 1 mol of H_2/mol) is 0.36 mM^{-1} cm^{-1}. The anaerobic reaction buffer (pH 8.5, 37°) contains 500 mM Tris-HCl and 3 mM Na_2S. The addition of 1 mM $Na_2S_4O_6$ causes the formation of polysulfide according to reaction (4).

$$nS_4O_6^{2-} + (n + 1)HS^- \rightarrow 2nS_2O_3^{2-} + S_{n+1}^{2-} + (n + 1)H^+ \qquad (4)$$

The solution is flushed with H_2, and the reaction is started by addition of the enzyme. One unit of activity corresponds to 1 μmol of hydrogen consumed per minute under the assay conditions.

Polysulfide Reductase. The reduction of 2,3-dimethyl-1,4-naphthoquinone (DMN) by HS^- is recorded at 270–290 nm using a dual-wavelength spectrophotometer ($\Delta\varepsilon$ = 15.0 mM^{-1} cm^{-1}).[11] The anaerobic reaction buffer (pH 7.2, 37°) contains 200 mM triethanolamine, 0.2 mM DMN, and 33 mM Na_2S. The reaction is started by addition of the enzyme. The unit of activity represents 1 μmol of DMN reduced per minute.

Hydrogenase. The reduction of DMN by H_2 is recorded at 270–290 nm using a dual-wavelength spectrophotometer ($\Delta\varepsilon$ = 15.0 mM^{-1} cm^{-1}).[13] The anaerobic reaction buffer (pH 8.1, 37°) contains 50 mM glycylglycine, 0.5 mM dithiothreitol, and 0.2 mM DMN. The buffer is flushed with H_2 and the reaction is started by addition of the enzyme. One hydrogenase unit corresponds to 1 μmol of DMN reduced per minute.

Isolation Procedures

Growth of Wolinella succinogenes (DSM 1740). The growth medium has the following composition: fumaric acid (90 mM), sodium formate (100 mM), K_2HPO_4 (20 mM), $(NH_4)_2SO_4$ (5 mM), NH_4Cl (5 mM), sodium acetate (20 mM), glutamate (1 mM), and Tris-HCl (50 mM). The pH of the solution is adjusted to pH 7.8 by the addition of KOH (about 100 mM). After the addition of $MgCl_2$ (0.25 mM), $CaCl_2$ (0.05 mM), and trace element solution (2 ml/liter), the medium is sterilized, inoculated (0.4%), flushed with N_2, and kept at 37° until the cell density is about 10^{12}/liter (about 17 hr). The trace element solution contains (mg/liter): disodium EDTA (5200), $FeCl_2 \cdot 4H_2O$ (1500), $ZnCl_2$ (70), $MnCl_2 \cdot 4H_2O$ (100), H_3BO_3

(62), $CoCl_2 \cdot 6H_2O$ (190), $CuCl_2 \cdot 2H_2O$ (17), $NiCl_2 \cdot 6H_2O$ (24), and $Na_2MoO_4 \cdot 2H_2O$ (36).

Isolation of Polysulfide Reductase

See Table II for a summary of the following procedures.

Preparation of Triton X-100 Extract. The bacteria are harvested from 40 liters of culture medium, using a filtration system. After the addition of dithiothreitol (1 m*M*) and DNase I (80 mg/liter) the concentrated bacterial suspension (about 0.3 liter) is passed once through a French press at 1500 psi. The "cell homogenate" is centrifuged for 40 min at 280,000 *g* to give the "membrane fraction" (sediment). The sediment is stirred for 90 min in an anaerobic buffer (50 ml, pH 7.8, 0°) containing potassium phosphate (35 m*M*), malonate (2 m*M*), N_3^- (2 m*M*), and dithiothreitol (1 m*M*). The mixture contains 1.5 g of Triton X-100 per gram of bacterial protein and is centrifuged as before to give the Triton X-100 extract (supernatant).

DEAE-Sepharose Cl-6B Chromatography. The Triton X-100 extract so obtained is then applied to a DEAE-Sepharose Cl-6B column (0.6 liter, 50-mm diameter) equilibrated with the buffer described above that contained Triton X-100 (0.3 g/liter). The enzyme elutes after about one column volume of the same buffer is passed through. More than 60% of the enzyme activity present in the Triton X-100 extract is recovered from the column. The fractions containing the enzyme with the highest specific activity are pooled and concentrated eightfold by pressure dialysis under N_2 at 0°.

Density Gradient Centrifugation. The centrifuge tubes (40 ml) contain the buffer described with Triton X-100 (0.3 g/liter) and a sucrose gradient (10–35%). The preparation obtained by DEAE-Sepharose Cl-6B chromatography is layered on top of this buffer. After centrifugation (5.5 hr) at 250,000 *g* in a vertical rotor, the contents of the tubes were fractionated

TABLE II
PURIFICATION OF POLYSULFIDE REDUCTASE FROM *Wolinella succinogenes*

Purification step	Specific activity (units/mg protein)	Total activity (units)	Recovery (%)	Purification (fold)
Cell homogenate	8.5	29,200	100	
Membrane fraction	21	26,800	91.8	2.5
Triton X-100 extract	36	19,300	66.1	4.2
DEAE-Sepharose Cl-6B column	326	4,300	14.7	38.4
Density gradient centrifugation	732	2,800	9.6	86.1

and analyzed for enzymatic activity. The recovery of activity is 9.6%. The fractions containing the enzyme with the highest specific activity are pooled and used for reconstitutional experiments.

Isolation of Hydrogenase

All buffers used, containing 2 m*M* dithiothreitol and 0.2 m*M* phenylmethylsulfonyl fluoride, are flushed with N_2 and kept at 0°. See Table III for a summary of these procedures.

Preparation of Triton X-100 Extract. Wolinella succinogenes [7.5 g of cells (wet weight)] grown with formate and fumarate as described above and stored at −80° is suspended in a buffer (0.1 liter, pH 7.8) containing 25 m*M* Tris-acetate, 10 m*M* ethylenediaminetetraacetic acid (EDTA), and lysozyme (0.5 g/liter). After stirring for 30 min, 15 m*M* $MgCl_2$ and DNase I (10 mg/liter) are added and stirring is continued for 10 min to give the "cell homogenate." The cell homogenate is then centrifuged for 20 min at 9000 *g*. The sediment is suspended in a buffer [50 ml of Tris-acetate (25 m*M*), pH 8.5] containing Triton X-100 (10 g/liter), and the suspension is stirred for 60 min. Centrifugation for 15 min at 160,000 *g* yields the Triton X-100 extract (supernatant).

Chromatofocusing. The Triton X-100 extract (0.3 g of protein) is applied to a chromatofocusing column (40 ml, PBE94; Pharmacia, Piscataway, NJ) equilibrated with a buffer (25 m*M* Tris-acetate, pH 8.5) containing Triton X-100 (0.5 g/liter). After two column volumes of equilibration buffer has been passed through, elution is carried out with 10 vol of a mixture (adjusted to pH 5.0 by addition of acetic acid) containing the polybuffers PB96 (3%) and PB74 (7%) as well as Triton X-100 (0.5 g/liter). Hydrogenase activity elutes as a symmetrical band with the peak at pH 7.6. The fractions containing hydrogenase with the highest specific activity are pooled, concentrated 20-fold by pressure dialysis, diluted with 50 m*M* Tris-acetate (pH 7.8), and stored in liquid N_2.

TABLE III
PURIFICATION OF HYDROGENASE FROM *Wolinella succinogenes*

Purification step	Specific activity (units/mg protein)	Total activity (units)	Purification (fold)	Recovery (%)
Cell homogenate	4.0	5000		100
Triton X-100 extract	24	3950	6.0	79
Chromatofocusing	505	1175	126.3	23.5

Properties of Isolated Enzymes

Properties of Polysulfide Reductase. The polysulfide reductase preparation obtained from the last step of the purification procedure catalyzed the reduction of DMN by sulfide at nearly 90 times the specific activity of the cell homogenate (Table II). The preparation contained molybdenum (6 μmol/g protein) but no cytochromes. The activities of hydrogenase, formate dehydrogenase, fumarate reductase,[18] and nitrite reductase[19] were absent. The subunit of formate dehydrogenase (FdhA) carrying the molybdenum of this enzyme was absent. An earlier preparation of polysulfide reductase contained 120 μmol/g protein of nonheme iron and acid-labile sulfide.[14]

After sodium dodecyl sulfate (SDS)–gel electrophoresis of the final preparation (Table II) and staining with Coomassie blue, five bands could be discriminated, the largest (M_r 85,000) and the smallest (M_r 23,000) of which were identified as PsrA and PsrB by N-terminal sequencing (Table I). Polysulfide reductase C has not been identified so far. As judged from the Coomassie stain, PsrA and PsrB amounted to approximately 70% of the total protein of the preparation and were present at nearly equimolar quantities.

The first 34 amino acid residues of the sequence predicted by *psrA* form a leader peptide that is similar to that encoded by the *fdhA* gene encoding the formate dehydrogenase of *W. succinogenes*.[20] Formate dehydrogenase is homologous to PsrA and likely carries the molybdenum cofactor present in the enzyme together with the substrate site, which is known to be exposed to the periplasmic side of the bacterial membrane.[21] The sequence homology suggests that PsrA carries the substrate site of polysulfide reductase together with the molybdenum cofactor, and is oriented toward the periplasmic side (Fig. 1).

Properties of Hydrogenase. The hydrogenase preparation obtained from the last step of the purification procedure (Table III) consisted of three polypeptides (M_r 30,000, 60,000 and 23,000) which were identified as HydA, HydB, and HydC by N-terminal sequencing[13] (Table I). The preparation contained 6.4 μmol of cytochrome *b* per gram protein. A content of 8.0 μmol of cytochrome *b* per gram protein would be expected if the enzyme were made up of one subunit each of HydA, B, and C

[18] G. Unden and A. Kröger, this series, Vol. 126, p. 387.

[19] I. Schröder, A. M. Roberton, M. Bokranz, G. Unden, R. Böcher, and A. Kröger, *Arch. Microbiol.* **140,** 380 (1985).

[20] M. Bokranz, M. Gutmann, C. Körtner, E. Kojro, F. Fahrenholz, F. Lauterbach, and A. Kröger, *Arch. Microbiol.* **156,** 119 (1991).

[21] A. Kröger, E. Dorrer, and E. Winkler, *Biochim. Biophys. Acta* **589,** 118 (1980).

(Table I) and contained 1 mol of heme b/mol enzyme. Hydrogenase C represents cytochrome *b* and carries the site of DMN reduction. This is concluded from comparison with a hydrogenase preparation that consisted merely of HydA and HydB, did not contain cytochrome *b*, and did not react with DMN.[22] This preparation contained 7.9 μmol of nickel per gram protein and 8 mol of acid-labile sulfide per mole nickel. The N-terminal 36 amino acid residues predicted by *hydA* represent a leader peptide that is similar to that derived from *W. succinogenes psrA* and *fdhA*, as well as to the leader peptides encoded by the *hydA* genes of the other membraneous and periplasmic [Ni-Fe] hydrogenases.

Restoration of Electron Transport Activity

Incorporation of Polysulfide Reductase and Hydrogenase into Liposomes. A suspension of sonic liposomes [10 g of phosphatidylcholine (from soybean) per liter] is prepared in an anaerobic buffer (pH 7.8, 0°) containing 35 m*M* potassium phosphate and 1 m*M* dithiothreitol. Polysulfide reductase (36 mg/g phospholipid), hydrogenase (65 mg/g phospholipid), and Amberlite XAD (Serva, Heidelberg, Germany) (20 g/g Triton X-100) are added to the suspension, and the mixture is gently shaken for 1.5 hr at 0°. After removal of the Amberlite XAD the suspension is frozen in liquid N_2 and thawed at room temperature. Freeze-thawing is repeated once.

Properties of Proteoliposomes. Gel filtration indicated that polysulfide reductase and hydrogenase had been incorporated into liposomes. The phospholipid eluted together with the enzymatic activities as coinciding bands at the void volume of a Sepharose Cl-4B column. When freeze-thawing was done in the presence of labeled glucose, part of the label eluted together with the proteoliposomes. The internal volume of the proteoliposomes calculated from the glucose content was 3.2 ml/g phospholipid, corresponding to 74-nm average inner diameter.

Proteoliposomes containing hydrogenase and polysulfide reductase catalyzed the electron transport from H_2 to polysulfide (Table IV), whereas those containing only one enzyme did not (not shown). When the enzymes were incorporated into liposomes containing vitamin K_1, the activity of electron transport was about the same as without vitamin K_1. Vitamin K_1 was found to be equally effective as the bacterial menaquinone in restoring the electron transport to fumarate.[18]

In intact *W. succinogenes* cells electron transport to polysulfide was half as fast as sulfide oxidation by DMN (Table IV). In the proteoliposomes the electron transport activity to polysulfide amounted to 2.5% of that of

[22] G. Unden, R. Böcher, J. Knecht, and A. Kröger, *FEBS Lett.* **145,** 230 (1982).

TABLE IV
COMPARISON OF ENZYMATIC ACTIVITIES OF PROTEOLIPOSOMES AND OF *Wolinella succinogenes*

Reaction	Activity (U/mg phospholipid)	
	Proteoliposomes	*W. succinogenes*[a]
$H_2 \rightarrow S_{n+1}{}^{2-}$	0.41	15
$HS^- \rightarrow$ DMN	16	30
$H_2 \rightarrow$ DMN	18	24

[a] The bacteria are assumed to contain 0.17 g of phospholipid per gram of protein.

sulfide oxidation by DMN at a hydrogenase activity that was commensurate with that in the bacteria. Thus the degree of restoration of electron transport activity was approximately 5%. The velocity of electron transport was possibly limited by a constituent of the chain, most of which had been removed by the isolation procedures. A likely candidate is PsrC, which may be present at less than 10% the molar amount of PsrA in the preparation of polysulfide reductase. The turnover number of the polysulfide reductase in the electron transport catalyzed by the proteoliposomes was nearly 2×10^3 min^{-1}. In earlier experiments[14] formate dehydrogenase was incorporated into liposomes together with polysulfide reductase and the turnover number of the latter enzyme in the electron transport from formate to polysulfide was 10^3 min^{-1}.

Sulfur Oxidoreductase from *Sulfurospirillum deleyianum*

A facultatively sulfur-reducing microaerophilic *Campylobacter*-like bacterium provisionally called "Spirillum" 5175 was isolated from an anaerobic enrichment culture for *Desulfuromonas* by Wolfe and Pfennig[9] and was reclassified as *S. deleyianum* DSM 6946.[23] *Sulfurospirillum deleyianum* is able to utilize nitrate, nitrite, fumarate, elemental sulfur, sulfite, thiosulfate, dimethyl sulfoxide, but not sulfate as terminal electron acceptor for growth with hydrogen or formate as electron donor.

We report here on the activation and reduction of elemental sulfur in *S. deleyianum* as well as on the purification and the partial characterization of the sulfur oxidoreductase from this microorganism.

[23] W. Schumacher, P. M. H. Kroneck, and N. Pfennig, *Arch. Microbiol.* **158,** 287 (1992).

Assay Methods

Preparation of Elemental Sulfur and Sulfur Compounds. Different preparations of elemental sulfur and sulfur compounds have been utilized to investigate the activation and reduction of elemental sulfur or sulfane-sulfur to hydrogen sulfide by *S. deleyianum*.[24] Hydrophilic sulfur has been prepared following the method described by Roy and Trudinger,[25] colloidal sulfur according to the procedure of Janek,[26] and sulfur suspension using the method of Odén.[27] The organic trisulfides, RS-S-SR (R = $-CH_2CH_2NH_2$; $-CH_2CH_2SO_3Na$), are prepared following the procedure described by Savige *et al.*[28] ^{35}S-Labeled compounds are synthesized utilizing sodium [^{35}S]sulfide (Amersham Buchler, Braunschweig, Germany).

Manometric Assay of Sulfur Oxidoreductase Activity. The enzymatic reduction of elemental sulfur and sulfur compounds of *S. deleyianum* is determined manometrically at 37° using small Warburg flasks (volume of liquid, 0.6 ml) as previously described by Fauque *et al.*[29] This manometric assay is based on the measurement of the consumption of dihydrogen, depending on the presence of an active hydrogenase. The sulfur oxidoreductase activity can also be determined by measuring the hydrogen sulfide produced during the reaction. For this purpose hydrogen sulfide is trapped with 0.1 ml of 12% NaOH in the central part of the Warburg vessel and it is determined using a modification of the methylene blue method described by Beinert.[30] This method requires only one-third the amount of protein utilized in the manometric assay.

Colorimetric Assay of Sulfur Oxidoreductase Activity. The manometric assay of dihydrogen uptake measurement and the method for the quantitative determination of hydrogen sulfide are laborious and time consuming. Consequently, Zöphel *et al.*[24] have developed a rapid and sensitive colorimetric assay using a redox-active dye that can be coupled to the sulfur oxidoreductase (E'_0 value of the $S°/HS^-$ couple equals −270 mV[31]). Phenosafranin, a diaminophenylphenazinium salt, is able to transfer electrons in its reduced form to the sulfur oxidoreductase and does not react chemi-

[24] A. Zöphel, M. C. Kennedy, H. Beinert, and P. M. H. Kroneck, *Arch. Microbiol.* **150,** 72 (1988).

[25] A. B. Roy and P. A. Trudinger, "The Biochemistry of Inorganic Compounds of Sulphur." Cambridge Univ. Press, Cambridge, 1970.

[26] A. Janek, *Kolloid. Z.* **64,** 31 (1933).

[27] E. Weitz, K. Gieles, J. Singer, and B. Alt, *Chem. Ber.* **10,** 2365 (1956).

[28] W. E. Savige, J. Eager, J. A. MacLaren, and C. M. Roxburgh, *Tetrahedron Lett.* **44,** 3289 (1964).

[29] G. Fauque, D. Hervé, and J. LeGall, *Arch. Microbiol.* **121,** 261 (1979).

[30] H. Beinert, *Anal. Biochem.* **131,** 373 (1983).

[31] R. K. Thauer, K. Jungermann, and K. Decker, *Bacteriol. Rev.* **41,** 100 (1977).

cally with the sulfur substrates, in contrast to methyl viologen and benzyl viologen. Phenosafranin is also able to work in the presence of detergents such as Triton X-100 and octyl β-D-glucopyranoside.[24] Phenosafranin can be reduced photochemically with oxalate and deazaflavin[32] under anaerobic conditions in a 1-cm Thunberg cell containing 50 m*M* Tris-HCl buffer, pH 8.7. The sulfur substrate (100 μl of cysteamine trisulfide, 26 m*M* in H_2O) is placed in one side arm and the second side arm contains the *S. deleyianum* membrane suspension having the sulfur oxidoreductase (approximately 0.125 mg). After five cycles of deaeration and regassing with purified argon the Thunberg cell is illuminated until the phenosafranin is reduced. Phenosafranin has a red color with an absorption maximum at 525 nm in the oxidized state, whereas it is colorless in the reduced form. After the photochemical reduction of the phenosafranin the sulfur and the sulfur oxidoreductase are added together into the main compartment of the Thunberg cell and the optical density at 525 nm is recorded. In contrast to the manometric assay the colorimetric procedure with phenosafrin does not require the addition of hydrogenase and permits measurement at much lower protein concentrations[24] (0.1 to 0.2 mg of protein per assay).

Activation and Elemental Sulfur Reduction in Sulfurospirillum deleyianum

Localization and Activity of Sulfur Oxidoreductase. Sulfur oxidoreductase activity is present in *S. deleyianum* cells grown with fumarate, nitrate, or elemental sulfur as terminal electron acceptor.[24] Consequently the sulfur oxidoreductase is a constitutive enzyme but the higher specific sulfur oxidoreductase activity is found with cells grown with elemental sulfur as the respiratory substrate. In contrast to the hydrogenase activity, which is present both in the membrane fraction ([Ni-Fe-S] hydrogenase) and in the soluble extract ([Fe-S] hydrogenase), the sulfur oxidoreductase activity is localized only in the membrane.[24] No activity was detected in the cytoplasmic fraction in contrast to *W. succinogenes*, *Desulfomicrobium* (*Dsm.*) *baculatum* strains Norway 4 and DSM 1743, and some thiophilic *Desulfovibrio* species.[33] The soluble extract from *S. deleyianum* did not contain the triheme cytochrome c_3 (previously called cytochrome $c_{551.5}$ or cytochrome c_7) or the tetraheme cytochrome c_3.[24,34]

The specific sulfur oxidoreductase activities of the crude extracts from

[32] V. Massey and P. Hemmerich, *Biochemistry* **17,** 9 (1978).

[33] G. D. Fauque, this volume [24].

[34] J. Le Gall and G. Fauque, *in* "Biology of Anaerobic Microorganisms" (A. J. B. Zehnder, ed.), p. 587. Wiley, New York, 1988.

mesophilic sulfur-reducing eubacteria and thiophilic sulfate reducers are reported in Table V. The crude extract from *S. deleyianum* is the most active with a specific activity of 0.270 μmol of hydrogen consumed per minute per milligram of protein.[24]

Activation and Reduction of Elemental Sulfur and Sulfur Compounds. The pH optimum of *S. deleyianum* sulfur oxidoreductase activity with elemental sulfur as electron acceptor is between pH 8.7 and 8.9 by the manometric test.[24] Using the colorimetric procedure with phenosafrin and cysteamine trisulfide as sulfur substrate, maximum reaction rates are also found at pH 8.7.[24] A linear increase in hydrogen consumption in the manometric test with Janek sulfur at pH 8.9 is obtained during the first 20 min for a crude extract of *S. deleyianum*. At pH 7.3 a lag phase of approximately 40 min is observed, then a milky turbidity appears in the Warburg flask coincident with the start of the reduction of elemental sulfur to hydrogen sulfide.[24] It seems that a sulfur solubilization step occurs in the metabolism of elemental sulfur, which is converted to a more active or "hydrophilic" form with a change in the colloidal sulfur particles. This solubilization step could be induced by a nucleophilic attack on the S—S bond of the substrate by a thiol group, such as glutathione, or by free HS^- in solution.[24] A milky turbidity also appeared in the Warburg vessels during the lag phase preceding the starting of the colloidal sulfur reduction

TABLE V
SULFUR OXIDOREDUCTASE ACTIVITY IN CRUDE EXTRACTS FROM MESOPHILIC SULFUR-REDUCING EUBACTERIA AND THIOPHILIC SULFATE REDUCERS[a]

Microorganism	Specific activity[b]
Sulfurospirillum deleyianum	0.270
Desulfomicrobium baculatum DSM 1743	0.069
Desulfomicrobium baculatum Norway 4	0.063
Desulfovibrio gigas	0.058
Desulfovibrio desulfuricans Berre-Eau	0.037
Desulfuromonas acetexigens DSM 1397	0.070
Desulfuromonas acetoxidans DSM 1675	0.026
Desulfuromonas succinoxidans Gö 20	0.040

[a] From G. Fauque, D. Hervé, and J. LeGall, *Arch. Microbiol.* **121,** 261 (1979); A. Zöphel, M. C. Kennedy, H. Beinert, and P. M. Kroneck, *Arch. Microbiol.* **150,** 72 (1988); G. D. Fauque, this volume [24].

[b] Specific sulfur oxidoreductase activity is expressed in micromoles of hydrogen consumed per minute per milligram of protein.

by tetraheme cytochromes c_3 from some *Desulfovibrio* and *Desulfomicrobium* species.[29] Furthermore, Cammack *et al.*[35] also showed that sodium sulfide decreased the lag phase observed during the reduction of colloidal sulfur by the tetraheme cytochrome c_3 from *Dsm. baculatum* Norway 4 and DSM 1743.

Among the different forms of elemental sulfur tested, colloidal sulfur prepared according to Janek gave the best results with respect to *S. deleyianum* sulfur oxidoreductase activity and reproducibility of activity measurements.[24] The sulfur suspension prepared according to Odén was not very active (approximately 5% compared to Janek sulfur) and the hydrophilic sulfur obtained following the method of Roy and Trudinger gave generally lowered activities with less reproducible results.[24] In contrast, the elemental sulfur prepared according to Roy and Trudinger has been reported to be more active than Janek sulfur with the tetraheme cytochromes c_3 from *Dsm. baculatum* DSM 1743 and Norway 4 and from some *Desulfovibrio* species.[33,36]

The elemental sulfur reduction catalyzed by the membrane fraction of *S. deleyianum* is facilitated by the addition of thiols, such as glutathione or hydrosulfide, which cleaved the S—S bond of the substrate by a nucleophilic attack.[24] Addition of glutathione shortened the lag phase and caused a slight increase in sulfur oxidoreductase activity (15%), confirming that the Janek sulfur was converted to a more active form. Organic trisulfides, RS-S-SR ($R = -CH_2CH_2NH_3^+$; $-CH_2CH_2SO_3^-$), are excellent sulfur substrates, showing more comparable reaction rates and activity values than colloidal sulfur.[24] The sulfane-sulfur moiety can be specifically marked using the ^{35}S isotope and 94% of the total radioactivity can be recovered as $H_2{}^{35}S$. Thus, in contrast to elemental sulfur, organic trisulfides are well-defined chemical compounds suitable for mechanistic and quantitative investigations.[24]

Purification Procedure of Sulfur Oxidoreductase from Sulfurospirillum deleyianum

Growth Conditions of Sulfurospirillum deleyianum. Sulfurospirillum deleyianum is grown under N_2/CO_2 (90/10%) with gentle stirring at 30° in a medium of the following composition[37]: 10 m*M* KH_2PO_4, 0.7 m*M* $CaCl_2$, 1.5 m*M* $MgSO_4$, 24 m*M* $NaHCO_3$, 2 m*M* NH_4Cl, 10 m*M* CH_3COONa,

[35] R. Cammack, G. Fauque, J. J. G. Moura, and J. LeGall, *Biochim. Biophys. Acta* **784,** 68 (1984).

[36] G. Fauque, Doctoral Thesis, University of Aix-Marseille I, France (1979).

[37] W. Schumacher and P. M. H. Kroneck, *Arch. Microbiol.* **156,** 70 (1991).

40 m*M* HCOONa, and 10 m*M* fumarate. Trace elements solution[38] (2 ml/liter) and 0.05% yeast extract are also added. L-Cysteine (0.7 m*M*) is utilized instead of sodium sulfide as the sulfur source and reductant. The pH is adjusted to 7.2 with 2 *M* H_2SO_4 or 2 *M* Na_2CO_3. The yield is approximately 1 g of cells/liter.

Preparation of Crude Extract and Membranes. Cells are harvested with the Pellicon system (Millipore, Eschborn, Germany) and are frozen in liquid nitrogen and stored at −80°.[39] The cells are broken either with a French press or by lysozyme treatment in the presence of EDTA, $MgCl_2$, and DNase. The homogenate in 50 m*M* Tris-HCl buffer, pH 7.6, is centrifuged at 4° for 20 mn at 10,000 *g*, and the supernatant constitutes the crude extract. The supernatant is then centrifuged at 2° for 60 min at 100,000 *g* and the sediment obtained is washed with cold 50 m*M* sodium phosphate, pH 7.3, and homogenized with the same buffer. This membrane suspension is further purified by sucrose density gradient centrifugation and solubilized by incubation with 1% octyl β-D-glucopyranoside (OcGlc) for 15 min at 20° and centrifuging it at 2° for 60 min at 100,000 *g*.

Purification Procedure

The purification procedure is taken from that described by Zöphel *et al.*[39] The sulfur oxidoreductase has been purified in four chromatographic steps.

Step 1: DEAE Column Chromatography. The solubilized membrane material is loaded onto a TSK DEAE-650 S column (1.6 × 10 cm) (Merck, Darmstadt, Germany) equilibrated with 1% OcGlc in 50 m*M* sodium phosphate buffer, pH 7.3. The column is then washed with 1% OcGlc in phosphate buffer and approximately 50% of the total proteins, including major portions of the yellow flexirubin-type pigment,[40] the hydrogenase, and the sulfur oxidoreductase, are eluted in the wash fraction.

Step 2: Hydroxylapatite Column Chromatography. The wash fraction from the previous step is concentrated by ultrafiltration with a PM30 membrane and diluted with 1% OcGlc in distilled water to a final 10 m*M* phosphate concentration. This fraction is adsorbed onto a hydroxylapatite column (1.6 × 10 cm, high resolution; Calbiochem, Frankfurt, Germany) and eluted with a linear 10–500 m*M* phosphate gradient, pH 7.3, containing 1% OcGlc. The yellow flexirubin-type pigment is in the wash fraction. The sulfur oxidoreductase activity is present in the second peak in the

[38] F. Widdel, *Appl. Environ. Microbiol.* **51,** 1056 (1986).

[39] A. Zöphel, M. C. Kennedy, H. Beinert, and P. M. H. Kroneck, *Eur. J. Biochem.* **195,** 849 (1991).

[40] H. Reichenbach and H. Kleinig, *Arch. Microbiol.* **101,** 131 (1974).

elution profile of the hydroxylapatite column and the main hydrogenase activity is eluted afterward.

Step 3: Mono Q HR 5/5 Column (FPLC System). The previous fraction containing the sulfur oxidoreductase activity is applied to an anion-exchange column Mono Q HR 5/5 (FPLC system; Pharmacia, Freiburg, Germany). Separation of hydrogenase and sulfur oxidoreductase is obtained at pH 8.3 and the [Ni-Fe] hydrogenase is of high purity as judged by SDS-polyacrylamide gel electrophoresis and by absorption and electron paramagnetic resonance (EPR) spectra.[39]

Step 4: Superose 12B Column (FPLC System). The sulfur oxidoreductase is subsequently applied to a gel-filtration column Superose 12 B (FPLC System, Pharmacia) in which both *b*-type and *c*-type cytochromes are separated. The fraction with the highest sulfur oxidoreductase activity does not contain *b*-type cytochromes but a *c*-type cytochrome in low concentration is still detectable.

Properties of Sulfurospirillum deleyianum Sulfur Oxidoreductase

The *S. deleyianum* sulfur oxidoreductase has been only partially characterized. Low-temperature EPR spectroscopy indicated that the sulfur oxidoreductase from *S. deleyianum* is an iron-sulfur protein containing [4Fe-4S] centers.[41] Furthermore, the presence of *b*-type cytochromes, as reported in the fumarate reductase of this strain, can be excluded. The fraction with the highest sulfur oxidoreductase activity contained a *c*-type cytochrome with absorption maxima at 553 and 521.5 nm in the reduced state.[39] It is not clear at this stage whether this membrane-bound *c*-type cytochrome is actually involved in the *S. deleyianum* activation and reduction of elemental sulfur.

Conclusion

The reduction of elemental sulfur to hydrogen sulfide (or sulfur respiration) is coupled to the phosphorylation of ADP with inorganic phosphate[42–44] in the catabolism of some anaerobic microorganisms, most of which belong to the group of extremely thermophilic archaebacteria. The

[41] P. M. H. Kroneck, J. Beuerle, and W. Schumacher, *in* "Metals Ions in Biological Systems" (H. Sigel and A. Sigel, eds.), Vol. 28, p. 455. Dekker, New York, 1992.

[42] G. D. Fauque, L. L. Barton, and J. LeGall, *in* "Sulphur in Biology," Ciba Foundation Symposium 72, p. 71. Excerpta Medica, Amsterdam, 1980.

[43] J. Paulsen, A. Kröger, and R. K. Thauer, *Arch. Microbiol.* **144,** 78 (1986).

[44] A. Kröger, J. Schröder, J. Paulsen, and A. Beilmann, *in* "The Nitrogen and Sulphur Cycles" (J. A. Cole and S. J. Ferguson, eds.), p. 133. Cambridge Univ. Press, Cambridge, 1988.

facultative spirilloid dissimilatory sulfur reducers are able to oxidize hydrogen or formate with nitrate, nitrite, elemental sulfur, fumarate, and malate as electron acceptor for growth and they contain menaquinone-6 as major quinone with low levels of thermoplasmaquinone-6.[1]

Enzymes involved in elemental sulfur reduction have been studied in the spirilloid sulfur reducers *S. deleyianum* and *W. succinogenes* and in the true sulfur-reducing bacterium *Desulfuromonas* (*Drm.*) *acetoxidans*. A sulfur oxidoreductase acting also as a sulfide dehydrogenase is present in the membrane fraction of *Drm. acetoxidans* and a *c*-type cytochrome with a low midpoint redox potential has been suggested to serve as electron donor for the sulfur reductase.[44] The polysulfide reductase of *W. succinogenes* is a component of the phosphorylative electron transport system with polysulfide as the terminal acceptor. This polysulfide reductase has been solubilized and purified in three steps; it is a molybdoenzyme containing [Fe-S] clusters but no cytochromes. Proteoliposomes containing hydrogenase and polysulfide reductase were able to catalyze the electron transport from dihydrogen to polysulfide. The membrane-bound sulfur oxidoreductase from *S. deleyianum* has been purified in four chromatographic steps. It is an iron-sulfur protein containing [4Fe-4S] clusters. The mechanism of electron transport from formate dehydrogenase to sulfur oxidoreductase is not yet understood.

Section IV

Oxidation of Reduced Sulfur Compounds

[26] Purification of Rusticyanin, a Blue Copper Protein from *Thiobacillus ferrooxidans*

By W. JOHN INGLEDEW and DAVID H. BOXER

Introduction

Rusticyanin is a high-potential blue copper protein expressed by *Thiobacillus ferrooxidans* when cells grow on the energy derived from the aerobic oxidation of ferrous iron at acid pH. The protein was discovered by Cobley and Haddock[1] and purified to homogeneity by Cox and Boxer.[2] It is a small, type 1 blue copper protein, M_r 16,300, containing a redox-active, single copper center. The protein is present in Fe(II)-grown cells at relatively high concentrations (up to 5% of total cell protein[2,3]) and is located in the periplasm.[4] It has been sequenced[5-8] and crystallized[9] and has been the subject of a number of biophysical studies.[10-14] Rusticyanin has an optical spectrum similar to those of other type 1 copper proteins (although with a lower extinction coefficient in the 590-nm region) and an electron paramagnetic resonance (EPR) spectrum similar to that of the type 1 protein stellacyanin.[15] Its midpoint potential is high and it is stable under acidic conditions. Nuclear magnetic resonance spectra show that the protein is intact and globular in structure even at pH 1.0. At 680 mV, the redox midpoint potential of rusticyanin is considerably higher than those of other blue copper proteins. This value is pH independent between

[1] J. G. Cobley and B. A. Haddock, *FEBS Lett.* **60,** 29 (1975).
[2] J. C. Cox and D. H. Boxer, *Biochem. J.* **174,** 497 (1978).
[3] W. J. Ingledew, *Biotechnol. Bioeng.* **16,** 23 (1986).
[4] W. J. Ingledew and A. Houston, *Appl. Biochem.* **8,** 1 (1986).
[5] R. C. Blake, E. A. Shute, M. M. Greenwood, G. H. Spencer, and W. J. Ingledew, *FEMS Microbiol. Rev.* **11,** 9 (1993).
[6] M. Ronk, J. E. Shively, E. A. Shute, and R. C. Blake, *Biochemistry* **30,** 9435 (1991).
[7] T. Yano, Y. Fukumori, and T. Yamanaka, *FEBS Lett.* **288,** 159 (1991).
[8] F. Nunzi, M. Woudstra, D. Campese, J. Bonicel, D. Morin, and M. Bruschi, *Biochim. Biophys. Acta* **1162,** 28 (1993).
[9] A. Djebli, P. Proctor, R. C. Blake, and M. Shoham, *J. Mol. Biol.* **227,** 581 (1992).
[10] S. D. Holt, B. Piggott, W. J. Ingledew, M. C. Feiters, and G. P. Diakun, *FEBS Lett.* **269,** 117 (1991).
[11] J. McGinnis, W. J. Ingledew, and A. G. Sykes, *Inorg. Chem.* **25,** 3730 (1986).
[12] A. G. Lappin, C. A. Lewis, and W. J. Ingledew, *Inorg. Chem.* **24,** 1446 (1985).
[13] R. C. Blake and E. A. Shute, *J. Biol. Chem.* **262,** 1483 (1987).
[14] R. C. Blake, K. J. White, and E. A. Shute, *Biochemistry* **30,** 9443 (1991).
[15] J. C. Cox, R. Aasa, and B. G. Malmstrom, *FEBS Lett.* **93,** 157 (1978).

pH 1.5 and 7.0, with an indication of pH dependency below pH 1.5 (determined by cyclic voltammetry and redox potentiometry[3,4,11]). For comparison the midpoint potentials, at pH 7.0, for plastocyanin, umacyanin, azurin, and stellacyanin are 370, 280, 330, and 184 mV, respectively.

The role of rusticyanin is unclear. The direct reduction of pure rusticyanin by Fe(II) has been studied and the reaction shown to be slow.[12,13] The reduction of rusticyanin by other reactants is usually faster.[11] Even though the conditions of the periplasm cannot be readily mimicked in *in vitro* experiments, it is clear that the rate of rusticyanin reduction by Fe(II) is orders of magnitude slower than that required for a role as the primary Fe(II) oxidant in Fe(II) oxidation. The reduction of rusticyanin in intact cells by Fe(II) has not been observed but reduced rusticyanin undergoes rapid oxidation, along with a cytochrome *c*, on exposure to oxygen.[16,17]

The amino acid sequence of rusticyanin from different strains has been determined by four groups.[5–8] The sequences show no unusual features when compared to those of other blue copper proteins. The protein does not contain detectable amounts of carbohydrate. Comparison of the rusticyanin sequence with that of other blue copper proteins indicates a conserved region near the C terminus comprising cysteine, methionine, and histidine residues. This is recognized as part of the copper-binding motif in plastocyanin but the role of methionine is controversial. A number of histidine and asparagine residues are present close to this region. The copper-binding motif, incorporating two sulfur ligands, has been confirmed by extended X-ray absorption fine structure (EXAFS) studies.[10] Rusticyanin has been crystallized by the hanging drop vapor diffusion technique and preliminary X-ray crystallographic studies have been reported.[9]

Growth of Cells

Rusticyanin is present in the bacterium only when it is grown aerobically, with vigorous aeration, on Fe(II) in acidic sulfate media. The cells are best grown in chemostat mode in a large vessel. In our laboratory *Tb. ferrooxidans* is grown in a chemostat (working volume, 50 liters) with a dilution rate of 0.02/hr. The medium consists of 180 m*M* $FeSO_4 \cdot 7H_2O$, 1 m*M* $(NH_4)_2SO_4$, 0.2 m*M* KH_2PO_4, 37.5 m*M* H_2SO_4, and salts solution (0.2 ml/liter).[18] Earlier growth media contained 20 m*M* H_2SO_4, but the present medium gives less precipitation of ferric hydroxysulfate (Jarosite). The $FeSO_4$ needs to be better than general reagent grade, otherwise the

[16] W. J. Ingledew, *Biochim. Biophys. Acta* **683,** 89 (1982).

[17] W. J. Ingledew and J. G. Cobley, *Biochim. Biophys. Acta* **590,** 141 (1980).

[18] B. Alexander, S. Leach, and W. J. Ingledew, *J. Gen. Microbiol.* **133,** 1171 (1987).

precipitation can present problems. The medium does not require autoclaving. The salts solution has the following composition: 31% (w/w) HCl (250 ml/liter), $MgCl_2$ (1.2 *M*), $CaCl_2$ (10 m*M*), $ZnSO_4$ (5 m*M*), $CoCl_2 \cdot 6H_2O$ (5 m*M*), H_3BO_3 (5 m*M*), and $Na_2MoO_4 \cdot H_2O$ (5 m*M*).

Cell Harvesting

Cells are harvested by continuous flow centrifugation. Cells have also been successfully harvested by other groups by cross-flow filtration. We have found that the amount of ferric iron precipitation tends to cause problems with this method, which may be averted by increasing the H_2SO_4 concentration in the growth medium.

Chemostat effluent (45 liters at a time) is harvested at 10° in a continuous flow rotor fitted to an MSE 18 centrifuge (Fisons Scientific Equipment, Loughborough, United Kingdom) operating at 17,000 rpm with a flow rate of 15–20 liters/hr. Cells are washed twice in 10 m*M* H_2SO_4, 50 m*M* Na_2SO_4 (to maintain ionic strength) and twice more in 50 m*M* Na_2SO_4 before storage at −20° as a pellet. Successful rusticyanin preparations have been carried out on cells stored in this manner for 6 months or more.

Cell Breakage

Rusticyanin is a soluble protein and is released from cells after breakage and centrifugation (to remove cell debris) at pH 2.0. At this pH most intracellular proteins are denatured and therefore easily removed by centrifugation. Rusticyanin is the predominant protein in this supernatant. Rusticyanin autoreduces at pH values much above 5.5, and therefore cannot be detected by its blue color above this pH. *Thiobacillus ferrooxidans* is resistant to breakage. In our laboratory, cells are broken by two or three passes through a French pressure cell. Other laboratories have successfully used sonication.[13]

Purification of Rusticyanin

The blue color of rusticyanin provides an easy way to monitor the progress of purification. The absorbance at 597 nm is a quick and sufficiently sensitive method for following rusticyanin. All chemicals used in the purification procedure are of Analar or Aristar (Merck, Rahway, NJ) quality. All steps are carried out at 4°. Approximately 25 g (wet weight) of *Tb. ferrooxidans* cell paste is thawed and resuspended (1 : 4, w/v) in 10 m*M* H_2SO_4. This step is usually said to be performed at pH 2.0, but the pH is more accurately 1.9. Cells are disrupted by passage three times through a French pressure cell at 1.4×10^5 kPa. The broken cell suspension is stirred, then centrifuged at 10,000 *g* for 10 min to remove debris. The blue-brown supernatant is removed and stored on ice. The pellet is resuspended in 40 ml of 10 m*M* H_2SO_4 and recentrifuged as described above. The two supernatants are combined and further centrifuged at 150,000 *g* for 1 hr. The resulting pellet consists of the membrane fragments. The supernatant is deep blue in color and contains the rusticyanin in a substantially purified state.

Ammonium Sulfate Fractionation

Solid, crushed $(NH_4)_2SO_4$ (Aristar grade; Merck, Rahway, NJ) is added slowly, while stirring the supernatant, to 65% saturation. The resultant solution is stirred on ice for a further 30 min and centrifuged at 30,000 g for 15 min to give a red pellet, which contains cytochrome *c*, and a blue supernatant. The supernatant is brought to 95% $(NH_4)_2SO_4$ saturation, while slowly stirring on ice, by the gradual addition of solid. The slow stirring is continued for an additional 30 min. The suspension is centrifuged at 30,000 *g* for 20 min. The pale yellow supernatant is discarded and the dark blue pellet resuspended in 10 ml of 10 m*M* sodium acetate, pH 5.4. The resuspended pellet is dialyzed against three changes of 2 liters of the same buffer, with a minimum of 4 hr between each change.

Ion-Exchange Chromatography

The dialyzed solution is layered on a DEAE-cellulose column (2 × 10 cm) equilibrated with 10 m*M* sodium acetate, pH 5.4. The blue protein passes through in the nonadsorbed fraction. The combined (blue) fractions from the DEAE-cellulose column are applied directly to a CM-cellulose column (2 × 8 cm) equilibrated with 10 m*M* sodium acetate, pH 5.4. The rusticyanin is adsorbed as a dark blue band at the top of the column. The column is washed with the equilibrating buffer until the A_{280} of the eluate falls to a minimum. Rusticyanin is eluted by the step addition of 100 m*M* NaCl in 10 m*M* sodium acetate, pH 5.4. The fractions containing the blue protein are combined.

Subsequent treatment depends on intended use. The protein can be dialyzed against 10 m*M* H_2SO_4 and stored at 4° for several weeks. For longer term storage under liquid N_2, we dialyze against 50 m*M* β-alanine-SO_4, pH 3.5, and concentrate. Concentration is done using Centriprep-10 and Centricon PM10 filtration concentrators (Amicon, Danvers, MA). The balance sheet for a typical purification of rusticyanin is shown in Table I.

TABLE I
BALANCE SHEET FOR RUSTICYANIN PURIFICATION

Fraction	Total protein[a] (mg)	Total A_{597}[b]	A_{597}/mg of protein	Purification factor	Yield A_{597} (%)	Total Cu[c] (μg)
Supernatant	715	63.7	0.089	1	100	1584
95% Saturated $(NH_4)_2SO_4$	365	34.4	0.094	1.06	54	867
DE-52 chromatography	312	30.2	0.096	1.08	48	853
CM-32 chromatography	248	26	0.108	1.21	41	739

[a] O. G. Lowry, N. J. Rosebrough, A. L. Farr, and R. Randall, *J. Biol. Chem.* **193,** 265 (1951).
[b] Obtained as volume (ml) × A_{597}.
[c] Assayed by atomic absorption spectroscopy.

Properties

The optical absorption spectrum provides a convenient confirmation of the preparation. The spectrum of the oxidized rusticyanin in Fig. 1 reveals a broad characteristic peak centered at 597 nm. The irregular feature at about 420 nm is due to the Soret absorption of a slight contamination by cytochrome *c*. This can be removed by repeating the CM-cellulose

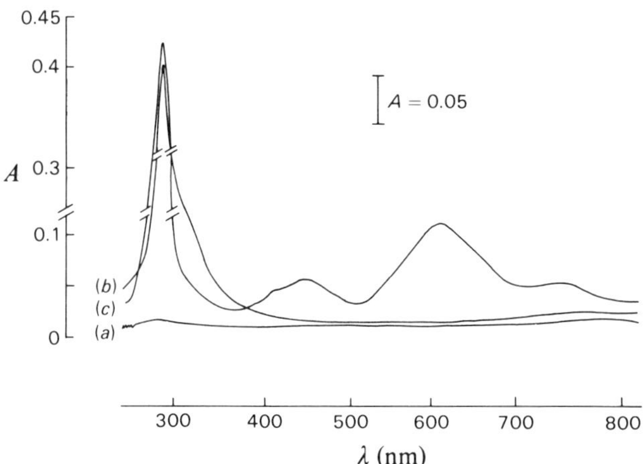

FIG. 1. Optical absorption of rusticyanin. The spectra were obtained from a solution of purified rusticyanin (1.10 mg of protein/ml; 3.40 μg of copper/mg of protein) in 10 m*M* H_2SO_4, in a 1-cm light path cuvette. Curve (a), baseline; curve (b), absolute spectrum of rusticyanin; curve (c), absolute spectrum of rusticyanin after reduction by 10 m*M* $FeSO_4$. [Adapted from J. C. Cox and D. H. Boxer, *Biochem. J.* **174,** 174 (1978).]

TABLE II
OPTICAL EXTINCTION COEFFICIENTS OF PURIFIED RUSTICYANIN

Wavelength	Molar absorption coefficient[a] (liters mol^{-1} cm^{-1})
287	4550
450	1060
597	1950
750	750

[a] Determined in 10 mM H_2SO_4.

ion-exchange step. The optical extinction coefficients for rusticyanin are given in Table II. Sodium dodecyl sulfate-polyacrylamide gel analysis of the final preparation should give a single readily staining polypeptide of apparent M_r ~16,500. The reduction of rusticyanin by Fe(II) salts can be readily monitored by absorption spectroscopy (Fig. 2).

Problems

Occasionally the blue color is lost during the preparative or subsequent procedures. This may be due to reduction of the blue cupric center to the colorless, cuprous form. In this case color can be restored by oxidation.

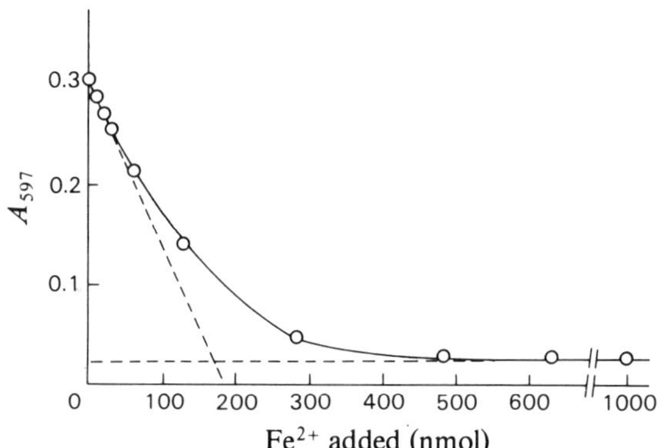

FIG. 2. Titration of rusticyanin with Fe(II). Purified rusticyanin (3.20 mg of protein/ml; 7.30 mg of copper/ml) in 10 mM H_2SO_4 was titrated with $Fe^{II}SO_4$. A_{597} was measured (1-cm light path cuvette) initially and 10 min after each Fe(II) addition. The end point was estimated by extrapolation, assuming that at low [Fe(II)] the change in A_{597} is directly proportional to the [Fe(II)]. [Adapted from J. C. Cox and D. H. Boxer, *Biochem. J.* **174,** 497 (1978).]

The high midpoint potential of rusticyanin requires the use of a high-potential oxidant and care must be taken to add only minimum amounts. We use sodium hexachloroiridate (K and K Laboratories, New York, NY) as oxidant and always remove excess oxidant by dialysis or gel filtration.

[27] Adenylylsulfate Reductases from Thiobacilli

By BARRIE F. TAYLOR

Adenylylsulfate (APS) reductase[1,2] (EC 1.8.99.2) catalyzes the following reversible reaction:

$$AMP + SO_3^{2-} + acceptor_{ox} \overset{2e}{\rightleftharpoons} APS + acceptor_{red}$$

where APS is adenylylsulfate or adenosine-5′-phosphosulfate. Adenylyl sulfate reductase occurs in sulfate-reducing bacteria,[3–5] and in some phototrophic bacteria[6,7] and thiobacilli[8–10] that oxidize reduced inorganic sulfur compounds. Adenylyl sulfate reductase has been purified from two species of *Thiobacillus, T. denitrificans*[9] and *T. thioparus*.[10]

Assay Methods

Principle

The activity of APS reductase is easily measured with ferricyanide as the electron acceptor and by following the decrease in absorbance at 420 nm.[9,11] The enzyme from *T. thioparus* may also be assayed with cytochrome *c* as the electron acceptor.[10]

[1] H. D. Peck, Jr., *Proc. Natl. Acad. Sci. U.S.A.* **45,** 701 (1959).
[2] M. Ishimoto and D. Fujimoto, *Proc. Jpn. Acad.* **35,** 243 (1959).
[3] H. D. Peck, Jr., T. E. Deacon, and J. T. Davidson, *Biochim. Biophys. Acta* **96,** 429 (1965).
[4] W. Stille and H. G. Trüper, *Arch. Microbiol.* **137,** 145 (1984).
[5] N. Speich and H. G. Trüper, *J. Gen. Microbiol.* **134,** 1419 (1988).
[6] H. G. Trüper and H. D. Peck, *Arch. Microbiol.* **73,** 125 (1970).
[7] H. G. Trüper and L. A. Rogers, *J. Bacteriol.* **108,** 1112 (1971).
[8] H. D. Peck, Jr., *Proc. Natl. Acad. Sci. U.S.A.* **46,** 1053 (1960).
[9] T. J. Bowen, F. C. Happold, and B. F. Taylor, *Biochim. Biophys. Acta* **118,** 516 (1965).
[10] R. M. Lyric and I. Suzuki, *Can. J. Biochem.* **48,** 344 (1970).
[11] H. D. Peck, Jr., *Biochim. Biophys. Acta* **49,** 621 (1961).

Reagents

Tris-HCl buffer, 0.1 *M*, pH 8.0
Na_2EDTA, 0.04 *M*
AMP, 0.04 *M*
$K_3Fe(CN)_6$, 5 m*M*
Na_2SO_3, 0.2 *M* (freshly prepared in 0.05 *M* Tris-HCl buffer, pH 8.0, containing 5 m*M* Na_2EDTA)

Procedure

The standard assay using ferricyanide as electron acceptor contains, in a final volume of 2.65 ml, 1.25 ml of Tris-HCl buffer, 0.5 ml of EDTA, 0.25 or 0.025 ml of AMP, and 0.25 ml of ferricyanide. Assays are performed at 25° in 3-ml silica cuvettes (1-cm light path) with a recording spectrophotometer, using a water blank and measurement of absorbance at 420 nm. Na_2SO_3 (0.05 ml) is added and, after the endogenous rate of ferricyanide reduction has been recorded (1 min), 0.10 ml of cell-free extract is added. The endogenous rate of ferricyanide reduction by sulfite is subtracted from the rate with extract present.

The cytochrome *c* assay[10] works for *T. thioparus* but not with *T. denitrificans*. The assay mixture contains 3.0 μmol of cytochrome *c*, 3.0 μmol of Na_2SO_3, 0.3 μmol of AMP, 20 μmol of Tris-HCl buffer (pH 9.5), cell extract, and water to 3.0 ml. After adding the cell extract or fraction the increase in absorbance at 550 nm is measured.

Definition of Unit

A 1 m*M* solution of ferricyanide has an absorbance of 1.0 at 420 nm; a decrease in absorbance of 1.0 corresponds to the reduction of 2.65 μmol of ferricyanide for an assay volume of 2.65 ml. A unit of enzyme catalyzes the reduction of 1 μmol of ferricyanide per minute.

Purification Procedures

Growth of Organisms

The medium[12] for *T. denitrificans* contains (per liter) $Na_2S_2O_3 \cdot 5H_2O$ (5 g), KNO_3 (2 g), KH_2PO_4 (2 g), $NaHCO_3$ (2 g), NH_4Cl (1 g), $MgSO_4 \cdot 7H_2O$ (0.8 g), $FeSO_4 \cdot 7H_2O$ [2% (w/v) solution in 1 *N* HCl] (0.02 g), and trace metal solution (1 ml). The trace metal solution[13] contains (per liter)

[12] B. F. Taylor, D. S. Hoare, and S. L. Hoare, *Arch. Microbiol.* **78**, 193 (1971).
[13] W. Vishniac and M. Santer, *Bacteriol. Rev.* **21**, 195 (1957).

Na_2EDTA (50 g), $ZnSO_4 \cdot 7H_2O$ (22 g), $CaCl_2 \cdot 2H_2O$ (7.34 g), $MnCl_2 \cdot 4H_2O$ (7.95 g), $FeSO_4 \cdot 7H_2O$ (5 g), $(NH_4)_2Mo_7O_{24} \cdot 4H_2O$ (1.1 g), $CuSO_4 \cdot 2H_2O$ (1.57 g), $CoCl_2 \cdot 6H_2O$ (1.61 g), and KOH (about 60 ml of a 10 *N* solution to adjust the pH to 6.0). The trace metals, ferrous sulfate, phosphate, and bicarbonate are autoclaved separately and the medium mixed after cooling. The final pH of the medium is 7.0. *Thiobacillus denitrificans* (ATCC 25259) is maintained in 18-ml tubes full of medium and closed with rubber-lined screw caps. Batch cultures are grown at 30° in 2-liter flasks filled with medium and sealed with rubber stoppers that are vented with Bunsen valves. Batches (each 100 liters) in a fermentor yield up to 0.7 g (dry weight)/liter of cells if a neutral pH is maintained by adding $NaHCO_3$ solution, and by periodically adding more thiosulfate and nitrate.[14] Cells are harvested by centrifugation at 10,000 *g* for 10 min at 5° and stored frozen.

The medium[15] for *T. thioparus* contains (per liter) $Na_2S_2O_3 \cdot 5H_2O$ (10 g), KH_2PO_4 (40 g), K_2HPO_4 (40 g), $(NH_4)_2SO_4$ (3 g), $CaCl_2$ (0.7 g), $MnSO_4 \cdot 2H_2O$ (0.2 g), $FeCl_3 \cdot 6H_2O$ (0.02 g), and phenol red [2% (v/v) aqueous solution] (0.3 ml). *Thiobacillus thioparus* (ATCC 8158) is maintained and grown[10] aerobically at 25° in 200-ml batches of medium in 500-ml Erlenmeyer flasks. Ten-liter batches are grown, with forced aeration through sintered glass spargers, in 15-liter carboys using 10% (v/v) inocula. The pH is maintained at 6.5 to 7.0 by adding 10% (w/v) K_2CO_3. After 3–5 days the cells are harvested by centrifugation, washed in 0.1 *M* phosphate buffer (pH 7.5), and elemental sulfur removed by differential centrifugation. Cell pellets are stored frozen.

Purification of Adenylylsulfate Reductase of Thiobacillus denitrificans

Step 1: Preparation of Cell-Free Extract. Frozen cells are disrupted in a Hughes press[16] and then mixed with three parts (w/v) of 0.05 *M* Tris-HCl buffer, pH 8.0.[9] After adding DNase and RNase, the mixture of buffer and crushed cells is stirred for 1 hr at 0–5°. Crude extract is obtained by centrifugation at 100,000 *g* for 3 hr at 0°. Cell disruption by sonication for 5 to 10 min with a sonic disintegrator (Measuring and Scientific Equipment, Ltd., England) and extraction with either Tris-HCl buffer or 0.05 *M* potassium phosphate buffer (pH 7.0)[17] produce crude extracts with less than half the specific activity of Hughes press extracts.

[14] K. Sargeant, P. W. Buck, J. W. S. Ford, and R. G. Yeo, *Appl. Microbiol.* **14,** 998 (1966).
[15] R. L. Starkey, *J. Bacteriol.* **28,** 365 (1935).
[16] D. E. Hughes, *Br. J. Exp. Pathol.* **32,** 97 (1951).
[17] B. F. Taylor and D. S. Hoare, *Arch. Microbiol.* **80,** 262 (1971).

Step 2: Heat Treatment of Crude Extract. Crude extract (60 ml) is heated, in 10-ml portions in test tubes, in a water bath for 15 min at 70°. After cooling in ice, denatured protein is removed by centrifugation at 14,000 *g* for 30 min at 0°.

Step 3: Ammonium Sulfate Fractionation. All steps are now carried out in a cold room at 5°. The enzyme precipitates between 50 and 70% saturation with ammonium sulfate. The supernatant from step 2 is diluted with 0.05 *M* Tris-HCl buffer (pH 8.0) to give an activity of 8.60 units/ml. Solid ammonium sulfate is added to 50% saturation with continuous stirring. After removing precipitated protein by centrifugation (14,000 *g*, 30 min, 0°) the supernatant is saturated to 70% with ammonium sulfate and the 50–70% fraction collected by centrifugation. The 50–70% ammonium sulfate fraction is dissolved in 6 ml of 0.01 *M* Tris-HCl buffer (pH 7.0) containing 0.11 *M* NaCl and then applied to a column of Sephadex G-25 (12.5-cm length, 2.4-cm diameter) equilibrated with the same buffer including 0.11 *M* NaCl.

Step 4: DEAE-Cellulose Chromatography. The protein eluant from the Sephadex G-25 column is applied to a column of DEAE-cellulose (15 × 2.4 cm) previously equilibrated with 0.01 *M* Tris-HCl buffer (pH 7.0) containing 0.11 *M* NaCl. After elution of nonadsorbed protein a linear gradient of NaCl (0.11 to 0.22 *M*) in 0.01 *M* Tris-HCl (pH 7.0) is used to elute APS reductase. The fractions eluting between 0.15 and 0.18 *M* NaCl are combined.

Table I summarizes the purification steps for APS reductase from *T. denitrificans*.

TABLE I
PURIFICATION OF ADENYLYLSULFATE REDUCTASE FROM *Thiobacillus denitrificans*[a]

Fraction	Total volume (ml)	Total protein (mg)	Total activity[b] (units)	Specific activity[b] (units/mg protein)	Purification (fold)	Yield (%)
Step 1: Cell-free extract	60	1512	750 (1878)	0.50 (1.24)	1.0	100
Step 2: Heat treatment, 70° for 15 min	51	433	592 (1480)	1.36 (3.21)	2.7 (2.6)	79 (79)
Step 3: 50–70% $(NH_4)_2SO_4$ fraction after Sephadex G-25 gel filtration	15	192	548 (1470)	2.86 (7.72)	5.8 (6.2)	73 (78)
Step 4: DEAE-cellulose fractions	45	46.3	445 (980)	9.60 (21.2)	19.2 (17.1)	59 (52)

[a] B. F. Taylor, unpublished data, 1965.
[b] Assayed with 3.8 m*M* AMP or, in parentheses, with 0.38 m*M* AMP.

Purification of Adenylylsulfate Reductase of Thiobacillus thioparus

Step 1: Preparation of Cell-Free Extract. Cells, 1 g (wet weight)/10 ml, are suspended in 0.1 *M* phosphate buffer (pH 7.5) and sonicated for 30 min at maximum power in a Raytheon 10KC sonic oscillator.[10] Cell-free extract is obtained by centrifugation at 30,000 *g* for 20 min.

Step 2: Ultracentrifugation. Cell-free extract (50–70 ml) from step 1 is centrifuged at 105,000 *g* for 2 hr.

Step 3: DEAE-Cellulose Chromatography. The supernatant from step 2 is applied to a DEAE-cellulose column (40-cm long by 2.5-cm diameter) equilibrated with 0.02 *M* phosphate buffer (pH 7.4). A linear gradient of 0.05–0.2 *M* phosphate (pH 7.4) is applied and APS reductase elutes at 0.15 *M* phosphate.

Step 4: Ammonium Sulfate Fractionation. The precipitate between 40 and 70% saturation is collected by centrifugation and dissolved in 0.1 *M* phosphate buffer (pH 7.4).

The purification procedure for APS reductase of *T. thioparus* is summarized in Table II.

Properties

Purity

A single, symmetrical peak is obtained during ultracentrifugation at 210,000 *g* at 20° of the purified enzyme from *T. denitrificans*.[9] A yellow color (flavin) is delineated by the peak during sedimentation. Analytical polyacrylamide electrophoresis at pH 8.3 and staining with naphthol black reveals one major protein band and some contamination with a much

TABLE II
PURIFICATION OF ADENYLYLSULFATE REDUCTASE FROM *Thiobacillus thioparus*[a]

Fraction	Total protein (mg)	Total activity (units)	Specific activity (units/mg protein)	Purification (fold)	Yield (%)
Step 1: Cell-free extract	—	—	0.14	1.0	—
Step 2: Supernatant from ultracentrifugation	400	240	0.60	3.1	100
Step 3: DEAE-cellulose eluate	40	144	3.60	7.0	60
Step 4: 40–70% $(NH_4)_2SO_4$ fraction	13.6	87	6.40	16.1	36

[a] Taken from Lyric and Suzuki.[10]

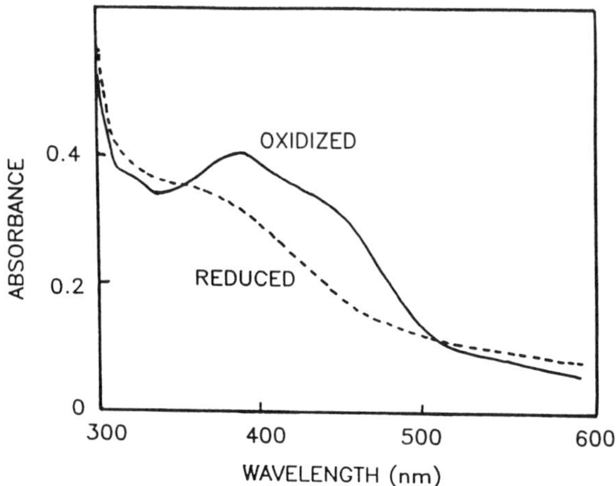

FIG. 1. Absorption spectra of oxidized and reduced APS reductase from *T. denitrificans*. Purified enzyme (specific activity, 8.38 units/ml) at a concentration of 1.05 mg of protein/ml in 0.01 *M* potassium phosphate buffer (pH 7.0). (– – –) Reduced ($NaBH_4$); (—) oxidized. (Taken from Taylor.[18])

fainter band. The enzyme from *T. thioparus* shows a major band, colored yellow (flavin) before staining, and three light bands of protein. Adenylylsulfate reductase constitutes about 4% of the soluble cell protein in both organisms.

Stability

Adenylylsulfate reductase from *T. denitrificans* is stable for months in frozen crude extracts. The unpurified enzyme retains nearly 100% activity after exposure to 75° for 15 min but is rapidly destroyed at 80°. The frozen purified enzyme from *T. thioparus,* at protein concentrations above 5 mg/ml, maintains full activity for several months. At 5° the purified enzyme of *T. thioparus* shows no decrease in activity for 6–8 hr but loses 20% activity after 5 days. The activity of the purified enzyme is unaffected by 5 min at 60° but shows an 80% loss after 5 min at 65°.

Molecular Weight and Cofactors

Adenylylsulfate reductases from thiobacilli contain FAD and nonheme iron, and exhibit characteristic absorbances in the regions of 390 and 445 nm, which are bleached by reducing agents (dithionite and borohydride) (Fig. 1).[10,18] The molecular weight of APS reductase from *T. thioparus* is

[18] B. F. Taylor, *FEMS Microbiol. Lett.* **59,** 351 (1989).

170,000 + 9000.[10] Assuming a molecular weight of 170,000, each mole of enzyme contains about 1 mol of FAD and several moles of nonheme iron (8–13 mol, *T. denitrificans;* 8–10 mol, *T. thioparus*). Acid-labile sulfide has not been determined for APS reductase from *T. denitrificans* but constitutes 4–5 mol/mol of enzyme in *T. thioparus*.

Electron Acceptors and pH Optima

Yeast cytochrome *c* functions with the enzyme from *T. thioparus* but is about 80-fold less effective as an electron acceptor than ferricyanide. Cytochrome *c* has not been tested as an electron acceptor for APS reductase from *T. denitrificans*. In the ferricyanide assay the pH optimum for the purified enzyme from *T. thioparus* is 7.0 (phosphate buffer) or 7.4 (Tris-HCl buffer); with cytochrome *c* as electron acceptor the pH optimum is 9.5 (Tris-HCl buffer). The pH optimum for APS reductase of *T. denitrificans* is 7.2 (Tris-maleate buffer).

Substrate Specificity

AMP and GMP are substrates for both enzymes. Apparent K_m values for nucleotides and sulfite are shown in Table III. Substrate inhibition by AMP is characteristic of both enzymes; a higher rate of ferricyanide reduction, and a more sensitive assay, is obtained by lowering the AMP concentration in the assay from 3.8 to 0.38 m*M* (Table I). The enzyme from *T. thioparus* is active with deoxyAMP, IMP, UMP, and CMP in either the ferricyanide or cytochrome *c* assays but the APS reductase of *T. denitrificans* is inactive with UMP and CMP.

TABLE III
APPARENT K_m VALUES (m*M*) FOR ADENYLYLSULFATE REDUCTASES FROM THIOBACILLI

Substrate	*T. denitrificans*[a] ferricyanide assay	*T. thioparus*[b] Ferricyanide assay	*T. thioparus*[b] Cytochrome *c* assay
Sulfite	1.5	2.5	0.017
AMP	0.041	0.1	0.0025
GMP	0.63	—	0.01

[a] Taken from Bowen *et al.*[9]
[b] Taken from Lyric and Suzuki.[10]

Inhibitors

Sulfhydryl groups are essential for the activity of APS reductases of thiobacilli. Complete inhibition is caused by 5 m*M* *N*-ethylmaleimide for *T. thioparus* enzyme and by a 10 m*M* concentration for the *T. denitrificans* enzyme. The enzymes are also sensitive to *p*-chloromercuribenzoate, which causes 100% inhibition at a 1 m*M* concentration for *T. thioparus*, and 30% inhibition at 0.1 m*M* (with a 15-min preincubation) for *T. denitrificans*. Vicinal dithiol groups are implicated in the interaction of cytochrome *c* with the enzyme from *T. thioparus* because of significant inhibition by 1 m*M* arsenite. Arsenite (10 m*M*) does not affect the ferricyanide activity for APS reductase from either organism. Even though the enzymes contain iron, phenanthroline, or bipyridyl, up to 10 m*M* levels have no significant effects on ferricyanide activity for either enzyme and do not affect cytochrome *c* activity for the *T. thioparus* enzyme. The general metal chelator EDTA (10 m*M*) does not inhibit ferricyanide reduction by the *T. thioparus* enzyme.

Acknowledgments

I thank Dr. Ulrich Fischer for his comments. Financial support for this chapter was provided by the National Science Foundation (OCE 9012157).

[28] Enzymes of Dissimilatory Sulfide Oxidation in Phototrophic Sulfur Bacteria

By CHRISTIANE DAHL and HANS G. TRÜPER

Introduction

The classic anaerobic phototrophic "purple" and "green" sulfur bacteria derive energy from light by using sulfide, thiosulfate, or elemental sulfur as photosynthetic electron donor. Some of these organisms can also thrive on hydrogen gas or simple carbon compounds as electron donors.

Enzymological studies have revealed that at least 10 enzymes and 2 nonenzymatic steps may be involved in dissimiliatory sulfur metabolism of different phototrophic bacteria (Fig. 1). The pathway of reduced sulfur oxidation in these organisms includes three essential steps: (1) formation of elemental sulfur that appears as globules, inside or outside the bacteria, during oxidation of sulfide or thiosulfate, (2) oxidation of sulfide or elemental sulfur to sulfite, and (3) formation of sulfate as the final product.

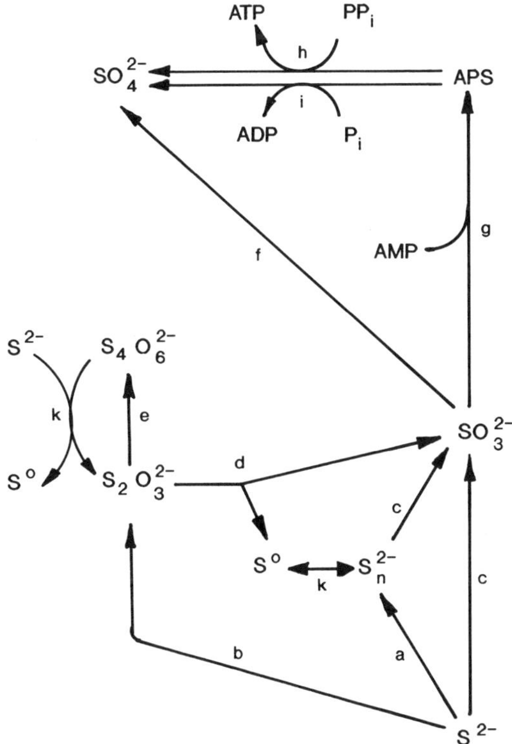

FIG. 1. Pathways of dissimilatory sulfur metabolism in phototrophic sulfur bacteria; (a), cytochromes *c* (and most probably sulfide quinone reductase); (b) flavocytochromes *c*; (c) reverse siroheme sulfite reductase; (d) thiosulfate sulfurtransferase; (e) thiosulfate : acceptor oxidoreductase; (f) sulfite : acceptor oxidoreductase; (g) adenylylsulfate reductase; (h) ATP-sulfurylase; (i) ADP-sulfurylase; (k) nonenzymatic reaction. S°, So-called elemental sulfur (at oxidation level zero), consisting of very long chain polysulfides and different elemental sulfur rings (S_8, S_6, etc.).

We first give a brief overview of the enzymes involved in photolithotrophic sulfur oxidation, focusing on those enzymes for which assays and purification procedures are described in this chapter. For more detailed information on phototrophic sulfur metabolism the reader is referred to a number of reviews.[1-5]

[1] H. G. Trüper and U. Fischer, *Philos. Trans. R. Soc. London B* **298,** 529 (1982).

[2] H. G. Trüper, *in* "Sulfur, Its Significance for Chemistry, for Geo-, Bio-, and Cosmosphere and Technology" (A. Müller and B. Krebs, eds.), p. 367. Elsevier, Amsterdam, 1984.

[3] D. C. Brune, *Biochim. Biophys. Acta* **975,** 189 (1989).

The splitting of thiosulfate to elemental sulfur and sulfite is catalyzed by an enzyme with rhodanese (thiosulfate sulfurtransferase, EC 2.8.1.1) activity. Sulfide is utilized by at least three different enzymes, leading to the production of polysulfides, thiosulfate, or sulfite, respectively. While the enzymatic step from sulfide to polysulfides and sulfur involves cytochromes of the *c* type, formation of thiosulfate from sulfide is catalyzed by flavocytochrome *c* in *Chromatium* and *Chlorobium* species. In *Chromatium vinosum* a "reverse" siroheme-containing sulfite reductase (EC 1.8.99.1) is responsible for sulfite formation from sulfide directly, as well as from polysulfides and/or elemental sulfur. A sulfide quinone reductase has been suggested to be involved not only in sulfide oxidation by *Chlorobium limicola* f. *thiosulfatophilum* but also to be of universal importance for sulfide oxidation by photoautotrophs.[6] Unfortunately, the oxidized product of this reaction has not yet been identified.

Sulfite is oxidized and reacts with AMP to form adenylylsulfate [adenosine-5′-phosphosulfate (APS)] catalyzed by the enzyme adenylylsulfate (APS) reductase (EC 1.8.99.2). This is an essential enzyme not only in species of the families Chromatiaceae and Chlorobiaceae[7,8] but has also been found to occur in dissimilatory sulfate-reducing prokaryotes[9] and in some *Thiobacillus* species.[10] Whereas APS reductases from the latter are soluble and reside in the cytoplasm, the enzyme from the purple sulfur bacteria appears to be associated with the membrane fraction ("chromatophores").[8] The degree of membrane binding can be strain specific as observed for different *Thiocapsa roseopersicina* and *Chr. vinosum* strains.[8,11,12] The APS reductase readily leaches from the chromatophores of *T. roseopersicina* 6311 (DSM 219) and EP2202 and from *Chr. vinosum* 1611 (DSM 182).[8,13] In contrast the enzyme from *T. rosoepersicina* M1,

---

[4] N. Pfennig and H. G. Trüper, *in* "The Prokaryotes" (A. Balows, H. G. Trüper, M. Dworkin, W. Harder, and K.-H. Schleifer, eds.), p. 3200. Springer-Verlag, New York, 1992.

[5] H. G. Trüper and N. Pfennig, *in* "The Prokaryotes" (A. Balows, H. G. Trüper, M. Dworkin, W. Harder, and K.-H. Schleifer, eds.), p. 3583. Springer-Verlag, New York, 1992.

[6] Y. Shahak, B. Arieli, and G. Hauska, *FEBS Lett.* **299,** 127 (1992).

[7] H. H. Thiele, *Antonie van Leeuwenhoek* **34,** 350 (1968).

[8] H. G. Trüper and H. D. Peck, Jr., *Arch. Microbiol.* **73,** 125 (1970).

[9] G. Fauque, J. LeGall, and L. L. Barton, *in* "Variations in Autotrophic Life" (J. M. Shively and L. L. Barton, eds.), p. 271. Academic Press, New York, 1991.

[10] S. Takakuwa, *in* "Organic Sulfur Chemistry: Biochemical Aspects" (S. Oae and T. Okuyama, eds.), p. 1. CRC Press, Boca Raton, Florida, 1992.

[11] C. Dahl and H. G. Trüper, *Z. Naturforsch. C: Biosci.* **44C,** 617 (1989).

[12] J. D. Schwenn and M. Biere, *FEMS Microbiol. Lett.* **6,** 19 (1979).

[13] J. Tschäpe, personal communication, (1993).

BBS, OP3, and 5811 is strictly membrane bound and little activity is found in the soluble fraction of cell extracts from *Chr. vinosum* D (DSM 180^{T}).[11-13] In *Chromatium warmingii* 6512 (ATCC 14959^{T},[14] once DSM 173) APS reductase is firmly membrane bound and cannot be solubilized.[15] In *Chromatium purpuratum, Chromatium gracile,* and *Ectothiorhodospira* species this enzyme could not be found.[1]

The intermediary formation of APS demands an additional enzyme to split off the sulfate moiety. This step allows conservation of the phosphate bond energy contained in APS. The enzyme ADP-sulfurylase (ADP-sulfate adenylyltransferase, EC 2.7.7.5) replaces the sulfate moiety by inorganic phosphate, thus producing adenosine diphosphate. ADP can then be disproportionated by the enzyme adenylate kinase, leading to the formation of one ATP and one AMP per two ADP. This pathway has been found in six species of the Chromatiaceae and in *Chlorobium vibrioforme* f. *thiosulfatophilum*.[1,16-18]

Another pathway that conserves the energy of APS is the replacement of its sulfate moiety by inorganic pyrophosphate, catalyzed by the action of ATP-sulfurylase (ATP-sulfate adenylyltransferase, EC 2.7.7.4). ATP-sulfurylase not only conserves the energy stored in APS but also conserves that of pyrophosphate, a common product in biosynthetic reactions, especially protein biosynthesis, and which is readily available during the exponential growth phase of cells. Some organisms contain both ATP- and ADP-sulfurylase, others only one or the other.[11,16,17]

An enzyme bypassing the formation of APS in phototrophic bacteria is sulfite : acceptor oxidoreductase (EC 1.8.2.1). This enzyme occurs in most members of the Chromatiaceae in addition to the APS pathway; in a few cases, instead of the latter.[1] In contrast to the enzyme found in members of the family Rhodospirillaceae, which is membrane bound in the four species so far examined,[19] sulfite : acceptor oxidoreductase from Chromatiaceae species is found in the soluble fraction.[16,20] The enzyme appears to be absent, or present only at low levels, in the Chlorobiaceae species. The *in vivo* acceptors of electrons from sulfite in the reactions catalyzed by sulfite : acceptor oxidoreductase and APS reductase are unknown.

[14] A superscript "T" indicates the type strain of a species.

[15] W. Leyendecker, unpublished results (Diploma Thesis, University of Bonn, Germany (1983).

[16] H. Ulbricht, Ph.D. Thesis, University of Bonn, Germany (1984).

[17] U. Bias and H. G. Trüper, *Arch. Microbiol.* **147,** 406 (1987).

[18] S. Khanna and D. J. D. Nicholas, *Arch. Microbiol.* **129,** 1365 (1983).

[19] O. Neutzling, C. Pfleiderer, and H. G. Trüper, *J. Gen. Microbiol.* **131,** 791 (1985).

[20] H. Ulbricht, unpublished results (Diploma Thesis, University of Bonn, Germany (1981).

TABLE I
OCCURRENCE OF SULFITE-OXIDIZING ENZYMES IN PHOTOTROPHIC SULFUR BACTERIA[a]

Species	APS pathway: APS reductase	APS pathway: ADP-sulfurylase	APS pathway: ATP-sulfurylase	Direct pathway (sulfite oxidoreductase)
Chlorobium limicola f. *thiosulfatophilum*	+(S)	−	+	(+)(M)
Chlorobium vibrioforme f. *thiosulfatophilum*	+(S)	+	−	N
Chromatium warmingii	+(M)	+	+	+(M)
Chromatium vinosum	+(M/S)	+	(+)	+(S)
Chromatium minutissimum	+(M)	+	(+)	+(S)
Thiocapsa roseopersicina	+(S)(M)	+	+	−/+(S)
Thiocystis violacea	+(M/S)	+	N	+
Chromatium gracile	−	+	(+)	+(S)
Chromatium purpuratum	−	−	(+)	+(S)
Ectothiorhodospira mobilis	−	(+)	−	+(M)
Ectothiorhodospira shaposnikovii	−	N	N	+(M)
Ectothiorhodospira halophila	−	N	N	+(M)

[a] +, Enzyme present; −, enzyme not found; (+) very low activity; N, not studied; M, membrane bound; S, soluble or solubilized (or solulizable).

From this short overview it is clear that the enzymatic steps described are not common to all phototrophic bacteria but that different pathways exist in different bacterial species. The occurrence of enzymes in different species of the Chromatiaceae and Chlorobiaceae is summarized in Table I.

Assay Method for Sulfite Reductase

Principle

In the assay described here sulfite reductase catalyzes sulfite reduction with enzymatically reduced methyl viologen as electron donor. The reaction mixture contains hydrogenase as an auxiliary enzyme, catalyzing methyl viologen reduction with hydrogen. Sulfite reductase activity can therefore be measured as H_2 consumption in a Warburg apparatus.

Neutzling *et al.*[19,21] developed a method to reduce methyl viologen electrochemically, thereby circumventing the necessity to purify a hydrogenase suited for the assay described above. This method has been used

[21] O. Neutzling, Ph.D. Thesis, University of Bonn, Germany (1981).

successfully to detect sulfite reductase in *Archaeoglobus fulgidus*[22] and can probably be adapted to assay the enzyme from phototrophic sulfur bacteria.

Procedure

Sulfite reductase activity is measured manometrically in Warburg flasks under hydrogen at 30°, following the method described by Schedel *et al.*[23] The main compartment contains (in a total volume of 2.5 ml) 1 ml of *Desulfovibrio gigas* hydrogenase (1–2 units) that is free from sulfite and thiosulfate reductase activities (prepared according to Trüper[24]), 0.1 ml of 75 m*M* methyl viologen, 0.1 ml of 1 *M* potassium phosphate (pH 7.0), and 0.1–1.3 ml of enzyme solution. The center well contains 0.2 ml of 5 *N* NaOH. The reaction is started by adding 0.3 ml of 50 m*M* Na_2SO_3 from the side arm. One enzyme unit is defined as that amount of protein that consumes 1 μmol of H_2 per minute.

Assay Methods for Sulfite : Acceptor Oxidoreductase

Ferricyanide-Dependent Assay

The enzyme is assayed by measuring oxidation of sulfite with ferricyanide ($\varepsilon_{420} = 1.09$ cm^2 μmol^{-1}) as electron acceptor at 25°. The assay contains 1 ml of 50 m*M* Tris-HCl, pH 9.0, with 0.1 m*M* $K_3[Fe(CN)_6]$, 1 m*M* Na_2SO_3 [the sulfite solution is always freshly prepared in 50 m*M* Tris-HCl, pH 8.0, containing 5 m*M* ethylenediaminetetraacetic acid (EDTA)], 8 m*M* EDTA, and an appropriate amount of enzyme extract.

As pointed out by Brune,[3] nonspecific reduction of intermediate electron carriers that in turn react with ferricyanide is a potential hazard in assaying sulfite : acceptor oxidoreductase activity. This artifact must be taken into account especially when trying to detect the enzyme in crude extracts.

Cytochrome c-Dependent Assay

The reaction mixture for the cytochrome *c*-coupled assay consists of 50 m*M* Tris-HCl, pH 7.8, 1.0 m*M* Na_2SO_3 (solution as given above), and 0.05 m*M* cytochrome *c*. The assay volume is 1 ml. Absorbance is measured in 1-cm glass cells at 550 nm and 25° (ε = 32,700 cm^2 mmol^{-1} in the reduced state).

[22] C. Dahl, N. Speich, and H. G. Trüper, this volume [23].

[23] M. Schedel, M. Vanselow, and H. G. Trüper, *Arch. Microbiol.* **121,** 29 (1979).

[24] H. G. Trüper, this volume [29].

Reaction rates with *c*-type cytochromes are an order of magnitude lower than those observed using $Fe(CN)_6^{3-}$ with the bacterial enzymes tested so far.[19]

Assay Methods for Adenylylsulfate Reductase

Adenylylsulfate reductase activity is measured in the direction of APS formation from sulfite and AMP with ferricyanide or cytochrome *c* as electron acceptor. Each molecule of SO_3^{2-} that reacts with AMP to form APS leads to the reduction of two molecules of ferricyanide or cytochrome *c*. Adenylylsulfate reductases from *T. roseopersicina* M1 and 6311 (DSM 219) function with ferricyanide and cytochrome *c* from *Candida crusei* but not with cytochrome *c* from horse heart, whereas APS reductase from *T. roseopersicina* BBS is active with both mammalian and fungal cytochrome *c*.[11] Adenylylsulfate reductases from *Chromatium minutissimum* and *Chr. vinosum* also do not react with cytochrome *c* from horse heart. Thus, the reactivity of APS reductases with the different electron acceptors can vary strain specifically and must be determined empirically for the enzyme under investigation.

Continuous Spectrophotometric Assays

Ferricyanide-Dependent Assay. The classic ferricyanide-dependent assay[25] is the easiest and least expensive method for monitoring column fractions during enzyme purification. The enzyme is assayed by measuring AMP-dependent reduction of ferricyanide (ε_{420} = 1.09 cm^2 μmol^{-1}) with sulfite. The assay contains 1 ml of 50 m*M* Tris-HCl, pH 8.0, with 0.5 m*M* $K_3[Fe(CN)_6]$, 4 m*M* Na_2SO_3 (the sulfite solution is always freshly prepared in 50 m*M* Tris-HCl, pH 8.0, containing 5 m*M* EDTA), 8 m*M* EDTA, and appropriately diluted extract. After determining the rate of unspecific ferricyanide reduction for 2–3 min the reaction is started by adding AMP to a final concentration of 0.4 m*M*.

Cytochrome c-Dependent Assay. The reaction mixture for the cytochrome *c*-coupled assay consists of 6.6 m*M* Tris-HCl, pH 9.0, 0.1 m*M* AMP, 1.0 m*M* Na_2SO_3 (solution is given above) and 0.1 m*M* cytochrome *c*.[26] The assay volume is 1 ml. Absorbance is measured in 1-cm glass cells at 550 nm and 25°.

Adenylylsulfate Formation Assay

In many strains of the phototrophic sulfur bacteria APS reductase cannot be detected via the continuous spectrophotometric assay. In all

[25] H. G. Trüper and L. A. Rogers, *J. Bacteriol.* **108,** 1112 (1971).

[26] R. M. Lyric and I. Suzuki, *Can. J. Biochem.* **48,** 344 (1970).

these strains APS reductase is strictly membrane bound and reaction with ferricyanide requires a 5- to 10-fold higher concentration of this electron acceptor in the assay.[11,13] This is probably because the membrane-bound enzymes cannot efficiently interact with water-soluble ferricyanide.[11] The assay described here could be used to search for APS reductase activity in organisms such as *Chr. gracile,* in which the presence of the enzyme has not been proved, but the enzyme has been presumed to exist because of the presence of ADP-sulfurylase.[1]

Principle. In a first step APS reductase catalyzes APS generation from sulfite and AMP with ferricyanide as electron acceptor. In a second step ATP-sulfurylase transforms the newly formed APS into ATP,[27] which is then determined via hexokinase and glucose-6-phosphate dehydrogenase-coupled NADP reduction in a spectrophotometric test system.

Procedure. The reaction mixture (1.0-ml total volume) contains 50 m*M* Tris-HCl, pH 7.0, 4.0 m*M* $K_3[Fe(CN)_6]$, 10 m*M* Na_2SO_3 (the sulfite solution is always freshly prepared in 50 m*M* Tris-HCl, pH 8.0, containing 5 m*M* EDTA), 8 m*M* EDTA, and enzyme extract. The reaction mixture is preincubated in small centrifuge tubes at 30° for 5 min. To minimize evaporation the tubes should be closed with glass marbles. The reaction is started by addition of appropriately diluted extract. After the desired incubation period, the reaction is terminated by boiling. Denatured protein is removed by centrifugation at 25,000 *g* for 10 min. An aliquot of supernatant is used for quantitative determination of generated APS: the reaction mixture contains 100 m*M* Tris-HCl, pH 8.0, 20 m*M* β-D-glucose, 10 m*M* $MgCl_2$, 0.5 m*M* NADP, 1 m*M* $Na_4P_2O_7$, 5 units of glucose-6-phosphate dehydrogenase, up to 120 nmol of APS, 5 units of hexokinase, 3 units of ATP-sulfurylase (obtained from Sigma, St. Louis, MO), and water to make up a total volume of 1 ml. The reaction is started by addition of ATP-sulfurylase after a preincubation time during which traces of ATP have been used up. The change in A_{340} is monitored and the APS concentration calculated from the total change in absorption and the extinction coefficient for NADPH (ε_{340} = 6.22 cm^2 μmol^{-1}).

Assay Methods for ATP-Sulfurylase

Coupled Spectrophotometric Assay

Principle. ATP-sulfurylase from phototrophic sulfur bacteria is measured in the physiological direction of ATP generation from the reaction of APS with pyrophosphate. In the coupled spectrophotometric assay the rate of ATP production is followed via the phosphorylation of glucose

[27] B. P. Cooper and H. G. Trüper, *Arch. Microbiol.* **141,** 384 (1985).

and the subsequent oxidation to D-6-P-glucono-δ-lactone by NADP.[28,29] The side reaction of pyrophosphatase, which rapidly depletes pyrophosphate from reaction mixtures for ATP-sulfurylase from dissimilatory sulfate reducers,[22] does not disturb the ATP-sulfurylase assay as severely in phototrophic bacteria. This is probably because ATP-sulfurylase from these organisms has a much higher affinity for pyrophosphate than has pyrophosphatase. In *Chr. warmingii* K_m values of 8 μM versus 0.15 mM have been observed for ATP-sulfurylase and pyrophosphatase, respectively.[16]

Procedure. In a total volume of 1.0 ml, combine 100 mM Tris-HCl, pH 8.0, 20 mM β-D-glucose, 10 mM $MgCl_2$, 0.5 mM Na-NADP, 1 mM $Na_4P_2O_7$ (or varied for kinetic studies), 3 units of glucose-6-phosphate dehydrogenase, 3 units of hexokinase, 0.05 mM APS (or varied for kinetics), and enzyme extract.[27] Reduction of NADP is followed at 340 nm (ε_{340} = 6.22 cm^2 μmol^{-1}) and 30°. The assay described is well suited to follow ATP-sulfurylase during purification steps. Kinetic studies of ATP-sulfurylase can best be performed using the method of Farley *et al.*,[30] by which the incorporation of $^{35}SO_4^{2-}$ into APS is determined. For an extensive review of this procedure and various other methods to assay, especially assimilatory ATP-sulfurylases, see Segel *et al.*[31]

ATP Formation Assay

Principle. Occasionally it is necessary to determine ATP-sulfurylase activity without the presence of auxiliary enzymes, for example, when determining temperature and pH optima of the enzyme. In a first step ATP-sulfurylase catalyzes ATP generation from APS with pyrophosphate. In a second step the ATP formed is spectrophotometrically determined.

Procedure. The reaction mixture (1.0-ml total volume) contains 100 mM Tris-HCl, pH 7.6, 1 mM APS (or varied for kinetic studies), 10 mM $MgCl_2$, 1 mM PP_i (or varied for kinetics), ATP-sulfurylase, and distilled water. The assay is incubated at 30° and the reaction is started by addition of appropriately diluted extract. After 10 min the ATP-sulfurylase reaction is terminated by boiling for 90 sec. Denatured protein is removed by centrifugation at 25,000 *g* for 10 min. An aliquot of supernatant is used for quantitative determination of generated ATP by a standard hexokinase

[28] R. J. Guillory and R. R. Fischer, *Biochem. J.* **129,** 471 (1972).

[29] J. W. Groenestijn, M. H. Deinema, and A. J. B. Zehnder, *Arch. Microbiol.* **148,** 14 (1987).

[30] J. R. Farley, D. F. Cryns, Y. H. Joy Yang, and I. H. Segel, *J. Biol. Chem.* **251,** 4389 (1976).

[31] I. Segel, F. Renosto, and P. A. Seubert, this series, Vol. 143, p. 334.

and glucose-6-phosphate dehydrogenase-coupled spectrophotometric test system.[32]

Assay Methods for ADP-Sulfurylase

ADP-sulfurylase combined with adenylate kinase is often cited as an alternative route for ATP synthesis from APS in sulfide- or thiosulfate-oxidizing bacteria.[33,34] In organisms such as the many Chromatiaceae species that contain both ADP- and ATP-sulfurylase, the former might allow energy conservation from APS even under pyrophosphate-limited conditions.

Coupled Spectrophotometric Assay

Principle. ADP-sulfurylase from phototrophic sulfur bacteria is measured in the physiological direction of ADP generation from the reaction of APS with phosphate. The rate of ADP production is followed via the reaction of ADP and phosphoenolpyruvate (PEP) yielding pyruvate, which is subsequently reduced to lactate by NADH.[35] Measurements of ADP-sulfurylase activity in partially purified preparations can be interfered with by other enzymatic activities, namely ATP-sulfurylase acting on contaminating pyrophosphate and ATPase splitting the resulting ATP into phosphate and ADP. As pointed out by Renosto *et al.*,[36] some of the "ADP-sulfurylases" described in the literature might not even represent independent entities. Therefore, microorganisms that oxidize sulfide or thiosulfate to sulfate but lack ATP-sulfurylase[17] are preferred sources for the detection, purification, and physiological evaluation of ADP-sulfurylase.

Procedure. In a total volume of 1.0 ml the reaction mixture contains 100 m*M* Tris-HCl buffer, pH 8.0, 1 m*M* APS (or varied for kinetic studies), 10 m*M* potassium phosphate (or varied for kinetics), 10 m*M* $MgCl_2$, 0.16 m*M* Na-NADH, 0.5 m*M* PEP, 20 units of pyruvate kinase, 20 units of

[32] W. Lamprecht and I. Trautschold, *in* "Methoden der enzymatischen Analyse" (H. U. Bergmeyer, ed.), p. 2024. Verlag Chemie, Weinheim, 1970.

[33] D. W. Smith and W. R. Strohl, *in* "Variations in Autotrophic Life" (J. M. Shively and L. L. Barton, eds.), p. 121. Academic Press, San Diego, 1991.

[34] D. P. Kelly, *in* "Autotrophic Bacteria" (H. G. Schlegel and B. Bowien, eds.), p. 193. Science Tech, Madison, Wisconsin/Springer-Verlag, Berlin, 1989.

[35] D. Jaworek and J. Welsch, *in* "Methods of Enzymatic Analysis" (H. U. Bergmeyer, ed.), Vol. 7, p. 364. VCH Publ. Weinheim, 1985.

[36] F. Renosto, R. L. Martin, J. L. Borrell, D. C. Nelson, and I. H. Segel, *Arch. Biochem.* **290,** 66 (1991).

lactate dehydrogenase, enzyme, and distilled water.[27] The reaction is started by adding the enzyme. The decrease in NADH, which is equivalent to ADP formation, is measured spectrophotometrically at 340 nm (ε_{340} = 6.22 cm^2 μmol^{-1}) and 30°.

ADP Formation Assay

Principle. In the ADP formation assay ADP synthesis from APS and phosphate is separated from the determination of generated ADP. The assay is especially useful when measuring temperature and pH optima of ADP-sulfurylase.

Procedure. The reaction mixture (1.0-ml total volume) contains 100 m*M* Tris-HCl, pH 7.6, 50 m*M* K_2HPO_4, pH 7.6, 1 m*M* APS, 10 m*M* $MgCl_2$, and enzyme extract. The assay is preincubated at the desired temperature and the reaction is started by addition of appropriately diluted extract. After 10 min the ADP-sulfurylase reaction is terminated by boiling for 90 sec. Denatured protein is removed by centrifugation at 25,000 *g* for 10 min. An aliquot of supernatant is used for quantitative determination of generated ADP in a coupled enzyme assay that contains (in 1 ml) 100 m*M* Tris-HCl, pH 8.0, 0.4 m*M* PEP, 2 m*M* $MgCl_2$, 0.12 m*M* Na-NADH, 11 units of lactate dehydrogenase, 10 units of pyruvate kinase, and up to 100 nmol of ADP in the sample. The reaction is started with pyruvate kinase. The change in A_{340} is monitored at 30° and the ADP concentration calculated from the total change in absorption and the extinction coefficient of NADH (ε_{340} = 6.22 cm^2 μmol^{-1}).

Synthesis of Adenosine-5′-phosphosulfate

Commercially available APS is rather impure (85%; impurity, AMP) and expensive. Therefore, a number of chemical methods[37,38] and biochemical methods[31,39] for the synthesis of unlabeled and radioactively labeled APS have been described. Most of these methods result in low yields or are unacceptably time consuming. In the following an optimized synthetic procedure, developed by Cooper and Trüper[40] and modified by Imhoff[41] that yields APS in millimole amounts for laboratory-scale use is described.

Adenylylsulfate synthesis is carried out with cell extracts of *Thiobacillus denitrificans* RT (DSM 807),[39,40] which contains APS reductase with

[37] J. Baddiley, J. G. Buchanan, and R. Letters, *J. Chem. Soc.*, 1067 (1957).

[38] R. Cherniak and E. A. Davidson, *J. Biol. Chem.* **239,** 2986 (1964).

[39] C. A. Adams, G. M. Warnes, and D. J. D. Nicholas, *Anal. Biochem.* **42,** 207 (1971).

[40] B. P. Cooper and H. G. Trüper, *Z. Naturforsch. C: Biosci.* **34C,** 346 (1979).

[41] J. F. Imhoff, *Arch. Microbiol.* **132,** 197 (1982).

high specific acitivity.[42] The organism is grown in the medium described by Schedel and Trüper.[43] Cells are suspended in about three times their volume of 50 m*M* Tris-HCl, pH 7.6, and disrupted by passing the suspension two or three times through a French pressure cell at about 138 MPa. The broken cell mass is then centrifuged at 17,000 *g* for 15 min to remove larger fragments. The synthesis reaction mixture, 100-ml total volume, contains 50 m*M* Tris-HCl, pH 7.6, 20 m*M* AMP, 5 m*M* Na_2SO_3 (dissolved in 5 m*M* EDTA, pH 8.0), 5 m*M* $K_3[Fe(CN)_6]$, and 25 ml of the crude extract. The reaction is started by adding Na_2SO_3 and incubated at 30°. After 5, 10, 15, 20, 25, and 30 min an additional 0.5 mmol of Na_2SO_3 is added. After 10, 20, 25, 35, and 45 min, 0.5 mmol of $K_3[Fe(CN)_6]$ is added. In this way a total of 3.5 mmol of Na_2SO_3 and 3.0 mmol of $K_3[Fe(CN)_6]$ is normally added in a synthesis of 100 ml. The reaction is terminated on ice after 55 min. The APS yield is usually more than 90% of the AMP in the reaction mixture. Proteins and ferrocyanide are removed on Sephadex G-25 (4.8 × 85 cm, equilibrated with 50 m*M* Tris-HCl, pH 7.6). The fractions containing APS are pooled and separated on DEAE-Sephadex A-25 (3 × 40 cm) equilibrated with 200 m*M* Tris-HCl, pH 7.6, and eluted with a linear gradient from 250–550 m*M* Tris-HCl, pH 7.6 (gradient volume, 2 liter). During chromatography APS is monitored at 259 nm and the concentration determined according to the extinction coefficient (ε_{259} = 15.4 cm^2 μmol^{-1}). Concentrations of APS in the fractions of the elution maximum are approximately 9 m*M*. Adenylylsulfate is identified and differentiated from impurities such as AMP by thin-layer chromatography.

High-Performance Thin-Layer Chromatography of Adenylylsulfate

High-performance thin-layer chromatography can be used to trace the distribution of APS in purification fractions. Five microliters of each fraction is spotted onto high-performance thin-layer chromatography (HPTLC) plates coated with silica gel F_{254} [Merck (Rahway, NJ) 5628], which are then developed using a mixture of 2-propanol, ammonia (33%), and water (6 : 3 : 1).[40]

Enzyme Purifications

Introductory Remarks

All enzyme purifications are performed at 4°. To date, most enzymes of dissimilatory sulfur metabolism in phototrophic sulfur bacteria have not been purified to homogeneity.

[42] T. J. Bowen, F. C. Happold, and B. F. Taylor, *Biochim. Biophys. Acta* **118,** 566 (1966).
[43] M. Schedel and H. G. Trüper, *Biochem. Biophys. Acta* **568,** 454 (1978).

Growth Conditions

All phototrophic sulfur bacteria are grown in Pfennig's medium[44] containing 0.036% (w/v) sulfide for purple sulfur bacteria and 0.06% (w/v) sulfide for green sulfur bacteria. The pH of the medium is adjusted to 7.3 and 6.8 for purple and green sulfur bacteria, respectively. Cultures are inoculated (10%, v/v) and grown anaerobically in completely filled 1-liter screw-cap bottles or 20-liter carboys at 30° and 2000–10,000 lx light intensity (depending on the organism). Carboys are stirred magnetically. When the cultures are free of intra- or extracellular sulfur globules they are fed with a neutral sulfide solution.[45] Cells are harvested by centrifugation and stored as cell paste at −20°.

Sulfite Reductase from Chromatium vinosum D (DSM 180^T)

Purification. The purification method described here was developed by Schedel *et al.*[23] Frozen cell paste is thawed and suspended in 50 m*M* potassium phosphate buffer, pH 7.0. The cells are broken by passing the suspension two or three times through a French pressure cell at about 138 MPa. The crude extract is centrifuged at 140,000 *g* for 2 hr. The supernatant, which contains the sulfite reductase, is fractionated with $(NH_4)_2SO_4$. The protein precipitates between 45 and 70% saturation and is collected by centrifugation, dissolved in 50 m*M* potassium phosphate, pH 7.0, and applied to Ecteola cellulose (Serva, Heidelberg, Germany) (3 × 15 cm) previously equilibrated with the same buffer. The protein is eluted with a linear gradient between 50 and 700 m*M* potassium phosphate, pH 7.0. Fractions with sulfite reductase activity (green colored, coming off at an elution molarity of about 400 m*M*) are combined, dialyzed against 50 vol of 50 m*M* potassium phosphate, pH 7.0, and then stirred into an appropriate volume of alumina Cγ-gel slurry (Serva, Heidelberg, Germany) suspended in water (~0.3 ml of gel slurry per milligram of protein). The gel is centrifuged, washed once with 50 m*M* potassium phosphate, pH 7.0, and then eluted with stepwise increasing concentrations of potassium phosphate, pH 7.0 (80, 110, 140, 170, and 210 m*M*). The eluates at 110, 140, and 170 m*M*, which contain sulfite reductase, are pooled. To lower the potassium phosphate concentration in the combined alumina Cγ-gel fractions 2 vol of distilled water is added. Sulfite reductase is then adsorbed on hydroxyapatite, washed once with 50 m*M* potassium phosphate, pH 7.0, and eluted with stepwise increasing concentrations of potassium phosphate, pH 7.0 (80, 110, 140, 170, and 210 m*M*) in the same way as from alumina Cγ-gel. The 80, 110, and 140 m*M* eluates are combined and

[44] Medium 2 in Ref. 4.
[45] E. Siefert and N. Pfennig, *Arch. Microbiol.* **139,** 100 (1984).

TABLE II
PARTIAL PURIFICATION OF SULFITE REDUCTASE FROM *Chromatium vinosum* D (DSM 180^T)

Step	Protein (mg)	Activity (units)	Specific activity (units/mg)
1. Soluble fraction	12,706	50.0	0.004
2. 45–70% $(NH_4)_2SO_4$	6,636	32.0	0.005
3. Ecteola eluate	551	34.0	0.061
4. Alumina Cγ eluate	269	23.0	0.085
5. Hydroxyapatite eluate	145	19.0	0.131
6. Sephadex G-200	55	9.0	0.163

brought to 80% saturation with solid $(NH_4)_2SO_4$. The precipitated protein is collected by centrifugation, dissolved in 50 m*M* potassium phosphate, pH 7.0, and applied to a Sephadex G-200 column (3 × 100 cm) equilibrated with the same buffer. Green fractions of the eluate are combined and frozen at −20°. Table II summarizes the purification of sulfite reductase that yields an 80% pure enzyme preparation.

Properties. Sulfite reductase occurs in cells grown photoautotrophically on sulfide and carbon dioxide, but not in cell grown photoheterotrophically on malate and sulfate, suggesting a function in sulfide oxidation rather than in sulfate assimilation. The *Chr. vinosum* sulfite reductase reduces sulfite but not di-, tri-, or tetrathionate or thiosulfate. The enzyme is active with reduced methyl or benzyl viologen as electron donors but not with $NADH_2$ or $NADPH_2$. Besides sulfide, trithionate (17%) and thiosulfate (28%) are formed as reaction products (as is typical of dissimilatory sulfite reductases). The pH optimum for activity is pH 6.0.

Chromatium vinosum sulfite reductase consists of two different subunits with molecular masses of 42,000 and 37,000 Da. Analytical gel filtration gives an apparent molecular weight of 280,000 suggesting an $\alpha_4\beta_4$ structure of the native enzyme. The subunit composition of the enzyme is, however, not clearly settled, because Kobayashi *et al.*[46] reported the isolation of a sulfite reductase with a molecular weight of 180,000 from photoautotrophically grown *Chr. vinosum*. Their enzyme is probably the same as that described by Schedel *et al.*[23] On the basis of a molecular weight of 180,000 a subunit composition of $\alpha_2\beta_2$ is suggested for dissimilatory sulfite reductase from *Chr. vinosum,* identical to the structure reported for most dissimilatory sulfite reductases from sulfate-reducing prokaryotes.

[46] K. Kobayashi, E. Katsura, T. Kondo, and M. Ishimoto, *J. Biochem.* (*Tokyo*) **84,** 1209 (1978).

The oxidizing enzyme has absorption maxima at 280, 392, 595, and 724 nm. No significant change occurs on reduction with dithionite or sulfide. The molar extinction coefficients of *Chr. vinosum* sulfite reductase at 392, 595, and 724 nm are estimated to be 302,000, 98,000, and 22,000 $cm^2\ mmol^{-1}$. The heme prosthetic group can be released from the enzyme by mixing 1 vol of enzyme solution in 50 m*M* potassium phosphate, pH 7.0, with 5 vol of 50 m*M* HCl in acetone at 0°. The spectrum obtained resembles the spectrum of siroheme prepared in an analogous way from *Escherichia coli* sulfite reductase.[47] On the basis of a molecular weight of 280,000 analyses for acid-labile sulfide[48] and iron content[49] give values of 47 mol of sulfide and 51 mol of iron per mol of sulfite reductase.

Sulfite : Acceptor Oxidoreductase from Chromatium purpuratum (DSM 1591^T)

Among the purple sulfur bacterial sulfite : acceptor oxidoreductases only the enzymes from *Chr. vinosum* (DSM 180^T), *Chr. gracile* (DSM 203^T), *Chr. minutissimum* (DSM 1376^T), and *Chromatium purpuratum* (DSM 1591^T) have been investigated in any detail.[16,20] Unlike the enzymes from Rhodospirillaceae species, all of these enzymes are soluble proteins.

Purification. Sulfite : acceptor oxidoreductase from *C. purpuratum* is a rather labile enzyme. It is especially important to note that recoveries of the enzyme after only one freeze/thaw cycle are extremely low. Furthermore, stability of the enzyme decreases in diluted solutions. Freezing and excessive dilution of the enzyme should therefore be avoided.

Frozen cell paste is thawed, suspended in about twice its volume of 50 m*M* Tris-HCl buffer, pH 7.6, and the cells are broken by ultrasonic treatment (20 sec ml^{-1} in 10-sec intervals at 4°). The broken cell mass is centrifuged at 17,000 *g* for 15 min and the supernatant subjected to ultracentrifugation (140,000 *g*, 3 hr). The supernatant, which contains the sulfite : acceptor oxidoreductase, is fractionated with $(NH_4)_2SO_4$. The protein precipitates between 45 and 60% saturation and is collected by centrifugation, dissolved in 50 m*M* Tris-HCl, pH 7.6, and desalted by gel filtration on Sephadex G-25. The extract is applied to a column of DEAE-cellulose (1.5 × 45 cm) equilibrated with 50 m*M* Tris-HCl, pH 7.6. Via a linear gradient from 0 to 600 m*M* NaCl sulfite : acceptor oxidoreductase is recovered at 500–540 m*M* NaCl. Active sulfite : acceptor oxidoreductase fractions are combined and concentrated by ultrafiltration (PM 10 mem-

[47] L. M. Siegel, M. J. Murphy, and H. Kamin, this series, Vol. 52, p. 436.

[48] T. E. King and R. O. Morris, this series, Vol. 10, p. 634.

[49] E. B. Sandell, "Colorimetric Determination of Trace Metals." Interscience Publ., New York, 1944.

TABLE III
PARTIAL PURIFICATION OF SULFITE : ACCEPTOR OXIDOREDUCTASE FROM *Chromatium purpuratum* DSM 1591

Step	Volume (ml)	Protein (mg)	Activity (units)	Specific activity (units/mg)
1. Soluble fraction	24.5	284.2	221.6	0.78
2. 45–60% $(NH_4)_2SO_4$	3.1	46.3	186.1	4.02
3. DE-52 eluate	11.5	6.1	74.1	12.2
4. Sephadex G-200 eluate	5.0	0.5	31.5	63.0

branes). The concentrated protein solution is subjected to gel filtration on Sephadex G-200 (1.5 × 90 cm) equilibrated with 50 m*M* Tris-HCl, pH 7.6. Table III summarizes the purification of sulfite : acceptor oxidoreductase.

Properties. Analytical gel filtration of sulfite : acceptor oxidoreductase from *Chr. purpuratum* gives an apparent molecular weight of 87,000. In contrast to sulfite : acceptor oxidoreductases from *Chr. vinosum* and *Chr. gracile,* which show activity with cytochrome *c* from *C. crusei,* the enzyme from *Chr. purpuratum,* like that from *Chr. minutissimum,* does not react with cytochrome *c* as electron acceptor. The pH optimum is 9.3, and the K_m values for sulfite and ferricyanide are 1.8 and 0.41 m*M*, respectively.

Nothing is known about prosthetic groups of phototrophic bacterial sulfite : acceptor oxidoreductases. Fischer[50] remarked that the absorption spectrum of the enzyme resembled that of cytochrome *c'*, but noted that cytochrome *c'* may have been present as a contaminant in the preparation. The enzyme from *Chr. minutissimum* does not contain cytochrome *c*.[20]

Soluble Adenylylsulfate Reductase from Thiocapsa roseopersicina (DSM 219)

Adenylylsulfate reductase from *T. roseopersicina* DSM 219 was first purified by Trüper and Rogers.[25] We improved this purification procedure and omitted some of the originally included, time-consuming ammonium sulfate precipitations. The reproducible procedure described yields APS reductase with a specific activity around 11.5 μmol of APS min^{-1} (mg protein)$^{-1}$, which is higher than the 8.7 μmol of ferricyanide min^{-1} (mg protein)$^{-1}$, which is higher than the 8.7 μmol of ferricyanide min^{-1} (mg protein)$^{-1}$ obtained with the original procedure.

[50] U. Fischer, *in* "Sulfur, Its Significance for Chemistry, for Geo-, Bio-, and Cosmosphere and Technology" (A. Müller and B. Krebs, eds.), p. 383. Elsevier, Amsterdam, 1984.

Preparation of Crude Extracts. Cells are taken up in about twice their volume of 50 m*M* Tris-HCl, pH 7.5, homogenized, and broken up by ultrasonic disintegration (60 sec ml^{-1} in intervals of 30 sec at 4°). The broken cell mass is then centrifuged at 25,000 *g* for 30 min at 4°. The pellet, containing cell wall debris and elemental sulfur globules, is discarded. The crude extract is subjected to further centrifugation at 140,000 *g* for 3 hr. Solid ammonium sulfate is added to the supernatant to 40% saturation at 0°. The precipitated protein is collected by centrifugation (17,000 *g*, 20 min) and discarded. The supernatant is applied to a column of Phenyl-Sepharose CL-4B (47-ml volume; flow rate, 30 ml/hr) equilibrated with 50 m*M* Tris-HCl, pH 7.5, containing ammonium sulfate at 40% saturation. The enzyme is eluted with a linear gradient (400 ml) of ammonium sulfate between 40 and 0% saturation. The active enzyme fractions are pooled and desalted by a Sephadex G-25 column. The eluate is applied to a column of DEAE-cellulose (4 × 10 cm) equilibrated with 50 m*M* Tris-HCl, pH 7.5, and eluted with a gradient (400 ml) of 0–500 m*M* NaCl in the same buffer. The flow rate employed is 30 ml/hr. Active APS reductase fractions are pooled and concentrated by ultrafiltration using an Amicon (Danvers, MA) PM30 filter. The concentrated protein solution is transferred into 50 m*M* Tris-HCl, pH 7.5, via a Pharmacia (Piscataway, NJ) PD-10 column and subjected to gel filtration on a Sephacryl S-300 column (1.5 × 90 cm; flow rate, 15 ml/hr), equilibrated with the same buffer. Table IV summarizes the purification of APS reductase.

Properties. The molecular weight of APS reductase from *T. roseopersicina* DSM 219 is estimated to be 180,000. Each molecule of APS reductase contains one FAD group, four atoms of nonheme iron, and six labile sulfide groups. When purified as described above, enzyme preparations contain *c*-type cytochrome, amounting to two heme groups per enzyme molecule. The enzyme from *Thiocystis violacea* is the only other known

TABLE IV
PARTIAL PURIFICATION OF SOLUBLE ADENYLYLSULFATE REDUCTASE FROM *Thiocapsa roseopersicina* DSM 219

Step	Volume (ml)	Protein (mg)	Activity (units)	Specific activity (units/mg)
1. Crude extract	200	388.0	155	0.4
2. Soluble fraction	145	178.5	105	0.6
3. 40% $(NH_4)_2SO_4$	135	130.5	88.5	0.7
4. Phenyl-Sepharose	35	23.5	54	2.3
5. DE-52 eluate	2	5.0	39	7.8
6. Sephacryl S-300	8	2.0	25	11.5

APS reductase that contains heme groups,[51] whereas the corresponding enzyme from *Chl. limicola* f. *thiosulfatophilum* lacks heme groups.[52]

Cytochrome *c* from *C. crusei* and ferricyanide serve as electron acceptors. With ferricyanide as the electron acceptor, the pH optimum of the enzyme is 8.0; with cytochrome *c*, the pH optimum is at 9.0. The K_m values for ferricyanide and cytochrome *c* are 0.13 and 0.033 m*M*, respectively. K_m values for sulfite and AMP are 1.5 and 0.073 m*M* in the ferricyanide-coupled assay, and 0.093 and 0.05 m*M* when the enzyme is measured with cytochrome *c*. Substrate inhibition with AMP in the ferricyanide and in the cytochrome *c*-dependent assay occurs above 0.5 and 0.08 m*M*, respectively.

Membrane-bound Adenylylsulfate Reductase from Thiocapsa roseopersicina M1

With the methods applied so far most attempts have failed to solubilize the strictly membrane-bound APS reductases from phototrophic sulfur bacteria. As an example, the enzyme from *Chr. warmingii* cannot be solubilized by treatment with detergents, ethylene glycol, or glycerol, nor is it found in the soluble fraction after incubation with high concentrations of various salts.[15] The enzyme from *T. roseopersicina* M1 presents an exception because it is solubilized by treatment of chromatophores with 1% Triton X-100 and can be enriched by column chromatography.

Procedure. Membrane-bound APS reductase from strain M1 can be enriched as follows[11]: crude extracts are prepared as described for *T. roseopersicina* DSM 219. To achieve solubilization of the strictly membrane-bound enzyme the supernatant of a first ultracentrifugation is incubated with 1% Triton X-100 for 15 min at room temperature. After a second ultracentrifugation (140,000 *g*, 2 hr) the detergent is removed from the supernatant obtained by chromatography on BioBeads SM2 (column size, 2 × 20 cm; flow rate, 18 ml/hr). It is important to remove the Triton as fast as possible because its presence diminishes enzyme activity. The resuspended pellet (50 m*M* Tris-HCl, pH 7.5) of a subsequent 30% ammonium sulfate precipitation is loaded on a phenyl-Sepharose CL-4B column (gel volume, 47 ml) equilibrated with 50 m*M* Tris-HCl, pH 7.5. A linear gradient from 0 to 0.8% Triton X-100 in the same buffer is applied and APS reductase elutes at a calculated detergent concentration of 0.5–0.7%. Active APS reductase fractions are pooled, freed from detergent, and concentrated by ultrafiltration using an Amicon PM30 filter. The concentrated protein solution is further purified by gel filtration on Sephacryl S-

[51] G. Tatzki, unpublished results (Diploma Thesis, University of Bonn, Germany (1979).

[52] J. Kirchhoff and H. G. Trüper, *Arch. Microbiol.* **100,** 115 (1974).

300 (column size, 1.5 × 90 cm). Table V summarizes the purification of adenylylsulfate reductase.

Properties. With the methods described APS reductase is enriched 23-fold. The enzyme seems to aggregate after removal of Triton X100 because it elutes within the void volumen of the gel-filtration column, indicating a molecular weight of more than 5×10^6. pH and temperature optima are 7.0 and 40°. K_m values for AMP, sulfite, and ferricyanide are 0.1, 0.34, and 1.0 m*M*, respectively. Substrate inhibition with AMP occurs above 0.5 m*M*. The enzyme functions with cytochrome *c* from *C. crusei* but not with cytochrome *c* from horse heart.

Partial Purification of ATP-Sulfurylase from Chlorobium limicola f. thiosulfatophilum (DSM 257)

Purification procedures have been described for the enzymes from the green sulfur bacterium *Chl. limicola* f. *thiosulfatophilum* (DSM 257)[17] and the purple sulfur bacterium *Chr. warmingii* (ATCC 14959^T).[16] Neither of the two enzymes has been enriched to homogeneity. This section describes the purification procedure of ATP-sulfurylase from *Chl. limicola* f. *thiosulfatophilum*.

Cells are suspended in 2 ml of 50 m*M* Tris-HCl, pH 7.6, per gram wet weight and ruptured by sonification (30 sec ml^{-1} in 20-sec intervals at 4°) in a Schöller sonifier. Heavy particles are removed by centrifugation at 27,000 *g* for 10 min. The supernatant, called crude extract, is further centrifuged at 118,000 *g* for 150 min, resulting in a soluble and a particulate fraction. The soluble fraction obtained after ultracentrifugation is fractionated with $(NH_4)_2SO_4$. The precipitate formed between 40 and 60% saturation is collected by centrifugation at 27,000 *g* for 15 min and is desalted

TABLE V
PARTIAL PURIFICATION OF ADENYLYLSULFATE REDUCTASE FROM *Thiocapsa roseopersicina* M1

Step	Volume (ml)	Protein (mg)	Activity (units)	Specific activity (units/mg)
1. Crude extract	160.0	2288.0	64.0	0.028
2. Pellet of first ultracentrifugation	94.0	1241.0	57.7	0.047
3. Supernatant of second ultracentrifugation	132.2	730.4	36.5	0.050
4. 30% $(NH_4)_2SO_4$	8.6	120.4	15.7	0.130
5. Phenyl-Sepharose	2.0	12.0	5.5	0.455
6. Sephacryl S-300	6.0	5.6	3.6	0.637

by gel filtration with Sephadex G-25 equilibrated in 50 m*M* Tris-HCl buffer, pH 7.6. The enzyme is then applied to a DE-52 cellulose column (2.0 × 10 cm), equilibrated with 50 m*M* Tris-HCl, pH 7.6. The column is washed with 2 vol of the starting buffer. ATP-sulfurylase is eluted from the column with a gradient between 0 and 250 m*M* NaCl (400-ml total volume). The enzyme is eluted from the DE-52 cellulose column with a twofold maximum between 0.18 and 0.2 *M* NaCl. It is at this stage free from adenylate kinase, which elutes at 250 m*M* NaCl. The pooled active ATP-sulfurylase fractions are also free of inorganic pyrophosphatase (EC 3.6.1.1), which is not eluted from the DE-52 column under the applied conditions. The pooled active fractions from DE-52 cellulose are concentrated by ultrafiltration (Diaflo PM10 membrane) to about 1 ml and passed through a Sephacryl S-200 gel-filtration column (2.0 × 85 cm). ATP-sulfurylase has been further purified by preparative isoelectric focusing (pH range, 3.5–9.5), resulting in a 60-fold-enriched enzyme preparation with a specific activity of 9.9 units (mg protein)$^{-1}$. Because recovery after this step is extremely low, it is usually omitted and characterization of catalytic properties is done with fractions eluted from Sephacryl S-200 that are free of interfering activities. Table VI summarizes the purification of ATP-sulfurylase.

Properties of ATP-Sulfurylase. The *Chl. limicola* f. *thiosulfatophilum* enzyme has a molecular weight of 230,000. The isoelectric point determined by isoelectric focusing is 4.8. The pH optima in 100 m*M* Tris-maleate and 100 m*M* Tris-HCl buffer are 7.2 and 8.6, respectively.

Hg^{2+}, Ag^{2+}, and *p*-chloromercuribenzoate (*p*-CMB) (0.1 m*M*, each) cause a decrease in ATP-sulfurylase activity of 100%, whereas Cs^{2+} (10 m*M*) and Cd^{2+} (10 m*M*) strongly inhibit the enzyme, suggesting unspecific denaturation and inactivation of thiol groups of the enzyme. The inhibitory effect of *p*-CMB and Cd^{2+} can be reduced to about 40% by prior treatment of the enzyme with dithioerythritol (10 and 1 m*M*), indicating the involvement of SH groups in the catalytic reaction. Under physiological condi-

TABLE VI
PARTIAL PURIFICATION OF ATP-SULFURYLASE FROM *Chlorobium limicola* f. *thiosulfatophilum*

Step	Volume (ml)	Protein (mg)	Activity (units)	Specific activity (units/mg)
1. Crude extract	74.0	532.8	186.4	0.35
2. Soluble fraction	54.2	183.6	97.4	0.53
3. DE-52 eluate	8.4	6.2	9.2	1.0
4. Sephacryl S-200	3.2	0.6	4.0	6.6

TABLE VII
PARTIAL PURIFICATION OF ADP-SULFURYLASE FROM *Thiocapsa roseopersicina* DSM 219

Step	Volume (ml)	Protein (mg)	Activity (units)	Specific activity (units/mg)
1. Crude extract	74.0	532.8	186.4	0.35
2. Ultracentrifugation	54.2	183.6	97.4	0.53
3. DE-52 eluate	8.4	6.2	9.2	1.0
4. Sephacryl S-300	3.2	0.6	4.0	6.6

tions a potent inhibitor of ATP-sulfurylase could not be found and enzyme activity does not depend on divalent cations. The sulfur compounds cysteine, methionine (2 m*M* each), mercaptoethanol, Na_2SO_3, Na_2SO_4, $Na_2S_2O_3$ (10 m*M* each) show no effect on the dissimilatory reaction of ATP-sulfurylase.

Partial Purification of ADP-Sulfurylase from Thiocapsa roseopersicina DSM 219

ADP-sulfurylase has been studied from several species of the Chromatiaceae and Chlorobiaceae.[16–18,20] So far none of these enzymes has been purified to homogeneity. The fact that most of the enriched preparations were free of contaminating activities of adenylate kinase, APS kinase, ATP-sulfurylase, and pyrophosphatase shows, however, that the enzyme ADP-sulfurylase does represent an independent entity in phototrophic sulfur bacteria.

Purification. Like all other ADP-sulfurylases from phototrophic sulfur bacteria studied so far, the enzyme from *T. roseopersicina* is labile and therefore difficult to purify.[53] Although the enzyme from *Chl. vibrioforme* f. *thiosulfatophilum* (DSM 263) has been reported to be partially stabilized by 5 m*M* EDTA and 1 m*M* mercaptoethanol[18] we found that glycerol is a suitable stabilizing agent for ADP-sulfurylase from *T. roseopersicina*. Nevertheless, most attempts to purify ADP-sulfurylase result in considerable loss of activity, and the purification factor obtained is low. For example, ammonium sulfate precipitation is inapplicable. The fraction between 30 and 70% saturation contains 80% of the recovered activity but the loss in total activity amounts to more than 80% and the purification factor achieved is only 1.2.

Crude extracts are prepared as described for purification of APS reductase from the same organism. For purification of ADP-sulfurylase low

[53] M. Algueró, C. Dahl, and H. G. Trüper, *Microbiologia (Madrid)* **4,** 149 (1988).

molecular weight constituents are removed from the supernatant obtained after ultracentrifugation by gel filtration on Sephadex G-25 equilibrated with 50 m*M* Tris-HCl, pH 7.5, containing 20% (v/v) glycerol. The eluate is loaded on a DEAE-cellulose column [Whatman (Clifton, NJ) DE-52; volume, 40 ml; flow rate, 30 ml/hr] equilibrated with 50 m*M* Tris-HCl, pH 7.5, containing 20% (v/v) glycerol. A linear KCl gradient (0–0.5 *M* in the aforementioned buffer) is applied and ADP-sulfurylase is eluted at a calculated KCl concentration of 120–150 m*M*. The combined fractions containing activity are concentrated by ultrafiltration in a Diaflo chamber with a PM30 filter and an N_2 pressure of 2 atm. The concentrated enzyme solution is desalted by gel filtration on Sephadex G-25 equilibrated with 50 m*M* Tris-HCl, pH 7.5. Kinetic characterization of the enzyme can be done with this fraction because it is free from interfering enzyme activities. For further purification ADP-sulfurylase is passed through a Sephacryl S-300 column (1.5 × 90 cm; flow rate, 15 ml/hr), equilibrated with 50 m*M* Tris-HCl, pH 7.5. Table VII summarizes the purification of ADP-sulfurylase. In polyacrylamide gels the preparation obtained with the procedure described migrates as one heavy and two rather weak bands.

Properties. ADP-sulfurylase from *T. roseopersicina* is a labile soluble enzyme with a molecular weight of 250,000. The optimum pH is 7.5 and the optimum temperature 35°. Under test conditions the apparent K_m values were determined to be 0.33 m*M* for adenylylsulfate and 13 m*M* for phosphate.

Acknowledgments

Our work was sponsored by the Deutsche Forschungsgemeinschaft (Grants Tr 133), the European Community (Contract No. EV4V-0207ES), and the Fonds der Chemischen Industrie.

[29] Reverse Siroheme Sulfite Reductase from *Thiobacillus denitrificans*

By HANS G. TRÜPER

Within the chemotrophic dissimilatory sulfur-oxidizing bacteria the Gram-negative, single-cell, short motile rods have been traditionally grouped into the genus *Thiobacillus*.[1] On the basis of 16S rRNA-based phylogenetic taxonomy this concept is no longer valid, as different species of "*Thiobacillus*" have been shown to belong to rather diverse branches within the two subclasses alpha and beta of the class Proteobacteria.[2,3] The type species, *Thiobacillus thioparus,* belongs to the beta subclass and is closely related only to *Thiobacillus denitrificans*. All other "*Thiobacillus*" species are not closely related to these two and will have to be assigned to other genera in the future.

The phylogenetic biodiversity of the thiobacilli is paralleled by an almost exotic diversity in their oxidation pathways for reduced sulfur compounds.[4]

Adenylylsulfate reductase (EC 1.8.99.2) has been found to occur only in the anaerobe *T. denitrificans*[5] and in the aerobes *T. thioparus*[6] and "*Thiobacillus*" *thiooxidans*[7]: however, it occurs in intracellular concentrations of around 3% of the total cell protein. These high concentrations, which are also typical of dissimilatory sulfate-reducing bacteria, indicate the active participation of the enzyme in dissimilatory sulfur oxidation in these thiobacilli.

[1] D. P. Kelly and A. P. Harrison *in* "Bergey's Manual of Systematic Bacteriology, (J. T. Staley, M. P. Bryant, N. Pfennig, and J. G. Holt, eds.), Vol. 3, p. 1842, Williams & Wilkins, Baltimore, Maryland, 1989.

[2] D. J. Lane, A. P. Harrison, Jr., D. Stahl, B. Pace, S. J. Giovannoni, G. J. Olsen, and N. R. Pace, *J. Bacteriol.* **174,** 269 (1992).

[3] S. Takakuwa, *in* "Organic Sulfur Chemistry: Biochemical Aspects" (S. Oae and T. Okuyama, eds.), p. 1. CRC Press, Boca Raton, Florida, 1992.

[4] For review cf. S. Takakuwa as cited above, and D. P. Kelly, *in* "The Nitrogen and Sulphur Cycles" (J. A. Cole and S. J. Ferguson, eds.), p. 65. Cambridge Univ. Press, Cambridge, 1988.

[5] T. J. Bowen, F. C. Happold, and B. F. Taylor, *Biochim. Biophys. Acta* **118,** 566 (1966); M. Aminuddin and D. J. D. Nicholas, *Biochim. Biophys. Acta* **325,** 81 (1973).

[6] R. M. Lyric and I. Suzuki, *Can. J. Biochem.* **48,** 334 (1970); K. Adachi and I. Suzuki, *Can. J. Biochem.* **55,** 91 (1977).

[7] H. D. Peck, Jr., *J. Bacteriol.* **82,** 933 (1961); H. D. Peck, Jr., T. E. Deacon, and J. T. Davidson, *Biochim. Biophys. Acta* **96,** 429 (1965).

A siroheme-containing sulfite reductase (EC 1.8.99.1) in similar high intracellular concentration was found in *T. denitrificans*.[8,9] In contrast to the sulfite reductases in sulfate-reducing bacteria, in *T. denitrificans* the enzyme must function in the oxidative—or "reverse"—direction, oxidizing sulfane sulfur to sulfite. So far, however, owing to a lack of electron acceptors that would not react with sulfide directly, no appropriate test system has been developed that would allow testing of the enzyme in the oxidative direction, that is, at the present state of the art, sulfite reductase can be tested only in the reductive direction.

The overall sulfur metabolism of *T. denitrificans* has been studied in thiosulfate-grown strain RT (DSM 807) and shown to proceed as follows[10]:

1. Thiosulfate is split to sulfite and sulfane sulfur by a thiosulfate sulfurtransferase (rhodanese, EC 2.8.1.1).
2. The sulfane moiety is transiently stored inside the cells in the form of polysulfides, most probably in the periplasm. No microscopically visible "sulfur" globules appear inside or outside the cells.
3. The polysulfide is oxidized to sulfite by reverse sulfite reductase.
4. Sulfone as well as sulfane-derived sulfite is oxidized to adenylylsulfate by adenylylsulfate reductase.
5. Sulfate is released from adenylylsulfate by ATP-sulfurylase or ADP-sulfurylase, both of which are present in this organism.

In addition to adenylylsulfate reductase, an AMP-independent sulfite oxidase has been found in a different strain ("Oslo") and studied in detail.[11]

Enzyme Assay

Principle. Sulfite reductase is measured in a manometric assay in the direction of sulfite reduction with enzymatically reduced methyl viologen as electron donor. The reduction of methyl viologen by hydrogen gas is catalyzed by purified hydrogenase (free from sulfite reductase and thiosulfate reductase activities) from *Desulfovibrio gigas*. The consumption of hydrogen is recorded manometrically.

Procedure. The reaction is carried out in Warburg flasks under hydrogen at 30°. The main compartment contains (in a total volume of 2.5 ml) 1 ml of hydrogenase solution (1–2 U), 0.1 ml of 75 m*M* methyl viologen, 0.1 ml of 1 *M* potassium phosphate (pH 7.0), and 0.1–1.3 ml of enzyme

[8] M. Schedel, J. LeGall, and J. Baldensperger, *Arch. Microbiol.* **105,** 339 (1975).

[9] M. Schedel and H. G. Trüper, *Biochim. Biophys. Acta* **568,** 454 (1979).

[10] M. Schedel and H. G. Trüper, *Arch. Microbiol.* **124,** 205 (1980); A. Schug, H. Ulbricht, and H. G. Trüper, unpublished results, 1982.

[11] M. Aminuddin and D. J. D. Nicholas, *J. Gen. Microbiol.* **82,** 103 (1974).

solution. The center well contains 0.2 ml of 5 *N* NaOH. The reaction is started by adding 0.3 ml of 50 m*M* sodium sulfite from the side arm. One enzyme unit (U) is defined as that amount of protein that consumes 1 μmol of hydrogen per minute.[12]

Preparation and Assay of Hydrogenase from Desulfovibrio gigas. Desulfovibrio gigas (DSM 496) is grown in a medium (pH 7.0) that contains (in 1 liter) 12 ml of 45% sodium lactate solution, 2 g of NH_4Cl, 2 g of $MgSO_4 \cdot 7\ H_2O$, 4.4 g of Na_2SO_4, 1 g of yeast extract, 2 ml of 10-fold concentrated trace element solution SL4,[13] and 20 ml of 1 *M* potassium phosphate. The latter is sterilized separately and added to the autoclaved medium after cooling. Cells are incubated at 30° under anoxic conditions for 2 days, harvested by centrifugation, and stored at −18° under hydrogen. For preparation of hydrogenase, aliquots of the cell paste are thawed, mixed with 1 vol of hydrogen-saturated 30 m*M* potassium phosphate (pH 7.0) (HSPP) at 0°, and stirred in an ice bath under hydrogen for 5 min. The suspension is then centrifuged and the dark-red supernatant is collected. This procedure for washing the cell pellet is repeated four times. The combined supernatants are supplied with HSPP-suspended alumina Cγ-gel (Serva, Heidelberg, Germany). After centrifugation the hydrogenase-containing yellow supernatant is concentrated under hydrogen in a Diaflo cell (membrane PM10; Amicon, Danvers, MA) and filtered through an HSPP-equilibrated Sephadex G-75 column. Hydrogenase-containing fractions of the eluate are combined, concentrated under hydrogen in a Diaflo cell, and stored at −18°. Hydrogenase activity is measured manometrically in Warburg flasks under hydrogen at 30°: the main compartment contains 1.55 ml of hydrogenase solution and 0.15 ml of 1 *M* potassium phosphate (pH 7.0). The reaction is started by the addition of 0.3 ml of 4% methyl viologen from the side arm. Hydrogenase activity is calculated from the initial rate of hydrogen consumption. One enzyme unit (U) is defined as the amount of protein that consumes 1 μmol of hydrogen per minute.

Purification

Frozen cell paste is thawed and suspended in 50 m*M* potassium phosphate (pH 7.0). The cells are broken by passing the suspension two or three times through a French pressure cell at about 138 MPa. The crude

[12] M. Schedel and H. G. Trüper, cf. Ref. 9; modified after A. Yoshimoto and R. Sato, *Biochim. Biophys. Acta* **153,** 555 (1968).

[13] N. Pfennig and K. D. Lippert, *Arch. Microbiol.* **55,** 245 (1966).

extract is adjusted to pH 4.5 by adding 1 vol of 0.3 *M* ammonium acetate (pH 4.1). After centrifugation the precipitated material is discarded.

This step is necessary to remove the dissimilatory nitrite reductase (cytochrome *cd*) of *T. denitrificans,* which like the siroheme sulfite reductase will reduce sulfite with viologen dyes as electron donors. In crude extracts it may contribute up to 50% of the total sulfite reductase measured.

The supernatant is adjusted to pH 5.5 with 1 *M* K_2HPO_4 and then fractionated with $(NH_4)_2SO_4$. The protein precipitating between 45 and 70% saturation is collected by centrifugation, dissolved in 50 m*M* potassium phosphate (pH 7.0), and applied to Ecteola cellulose (column size, 3 × 100 cm, Serva, Heidelberg, Germany) previously equilibrated with the same buffer. The protein that adsorbs to the gel is eluted with a linear gradient between 50 and 350 m*M* potassium phosphate (pH 7.0). The fractions containing sulfite reductase are eluted at about 220 m*M* and show an intense green color; they are combined, dialyzed against 50 vol of 50 m*M* potassium phosphate, and then stirred into an appropriate volume of alumina Cγ-gel slurry suspended in water. The gel to which sulfite reductase adsorbs is centrifuged, washed once with 50 m*M* potassium phosphate, and then eluted with stepwise increasing concentrations of potassium phosphate (pH 7.0; 80, 110, 140, 170, and 210 m*M*). The eluates at 110, 140, and 170 m*M* are combined and brought to 80% saturation with solid $(NH_4)_2SO_4$. The precipitated protein is collected by centrifugation, dissolved in 50 m*M* potassium phosphate (pH 7.0) and applied to a Sephadex G-200 column (diameter, 3 cm; length, 100 cm), that has been previously equilibrated with the same buffer. The green fractions of the eluate are combined and frozen at −20°. Under these conditions the enzyme is stable for 3 months. A typical purification is summarized in Table I.

TABLE I
PURIFICATION OF SULFITE REDUCTASE FROM *Thiobacillus denitrificans* RT

Step	Protein (mg)	Activity (units)	Specific activity (mU/mg)
1. Crude extract	18,400	258	14
2. Ammonium acetate fraction	5,280	121	31
3. 45–70% $(NH_4)_2SO_4$	3,540	71	20
4. Ecteola eluate	386	113	293
5. Alumina Cγ eluate	312	106	339
6. Sephadex G-200 filtrate	254	84	331

Properties

Molecular Properties. The enzyme obtained by the procedure given above is electrophoretically homogeneous.[9] Determined by gel filtration with protein standards as markers, the molecular mass of the purified enzyme is 160 kDa. The isoelectric point is at pH 4.8.

The enzyme has an $\alpha_2\beta_2$ subunit structure; the α subunit has a molecular mass of 38 kDa, and the molecular mass of the β subunit is 43 kDa.

Absorption spectra. The oxidized and reduced spectra are almost identical. The oxidized spectrum shows absorption peaks at 274, 393, and 594 nm. The molar extinction coefficients at these wavelengths are 280×10^3, 181×10^3, and 60×10^3 cm^2 $mmol^{-1}$, respectively. The siroheme content indicated by these spectral properties may be extracted with HCl in acetone[15]; in this extract it shows typical absorption peaks at 370 and 594 nm.

Iron-Sulfur Clusters. The enzyme contains 24 mol of iron and 20 mol of (acid-labile) sulfur per mole of enzyme. These data indicate that the enzyme contains six [4Fe-4S] clusters per $\alpha_2\beta_2$ molecule. Unlike assimilatory sulfite reductases the enzyme does not contain flavin groups.

Catalytic Properties. The enzyme reduces sulfite, but not thiosulfate, dithionate, trithionate, or tetrathionate. As electron donors in the test system methyl or benzyl viologen is suitable, whereas $NADH_2$ and $NADPH_2$ are inactive. The pH optimum of the enzyme is about 6.0.[14]

[14] M. Schedel, Doctoral Thesis, University of Bonn, Germany (1977).

[15] M. J. Murphy and L. M. Siegel, *J. Biol. Chem.* **248,** 6911 (1973).

[30] Purification and Properties of Cytochrome *c*-555 from Phototrophic Green Sulfur Bacteria

By T. E. MEYER

Background

The green sulfur bacteria are strictly anaerobic, obligately photosynthetic, and autotrophic bacteria not closely related to any other group.[1] They utilize reduced sulfur compounds such as sulfide, sulfur, and less commonly thiosulfate as electron donors and carbon dioxide as the primary carbon source. Acetate, but no other organic compound, is assimi-

[1] J. F. Imhoff, *in* "Photosynthetic Prokaryotes" (N. H. Mann and N. G. Carr, eds.), p. 53. Plenum, New York, 1992.

lated in the presence of sulfide and carbon dioxide. The green bacteria have a unique pathway for carbon assimilation via a reversed tricarboxylic acid (TCA) cycle. They are not as easily cultivated as are other photosynthetic bacteria, which may account for the relatively few biochemical studies. The photosynthetic reaction center is related to plant, algal, and cyanobacterial photosystem I.[2] The green bacteria contain bacteriochlorophyll *a*, as do the purple bacteria, but also synthesize unique light-harvesting species known as *Chlorobium* chlorophylls or bacteriochlorophylls *c, d,* and *e*.

Green bacterial taxonomy is not highly developed in the sense that ill-defined classic criteria such as cellular morphology have been used to differentiate the species and genera. Application of the more reliable criteria (the CG content of DNA, the salt requirement for growth, and the cytochrome *c*-555 sequences[3]) reveal that some closely related strains have been placed in different species and some widely differing strains have been placed in the same species. Partial ribosomal RNA sequences indicate that green bacteria are distantly related to all other bacteria,[4] but insufficient data in the form of complete sequences have been collected to improve the definition of genera and species. The practical result of the outdated taxonomy is that strain designations are more important than species and should always be reported to ensure reproducibility. To date, this practice has not been followed in the literature and it is sometimes difficult to tell what organism has been examined.

Cytochrome Content of Green Bacteria

The green sulfur bacteria are a rich source of cytochromes and other electron transfer proteins as reviewed by Fischer[5] and shown in Table I. Cytochrome *c*-555 is the most commonly occurring soluble cytochrome and is present in all strains examined to date. Cytochrome *c*-555 is presumably the soluble mediator between a cytochrome bc_1 complex and the photosynthetic reaction center, although this remains to be proved. It is not to be confused with the membrane-bound reaction center cytochrome

[2] M. Büttner, D. L. Xie, H. Nelson, W. Pinther, G. Hauska, and N. Nelson, *Proc. Natl. Acad. Sci. U.S.A.* **89,** 8135 (1992).

[3] J. Van Beeumen, R. P. Ambler, T. E. Meyer, M. D. Kamen, J. M. Olson, and E. K. Shaw, *Biochem. J.* **159,** 757 (1976).

[4] J. Gibson, W. Ludwig, E. Stackebrandt, and C. R. Woese, *Syst. Appl. Microbiol.* **6,** 152 (1985).

[5] U. Fischer, *in* "Green Photosynthetic Bacteria" (J. M. Olson, J. G. Ormerod, J. Amesz, and E. Stackebrandt, eds.), p. 127. Plenum, New York, 1988.

TABLE I
ELECTRON TRANSFER PROTEINS FOUND IN GREEN BACTERIA AND OTHER DIAGNOSTIC PROPERTIES OF SPECIES

Species	CG[a]	S_2O_3[b]	NaCl[c]	c-555[d]	FC[e]	c-551[f]	Rd[g]	Fd[h]
Chlorobium limicola Tassajara (DSM 249)[i]	58	+	0	+	+	+	+	+
Chlorobium vibrioforme PM (DSM 264)[i]	56	+	0	+	+	+	+	+
Chlorobium vibrioforme L (DSM 263)[i,j]	57	+	0	+	+	+	+	+
Chlorobium limicola GHS (DSM 245)[k]	51	−	0	+	+		+	+
Chlorobium vibrioforme ML (DSM 260)[l]	53	−	1	+			+	+
Chlorobium phaeobacteriodes LJS[m]		−		+	+			
Pelodictyon luteolum (DSM 273)[n]	58	−	0	+			+	+
Prosthecochloris aestuarii 2K[o,p]	53	−	1	+			+	+

[a] %CG, DNA base composition.
[b] S_2O_3, Thiosulfate utilization.
[c] %NaCl, Salt concentration required for growth.
[d] *c*-555, Presence of cytochrome *c*-555.
[e] FC, Presence of flavocytochrome *c*.
[f] *c*-551, Presence of soluble cytochrome *c*-551.
[g] Rd, Presence of rubredoxin.
[h] Fd, Presence of bacterial ferredoxin.
[i] T. E. Meyer, R. G. Bartsch, M. A. Cusanovich, and J. H. Mathewson, *Biochim. Biophys. Acta* **153,** 854 (1968).
[j] M. A. Steinmetz and U. Fischer, *Arch. Microbiol.* **131,** 19 (1982).
[k] M. A. Steinmetz and U. Fischer, *Arch. Microbiol.* **130,** 31 (1981).
[l] M. A. Steinmetz, H. G. Truper, and U. Fischer, *Arch. Microbiol.* **135,** 186 (1983).
[m] U. Fischer, *Z. Naturforsch. C: Biosci.* **44C,** 71 (1989).
[n] M. A. Steinmetz and U. Fischer, *Arch. Microbiol.* **132,** 204 (1982).
[o] J. M. Olson and E. K. Shaw, *Photosynthetica* **3,** 288 (1969).
[p] Y. Shioi, K. Takamiya, and M. Nishimura, *J. Biochem.* (*Tokyo*) **71,** 285 (1972).

c-551, which is the immediate electron donor to the reaction center.[6] Cytochrome *c*-555 is presumably the terminal electron acceptor for enzymes involved in oxidation of reduced sulfur compounds, such as sulfide-cytochrome *c* oxidoreductase (no EC numbers assigned).[7] Cytochrome *c*-555 also forms a complex with sulfide-cytochrome *c* oxidoreductase (flavocytochrome *c*),[8] consistent with such a role.

Cytochrome *c*-555 as Species Marker

The most thoroughly characterized cytochrome *c*-555 is from *Chlorobium* strain PM (NCIB 8346, DSM 264). This strain is officially assigned

[6] J. S. Okkels, B. Kjaer, O. Hansson, I. Svendsen, B. L. Möller, and H. V. Scheller, *J. Biol. Chem.* **267,** 21139 (1992).
[7] A. Kusai and T. Yamanaka, *Biochim. Biophys. Acta* **325,** 304 (1973).
[8] M. W. Davidson, T. E. Meyer, M. A. Cusanovich, and D. B. Knaff, *in* "Green Photosynthetic Bacteria" (J. M. Olson, J. G. Ormerod, J. Amesz, and E. Stackebrandt, eds.), p. 133. Plenum, New York, 1988.

TABLE II
PROPERTIES OF CYTOCHROMES *c*-555

Species	p*I*[a]	E_{m7}[b]	SEQ[c]	Charge[d]	Elution[e]
Chlorobium limicola Tassajara[f]	10.5	145	+	+6	CMC (30 m*M*)
Chlorobium vibrioforme PM, L[f-i]	10.5	145	+	+6	CMC (30 m*M*)
Chlorobium limicola GHS[j]	10	140			CMC
Chlorobium vibrioforme ML[k]	7.3	80			DEAE (5 m*M*)
Chlorobium phaeobacteroides LJS[l]	10	105			CMC
Pelodictyon luteolum DSM 273[m]	10.5	160			CMC
Prosthecochloris aestuarii 2K[n,o]	4.7	103	+	−1	DEAE (5–60 m*M*)

[a] p*I*, Isoelectric point.
[b] E_{m7}, Redox potential at pH 7.
[c] SEQ, Amino acid sequence determined.
[d] Charge, Net charge of the protein based on sequence.
[e] Elution, Adsorbant followed by ionic strength.
[f] T. E. Meyer, R. G. Bartsch, M. A. Cusanovich, and J. H. Mathewson, *Biochim. Biophys. Acta* **153,** 854 (1968).
[g] J. Gibson, *Biochem. J.* **79,** 151 (1961).
[h] T. Yamanaka and K. Okunuki, *J. Biochem.* (*Tokyo*) **63,** 341 (1968).
[i] M. A. Steinmetz and U. Fischer, *Arch. Microbiol.* **131,** 19 (1982).
[j] M. A. Steinmetz and U. Fischer, *Arch. Microbiol.* **130,** 31 (1981).
[k] M. A. Steinmetz, H. G. Truper, and U. Fischer, *Arch. Microbiol.* **135,** 186 (1983).
[l] U. Fischer, *Z. Naturforsch. C: Biosci.* **44C,** 71 (1989).
[m] M. A. Steinmetz and U. Fischer, *Arch. Microbiol.* **132,** 204 (1982).
[n] J. M. Olson and E. K. Shaw, *Photosynthetica* **3,** 288 (1969).
[o] Y. Shioi, K. Takamiya, and M. Nishimura, *J. Biochem.* (*Tokyo*) **71,** 285 (1972).

to the species *Chlorobium vibrioforme* f. *thiosulfatophilum*. The amino acid sequences of cytochromes *c*-555 from strain PM (Pond Mud, NCIB 8346, DSM 264), strain L (Larsen, NCIB 8327, DSM 263), and strain Tassajara (DSM 249) appear to be identical to one another,[3,9] which suggests along with other criteria such as those mentioned above that they should be members of the same species. Nevertheless, strain Tassajara is currently designated *Chlorobium limicola* f. *thiosulfatophilum*. It can be seen in Table I that *C. vibrioforme* ML (Moss Landing, DSM 260) has a significantly lower CG content than do strains PM and L; it is from a brackish water habitat, and cannot be a member of the same species as strains PM and L. The cytochrome *c*-555 also has significantly different properties, as shown in Table II, although the sequence has not yet been determined. *Chlorobium limicola* GHS (Gilroy Hot Spring, DSM 245) has a significantly lower CG content, which shows that it cannot be a member of the same species as strain Tassajara. Cellular morphology, on which the current names are based, may well be a poor indicator of relatedness in this family of bacteria. Cytochrome *c*-555, on the other hand, appears

[9] R. P. Ambler and T. E. Meyer, unpublished, 1976.

to be an important molecular evolutionary marker that has led us to use the older nomenclature, emphasizing utilization of thiosulfate as an important diagnostic characteristic. Thus, fresh-water strains having a CG content of 56–58%, which utilize thiosulfate, and which have virtually identical cytochrome *c*-555 sequences, are called *Chlorobium thiosulfatophilum*. It is possible that all strains that use thiosulfate are members of a single species; however, this has yet to be proved. The bacteria that do not utilize thiosulfate appear to be more heterogeneous than those described above and may represent several species. This must be tested by further sequence determination. In this regard, the sequence of *Prosthecochloris aestuarii* 2K (Pfennig 5030) cytochrome *c*-555 is only 55% identical to those described above and obviously represents a distinct species.

Cytochrome *c*-555 Properties

Cytochrome *c*-555 is a small (10 kDa), soluble protein containing 86–99 amino acid residues and a single heme (Table II). It has a relatively low redox potential of 80–160 mV at pH 7, which may reflect the anaerobic environment in which the species lives. The amino acid sequences[3] and a low-resolution three-dimensional structure[10] show that cytochrome *c*-555 is a class I cytochrome related to mitochondrial cytochrome *c*, although not closely. There is a greater similarity to cyanobacterial and algal cytochromes *c*-553 or c_6 and to *Pseudomonas* and *Azotobacter* cytochromes c_5 than there is to the mitochondrial cytochromes *c* or to purple bacterial cytochromes c_2. This is consistent with the relatedness between the reaction centers of green bacteria and plant, algal, and cyanobacterial photosystem I.[2] Although some are acidic, most cytochromes *c*-555, like mitochondrial cytochromes *c*, are strongly basic proteins. The acidic proteins come from marine or moderately halophilic species, which are known to contain more acidic proteins in general. Because of the relative rarity of strongly basic proteins, the fresh-water species of cytochrome *c*-555 are often used in model systems in which a basic cytochrome is called for, but that has other properties much different from those of mitochondrial cytochrome *c*. The relatively low redox potential is such a property, and has been exploited in model studies. The red-shifted α peak is another such characteristic. Despite these useful properties, the protein is not commercially available, nor has the gene been cloned or expressed in a more amenable host. Therefore, a description of bacterial cultivation and protein purification is in order.

[10] Z. R. Korszun and F. R. Salemme, *Proc. Natl. Acad. Sci. U.S.A.* **74,** 5244 (1977).

Cultivation of Chlorobium Strains

Chlorobium strains that utilize thiosulfate are easier to cultivate than are those that use sulfide alone. Only small quantities of sulfide are necessary to make the culture anaerobic and sufficiently large amounts of thiosulfate (in the form of an innocuous salt) can be added to the medium to obtain full growth.[11] Even then, cell yields are only about 0.5 to 1 g (wet weight) per liter. If acetate is added to the medium, cell yield can be doubled, but is still far less than the approximately 8 g/liter obtained with organotrophic purple bacteria grown in the light. With acetate, one takes the risk of having the culture overrun with more rapidly growing purple bacterial contaminants, which are relatively sulfide tolerant. *Rhodopseudomonas palustris* is such an opportunistic contaminant we have encountered in *Chlorobium* cultures.

For *Chlorobium* strains that use sulfide only, there is a limit to the quantity of sulfide that can be added owing to its toxicity (which varies with the strain). In this case small quantities of sulfide must be added periodically to the culture until it is fully grown. To increase cell yield and to avoid feeding the culture periodically, it is possible to grow *Chlorobium* strains in mixed culture with *Desulfuromonas acetoxidans*[12,13] or with Spirillum 5175,[14] now known as *Sulfurospirillum deleyianum,*[15] both of which utilize acetate and continuously recycle into sulfide the elemental sulfur externally produced by *Chlorobium* strains. *Desulfuromonas acetoxidans* Pfennig 5071 is the same organism found growing in consortium with *P. aestuarii* 2K (Pfennig 5030); the mixture was originally described as *Chloropseudomonas ethylica* 2K.[16] The different strains of "*Cps. ethylica*" contain various green bacterial strains or species. When growing two species in consortium, there is a risk of misidentifying proteins isolated from the mixed culture. Because of the unusually red-shifted α peak and typically basic isoelectric point, cytochrome *c*-555 is relatively easy to identify. *Note:* The α peak is sensitive to pH and can shift up to 1 nm toward the blue end of the spectrum at pH 5 and toward the red at pH 9[17]; thus this protein has also been called cytochrome *c*-554.

[11] S. K. Bose, *in* "Bacterial Photosynthesis" (H. Gest, A. San Pietro, and L. P. Vernon, eds.), p. 501. Antioch Press, Yellow Springs, Ohio, 1963.

[12] N. Pfennig and H. Biebl, *Arch. Microbiol.* **110,** 3 (1976).

[13] H. Biebl and N. Pfennig, *Arch. Microbiol.* **117,** 9 (1978).

[14] R. S. Wolfe and N. Pfennig, *Appl. Environ. Microbiol.* **33,** 427 (1977).

[15] W. Schumacher, P. M. H. Kroneck, and N. Pfennig, *Arch. Microbiol.* **158,** 287 (1992).

[16] V. V. Shaposhnikov, E. N. Kondratieva, and V. D. Federov, *Nature (London)* **187,** 167 (1960).

[17] G. W. Pettigrew and G. R. Moore, "Cytochromes c. Biological Aspects." Springer-Verlag, Berlin and New York, 1987.

Purification of Cytochrome *c*-555

Many of the published procedures for isolation and purification of cytochrome *c*-555 are either too brief or are too out of date to be useful. However, an adaptation of the procedure of Meyer *et al.*[18] for *Chlorobium* strains PM, L, and Tassajara is described below. Appropriate modifications will be required for the more acidic strains of cytochrome *c*-555 such as *Chlorobium* strains ML and 2K. The accumulated frozen cell paste (1 kg) is suspended to 20–25% (w/v) in 50 m*M* phosphate buffer, pH 7, or in 100 m*M* Tris-HCl buffer, pH 8 at 4°C. The cells can be broken by repeated cycles of freezing and thawing, by sonication, or by French press. The latter is preferred, although the sulfur granules can damage the apparatus by "sand-blasting" the needle valve. Membranes and sulfur should be removed by high-speed centrifugation (4 hr in a Beckman Ti45 rotor at 45,000 RPM at 4°C) or by 40% saturation ammonium sulfate precipitation, followed by a short period of low-speed centrifugation (10 min in a Beckman JA20 Rotor at 16,000 RPM at 4°C). There are smaller losses in the former than in the latter procedure and one does not have to remove the ammonium sulfate afterward.

If it is not desired to purify other cytochromes or ferredoxins in addition to cytochrome *c*-555, then the extract can be desalted with a buffer change to 1 m*M* phosphate, pH 7, by dialysis or Sephadex G-25 gel filtration. The desalted extract is loaded onto a CM-cellulose column [such as Whatman (Clifton, NJ) CM 52] or on to an ion exchanger such as CM-Sepharose. A short, fat column (4 × 4 cm) is preferred over a long, narrow column for rapid recovery of protein. The theoretical reduction in resolution of such a short column is minor in comparison to the losses of resolution due to uneven loading, wall effects, failure to buffer the column properly, and gradients that are too steep. If the column is too small, then it can be coupled in series to another or the contents can be removed and added to the top of a larger column already packed with adsorbant. The column can be developed by washing with several column volumes of phosphate (1, 5, 10 m*M*, etc.) until the red-colored band of cytochrome begins to move and then the column is washed with a single concentration of buffer until the protein has been eluted. In previous experiments, cytochrome *c*-555 from *Chlorobium* strain PM eluted at about 30 m*M* ionic strength. The actual concentration for elution may vary depending on the dimensions of the column, the pH, the temperature, and the relative volumes of column and elution buffer. The fractions are assayed for the ratio of absorbance at 280 and 412 nm (A_{280}/A_{412}). The percentage of cytochrome with the best purity (~80%) is combined and concentrated by pressure dialysis via Diaflo YM5 (Amicon, Danvers, MA) or other appropriate

[18] T. E. Meyer, R. G. Bartsch, M. A. Cusanovich, and J. H. Mathewson, *Biochim. Biophys. Acta* **153,** 854 (1968).

membrane. It is also possible to dialyze the sample against 1 m*M* phosphate, concentrate it on a small CM cellulose column (1 × 1 cm), and elute it with a minimal volume of 100 m*M* phosphate.

It is convenient at this point to precipitate cytochrome *c*-555 with enzyme-grade ammonium sulfate[19] in increments of 10% saturation, starting with about 30%. The cytochrome should precipitate between about 50 and 80% saturation. Concentrated solutions of cytochrome will precipitate at a lower percentage of ammonium sulfate than will more dilute solutions, and the buffer composition and pH will also affect the outcome. The protein precipitates at a higher concentration of ammonium sulfate in the presence of NaCl. The cytochrome solution is centrifuged (10 min in a Beckman JA20 rotor at 16,000 RPM at 4°C) after each addition of ammonium sulfate. The supernatant is poured off and the remaining liquid removed with a Pasteur pipette and saved for the next addition of ammonium sulfate. Each of the pellets is dissolved in a minimal volume of buffer and assayed. The 80% of the cytochrome having the best purity is combined and applied directly to Sephadex G-75 (4 × 80 cm) or appropriate gel-filtration column equilibrated with buffer and 100 m*M* NaCl. Resolution will depend on the sample volume relative to column volume and the skill with which the sample is applied. A small sample volume may also be more dense than the column buffer, in which case the sample may load unevenly and the top of the column will have to be stirred. The final step of purification is chromatography on CM-cellulose, using a linear gradient and starting with a concentration of buffer that just causes the cytochrome to move on the column and ending with twice that concentration. When the fractions are assayed, the purity ratio should be constant, or nearly so, throughout the elution. If the ratio is not small and constant, then additional purification may be necessary. Pure cytochrome *c*-555 should have a ratio of about 0.18 in the completely oxidized form. The yield of pure protein should be ~15 μmoles/kg cells. The extinction coefficient in Meyer *et al.*[18] is too large, but should be closer to the 21.6 mM^{-1} cm^{-1} at 555 nm reported by Yamanaka and Okunuki.[20] If additional purification is necessary, then hydroxylapatite or reverse ammonium sulfate gradient chromatography on phenyl-Sepharose may be effective, although we have never used these adsorbants for cytochrome *c*-555.

Purification of Other Electron Transfer Proteins

If one desires to purify other electron transfer proteins from *Chlorobium* strains, such as ferredoxin and rubredoxin,[21,22] then the buffer of

[19] A. A. Green and W. L. Hughes, this series, Vol. 1, p. 67.
[20] T. Yamanaka and K. Okunuki, *Biochem. J.* **63,** 341 (1968).
[21] T. E. Meyer, J. J. Sharp, and R. G. Bartsch, *Biochim. Biophys. Acta* **234,** 266 (1971).
[22] K. J. Woolley and T. E. Meyer, *Eur. J. Biochem.* **163,** 161 (1987).

choice for cell disintegration is 100 m*M* Tris-HCl, pH 8. The clarified extract from centrifugation (4 hr in a Beckman Ti45 rotor at 45,000 RPM at 4°C) is passed directly through DEAE-cellulose (8 × 8 cm) or appropriate substitute to minimize denaturation of ferredoxin. After washing this column with the application buffer, the ferredoxin and rubredoxin are eluted with a minimal volume of 500 m*M* NaCl in buffer without chromatography, again to minimize losses. This concentrated protein solution (further concentrated by Diaflo treatment) is then directly chromatographed on Sephadex G-75 (4 × 80 cm) in 20 m*M* Tris-HCl, pH 8, 100 m*M* NaCl. After assaying the ferredoxin–rubredoxin eluate at 280 and 380 nm, the fractions with the best purity are combined and loaded onto DEAE-cellulose (4 × 4 cm) in the same buffer. A stepwise gradient is preferred to separate rubredoxin from ferredoxin. This is accomplished by adding NaCl in 20 m*M* increments to the loading buffer until rubredoxin begins to elute (this should begin at about 150 m*M* ionic strength), then holding it constant until the reddish-purple color has all eluted. The NaCl should then be increased in 50 m*M* increments until the green color of ferredoxin begins to elute, then held constant until all the color is off the column (this should be at about 300 m*M* ionic strength). The usual contaminant of ferredoxin and rubredoxin is nucleic acid, which is most effectively removed by fractionation with ammonium sulfate either via simple precipitation or by a reverse gradient on phenyl-Sepharose. The best purity achieved for rubredoxin, measured as the A_{280}/A_{492} ratio, is 2.6. *Chlorobium* ferredoxin is extremely labile; therefore procedures should be minimized to complete the purification as quickly as possible. The best purity, measured as the A_{280}/A_{390} absorbance ratio, is 1.4.

The other soluble acidic cytochromes of *Chlorobium* strains can be isolated by desalting the unadsorbed protein from the initial DEAE-cellulose column used to isolate ferredoxin and rubredoxin, and passing it through a larger DEAE-cellulose column (8 × 15 cm in 5 m*M* buffer). The column is developed stepwise by increasing buffer to 20 m*M* in 5 m*M* increments, then adding NaCl in 10 m*M* increments. One should be vigilant and not increase the ionic strength too rapidly, as this purification step provides the best opportunity for separating flavocytochrome *c* from cytochrome *c*-551. Flavocytochrome *c* should elute at about 30 m*M* and cytochrome *c*-551 at 50 m*M* ionic strength. Further purification steps are the same as for cytochrome c-555 but with substitution of DEAE for cm cellulose. The best A_{280}/A_{410} ratio for oxidized flavocytochrome *c* is 0.9. The best A_{280}/A_{411} ratio for oxidized cytochrome *c*-551 is 0.48.

Basic cytochrome *c*-555 can be adsorbed on CM-cellulose (4 × 4 cm) from the extract not adsorbed to DEAE-cellulose. For those strains of *Chlorobium* with an acidic cytochrome *c*-555, DEAE-cellulose (8 × 15 cm) is the mandatory adsorbant. Further purification of acidic cytochrome *c*-555 involves ammonium sulfate treatment, Sephadex G-75 (4 × 80 cm), and a repetition of the DEAE-cellulose (4 × 4 cm) step with a linear gradient.

Prosthecochloris 2K cytochrome *c*-555 elutes at about 5–50 m*M* ionic strength. *Chlorobium* strain ML cytochrome *c*-555 elutes at 5 m*M* ionic strength; thus, the extracts should be loaded onto the column from 1 m*M* buffer. Because most of the proteins in bacteria are acidic, the cytochromes *c*-555, which are also acidic, may need additional treatments with hydroxylapatite or phenyl-Sepharose to remove contaminants.

If *Chlorobium* strains are grown in consortium with *Desulfuromonas*, be aware that purification of cytochrome *c*-555 may now be complicated by the presence of additional protein species. *Desulfuromonas* is known to produce ferredoxin and rubredoxin[23] and both large and small cytochromes c_3[24,25] plus a small cytochrome *c*-553, which may be confused with cytochrome *c*-555. Because the cytochromes c_3 are multiheme proteins, they will appear to be more abundant than any other cytochrome in the extract. The small cytochrome c_3 and *c*-553 are acidic and adsorb to DEAE-cellulose, whereas the large cytochrome c_3 is neutral and does not readily adsorb to either DEAE- or CM-cellulose. The small cytochrome c_3 elutes from DEAE cellulose at 100–200 m*M* ionic strength, that is, much later than cytochrome *c*-555.

Acknowledgment

This work was supported in part by a grant from the National Institutes of Health (GM 21277).

[23] I. Probst, J. J. G. Moura, I. Moura, M. Bruschi, and J. Le Gall, *Biochim. Biophys. Acta* **502,** 38 (1978).
[24] R. P. Ambler, *FEBS Letts.* **18,** 351 (1971).
[25] I. Probst, M. Bruschi, N. Pfennig, and J. Le Gall, *Biochim. Biophys. Acta* **460,** 58 (1977).

[31] Purification and Properties of High-Potential Iron-Sulfur Proteins

By T. E. MEYER

Background

High-potential iron-sulfur proteins (HiPIPs) comprise a class of proteins (also known as high redox potential ferredoxins) usually found in purple phototrophic bacteria, where they presumably function to mediate electron transfer between the cytochrome bc_1 complex and the photosynthetic reaction center. That the concentration in aerobically grown purple bacteria is low relative to that in anaerobically grown cells supports such a role. As expected, HiPIP was found to interact effectively with the reaction center in *Chromatium vinosum*.[1] High-potential iron-sulfur protein could also be involved more directly in sulfur metabolism in the purple sulfur bacteria as electron acceptors for sulfide dehydrogenase[1] and thiosulfate dehydrogenase.[2] However, hard evidence in the form of genetic data linking HiPIP to any of these processes has not yet been obtained. The HiPIP found in a halophilic *Paracoccus* species is thought to be involved in denitrification, perhaps as electron donor to the cytochrome *cd* type of nitrate reductase.[3] The higher level of HiPIP in anaerobically grown cells is consistent with this role. High-potential iron-sulfur protein from *Thiobacillus ferrooxidans* has the most clearly defined function as an iron-oxidizing enzyme.[4] Nonphysiological reactions with flavodoxin[5] and with ferredoxin-NADP$^+$ reductase[6] illustrate the versatility of HiPIP in studies of electron transfer. In general, HiPIPs function in much the same capacity as do cytochromes and copper proteins, by virtue of their high redox potentials. The average redox potential of HiPIP is about 330 mv, although those from halophilic species can have potentials as low as 50 mV and that from *Rhodopila globiformis* is in the vicinity of 450 mV.[7]

The purification and properties of HiPIP were previously reviewed by Bartsch.[8] However, the number of species of HiPIP has more than doubled since then. Also, there has been a corresponding increase in characterization, including nuclear magnetic resonance (NMR), X-ray crystallography, amino acid sequence, and even some genetics. The currently known species of HiPIP are shown in Table I. Equally abundant HiPIP isozymes have been found in the *Ectothiorhodospira* species and in *Rhodospirillum salinarum*. High-potential iron-sulfur protein is generally an abundant protein except in *Rhodomicrobium vannielii*. With one exception (*R. salinarum* iso-2), HiPIP is remarkably stable in comparison with other ferredoxins.

[1] S. J. Kennel, R. G. Bartsch, and M. D. Kamen, *Biophys. J.* **12,** 882 (1972).

[2] Y. Fukumori and T. Yamanaka, *Curr. Microbiol.* **3,** 117 (1979).

[3] K. Hori, *J. Biochem. (Tokyo)* **50,** 481 (1961).

[4] Y. Fukumori, T. Yano, A. Sato, and T. Yamanaka, *FEMS Microbiol. Lett.* **50,** 169 (1988).

[5] C. T. Przysiecki, G. Cheddar, T. E. Meyer, G. Tollin, and M. A. Cusanovich, *Biochemistry* **24,** 5647 (1985).

[6] A. K. Bhattacharyya, T. E. Meyer, M. A. Cusanovich, and G. Tollin, *Biochemistry* **26,** 758 (1987).

[7] C. T. Przysiecki, T. E. Meyer, and M. A. Cusanovich, *Biochemistry* **24,** 2542 (1985).

[8] R. G. Bartsch, this series. Vol. 53, p. 329.

The ultraviolet (UV)–visible absorption spectra of HiPIP are distinctive and provide a means for ready identification.[8] Because HiPIPs generally have such a high redox potential, they are usually isolated in the green-colored reduced form and the spectrum is suggestive of bacterial ferredoxin with a maximum at about 380–390 nm. High-potential iron-sulfur protein is unique in showing a large increase in absorption in the visible region on addition of a crystal of potassium ferricyanide, which is diagnostic. On the other hand, bacterial ferredoxins are bleached by ferricyanide to varying extents, due to partial destruction of their iron-sulfur clusters. High-potential iron-sulfur proteins with the lowest redox potential are isolated in the purplish-brown oxidized form, which has three broad maxima at about 320, 380, and 450 nm in fresh protein.[8] Aged protein loses this fine structure. There is no change in the absorption spectrum of most HiPIPs on addition of a few crystals of sodium dithionite, whereas the bacterial ferredoxins are partially bleached (one must take care to have the solution well buffered to prevent formation of elemental sulfur, which would result in turbidity). The HiPIPs with the lowest redox potentials, which are isolated in the oxidized form, also bleach on reduction, but in this case there is a maximum in the oxidized minus reduced difference spectrum at about 480 nm, which is also characteristic of HiPIP.[8] One may quantify the amount of HiPIP by using the absolute extinction coefficients for the reduced protein at about 390 nm (16 mM^{-1} cm^{-1}), for the oxidized protein at 390 nm (20 mM^{-1} cm^{-1}), or the difference extinction at 480 nm (~9 mM^{-1} cm^{-1}).

High-potential iron-sulfur protein is remarkably stable to storage, heat, and pH extremes, as contrasted with the bacterial ferredoxins. For example, *Thiocapsa roseopersicina* HiPIP has a half-life of about 40 min at 80°.[9] The periplasmic HiPIP from *T. ferrooxidans* functions at a pH of 3.5.[4] The iso-2 HiPIP from *R. salinarum* is an exception in that it is stable only in the reduced state as isolated.[10] There is a transient increase in absorbance on addition of ferricyanide as expected for an HiPIP, but this is followed by the complete loss of absorbance in a matter of minutes, due to destruction of the iron-sulfur cluster. It does not appear to be affected by dithionite, suggesting that the 1+, 2+ redox transition has a potential much lower than in bacterial ferredoxins. High-potential iron-sulfur proteins stored in the oxidized state at −20° slowly decompose over a period of months, releasing hydrogen sulfide, which reduces the remainder of the protein. The exceptions to this rule are the low-potential

[9] N. A. Zorin and I. N. Gogotov, *Biokhimiya (Engl. Transl.)* **48,** 1011 (1984).

[10] T. E. Meyer, J. Fitch, R. G. Bartsch, D. Tollin, and M. A. Cusanovich, *Biochim. Biophys. Acta* **1017,** 118 (1990).

TABLE I
PROPERTIES OF HIGH-POTENTIAL IRON-SULFUR PROTEINS

	E_m[a]	Res.[b]	R[c]	Charge[d]	Elution conditions[e]
Chromatiaceae					
Chromatium vinosum ATCC 17899[f–h]	356	85	2.6	−5	DEAE 20T + 40 NaCl
Chromatium gracile Hol1[i,j]	350	83	2.6	−7	DEAE 20T + 80 NaCl
Chromatium warmingii DSM 173[k,l]	355	85	2.5	−4	DEAE 20T + 60 NaCl
Thiocapsa roseopersicina DSM 219[j–n]	346	85	2.6	−6	DEAE 20T + 60 NaCl
Thiocapsa pfennigii DSM 227[f,o]	352	81	2.8	−9	DEAE 20T + 100 NaCl
Thiocapsa sp. strain CAU1[p]			2.5		DEAE 20T + 80 NaCl
Thiocystis violacae[k]			2.5		DEAE 20T + 50 NaCl
Ectothiorhodospiraceae					
E. halophila BN 9626 iso-1[q–s]	120	71	2.6	−12	DEAE 20T + 210 NaCl
E. halophila BN 9626 iso-2[q,r]	50	76	2.9	−15	DEAE 20T + 240 NaCl
E. vacuolata DSM 2111 iso-1[p,t]	260	72	2.4	−5	DEAE 20T + 100 NaCl
E. vacuolata DSM 2111 iso-2[p,t]	150	71	2.6	−8	DEAE 20T + 180 NaCl
E. shaposhnikovii DSM 234 iso-1[l,u]	270	72	2.4	−6	DEAE 40T + 80 NaCl
E. shaposhnikovii DSM 234 iso-2[l,u]	155	71	2.6	−8	DEAE 40T + 180 NaCl
Rhodospirillaceae					
Rhodocyclus gelatinosus ATCC 17011[f,v]	332	74	2.3	+3	CM 10P
Rhodocyclus tenuis ATCC 25093[f,w,x]	300	62	1.5	+2	CM 7P
Rhodomicrobium vannielii ATCC 17100[l,p]		53	1.6	+4	CM 15P
Rhodopila globiformis DSM 161[t,y]	450	57	1.8	−3	DEAE 20T + 40 NaCl
R. salinarum ATCC 35394 iso-1[t,z]	265	57	2.2	−5	DEAE 20T + 60 NaCl
R. salinarum ATCC 35394 iso-2[t,z]		54	1.9	−1	DEAE 20T + 60 NaCl
Rhodopseudomonas marina DSM 2698[t,aa]	345	53	1.8	+5	CM 10P + 100 NaCl
Denitrifying bacteria					
Paracoccus sp. strain ATCC 12084[bb,cc]	282	71	2.2	−13	DEAE 20T + 180 NaCl
Aerobic sulfur bacteria					
Thiobacillus ferrooxidans Fe1[dd,ee]		53	1.7	+1	

[a] E_m, Midpoint redox potential at pH 7.
[b] Res., Total number of amino acid residues.

[c] R, Absorbance ratio (A_{280}/A_{390}) for reduced protein.
[d] Charge: Net protein charge based on the sequence.
[e] Elution conditions: Adsorbant followed by buffer (T, Tris-HCl; P, phosphate) and salt concentration (mM) required to elute the protein.
[f] R. G. Bartsch, this series, Vol. 53, p. 329.
[g] K. Dus, S. M. Tedro, and R. G. Bartsch, *J. Biol. Chem.* **248,** 7318 (1973).
[h] C. W. Carter, Jr., J. Kraut, S. T. Freer, N. H. Xuong, R. A. Alden, and R. G. Bartsch, *J. Biol. Chem.* **249,** 4212 (1974).
[i] R. G. Bartsch, unpublished, 1994.
[j] S. M. Tedro, T. E. Meyer, R. G. Bartsch, and M. D. Kamen, *J. Biol. Chem.* **256,** 731 (1981).
[k] U. Wermter and U. Fischer, *Z. Naturforsch.* **38C,** 968 (1983).
[l] J. Van Beeumen, unpublished, 1994.
[m] U. Fischer, *Z. Naturforsch.* **35C,** 150 (1980).
[n] N. A. Zorin and I. N. Gogotov, *Biokhimiya (Engl. Transl.)* **48,** 1011 (1984).
[o] S. M. Tedro, T. E. Meyer, and M. D. Kamen, *J. Biol. Chem.* **249,** 1182 (1974).
[p] T. E. Meyer, unpublished, 1994.
[q] T. E. Meyer, *Biochim. Biophys. Acta* **806,** 175 (1985).
[r] S. M. Tedro, T. E. Meyer, and M. D. Kamen, *Arch. Biochem. Biophys.* **241,** 656 (1985).
[s] D. R. Breiter, T. E. Meyer, I. Rayment, and H. M. Holden, *J. Biol. Chem.* **266,** 18660 (1991).
[t] R. P. Ambler, unpublished (1994).
[u] W. H. Kusche and H. G. Truper, *Arch. Microbiol.* **137,** 266 (1984).
[v] S. M. Tedro, T. E. Meyer, and M. D. Kamen, *J. Biol. Chem.* **251,** 129 (1976).
[w] S. M. Tedro, T. E. Meyer, and M. D. Kamen, *Arch. Biochem. Biophys.* **239,** 94 (1985).
[x] I. Rayment, G. Wesenberg, T. E. Meyer, M. A. Cusanovich, and H. M. Holden, *J. Mol. Biol.* **228,** 672 (1992).
[y] R. P. Ambler, T. E. Meyer, M. A. Cusanovich, and M. D. Kamen, *Biochem. J.* **246,** 115 (1987).
[z] T. E. Meyer, J. Fitch, R. G. Bartsch, D. Tollin, and M. A. Cusanovich, *Biochim. Biophys. Acta* **1017,** 118 (1990).
[aa] T. E. Meyer, V. Cannac, J. Fitch, R. G. Bartsch, D. Tollin, G. Tollin, and M. A. Cusanovich, *Biochim. Biophys. Acta* **1017,** 125 (1990).
[bb] K. Hori, *J. Biochem. (Tokyo)* **50,** 481 (1961).
[cc] S. M. Tedro, T. E. Meyer, and M. D. Kamen, *J. Biol. Chem.* **252,** 7826 (1977).
[dd] Y. Fukumori, T. Yano, A. Sato, and T. Yamanaka, *FEMS Microbiol. Lett.* **50,** 169 (1988).
[ee] T. Kusano, T. Takeshima, K. Sugawara, C. Inoue, T. Shiratori, T. Yano, Y. Fukumori, and T. Yamanaka, *J. Biol. Chem.* **267,** 11242 (1992).

HiPIPs from *Ectothiorhodospira halophila,* which appear to be more stable in the oxidized state than are other species.

High-potential iron-sulfur proteins generally appear to be monomeric, although evidence from electron paramagnetic resonance (EPR)[11] and X-ray crystallography[12,13] suggests that dimerization may be significant under some conditions. The native HiPIP from *T. ferrooxidans* has a mass of 63 kDa and, based on the sequence, may be a decamer.[4,14] The liable iso-2 HiPIP from *R. salinarum* discussed above is 45 kDa in size and based on the sequence may be at least a hexamer.[10,15] Results of sodium dodecyl sulfate-polyacrylamide gel electrophoresis (SDS–PAGE) must be viewed with caution because the relatively high cysteine content of HiPIP (and bacterial ferredoxin as well) often results in aggregation following denaturation even in the presence of mercaptoethanol. Generally one sees a ladder of monomer, dimer, trimer, and larger species, or only the larger forms, which can actually aid in determining an average monomer size provided the artifact is recognized as such.

The amino acid sequences of the HiPIPs are highly variable; they differ in length from 53 to 85 residues and there are few conserved residues (see Tedro *et al.*[16] for a summary and references in Table I). The largest HiPIPs are found in the family Chromatiaceae and the smallest in the family Rhodospirillaceae. It would be nearly impossible to align them without the help of the three-dimensional structures, which show where insertions and deletions may have occurred (see Rayment *et al.*[13] and references therein). An HiPIP sequence may be recognized by a pair of cysteines separated by two amino acid residues approximately in the center of the protein (positions 43 and 46 in *C. vinosum*), followed by another cysteine about 15–17 residues downstream, and then by a cysteine 14–15 residues further downstream, not far from the C terminus. There is a conserved tyrosine at positions 12–19 and a glycine at positions 53–75, which are spatially adjacent to one another. Other than the cysteines, these are the only highly conserved residues in the sequences. The labile iso-2 HiPIP from *R. salinarum* is lacking nearly the whole N terminus, including the

[11] W. R. Dunham, W. R. Hagen, J. A. Fee, R. H. Sands, J. B. Dunbar, and C. Humblet, *Biochim. Biophys. Acta* **1079,** 253 (1991).

[12] D. R. Breiter, T. E. Meyer, I. Rayment, and H. M. Holden, *J. Biol. Chem.* **266,** 18660 (1991).

[13] I. Rayment, G. Wesenberg, T. E. Meyer, M. A. Cusanovich, and H. M. Holden, *J. Mol. Biol.* **228,** 672 (1992).

[14] T. Kusano, T. Takeshima, K. Sugawara, C. Inoue, T. Shiratori, T. Yano, Y. Fukumori, and T. Yamanaka, *J. Biol. Chem.* **267,** 11242 (1992).

[15] R. P. Ambler, T. E. Meyer, and M. A. Cusanovich, unpublished, 1994.

[16] S. M. Tedro, T. E. Meyer, and M. D. Kamen, *Arch. Biochem. Biophys.* **241,** 656 (1985).

conserved tyrosine prior to the first cysteine, which may partially explain both aggregation and lability.[15] Tryptophans commonly occur in HiPIP, but are rare in the bacterial ferredoxins, thus resulting in a large absorbance at 280 nm, which may aid in their recognition. The HiPIPs from Chromatiaceae species generally have tryptophans at positions 60, 76, and 80, but these are substituted to varying extents by other aromatic residues in the HiPIPs from members of the family Rhodospirillaceae; thus the latter often have ultraviolet (UV) maxima similar to those of the bacterial ferredoxins. The gene sequence of *T.* ferrooxidans HiPIP shows that there is a leader sequence that directs the protein to the periplasmic space.[14] There are no data to show whether the phototrophic HiPIPs are cytoplasmic or periplasmic, which would have important implications for their supposed roles as electron acceptors for sulfur-oxidizing enzymes, which are presumably cytoplasmic in Chromatiaceae species and periplasmic in members of the family Ectothiorhodospiraceae, on the basis of the location of transiently accumulated sulfur granules.

Three-dimensional structures are now known for three HiPIPs from *C. vinosum, E. halophila,* and *Rhodocyclus tenuis* (Rayment *et al.*[13] and references therein). In spite of the low sequence homology in this family of proteins, there is a remarkable conservation of three-dimensional structure. There is little secondary structure in the HiPIPs, with only one small helix near the N terminus. The [4Fe-4S] cluster is buried in the protein interior and is inaccessible to solvent. This may account for the remarkable stability and may also be responsible for the use of the $2+$, $3+$ oxidation states and high redox potential of the iron sulfur cluster. Because the reduced state of the iron sulfur cluster and its cysteine ligands has an overall charge of $2-$, the more hydrophobic environment of the cluster in HiPIP favors the oxidized form. The more hydrophilic environment and greater numbers of hydrogen bonds to the iron-sulfur clusters in bacterial ferredoxins allow them to be reduced to the $1+$ state with a larger overall charge on the cluster.

Purification of High-Potential Iron-Sulfur Proteins

Bartsch[8] described the purification of *C. vinosum, Thiocapsa pfennigii, Rhodocyclus gelatinosus, Paracoccus* species, and *Rc. tenuis* HiPIPs from 1 kg cells. Those procedures remain valid today and may be applied to additional related species shown in Table I. For example, *Chromatium gracile, Chromatium warmingii, Thiocapsa roseopersicina, Thiocystis violaceae,* and *Thiocapsa* sp. strain CAU1 and their HiPIPs are similar enough to that of *C. vinosum* that purification follows the same protocol. Purification of other HiPIPs is described below. However, a few general comments on purification are in order. Small cytochromes are often persistent contaminants of HiPIP preparations. In *C. vinosum* and related species, these can be removed by hydroxylapatite (4 × 4 cm column) chromatography at the latter stages of purification. The HiPIPs usually do not adsorb or do so only weakly from 100 m*M* NaCl solutions, whereas the cytochromes may be bound more tightly. In general, a linear phosphate gradient of 0–50 m*M* in 100 m*M* NaCl has proved sufficient for separation[17] provided that the HiPIP adsorbs to the column.

With HiPIP from *Paracoccus* sp., gel filtration is the single most effective procedure for separation from cytochromes and may be repeated for effect with high-resolution media. Generally speaking, however, ion exchange on DEAE- or CM-cellulose, Sepharose, or Sephacel provides the most effective purification of HiPIPs. For adsorption of acidic HiPIPs onto columns of ion-exchange resin from crude extracts, a much larger DEAE-cellulose column (20 cm high × 8 cm diameter) may be required than for basic HiPIPs on CM-cellulose (10 cm high × 4 cm diameter) because most cellular proteins are acidic and because CM-cellulose has a greater exchange capacity. It follows that the basic HiPIPs will be easier to purify because of the fewer contaminating proteins adsorbed on CM-cellulose. The crude extract should first be passed through DEAE-cellulose in 1 m*M* Tris-HCl buffer, pH 8, to remove acidic proteins before passing it through CM-cellulose. The columns should be developed with a stepwise gradient tailored to suit the species in question. A typical protocol would be to wash with several column volumes of 1, 5, 10, and 20 m*M* buffer. Unadsorbed protein will be washed off and if there are no colored bands moving on the column, then the elution should be continued with 10, 20, 40, 60, 80, 100, 150, 200, 300, and 500 m*M* NaCl in buffer until some movement occurs. Once color begins to move down the column, then the concentration of the elution buffer should not be changed until that band of protein is off the column. A linear gradient does not provide the resolution needed at this stage of the purification. Reduced HiPIP is recognized by its green color and increase in absorption throughout the visible region on addition of a crystal of ferricyanide. The oxidized protein is purple to brown in color and sometimes precedes the green reduced protein on DEAE-cellulose (or follows it on CM-cellulose). If no HiPIP is adsorbed to DEAE-cellulose, then that portion of the extract that does not adsorb is adjusted to pH 6 and passed through a CM-cellulose column equilibrated with 1 m*M* phosphate buffer, pH 6. Chromatography is repeated in the same manner as for DEAE-cellulose except for the substitution of phosphate buffer for Tris.

Those bacterial species that produce a carbohydrate capsule can make

[17] R. G. Bartsch, unpublished, 1994.

it difficult to purify HiPIP and other electron transfer proteins. This capsule is presumably composed of polygluconate or polyglucuronate, which is partly water soluble and interferes with column chromatography by increasing the viscosity of solutions; it may also displace protein from anion-exchange columns or prevent their binding. Such problems have been recognized with *Thiocapsa pfennigii, Rc. tenuis,* and *Rhodopila globiformis*. Several short, fat columns of large volume (4 cm high × 8 cm diameter) may be required in these cases in order to cope with the slow flow rate as well as high occupancy by acidic polysaccharides. If the adsorbant shrinks during loading of sample, the protein may load unevenly, in which case the top layer will have to be stirred. We have not found a satisfactory solution to this problem, but it may require selection of mutants lacking a polysaccharide capsule.

High-potential iron-sulfur protein is assayed by measuring the absorbance of fractions at 390 and 280 nm. The ratio of absorbance at these wavelengths provides a measure of purity. Once the HiPIP fractions are identified and the purest 80% or so are pooled, then the volume may be reduced by Amicon (Danvers, MA) Diaflo pressure dialysis using a YM5 membrane or its equivalent. Because of the small size of most HiPIPs, they easily diffuse through dialysis membranes of larger size cutoff. Ammonium sulfate precipitation is the next step in purification. Solid enzyme-grade ammonium sulfate is added to a well-buffered solution, usually in 10% saturation increments.[18] The HiPIPs will generally precipitate over a broader range when impure and will come down at higher ammonium sulfate concentrations when dilute. The buffer composition, pH, ionic strength, and temperature will affect the outcome. Normally 60–80% saturated ammonium sulfate is required to precipitate HiPIP, which can be collected by a short centrifugation at low speed (10 min in a Beckman JA20 rotor at 16,000 RPM at 4°C), but there will be cases in which considerable color remains in solution in saturated ammonium sulfate. In these cases, the solution is filtered through a small DEAE- or CM-cellulose column that will adsorb the protein as in hydrophobic chromatography (usually performed with phenyl-Sepharose). The HiPIP may be removed from the column with water or buffer. Each of the ammonium sulfate precipitates is dissolved in minimal buffer and assayed. The purest 80% or so of the protein is pooled and chromatographed by gel filtration without the necessity of removing residual ammonium sulfate. Because HiPIP is usually a small protein, a well-buffered column of Sephadex G-75 or G-50 (4 × 80 cm) or the equivalent is appropriate. If the dense protein solution does not load evenly, the top may have to be stirred. At the final stage of purification, ion-exchange chromatography is usually repeated

[18] A. A. Green and W. L. Hughes, this series, Vol. 1, p. 67.

with a linear gradient. The protein is adsorbed at low ionic strength, the buffer and salt concentration are increased until the band begins to move, and then the gradient is started at that concentration and ends with about double the concentration. The HiPIP fractions from the column should have a small and constant ratio of absorbance at 280 to 390 nm if pure, otherwise additional procedures may be necessary. Chromatography on hydroxylapatite is one such useful technique, although as stated above HiPIP does not adsorb strongly if at all. Nevertheless, it may remove contaminants such as cytochromes. If the sample loads unevenly on hydroxylapatite, stirring in the same manner as for cellulose or Sephadex may crush the fragile crystals, resulting in a slow flow rate. In this case, a palette knife may be used to loosen the top layer of crystals and to stir them gently.

Hydrophobic interaction chromatography is another procedure that can be effective. If it is known at which point the protein precipitates with ammonium sulfate, then an amount of ammonium sulfate just below that required for precipitation is added to the solution, which is then passed through a phenyl-Sepharose (Pharmacia, Piscataway, NJ) column. The protein should adsorb to the top of the column; if it does not, then more ammonium sulfate may have to be added and the procedure repeated. If the protein runs down the side of the column or loads unevenly, then one should gently stir the top of the column until homogeneous. Once the phenyl-Sepharose column is loaded, then a linear gradient starting with the initial ammonium sulfate concentration and ending with water or dilute buffer should be applied.

Purification of *Ectothiorhodospira halophila* High-Potential Iron-Sulfur Proteins

The purification of *E. halophila* iso-HiPIPs was described by Meyer.[19] Strain BN 9626 was used, but similar results were obtained with the type strain SL1 (DSM 244). There are a number of brightly colored proteins that chromatograph on DEAE-cellulose from extracts of *E. halophila* and the salt concentration for elution of HiPIP will vary from one preparation to the next. Therefore it is important to know the order in which the proteins elute. The first colored protein to elute from DEAE of *E. halophila* extracts is the photoactive yellow protein, followed by cytochrome *c*-551, a flavoprotein, iso-1 HiPIP (oxidized and reduced forms, respectively), iso-2 HiPIP (oxidized and reduced forms), cytochrome *c′*, and a mixture of cytochrome *c′*, bacterial ferredoxin, and a purple protein. Provided that this chromatography gives well-resolved peaks, then subsequent puri-

[19] T. E. Meyer, *Biochim. Biophys. Acta* **806,** 175 (1985).

fication of the two HiPIPs by ammonium sulfate precipitation, Sephadex G-75, and repetition of DEAE-cellulose will be straightforward.

Purification of *Ectothiorhodospira vacuolata* and *Ectothiorhodospira shaposhnikovii* High-Potential Iron-Sulfur Proteins

The purification of *Ectothiorhodospira vacuolata* HiPIPs is similar to that reported for *Ectothiorhodospira shaposhnikovii*.[20] Electron transfer proteins were eluted from DEAE-cellulose in the following order: cytochrome c_4, iso-1 HiPIP (oxidized and reduced), cytochrome c', iso-2 HiPIP (oxidized and reduced), cytochrome *b*-558, and bacterial ferredoxin. The two HiPIPs were purified by the usual ammonium sulfate, Sephadex G-75, and repetition of DEAE-cellulose chromatography.

Purification of *Rhodospirillum salinarum*

The isolation of HiPIPs from *R. salinarum* was described by Meyer *et al.*[10] Although the HiPIP isozymes differ in charge, they elute in a single green band, followed by cytochrome *c*-551, cytochrome c', and a large amount of unidentified orange and green pigments unique to this species. There was partial resolution of the HiPIP isozymes by ammonium sulfate precipitation (iso-2 HiPIP precipitated at 40–60% saturation and iso-1 HiPIP at 70–90% saturation ammonium sulfate) and there was complete separation on Sephadex G-75. Subsequent DEAE-cellulose chromatography was sufficient to complete the purification. The amino acid sequences of the HiPIPs show that the subunit of iso-2 HiPIP is actually smaller than iso-1 HiPIP.[15] Native iso-2 HiPIP may thus be a hexamer rather than a tetramer as previously reported.[10]

Purification of *Rhodopseudomonas marina* High-Potential Iron-Sulfur Protein

HiPIP from *Rhodopseudomonas marina* was characterized by Meyer *et al.*[21] Two strains were studied, the type strain DSM 2698 and strain *agilis* (ATCC 35601). There are many soluble electron transfer proteins in *Rps. marina,* but only cytochrome c_2 and HiPIP are basic and adsorb to CM-cellulose (some or all of the cytochrome c_2 also adsorbs to DEAE-cellulose). Two HiPIPs were separated on CM-cellulose from *Rhodopseu-*

[20] W. H. Kusche and H. G. Truper, *Arch. Microbiol.* **137,** 266 (1984).

[21] T. E. Meyer, V. Cannac, J. Fitch, R. G. Bartsch, D. Tollin, G. Tollin, and M. A. Cusanovich, *Biochim. Biophys. Acta* **1017,** 125 (1990).

domonas agilis. The N-terminal sequences of *Rps. agilis* iso-1 HiPIP (the more basic form) and that of the type strain appear to be identical.[15,21] The N terminus of iso-2 HiPIP is blocked. It is possible that iso-2 HiPIP may be a proteolytically degraded form of iso-1 because the first six residues of iso-1 HiPIP contain two lysine residues followed by a glutamine, which could cyclize following cleavage of the Lys-Gln bond.

Purification of *Rhodophila globiformis* High-Potential Iron-Sulfur Protein

The purification of *Rp. globiformis* HiPIP was described by Ambler *et al.*[22] This bacterium is difficult to culture and polysaccharide interferes with chromatography of the HiPIP on DEAE-cellulose. This problem was overcome by adsorbing HiPIP to CM-cellulose at pH 5 and chromatographing it with a 1–20 m*M* phosphate gradient, by ammonium sulfate precipitation, Sephadex G-75 gel filtration, and repetition of CM-cellulose chromatography, resulting in pure protein.

Purification of *Rhodocyclus tenuis* High-Potential Iron-Sulfur Protein

Problems with polysaccharide were experienced during purification of *Rc. tenuis* HiPIP previously reported by Bartsch.[8] However, it has been much more difficult to adsorb the HiPIP to initial DEAE- or CM-cellulose columns from crude extracts of this species than for others. In those instances when difficulties were encountered, larger columns were employed. Reduced HiPIP would adsorb weakly to DEAE-cellulose at pH 8 and oxidized HiPIP would adsorb weakly to CM-cellulose at pH 5. It was difficult to maintain the HiPIP in completely oxidized or reduced forms in crude extracts and several columns have occasionally been necessary. Generally, the HiPIP that adsorbs to CM-cellulose is more easily purified than the one that adsorbs to DEAE-cellulose. It is possible that there are two similar HiPIP isozymes in *Rc. tenuis*[13] because only a portion of the protein has been purified from each extract.

Purification of *Rhodomicrobium vannielii* High-Potential Iron-Sulfur Protein

The purification of *Rm. vannielii* HiPIP is similar to that for *Rc. gelatinosus*.[8] Cytochrome c_2 elutes from CM-cellulose before the HiPIP does. The HiPIP is more basic than other species and requires higher ionic

[22] R. P. Ambler, T. E. Meyer, M. A. Cusanovich, and M. D. Kamen, *Biochem. J.* **246,** 115 (1987).

strength to elute. Because there is only a small quantity of HiPIP in *Rm. vannielii,* the purification results in greater losses. The final CM-cellulose chromatography was repeated twice to obtain reasonably pure protein, but in retrospect, chromatography on hydroxylapatite or phenyl-Sepharose may be more effective in obtaining a larger quantity of pure protein.

Purification of *Thiobacillus ferrooxidans* High-Potential Iron-Sulfur Protein

The purification of *Thiobacillus* HiPIP was described by Fukumori *et al.*[4] However, the availability of the cloned gene[14] may allow it to be overproduced in a host organism that could be more easily cultivated than is *T. ferrooxidans*.

Acknowledgment

This work was supported in part by a grant from the National Institutes of Health, GM 21277.

[32] Sulfite : Cytochrome c Oxidoreductase of Thiobacilli

By ISAMU SUZUKI

Introduction

Sulfite : cytochrome *c* oxidoreductase, often called sulfite oxidase, catalyzes the oxidation of sulfite to sulfate with the transfer of electrons to cytochrome *c* or other electron acceptors:

$$SO_3^{2-} + 2\,\text{cytochrome}\,c(Fe^{3+}) + H_2O \rightarrow SO_4^{2-} + 2\,\text{cytochrome}\,c(Fe^{2+}) + 2H^+ \qquad (1)$$

The enzyme is considered to be responsible for the oxidation of sulfite to sulfate in many thiobacilli[1] through the electron transport systems. Some thiobacilli also possess adenylylsulfate reductase (APS reductase), which requires AMP for the oxidation of sulfite.[1,2] The former enzyme, therefore, is sometimes called AMP-independent sulfite oxidase. The enzyme in thiobacilli is similar to sulfite oxidase in animal livers[3] except in its inability

[1] I. Suzuki, *Annu. Rev. Microbiol.* **28,** 85 (1974).

[2] B. F. Taylor, this volume [27].

[3] R. M. MacLeod, W. Farkas, I. Fridovich, and P. Handler, *J. Biol. Chem.* **236,** 1841 (1961).

to use O_2 as electron acceptor.[4] The enzyme has been purified from *Thiobacillus novellus*,[4–7] *Thiobacillus thioparus*,[8] and sulfur-grown *Thiobacillus ferrooxidans*.[9]

Assay Methods

The enzyme is assayed by following the reduction of cytochrome *c* or ferricyanide in the presence of sulfite and the reduction rates are corrected for nonenzymatic rates.

Ferricyanide as Electron Acceptor

The original assay procedure[4] is a modification of the method of Peck[10] for APS reductase. The reaction mixture in a total volume of 3.0 ml contains 5 μmol of Tris-HCl (pH 8.0), 1.5 μmol of potassium ferricyanide, 5 μmol of sodium sulfite [0.1 *M* stock solution in 5 m*M* ethylenediaminetetraacetic acid (EDTA)], and enzyme. The reaction is initiated by the addition of sulfite or enzyme and the reduction of ferricyanide is followed at 420 nm (molar extinction coefficient, 1×10^3 cm^{-1}). The low buffer concentration used is to minimize the inhibition of enzyme activity by anions.[4,5,8,9]

Cytochrome c as Electron Acceptor

The enzyme is assayed[5] as above except for the replacement of potassium ferricyanide with 0.15 μmol of cytochrome *c* (horse heart) and following the reduction of cytochrome *c* in terms of absorbance increase at 550 nm (molar extinction coefficient, 19.2×10^3 cm^{-1}). In a modification of the method the Tris buffer may be replaced with 20 μmol of Tricine-NaOH buffer (pH 8.5)[8] and the sulfite concentration may be lowered to 0.3–0.6 μmol.[6,8] Because of the 20-fold higher molar extinction coefficient of reduced cytochrome *c* compared to ferricyanide the cytochrome *c* method is much more sensitive.

[4] A. M. Charles and I. Suzuki, *Biochem. Biophys. Res. Commun.* **19,** 686 (1965).
[5] A. M. Charles and I. Suzuki, *Biochim. Biophys. Acta* **128,** 522 (1966).
[6] T. Yamanaka, T. Yoshioka, and K. Kimura, *Plant Cell Physiol.* **22,** 613 (1981).
[7] F. Toghrol and W. M. Southerland, *J. Biol. Chem.* **258,** 6762 (1983).
[8] R. M. Lyric and I. Suzuki, *Can. J. Biochem.* **48,** 334 (1970).
[9] J. R. Vestal and D. G. Lundgren, *Can. J. Biochem.* **49,** 1125 (1971).
[10] H. D. Peck, Jr., *Biochim. Biophys. Acta* **49,** 621, (1961).

Purification Methods

Thiobacillus novellus Enzyme Purification

Thiobacillus novellus (ATCC 8093), a facultative chemoautotroph, is adapted to and grown in thiosulfate as energy source under autotrophic conditions.[11] A large-scale culture is grown in glass carboys with forced aeration with air or a mixture of 5% CO_2 and 95% air[11] at 30° for 5 to 6 days with a 10% inoculum. The pH of the medium is maintained between pH 7 and 8 by the addition of a 10% Na_2CO_3 solution with the aid of phenol red in the medium (0.2 ml of 0.2% phenol red per liter).

The cells are collected by centrifugation, washed in 20 m*M* potassium phosphate buffer (pH 7.0), resuspended in the same buffer at 40 m*M* [50 mg of cells (wet weight) per milliliter], and are stirred gently at 4° overnight. The cells are collected by centrifugation and after washing in the 40 m*M* buffer either stored at −20° or used for making extracts.

The cell-free extracts[11] are prepared from a cell suspension [200 mg of cells (wet weight) per milliliter of 40 m*M* potassium phosphate buffer of pH 7.0] by sonication for 15 min under a stream of N_2 in a 10-kHz Raytheon (Waltham, Mass.) sonic disintegrator at 5°. All the following procedures are carried out at 4° unless otherwise stated. The crude sonicate is centrifuged at 12,100 *g* for 20 min to obtain the extracts.

The crude extracts are diluted with an equal volume of 0.2 *M* Tris-HCl buffer (pH 8.0) and solid ammonium sulfate is added to a concentration of 40% saturation.[5] After centrifugation for 20 min at 17,300 *g* the supernatant is dialyzed extensively against 2 m*M* Tris-HCl buffer (pH 8.0). The dialyzed supernatant is placed on a DEAE-cellulose column (1 × 15 cm) and washed successively with 2, 10, 20, 40, 200, and 500 m*M* potassium phosphate (pH 7.0). A red band of cytochrome in the column is eluted in two fractions at 2 and 200 m*M* phosphate. The enzyme is eluted in 20 m*M* phosphate.[5] After concentration in a dialysis bag placed in crystalline sucrose the enzyme is dialyzed extensively against 2 m*M* Tris-HCl (pH 8.0). Treatment with alumina Cγ gel (40 mg/mg protein) and centrifugation removes contaminating proteins, leaving the enzyme in the supernatant. The enzyme is 74-fold purified with a high recovery of 53%.

A further purification is achieved by the method of Toghrol and Southerland.[7] The supernatant from the 40% ammonium sulfate saturation is heated at 58° for 30 min in the presence of 10 m*M* sodium sulfite, which stabilizes the enzyme. The enzyme is then precipitated with ammonium sulfate at 70% saturation. The enzyme after dialysis is chromatographed

[11] A. M. Charles and I. Suzuki, *Biochim. Biophys. Acta* **128,** 510 (1966).

on DEAE-cellulose with a linear gradient of potassium phosphate (pH 7.8, 5–50 m*M*), the enzyme being eluted betwen 20 and 30 m*M*. The fractions are concentrated by ultrafiltration and after dilution to 5 m*M* potassium phosphate (based on conductivity) the enzyme is chromatographed on hydroxylapatite at the same linear gradient as above. The active fractions are applied to a column of Affi-Gel blue at 5 m*M* potassium phosphate. The enzyme is not adsorbed and collected fractions are concentrated. The purification achieved is 206-fold with 18% recovery. The heme content of enzyme preparations increases during the purification and purified enzyme has a spectrum characteristic of cytochrome *c* (reduced peaks at 414.5, 521, and 550 nm).

An alternative method developed by Yamanaka *et al.*[6] is based on their earlier purification of cytochrome *c*-550 and cytochrome *c*-551 from *T. novellus*.[12] The crude sonic extract in 10 m*M* Tris-HCl (pH 8.5) is passed through an Amberlite CG-50 column to remove cytochrome *c*-550 and then loaded onto a DEAE-cellulose column. The enzyme and cytochrome *c*-551 are eluted with 0.1 *M* NaCl, and also move together on a Sephadex G-100 column. In a linear gradient DEAE-cellulose chromatography (20–200 m*M* NaCl) in 10 m*M* Tris-HCl (pH 8.5) a major portion of cytochrome *c*-551 moves ahead of enzyme activity, but the enzyme fraction always contains the cytochrome and the total activity is reduced by the removal of *c*-551 from the enzyme fraction. Isoelectric focusing (pH 3–10), polyacrylamide gel electrophoresis, and starch electrophoresis all indicate the close association of enzyme activity and cytochrome *c*-551. Yamanaka *et al.* conclude that the enzyme contains cytochrome *c*-551 as its integral part.[6]

Thiobacillus thioparus Enzyme Purification

Thiobacillus thioparus (ATCC 8158), a strict autotroph, is grown on Starkey's medium No. 2.[13] Large-scale cultivation[8] is performed as described for *T. novellus* and aerated with air. The pH is maintained at 6.5–7.0. After 3–5 days the cells are collected and washed in 0.1 *M* potassium phosphate buffer (pH 7.5). Sulfur is removed by differential centrifugation. A cell suspension [1 g of cells (wet weight) per 10 ml of the phosphate buffer] is sonicated in a Raytheon 10-kHz oscillator for 30 min and is centrifuged at 30,000 *g* for 20 min to obtain the cell-free extract. The extract is centrifuged further at 105,000 *g* for 2 hr and the supernatant (50–70 ml) is applied to a column (2.5 × 40 cm) of DEAE-cellulose in 20

[12] T. Yamanaka, S. Takenami, N. Akiyama, and K. Okunuki, *J. Biochem.* (*Tokyo*) **70,** 349 (1971).

[13] R. L. Starkey, *J. Bacteriol.* **28,** 365 (1934).

mM potassium phosphate (pH 7.4). Sulfite oxidase is eluted with 50 mM potassium phosphate (pH 7.4) and is separated from thiosulfate-oxidizing enzyme and APS reductase, which remain bound to the column.[8] The eluate is mixed with calcium phosphate gel (2.5 mg/mg protein) and stirred for 30 min. The gel with adsorbed enzyme is collected by centrifugation and sulfite oxidase is eluted with 0.15 M potassium phosphate (pH 7.4). The crude extract shows only 10–20% of the total sulfite oxidase activity of the DEAE-cellulose eluate, due to interfering inhibitors including cytochrome oxidase.[8] The calcium phosphate step increases the purification by 15-fold with 50% recovery of activity. Assuming the presence of total enzyme activity in the crude extract equals that in the DEAE-cellulose eluate, the overall purification achieved is 163-fold.

Thiobacillus ferrooxidans (Sulfur-Grown) Enzyme Purification

Thiobacillus ferrooxidans (ATCC 13661) is grown on sulfur[9] and cells are collected by centrifugation, washed with water adjusted to pH 3 with sulfuric acid, and suspended in 0.1 M Tricine-NaOH buffer (pH 8.0) at a concentration of 1 g of cells (wet weight) per 10 ml for sonication. Three 5-min treatments in a 10-kHz Raytheon sonic oscillator result in breakage of 90% of the cells.[9] Sulfite oxidase is purified essentially according to the method described for *T. thioparus* enzyme above, except for the use of Tricine-NaOH (pH 8.0) buffer. Purification of 7.3-fold is reported with recovery of 17% activity.[9]

Summary of Purification

A summary of purification results is presented in Table I. *Thiobacillus novellus* has a much higher level of enzyme activity than other thiobacilli and is considered an excellent source for the purification.

Properties

Electron Acceptors

Sulfite : cytochrome *c* oxidoreductase of thiobacilli is similar to sulfite oxidase from animal livers[3] in many respects, but differs in the electron acceptor specificity. The hepatic enzyme[3,14] can use ferricyanide 10 to 14 times faster than cytochrome *c* and can use O_2 as fast as cytochrome *c*, whereas the enzyme from thiobacilli can use ferricyanide only at half the rate of cytochrome *c*[5,7,8] and cannot use O_2.[5,8] The difference is probably

[14] H. J. Cohen and I. Fridovich, *J. Biol. Chem.* **246**, 359 (1971).

TABLE I
SUMMARY OF SULFITE OXIDASE PURIFICATION RESULTS

Organism	Specific activity (units/mg)[a]		Purification (fold)	Assay
	Cell-free extract	Enzyme		
T. novellus[b]	1.9	140	74	$Fe(CN)_6^{3-}$
T. novellus[c]	1.4	282	206	Cytochrome *c*
T. thioparus[d]	0.004	0.69	163	Cytochrome *c*
T. ferrooxidans[e]	0.02	0.12	7	$Fe(CN)_6^{3-}$

[a] One unit: reduction of 1 μmol of $Fe(CN)_6^{3-}$ or cytochrome *c* per minute. The *T. thioparus* cell-free extract has only 10–20% of the total activity, due to some inhibitors. The figure cited here assumes an absence of inhibitors.

[b] A. M. Charles and I. Suzuki, *Biochim. Biophys. Acta* **128,** 522 (1966).

[c] F. Toghrol and W. M. Southerland, *J. Biol. Chem.* **258,** 6762 (1983).

[d] R. M. Lyric and I. Suzuki, *Can. J. Biochem.* **48,** 334 (1970).

[e] J. R. Vestal and D. G. Lundgren, *Can. J. Biochem.* **49,** 1125 (1971).

due to the presence of a *b*-type cytochrome in the former[3] and a *c*-type cytochrome in the latter.[6,7]

The *T. novellus* enzyme in the presence of sulfite reduces the native cytochrome *c*-550, which is reoxidized in the presence of native cytochrome oxidase.[5,6] Thus the enzyme, cytochrome *c*-550, and cytochrome oxidase reconstitute the sulfite oxidase system.[6]

Molecular Weight

The molecular weight of enzyme from *T. novellus*,[7] M_r 40,000, agrees with the value for *T. ferrooxidans* enzyme[9] (M_r 41,500), but is smaller than the value for *T. thioparus* enzyme[8] (M_r 54,000). The value is close to half the value of M_r 115,000–120,000 for the hepatic sulfite oxidase, which is composed of two subunits and contains one cytochrome (*b*-type) and one molybdenum per subunit.[15,16] The *T. novellus* enzyme also contains a heme (*c* type) and, in addition, molybdenum and molybdopterin similar to the hepatic enzyme.[7] Molybdenum is also found in the *T. thioparus* enzyme preparation[17] in addition to iron and *c*-type heme.[8]

[15] H. J. Cohen and I. Fridovich, *J. Biol. Chem.* **246,** 367 (1971).

[16] W. M. Southerland and K. V. Rajagopalan, *J. Biol. Chem.* **253,** 8753 (1978).

[17] D. L. Kessler and K. V. Rajagopalan, *J. Biol. Chem.* **247,** 6566 (1972).

Anionic Inhibition

The *T. novellus* enzyme is inhibited by increasing concentrations of Tris-HCl, Tris-acetate, NaCl, KCl, and potassium phosphate.[4] In the ferricyanide assay system the inhibition by NaCl is competitive with respect to sulfite concentration, with a K_i value of 4.5 m*M*.[5] The inhibition seems to be by monovalent anions because phosphate inhibition is 5–10 times stronger at pH 6 than at pH 8.[4] The *T. thioparus* enzyme (cytochrome *c* assay), on the other hand, is inhibited by chloride ion only weakly and noncompetitively with sulfite, with a higher K_i of 38 m*M*. Phosphate is a stronger inhibitor uncompetitive with sulfite (K_i 10 m*M*), similar to the product inhibitor sulfate (K_i 6.3 m*M*). The *T. ferrooxidans* enzyme is also strongly inhibited by phosphate.[9] The hepatic sulfite oxidase is inhibited by anions, but only when one-electron acceptors are used and not when O_2 is used.[14]

Effect of pH

The study of pH effect on the enzyme activity is complicated by the anionic inhibition. The *T. novellus* enzyme in the ferricyanide assay shows a maximal activity around pH 8.0 in a potassium phosphate buffer, whereas a broad plateau from pH 7 to 8 is observed in a Tris-HCl buffer.[5] The *T. thioparus* enzyme in the cytochrome *c* assay shows an optimal pH around pH 8 either in Tris-HCl or Tricine-NaOH at sulfite concentrations of 0.1 to 0.4 m*M*, but the extrapolation to the rates at saturating sulfite concentrations (V_{max}) shows a plateau from pH 7 to 9.5.[8] The hepatic sulfite oxidase has an optimum at pH 8.6.[14]

Effect of Substrate Concentrations

The *T. novellus* enzyme[5] shows an apparent K_m value for sulfite of 40 μM at pH 8 and 2 μM at pH 6.5, although the maximal activity is over 10 times higher at pH 8 in the cytochrome *c* assay method in 8 m*M* potassium phosphate. In the ferricyanide assay at pH 8 the value is 20 μM sulfite. With the *T. thioparus* enzyme[8] the apparent K_m for sulfite is 88 μM in 3 m*M* Tris-HCl (pH 8.0) and the value for cytochrome *c* is 13 μM. The *T. ferrooxidans* enzyme[9] has a much higher apparent K_m for sulfite of 0.54 m*M* with ferricyanide and 0.58 m*M* with cytochrome *c* as electron acceptor in 33 m*M* Tricine-NaOH (pH 8.0). The apparent K_m value for ferricyanide is reported as 0.25 m*M*. The hepatic sulfite oxidases[14] show apparent K_m values for sulfite of 10–30 μM with cytochrome *c*.

Inhibitors

The *T. novellus* and *T. thioparus* enzymes are inhibited by thiol-binding agents: mercuric chloride, *p*-chloromercuribenzoate, sodium arsenite, and *N*-ethylmaleimide.[5,8] The *T. thioparus* enzyme[8] is shown to be inhibited by metal-chelating agents; 2,2′-bipyridyl, *o*-phenanthroline, and EDTA.[8] *Thiobacillus novellus* enzyme[7] is also inhibited by cyanide.

Kinetics

The initial velocity and product inhibition study[18] of *T. thioparus* enzyme gives 0.1 m*M* sulfite and 14 μM cytochrome *c* as K_m values when the other substrate is saturating. Sulfate inhibits the reaction competitively with cytochrome *c* concentration (K_i: 2 m*M* sulfate), and uncompetitively with respect to sulfite concentration (K_i 6 m*M* sulfate).

Stability

The enzyme is stable when stored at −20°.[5,8,9] The *T. novellus* enzyme activity is destroyed at 60° for 1 min or at 55° for 5 min, but sulfite at 50 m*M* protects the enzyme during the heating.[5] In fact the protection is used for purification of the enzyme.[7] The *T. thioparus* enzyme is also destroyed by heating at similar temperatures.[8]

Other Types of Sulfite Oxidase

Sulfite oxidase purified (50-fold) from *Thiobacillus denitrificans* (anaerobically grown with nitrate) cell membranes[19] after solubilization with sodium deoxycholate reduces ferricyanide with sulfite but not cytochrome *c* (only 10% the rate) and is not inhibited by buffers. The apparent K_m for sulfite is 0.5 m*M* and the optimal pH is 8.3. The activity is inhibited by thiol-binding reagents. The enzyme is excluded from Sepharose-6B.

Sulfite oxidase purified (34-fold) from iron-grown *T. ferrooxidans* (AP 19-3) membranes[20,21] after solubilization with Nonidet P-40 reduces only ferric iron with sulfite and not ferricyanide or cytochrome *c*. The enzyme is reported to have an apparent molecular weight of 650,000, and consists of two subunits (m_r 61,000 and 59,000). The optimum pH is 6.5 with apparent K_m values of 71 μM sulfite and 1.0 m*M* Fe^{3+}. The activity is inhibited by Fe^{2+} and Hg^{2+}.

[18] R. M. Lyric and I. Suzuki, *Can. J. Biochem.* **48,** 594 (1970).

[19] M. Aminuddin and D. J. D. Nicholas, *J. Gen. Microbiol.* **82,** 103 (1974).

[20] T. Sugio, T. Katagiri, M. Moriyama, Y. L. Zhen, K. Inagaki, and T. Tano, *Appl. Environ. Microbiol.* **54,** 153 (1988).

[21] T. Sugio, T. Hirose, Y. L. Zhen, and T. Tano, *J. Bacteriol.* **174,** 4189 (1992).

[33] Sulfur-Oxidizing Enzymes

By ISAMU SUZUKI

Introduction

Sulfur-oxidizing enzymes oxidize elemental sulfur to sulfite with molecular oxygen:

$$S^0 + O_2 + H_2O \rightarrow H_2SO_3 \quad (1)$$

The sulfur-oxidizing enzyme of thiobacilli (*Thiobacillus thiooxidans,*[1,2] *Thiobacillus thioparus,*[2] and sulfur-grown *Thiobacillus ferrooxidans*[3]) requires reduced glutathione (GSH) for the reaction, whereas the enzyme from archaebacteria (*Sulfolobus brierleyi*[4] or *Acidianus brierley;* and *Desulfurolobus ambivalens*[5]) oxidizes sulfur without GSH, although at higher temperatures. Thiosulfate is often produced during the enzyme assay by a chemical reaction between sulfur and sulfite[2]:

$$S^0 + SO_3^{2-} \rightarrow S_2O_3^{2-} \quad (2)$$

Because of the requirement for O_2 in Eq. (1) and the incorporation of some ^{18}O from $^{18}O_2$ into the product[4,6] the enzyme has been considered as sulfur oxygenase.[6] The *Desulfurolobus* enzyme, however, produced (in addition to sulfite and thiosulfate) sulfide from elemental sulfur and was called sulfur oxygenase reductase.[5] An enzyme isolated from *T. ferrooxidans* grown on ferrous iron is unique in that aerobically it carries out Eq. (1) but anaerobically ferric iron can replace O_2 as electron acceptor.[7] The enzyme required GSH and was called sulfur : ferric ion oxidoreductase.[7] Later the name was changed to hydrogen sulfide : ferric ion oxidoreductase although GSH was still required.[8]

[1] I. Suzuki, *Biochim. Biophys. Acta* **104,** 359 (1965).
[2] I. Suzuki and M. Silver, *Biochim. Biophys. Acta* **122,** 22 (1966).
[3] M. Silver and D. G. Lundgren, *Can. J. Biochem.* **46,** 457 (1968).
[4] T. Emmel, W. Sand, W. A. König, and E. Bock, *J. Gen. Microbiol.* **132,** 3415 (1986).
[5] A. Kletzin, *J. Bacteriol.* **171,** 1638 (1989).
[6] I. Suzuki, *Biochim. Biophys. Acta* **110,** 97 (1965).
[7] T. Sugio, W. Mizunashi, K. Inagaki, and T. Tano, *J. Bacteriol.* **169,** 4916 (1987).
[8] T. Sugio, T. Katagiri, K. Inagaki, and T. Tano, *Biochim. Biophys. Acta* **973,** 250 (1989).

Assay Methods

Manometric Assay for Thiobacillus Enzymes

The sulfur-oxidizing enzyme activity can be determined in a Warburg respirometer at 30° by following the rate of O_2 consumption and also by measuring the amount of thiosulfate formed.[1-3] The standard reaction mixture in a total volume of 2 ml contains 500 μmol of Tris-HCl (pH 7.8), 48 mg of sulfur, 5 μmol of GSH, and enzyme. Precipitated sulfur powder as substrate is prepared by sonication (10-kHz; Raytheon, Waltham, Mass.) for 30 min at 4° of a sulfur suspension (32 g/100 ml) in water containing 0.05% (w/v) Tween 80 and by extensive dialysis against the 0.05% (w/v) Tween 80 solution. The sulfur suspension without sonication may be used as substrate more conveniently after homogenization with a magnetic stirrer for several hours. Because the sulfur oxidation rate is a function of the surface area of sulfur rather than the weight, more sulfur (in weight) is required (normally two to three times more) when the sulfur suspension is used without sonication.

The reaction is started by tipping a GSH solution (0.1 *M*) from the side arm of Warburg vessels and the O_2 consumption is followed for one to several hours. The reaction is stopped by the addition of 0.3 ml of 1 *M* cadmium acetate and after removal of sulfur, GSH, and proteins by centrifugation the supernatant is used for thiosulfate determination by the cyanolysis method of Sörbo.[9] The rates of O_2 consumption and of thiosulfate formation are equimolar.[1-3] Control experiments without the enzyme are required to measure the nonenzymatic rates of O_2 consumption and thiosulfate formation.

For the assay of partially purified sulfur-oxidizing enzyme preparation the standard assay mixture is supplemented with 250 μg of catalase (liver, 2× crystallized) and 0.2 μmol of 2,2′-dipyridyl, mainly to protect GSH from chemical oxidation[2] during long incubation periods.

Oxygen Electrode Assay for Thiobacillus Enzymes

The use of a Clark oxygen electrode to follow the O_2 concentration in a solution polarographically is a convenient assay of the sulfur-oxidizing enzyme; the reaction time is shorter than with the manometric method. A Gilson (Middleton, WI) oxygraph at 25 or 30° or a similar instrument is suitable.

The reaction mixture in a total volume of 1.2 ml contains 120 μmol of Tris-HCl (pH 7.5) or 60 μmol of potassium phosphate (pH 7.5), 0.5 to 1.0

[9] B. Sörbo, *Biochim. Biophys. Acta* **23,** 412 (1957).

mg of colloidal sulfur, 5 μmol of GSH, and enzyme.[10] The colloidal sulfur is prepared by sonication of sulfur in 0.05% (w/v) Tween 80 as described above, but the milky white colloidal sulfur is collected by decantation of larger particles before dialysis. Sonication may be carried out in 20-kHz ultrasonic oscillators. If sulfur is used as substrate without sonication 30 mg of homogenized sulfur is required. The reaction is started by the addition of GSH with a microsyringe and the consumption of O_2 is followed to obtain the rate of oxidation in nanomoles of O_2 per minute. The assay is limited, in terms of the total O_2 available, by the solubility of O_2 (approximately 0.3 μmol) and therefore is suitable for measurement for only a few minutes (normally less than 0.5 hr). The rate of nonenzymatic O_2 consumption by the sulfur–GSH system must be determined for correction. The O_2 electrode assay and manometric assay methods give similar results when the activity is expressed in the rate of O_2 consumption (micromoles of O_2 per minute). The former assay is not suitable for thiosulfate determination because of the limiting O_2 and short incubation period.

Thermophilic Archaebacterial Enzyme Assay

The enzyme from *S. brierleyi* is assayed[4] at 65° in test tubes shaken at 300 rpm. The reaction mixture in a total volume of 1 ml contains 70 μmol of Tris-H_2SO_4 (pH 7.0), 10 mg of sulfur, 0.005% (w/v) Tween 20, and enzyme. Sulfur is powdered by grinding with a mortar and pestle. The reaction is started by the addition of enzyme and the increase in sulfite concentration is measured at intervals for 1 hr by the fuchsin method of Grant.[11]

The *D. ambivalens* enzyme is assayed[5] at 85° in stoppered Hungate tubes shaken at 120 rpm. The reaction mixture contains 70 μmol of Tris-acetate (pH 7.4), 20 mg of sulfur, 0.005% (w/v) Tween 20, and enzyme per milliliter. Sulfur is sonicated with a Branson (Danbury, CT) sonifier for 5 min. Sulfite formation is followed by a fuchsin method,[5] thiosulfate by discoloration of methylene blue,[5] and H_2S by methylene blue formation.[12] Sulfite is the major product (80–90%) after 20 min of reaction.

Fe^{2+}-Grown Thiobacillus ferrooxidans Enzyme Assay

The enzyme from Fe^{2+}-grown *T. ferrooxidans* may be assayed[7] either anaerobically by the reduction of Fe^{3+} to Fe^{2+} or aerobically by the

[10] R. S. Bhella, M.Sc. Thesis, University of Manitoba, Winnipeg (1981); B. S. Chahal, M.Sc. Thesis, University of Manitoba, Winnipeg (1986).

[11] W. M. Grant, *Ind. Eng. Chem., Anal. Ed.* **19,** 345 (1947).

[12] T. E. King and R. O. Morris, this series, Vol. 10, p. 634.

formation of sulfite. The latter reaction is similar to the reaction catalyzed by other thiobacilli enzymes. The reaction mixture contains 80 μmol of sodium phosphate (pH 6.5), 20 mg of sulfur, 40 μg of bovine serum albumin, enzyme, and 4 μmol of GSH (pH 6.5 with NaOH) per milliliter. The reaction is started with the addition of GSH and the formation of sulfite at 30° with shaking is followed by the *p*-rosaniline method.[13] For the initial 10–15 min the major product is sulfite (90%), but a longer reaction time leads to increased thiosulfate formation.[2]

Purification Methods

Thiobacillus Enzymes

The sulfur-oxidizing enzyme is partially purified from sulfur-grown *T. thiooxidans* (ATCC 8085)[1] and *T. ferrooxidans* (*Ferrobacillus ferrooxidans,* ATCC 13661)[3] and thiosulfate-grown *T. thioparus* (ATCC 8158)[2] cells by similar procedures. The crude extracts are prepared by sonication at 4° for 20 min under N_2 atmosphere in a 10-kHz Raytheon oscillator of a 15% (w/v) cell suspension in 0.2 *M* Tris-HCl buffer of pH 7.5.[1] *Thiobacillus thiooxidans* cells are pretreated[1] in distilled water [5 g of cells (wet weight) in 20 ml] with 5 g each of a cationic resin (e.g., Dowex-50, H^+ form) and an anionic resin (e.g., Dowex-1, OH^- form) by vigorous rotary shaking for 10 min at room temperature before the removal of resin by filtration and collection of cells by centrifugation. *Thiobacillus ferrooxidans* cells[3] are pretreated instead in 0.5 *M* Tris-HCl (pH 7.8) for 12 hr at 4° before the collection by centrifugation and suspension in the same buffer for sonication. All of the following procedures are at 4°, unless otherwise indicated.

The cell-free extracts after removal of cell debris by centrifugation for 20 min at 12,000 *g* (23,500 *g* for *T. thioparus* extracts) are acidified to pH 5.0 with 1 *M* acetic acid and the precipitate is removed by centrifugation at 23,000 *g* for 20 min. The supernatant is adjusted to pH 7.5 with 1 *M* Tris-HCl (pH 9.0)[1] and absolute ethanol is added to a concentration of 40%, while maintaining the temperature below 0°. The precipitate is removed by centrifugation at −10° and the enzyme is precipitated from the supernatant by making it 50% ethanol and leaving it overnight at −20°. The enzyme is collected by centrifugation at −10° and is suspended in an aliquot of 0.2 *M* Tris-HCl buffer for storage at −20°. The *T. thioparus* and *T. ferrooxidans* enzyme is precipitated from the pH 5 supernatant without pH adjustment between 15 and 30% ethanol with similar results.[2,3] The

[13] P. W. West and G. C. Gaeke, *Anal. Chem.* **28,** 1816 (1956).

T. thioparus extract increases in total activity nearly three times in the pH 5 supernatant and five times in the ethanol fraction, presumably due to removal of unknown inhibitory components. The enzyme is purified 12- to 24-fold by these procedures with a high recovery of 53% to more than 100% (*T. thioparus*).

A further attempt to purify the above enzyme with DEAE-cellulose chromatography leads to considerable inactivation, although the enzyme is adsorbed at 50 m*M* Tris-HCl (pH 7.5) and elutes at 0.2–0.25 *M* Tris-HCl (pH 7.5) from a small volume of DEAE-cellulose.[1,2]

Thiobacillus thiooxidans Enzyme Purification

The enzyme activity (10–20% of intact cells) can be recovered in cell-free extracts by breaking the cells by sonication or by passage through a French pressure cell with or without previous treatment of cells. The purification, however, is simpler when more enzyme is released in a soluble form. A treatment with trypsin[10] leads to such extracts. Cells are washed with 0.2 *M* Tris-HCl (pH 7.5), suspended in 50 m*M* Tris-HCl (pH 7.5), and treated with trypsin [bovine, pancreatic, crystalline; 1.5 μg per mg cells (wet weight)] for 15 min at room temperature with gentle stirring. The reaction is stopped by adding trypsin inhibitor (soybean). All of the following procedures are at 4°. The cells are easily disrupted after a 10-min sonication in a 10-kHz Raytheon oscillator. The cell-free extract (80 ml, 1 g of protein) is obtained after centrifugation at 22,000 *g* for 15 min. The activity remains in the supernatant after centrifugation at 150,000 *g* for 2 hr, acidification to pH 5.0 with 1 *M* acetic acid, and centrifugation at 20,000 *g* for 20 min. The enzyme can be adsorbed on a small DEAE-cellulose column (2.5 × 15 cm) with a fast flow rate of 20–25 ml/hr in 50 m*M* Tris-HCl (pH 7.5). After washing with 50 m*M*, 0.1 *M*, and 0.15 *M* Tris-HCl (pH 7.5) the enzyme is eluted with 0.3 *M* Tris-HCl (pH 7.5) as a single yellow band. The fraction (8–10 ml) is passed through a Sephadex G-100 column (2.5 × 55 cm) in 50 m*M* Tris-HCl (pH 7.5) at a fast flow rate (12–15 ml/hr). Active fractions are concentrated by passage through a DEAE-cellulose column as described above before a second round of Sephadex G-100 column chromatography as described above, followed by DEAE-cellulose column concentration. Purification of 26.5-fold with 10% recovery of activity is achieved.

A modification[10] of the DEAE-cellulose chromatography procedure to include elution with 0.2 *M* Tris-HCl (pH 7.5) results in removal of a large portion of yellow color, still adsorbed in the column, but with the activity spreading into 35 ml. The enzyme is concentrated to 4–5 ml by a 4× dilution of the solution with water and passage through a DEAE-cellulose column (2 × 12 cm) as above. After chromatography on Sephadex G-100 the enzyme is concentrated by precipitation with solid ammonium sulfate (100% saturation), centrifugation, suspension in 50 m*M* Tris-HCl (pH 7.5) buffer, and dialysis in the same buffer. Purification of 45-fold with 12% recovery of activity is achieved.

Thiobacillus ferrooxidans (Fe^{2+}-Grown) Enzyme Purification

The purification procedures of Sugio *et al.*[7] are carried out at 4°. The Fe^{2+}-grown cells of *T. ferrooxidans* (AP19-3) are washed three times in 0.1 *M* sodium phosphate (pH 7.5) buffer and disrupted by passage through a French pressure cell twice. The supernatant after centrifugation at 12,000 *g* for 20 min and 105,000 *g* for 60 min is brought to 40% saturation with solid ammonium sulfate. The precipitate collected by centrifugation at 10,000 *g* for 20 min is dissolved in the same buffer and after dialysis applied to a Sephadex G-100 column (1.6 × 100 cm). Active fractions, after dialysis against 20 m*M* Tris-HCl (pH 7.5), are applied to a strong anion-exchange column (Mono Q; 0.5 × 5 cm) in a Pharmacia (Piscataway, NJ) fast protein liquid chromatography (FPLC) system and eluted with a linear gradient of NaCl from 0 to 0.25 *M* (33 ml). The active fraction (1.1 mg of protein from 744 mg of protein in a cell-free extract with 102-fold purification and 15% recovery) contains four protein bands in polyacrylamide disk gel electrophoresis [7.5% (w/v) polyacrylamide, pH 9.4, 3 mA/tube for 1.5 hr]. The enzyme band is cut, frozen at −20°, disrupted in 0.1 *M* sodium phosphate (pH 6.5), and eluted by centrifugation to obtain a pure enzyme preparation (154-fold purification, 2% yield).

Thermophilic Archaebacterial Enzyme Purification

Sulfolobus brierleyi (*A. brierleyi*)[4] and *D. ambivalens* (DSM 3772)[5] cells grown aerobically on sulfur yield purified enzyme preparations by similar procedures: sucrose density gradient centrifugation, taking advantage of their large molecular size (near 0.5 million) and electrophoresis in nondenaturing polyacrylamide gradient gel.

In the purification procedures of Emmel *et al.*[4] *S. brierleyi* cells (1 g dry weight) suspended in 10 ml of a mineral salts solution[14] adjusted to pH 7.0 are disrupted in a French pressure cell and centrifuged at 25,000 *g* at 4° for 1 hr. The supernatant is fractionated in a linear (25–45%, w/v) sucrose gradient centrifugation at 300,000 *g* at 4° for 4 hr. The active fraction is further purified by nondenaturing polyacrylamide gel electrophoresis (PAGE) [3–15% (w/v) polyacrylamide gradient; acrylamide : bis-

[14] M. E. Mackintosh, *J. Gen. Microbiol.* **105,** 215 (1978).

TABLE I
RESULTS OF SULFUR-OXIDIZING ENZYME PURIFICATION

Organism	Specific activity (units/mg)[a] Cell-free extract	Enzyme	Purification (fold)	Assay
T. thiooxidans[b]	0.011	0.13	12	S^0 + GSH
T. thiooxidans[c]	0.016	0.73	46	S^0 + GSH
T. thioparus[d]	0.010	0.25	25	S^0 + GSH
T. ferrooxidans (S^0)[e]	0.009	0.15	17	S^0 + GSH
T. ferrooxidans (Fe^{2+})[f]	0.010	1.54	154	S^0 + GSH
S. brierleyi[g]	0.030	0.91	30	S^0, 65°
D. ambivalens[h]	1.89	10.6	6	S^0, 85°

[a] One unit: oxidation of 1 μmol of sulfur per minute. The *T. thioparus* cell-free extract has only 20% of the total activity of the purified enzyme fraction, due to the presence of some inhibitors. The figure listed here assumes no inhibitor is present.
[b] I. Suzuki, *Biochim. Biophys. Acta* **104,** 359 (1965).
[c] R. S. Bhella, M.Sc. Thesis, University of Manitoba, Winnipeg (1981); B. S. Chahal, M.Sc. Thesis, University of Manitoba, Winnipeg (1986).
[d] I. Suzuki and M. Silver, *Biochim. Biophys. Acta* **122,** 22 (1966).
[e] M. Silver and D. G. Lundgren, *Can. J. Biochem.* **46,** 457 (1968).
[f] T. Sugio, W. Mizunashi, K. Inagaki, and T. Tano, *J. Bacteriol.* **169,** 4916 (1987).
[g] T. Emmel, W. Sand, W. A. König, and E. Bock, *J. Gen. Microbiol.* **132,** 3415 (1986).
[h] A. Kletzin, *J. Bacteriol.* **171,** 1638 (1989).

acrylamide ratio, 30 : 0.8) containing deoxycholate [0.5 *M* Tris-HCl, 0.25% (w/v) sodium deoxycholate] for 12 hr at 120 V. The gel area with the enzyme (located with a gel scanner at 280 nm) is excised and placed in a dialysis tube with 2 ml of 0.0025 *M* Tris–0.5 *M* glycine buffer (pH 8.4) containing 0.25% (w/v) sodium deoxycholate for electroelution (15 V/cm for 3 hr) in the same buffer. The enzyme is purified 31-fold with 7% recovery of activity.

In the procedures of Kletzin[5] *D. ambivalens* cells (1 g of cells suspended in 9 ml of 20 m*M* Tris-acetate, pH 8.0) are disrupted by sonication and after readjustment of the pH to 8.0 with Tris and DNase treatment for 30 min the crude extract is centrifuged for 60 min at 100,000 *g*. The supernatant is fractionated in a linear sucrose gradient [28–40% (w/w) in the Tris-acetate buffer] by centrifugation at 290,000 *g* for 48 hr. Active fractions are further fractionated by horizontally run nondenaturing PAGE (4–15% gradient gel) at pH 8.8 as described in detail by Kletzin.[5] The enzyme protein band is located by Coomassie blue staining of a narrow strip of gel and is electroeluted at pH 8.8. No deoxycholate is used during electrophoresis or electroelution. The enzyme is lyophilized and dialyzed against 20 m*M* Tris-acetate (pH 8.0). The enzyme is purified 5.6-fold with 7.7% recovery of activity.

A summary of purification results is shown in Table I. The cell-free extracts of thiobacilli seem to show similar specific activities. The *D. ambivalens* extract shows a specific activity more than 100 times that of thiobacilli (although the assay temperature at 85° is much higher than 25–30° used for thiobacilli), but the enzyme can be purified only 6-fold, implying high content in the cells.

Properties

All the thiobacilli enzymes require GSH for elemental sulfur oxidation.[1-3,7] All the sulfur oxidizing enzymes require molecular oxygen, O_2, except the Fe^{2+}-grown *T. ferrooxidans* enzyme,[7] which can use either Fe^{3+} or O_2 as electron acceptor. The archaebacterial enzymes oxidize sulfur without GSH, but only at high temperatures; the optima are 65° for *S. brierley*[4] and 85° for *D. ambivalens,*[5] temperatures destructive to thiobacilli enzymes.[1-3] The optimal pH for thiobacilli enzymes, pH 7.5–8.0,[1-3,10] is only slightly higher than the value for archaebacterial enzymes, pH 7–7.5.[4,5] The enzyme from Fe^{2+}-grown *T. ferrooxidans*[7] has a slightly lower optimal pH of 6.5 when measured for sulfite formation.

The archaebacterial enzymes are similar in their large sizes: molecular weights of 560,000 composed of subunits of 35,000 in *S. brierleyi*[4] and molecular weights of 550,000 composed of subunits of 40,000 in *D. ambivalens.*[5] The enzyme from Fe^{2+}-grown *T. ferrooxidans* is smaller, with a molecular weight of 46,000 consisting of subunits of 23,000.[7] The enzyme from *T. thiooxidans* has a molecular weight of 40,000 in Sephadex G-100 column chromatography.[10] Enzyme preparations from thiobacilli contain iron and labile sulfide.[2,3,10]

The *D. ambivalens* enzyme, which does not require GSH, is nevertheless inhibited by thiol-binding agents *p*-chloromercuribenzoate, iodoacetate, and *N*-ethylmaleimide.[5] Cu^{2+}, a thiol-binding ion, inhibits the enzyme in *T. thioparus,*[2] *T. ferrooxidans,*[3] and Fe^{2+}-grown *T. ferrooxidans.*[8] The K_m value for GSH in thiobacilli is similar (2–6 m*M*),[1-3,7,10] including the Fe^{2+}-grown *T. ferrooxidans* enzyme. The K_m value for sulfur is difficult to determine because of the size difference of elemental sulfur powder. It can be as low as 6 m*M*[10] for the milky white colloidal sulfur obtained by decantation of sonicated sulfur but for commercial powdered sulfur it is many times higher although not determined accurately. The K_m for sulfur (powdered by grinding) in the *S. brierley* system is reported as 50 m*M*.

[34] Sulfide-Cytochrome *c* Reductase (Flavocytochrome *c*)

By Tateo Yamanaka

Introduction

Flavocytochromes *c*, which have sulfide-cytochrome *c* reductase activity, occur in two genera of photosynthetic sulfur bacteria, *Chlorobium* and *Chromatium*. Bartsch and co-workers have shown that *Chlorobium limicola* f. *thiosulfatophilum* cytochrome *c*-553 and *Chromatium vinosum* cytochrome *c*-552 have covalently bound flavin as well as heme *c* and characterized them to some extent.[1] However, the function of the flavocytochromes *c* was not known until the studies by Kusai and Yamanaka, who found that *Chl. limicola* f. *thiosulfatophilum* cytochrome *c*-553 has sulfide–cytochrome *c* reductase activity.[2] *Chromatium* cytochrome *c*-552 has also been found to have sulfide–cytochrome *c* reductase activity.[3] Yamanaka found that *Chl. limicola* f. *thiosulfatophilum* cytochrome *c*-553 is composed of flavoprotein and cytochrome subunits.[4] After that, Fukumori and Yamanaka showed that *Chr. vinosum* cytochrome *c*-552 is also composed of two subunits.[3] Van Beeumen and co-workers determined the amino acid sequences of the cytochrome subunits of *Chl. limicola* f. *thiosulfatophilum* cytochrome *c*-553[5] and *Chr. vinosum* cytochrome *c*-552[6,7] and the partial sequences of the flavoprotein subunits of both cytochromes.[5,6] Flavocytochromes *c* of the two organisms also show sulfur-reducing activity with benzyl viologen radical[3,8] as an electron donor.

[1] R. G. Bartsch, T. E. Meyer, and A. B. Robinson, *in* "Structure and Function of Cytochromes" (K. Okunuki, M. D. Kamen, and I. Sekuzu, eds.), p. 443. Univ. of Tokyo Press, Tokyo/Univ. Park Press, Baltimore, Maryland, 1968.

[2] A. Kusai and T. Yamanaka, *FEBS Lett.* **34,** 235 (1973).

[3] Y. Fukumori and T. Yamanaka, *J. Biochem. (Tokyo)* **85,** 1405 (1979).

[4] T. Yamanaka, *J. Biochem. (Tokyo)* **79,** 655 (1976).

[5] J. J. Van Beeumen, S. Van Bun, T. E. Meyer, R. G. Bartsch, and M. A. Cusanovich, *J. Biol. Chem.* **265,** 9793 (1990).

[6] J. J. Van Beeumen, H. Demol, B. Samyn, R. G. Bartsch, T. E. Meyer, M. M. Dolata, and M. A. Cusanovich, *J. Biol. Chem.* **266,** 12921 (1991).

[7] J. J. Van Beeumen, *Biochim. Biophys. Acta* **1058,** 56 (1991).

[8] T. Yamanaka and Y. Fukumori, *in* "Flavins and Flavoproteins" (K. Yagi and T. Yamano, eds.), p. 631. Japan Scientific Societies Press, Tokyo, 1980.

p-Cresol methylhydroxylase of *Pseudomonas putida*[9,10] and adenosine-5′-phosphosulfate (APS) reductase of *Thiocapsa roseopersicina*[11] are also flavocytochromes *c*. The former enzyme converts *p*-cresol first to *p*-hydroxybenzyl alcohol with *N*-methylphenazonium methosulfate (PMS) as the hydrogen acceptor and then to *p*-hydroxybenzylaldehyde. The enzyme of *T. roseopersicina* seems to participate in the oxidation of sulfite in the bacterium; it catalyzes production of APS from sulfite and AMP with ferricytochrome c. A flavocytochrome *c* has been isolated from *Beggiatoa alba,* but its function has not been determined.[12] It has flavin mononucleotide (FMN) as well as heme *c*.

Chlorobium limicola f. *thiosulfatophilum* Flavocytochrome c (Cytochrome *c*-553)

Assay Method for Sulfide–Cytochrome c Reductase

Reagents

Tris-HCl buffer (0.1 *M*), pH 7.4
Na_2S, 2.5 m*M*
Yeast ferricytochrome *c*, 2.5 m*M*
Chlorobium flavocytochrome *c*, 3 μM

Assay Procedure. The reaction is carried out at room temperature in a cuvette (1-cm light path) of a spectrophotometer. The reaction mixture is prepared by mixing 0.95 ml of 0.1 *M* Tris-HCl buffer, pH 7.4, 0.02 ml of 2.5 m*M* yeast ferricytochrome *c*, 0.01 ml of 3 μM flavocytochrome *c*, and 0.02 ml of 2.5 m*M* Na_2S. The reaction is started by the addition of Na_2S and the increase of the absorbance at 550 nm is spectrophotometrically followed. The solution of Na_2S should be prepared just before the start of the experiment.

Purification

Acetone-dried cells (~200 g) of *Chl. limicola* f. *thiosulfatophilum* are suspended at 4°C in 3 liters of 10 m*M* Tris-HCl buffer, pH 8.5, containing 10% saturated $(NH_4)_2SO_4$, and the resulting suspension is allowed to stand overnight with agitation by a magnetic stirrer. The suspension thus treated

[9] D. J. Hopper and D. G. Taylor, *Biochem. J.* **167,** 157 (1977).

[10] N. Shamala, L. W. Lim, F. S. Mathews, W. McIntire, T. P. Singer, and D. J. Hopper, *J. Mol. Biol.* **183,** 517 (1985).

[11] H. G. Trüper and L. A. Rogers, *J. Bacteriol.* **108,** 1112 (1971).

[12] T. M. Schmidt and A. A. DiSpirito, *Arch. Microbiol.* **154,** 453 (1990).

is centrifuged at 10,000 *g* for 25 min. The resulting supernatant is 40% saturated with $(NH_4)_2SO_4$ and centrifuged at 10,000 *g* for 15 min. The supernatant obtained is 90% saturated with $(NH_4)_2SO_4$, centrifuged at 10,000 *g* for 25 min, and the resulting precipitate is dissolved in a minimal volume of 10 m*M* Tris-HCl buffer, pH 8.0. After centrifugation at 10,000 *g* for 15 min, the resulting supernatant is dialyzed against 20 liters of 10 m*M* Tris-HCl buffer, pH 8.5. When the dialyzed protein solution is loaded on a DEAE-cellulose column, flavocytochrome *c* passes through the column together with cytochrome *c*-555. The passed solution containing the cytochromes is dialyzed against 10 m*M* phosphate buffer, pH 7.0, for 2 days, changing the outer solution two or three times. The dialyzed cytochrome solution is loaded on an Amberlite CG-50 column. Flavocytochrome *c* is adsorbed under the band of cytochrome *c*-555. When the column is washed with 0.1 *M* phosphate buffer, pH 7.0, flavocytochrome *c* is eluted whereas cytochrome *c*-555 is eluted with 0.5 *M* phosphate buffer, pH 7.0. The eluate containing flavocytochrome *c* is further purified by passing the DEAE-cellulose column to remove contaminant proteins and by adsorption on and elution from the Amberlite CG-50 column.

Properties

Flavocytochrome *c* of *Chl. limicola* f. *thiosulfatophilum* has a molecular weight of 58,000, and contains one molecule each of heme *c* and FAD per molecule.[4,13] Its molecule is composed of one molecule each of cytochrome *c* (M_r 11,000) and flavoprotein (M_r 47,000) subunits. The absorption spectrum of the flavocytochrome shows a peak at 410 nm and shoulders at 450 and 480 nm in the oxidized form, and peaks at 417, 523, and 553 nm in the reduced form. When cyanide is added to flavoferricytochrome *c*, the shoulders at 450 and 480 nm disappear, and the absorbance in the wavelength region from 600 to 700 nm increases.[14]

The affinity of the ferricytochrome for cyanide is so high that the spectral change reaches its maximum in the presence of cyanide equimolar to the cytochrome.[15] Cyanide bound to the ferricytochrome is deprived of by treating the complex with Hg^{2+} or by gel filtration of the complex after the cytochrome has been reduced. Sulfite in place of cyanide causes a similar spectral change in the ferricytochrome.[16] The flavin of the flavo-

[13] T. Yamanaka, "The Biochemistry of Bacterial Cytochromes." Japan Scientific Societies Press, Tokyo/Springer-Verlag, Berlin, 1992.

[14] A. Kusai and T. Yamanaka, *Biochim. Biophys. Acta* **325,** 304 (1973).

[15] T. Yamanaka and A. Kusai, *in* "Flavins and Flavoproteins" (T. P. Singer, ed.), p. 292. Elsevier, Amsterdam, 1976.

[16] T. E. Meyer and R. G. Bartsch, *in* "Flavin and Flavoproteins" (T. P. Singer, ed.), p. 312. Elsevier, Amsterdam, 1976.

cytochrome is FAD, which binds to the cytochrome through thioether linkage; the 8α-methyl group of FAD links with Cys-42 of the cytochrome protein through the sulfur atom.[5,17]

Flavocytochrome *c* rapidly reduces *Chl. limicola* f. *thiosulfatophilum* ferricytochrome *c*-555 and eukaryotic ferricytochrome *c* in the presence of sulfide even at a concentration lower than 10 μM.[2,14] Flavocytochrome *c* is also obtained from *Chlorobium limicola,* which does not utilize thiosulfate, and from two strains of *Chlorobium vibrioforme;* one utilizes thiosulfate and the other does not.[18,19] *Chlorobium limicola* f. *thiosulfatophilum* flavocytochrome *c* reduces elemental sulfur to sulfide with benzyl viologen radical[8] (see below). This is related to the findings that elemental sulfur is reduced to sulfide by the whole cells of the organism under illumination.[20] In Table I, the properties of *Chl. limicola* f. *thiosulfatophilum* flavocytochrome *c* are shown together with those of *Chr. vinosum* flavocytochrome *c*.

Subunit

Preparation. Trichloroacetic acid (10%, w/v) is added to 5–10 ml of 230–420 μM flavocytochrome *c* dissolved in 10 mM Tris-HCl buffer at pH 8.0 to a final concentration of 2%. The resulting precipitate is collected by centrifugation at 10,000 *g* for 5 min and the precipitate obtained is dissolved in 5.0 ml of 0.1 M Tris-HCl buffer, pH 8.5. When the turbid solution obtained is centrifuged at 10,000 *g* for 5 min, the supernatant is red in color whereas the precipitate is yellow. The supernatant shows the absorption spectrum of a cytochrome *c*. However, as it is still contaminated with the original flavocytochrome *c*, it is once more subjected to successive treatment with 2% trichloroacetic acid and dissolution in 0.1 M Tris-HCl buffer, pH 8.5. In some cases, further purification is performed by chromatography on a DEAE-cellulose column. The cytochrome subunit is adsorbed tightly on the column in 10 mM Tris-HCl buffer, pH 8.5, and eluted with 0.1 M Tris-HCl buffer, pH 8.5. However, the amino acid composition of the subunit preparation purified by the chromatography differs slightly from that of the preparation obtained by trichloroacetic acid treatment alone.[4]

The insoluble yellow fraction is suspended in 3.0 ml of 0.1 M Tris-HCl buffer, pH 8.5, is then dissolved at pH 11.5 by the addition of 0.1 N NaOH, and again precipited by gradual lowering of the pH to 8.0 with

[17] W. C. Kenney, W. McIntire, and T. Yamanaka, *Biochim. Biophys. Acta* **483,** 467 (1977).
[18] M. A. Steinmetz and U. Fischer, *Arch. Microbiol.* **130,** 31 (1981).
[19] M. A. Steinmetz and U. Fischer, *Arch. Microbiol.* **131,** 19 (1982).
[20] H. Paschinger, J. Paschinger, and H. Gaffron, *Arch. Microbiol.* **96,** 341 (1974).

TABLE I

SOME PROPERTIES OF FLAVOCYTOCHROMES *c* FROM *Chlorobium limicola* f. *thiosulfatophilum* and *Chromatium vinosum*

Property	*Chl. limicola* f. *thiosulfatophilum*	*Chr. vinosum*
Absorption spectrum (maxima, nm)		
Native cytochrome (oxidized)[a,b]	(450),[c] (480)[c]	(450),[c] (475)[c]
Native cytochrome (reduced)[a,b]	417, 523, 553	416, 523, 552
Flavoprotein subunit (oxidized)[a,b]	350, 452	360, 453
Cytochrome subunit (reduced)[a,b]	417, 523, 553	416, 523, 552
A_{protein} (oxidized)/A_α (reduced)		
Native cytochrome[a,b,d]	3.90	3.73
Cytochrome subunit[a,b]	0.86	1.20
Molecular weight		
Native cytochrome[a,b]	58,000	72,000[e] 67,000
Flavoprotein subunit[a,b]	47,000	46,000
Cytochrome subunit[a,b]	11,000	21,000
$E_{m,7}$ (volt)		
Native cytochrome[f]	0.098	0.032
Heme in native cytochrome[g]	ND[h]	0.016 −0.033
Heme *c* (mol/mol)[e]	1	2
Flavin (mol/mol)[e,i]	One covalently bound FAD	One covalently bound FAD
Isoelectric point		
Native cytochrome[e]	6.7	5.1
Flavoprotein subunit[b]	ND[h]	5.6
Cytochrome subunit[b]	ND[h]	5.2
Sulfide-cytochrome *c* reductase activity[a,b,j]	Yes	Yes
S^0 reductase activity[b,k]	Yes	Yes

[a] T. Yamanaka, *J. Biochem.* (*Tokyo*) **79,** 655 (1976).

[b] Y. Fukumori and T. Yamanaka, *J. Biochem.* (*Tokyo*) **85,** 1405 (1979).

[c] Shoulder.

[d] T. E. Meyer, R. G. Bartsch, M. A. Cusanovich, and J. H. Mathewson, *Biochim. Biophys. Acta* **153,** 854 (1968).

[e] R. G. Bartsch, T. E. Meyer, and A. B. Robinson, *in* "Structure and Function of Cytochromes" (K. Okunuki, M. D. Kamen, and I. Sekuzu, eds.), p. 443. Univ. of Toyko Press, Tokyo, 1968.

[f] M. A. Cusanovich, T. E. Meyer, and R. G. Bartsch, *in* "Chemistry and Biochemistry of Flavoenzymes" (F. Muller, ed.), Vol. 2, p. 377. CRC Press, Boca Raton, Florida, 1991.

[g] T. Yamanaka and Y. Fukumori, *in* "Photosynthesis II. Electron Transport and Photophosphorylation" (G. Akoyunoglou, ed.), p. 577. Balaban International Science Series, Philadelphia, Pennsylvania, 1981.

[h] ND, Not determined.

[i] W. C. Kenney, W. McIntire, and T. Yamanaka, *Biochim. Biophys. Acta* **483,** 467 (1977).

[j] A. Kusai and T. Yamanaka, *FEBS Lett.* **34,** 235 (1973).

[k] T. Yamanaka and Y. Fukumori, *in* "Flavins and Flavoproteins" (K. Yagi and T. Yamano, eds.), p. 631. Japan Scientific Societies Press, Tokyo, 1980.

0.2 *M* NaH_2PO_4. The resulting turbid solution is centrifuged at 10,000 *g* for 5 min. The yellow precipitate thus obtained is dissolved at pH 11.5 and the resulting yellow solution is used as the flavoprotein subunit. If the flavoprotein subunit is still contaminated with cytochrome as judged from the absorption spectrum, the above procedures of dissolution at alkaline pH and precipitation at pH 8.0 are repeated. The resulting alkaline solution of the flavoprotein subunit is yellow in color.

Properties. The cytochrome subunit shows the absorption spectrum of a usual *c*-type cytochrome; there is no peak or shoulder around 450 and 480 nm in the oxidized form. The oxidized form of the cytochrome subunit shows absorption peaks at 276 and 409 nm, and its reduced form shows peaks at 417, 523, and 553 nm. The ratio of A_{276}(oxidized)/A_{553}(reduced) is 0.86. The reduced form of cytochrome subunit is oxidized by *Pseudomonas aeruginosa* nitrite reductase fairly rapidly, whereas it does not react with cow cytochrome *c* oxidase.

The flavoprotein subunit shows absorption peaks at 272, 350, and 452 nm in the oxidized form. Although the ratio of A_{272}(oxidized)/A_{452}(oxidized) is 6.2, this value is uncertain because the protein is dissolved at pH 11.5. A small peak at 408 nm is observed in the reduced form and this is attributable to the remaining cytochrome subunit. The amount of the remaining cytochrome subunit, however, seems to be small, because the amino acid composition of the flavoprotein subunit is close to that calculated by subtracting the values for the cytochrome subunit from those for the original flavocytochrome *c*.[4] The original flavocytochrome *c* shows a fluorescence spectrum with a peak at 450 nm on excitation at 375 nm, whereas the flavoprotein subunit shows two emission peaks at 445 and 510 nm in the fluorescence spectrum. The molecular weights of cytochrome and flavoprotein subunits are determined to be 11,000 and 47,000, respectively, by sodium dodecyl sulfate-polyacrylamide gel electrophoresis (SDS–PAGE).

The yields of the cytochrome subunit and flavoprotein subunit are usually about 65%, on the basis of the amount of the original flavocytochrome *c* used. The fairly low yield may be attributable to decomposition of the subunits and/or to the flavocytochrome *c* left undissociated during the treatment of the cytochrome with trichloroacetic acid and alkali. Further, it is difficult to estimate the amount of the flavoprotein subunit, because it is soluble only at alkaline pH.

Neither of the subunits obtained shows sulfide–cytochrome *c* reductase activity or reactivity with cyanide. When the subunits are separately excited with a laser ray at 457.9 nm in the presence of ethylenediaminetetraacetic acid (EDTA), the cytochrome subunit is not reduced whereas the flavoprotein subunit is reduced.[21] As flavoferricytochrome *c* is completely reduced by the laser illumination, the electron transfer in the flavocytochrome *c* molecule in the presence of sulfide seems to occur from flavin to heme c. Tollin *et al.*[22] have obtained similar results.

Chromatium vinosum Flavocytochrome c

The assay for sulfide–cytochrome *c* reductase activity of *Chromatium* flavocytochrome *c* (cytochrome *c*-552) is carried out by a method similar to the assay for the activity of *Chlorobium* flavocytochrome *c*.

Purification

The acetone-dried cells (200 g) of *Chr. vinosum* are suspended at 4°C in 1.5 liters of 10 m*M* Tris-HCl buffer, pH 8.5, and the resulting suspension is allowed to stand overnight and centrifuged at 10,000 *g* for 40 min. The supernatant obtained is dialyzed against 10 liters of 10 m*M* Tris-HCl buffer, pH 8.5. The dialyzed cytochrome solution is 50% saturated with $(NH_4)_2SO_4$ and the precipitate is collected by centrifugation at 10,000 *g* for 40 min. The supernatant obtained is dialyzed overnight against 10 m*M* Tris-HCl buffer, pH 8.5, and loaded on a DEAE-cellulose column that has been equilibrated with the same buffer as used for the dialysis described above. *Chromatium* flavocytochrome *c* adsorbed on the column is eluted by a linear gradient of 0.1–0.2 *M* NaCl produced in 10 m*M* Tris-HCl buffer, pH 8.5. The eluates containing *Chromatium* flavocytochrome *c* are slightly contaminated with cytochrome *c'*. The eluates containing flavocytochrome *c* are combined and dialyzed against 10 m*M* Tris-HCl buffer, pH 7.5. The dialyzed cytochrome solution is loaded on a DEAE-cellulose column that has been equilibrated with 10 m*M* Tris-HCl buffer, pH 7.0. Flavocytochrome *c* adsorbed on the column is eluted with a linear gradient of 0–0.4 *M* NaCl produced in 0.1 *M* Tris-HCl buffer, pH 7.5. The eluates containing flavocytochrome *c* are combined and concentrated with the aid of a small DEAE-cellulose column. *Chromatium* flavocytochrome *c* thus concentrated is loaded on a Sephadex G-100 column and the eluate containing flavocytochrome *c* is used as its purified preparation.

Properties

The molecular weight of *Chromatium* flavocytochrome *c* is 67,000 (or 72,000[1]) and the protein is composed of one molecule each of the cyto-

[21] T. Kitagawa, Y. Fukumori, and T. Yamanaka, *Biochemistry* **19,** 5721 (1980).
[22] G. Tollin, T. E. Meyer, and M. A. Cusanovich, *Biochemistry* **21,** 3849 (1982).

chrome and flavoprotein subunits.[3] Molecular weights of the cytochrome and flavoprotein subunits are 21,000 and 46,000, respectively, when they are estimated by SDS–PAGE, whereas the molecular weight of cytochrome subunit per heme is 10,500. Therefore, each *C. vinosum* flavocytochrome *c* and its cytochrome subunit contains two heme *c* molecules per molecule, as determined by Meyer *et al.*[1] The complete amino acid sequence of the cytochrome subunit of *Chromatium* flavocytochrome *c* has been determined, and the occurrence of two heme *c*-binding sites has been confirmed.[6]

Chromatium flavocytochrome *c* shows sulfide–cytochrome *c* reductase activity; it reduces horse ferricytochrome *c* and *Nitrosomonas europaea* ferricytochrome *c*-552 in the presence of 10 μM Na_2S. However, *Chr. vinosum* ferricytochrome *c*-553(550) or *Chr. vinosum* ferricytochrome *c'* is not reduced by the flavocytochrome.[3] Until now, the electron acceptor for the flavocytochrome in the sulfide oxidation by the bacterium has not been found. The flavocytochrome also reduces S^0 with benzyl viologen radical (see Assay for Elemental Sulfur Reductase, below).[8] Therefore, the flavocytochrome will be responsible for reduction of S^0 by *Chr. vinosum*.[23]

Subunit

Preparation. The subunits of *Chromatium* flavocytochrome *c* are not obtained by the treatment with trichloroacetic acid of the cytochrome as used for *Chlorobium* flavocytochrome *c*, because its flavoprotein subunit is soluble at pH values at which its cytochrome subunit is soluble.

The flavoprotein subunit of *Chromatium* flavocytochrome *c* is obtained by isoelectric focusing at 1.7 mA and 800 V in the presence of 6 *M* urea and 1% 2-mercaptoethanol with 1% carrier ampholyte of pH range 3–6. About 70 nmol of flavocytochrome *c* is applied to the electrophoresis apparatus (110 ml in volume). The electrophoresis is performed at 4° for 2 days. The cytochrome subunit is not obtained by the isoelectric focusing, because the heme of the cytochrome moiety is quickly destroyed in the presence of 2-mercaptoethanol. The cytochrome subunit is obtained by gel filtration with Sephacryl S-200 in 0.1 *M* Tris-HCl buffer, pH 8.5, containing 6 *M* urea and 0.1 *M* KCl. Before being subjected to the gel filtration, flavocytochrome *c* is dialyzed against 0.1 *M* Tris-HCl buffer, pH 8.5, containing 6 *M* urea and 0.1 *M* KCl for 12 hr. In the gel filtration, the cytochrome subunit is obtained in a pure state, whereas the flavoprotein moiety is eluted together with the undissociated cytochrome.

[23] H. Van Gemerden, *Arch. Microbiol.* **64,** 118 (1968).

Properties. The cytochrome subunit shows peaks at 275 and 407 nm in the oxidized form, and peaks at 416, 523, and 552 nm in the reduced form. The flavoprotein subunit shows peaks at 276, 360, and 453 nm in the oxidized form. The ratio of A_{275}/A_{552} of the cytochrome subunit is 1.20, whereas the ratio of the original flavocytochrome is 3.73. Molecular weights of the cytochrome and flavoprotein subunits are estimated to be 21,000 and 46,000, respectively, by SDS–PAGE. The cytochrome subunit has two heme *c* molecules per molecule[24] and $E_{m,7}$ values of the two hemes are 0.016 and −0.033 V, respectively.[25] The flavoprotein subunit has one molecule of covalently bound FAD per molecule.[1] Neither of the subunits has sulfide–cytochrome *c* reductase activity.

Assay for Elemental Sulfur Reductase

Reagents

Tris-HCl buffer (3.0 *M*), pH 8.5
Benzyl viologen, 100 μM
Chromatium flavocytochrome *c*, 30 μM
A mixture of $Na_2S_2O_4$ and glucose (1 : 50, in moles)
Elemental sulfur (powder)

Procedure. The reaction is performed using a Thunberg-type cuvette. The main chamber of the cuvette contains 0.1 ml of 3 *M* Tris-HCl buffer, pH 8.5, 0.1 ml of 100 μM benzyl viologen, 0.1 ml of 30 μM flavocytochrome *c*, 100 μmol of elemental sulfur, and 2.7 ml of deionized water, and the side arm contains 7 μmol of $Na_2S_2O_4$ as the mixture with glucose. Benzyl viologen is half-reduced (radical) instantly when $Na_2S_2O_4$ is mixed with the reaction mixture in the main chamber, and then the radical of the pigment is oxidized in the presence of elemental sulfur and flavocytochrome *c*. In the absence of the cytochrome, the purple color of the radical of benzyl viologen radical is unchanged for more than 30 min. Although elemental sulfur is easily precipitated in the absence of flavocytochrome *c*, it remains in the suspension for more than a few hours in the presence of the flavocytochrome. The reference cuvette contains all of the components but $Na_2S_2O_4$.[8]

[24] M. M. Dolata, J. J. Van Beeumen, R. P. Ambler, T. E. Meyer, and M. A. Cusanovich, *J. Biol. Chem.* **268,** 14426 (1993).

[25] T. Yamanaka and Y. Fukumori, *in* "Photosynthesis II. Electron Transport and Photophosphorylation" (G. Akoyunoglou, ed.), p. 577. Balaban International Science Services, Philadelphia, Pennsylvania, 1981.

Although it may seem difficult for the cytochrome to reduce S^0 to S^{2-} as the $E_{m,7}$ of *Chromatium* flavocytochrome *c* is about 0.032 V[26] and that of the S^0/S^{2-} couple is −0.23 V, the benzyl viologen radical ($E_{m,7}$ −0.36 V) pushes electrons powerfully to elemental sulfur, overcoming the energetic disadvantage.

Acknowledgment

The author wishes to thank Dr. Y. Fukumori for his collaboration in the study of *Chromatium vinosum* flavocytochrome *c*.

[26] M. A. Cusanovich, T. E. Meyer, and R. G. Bartsch, *in* "Chemistry and Biochemistry of Flavoenzymes" (F. Muller, ed.), Vol. 2, p. 377. CRC Press, Boca Raton, Florida, 1991.

Section V

Metabolism of Polythionates

[35] Synthesis and Determination of Thiosulfate and Polythionates

By DON P. KELLY and ANN P. WOOD

Introduction

Although many inorganic and organic sulfur compounds are commercially available in high purity and often at low cost, most polythionates are not normally available from chemical manufacturers. As polythionates, such as tri-, tetra-, penta-, and hexathionate, have been implicated as intermediates in the oxidation of inorganic sulfur substrates (e.g., sulfur, sulfide, and thiosulfate) since the earliest studies of thiobacilli almost a century ago,[1,2] and more recently trithionate has been a contender as an intermediate in dissimilatory sulfate reduction (see [17] in this volume), there has long been a keen interest in the detection, identification, and quantitation of polythionates. Such work has undoubtedly been hampered by the unavailability of pure specimen compounds, both as chemical standards and as potential substrates.

Our purpose in this chapter is to provide sufficient detail of reliable and relatively simple procedures for the synthesis of the shorter chain polythionates to enable the biochemist and microbiologist to prepare these materials in sufficient quantity and purity for their research. We also describe the more basic procedures for the detection and analysis of polythionates (and other low molecular weight sulfur species) both individually and in mixture with each other and with thiosfulate and thiocyanate.

Commercial Sources and Synthetic Procedures for Thiosulfate and Tetrathionate

Commercial Availability

Thiosulfate and tetrathionate are easily available at high purity and relatively low cost. All major chemical suppliers market $Na_2S_2O_3 \cdot 5H_2O$ at analytical reagent quality (99.5% or better) at less than \$20 per kilogram and $K_2S_2O_3$ (and its pentahydrate) are available at purities of exceeding 98 and 99.5% at around \$120 and \$300 per kilogram, respectively. $K_2S_4O_6$

[1] D. P. Kelly, *Philos. Trans. R. Soc. London B* **298,** 499 (1982).
[2] D. P. Kelly, *Soc. Gen. Microbiol. Symp.* **42,** 65 (1987).

and $Na_2S_4O_6 \cdot 2H_2O$ are available from major and specialist suppliers at high purity (>99.5%) at around $200 and $500 per kilogram, respectively.

Synthesis of Potassium Tetrathionate

Several procedures are available, the simplest of which is the oxidation of thiosulfate with iodine. Two common methods for the bulk synthesis of $K_2S_4O_6$ are described.

From $Na_2S_2O_3$ and Iodine

$$2S_2O_3^{2-} + I_2 \rightleftharpoons S_4O_6^{2-} + 2I^-$$

A solution of 50 g of $Na_2S_2O_3 \cdot 5H_2O$ in 100 ml of water is chilled on ice and stirred vigorously while adding enough powdered iodine (about 50 g) to it to produce a faint yellow color.[3,4] Add 100 ml of saturated potassium acetate solution, mix, and add 800 ml of ethanol. Crystalline $K_2S_4O_6$ is deposited and is recovered by filtration. To purify the $K_2S_4O_6$, the product can be redissolved with vigorous stirring in about 60 ml water at 70°, then hot-filtered (60°) into a beaker chilled in ice–water. $K_2S_4O_6$ (about 50 g) at 99–100% purity crystallizes and is filtered off, washed with ethanol, and dried over P_2O_5. This recrystallization step may better be carried out using 0.5 *M* HCl, rather than water, as recommended by Foss.[3-5] This takes account of the fact that tetrathionate is less stable in aqueous solution than in acid.[6]

A similar procedure is described by Feher,[7] in which 39.5 g of $3K_2S_2O_3 \cdot 5H_2O$ (as a saturated solution) is added dropwise with intense stirring to 26 g of iodine in ethanol (containing a few milliliters of water). $K_2S_4O_6$ is deposited and recovered by filtration. It is washed on the filter with ethanol until free of iodine and iodide, then purified by dissolving in a minimum volume of water at room temperature and reprecipitating with ethanol. The crystalline $K_2S_4O_6$ is dried on filter paper over H_2SO_4.

From Sulfurous Acid and Disulfur Dichloride

$$2H_2SO_3 + S_2Cl_2 \rightleftharpoons H_2S_4O_6 + 2HCl$$
$$H_2S_4O_6 + 2KOH \rightleftharpoons K_2S_4O_6 + 2H_2O$$

[3] A. B. Roy and P. A. Trudinger, "The Biochemistry of Inorganic Compounds." Cambridge Univ. Press, Cambridge, 1970.

[4] P. A. Trudinger, *Biochem. J.* **780,** 680 (1961).

[5] O. Foss, *K. Norske Videnskab. Selksabs Skrifter.* 2 (1945).

[6] A. P. Wood, D. P. Kelly, and P. R. Norris, *Arch. Microbiol.* **146,** 382 (1987).

[7] F. Feher, *in* "Handbuch der Präparativen Anorganischen Chemie" (G. Brauer, ed.), p. 398. Enke, Stuttgart, 1975.

This is the method of Stamm and Goehring.[8] Water (750 ml) cooled to 0° in a 2- to 3-liter stoppered flask is saturated by gassing with SO_2.[5,7,8] The solution produces crystals of $SO_2 \cdot 6H_2O$. To this solution at 0° are added successive 100-ml portions of S_2Cl_2 (75 g in 500 ml of petroleum ether, precooled to − 15°). After each addition the mixture becomes yellow and must be decolorized with vigorous shaking (and cooling to 0°) before adding more S_2Cl_2. The mixture should still smell of SO_2 when all the S_2Cl_2 has been added. The aqueous layer (containing the $K_2S_4O_6$) is separated from the petroleum layer with a separating funnel, then SO_2 is expelled by flushing with air for several hours. After cooling to 0°, the solution is brought to pH 6–7 with about 1 liter of 15% (w/v) KOH in ethanol. $K_2S_4O_6$ containing about 10% KCl is deposited (165 g). This can be purified by dissolving in 120 ml of water at 70° and recrystallizing at 0° as described above. About 120 g of 100% $K_2S_4O_6$ is produced, washed with ethanol, and dried at room temperature. Adding an equal volume of ethanol to the aqueous filtrate from the recrystallization will precipitate about a further 20 g of 99% $K_2S_4O_6$.

[^{35}S]Thiolsulfate

Considerable use has been made of [^{35}S]thiosulfate in studying sulfur oxidation by thiobacilli and phototrophic sulfur bacteria (see [37], this volume). Its synthesis has been described,[9] but most researchers requiring it as a biological tracer would prefer a high-purity commercial source. [^{35}S]Thiosulfate is available as a custom-synthesized item from Amersham International (Amersham, England) labeled in either the outer (sulfane) sulfur ($^{35}S{-}SO_3^{2-}$), or the inner (sulfonate) sulfur atom ($S{-}^{35}SO_3^{2-}$). Synthesis by Amersham[10] is from sodium sulfite and elemental sulfur, with ^{35}S being provided in one or the other compound to produce labeling in the inner or outer position. At least 95% of the ^{35}S is present in the "correct" position, and no internal chemical exchange occurs to randomize the label position.[9] The synthesis of both compounds is conducted in 50-ml bulbs with or without a side arm, and a constriction and B14 cone (Fig. 1a and b).

$Na_2S\,^{35}SO_3$ (Sodium [inner-^{35}S]Thiosulfate)

One molar sodium hydroxide (6.0 ml) is degassed in a bulb (Fig. 1a) on a vacuum manifold and $^{35}SO_2$ (100 mCi) is condensed into the bulb.

[8] H. Stamm and M. Goehring, *Z. Anorg. Allg. Chem.* **250,** 226 (1942).

[9] D. P. Ames and J. E. Willard, *J. Am. Chem. Soc.* **73,** 164 (1951).

[10] Information provided by the Life Sciences Manufacturing Manager, Amersham International plc, Amersham, UK. The original method (cited in Ref. 9) may be found in Watson and Rajagopalan, *J. Ind. Inst. Sci.* 8A, 275 (1925).

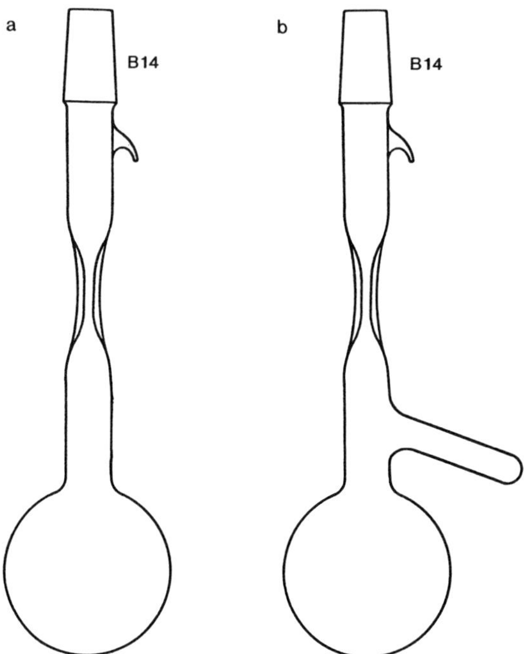

FIG. 1. Flasks for the preparation of (a) inner-labeled and (b) outer-labeled [^{35}S]thiosulfate. (Kindly provided by Amersham International, plc.)

After isolation of the bulb from the manifold, its contents are allowed to warm to room temperature. Nonradioactive sulfur dioxide is measured out to make the total amount of active plus inactive gas equal to 3 mmol. The measured inactive sulfur dioxide is then condensed into the bulb, and the bulb allowed to warm to room temperature and left for 1 hr for the gas to absorb.

The bulb is then flooded with nitrogen, detached from the manifold, and finely ground elemental sulfur (192 mg, 6 mmol) added. The bulb is again flushed with nitrogen, flame sealed, and heated in an oven for 4 hr.

After cooling, the bulb is opened and decolorizing charcoal added, and the mixture gently swirled for a few minutes before filtering through a 0.45-μm pore size syringe filter into a centrifuge tube.

Excess sulfate in the solution is removed by barium precipitation: barium thiosulfate solution is added until the solution gives a positive test for barium. The presence of barium ions is indicated by a color change from orange-yellow to red when a drop of the solution is added to a drop of sodium rhodizonate solution on a filter paper. Excess barium ions are then removed by the addition of a dilute solution of sodium sulfate, taking care not to add excess, and retesting for barium. The barium sulfate precipitate is removed by filtration, as described above, into a tared flask. The solution is freeze-dried overnight and finally heated at 50° for 30 min to remove water of crystallization. The dried solid is weighed and a sample subjected to infrared analysis. The remainder is dissolved in deoxygenated water to give a radioactive concentration of about 10 mCi/ml, and dispensed into nitrogen-flushed plastic bottles for storage at −140°.

The radiochemical purity can be determined by thin-layer chromatography (TLC) on cellulose in pyridine–propan-2-ol–water (60/100/100) and butan-1-ol–acetone–water (40/40/20). Degradation procedures for the determination of the position of the ^{35}S in the molecule are described below.

$Na_2\,^{35}SSO_3$ (Sodium [outer-^{35}S]Thiosulfate)

Elemental ^{35}S in toluene (100 mCi) is placed in a bulb with a side arm (Fig. 1b) and the toluene removed in a stream of nitrogen. The bulb is then pumped to a hard vacuum to remove all traces of toluene. Finely ground inactive sulfur (128 mg, 4 mmol) is carefully placed in the side arm and sodium sulfite heptahydrate (514 mg, 2 mmol, in 5 ml of deoxygenated water) is placed in the main bulb. The bulb is then flushed with nitrogen, flame sealed, and heated at 100° for 1.5 hr. After cooling, the contents of the side arm are tipped into the solution and heating continued for a further 3 hr.

After opening the ampoule the thiosulfate product is recovered and purified as described above.

[^{35}S]Tetrathionate

Trudinger[3,11] described a modification of the thiosulfate/iodine method for the synthesis of milligram quantities of [^{35}S]tetrathionate. Sodium [^{35}S]thiosulfate in a small volume of ice-cold water is titrated with a concentrated iodine solution until it is faintly yellow. A slight excess of barium acetate is added, followed by ethanol to a total of 50% (v/v). This precipitates BaS_4O_6, which is recovered by centrifugation, washed with ethanol, and dissolved in a small volume of water. The BaS_4O_6 is reprecipitated with ethanol and this procedure repeated three or four times. This produces about 85% of the theoretical amount of BaS_4O_6. $K_2S_4O_6$ can be recovered by double decomposition and precipitation (of $BaSO_4$) using K_2SO_4. Alternatively the barium salt can be treated with Dowex-50 (K^+), yielding a solution of $K_2S_4O_6$.

[11] P. A. Trudinger, *Biochem. J.* **90,** 640 (1964).

An alternative way of producing [^{35}S]tetrathionate labeled in either the two central (sulfane) atoms or the two outer (sulfonate) atoms is simply by mixing unlabeled $K_2S_4O_6$ with [^{35}S]thiosulfate. A rapid exchange reaction occurs in which the tetrathionate adopts the same labeling distribution as the thiosulfate.[3,12,13] The [^{35}S]thiosulfate available from Amersham is of specific activities up to 10 mCi/mmol, and therefore mixing 0.01 mCi with 1 mmol of $K_2S_4O_6$ produces [^{35}S]tetrathionate with an activity of around 22,000 dpm/μmol but containing only 0.1% [^{35}S]thiosulfate as an impurity.

Synthesis of Trithionate (Sodium or Potassium Salt)

Four procedures have been used to prepare trithionate of 97–100% purity.[3,8,14–19] Three of these are described below, and in the authors' hands the method employing the oxidation of thiosulfate with hydrogen peroxide has proved safe and effective. The sulfur dichloride/bisulfite procedure also yielded good-quality material for studies with sulfate-reducing bacteria.[20]

From Reaction of Thiosulfate with Hydrogen Peroxide

$$2Na_2S_2O_3 + 4H_2O_2 \rightleftharpoons Na_2S_3O_6 + Na_2SO_4 + 4H_2O$$

Dissolve 150 g of sodium thiosulfate pentahydrate in 90 ml of distilled water in a stainless steel beaker.[14–16] Pack the beaker in ice and stir continuously on a magnetic stirrer to reduce the temperature to about 1°. Maintaining continuous stirring, and not allowing the temperature to rise above 20°, gradually add 140 ml of 30% (w/v) hydrogen peroxide ("100 vol") solution dropwise by means of a funnel with a dropping pipette attached. Cease stirring and maintain the beaker in ice for 1–2 hr. Sodium sulfate crystallizes during this period (~38 g). Failure to crystallize should be remedied by adding a few small crystals of sodium sulfate and stirring vigorously with a glass rod. Remove the sulfate by filtration under suction

[12] A. Fava, *Gazz. Chim. Ital.* **83,** 87 (1953).

[13] A. Fava and S. Bresadola, *J. Am. Chem. Soc.* **77,** 5792 (1955).

[14] R. Willstätter, *Ber. dtsch. chem. Ges.* **36,** 1831 (1903).

[15] F. Auerbach and I. Koppel, *in* "Handbuch der Anorganischen Chemie" (R. Abegg, F. Auerbach, and F. Koppel, eds.), Vol. 4, p. 554. Hirzel, Leipzig, 1927.

[16] A. P. Wood and D. P. Kelly, *Arch. Microbiol.* **144,** 71 (1986).

[17] F. Feher, *in* "Handbuch der Präparativen Anorganischen Chemie" (G. Brauer, ed.), p. 397. Enke, Stuttgart, 1975.

[18] F. Martin and L. Metz, *Z. Anorg. Allg. Chem.* **127,** 83 (1923).

[19] A. Fava and D. Divo, *Gazz. Chim. Ital.* **82,** 558 (1952).

[20] H. Sass, J. Steuber, M. Kroder, P. M. H. Kroneck, and H. Cypionka, *Arch. Microbiol.* **158,** 418 (1992); and personal communication from H. Cypionka, 1993.

(Büchner flask) through a Whatman (Clifton, NJ) No. 1 filter paper. Wash the sulfate on the filter with 100 ml of ethanol, which is allowed to join the filtrate. Discard the sulfate crystals, which contain insignificant thiosulfate or trithionate. Transfer the filtrate to a 2-liter beaker at 3°, mix with 250 ml of ice-cold ethanol, and leave at 0–3° for 1 hr. A white crystalline material is deposited (~6.5 g) which is predominantly sulfate and contains less than 0.8 g of $Na_2S_3O_6$. This is removed by filtration and washed on the filter with 200 ml of ice-cold ethanol, which mixes with the filtrate. This material is discarded. The filtrate (together with 100 ml of ethanol used to rinse the flask) is transferred to a 5-liter beaker containing 1 liter of cold ethanol, the mixture stirred thoroughly, and left at 3° for 1–2 hr. Sodium trithionate crystallizes and is separated by filtration under suction, washed with 50 ml of ethanol, 50 ml of acetone, and dried over silica gel *in vacuo*. The yield is about 62 g of $Na_2S_3O_6$ at a purity of 96–100%. This represents about 87% of the theoretical yield for this product. Analysis of the three crystalline products from each stage of the purification indicates production of $Na_2S_3O_6$ and Na_2SO_4 at 91 and 108%, respectively, of the amounts predicted by the equation.

From Reaction of Sulfur Dichloride and Potassium Bisulfite

$$SCl_2 + 2KHSO_3 \rightleftharpoons K_2S_3O_6 + 2HCl$$

Cool 800 ml of 5 *M* KOH to $-5°$ in a large (3- to 4-liter) stoppered flask, then bubble gaseous SO_2 through it until it reaches about pH 7, indicating production of $KHSO_3$.[3,8,17] The $KHSO_3$ solution is maintained at $-5°$ and successive additions made to it of 200-ml amounts of a solution of 100 g of SCl_2 in 1500 ml of petroleum ether. After each addition the mixture becomes yellow and must be vigorously shaken to discharge color. The temperature must be kept below 10° during SCl_2 additions. The mixture is then maintained at 0° for about 1 hr, during which a coarse crystalline slurry separates. This solid is separated by filtration, washed with acetone, and dried at room temperature (on a clay tile in the original method[13]). This yields about 120 g of crystalline material containing about 86% $K_2S_3O_6$, containing KCl and sulfur as impurities. Further purification is achieved by dissolving in 350 ml of distilled water at 35°, filtering through a warmed filter, then cooling rapidly to 0°. $K_2S_3O_6$ reportedly at 100% purity separates.[13] After filtering off the $K_2S_3O_6$, the addition of an equal volume of acetone to the filtrate and cooling to 0° produces a further crop of $K_2S_3O_6$ of the same purity. The pure $K_2S_3O_6$ is washed with acetone and dried at room temperature. The yield is about 85 g, which indicates recovery as $K_2S_3O_6$ of about 30% of the added SCl_2.

Potassium trithionate prepared in this way by Kroder and co-workers[20] assayed at 96–97% purity (both cold crystallized and acetone precipitated), contained 2.1–2.5% sulfate and thiosulfate, but no tetrathionate, as impurities, and elemental analyses gave 32.5–34.7% S (expected, 35.6%) and 27.6–27.9% K (expected, 28.9%).

A modification of this procedure was described by Akagi *et al.*,[21] using 11 g of $K_2S_2O_5$ dissolved in 30 ml of water and adjusted to pH 7.0 with 4 *M* KOH. This forms an equivocal mixture of $KHSO_3$ and K_2SO_3. Reaction with SCl_2 (5 g in 65 ml of petroleum spirit) was carried out in a 500-ml stoppered flask chilled to −5° in an ice–salt slurry, and purification conducted as above. In the authors' hands this modification (using 140 g of $K_2S_2O_5$ in 500 ml of 1.23 *M* KOH with 64 g of SCl_2 in 800 ml of petroleum spirit) yielded 76 g of "first crop" crystals containing 86% $K_2S_3O_6$ and about 5% $K_2S_4O_6$. This was exactly comparable with the amount expected in the unmodified method (120 g).

From Reaction of Sulfur Dioxide with Thiosulfate

A solution (200 ml) of $Na_2S_2O_3$ (saturated at 30°) is cooled under running water and mixed with 20 ml of saturated sulfur dioxide solution (sulfurous acid).[17,18] The resultant yellow color is fairly rapidly discharged. The mixture is then gassed with a stream of SO_2, producing a strong yellow color. Gassing is stopped and the mixture allowed to stand until the color discharges. These two steps are then repeated until the yellow color persists indefinitely. The mixture is cooled to 10° and held at this temperature for several hours, during which a faintly yellow product is deposited. This is dissolved in a minimum of water, filtered to remove suspended sulfur, and pure $Na_2S_3O_6$ precipitated by the addition of about an equal volume of ethanol. The $Na_2S_3O_6$ is rapidly filtered, washed with ethanol, and dried on a tile at room temperature. We have no direct experience with this method.

From Reaction of Thiosulfate with Sulfite

$$K_2S_4O_6 + K_2SO_3 \rightleftharpoons K_2S_3O_6 + K_2S_2O_3$$

This procedure can make use of commercially available tetrathionate and requires differential crystallization to separate the products.[19] We have no experience with the procedure but would suspect that a negative factor in its use would be the losses likely to accompany separating the trithionate product from an excess of sulfite and from the coproduct thiosulfate, which is also soluble in water.

[21] J. M. Akagi, M. Chan, and V. Adams, *J. Bacteriol.* **120,** 240 (1974).

Synthesis of Potassium Pentathionate

From Sodium Thiosulfate and Sulfur Dichloride ($K_2S_5O_6 \cdot 1\frac{1}{2}H_2O$)

$$SCl_2 + 2H_2S_2O_3 \xrightarrow{HCl} H_2S_5O_6 + 2NaCl$$

Solution 1: 51 g of SCl_2 dissolved in 200 ml of CCl_4 in a 2-liter glass-stoppered flask and cooled to $-15°$

Solution 2: 250 g of $Na_2S_2O_3 \cdot 5H_2O$ dissolved in 400 ml of water and chilled on ice

Solution 3: 200 ml of concentrated HCl (11.6 *N*) mixed with 200 ml of water, chilled on ice

Solutions 2 and 3 are rapidly and simultaneously poured into solution 1 and the flask stoppered and immediately shaken vigorously.[3,17,22,23] The temperature of the mixture must not rise above 0°. The mixture becomes colored but the color discharges within 20 sec, and the aqueous phase should not show more than a slight turbidity due to sulfur. Rapidly add about 120 ml of 0.3 *M* $FeCl_3$, prechilled on ice, until the aqueous phase is a bright yellow: there is an intermediate formation of the dark color of the Fe(III)–thiosulfate intermediate, which disappears rapidly. The aqueous fraction is separated and immediately concentrated to about 170 ml at reduced pressure (12 mmHg) in a bath at 35–40°. The NaCl that separates is filtered off (under suction). The filtrate is chilled to 0° and, while stirring vigorously, about 20 g of KOH in 100 ml of methanol is added dropwise or in a slow trickle, not allowing the temperature to rise above 10°. Brown hydrated iron oxide appears momentarily with each drop added but the end point for the titration is indicated by a greenish-black precipitate of ferric hydroxide beginning to separate when the mixture reaches about pH 3. The mixture is then cooled to 0° and the crystalline slurry filtered and washed on the filter with acetone to discharge its initial yellow color. After drying at room temperature on a tile about 102 g of 85% $K_2S_5O_6 \cdot 1\frac{1}{2}H_2O$ is obtained (with KCl as impurity).[4,17,23] Purification is effected by adding 50 g of the impure $K_2S_5O_6 \cdot 1\frac{1}{2}H_2O$ to 100 ml of 0.5 *M* HCl at 60°, thereby cooling it considerably. The solution is quickly reheated to 50° and filtered through a warmed funnel into a beaker chilled in ice. Pure $K_2S_5O_6 \cdot 1\frac{1}{2}H_2O$ is deposited as star-shaped crystals and is recovered by filtration, washed with ethanol, and can be dried *in vacuo* over P_2O_5.[14] The yield is about 23 g, meaning that the whole synthesis yields

[22] H. Stamm, O. Seipold, and M. Goehring, *Z. Anorg. Allg. Chemie* **247,** 277 (1941).
[23] M. Goehring and U. Feldmann, *Z. Anorg. Chem.* **257,** 223 (1948).

about 46 g of 100% $K_2S_5O_6 \cdot 1\frac{1}{2}H_2O$, which contains about 46% of the sulfur used in the initial reactants.

Foss[5] recommends a more cautious recrystallization procedure, owing to the instability of pentathionate. A little less than double the volume of 0.5 *M* HCl is heated to 50° and poured into a beaker containing the pentathionate. This dissolves rapidly and should not exceed 35° in temperature. It is immediately filtered under suction through a warm filter into a beaker chilled in ice. The whole procedure from adding the acid to completing filtration should take no more than 1 min.

Adding methanol to the filtrate from the recrystallization precipitates about a further 13 g of $K_2S_5O_6$ of around 80% purity.

From Sodium Thiosulfate Using HCl and Arsenious Oxide

$Na_2S_2O_3$ (500 g) is dissolved in 600 ml of water in a 5-liter beaker and 8–10 g of As_2O_3 in 50% NaOH is added. The mixture is then vigorously stirred and cooled to −10° until crystallization begins.[17,22] With rapid mixing, 800 ml of concentrated HCl, precooled to −15°, is poured into the beaker. This results in NaCl precipitation and this is removed by filtration, to yield a clear filtrate. This is left at 25° for 3–4 days. During this time arsenic sulfide and sulfur precipitate and are removed by filtration. The filtrate is concentrated under vacuum at 38–40° to about 200 ml and filtered to remove NaCl, yielding a filtrate of *D*, 1.6. This is supplemented with 100 ml of acetic acid and cooled in a tall 1-liter beaker with vigorous stirring to −10°. To this is added, stepwise, a slurry of finely crystalline potassium acetate: this is prepared by adding 80 g of potassium acetate to 250 ml of hot ethanol, shaking to bring to room temperature, then adding (while shaking vigorously) a thin stream of 50 ml of acetic acid. The temperature is held below −2° and spontaneous crystallization of $K_2S_5O_6$ begins within 1 min. This is immediately filtered and washed successively with a minimum amount of dilute acetic acid, followed by 96% ethanol, and finally absolute ethanol, before drying at room temperature. The yield is 80–100 g of pure $K_2S_5O_6 \cdot 1\frac{1}{2}H_2O$, which gives a clear aqueous solution. This can be recrystallized from a saturated solution in warm 0.5 *M* HCl after filtering into a chilled vessel.

Synthesis of Potassium Hexathionate

From Disulfur Dichloride and Thiosulfate

$$S_2Cl_2 + 2Na^+ + 2S_2O_3^{2-} \xrightarrow{HCl} S_6O_6^{2-} + 2NaCl$$

S_2Cl_2 (27 g) is dissolved in 100 ml of CCl_4 and cooled to $-15°$ in a wide-bottomed 1-liter flask.[3,17,23,24] To this are simultaneously and rapidly added 150 ml of water containing 100 g of $Na_2S_2O_3 \cdot 5H_2O$ and 160 ml of concentrated HCl : water (1 : 1), both of which are prechilled on ice. The mixture is vigorously shaken and its color is discharged within about 20 sec. About 15 ml of 0.6 *M* $FeCl_3$ is added until the aqueous layer becomes faintly yellow. The aqueous phase is separated and concentrated to about 50 ml at 35° under reduced pressure (12 mm). NaCl is deposited and is filtered off. The filtrate is chilled on ice and titrated to pH 1–2 with about 40 ml of a solution of 20 g of KOH in 100 ml of methanol. $K_2S_6O_6$ separates and is recovered by filtration under suction, washed twice with about 40 ml of acetone, and dried at room temperature. This yields about 42 g of 81% $K_2S_6O_6$. Purification is achieved by dissolving 20 g in 30 ml of 2 *M* HCl, quickly warming to 60°, filtering, and immediately chilling to 0°. About 11 g of 96% $K_2S_6O_6$ crystallizes.

From Thiosulfate and Nitrite in Acid Solution

Concentrated HCl (200 ml) and water (100 ml) are cooled to $-35°$ in a 3-liter wide-necked, round-bottomed flask.[3,17,22,24] To this is rapidly added with vigorous mixing a solution in 80 ml of water of 80 g of crystalline $K_2S_2O_3 \cdot \frac{5}{3}H_2O$ and 12 g of KNO_2. In the space of a few minutes the mixture becomes first dark brown, then dark green with vigorous gas production, then green yellow, and finally white with precipitated KCl. Up to this point the mixture must be vigorously shaken. Nitrogen oxides are then expelled from the mixture with a stream of nitrogen gas (there is a strong smell of SO_2), it is stood in a chilling mixture for about 0.5 hr, then the KCl is removed by filtration. The clear filtrate from two preparations is concentrated to give a thick crystalline slurry under reduced pressure (15–18 mm) at 25–30°. This is filtered through a glass filter, washed with 96% alcohol and then with absolute ethanol, and dried at room temperature to give 60–70 g of 60% $K_2S_6O_6$ contaminated with KCl. Purification is effected by dissolving 50 g in 75 ml of 2 *M* HCl, rapidly heating to 80°, then rapidly cooling with constant shaking. $K_2S_6O_6$ deposits and is recovered by filtration and washed with ethanol and ether. This produces about 40–44 g of 97.5% $K_2S_6O_6$ from the two preparations.

[24] E. Weitz and F. Achterberg, *Ber. Dtsch. Chem. Ges.* **61,** 399 (1928).

Determination of Thiosulfate and Polythionates

Titrimetric

The older literature contains detailed procedures for the estimation of thiosulfate in mixture with polythionates by means of iodine titration of thiosulfate and of thiosulfate released from the scission of polythionates by cyanide, sulfite, and alkaline hydrolysis.[25,26] These methods have been supplanted by more rapid and sensitive colorimetric procedures, but a detailed description of the titration protocols is given by Starkey.[27]

Cyanolytic Colorimetric Methods

Thiosulfate and polythionates react with cyanide to form thiocyanate, which can be determined colorimetrically as ferric thiocyanate.[28–30] Differences in the reactivity of the thionates with cyanide enable the quantitative characterization and determination of mixtures of several compounds: trithionate is stable at high pH values and reacts with cyanide only at elevated temperatures, thiosulfate reacts with cyanide at room temperature only in the presence of copper(II) as a catalyst, whereas the higher polythionates ($S_nO_6^{2-}$, where $n = 4$ or more) react rapidly with cyanide at room temperature to form thiosulfate and sulfite:

$$S_2O_3^{2-} + CN^-(+ Cu^{2+}) \rightleftharpoons SCN^- + SO_3^{2-}$$
$$S_3O_6^{2-} + 3CN^- + H_2O\ (100°) \rightleftharpoons SCN^- + SO_3^{2-} + SO_4^{2-} + 2HCN$$
$$S_4O_6^{2-} + 3CN^- + H_2O \rightleftharpoons SCN^- + S_2O_3^{2-} + SO_4^{2-} + 2HCN$$
$$S_5O_6^{2-} + 4CN^- + H_2O \rightleftharpoons 2SCN^- + S_2O_3^{2-} + SO_4^{2-} + 2HCN$$
$$S_6O_6^{2-} + 5CN^- + H_2O \rightleftharpoons 3SCN^- + S_2O_3^{2-} + SO_4^{2-} + 2HCN$$

Among the earliest procedures described for thiosulfate and tetrathionate were those of Sörbo[28] and of Nietzel and DeSesa,[31] who recognized the usefulness of the catalysis by copper(II) of the reaction of thiosulfate with cyanide. In the absence of copper(II), thiosulfate reacts only slowly unless heated to 100°. In the absence of cyanide, copper(II) catalyzes the

[25] R. R. Jay, *Anal. Chem.* **25,** 288 (1953).
[26] A. Kurtenacker, *in* "Handbuch der Anorganischen Chemie" (R. Abegg, F. Auerbach, and I. Koppel, eds.), Vol. 4, p. 449. Hirzel, Leipzig, 1927.
[27] R. L. Starkey, *J. Gen. Physiol.* **18,** 325 (1935).
[28] B. Sörbo, *Biochem. Biophys. Acta* **23,** 412 (1957).
[29] T. Koh and I. Iwasaki, *Bull. Chem. Soc. Jpn.* **39,** 352 (1966).
[30] D. P. Kelly, L. A. Chambers, and P. A. Trudinger, *Anal. Chem.* **41,** 898 (1969).
[31] O. A. Nietzel and M. A. DeSesa, *Anal. Chem.* **27,** 1839 (1955).

oxidation of thiosulfate to tetrathionate,[32–34] but this is not the mechanism for production of thiocyanate as the copper-catalyzed cyanolysis reaction is essentially instantaneous while the oxidation to tetrathionate is slow.[28,32,33] Procedures are described below for the determination of tetrathionate, pentathionate, and hexathionate, separately and in mixture,[35–38] and for thiosulfate, trithionate, and tetrathionate, separately or in mixture.[30]

Tetra-, Penta-, and Hexathionate

Reagents

Buffer, pH 7.0: 50 ml of 0.2 *M* NaH_2PO_4 + 29 ml of 0.2 *M* NaOH

Ferric nitrate reagent: 303 g of $Fe(NO_3)_3 \cdot 9H_2O$ in 217 ml of 72% (w/v) perchloric acid, made up to 500 ml with distilled water

Sodium cyanide, 0.1 *M*

$CuCl_2$, 0.05 *M*

Procedures. When present separately, these may be determined as follows: samples (10 ml containing up to 0.6, 0.3, or 0.2 m*M* tetra-, penta-, or hexathionate, respectively) are placed in 25-ml volumetric flasks.[35–38] To these are added 4 ml of buffer and 2.5 ml of NaCN and mixed, thereby bringing them to pH 8.7. The flasks are held in a water bath at 40° for 30 min, by which time conversion to 1 mol of thiosulfate per mole of polythionate, and to different amounts of thiocyanate, occurs as follows (procedure A):

$$S_nO_6^{2-} + (n-1)CN^- + H_2O \rightleftharpoons S_2O_3^{2-} + SO_4^{2-} + 2HCN + (n-3)SCN^-$$

(where $n = 4$, 5, or 6).

To each flask is then added 3 ml of ferric nitrate reagent, the volume made up to 25 ml with water and thoroughly mixed. The optical density of the color due to ferric thiocyanate is read at once at 460 nm, using a reagent blank as the reference zero. The amount of polythionate present is indicated by the molar conversion of tetra-, penta-, and hexathionate to one, two, and three thiocyanates, respectively.

[32] A. Kurtenacker, *in* "Handbuch der Anorganischen Chemie" (R. Abegg, F. Auerbach, and I. Koppel, eds.), Vol. 4, p. 553. Hirzel, Leipzig, 1927.

[33] D. P. Kelly, *J. Chromatogr.* **66,** 185 (1972).

[34] J. J. Byerley, S. A. Fouda, and G. L. Rempel, *J. Chem. Soc. Dalton Trans.*, 889 (1973).

[35] T. Koh and I. Iwasaki, *Bull. Chem. Soc. Jpn.* **39,** 352 (1966).

[36] T. Koh, *Bull. Chem. Soc. Jpn.* **38,** 1510 (1965).

[37] T. Koh and I. Iwasaki, *Bull. Chem. Soc. Jpn.* **38,** 2135 (1965).

[38] T. Koh and I. Iwasaki, *Bull. Chem. Soc. Jpn.* **39,** 703 (1966).

When present in mixture, two replicate sets of flasks are set up as above, and one set is treated exactly as before. To the other set, after the incubation with NaCN, 1.5 ml of $CuCl_2$ is rapidly added with vigorous shaking, bringing the solution to pH 7.1. This catalyzes the conversion into thiocyanate of the thiosulfate formed in the initial cyanolysis, to give this overall result (procedure B):

$$S_nO_6^{2-} + nCN^- + H_2O \rightleftharpoons SO_3^{2-} + SO_4^{2-} + 2HCN + (n - 2)SCN^-$$

(where $n = 4, 5,$ or 6).

The molar yield of thiocyanate from each polythionate becomes 2, 3, and 4, or one more than produced in the absence of Cu(II). The relative concentrations of pairs of polythionates in mixture can be calculated as follows, from the thiocyanate formed in each procedure (*A* and *B*):

Tetrathionate plus pentathionate:

$$S_4O_6^{2-} = 2B - 3A$$
$$S_5O_6^{2-} = 2A - B$$

Tetrathionate plus hexathionate:

$$S_4O_6^{2-} = (3B - 4A)/2$$
$$S_6O_6^{2-} = (2A - B)/2$$

Pentathionate plus hexathionate:

$$S_5O_6^{2-} = 3B - 4A$$
$$S_6O_6^{2-} = 3A - 2B$$

Thiosulfate, Trithionate, and Tetrathionate

Kelly *et al.*[30] developed the cyanolysis procedures of Koh and Iwasaki,[29] so that thiosulfate, trithionate, and tetrathionate could be determined either singly or in mixture with each other. The modification by Kelly *et al.*[30] has come into relatively wide use in microbiological studies and is described below.

Reagents

Buffer, pH 7.4: 50 ml of 0.2 *M* NaH_2PO_4 plus 39 ml of 0.2 *M* NaOH
Ferric nitrate reagent: As described above
Potassium cyanide, 0.1 *M*
Copper sulfate, 0.1 *M*
Potassium thiocyanate standard, 0.001 *M*

Procedure. All procedures are carried out in 25-ml volumetric flasks. The sample to be analyzed (containing up to 8 μmol of total thionate) is added to 4 ml of buffer in a flask and water added to give a total volume of about 10 ml. Three replicated sets of samples are set up in this way. The replicates are treated separately as follows.

I. Cool to 5–10° and add 5 ml of chilled KCN and mix rapidly. Leave at 5–10° for 20 min.
II. As in treatment I, but rapidly mix in 1.5 ml of $CuSO_4$ after 5 min. Leave at 5–10° for 10–15 min.
III. Add 5 ml of KCN, mix, then heat in a boiling water bath for 45 min. After cooling, rapidly mix in 1.5 ml of $CuSO_4$ and leave for 10–15 min.

Finally, add 3 ml of ferric nitrate reagent to each flask with continuous agitation, allowing to warm to room temperature and ensuring that any (white) precipitate redissolves. Make up to 25 ml with distilled water and read optical density at 460 nm against a reagent blank (as in treatment I) lacking any sulfur compound. Thiocyanate formation can be quantified by reference to a calibration curve using thiocyanate or thiosulfate standards (0–10 μmol in treatment I or II, respectively). The millimolar extinction coefficient under the assay mixture conditions is about 4.4 $mM^{-1}cm^{-1}$, meaning that 1 μmol of ferric thiocyanate in the 25-ml assay has an optical density (1-cm light path) of about 0.175.

From the equations above it can be seen that thiosulfate, trithionate, and tetrathionate react to give the following amounts of thiocyanate in the three treatments:

Treatment I:

$$S_4O_6^{2-} \rightarrow 1SCN^- \quad (S_2O_3^{2-} \text{ and } S_3O_6^{2-} \rightarrow 0SCN^-)$$

Treatment II:

$$S_4O_6^{2-} \rightarrow 2SCN^- \quad (S_3O_6^{2-} \rightarrow 0SCN^-)$$
$$S_2O_3^{2-} \rightarrow 1SCN^-$$

Treatment III:

$$S_4O_6^{2-} \rightarrow 2SCN^-$$
$$S_2O_3^{2-} \rightarrow 1SCN^-$$
$$S_3O_6^{2-} \rightarrow 1SCN^-$$

If pure samples of only tetrathionate, thiosulfate, or trithionate are to be assayed, then treatment I, II, or III only is required, respectively, and the molar amount of thiocyanate formed will be equivalent to the molar amount of the thionate present. If these three compounds are present in mixture with each other, their relative concentrations are calculated as follows:

1. Tetrathionate (A) is given directly by treatment I = A(mol).
2. Thiosulfate (B) is given by subtracting twice the value for treatment I ($2A$) from the value for treatment II (x), $x - 2A = B$ (mol).
3. Trithionate (C) is given directly as the molar difference between treatments III (y) and II (x), $y - x = C$ (mol).

Nor and Tabatabai[39] have described a method for measurement of thiosulfate and tetrathionate that is essentially the same as described above (steps I and II), except that lower concentrations of KCN, Cu^{2+}, and ferric reagent are used, and the reactions are conducted at room temperature. They confirmed that slightly more thiocyanate was formed from tetrathionate (in the absence of Cu^{2+}) than the 1 : 1 ratio predicted by the equation (57% of the sulfane sulfur of tetrathionate converted to thiocyanate rather than 50%), and recommended the use of a factor of 1.75 rather than 2.0 in the calculation of thiocyanate formation from tetrathionate.

In all these procedures in which ferric thiocyanate is measured, caution needs to be exercised, as the color is light sensitive and readings should be made at once or samples stored in darkness after addition of ferric reagent. The sensitivity of the procedure can be enhanced by using small volumes of more concentrated reagents and reading optical density in long light path (4 cm) cuvettes. In this way the lower limit for detection of thiosulfate or trithionate can be reduced to about 1 μM.

Spectrophotometric Determination of Thiosulfate and Polythionates by Their Ultraviolet Absorption Spectra

The thionates have ultraviolet absorption spectra with maxima around 210–230 nm.[40,41] Meulenberg *et al.*[41] have developed a continuous ultraviolet (UV) assay for thiosulfate formation from trithionate, based on the difference in absorbance at 220 nm. In 25 mM potassium phosphate, pH 3.0, with 1 M $(NH_4)_2SO_4$, the respective molar extinction coefficients for thiosulfate and trithionate at 220 nm were 3.28×10^3 and 0.15×10^3 M^{-1} cm^{-1}.[41] The absorption maximum for trithionate is at 205 nm, but its molar extinction coefficient is still only about 0.8×10^3 M^{-1} cm^{-1}. Thiosulfate and tetrathionate have similar UV absorption spectra, both with α peaks around 220 nm, and in aqueous solution their molar extinction coefficients

[39] Y. M. Nor and M. A. Tabatabai, *Anal. Lett.* **8,** 537 (1975).

[40] D. P. Kelly, unpublished data, 1987.

[41] R. Meulenberg, J. T. Pronk, J. Frank, W. Hazeu, P. Bos, and J. G. Kuenen, *Eur. J. Biochem.* **209,** 367 (1992).

are about 3.7 and 9.0 × 10^3 M^{-1} cm^{-1} at 220 nm.[39] Sulfate also absorbs strongly in the UV, with a molar extinction coefficient of about 4.5 × 10^3 M^{-1} cm^{-1} at 220 nm and about twice that value at 200 nm.[39] Ultraviolet spectroscopy is thus of limited usefulness, and is of course subject to interference by many other inorganic ions with strong UV absorbance, such as nitrite and nitrate.[39] The molar extinction coefficient for thiosulfate is progressively depressed by increasing concentrations of Li^+, Na^+, K^+, and Mg^+ salts,[42] so it is important to ensure standardization of all conditions when using UV spectra to assay thiosulfate.

Spectrophotometric Iodometric Determination of Sulfite, Thiosulfate, and Tetrathionate, Individually or in Mixtures, after Polythionate Degradation with Sulfite or Cyanide

Sulfite and Thiosulfate Estimation. An exact amount of iodine is generated in the assay mixture by adding acetic acid, iodide, and iodate.[43,44] Sulfite and thiosulfate (but not polythionates) oxidize the iodine to iodate, discharging the iodine color. Comparing the reduction in color with a sulfur-free control and standards prepared with sulfite or thiosulfate enables estimation of these compounds in unknown samples. When both are present in mixture, sulfite can be masked by complexing with formaldehyde so that sulfite plus thiosulfate or thiosulfate alone can be measured separately. On a molar basis, sulfite oxidizes twice as much iodine as thiosulfate:

$$IO_3^- + 6H^+ + 5I^- \rightleftharpoons 3I_2 + 3H_2O$$
$$SO_3^{2-} + I_2 + H_2O \rightleftharpoons SO_4^{2-} + 2H^+ + 2I^-$$
$$2S_2O_3^{2-} + I_2 \rightleftharpoons S_4O_6^{2-} + 2I_2$$

Reagents

Iodate–iodide reagent (in a 250-ml volumetric flask): Dissolve 36.1 g of KI in 150 ml of water; add 0.1 g of Na_2CO_3 and allow to dissolve; add 10 ml of stock 0.025 *N* KIO_3 (0.1338 g/100 ml); make up to 250 ml with water

Phosphate buffer (pH 7.0): 100 ml of 0.2 *M* NaH_2PO_4 plus 59.3 ml of 0.2 *M* NaOH

Acetic acid (5 *N*): 30% (v/v) glacial acetic acid in water

Sulfite standard: 20 m*M* sulfite is prepared by dissolving 0.51 g of Na_2SO_3 in 100 ml of 5 m*M* ethylenediaminetetraacetic acid (EDTA) (1.86 g

[42] D. P. Ames and J. E. Willard, *J. Am. Chem. Soc.* **75,** 3267 (1953).

[43] T. Koh and K. Taniguchi, *Anal. Chem.* **45,** 2018 (1973).

[44] I. Iwasaki and S. Suzuki, *Bull. Chem. Soc. Jpn.* **39,** 576 (1966).

of disodium EDTA per liter), and a dilution of 1 ml made up to 100 ml with 5 m*M* EDTA gives 0.2 μmol/ml, which must be used soon after preparation

Thiosulfate standard: Dilute stock 0.1 *M* $Na_2S_2O_3$ to 0.2 m*M* in water

Formaldehyde (0.5 *M*): 4 ml of 38% (w/v) formaldehyde in 96 ml of water

All procedures are conducted in 25-ml volumetric flasks. Buffer (3.2 ml) is dispensed into a series of flasks, and 0–10 ml of samples or standards (with water to make a total of 10 ml) added. Samples should contain less than 2 μmol of sulfite or 4 μmol of thiosulfate (or equivalent concentrations for mixtures). If samples containing sulfite need dilution, this should be done with 5 m*M* EDTA to minimize autooxidation. Add 3 ml of acetic acid, mix, and immediately add 2.3 ml of iodate–iodide reagent. Make up to 25 ml with water, and immediately stopper and mix thoroughly. Absorbance (OD at 350 nm) should be read within 30 min against water as a blank. To determine thiosulfate in the presence of sulfite, a parallel series of flasks is prepared as above, but 1.5 ml of formaldehyde is added and thoroughly mixed before addition of the acetic acid. The reagent blank typically gives an OD of 1.76, which is decreased linearly with increasing amounts of sulfite and thiosulfate. Iodine consumption by 1 μmol of sulfite or thiosulfate decreases the OD_{350} by 0.92 and 0.46, respectively. Thus a mixture of 1 μmol each of sulfite and thiosulfate gives a decrease in OD of 1.38 without formaldehyde and of 0.46 with formaldehyde, thereby enabling calculation of the relative concentration of each in unknown mixtures. This procedure will detect sulfite or thiosulfate at lower limits of about 5–10 μM.

Sulfite Degradation of Tetrathionate. Polythionates react with sulfite in alkaline solution to produce thiosulfate:

$$S_nO_6^{2-} + (n - 3)SO_3^{2-} \rightleftharpoons (n - 3)S_2O_3^{2-} + S_3O_6^{2-}$$

Tetrathionate thus produces an equimolar amount of thiosulfate on sulfitolysis. The procedure described here is derived from Refs. 43 and 44, and uses the reagents described above.

The sample (10 ml) in a 25-ml flask, as above, is mixed with 0.5 ml of 0.5 *M* sodium sulfite and a drop of phenolphthalein is added. If there is no pink color, the solution is supplemented dropwise with 1 *M* NaOH until faint pink. After at least 5 min, 6 ml of formaldehyde is added, followed by 3 ml of acetic acid and 2.3 ml of iodate–iodide reagent. After making up to 25 ml with water, thiosulfate is determined from the OD at 350 nm, as described above.

When sulfite and thiosulfate are also present, triplicate series of assays are set up and one series treated as above: this will reveal both thiosulfate and tetrathionate. The second and third sets are treated as for sulfite plus thiosulfate mixtures (above) without pretreatment with sulfite. The tetrathionate content is thus obtained by difference.

Cyanide Degradation of Polythionates. As described in an earlier section, tetra-, penta-, and hexathionate all yield equimolar thiosulfate under cyanolysis, regardless of polythionate sulfur chain length (where $n = 4$, 5, or 6)[44]:

$$S_nO_6^{2-} + (n - 1)CN^- + H_2O \rightleftharpoons S_2O_3^{2-} + 2HCN + SO_4^{2-} + (n - 3)SCN^-$$

The procedure is based on that for sulfite and thiosulfate (described above), with 0.1 *M* NaCN as an additional reagent. Samples (10 ml) are mixed in 25-ml volumetric flasks with 3.2 ml of buffer and 2 ml of NaCN, then incubated in a bath at 40° for 30 min to effect cyanolysis. Allow to cool, add 1.5 ml of formaldehyde, and let stand for a few minutes (to ensure masking of cyanide); then add 3 ml of acetic acid and 2.3 ml of iodate–iodide reagent, mix, and make up to 25 ml. The OD at 350 nm is read as described above, and total polythionate concentration is calculated from the thiosulfate standard calibration. If sulfite and thiosulfate are also present, they are determined separately in samples not treated with NaCN.

Whereas tetra- and hexathionate react precisely as expected from the equation, pentathionate produced a greater reduction in OD in the iodine assay than predicted. This can be corrected[43] by incubating a replicate sample (10 ml) without NaCN with 1 ml of buffer at 40°, then treating the replicate with formaldehyde and the other reagents exactly as for the cyanide-treated samples. The decrease in OD value for the replicate without cyanide is added to the actual OD reading obtained after cyanide treatment and this is found to give the value expected for stoichiometric production of thiosulfate from pentathionate. Tetra- and hexathionate do not produce any change in OD in the iodine assay when pretreated without cyanide.[43]

Degradation Procedures for Thiosulfate and Polythionates

Determination of Polythionate Sulfur Chain Length by Cyanolysis

The differential reactions of polythionates and thiosulfate with cyanide in the presence or absence of copper can be exploited not only in their quantitative analysis (see above), but also to determine the sulfur chain length of pure compounds.[16,38]

Using hexathionate as an example, the following reactions with cyanide occur:

Reaction A:

$$S_6O_6^{2-} + 5CN^- \rightleftharpoons S_2O_3^{2-} + SO_4^{2-} + 3SCN^-$$

Reaction B:

$$S_6O_6^{2-} + 6CN^-(+Cu^{2+}) \rightleftharpoons SO_3^{2-} + SO_4^{2-} + 4SCN^-$$

Thus 4 and 3 mol of thiocyanate are formed, respectively, with (reaction B) and without (reaction A) copper(II). The relationship between chain length and thiocyanate formed is expressed as follows: $S_nO_6^{2-}$ gives thiocyanate in the ratio of $A/B = (n - 3)/(n - 2)$. Therefore, $n = A/(B - A) + 3$. For hexathionate, the value of $n = [3/(4 - 3)] + 3 = 3 + 3 = 6$ sulfur atoms.

Example Analyses. Analysis of three "unknowns" produced the following absorbance readings for thiocyanate after cyanolysis without (A) and with (B) copper(II)[16]:

Reading I:
$A = 1.000$, $B = 1.772$ $n = [1.000/(1.772 - 1.000)] + 3 = 4.30$
Reading II:
$A = 0.701$, $B = 0.934$ $n = [0.701/(0.934 - 0.701)] + 3 = 6.01$
Reading III:
$A = 1.418$, $B = 1.772$ $n = [1.418/(1.772 - 1.418)] + 3 = 7.01$

Compound I was thus predominantly tetrathionate, whereas compounds II and III were pure hexathionate and heptathionate.

Degradation of Thiosulfate and Polythionates Using Ag^+, Hg^{2+}, and Cyanide

Silver, mercury, and cyanide ions all decompose thiosulfate and polythionates by reactions that maintain the identity of the sulfane and sulfonate groups within the original molecules. Copper(II) will also specifically degrade trithionate to CuS (sulfane sulfur) and sulfate (sulfonate groups). These reactions may be used to determine specific ^{35}S-labeling patterns within the molecules, for example, after metabolism of sulfane- or sulfonate-labeled thiosulfate.

Silver Degradation of Thiosulfate

$$^-S\text{–}SO_3^- + 2Ag^+ + H_2O \rightleftharpoons Ag_2S + H_2SO_4$$

An aqueous solution of thiosulfate is heated with an excess of $AgNO_3$ at 95° for 45 min.[44,45] The outer (sulfane) sulfur atom precipitates as Ag_2S and can be recovered by centrifugation. The washed precipitate can be assayed for its ^{35}S content either as a solid or after dissolving in 2% (w/v) KCN.

Mercury Degradation of Thiosulfate and Polythionates

$$2Na_2(O_3S{-}S^*) + 3HgCl_2 + 2H_2O \rightarrow HgCl_2 \cdot 2HgS^* + 2Na_2SO_4 + 4HCl$$

$$2K_2(O_3S{-}S^*{-}SO_3) + 3HgCl_2 + 4H_2O \rightarrow HgCl_2 \cdot 2HgS^* + 2K_2SO_4 + 2H_2SO_4 + 4HCl$$

$$2K_2(O_3S{-}S^*{-}S^*{-}SO_3) + 3HgCl_2 + 4H_2O \rightarrow HgCl_2 \cdot 2HgS^* + 2S^* + 2K_2SO_4 + 2H_2SO_4 + 4HCl$$

The asterisks indicate the position of the outer (sulfane) atoms of the three compounds in order to demonstrate that these are always converted to the insoluble $HgCl_2 \cdot 2HgS$ complex, while the inner (sulfonate) atoms remain in solution as sulfate.[5,27,45–49]

To a solution (1 ml) of the labeled compound (1–3 mg) in a small centrifuge tube are added 0.5 ml of 38% (w/v) formaldehyde solution and 0.5 ml of 5% (w/v) $HgCl_2$ in 2% (w/v) sodium acetate. After allowing it to stand for 15 min at room temperature the mixture is supplemented with 2 mg (0.1 ml) of an unlabeled sample of the material being degraded and allowed to stand for a further hour. The precipitate is recovered by centrifugation, washed with a 1:20 dilution of the HCHO–$HgCl_2$–acetate mixture, and recentrifuged. The supernatants are combined and diluted to a standard volume with water. The precipitate is dissolved in 1 ml of concentrated HNO_3 saturated with bromine by gentle boiling for 30 min on a sand bath, then also made up to standard volume for determination of ^{35}S.

Cyanide Degradation of Thiosulfate and Polythionates

In the reactions with cyanide and copper(II) (see above) the outer (sulfane) sulfur atoms of both thiosulfate and tetrathionate are converted to thiocyanate and the inner (sulfonate) groups to sulfate. These may be separated by chromatography for the determination of the position of ^{35}S labeling within the original molecule.

[45] A. I. Brodskii and R. K. Eremenko, *Zh. Obshch. Khim.* **24,** 1142 (1954).

[46] D. P. Kelly and P. J. Syrett, *Biochem. J.* **98,** 537 (1966).

[47] H. B. van der Heijde and A. H. W. Aten *J. Am. Chem. Soc.* **74,** 3706 (1952).

[48] P. A. Trudinger, *Biochem. J.* **78,** 680 (1961).

[49] R. Abegg, F. Auerbach, and I. Koppel (eds.), "Handbuch der Anorganischen Chemie," Vol. 4, 1er Abt., Leipzig, Hirzel, 1927.

Tetrathionate reacts with cyanide (without copper) to give the following distribution of ^{35}S from tetrathionate specifically labeled in the sulfane atoms[48,50]:

$$Na_2(O_3S\text{–}S^*\text{–}S^*\text{–}SO_3) + NaCN + H_2O \rightarrow Na_2(S^*\text{–}SO_3) + NaS^*CN + H_2SO_4$$

The sample (1 ml, 10 m*M*) is made alkaline to phenolphthalein, mixed with 1 ml of 0.1 *M* KCN, and the products separated after 5 min at room temperature by chromatography on a Dowex-2 × 8 (acetate) ion-exchange column: sulfate and thiosulfate are eluted with 2 and 5 *M* ammonium acetate (pH 5.0), respectively, and thiocyanate with 2 *M* HNO_3.

Degradation of Trithionate with Copper

$$^-O_3S\text{–}S^*\text{–}SO_3^- + Cu^{2+} + 2H_2O \rightleftharpoons CuS^* + 2SO_4^{2-} + 4H^+$$

Trithionate (20 ml, 10 m*M*) is mixed with 5 ml of 1 *M* $CuSO_4$ and incubated in a covered beaker at 70° for 15–20 hr.[51] Recovery of the S* atom as a black precipitate is approximately 100% of that expected. Tetrathionate does not react significantly in this time period, but gives about 30% of the theoretical CuS after 190 hr.[49] Thiosulfate reacts at about the same rate as trithionate, giving about 92% recovery of CuS after 15 hr (D. P. Kelly, unpublished results, 1968).

Chromatographic Separation and Identification of Thiosulfate and Polythionates

High-Performance Liquid Chromatography

Thiosulfate has been determined in water and blood samples by several high-performance liquid chromatography (HPLC) procedures, which use fluorescence methods to detect the thiosulfate.[52–56]

[50] A. I. Brodskii and R. K. Eremenko, *Zh. Obshch. Khim.* **25,** 1189 (1955).

[51] E. H. Riesenfeld, E. Josephy, and E. Grunthal, cited in Ref. 49 (p. 582).

[52] P. R. Dando, A. J. Southward, and E. C. Southward, *Proc. R. Soc. London B* **227,** 227 (1986).

[53] S. H. Lee and L. R. Field, *Anal. Chem.* **56,** 2647 (1984).

[54] R. C. Fahey, R. Dorian, G. L. Newton, and J. Utley, *in* "Radioprotectors and Anticarcinogens" (O. F. Nygaard and M. G. Simie, eds), p. 103. Academic Press, New York, 1983.

[55] R. D. Vetter, P. A. Matrai, B. Javor, and J. O'Brien, *in* "Biogenic Sulfur in the Environment" (E. S. Saltaman and W. J. Cooper, eds), p. 243. American Chemical Society, Washington, D.C., 1989.

[56] J. A. Childress, C. R. Fisher, J. A. Favuzzi, and N. K. Sanders, *Physiol. Zool.* **64,** 1444 (1991).

High-Performance Liquid Chromatography with Cerium(III) Fluorescence Detection. Dando *et al.*[52] and Lee and Field[53] eluted thiosulfate from a 4.6 mm (i.d.) × 25 cm Vydac (Separations Group, Hesperia, CA) 302 IC anion-exchange resin column using 2 m*M* succinic acid adjusted to pH 7.0 with sodium borate. Eluate from the column is pumped into a packed bed reactor (a Teflon tube, 6 mm × 20 cm, filled with 100/120 mesh glass beads), where it reacts with 0.1 m*M* cerium(IV) sulfate (in 0.5 *M* sulfuric acid and previously boiled with 200 mg of sodium bismuthate per liter). Thiosulfate reduces Ce(IV) to Ce(III):

$$2S_2O_3^{2-} + 2Ce(IV) \rightleftharpoons S_4O_6^{2-} + 2Ce(III)$$

Ce(III) production is equivalent to thiosulfate concentration and is detected using a postcolumn cerium fluorescence detector, with the eluate being pumped through a fluorescence spectrophotometer cell at about 0.7 ml/min. Fluorescence peak height is proportional to thiosulfate concentration over the range 1–250 μM, and the detection limit for the cerium fluorescence detector is around 3.5 pmol.[53] P. R. Dando (personal communication, 1993) found this procedure also to be applicable to trithionate but not tetrathionate.

High-Performance Liquid Chromatography of Fluorescent Monobromobimane Derivatives. Thiosulfate and numerous other thiols can be separated and detected in this way.[54] Monobromobimane (mBBr) is not significantly fluorescent, but reacts with thiols optimally at pH 8.0 to produce highly fluorescent derivatives.

Reagents

Stock (100 m*M*) monobromobimane: Prepare in acetonitrile, store frozen, and use within 1 month. Dilutions are prepared in acetonitrile
Methanesulfonic acid (25 m*M*)
Acetonitrile
N-2-Hydroxyethylpiperazine-*N*′-2-ethanesulfonic acid (HEPES) (0.2 *M*), containing 5 m*M* EDTA

For samples such as blood or sea water no pH adjustment of the sample is necessary, otherwise the sample should be brought to pH 8.0 using HEPES buffer. Samples with 0–0.5 or 0.5–5 m*M* total reduced sulfur are derivitized with 2 and 10 m*M* mBBr, respectively. Rapid sample derivitization needs at least a twofold excess of mBBr over total thiol, and is then complete in about 2 min[54] (pH 8.0, 20°). Typically[55] a sample (e.g., water or blood, 0.1 ml) in a 1.5-ml disposable microfuge tube is supplemented with 0.01 ml of mBBr in acetonitrile and incubated for 10–15 min in

TABLE I
PAPER CHROMATOGRAPHY OF SOME INORGANIC SULFUR COMPOUNDS

Compound	R_f with the following solvent systems[a]				
	1	2	3	4	5
Sulfate	0.02	0.18			
Thiosulfate	0.03	0.21	0.43	0.37	0.26
Trithionate	0.13	0.43	0.52	0.60	0.50
Tetrathionate	0.21	0.57	0.53	0.67	0.66
Pentathionate				0.73	0.67
Hexathionate					0.77
Thiocyanate	0.71	0.81			

[a] Solvents: 1, butan-1-ol–acetone–water (2/2/1); 2, butan-1-ol–acetic acid–pyridine–water (30/6/20/24); 3, butan-1-ol–methanol–water (1/1/1); 4, pyridine–propan-1-ol–water (7/10/10); 5, butan-1-ol–ethylene glycol monomethyl ether (35/65).

subdued light; 0.1 ml of acetonitrile is added (then, if necessary, heated at 60° for 10 min to precipitate protein), followed by 0.3 ml of 25 mM methanesulfonic acid and centrifuged to compact any protein precipitate.

The sample can normally be subjected to HPLC without removal of excess mBBr. If the latter (or other reaction products) interfere they can be removed by extraction with ethyl acetate.[54] This procedure has been employed in determining thiosulfate in blood, using HPLC.[55,56] Derivatives can be separated on a 15-cm (or longer) reversed-phase column (e.g.,

TABLE II
THIN-LAYER CHROMATOGRAPHY OF INORGANIC SULFUR COMPOUNDS

Compound	R_f using the following solvent systems[a]				
	1	2	3	4	5
Sulfate	0	0	0		0
Thiosulfate	0.02	0.49	0.05	0	0
Trithionate	0.35	0.73	0.05	0.03	0
Tetrathionate	0.46	0.77	0.05	0.04	0
Thiocyanate	0.64	0.77	0.22	0.36	0.78

[a] Solvents: 1, butan-1-ol containing 5% (v/v) water (ITLC SA); 2, methanol–propan-1-ol (1/1) (ITLC SA); 3, heptan-2-ol–methanol–water (85/10/5) (ITLC SA); 4, octan-1-ol saturated with water (ITLC SA); 5, octan-1-ol saturated with water (ITLC SG).

TABLE III
ELECTROPHORESIS OF INORGANIC SULFUR COMPOUNDS

Compound	Distance (mm) migrated relative to thiosulfate[a] under condition:					
	1	2	3a	3b	3c	3d
Sulfate	0.91					
Thiosulfate	1.00	1.00	1.00	1.00	1.00	1.00
Trithionate		0.91	0.59	0.47	0.94	
Tetrathionate		0.81	0.44	0.31	0.85	0.90
Pentathionate		0.69				
Hexathionate		0.60				
Thiocyanate	0.88		0.78			

[a] Conditions: 1, 0.1 *M* $(NH_4)_2CO_3$, Whatman 3MM paper, 100 V cm^{-1}; 2, 0.05 *M* potassium hydrogen phthalate, Whatman No. 1, 19.6 V cm^{-1}; 3, 0.1 *M* citrate buffer (pH 4.9), 500 V, 30–50 mA: (a) Whatman anion-exchange paper DE20 (45 min); (b) Whatman anion-exchange paper ET20 (60 min); (c) Gelman glass microfiber sheets (45 min); (d) Gelman Sepraphore III strips (30 min). Thiosulfate migrated 60–90 mm in 3(a)–(d).

Gilson apparatus with an Altex C_{18} column[55]) using a flow rate of 1.5 ml/min. Gradient elution is preferred and uses methanol (solvent A) and 2% acetic acid adjusted to pH 3.5 with 10 *M* NaOH (solvent B). For the first 5 min the eluant is 90 : 10 (solvent A : solvent B), which is linearly increased to 35% solvent B at 20 min and to 40% solvent B at 40 min. Eluate is passed through a fluorometer with a 235-nm filter for excitation and a 442-nm filter for detection of fluorescence.[56] Vetter *et al.*[55] recommended excitation using a 305- to 395-nm filter, and a narrow-band emission filter centered at 480 nm. Other HPLC solvent systems have also been used.[54,56]

Paper Chromatography

There are numerous solvents for the separation of thionates in this way, including the reversed phase technique developed by Pollard *et al.*[57,58] Convenient and reproducible simple solvent procedures using overnight runs of descending chromatography on Whatman No. 1 chromatography paper have been described.[3,58] Some of these are summarized in Table I. Further detail and source literature can be found in Refs. 3 and 59.

[57] F. H. Pollard, J. F. W. McOmie, and D. J. Jones, *J. Chem. Soc.*, 4337 (1955).
[58] F. H. Pollard, D. J. Jones, and G. Nickless, *J. Chromatogr.* **15,** 393 (1964).
[59] D. P. Kelly, *J. Chromatogr.* **66,** 185 (1972).

Thin-Layer Chromatography

Sixteen possible solvents for TLC were tested by Kelly,[60] but few were suitable. Using Gelman (Ann Arbor, Michigan) instant thin-layer chromatography (ITLC) media (silicic acid and silica gel on glass fiber supports), types SA and SG, complete separation of sulfate, thiosulfate, tri-, and tetrathionates and thiocyanate could be achieved in about 5 hr (Table II).

Ion-Exchange Chromatography

Roy and Trudinger[3] have described several procedures for the separation of thiosulfate and other inorganic sulfur compounds on standard ion-exchange columns. They concluded that these were unsuitable for routine analysis or preparation, because of variability between runs and the fact that the polythionates had to be eluted in high concentrations of HCl (3–9 *M*) under the conditions they employed.

Electrophoresis

Several procedures for the paper electrophoresis of inorganic sulfur compounds have proved effective[3,59] For best results high voltages and short duration give the best reproducibility and sharpest resolution of bands. Paper strips can be cooled during electrophoresis by immersing in chlorobenzene or tetrachloromethane.[3] Table III describes satisfactory methods.

Detection of Thionates on Chromato- and Electrophoretograms

Thiosulfate and polythionates can be visualized on paper and thin-layer media by spraying with 8% (w/v) $AgNO_3$ in acetone containing 10% (v/v) water, when yellow spots appear. Alternatively, spraying with 0.5% (w/v) $AgNO_3$ in dilute ammonia solution [5 vol of aqueous ammonia (specific gravity, 0.88) plus 95 vol of water], and heating at 100–110° for 2–5 min, produces black-gray spots of Ag_2S. The sensitivity is of the order of 1 μg of sulfane sulfur on a 1-cm^2 spot. After chromatography of ^{35}S-labeled mixtures, the position of radiolabeled compounds can be detected by standard autoradiographic or scanning procedures. When [^{35}S]sulfate and [^{35}S]thiosulfate are present in a mixture, separate quantitation of their ^{35}S content is difficult because few solvents separate them sufficiently from each other. This problem can be overcome by treating one set of

[60] D. P. Kelly, *J. Chromatogr.* **51,** 343 (1970).

replicate samples with a slight excess of ethanolic iodine, which converts thiosulfate to tetrathionate. ^{35}S in sulfate is then determined directly and the amount of [^{35}S]thiosulfate present is given by the increase in the amount of [^{35}S]tetrathionate found in the presence of iodine.

[36] Enzymes Involved in Microbiological Oxidation of Thiosulfate and Polythionates

By DON P. KELLY and ANN P. WOOD

Introduction

Thiosulfate and polythionates ($S_nO_6^{2-}$) serve as energy-yielding or electron-donating substrates in the metabolism of a wide range of chemolithotrophic and photolithotrophic bacteria, as well as some heterotrophs.[1-3] Relatively few enzymes have been positively implicated in these oxidative processes and only some of these have been highly purified and characterized. In this chapter we describe effective assay procedures for a variety of enzymes occurring in lithotrophs and some heterotrophs. Some of these enzymes [adenylylsulfate (APS) reductase and thiosulfate reductase] also occur in sulfate-reducing bacteria.

Methods

Thiosulfate Dehydrogenase [Thiosulfate : Cytochrome-c Oxidoreductase (Tetrathionate Synthesizing), Tetrathionate Synthase, Thiosulfate-Oxidizing Enzyme]: EC 1.8.2.2

Thiosulfate dehydrogenase occurs in several thiobacilli and some heterotrophs and catalyzes the oxidation

$$2S_2O_3^{2-} \rightarrow S_4O_6^{2-} + 2e^-$$

$$2S_2O_3^{2-} + (\text{oxidized acceptor}) \rightarrow S_4O_6^{2-} + (\text{reduced acceptor})$$

[1] D. P. Kelly, *in* "Autotrophic Bacteria" (H. G. Schlegel and B. Bowien, eds.), p. 193. Science Tech Publ., Madison, Wisconsin, 1989.

[2] D. P. Kelly, *in* "The Nitrogen and Sulphur Cycles" (J. A. Cole and S. J. Ferguson, eds.), p. 65. Cambridge Univ. Press, Cambridge, 1988.

[3] D. P. Kelly, *in* "Bacterial Energetics" (T. A. Krulwich, ed.), p. 479. Academic Press, San Diego, 1990.

The physiological acceptor is generally a cytochrome *c*, but the enzyme is commonly assayed with ferricyanide as the artificial acceptor:

$$2S_2O_3^{2-} + 2Fe(CN)_6^{3-} \rightarrow S_4O_6^{2-} + 2Fe(CN)_6^{4-}$$

Although intact cells of some thiobacilli, phototrophs and heterotrophs, as well as cell-free extracts and the purified enzyme, produce tetrathionate, there is some doubt that tetrathionate is necessarily always a normal free intermediate in thiosulfate or sulfide oxidation by normal cells.

Verification of tetrathionate as the sole product of the reaction can be achieved either with [^{35}]thiosulfate as substrate[4] followed by paper or thin-layer chromatography (see [35] in this volume) or by chemical analysis[5] (see [35] in this volume).

Reagents

Potassium hydrogen phthalate-NaOH buffers (0.3 *M*), pH 4.5–6.0
Potassium phosphate buffers (0.3 *M*), pH 6.0–7.0
Potassium ferricyanide, 0.03 *M*
Sodium thiosulfate, 0.1 *M*

Procedure. The enzyme assay is based on that of Trudinger[4,6] in which ferricyanide reduction is measured spectrophotometrically at 420 nm in a total volume of 3 ml in a 1-cm cuvette containing phthalate or phosphate buffer, pH 4.5–7.0 (300 μmol), $Na_2S_2O_3$ (30 μmol), $K_3Fe(CN)_6$ (3 μmol). The reaction is started by addition of cell-free extract or enzyme preparation. Decrease in absorbance at 420 nm is recorded and ferricyanide reduction estimated from a calibration curve or by using the millimolar extinction coefficient of 1.0. Enzyme activity is expressed as nanomoles of ferricyanide reduced per minute per milligram protein.

In *Thiobacillus tepidarius* the specific activity measured between pH 4.5 and 7.0 showed the rate to be highest at the lowest pH.[7] Below pH 4.5 there is a chemical reaction between thiosulfate and ferricyanide that interferes with the assay. A similar assay procedure using one-third concentrations of reagents has been used with extracts of *Thiobacillus neapolitanus,* at pH values of 4.5–8.5, and again greatest activity was observed at the lowest pH.[8] The enzyme has also been successfully assayed using

[4] P. A. Trudinger, *Biochem. J.* **78,** 680 (1961).
[5] W. P. Lu and D. P. Kelly, *J. Gen. Microbiol.* **134,** 877 (1988).
[6] P. A. Trudinger, *Biochem. J.* **78,** 673 (1961).
[7] A. P. Wood and D. P. Kelly, *Arch. Microbiol.* **144,** 71 (1986).
[8] J. Mason, D. P. Kelly, and A. P. Wood, *J. Gen. Microbiol.* **133,** 1249 (1987).

acetate buffer.[9] More recently the assay has been carried out in unbuffered ammonium sulfate solution with pH adjusted by addition of sulfuric acid.[10]

It may also be assayed using horse heart cytochrome *c* as electron acceptor. Here the reaction mixture (1 ml) contains 0.1 *M* potassium phosphate (pH 7.0), 1 μmol of $Na_2S_2O_3$; 0.15 mg of cytochrome *c*, and 0.5–5.0 mg of protein. Cytochrome reduction is monitored as increase in absorbance at 550 nm.

Purification of Thiosulfate Dehydrogenase. This enzyme has been purified from a number of different *Thiobacillus* spp. and found to vary greatly in structural and catalytic properties (for summary see Refs. 5 and 10). Thiosulfate dehydrogenase preparations from different thiobacilli have been shown to have molecular weight values of 102,000–138,000, and to be made up either of identical subunits (M_r 45,000, *T. tepidarius*[5]) or unequal subunits (M_r 24,000 and 20,000, *Thiobacillus acidophilus*[10]). Most preparations were reported to contain no detectable heme[5] but that purified 1026-fold from *T. acidophilus* contained about 5.3 mol of *c*-type heme per mole of native enzyme, made up of two different *c*-553 hemes, and is present in both subunit types.[10]

Thiosulfate dehydrogenase from *T. acidophilus* was purified to homogeneity in the following way.[10] All procedures are carried out at pH 7.0 and room temperature, except for the ammonium sulfate precipitation, which is done on ice. Cells are disrupted by a French pressure cell, extracted three times with culture supernatant, centrifuged (48,000 *g*, 20 min), and pooled. Ammonium sulfate to 3.0 *M* is added to the supernatant, precipitated protein centrifuged, redissolved in sodium citrate (25 m*M*, pH 7.0), recentrifuged, then subjected to hydrophobic interaction chromatography on a phenyl-Sepharose column combined with a fast protein liquid chromatography (FPLC) system. Ammonium sulfate is added (to 1.5 *M*) to the enzyme preparation, centrifuged (48,000 *g*, 30 min), then enzyme solution is loaded onto the column, equilibrated with 25 m*M* sodium citrate plus 1.5 *M* ammonium sulfate, pH 7.0, and the column eluted with equilibration buffer at 3 ml/min until the absorbance of the eluant at 280 nm is less than 0.2. A linear gradient of 1.5 to 0 *M* ammonium sulfate in citrate is then applied. Fractions containing enzyme activity are pooled, concentrated, and desalted to a final concentration of 25 m*M* ammonium sulfate (Centriprep-30; Amicon, Danvers, MA). This is further purified by anion-exchange FPLC. After loading on a Mono Q column

[9] A. J. Smith, *J. Gen. Microbiol.* **42,** 371 (1966).

[10] R. Meulenberg, J. T. Pronk, W. Hazeu, J. P. van Dijken, J. Frank, P. Bos, and J. G. Kuenen, *J. Gen. Microbiol.* **139,** 2033–2039 (1993).

equilibrated with sodium citrate (25 m*M*, pH 7.0), and washing with buffer until the eluant shows an absorbance at 280 nm of less than 0.0005, a linear gradient of 0 to 1.0 *M* NaCl in citrate buffer is applied. Enzyme-containing fractions are further purified by gel filtration on a Superose-6 column equilibrated with 0.5 *M* ammonium sulfate. Protein elution is monitored at 280 nm, and active fractions are pooled, concentrated, and stored in liquid nitrogen. This produces about 1000-fold purification, giving a product that is homogeneous in gel filtration.[10]

The enzyme in *T. tepidarius* was located in the periplasm of the cells and could be released by lysozyme treatment and osmotic shock.[5].

Trithionate Hydrolase (Trithionate Thiosulfohydrolase): EC 3.12.1.1

$$^{-}O_3S{-}S{-}SO_3^{-} + H_2O \rightleftharpoons {}^{-}S{-}SO_3^{-} + H_2SO_4$$

The activity of trithionate hydrolase was first demonstrated in *T. neapolitanus* and the enzyme subsequently was recovered from both *T. tepidarius* and *T. acidophilus*.[5,11]

Procedure 1. The enzyme can be determined in a coupled assay with thiosulfate dehydrogenase (see above), using 0.1 m*M* sodium trithionate instead of thiosulfate, and measuring thiosulfate-dependent cytochrome *c* reduction at pH 7.0.[5] Thiosulfate produced by the hydrolase becomes the substrate for an excess of thiosulfate dehydrogenase and the rate of reduction of cytochrome *c* equates to the specific activity of the hydrolase. Trithionate solutions are prepared in 0.1 *M* potassium phosphate, pH 7.0, and used within 5 hr to minimize chemical hydrolysis.

Procedure 2. Trithionate hydrolase can also be measured in a discontinuous assay[11] in a well-mixed, thermostatted reaction chamber (10 ml) containing cell-free extract in 25 m*M* potassium phosphate and 1 *M* ammonium sulfate, pH 3.0. The reaction is started by adding 1 m*M* trithionate and 0.5-ml samples removed at intervals into 0.1 ml of 0.125 *M* potassium cyanide for thiosulfate determination by a modification of the cyanolysis method (see [35] in this volume). Following addition of 0.1 ml of 0.075 *M* copper chloride, 1.0 ml of 0.3 *M* ferric nitrate in 3 *M* HNO_3 is added and ferric thiocyanate color measured at 460 nm. Ferric nitrate is used in this assay at a concentration 10-fold higher than in the standard method (see [35] in this volume) to overcome inhibition of color development by the high ammonium sulfate concentrations present in the samples.

[11] R. Meulenberg, J. T. Pronk, J. Frank, W. Hazeu, P. Bos, and J. G. Kuenen, *Eur. J. Biochem.* **209**, 367 (1992).

Rhodanese (Thiosulfate:Cyanide Sulfurtransferase): EC 2.8.1.1

Rhodanese catalyzes transfer of the sulfane sulfur of thiosulfate to an acceptor, which is normally cyanide in the standard assay, and is likely to be cyanide under some physiological conditions.

$$^{-}S\text{-}SO_3^{-} + CN^{-} \rightleftharpoons SCN^{-} + SO_3^{2-}$$

Rhodanese has been found in representatives of all the kingdoms. It can transfer the sulfane sulfur to other acceptors, including lipoic acid.[12]

Procedure 1: Discontinuous Determination of Thiocyanate Product

Reagents

Tris-HCl buffers (1.0 *M*), pH 7.6–10.6
Sodium thiosulfate, 1.0 *M*
Potassium cyanide, 1.0 *M*
Formaldehyde solution (formalin), 38% (w/v)
Ferric reagent: 16% (w/v) ferric nitrate (nonahydrate) in 1.0 *M* nitric acid

Reaction mixtures in replicate small test tubes contain 0.25 ml of Tris buffer, 0.05 ml of thiosulfate, 2.15 ml of water, and cell-free extract.[13,14] After incubation for 2–3 min, 0.05 ml of KCN is added and mixed rapidly. The reaction is stopped at timed intervals by addition of 0.2 ml of 38% (w/v) formaldehyde to each tube, followed by immediate mixing. Addition of 1.3 ml of 16% (w/v) ferric nitrate in 1 *M* nitric acid to each tube produces ferric thiocyanate, which is measured by absorbance at 470 nm and its concentration calculated by reference to a standard curve. Activity is expressed as nanomoles of thiocyanate formed per minute per milligram protein.

Procedure 2: Continuous Spectrophotometric Determination of Sulfite Production by Coupled Oxidation of 2,6-Dichlorophenol-indophenol

Reagents

Tris-HCl buffer (0.2 *M*) pH 8.6
Sodium thiosulfate, 0.3 *M*
2,6-Dichlorophenolindophenol (DCPIP), 0.004 *M*

[12] M. Silver and D. P. Kelly, *J. Gen. Microbiol.* **97,** 277 (1976).
[13] B. Sörbo, *Acta Chem. Scand.* **7,** 1129 (1953).
[14] T. J. Bowen, P. J. Butler, and F. C. Happold, *Biochem. J.* **97,** 651 (1965).

N-Methylphenazonium methosulfate (PMS), 5 mg/ml
KCN, 2.0 *M*

Typically a final reaction mixture volume of 3 ml in a 1-cm cuvette will contain 1.5 ml of Tris buffer (300 μmol), 0.5 ml of thiosulfate (150 μmol), 0.05 ml of DCPIP (0.2 μmol), 0.1 ml of PMS, cell-free extract, and water to a total of 2.95 ml.[15–17] A blank treatment is prepared without DCPIP and extract. Assays can be run at the temperature within the spectrophotometer cell holder or at 30° in a controlled temperature holder. After incubation for 2–3 min, reaction is initiated by rapid addition and mixing of 0.05 ml of KCN (100 μmol). Decrease in absorbance at 600 nm measures DCPIP reduction and enzyme activity is expressed as nanomoles of DCPIP reduced per minute per milliliter protein. The 600-nm millimolar extinction coefficient for DCPIP is 20.6.[18]

It should be noted that the addition of cyanide raises the actual pH of the mixtures to about pH 9.3. For determination of pH optima or assay of rhodanese enzymes with pH optima higher than pH 9.3, different buffers (e.g., glycine-NaOH) including Tris (without acid, pH 10.6) can be used.[19]

Thiocyanate Hydrolase

$$SCN^- + 2H_2O \rightleftharpoons COS + NH_3 + OH^-$$

Thiocyanate hydrolase has to date been demonstrated only in *Thiobacillus thioparus,* in which it is induced by growth on thiocyanate, and was purified 52-fold by ammonium sulfate precipitation, and DEAE-Sephacel and hydroxylapatite column chromatography.[20] It is made up of three different subunits (M_r 19,000, 23,000, and 32,000) and the native enzyme has a molecular weight of 126,000. Activity is optimal at pH 7.5–8.0, 30–40°, and has a K_m of about 11 m*M*.

Assay Procedure. Activity can be determined by incubating a cell-free extract in 0.1 *M* phosphate buffer, pH 7.5, with 20 m*M* KSCN at 30° and measuring the disappearance of either or both thiocyanates (see [35] in this volume) and ammonia formation (by standard Nessler procedure) either after 20 min[20] or at suitable time intervals to obtain a reaction time course. The production of COS can be demonstrated by conducting the

[15] A. J. Smith and J. Lascelles, *J. Gen. Microbiol.* **42,** 357 (1966).
[16] D. P. Kelly, *Arch. Mikrobiol.* **61,** 59 (1968).
[17] W. P. Lu and D. P. Kelly, *FEMS Microbiol. Lett.* **18,** 289 (1983).
[18] M. A. Steinmetz and U. Fischer, *Arch. Microbiol.* **142,** 253 (1985).
[19] A. P. Wood and D. P. Kelly, *J. Gen. Microbiol.* **125,** 55 (1981).
[20] Y. Katayama, Y. Narahara, Y. Inoue, F. Amano, T. Kanagawa,and H. Kuraishi, *J. Biol. Chem.* **267,** 9170 (1992).

assay in a sealed vessel and sampling the head space for analysis by gas chromatography.

Thiosulfate Reductase (Thiosulfate–Thiol Transferase or Sulfurtransferase): Glutathione Dependent (EC 2.8.1.3) and Dithiothreitol Dependent (EC 2.8.1.5)

$$^{-}S{-}SO_3^{-} + 2e^{-} \rightleftharpoons S^{2-} + SO_3^{2-}$$

Thiosulfate reductase can be assayed using reduced methyl viologen as the reductant, with continuous spectrophotometric measurement of dye oxidation.[18]

Reagents

Tris-acetate buffer (pH 8.7), 1.0 *M*
Reduced methyl viologen, 2 m*M* in 0.05 *M* Tris-acetate, pH 8.7
Sodium thiosulfate, 0.05 *M*

Procedure. Reaction is carried out in a cuvette that can be stoppered with a serum cap: 0.05 ml of Tris buffer and 0.05–0.2 ml of cell-free extract are mixed and made up to 0.4 ml with water. The cuvette is stoppered, deaerated under vacuum for 10 min, and regassed with oxygen-free nitrogen, then 0.5 ml of reduced methyl viologen is added by Hamilton syringe. Reaction is started by injecting 0.1 ml of thiosulfate and the oxidation of reduced methyl viologen followed at 600 nm, using a millimolar extinction coefficient of 113.[18]

AMP-Dependent and AMP-Independent Oxidation of Sulfite

A number of enzymes whose activities result in the conversion of sulfite to sulfate are known. We describe assay of (1) "sulfite dehydrogenase" and (2) "APS reductase."

1. $SO_3^{2-} + H_2O \rightleftharpoons SO_4^{2-} + 2H^{+} + 2e^{-}$
2. $SO_3^{2-} + AMP \rightleftharpoons APS + 2e^{-}$

Adenylylsulfate can be converted to sulfate by the action of ADP- or ATP-sulfurylase:

$$APS + PO_4^{3-} \text{ (or } P_2O_7^{4-}) \rightleftharpoons SO_4^{2-} + ADP \text{ (or ATP)}$$

Reagents

Tris-HCl or Tris-H_2SO_4 buffers (0.3 *M*), pH 7.4, 8.0, or 8.4
Potassium ferricyanide, 0.03 *M*

Sodium sulfite, 0.18 *M*, freshly prepared in 0.1 *M* Tris-HCl, pH 7.8, containing 10 m*M* ethylenediaminetetraacetic acid (EDTA)

AMP, 0.045 *M*

Sulfite Dehydrogenase (Sulfite:Ferricytochrome-c Oxidoreductase, Sulfite Oxidase): EC 1.8.2.1

$$SO_3^{2-} + 2Fe(CN)_6^{3-} + H_2O \rightleftharpoons SO_4^{2-} + 2Fe(CN)_6^{4-} + 2H^+$$

Sulfite-dependent reduction of ferricyanide can be measured in reaction mixtures (3 ml) containing 1 ml of Tris buffer (300 μmol), 0.1 ml of potassium ferricyanide (3 μmol), 0.05 ml of Na_2SO_3 (9 μmol), and protein (1–3 mg). The reaction is started by adding enzyme or sulfite, and decrease in absorbance at 420 nm recorded, using buffer plus water as a blank. The millimolar extinction coefficient is 1.0. The rates of sulfite autooxidation (before enzyme addition) and any endogenous reduction of ferricyanide by the enzyme preparation (in the absence of sulfite) need to be assayed as controls.

The enzyme can also be assayed using horse heart cytochrome *c* as electron acceptor, using the same procedure as for thiosulfate dehydrogenase (above) but with sulfite as substrate.[21] Preparations of this enzyme purified 2000-fold from *Thiobacillus versutus* showed it to be a monomer (M_r 44,000) and to contain intimately associated cytochrome *c*-551.[21]

Adenylylsulfate Reductase: EC 1.8.99.2

Adenylylsulfate reductase activity is estimated from the difference in rate of reduction of ferricyanide (A_{420}) in the assay described above, in the presence of sulfite with and without AMP.[22] Assays (3 ml in 1-cm cuvettes) contain 0.1 *M* Tris-HCl (pH 7.4), 1 m*M* potassium ferricyanide, 3 m*M* sodium sulfite, and cell-free extract. The rate of ferricyanide reduction due to AMP-independent sulfite oxidation is determined for a few minutes before 0.05 ml of AMP (0.75 m*M*) is added to the cuvette and the rate of ferricyanide reduction measured for a further 5–10 min. Any stimulation of the rate of ferricyanide reduction by AMP is a measure of sulfite oxidation that can be ascribed to APS reductase. Controls should omit sulfite or AMP. The sequence of additions of AMP, sulfite, and extract should be tested to ensure that results obtained are independent of the order of exposure of the enzymes to the substrates.

[21] W.-P. Lu and D. P. Kelly, *J. Gen. Microbiol.* **130,** 1683 (1984).

[22] T. J. Bowen, F. C.Happold, and B. F. Taylor, *Biochim. Biophys. Acta* **118,** 566 (1966).

Sulfur Oxygenase (Sulfur Dioxygenase, Sulfur:Oxygen Oxidoreductase, Sulfur-Oxidizing Enzyme): EC 1.13.11.18

$$S_8 + 8H_2O + 8O_2 \rightleftharpoons 8H_2SO_3$$

Sulfur oxygenase is assayed by determination of oxygen uptake and thiosulfate production in the presence of powdered sulfur and reduced glutathione.[23,24] Respirometric assays are conducted in an oxygen electrode cell using reaction mixtures (2 ml) containing 0.25 *M* Tris-HCl (pH 7.8), sulfur (48 mg), catalase (0.25 mg), 2,2′-bipyridyl (0.1 m*M*), reduced glutathione (2.5 m*M*), and crude cell-free extract (0.05–0.25 ml). Oxygen uptake rates should be corrected for (low) chemical control oxidation rates. Activities measured at 30° with cell-free extracts of thiobacilli are generally low, requiring incubation periods of an hour or longer. The time course of thiosulfate formation in an identical assay mixture can be followed in small flasks shaken in air and sampled at intervals up to 3 hr. The samples (0.1 ml) are mixed with 0.1 ml of 1 *M* cadmium acetate and then assayed cyanolytically for thiosulfate (see [35] in this volume).

Glutathione-Independent Sulfur "Oxidase"

Sulfur oxidation in cell-free extracts of some thiobacilli is seen in mixtures lacking glutathione, and is apparently due to an uncharacterized enzyme(s) that might link oxidation to electron transport, rather than to being an oxygenase[25]:

$$S_8 + 24H_2O \rightleftharpoons 8H_2SO_3 + 32H^+ + 32e^-$$

Activity can be measured in an oxygen electrode cell using an assay mixture (2 ml) containing 0.1 *M* Tris-HCl (pH 8.0), powdered sulfur (30 mg), with and without Tween 80 (1%, w/v). Taylor[25] used about 25 mg of protein per assay and found that at least 15 mg was required to initiate the reaction.

Proteins A and B and Thiosulfate-Oxidizing Multienzyme System of Thiobacillus versutus

An enzyme system located in the periplasmic space of *T. versutus* is capable of the complete oxidation of thiosulfate to sulfate, without the intermediate formation or accumulation of polythionates, with the coupled reduction of several cytochromes.[26] This can be assayed only with a

[23] I. Suzuki, *Biochim. Biophys. Acta* **104,** 359 (1965).
[24] I. Suzuki and M. Silver, *Biochim. Biophys. Acta* **122,** 22 (1966).
[25] B. F. Taylor, *Biochim. Biophys. Acta* **170,** 112 (1968).
[26] W.-P. Lu, B. E. P. Swoboda, and D. P. Kelly, *Biochim. Biophys. Acta* **828,** 116 (1985).

complete system comprising at least two proteins (M_r 16,000 and 64,000) and two unusual *c*-type cytochromes. One protein (A) is a thiosulfate-binding enzyme and the other (B) contains an unusual binuclear manganese cluster. The mechanisms of action of these enzymes is unknown and the activity of the components of the system cannot be assayed independently. Their assay and properties are therefore not further described in this chapter, but the specialist reader is referred to the original literature.[26–31]

[27] W.-P. Lu and D. P. Kelly, *J. Gen. Microbiol.* **129,** 1673 (1983).
[28] W.-P. Lu and D. P. Kelly, *J. Gen. Microbiol.* **129,** 3549 (1983).
[29] W.-P. Lu and D. P. Kelly, *Biochim. Biophys. Acta* **765,** 106 (1984).
[30] W.-P. Lu, *FEMS Microbiol. Lett.* **34,** 313 (1986).
[31] R. Cammack, A. Chapman, W.-P. Lu, A. Karagouni, and D. P. Kelly, *FEBS Lett.* **253,** 239 (1989).

[37] Whole-Organism Methods for Inorganic Sulfur Oxidation by Chemolithotrophs and Photolithotrophs

By DON P. KELLY and ANN P. WOOD

Introduction

This chapter describes some of the techniques that have been used in studies of sulfur compound oxidation and energy coupling using suspensions of intact organisms and for the separate identification and characterization of enzymes that are located in the periplasm, rather than the cytoplasm, of thiobacilli. In some cases the techniques are derived from those developed for studies on other microorganisms or using mitochondria, and are outlined primarily to demonstrate that the same principles apply to the less routinely studied lithotrophs. Some of the procedures provide examples of the use of the analytical methods described in [35] in this volume.

Demonstration of Periplasmic Location of Some Enzymes Involved in Sulfur Compound Oxidation in Thiobacilli

Separation of the periplasmic fractions from both *Thiobacillus versutus* and *Thiobacillus tepidarius* has been achieved by methods based on a procedure developed for *Escherichia coli*.[1-3] The whole of the thiosulfate-oxidizing multienzyme system of *T. versutus* is periplasmic,[1,4-7] and the thiosulfate dehydrogenase (and probably trithionate hydrolase) enzyme of *T. tepidarius* is also in the periplasmic space.[2]

Periplasmic Enzymes of Thiobacillus versutus

Reagents

Suspension mixture: 1 *M* sucrose, 0.2 *M* Tris-HCl (pH 8.0), 5 m*M* ethylenediaminetetraacetic acid (EDTA)
Tris-HCl (pH 7.5), 0.025 *M*
Magnesium chloride, 0.1 *M*
Lysozyme
DNase

Procedure. Washed bacteria, harvested from autotrophic culture on thiosulfate, are taken up in suspension mixture [130 mg (dry weight)/10 ml], and 8.5 mg of lysozyme is added. The mixture is subjected to mild osmotic shock by mixing in 10 ml of water, then incubating at 26° for 25 min. This treatment lyses the bacterial cell walls but leaves the cytoplasmic membrane intact. The suspension is centrifuged (12,000 g, 20 min, 4°) to give a clear pink-orange supernatant liquid (periplasmic fraction) containing all the periplasmic proteins, including most of the cell content of cytochrome *c*, and a pale-colored pellet (spheroplasts). The periplasmic fraction is dialyzed against Tris-HCl (2 hr, 4°) and then concentrated with polyethylene glycol to about 2 ml. This fraction can then be assayed for cytochrome content and enzyme activities (Table I). The spheroplast fraction can be lysed to solubilize its enzymes by resuspending in 20 ml of Tris-HCl and adding 50 mg of DNase and 0.02 mg of $MgCl_2$ (to reduce viscosity due to DNA). After lysis, the mixture is centrifuged (150,000 *g*, 1 hr) to yield a straw-colored supernatant (cytoplasmic fraction) and a pellet (membrane fraction) that can be resuspended in Tris-HCl. The distribution of enzymes found by Lu[1] is summarized in Table I.

[1] W.-P. Lu, *FEMS Microbiol. Lett.* **34,** 313 (1986).
[2] W.-P. Lu and D. P. Kelly, *J. Gen. Microbiol.* **134,** 877 (1988).
[3] B. Withilt, M. Bodkhout, M. Brock, J. Kingma, H. van Heerikhuizen, and L. de Leij, *Anal. Biochem.* **74,** 160 (1976).
[4] W.-P. Lu and D. P. Kelly, *J. Gen. Microbiol.* **129,** 1673 (1983).
[5] W.-P. Lu and D. P. Kelly, *J. Gen. Microbiol.* **129,** 3549 (1983).
[6] W.-P. Lu and D. P. Kelly, *Biochim. Biophys. Acta* **765,** 106 (1984).
[7] W.-P. Lu and D. P. Kelly, *Biochim. Biophys. Acta* **828,** 116 (1985).

TABLE I
DISTRIBUTION (%) OF PROTEIN AND ENZYME ACTIVITIES BETWEEN PERIPLASMIC, CYTOPLASMIC, AND MEMBRANE FRACTIONS OF *Thiobacillus versutus*[a]

Fraction	Protein	Thiosulfate-oxidizing multienzyme system[b]	Sulfite dehydrogenase	Malate dehydrogenase	*c*-type cytochromes
Periplasm	36	95	96	7	81
Cytoplasm	47	5	4	93	3
Membrane	17	0	0	0	16

[a] Taken from Lu (1986).[1]
[b] Measured as thiosulfate-dependent reduction of horse heart cytochrome *c*.[1,5]

Periplasmic Location of Thiosulfate Dehydrogenase (Tetrathionate Synthase) of Thiobacillus tepidarius

Reagents

Suspension mixture: 0.5 *M* sucrose, 0.05 *M* potassium phosphate buffer (pH 7.0), 0.05 *M* Tris-HCl (pH 7.0), 5 m*M* EDTA
Lysozyme (1.5%, w/v)
Potassium phosphate buffer (pH 7.0), 0.1 *M*
DNase

Procedure. Bacteria harvested by centrifugation from cultures grown autotrophically on thiosulfate are resuspended in 5 ml of suspension mixture [80 mg (dry weight)/ml], supplemented with 1 ml of lysozyme and incubated for 2 hr at 30°.[2] After centrifuging (10,000 *g*, 18 min) the supernatant is recovered (periplasmic fraction) and the pellet resuspended in 2 ml of phosphate with 0.5 mg of DNase and centrifuged (10,000 *g*, 15 min). The pellet from this stage is discarded as unlysed spheroplasts and large debris, and the supernatant centrifuged at 45,000 *g* (30 min) to give a supernatant (cytoplasmic fraction) and a pellet (membrane fraction).

In preparations of the kind described above the relative distribution of protein and thiosulfate dehydrogenase activity is as shown in Table II.

Differential Radiolabeling to Demonstrate Enzymes of Thiosulfate-Oxidizing System of Thiobacillus versutus in Periplasm and Cytoplasm of Intact Organisms

Techniques developed for the assessment of amidination of the inner and outer membrane surfaces of human erythrocytes can be adapted for use with gram-negative bacteria, to determine whether proteins are located

TABLE II
DISTRIBUTION (%) OF PROTEIN AND THIOSULFATE DEHYDROGENASE IN PERIPLASMIC, CYTOPLASMIC, AND MEMBRANE FRACTIONS OF *Thiobacillus tepidarius*[2]

Fraction	Protein	Thiosulfate dehydrogenase
Periplasm	27	76
Cytoplasm	53	8
Membrane	20	16

in the periplasm or inside the cytoplasmic membrane of the cell.[8–10] The example of this technique in thiobacilli is described below: it could be applied to other enzymes and in other sulfur bacteria.

Reagents

BufferA: 0.05 *M* Tris-HCl (pH 7.2), containing 0.2 *M* KCl
Buffer B: 0.1 *M* phosphate (pH 8.0), containing 0.1 *M* KCl
Ethyl [1-^{14}C]acetimidate hydrochloride (EAI; about 57 mCi/mmol)
Isethionyl [1-^{14}C]acetimidate (IAI; about 28 mCi/mmol)
Unlabeled IAI
Buffer C: 0.1 *M* phosphate (pH 8.0), containing 0.2 *M* KCl
Buffer D: 0.1 *M* Tris-HCl (pH 7.8), containing 0.01 *M* EDTA
Buffer E: 0.02 *M* phosphate (pH 7.0)
Sodium dodecyl sulfate (SDS) buffer: 60 m*M* Tris-HCl (pH 6.8), containing (per liter) 30 g of SDS, 50 ml of 2-mercaptoethanol, 200 ml of glycerol, 20 mg of bromothymol blue

Principle and Procedure. EAI penetrates the cytoplasmic membrane of intact cells and binds to both periplasmic and cytoplasmic proteins. IAI does not penetrate the cytoplasmic membrane and binds only to periplasmic proteins. Pretreatment of cells with unlabeled IAI followed by treatment with [^{14}C]EAI results in labeling of the cytoplasmic proteins but not those in the periplasm, which are already "masked" by IAI. ^{14}C-Labeled EAI and IAI are supplied by Amersham International (Amersham, England) and the unlabeled compounds are available from Sigma (St. Louis, MO).

Using *T. versutus* grown on thiosulfate, organisms are harvested, washed with buffer A, then with buffer B, and finally resuspended in

[8] W.-P. Lu and D. P. Kelly, *Arch. Microbiol.* **149,** 297 (1988).
[9] E. R. Kasprzak and D. J. Steenkamp, *in* "Microbial Growth on C_1 Compounds" (R. L. Crawford and R. S. Hanson, eds.), p. 147. American Society for Microbiology, Washington, D.C., 1984.
[10] N. M. Whiteley and H. C. Berg, *J. Mol. Biol.* **87,** 541 (1974).

buffer B [50 mg (dry weight)/ml]. Cell suspension (0.1 ml) is incubated (20°, 30 min) with either 2.9 m*M* EAI or 4 m*M* IAI, then centrifuged and washed with buffer C, twice with buffer D, and finally suspended in 0.15 ml of buffer E. Samples are then mixed with an equal volume of SDS buffer, boiled for 2–5 min, then subjected to polyacrylamide gel electrophoresis (PAGE).[8] Comparison of autoradiographs of gels carrying EAI- or IAI-labeled proteins, respectively, will show that the proteins A and B (M_r 16,000 and 64,000) are much more heavily labeled by IAI than by EAI, relative to other cell proteins, consistent with their periplasmic location. A better demonstration is given if cell suspension is preincubated (20°, 30 min) with 9 m*M* unlabeled IAI, to "mask" periplasmic proteins so that they do not subsequently bind EAI. Cells are then centrifuged, washed once with buffer B, suspended in 0.1 ml of buffer B, then treated as before with [^{14}C]EAI. It is then found that proteins A and B on the gels are not ^{14}C labeled, because they are located in the periplasm. It can be demonstrated separately that pure samples of each of proteins A and B will readily bind [^{14}C]EAI or [^{14}C]IAI, showing the validity of the experimental approach.

Equilibrium Dialysis to Demonstrate Binding of [^{35}S]Thiosulfate to Protein A of Multienzyme System of *Thiobacillus versutus*

The technique of establishing an equilibrium between two dialysis chambers separated by a semipermeable membrane[11] can be applied to the binding of thiosulfate by the protein A component of the *Thiobacillus versutus* thiosulfate-oxidizing multienzyme system.[7]

One chamber of the dialysis system contains 0.4 ml of 50 m*M* Tris-HCl (pH 7.2) containing about 1 mg of protein A, and the other contains 0.4 ml of buffer with $^{-35}S{-}SO_3^-$ or $^-S{-}^{35}SO_3^-$ (10–200 μM). The membrane used in this case is one not allowing diffusion of small proteins: the molecular weight of protein A is 16,000. The dialysis device is rotated at 100 rpm on a rotary shaker (30°) for 5–6 hr. This is the time taken for equilibrium to be reached, after which no further change in the ^{35}S content of either chamber occurs. Applying this technique to other enzymes or organisms would necessitate monitoring of the time taken to achieve equilibrium. From the ^{35}S content of each chamber at equilibrium the difference in disintegrations per minute (dpm) is a measure of the amount of [^{35}S]thiosulfate bound to the enzyme, and the number of moles of thiosulfate bound per mole of enzyme can be calculated. The number of binding sites for thiosulfate on the enzyme (protein A) can be determined graphically from

[11] K. Cain and D. E. Griffiths, *Biochem. J.* **162,** 572 (1977).

a Scatchard plot. From the equilibrium values of bound and free thiosulfate, $[T_b]$ and $[T_f]$, and total enzyme, [A] (μM), a graph is plotted of ($[T_b]/[A][T_f] \times 10^3\ \mu M^{-1}$) against ($[T_b]/[A]$). The number of binding sites is given by the extrapolation value on the latter (x) axis. In the case of the *T. versutus* protein A, the value is one,[7] but this technique could be applicable to other enzymes capable of binding thiosulfate or polythionates in this or other organisms.

Oxidation of ^{35}S-Labeled Thiosulfate by Lithotrophs

Discriminative Oxidation of the Two Sulfur Atoms of Thiosulfate

Monitoring of the time course of the oxidation of thiosulfate labeled in either the sulfane (S–) or sulfonate ($-SO_3$) atom can be informative of the mechanism of initial attack on thiosulfate and the possible formation of intermediates of the process.[12–16] For example, if there is initial thiosulfate cleavage to form elemental sulfur and rapid conversion of the sulfonate group to sulfate, it is simple to monitor the separate metabolism of the two atoms:

$$^{-}O_3S{-}{*}S^{-} \xrightarrow{O} SO_4^{2-} + [{*}S]$$

$$[{*}S] + H_2O + 1.5O_2 \longrightarrow H_2{*}SO_4$$

Such two-stage oxidation has been shown in anaerobic *Chromatium*,[13,14] and in *Thiocapsa*[14] and aerobic *Thiothrix*,[15] and can be quantified by direct chemical analysis of the disappearance of thiosulfate, formation of sulfur and sulfate, as well as by the measurement of ^{35}S in these compounds, as described in [35] in this volume. Rapid separation of organisms and sulfur can be achieved either by the filtration of suspensions through membranes (0.2- or 0.45-μm pore sizes) or by rapid centrifugation (e.g., high-speed microfuge centrifugation of small-volume samples). The measurement of the chemical and ^{35}S content of sulfate in samples also gives the specific activity of the sulfate formed: this is particularly important when sulfate formation is concurrent from both atoms of thiosulfate but the rate from one is faster than the other. For example, when *Chromatium* oxidizes thiosulfate phototrophically, sulfate is formed at reasonably con-

[12] D. P. Kelly and P. J. Syrett, *Biochem. J.* **98,** 537 (1966).
[13] A. J. Smith and J. Lascelles, *J. Gen. Microbiol.* **42,** 357 (1966).
[14] H. G. Trüper and N. Pfennig, *Antonie van Leeuwenhoek J. Microbiol. Serol.* **32,** 261 (1966).
[15] E. V. Odintsova, A. P. Wood, and D. P. Kelly, *Arch. Microbiol.* **160,** (1993).
[16] O. H. Tuovinen and D. P. Kelly, *Plant Soil* **43,** 77 (1975).

stant rates throughout oxidation of both $^{-}S\text{–}^{35}SO_3^{-}$ and $^{-35}S\text{–}SO_3^{-}$, but the rate of [^{35}S]sulfate formation from the S– label is only 15% of that from the $\text{–}SO_3$ position.[13] Meanwhile the rate of intracellular accumulation of elemental ^{35}S from the $^{-35}S\text{–}SO_3^{-}$ is similar to the rate of $^{35}SO_4^{2-}$ production from $^{-}S\text{–}^{35}SO_3^{-}$.[13]

If sulfate is being determined by high-performance liquid chromatography (HPLC) it is possible simultaneously to determine the ^{35}S content of the sulfate with an in-line β counter and thus obtain a direct read-out of sulfate-specific activity. This will, for example, demonstrate if sulfate is initially formed only from $\text{—}SO_3$ and if there is a progressive or step increase in specific activity as the S– atom is oxidized. Given the cost (or preparation times) of specifically labeled thiosulfate, this technique means that thiosulfate labeled in only one atom can provide as much information on sulfate production kinetics (or sulfur formation) as an experiment in which both labeled thiosulfates are used.

Oxidation of [^{35}S]Thiosulfate by Thiobacillus neapolitanus

The analytical procedures of [35] (in this volume) can be applied to the products of oxidation of labeled thiosulfate by thiobacilli or other chemolithotrophs, and used to demonstrate the origins of sulfur atoms in intermediate polythionates and product sulfate.[8] Using shaken suspensions of *T. neapolitanus,* samples are taken at frequent intervals after addition of [^{35}S]thiosulfate (e.g., 0.25- to 1.0-min intervals for 8–10 min) and the reaction is terminated either by separating the bacteria by membrane filtration, or by killing the organisms by mixing with a one- or twofold excess of ethanol and separating cells by filtration.[12–16] Cell-free solutions can be analyzed chemically for sulfur compounds and subjected to chromatographic analysis. Residual thiosulfate (in kinetic experiments) and any polythionates formed can be purified by paper chromatography (or HPLC) and both their total ^{35}S content and the distribution of ^{35}S within their molecules determined (see [35] in this volume). If paper or thin-layer chromatography is used, $^{35}SO_4^{2-}$ is determined separately from thiosulfate by treatment of one set of replicate samples with iodine, which oxidizes thiosulfate to tetrathionate, which in turn separates completely from sulfate on chromatography. If kinetic experiments are conducted in Warburg or Gilson respirometers, oxygen uptake can be monitored simultaneously with sulfur compound turnover.

Simultaneous chemical and ^{35}S measurements of the same compounds show that chemically detectable thiosulfate is still present when thiosulfate provided initially as $^{-}S\text{–}^{35}SO_3^{-}$ has disappeared, whereas that provided as $^{-35}S–SO_3^-$ is still present, and that the specific activity of sulfate produced from $^{-35}S–SO_3^-$ increases progressively during oxidation. Such a result shows that sulfate is produced faster from the $–SO_3$ sulfur atom than from the S– atom of thiosulfate. It also shows that thiosulfate labeled in both sulfur atoms must be formed from the S– atom during oxidation. This can be demonstrated by mercury and silver degradation of the labeled thiosulfate (see [35] in this volume). Similarly, analysis by mercury degradation of trithionate accumulated during $^{-35}S–SO_3^-$ oxidation shows that ^{35}S is present in both the S– and $–SO_3$ atoms of trithionate. The data obtainable from these procedures are illustrated in Table III.[12] Identical experiments using $^-S–^{35}SO_3^-$ show that *T. neapolitanus* does not convert the $–SO_3$ sulfur atom into the S– atom of either thiosulfate or trithionate, but converts it only to sulfate or the $–SO_3$ groups of trithionate.[12] While these examples are for a thiobacillus, the techniques described can be applied to any bacteria that oxidize or reduce thiosulfate and polythionates.

TABLE III
INFORMATION THAT CAN BE DERIVED FROM ANALYSIS OF PRODUCTS OF [*sulfane*-^{35}S]THIOSULFATE ($^{-35}S–SO_3^-$) OXIDATION BY *Thiobacillus neapolitanus*[a]

Parameter measured	Data obtained
Amount of thiosulfate unoxidized	42% (of initial 6 m*M*)
^{35}S recovered as thiosulfate	47%
^{35}S recovered as trithionate	15%
^{35}S recovered as sulfate	29%
^{35}S in S– position in thiosulfate	64%
^{35}S in $–SO_3$ position in thiosulfate	36%
^{35}S in S– position in trithionate	48%
^{35}S in $–SO_3$ positions in trithionate	52%
Relative specific activity of [^{35}S]sulfate	59% of theoretical[b]

[a] *Thiobacillus neapolitanus* [1.85 mg (dry weight)/ml in 0.1 *M* phosphate, pH 7.0] was shaken at 23°. Data are given for analyses made 5 min after addition of 6 m*M* $^{-35}S–SO_3^-$ (4×10^6 dpm ^{35}S/ml). The data are exemplary only and further details are in Ref. 8. The data illustrate the application of analytical procedures given in [35] in this volume.

[b] Calculated as that predicted if the two atoms of thiosulfate were oxidized at the same rate rather than the $–SO_3$ group at a faster rate.

Respiration-Driven Proton Translocation in *Thiobacillus versutus* and *Thiobacillus tepidarius*

The classic techniques developed for mitochondria[17] and heterotrophic bacteria[18] can be applied to bacteria oxidizing inorganic sulfur compounds.[8,19]

Procedure: Oxygen Pulse Method

Acidification (i.e., proton extrusion) during sulfur compound oxidation dependent on small pulses of oxygen is measured using an oxygen electrode cell (3.0- or 3.5-ml reaction volumes) as reaction chamber. Suspensions of *T. versutus* or *T. tepidarius* [8–12 mg (dry weight)] are stirred in 0.15 *M* KCl, 1.5 m*M* glycylglycine, 20 μg of valinomycin or 50 m*M* KSCN, 35 μg of carbonate dehydratase, and various substrates. Initial pH is 6.4–6.6 and the temperatures are 25 and 40° for the two organisms, respectively. Suspensions are made anaerobic by continuous passage of oxygen-free nitrogen over their surfaces, and substrates provided can be thiosulfate (57 or 100 μM), tetrathionate (167 μM, for *T. tepidarius*), sulfite (14 μM), ascorbate plus TMPD (*N,N,N′,N′*-tetramethyl p-phenylenediamine) (125 plus 29 μM), or ferrocyanide (125 μM). Suspensions are preincubated for 1 hr when using valinomycin or 15 min with KSCN (which is not an oxidation substrate for these two thiobacilli). Proton translocation is induced by injecting air-saturated 0.15 *M* KCl, which contains 4.8 ng-atoms oxygen in 10 μl at 25° and 7.8 ng-atoms oxygen in 20 μl at 40°. Acidification is measured with a pH-measuring and recording system with a fast reaction time. Lu and Kelly[14,19] used a CMAWL pH probe (Russell pH, Ltd., Fife, Scotland) linked to a Pye Unicam PW 9409 pH meter and recorded on a JJCR 100 potentiometric recorder (Lloyd Instruments, Ltd., Southampton, England) giving full-scale deflection equivalent to a ΔpH of 0.12 unit. The system can be calibrated with 1–5 μl of oxygen-free 10 m*M* HCl.

Following addition of the oxygen pulse acidification is complete within 3–5 sec. The maximum concentration of H^+ produced by a pulse is estimated from the decay over time of the ΔpH. The extrapolation to time zero on a graphical plot of log ng-ions H^+ against time gives the maximum H^+ produced, and the $\rightarrow H^+/O$ quotient is calculated as H^+ generated divided by ng-atoms O added. Typical examples of the semilogarithmic plots derived in this way are shown in Fig. 1. For *T. versutus*, the oxidation

[17] P. Mitchell and J. Moyle, *Nature (London)* **208,** 147 (1965).

[18] P. Scholes and P. Mitchell, *Bioenergetics* **1,** 309 (1970).

[19] W.-P. Lu and D. P. Kelly, *J. Gen. Microbiol.* **134,** 865 (1988).

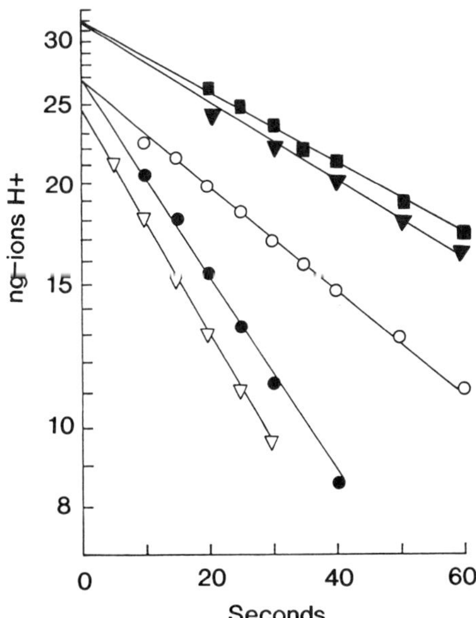

FIG. 1. Determination of $\rightarrow H^+/O$ quotients from extrapolation to zero time of the logarithmic decay of the proton gradient generated by oxygen pulses added to anaerobic suspensions of aerobic thiobacilli, provided with inorganic sulfur compounds as respiratory substrates. Mean values are given from several experiments for the translocated H^+ concentrations, 5–60 sec following the oxygen pulse (see text). Data shown are for *Thiobacillus tepidarius* given an oxygen pulse of 7.8 ng-atoms oxygen in the presence of thiosulfate and rotenone (■) or tetrathionate (▼), and for *Thiobacillus versutus* given 4.8 ng-atoms O (endogenous respiration, without sulfur substrate, ○) or 9.7 ng-atoms O in the presence of thiosulfate (●) or sulfite (▽).[8,19]

of thiosulfate or sulfite produces a $\rightarrow H^+/O$ quotient around 2.5, and for *T. tepidarius* a quotient of about 4.0 results for each of thiosulfate, tetrathionate, and sulfite.

Other Techniques Applicable to Whole-Cell Suspensions

Techniques developed for the study of the energetics of mitochondria and heterotrophs have also been applied to sulfur-oxidizing bacteria. Standard procedures for the measurement of $^{14}CO_2$ fixation, ATP synthesis, and the reduction of NAD^+ and $NADP^+$ have been used for various thiobacilli, enabling, for example, the demonstration of the dependence of carbon dioxide fixation on thiosulfate or polythionate oxidation, the presence of both uncoupler (e.g., dinitrophenol)-sensitive and -insensitive ATP synthesis in *T. neapolitanus*, ATP synthesis and $NAD(P)^+$ reduction during thiosulfate and tetrathionate oxidation in other thiobacilli, and their differential inhibition by FCCP (carbonyl cyanide *p*-[trifluoromethoxy]-phenylhydrazone), DCCD (*N,N'*-dicyclohexylcarbodiimide), and oligomycin.[19–22] Standard chemostat culture methods can also be applied to cultures of aerobic chemolithotrophs capable of growth on thiosulfate and polythionates, and can be used to provide data on the fundamental growth constants applicable to those compounds. These constants are specific for different species and are of use in calculating yields of ATP actually achieved by growing cultures.[22–25]

Details of these procedures are given in the literature cited, but are not detailed here as they are mainly derived from well-known methods applicable to microorganisms in general. Their use has helped demonstrate the fundamental similarity of the biochemistry of heterotrophs and those bacteria deriving energy from the oxidation of inorganic sulfur compounds.

[20] W.-P. Lu and D. P. Kelly, *Arch. Microbiol.* **149,** 303 (1988).

[21] D. P. Kelly and P. J. Syrett, *J. Gen. Microbiol.* **43,** 109 (1966).

[22] D. P. Kelly, *in* "The Bacteria" (T. A. Krulwich, ed.), Vol. 12, p. 479. Academic Press, New York, 1990.

[23] D. P. Kelly, *Proc. R. Soc. London Ser. B* **298,** 499 (1982).

[24] A. P. Wood and D. P. Kelly, *Arch. Microbiol.* **144,** 71 (1986).

[25] J. Mason, D. P. Kelly and A. P. Wood, *J. Gen. Microbiol.* **133,** 1249 (1987).

Section VI

Special Techniques

[38] Mössbauer Spectroscopy in Study of Cytochrome cd_1 from *Thiobacillus denitrificans*, Desulfoviridin, and Iron Hydrogenase

By Boi Hanh Huynh

Introduction

Mössbauer spectroscopy is a nuclear resonance technique involving recoilless emission and absorption of γ photons in low-energy nuclear transitions (10^2 keV). Because the most easily detectable Mössbauer transition is observed in ^{57}Fe, an isotope of a biologically important trace element, the Mössbauer effect has been used for biological studies almost immediately after its discovery in 1958.[1] Because of the high-resolution nature of the technique, Mössbauer measurements can provide electronic and physical information concerning the iron centers in proteins in details that no other spectroscopy can match. Over the past decades the technique has become an indispensable tool for the study of iron-containing proteins. In Volume 27 of this series, T. H. Moss described in detail the basic principles and experimental techniques of Mössbauer spectroscopy.[2] In Volume 54, E. Münck discussed at length biological applications of the technique with particular emphasis on iron-containing electron carriers.[3] Theoretical and technical considerations pertinent to studies of biological samples were also discussed in Münck's chapter. More recent reviews on similar subject can be found elsewhere.[4–7]

This chapter is written with the purpose of helping the nonspecialist to understand Mössbauer spectra. Emphasis is focused on information contained in the spectra. The theoretical background necessary for understanding Mössbauer spectra of biological samples is reviewed. Hyperfine interactions and their effect on Mössbauer spectra are described. The nuclear and electronic spin Hamiltonian commonly used for data analysis

[1] R. L. Mössbauer, *Z. Phys.* **151,** 124 (1958).
[2] T. H. Moss, this series, Vol. 27 [35].
[3] E. Münck, this series, Vol. 54 [20].
[4] B. H. Huynh and T. A. Kent, *in* "Advances in Inorganic Biochemistry" (G. L. Eichhorn and L. G. Marzilli, eds.), Vol. 6, p. 163. Elsevier, New York, 1984.
[5] P. G. Debrunner, *Hyperfine Interact.* **53,** 21 (1990).
[6] P. G. Debrunner *in* "Iron Porphyrins, Part 3" (A. B. Lever and H. B. Gray, eds.), p. 139. VCH Publ., New York, 1990.
[7] A. X. Trautwein, E. Bill, E. L. Bominar, and H. Winkler, *Struct. Bonding (Berlin)* **78,** 1 (1991).

are introduced. The correlation between Mössbauer spectra and electronic structure is discussed. Mössbauer investigations of cytochrome cd_1 from *Thiobacillus denitrificans,* iron hydrogenase from *Desulfovibrio vulgaris,* and desulfoviridins are used as examples to demonstrate the variety of information obtainable by Mössbauer spectroscopy. Contributions of Mössbauer spectroscopy to other iron-containing proteins involved in sulfur metabolism are presented in other chapters of this volume.

Theoretical Background

For biological applications, the Mössbauer transition involves the ground and first excited states of the ^{57}Fe nucleus. The ground state has a nuclear spin $I = 1/2$ and possesses a magnetic moment $\mu_n = g_n \beta_n I$, where the nuclear gyromagnetic ratio $g_n = 0.1806$ and β_n is the nuclear Bohr magneton. The excited state has $I = 3/2$, $g_n = -0.1033$, and, in addition, possesses an electric quadrupole moment $Q = 8 \times 10^{-26}$ cm^2. The transition energy is 14.4 keV with an uncertainty approaching the Heisenberg uncertainty limit of 4.5×10^{-9} eV. Such an extreme resolution allows accurate measurement of hyperfine interactions between the iron nucleus and its surrounding electrons, which are on the order of 10^{-8} to 10^{-6} eV. In practice, to obtain an energy absorption spectrum representing the hyperfine interactions, the energy of the source radiation is brought into resonance with the transition energy of the absorber by moving the source relative to the absorber. The energy unit of Mössbauer spectra is therefore generally expressed in terms of source velocity. A source velocity of 1 mm/sec corresponds to a Doppler shift in energy of 4.8×10^{-8} eV. Because the nuclear properties of ^{57}Fe are known, analysis of the Mössbauer spectra yields information concerning the electronic properties. For ^{57}Fe, there are two kinds of hyperfine interactions: the electric and the magnetic hyperfine interactions. These two kinds of interactions basically result in two different types of Mössbauer spectra, reflecting fundamentally different electronic systems.

Electric Hyperfine Interactions

The ^{57}Fe nucleus is a positively charged particle. It will interact with the surrounding negatively charged electrons. This electrostatic interaction can be separated into two terms: (1) interaction of the nuclear monopole with the electronic charge density at the nucleus and (2) interaction of the nuclear excited-state quadrupole moment with the electric field gradient (EFG) produced by the surrounding electrons. The former shifts the nuclear energy levels and gives rise to a shift of the center of gravity of the spectrum, called the isomer shift δ.[8]

$$\delta = K(|\Psi(0)|^2_{\mathrm{A}} - |\Psi(0)|^2_{\mathrm{S}}) \tag{1}$$

where K is a nuclear constant and $|\Psi(0)|^2_{\mathrm{A}}$ and $|\Psi(0)|^2_{\mathrm{S}}$ are electronic charge densities at the absorber and source nuclei, respectively. To eliminate the effect of source material differences and to obtain a standard measurement for isomer shift, the centroid of the room temperature Mössbauer spectrum of metallic α iron is adopted as zero velocity.

The electric quadrupole interaction can be described by the following nuclear Hamiltonian,[8]

$$\hat{\mathcal{H}}_{\mathrm{Q}} = \frac{eQV_{zz}}{4I(2I-1)}[3\mathbf{I}_z^2 - \mathbf{I}^2 + \eta(\mathbf{I}_x^2 - \mathbf{I}_y^2)] \tag{2}$$

where e is the magnitude of electron charge, V_{ii} values are the principal components of the EFG tensor ($i = x, y$, and z), and $\eta = (V_{xx} - V_{yy})/V_{zz}$ is the asymmetry parameter. According to the convention for EFG, the x, y, z are chosen such that $|V_{xx}| \leq |V_{yy}| \leq |V_{zz}|$. This convention forces $0 \leq \eta \leq 1$. For biological applications, however, it is more common to choose the EFG axes to coincide with the axes defined by the electronic spin Hamiltonian described in the next section. Consequently, V_{zz} is not necessarily the largest principal component and η is not confined to between 0 and 1.

For biological molecules, the electronic charge distribution surrounding the iron nucleus is generally asymmetric, producing a nonzero EFG at the nucleus and lifting the degeneracy of the nuclear excited state. However, reversal of nuclear spin orientation does not change the nuclear charge distribution and quadrupole interaction cannot completely lift the fourfold degeneracy of the excited state. It can split the state into only two doubly degenerate levels (Fig. 1A). Because the ground state possesses no quadrupole moment, it does not interact with the EFG and its energy remains unaffected. In such a situation, there are two possible Mössbauer transitions from the unsplit ground state to the two doubly degenerate excited states. The resulting Mössbauer spectrum consists of two equal-intensity absorption lines (Fig. 1B). Such a spectrum is called a *quadrupole doublet* and is a result of the electric hyperfine interactions. The center of the spectrum yields the parameter isomer shift δ. The energy separation between the two lines, called *quadrupole splitting* (ΔE_{Q}), is a measure of

[8] G. K. Wertheim, "Mössbauer Effect: Principles and Applications." Academic Press, New York (1964).

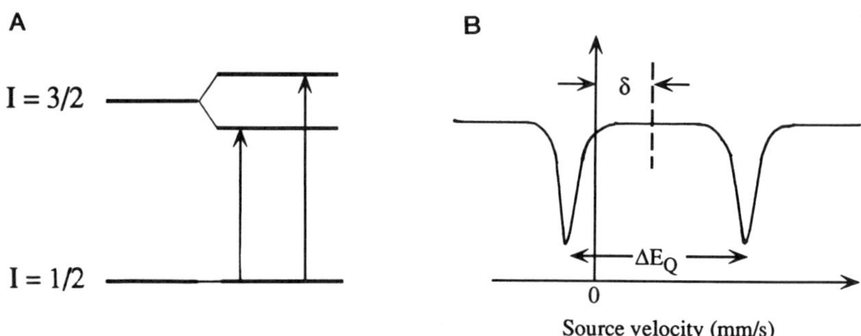

FIG. 1. Effect of the electric quadrupole interaction. The electric quadrupole interaction splits the ^{57}Fe nuclear excited state ($I = 3/2$) into two levels (A) and results in a quadrupole doublet (B).

the strength of the quadrupole interaction (and therefore, the EFG) and can be calculated by diagonalizing the matrix of $\hat{\mathscr{H}}_Q$.

$$\Delta E_Q = \frac{eQV_{zz}}{2}\sqrt{1 + (\eta^2/3)} \tag{3}$$

Both parameters ΔE_Q and δ depend not only on the $3d$ electronic population of the iron atom but also on the coordinated ligands. They are sensitive to the iron oxidation and spin states, as well as to the coordination number and types of ligand. Owing to the complexity and variety of contributions involved, quantitative theoretical approach to calculate ΔE_Q and δ are difficult and have met with little success. Nevertheless, qualitative understanding of the contributions has been achieved. These parameters are therefore generally used as empirical parameters for characterization of oxidation and spin states, degree of covalency, and coordination environment.

Table I lists the ranges of values observed for the four most commonly observed iron oxidation and spin states. The parameter δ follows a general

TABLE I
RANGES OF MÖSSBAUER PARAMETERS AT 4.2 K FOUND FOR FOUR COMMONLY OBSERVED IRON OXIDATION STATES

States	ΔE_Q (mm/sec)	δ (mm/sec)
Ferrous high-spin	2.0–4.3	0.7–1.3
Ferrous low-spin	0.3–1.4	0.2–0.5
Ferric high-spin	0.4–1.9	0.3–0.6
Ferric low-spin	1.5–2.8	0.1–0.4

trend that $\delta_{\mathrm{Fe(III)}} < \delta_{\mathrm{Fe(II)}}$ (i.e., δ decreases with increasing oxidation). For the same oxidation state, $\delta_{\text{low-spin}} < \delta_{\text{high-spin}}$. Within the same oxidation and spin state, $\delta_{\text{tetrahedral}} < \delta_{\text{octahedral}}$.[9] For example, in the case of high-spin Fe(III), δ varies from 0.26 mm/sec for tetrahedral sulfur[10] (or oxygen[11]) coordination to 0.55 mm/sec for a six-coordinated iron site in *Escherichia coli* ribonucleotide reductase.[12] The value of ΔE_Q is determined by the EFG, and in most cases the EFG is dominated by contributions from the valence $3d$ electrons. Consequently, ΔE_Q values for high-spin Fe(II) and low-spin Fe(III), in which the charge distributions of the $3d$ electrons are asymmetric, are generally large in comparison with those of low-spin Fe(II) and high-spin Fe(III), in which the $3d$ electronic distributions are nearly spherical.

Magnetic Hyperfine Interactions

Both the ground and excited states of the ^{57}Fe nucleus possess magnetic moments and can interact with a local magnetic field, $\mathbf{H}_{\text{eff}}$. This interaction can be written as

$$\hat{\mathscr{H}}_M = -g_n\beta_n\mathbf{I}\cdot\mathbf{H}_{\text{eff}} \tag{4}$$

Because the magnetic field $\mathbf{H}_{\text{eff}}$ is the vectorial sum of any external applied field $\mathbf{H}_{\text{app}}$ and the internal field $\mathbf{H}_{\text{int}}$, produced by unpaired $3d$ electrons, the magnetic hyperfine interaction can be rewritten as

$$\hat{\mathscr{H}}_M = -g_n\beta_n\mathbf{I}\cdot(\mathbf{H}_{\text{app}} + \mathbf{H}_{\text{int}}) = -g_n\beta_n\mathbf{I}\cdot\left(\mathbf{H}_{\text{app}} - \frac{\tilde{\mathbf{A}}}{g_n\beta_n}\cdot\mathbf{S}\right) \tag{5}$$

The tensor $\tilde{\mathbf{A}}$ is called the magnetic hyperfine coupling tensor. It represents the coupling strength between the nuclear spin I and the electronic spin S, and consists of three terms: the Fermi contact term, the dipole–dipole term, and the orbital interactions.[13] The Fermi contact term is isotropic and has a value of ~ -20 T. The dipole and orbital terms are anisotropic and depend strongly on the distribution of the unpaired $3d$ electrons. For high-spin Fe(III), the ground electronic configuration is the spherical ^6S (or $^6A_1$ for cubic symmetry). The anisotropic contributions are small and the $\tilde{\mathbf{A}}$ tensor approaches the isotropic Fermi contact term. For high-

[9] N. N. Greenwood and T. C. Gibb, "Mössbauer Spectroscopy." Champan & Hall, London, 1971.

[10] I. Moura, B. H. Huynh, R. P. Hausinger, J. LeGall, A. V. Xavier, and E. Münck, *J. Biol. Chem.* **255,** 2493 (1980).

[11] N. Ravi and R. Jagannathan, *Phys. Chem. Solids* **41,** 501 (1980).

[12] J. B. Lynch, C. Juarez-Garcia, E. Münck, and L. Que, Jr., *J. Biol. Chem.* **264,** 8091 (1989).

[13] G. Lang, *Q. Rev. Biophys.* **3,** 1 (1970).

spin Fe(II) or low-spin Fe(III), the anisotropic terms dominate and the components of the $\tilde{\mathbf{A}}$ tensor can vary from -50 T to 100 T, with the low-spin Fe(III) showing larger anisotropy.

The magnetic hyperfine interactions lift completely the degeneracy of the nuclear energy levels, resulting in six allowed Mössbauer transitions (Fig. 2). The corresponding Mössbauer spectrum consists of six absorption lines with intensities governed by magnetic dipole transition rules.[8] Such a spectrum is called a *magnetic spectrum*. In the presence of a small applied field (e.g., 0.05 T), the magnetic spectrum is determined by $\mathbf{H}_{\text{int}}$ because $\mathbf{A}$ is on the order of 10 to 10^2 T. Because $\mathbf{H}_{\text{int}} = -[\tilde{\mathbf{A}}/(g_n\beta_n)]\cdot\mathbf{S}$, which contains the electronic spin S, the magnetic spectrum is in turn strongly dependent on the electronic properties of the iron site.

Correlation with Electronic Properties

In the preceding section we introduced the hyperfine interactions and their accompanying characteristic parameters. Two types of Mössbauer spectra, namely, the quadrupole doublet and the magnetic spectrum, can result from these interactions. In this and the following sections, we discuss the effect of the electronic properties in relation to the type of observed spectra.

In most cases the iron centers in proteins have a well-isolated ground electronic multiplet, whose energy and magnetic properties can be described by the following spin Hamiltonian, $\hat{\mathscr{H}}_e$:

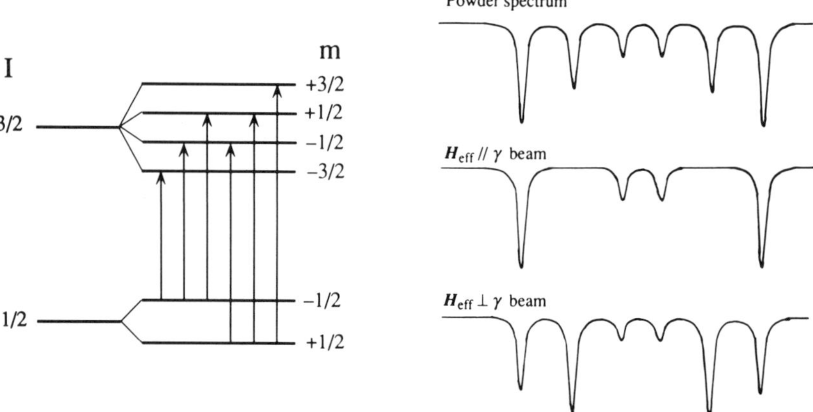

FIG. 2. Effect of the magnetic hyperfine interaction. The magnetic hyperfine interaction (left) splits the ^{57}Fe nuclear excited state ($I = 3/2$) into four levels and the ground state ($I = 1/2$) into two levels, resulting in six allowed transitions (indicated by arrows). Also shown (right) are the schematic drawings of resultant Mössbauer spectra.

$$\hat{\mathscr{H}}_e = D\left[\mathbf{S}_z^2 - \frac{S(S+1)}{3} + \frac{E}{D}(S_x^2 - \mathbf{S}_y^2)\right] + \beta\mathbf{H}_{app}\cdot\tilde{\mathbf{g}}_0\cdot\mathbf{S} \tag{6}$$

where S is the electronic spin of the system. The first term describes the energy separations of levels within the ground spin multiplet in the absence of an external field, called *zero-field splitting*. D and E/D are, respectively, the axial and rhombic zero-field splitting parameters. The second term represents the interaction of the electronic moment with the external field (Zeeman interaction). The tensor $\tilde{\mathbf{g}}_0$ takes into account both the spin and unquenched orbital moments. The complete Hamiltonian for analysis of Mössbauer spectra includes both the nuclear ($\hat{\mathscr{H}}_Q + \hat{\mathscr{H}}_M$) and the electronic spin Hamiltonian ($\hat{\mathscr{H}}_e$). Because the zero-field splitting and the electron Zeeman interaction (1–10 cm^{-1}) are much larger than the magnetic hyperfine interactions (10^{-4}–10^{-3} cm^{-1}), the electronic Hamiltonian is solved first and the resulting spin expectation value $\langle S\rangle$ is then substituted into the nuclear Hamiltonian for the calculation of the internal field,

$$\mathbf{H}_{int} = -\frac{\tilde{\mathbf{A}}}{g_n\beta_n}\cdot\langle\mathbf{S}\rangle \tag{7}$$

In principle, electronic systems can be separated into two distinct categories: those with even numbers of electrons and those with odd numbers of electrons. Each categorical system yields characteristics $\langle S\rangle$. In the following, we examine the expectation value $\langle S\rangle$ resulting from these two electronic systems and discuss its role in determining the Mössbauer spectral properties.

Systems with Even Numbers of Electrons

For systems with even numbers of electrons, the electronic spin S is an integer. When $E/D \neq 0$ (which is generally the case), the degeneracy of the spin multiplet is completely lifted and the expectation value $\langle S\rangle$ equals zero in the absence of an applied field. Consequently, $\mathbf{H}_{int}$ is also zero and the resulting Mössbauer spectrum is a quadrupole doublet. In the presence of an applied field, $\langle S\rangle$ increases as a function of the ratio $(\beta H_{app})/\Delta$, where Δ is the energy separation between the two lowest energy levels of the spin multiplet. $\langle S\rangle$ reaches its saturation value as $(\beta H_{app})/\Delta$ approaches unity.[14] For most of the integer-spin iron proteins studied so far, Δ is on the order of 1–10 cm^{-1} and quadrupole doublet is commonly observed even in the presence of a moderate applied field (<1 T;

[14] K. K. Surerus, M. P. Hendrich, P. D. Christie, D. Rottgardt, W. H. Orme-Johnson, and E. Münck, *J. Am. Chem. Soc.* **114,** 8579 (1992).

$\beta H_{app} = 0.47\ cm^{-1}$ for $H_{app} = 1$ T). Exceptions, however, have been found for proteins containing a [3Fe-4S] cluster, which, in its reduced form ($S = 2$), has a $\Delta \approx 0.2\ cm^{-1}$ and exhibits magnetic spectrum in the presence of a field of 0.05 T.[15] A Δ value of less than $10^{-3}\ cm^{-1}$ was suggested for the thionine-oxidized P clusters in the molybdenum-iron proteins of *Azotobacter vinelandii* and *Clostridium pasteurianum*.[14]

Systems with Odd Number of Electrons

For systems with odd number of electrons, the electronic spin S is a half-integer. According to the Kramers theorem, electronic states of half-integer spin system are at least doubly degenerate in the absence of a magnetic field.[16] Consequently, the zero-field splitting cannot lift the degeneracy completely, and the ground spin multiplet consists of doubly degenerate states called *Kramers doublets* (i.e., $\Delta = 0$). Assuming that $\tilde{\mathbf{g}}_0$ is isotropic and has the free electron value of 2.0, the expectation value $\langle S \rangle_k$ for the kth Kramers doublet can be written as

$$\langle S_i \rangle_k = g_i/4 \qquad (i = x, y, z) \tag{8}$$

where g_x, g_y, and g_z are the principal components of the apparent $\tilde{\mathbf{g}}$ tensor for the kth doublet and can be measured by electron paramagnetic resonance (EPR). They can also be derived from Eq. (6), and theoretical g values for Kramers doublets of $S = 3/2$ and $5/2$ systems as a function of E/D can be found elsewhere.[4] Because of the degenerate nature of their electronic states, half-integer spin systems always produce a nonzero $\langle S \rangle$ and therefore significant $\mathbf{H}_{int}$ at the nucleus and a magnetic Mössbauer spectrum is always observed even in the absence of a magnetic field. According to Eqs. (7) and (8) $\mathbf{H}_{int}$ is directly proportional to the apparent g values. The magnetic Mössbauer spectral properties are therefore closely related to the g values measured by EPR.

For Kramers doublets with isotropic $\tilde{\mathbf{g}}$ tensor (e.g., $g_x = g_y = g_z = 4.3$ for the middle doublet of an $S = 5/2$ system with $E/D = 1/3$), a small applied field is capable of polarizing the electronic spin along the direction of the applied field. Even though the molecules in a frozen protein sample are randomly oriented, their internal fields, which have the direction of $\mathbf{S}$, are aligned parallel to the applied field (isotropic $\tilde{\mathbf{A}}$ is also assumed in this argument). Consequently, applying a small magnetic field parallel or perpendicular to the observed γ radiation would align the internal field accordingly, regardless of the molecular orientation. Because the transi-

[15] B. H. Huynh, J. J. G. Moura, I. Moura, T. A. Kent, J. LeGall, A. V. Xavier, and E. Münck, *J. Biol. Chem.* **255,** 2493 (1980).

[16] K. Kramers, *Akad. van Wetenschappen, Amsterdam* **33,** 959 (1930).

tion probabilities for the six allowed Mössbauer transitions (indicated in Fig. 2) depend on the angle between the directions of the γ radiation and of the local magnetic field at the nucleus,[8] the relative intensities for the six absorption lines vary with the direction of the applied field (parallel, 6 : 0 : 2 : 2 : 0 : 6; perpendicular, 3 : 4 : 1 : 1 : 4 : 3). For Kramers doublets with comparable principal g components (e.g., $g_x = g_y = 6$ and $g_z = 2$ for the ground doublet of an $S = 5/2$ system with $E/D = 0$ and $D > 0$), a certain degree of spin polarization can still be achieved by the application of a small magnetic field. Field direction-dependent Mössbauer spectra can still be observed, although the intensity pattern of the spectrum would not reach the ratios given for the isotropic situation.[6] For Kramers doublets with extreme anisotropy (e.g., $g_x = g_y = 0$ and $g_z = 10$ for the highest excited state of an $S = 5/2$ system with $E/D = 0$ and $D > 0$), the magnetic moment is aligned preferentially along a particular molecular axis. A small applied field does not have the strength to polarize the magnetic moment and has an insignificant effect on the direction of the internal field. Consequently, a powder spectrum with relative intensities of 3 : 2 : 1 : 1 : 2 : 3 for the six absorption lines is observed independent of the direction of the applied field.

Systems with Fast Electronic Relaxation

In the above discussion, the electronic states are assumed to be stationary with respect to the nuclear precession time ($\sim 10^{-7}$ sec). In the other extreme case, in which the electronic relaxation is much faster than the nuclear precession, the expectation value $\langle S \rangle$ in Eq. (7) is replaced by the thermal average of the spin expectation values for all the thermally populated states. For biological molecules, this situation generally occurs at temperatures above 77 K. At these temperatures, the thermal average of $\langle S \rangle$ approaches zero even for paramagnetic systems and the resulting Mössbauer spectrum consists only of quadrupole doublets. It is therefore a common practice to perform measurements at high temperatures in order to determine the parameters ΔE_Q and δ for iron atoms associated with half-integer spin systems.

Multiiron Containing Proteins

It is important to realize that in biological applications of Mössbauer spectroscopy, only one transition is involved, namely, the transition between the ground and first excited state of the ^{57}Fe nucleus. The transition probability is therefore a constant. Even though hyperfine interactions may split the nuclear energy levels, resulting in multiple allowed transitions, the total transition probability of the allowed transitions remains invariant, independent of the electronic environment. Another factor that determines the observed Mössbauer absorption is the recoilless fraction, which is a function of the temperature and thermal properties of the matrix containing the iron nucleus.[8] For ^{57}Fe nuclei within the same frozen sample, a similar recoilless fraction is expected because the temperature and matrix are the same. Consequently,the relative Mössbauer absorption for each distinct iron site within the same sample is expected to be directly proportional to the relative concentration of the iron site. This expectation appears to be true for low-temperature measurements (below 77 K) and is extremely important for the study of proteins containing multiple iron centers and/or multinuclear iron clusters.

As discussed in the preceding sections, Mössbauer spectra are strongly correlated with the electronic properties. An iron center (containing either a single iron or a cluster of iron atoms) with an odd number of electrons generally exhibits a Mössbauer spectrum magnetically split by the internal field generated by the unpaired electrons. An iron center with even number of electrons, on the other hand, exhibits quadrupole doublets. Consequently, for proteins containing multiple iron centers, it is possible to distinguish and identify Mössbauer spectral components originating from metal centers with even or odd numbers of electrons. Because Mössbauer absorption is diretly proportional to the iron concentration, the number of iron atoms associated with each center can be deduced by correlating the relative absorption intensities of the Mössbauer spectral components with the iron content determined for the protein. Moreover, characteristic hyperfine parameters can be obtained from each Mössbauer spectral component. The intrinsic oxidation and spin state of each distinct iron site can also be determined. Owing to these unique properties, Mössbauer spectroscopy has been used to reveal the organization of the iron atoms in the molybdenum-iron protein of nitrogenases,[17] establish the presence of an exchange-coupled siroheme–[4Fe-4S] center in sulfite reductases,[18] and discover the [3Fe-4S] cluster in ferredoxins.[15] More recently, Mössbauer spectroscopy in conjunction with a rapid freeze-quench kinetic technique has been used successfully in trapping and structurally characterizing a kinetically competent iron intermediate during the assembly of the tyrosyl radical–diiron cluster cofactor in the R2 subunit of *E. coli* ribonucleotide reductase.[19]

[17] B. H. Huynh, E.Münck, and W. H. Orme-Johnson, *Biochim. Biophys. Acta* **623,** 124 (1980).

[18] J. A. Christner, E. Münck, P. A. Janick, and L. M. Siegel, *J. Biol. Chem.* **258,** 11147 (1983).

[19] J. M. Bollinger, Jr., J. Stubbe, B. H. Huynh, and D. E. Edmondson, *J. Am. Chem. Soc.* **113,** 6289 (1991).

Case Studies

Cytochrome cd_1 from Thiobacillus denitrificans

Dissimilatory nitrite reductases of the cytochrome cd_1 type have been found in many denitrifying bacteria including *T. denitrificans,*[20] *Paracoccus denitrificans,*[21] *Pseudomonas aeruginosa,*[22] *Pseudomonas perfectomarina,*[23] and *Alcaligenes faecalis.*[24] They are a group of enzymes that share common structural features such as a molecular mass of about 120 kDa and two identical (or equal-sized) subunits. Each subunit contains one *c* heme and one d_1 heme. Heme d_1 has a dioxoisobacteriochlorin structure,[25] unique for hemeproteins. Detailed Mössbauer and EPR studies have been performed on the reductase isolated from *T. denitrificans*[26,27] and the results are reviewed here.

As isolated the *T. denitrificans* cytochrome cd_1 exhibits an intense axial EPR signal at $g_z = 2.50$, $g_y = 2.43$ and $g_x = 1.70$, and a weak resonance at $g = 3.60$.[26] The intense signal is similar to those observed for diimidazole low-spin ferric chlorin complexes[28] and was attributed to heme d_1. An almost identical signal has also been observed for the *P. aeruginosa* cytochrome cd_1, indicating a similar iron environment for heme d_1 in both proteins. Cytochromes *c* are generally identified by their histidine/methionine axial ligand combination and display maximum *g* values in the range of 3.0–3.2.[29] The large $g = 3.60$ value observed for the *T. denitrificans* cytochrome cd_1 is unique for *c*-type cytochromes and resembles more closely those of diimidazole heme complexes with

[20] J. LeGall, W. J. Payne, T. V. Morgan, and D. V. DerVartanian, *Biochem. Biophys. Res. Commun.* **87,** 355 (1979).

[21] N. Newton, *Biochim. Biophys. Acta* **185,** 316 (1969).

[22] T. Horio, T. Higashi, T. Yamanaka, H. Matsubara, and K. Okunuki, *J. Biol. Chem.* **236,** 944 (1961).

[23] W. G. Zumft, B. F. Sherr, and W. J. Payne, *Biochem. Biophys. Res. Commun.* **88,** 1230 (1979).

[24] H. Iwasaki and T. Matsubara, *J. Biochem.* (*Tokyo*) **69,** 847 (1971).

[25] E. Weeg-Aerssens, W. Wu, R. W. Ye, J. M. Tiedje, and C. K. Chang, *J. Biol. Chem.* **266,** 7496 (1991).

[26] B. H. Huynh, M. C. Liu, J. J. G. Moura, I. Moura, P. O. Ljungdahl, E. Münck, W. J. Payne, H. D. Peck, Jr., D. V. DerVartanian, and J. LeGall, *J. Biol. Chem.* **257,** 9576 (1982).

[27] M.-C. Liu, B. H. Huynh, W. J. Payne, H. D. Peck, Jr., D. V. DerVartanian, and J. Le Gall, *Eur. J. Biochem.* **169,** 253 (1987).

[28] A. M. Stolzenbers, S. H. Strauss, and R. H. Holm, *J. Am. Chem. Soc.* **103,** 4763 (1981).

[29] D. L. Brautigan, B. A. Feinberg, B. M. Hoffman, E/ Margoliash, J. Peisach, and W. E. Blumberg, *J. Biol. Chem.* **252,** 574 (1977).

perpendicular ligand plane orientation.[30] For the *P. aeruginosa* cytochrome cd_1, imidazole was found to bind to both c and d_1 hemes at pH 7.0, shifting the heme c EPR signal from 3.02 to 3.55.[31] These observations strongly suggest a perpendicular dihistidine structure for the heme c in *T. denitrificans* cytochrome cd_1 .

Mössbauer spectra of the as-isolated cytochrome cd_1 recorded at high temperatures (e.g., 200 K) show two equal-intensity quadrupole doublets with parameters characteristic of low-spin ferric hemes. Doublet I with parameters $\Delta E_Q = 2.22 \pm 0.02$ mm/sec and $\delta = 0.22 \pm 0.02$ mm/sec, typical of c-type cytochromes, was attributed to the c heme and doublet II ($\Delta E_Q = 1.66 \pm 0.02$ mm/sec and $\delta = 0.20 \pm 0.02$ mm/sec) was attributed to the d_1 heme.[26] At low temperatures two resolved magnetic spectral components were observed with equal absorption areas (Fig. 3). The c heme with a large magnetic moment ($g = 3.6$), and therefore large $\mathbf{H}_{int}$, exhibits a magnetic spectrum with a total splitting of 15 mm/sec. The d_1 heme with a smaller magnetic moment displays a splitting of approximately 6 mm/sec. Analysis of the Mössbauer spectra using the spin Hamiltonians given in Eqs. (2), (5), and (6) yielded characteristic hyperfine parameters for both low-spin ferric hemes (Table II).[26] Extreme anisotropy for the $\tilde{\mathbf{A}}$ tensors was observed, as expected for low-spin ferric heme due to the substantial unquenched orbital angular momentum of the ground Kramers doublet.

Thiobacillus denitrificans cytochrome cd_1 can be reduced by sodium ascorbate plus phenazine methosulfate. In the reduced enzyme, both hemes are in the ferrous form (i.e., integer spin system). Two quadrupole doublets with equal absorption were observed at all temperatures in the absence of a strong applied field.[26] At 4.2 K, one quadrupole doublet exhibits $\Delta E_Q = 1.14 \pm 0.04$ mm/sec and $\delta = 0.45 \pm 0.04$ mm/sec, which are commonly observed for low-spin ferrous c-type cytochromes ($S = 0$), and was attributed to the c heme. The other doublet attributed to the d_1 heme shows $\Delta E_Q = 2.24 \pm 0.04$ mm/sec and $\delta = 0.97 \pm 0.04$ mm/sec, characteristic of high-spin ferrous heme complexes ($S = 2$). Because high-spin ferrous hemes are generally five coordinated, this observation indicates that the activation of the d heme involves transfer of one electron to the d heme with displacement of one axial ligand. The d heme would then be ready for substrate binding and reduction.

When the fully reduced *T. denitrificans* cytochrome cd_1 is reacted

[30] F. A. Walker, B. H. Huynh, W. R. Scheidt, and S. R. Osvath, *J. Am. Chem. Soc.* **108,** 5288 (1986).

[31] T. A. Walsh, M. K. Johnson, A. J. Thomson, D. Barber, and Greenwood, *J. Inorg. Biochem.* **14,** 1 (1981).

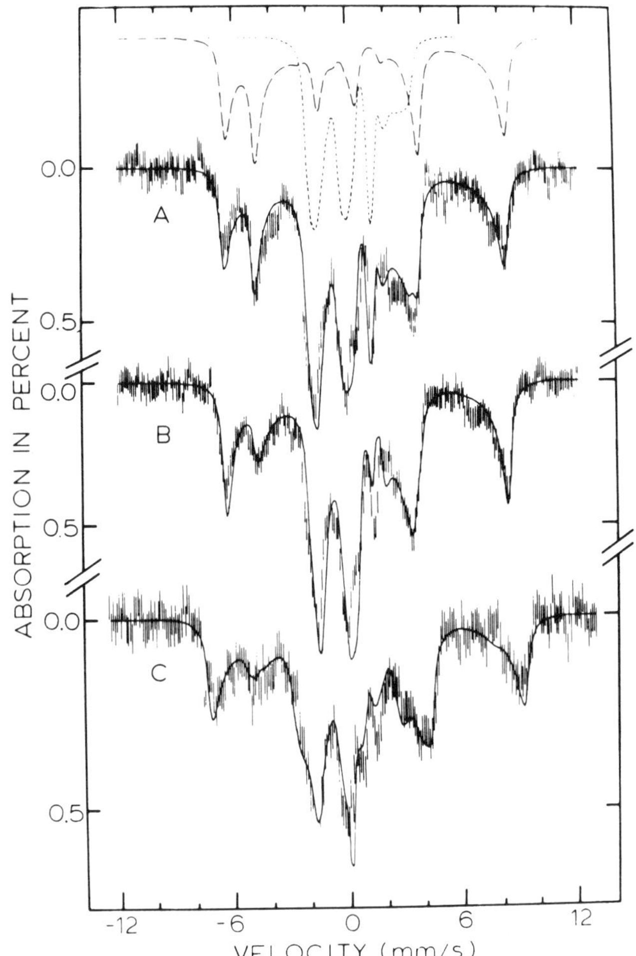

FIG. 3. Mössbauer spectra of ferricytochrome cd_1. The data were recorded at 4.2 K with a magnetic field of 60 mT applied perpendicular (A) and parallel (B) to the γ beam, and with a parallel applied field of 6 T (C). The solid lines are theoretical simulations using the parameters listed in Table II, and assuming equal absorption for each heme. In (A), the theoretical spectra corresponding to heme c (dashed line) and heme d_1 (dotted line) are also shown. [Taken from B. H. Huynh, M. C. Liu, J. J. G. Moura, I. Moura, P. O. Ljungdahl, E. Münck, W. J. Payne, H. D. Peck, Jr., D. V. DerVartanian, and J. LeGall, *J. Biol. Chem.* **257,** 9576 (1982).]

TABLE II
HYPERFINE PARAMETERS FOR FERRICYTOCHROME cd_1 AND NO-BOUND d_1 HEME[a]

Parameter	Ferric heme d_1	Ferric heme c	NO-bound heme d_1
g_x	1.70	—	1.95
g_y	2.43	—	2.00
g_z	2.50	3.60	2.05
$A_{xx}/g_n\beta_n$ (T)	−12.5	−30.0	−15.0
$A_{yy}/g_n\beta_n$ (T)	−12.0	24.0	−15.0
$A_{zz}/g_n\beta_n$ (T)	32.0	91.0	12.0
ΔE_Q (mm/sec)	1.70	2.22	0.8
η	−2.0	−1.8	0
δ (mm/sec)	0.26	0.28	0.34

[a] Data taken from B. H. Huynh, M. C. Liu, J. J. G. Moura, I. Moura, P. O. Ojungdahl, E. Münck, W. J. Payne, H. D. Peck, Jr., D. V. DerVartanian, and J. LeGall, *J. Biol. Chem.* **257,** 9576 (1982) and M.-C. Liu, B. H. Huynh, W. J. Payne, H. D. Peck, Jr., D. V. DerVartanian, and J. LeGall, *Eur. J. Biochem.* **169,** 253 (1987).

anaerobically with nitrite at pH 7.6, characteristic ferrous heme–NO EPR signal with hyperfine splitting caused by the $I = 1$ ^{14}N nucleus was observed at the $g = 2$ region.[27] The corresponding Mössbauer spectra recorded under different experimental conditions revealed a diamagnetic and a paramagnetic spectral component, again, of equal absorption area (Fig. 4). Because the diamagnetic component was found to be identical to that of the low-spin ferrous c heme, the paramagnetic species was assigned to the d_1 heme. Correlating these Mössbauer results with the EPR finding that the paramagnetic species is an NO-bound ferrous heme, it was concluded that NO was bound mainly to the d_1 heme.[27] Detailed analysis of the Mössbauer spectra yielded an axial $\tilde{\mathbf{A}}$ tensor for the NO-bound d_1 heme (Table II), consistent with the unpaired electron residing on the iron $3z^2 - r^2$ orbital. Examination of the hyperfine parameters suggested that in comparison with low-spin ferrous heme, the effect of NO binding is to increase the iron $3z^2 - r^2$ electronic population through σ-electron donation from the NO ligand and to reduce the iron xz and yz population through the π-electron back-donation from the iron to NO.[27]

Desulfoviridins

Desulfoviridins are a class of dissimilatory sulfite reductases found in many sulfate-reducing bacteria including *Desulfovibrio gigas, Desulfovibrio salexigens,* and *D. vulgaris* (Hildenborough).[32–34] They catalyze the

[32] J. P. Lee and H. D. Peck, Jr., *Biochem. Biophys. Res. Commun.* **45,** 583 (1971).
[33] J. P. Lee, J. LeGall, and H. D. Peck, Jr., *J. Bacteriol.* **115,** 529 (1973).
[34] I. Moura, J. LeGall, A. R. Lino, H. D. Peck, Jr., G. Fauque, A. V. Xavier, D. V. DerVartanian, J. J. G. Moura, and B. H. Huynh, *J. Am. Chem. Soc.* **110,** 1075 (1988).

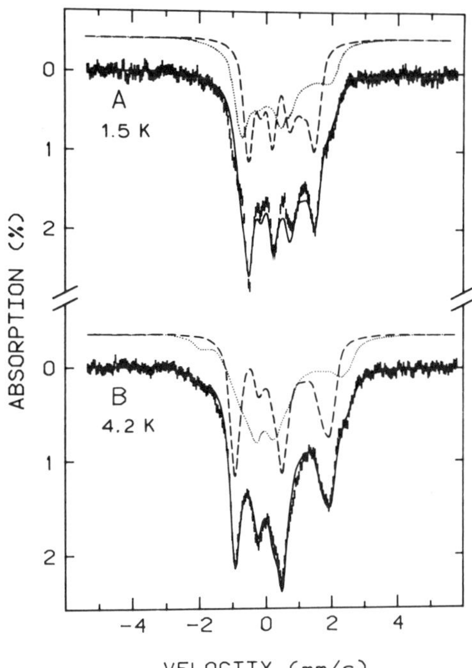

FIG. 4. Mössbauer spectra of nitrite-reacted ferrocytochrome cd_1. The data were recorded at temperatures as indicated with a magnetic field of 3 T (A) or 6 T (B) applied parallel to the γ beam. Theoretical simulations for the c heme (dashed lines) and the d_1 heme (dotted lines) are shown and their sum are plotted as solid lines over the experimental data.

reduction of sulfite to sulfide and are involved in the pathway of respiratory sulfate reduction. They are large oligomers with molecular mass on the order of 200 kDa and are composed of three different subunits organized in an $\alpha_2\beta_2\gamma_2$ configuration with subunit molecular masses of 50 kDa (α), 40 kDa (β), and 11 kDa (γ).[35] Each molecule was reported to contain 16–18 iron atoms and comparable amounts of labile sulfide. Their optical spectra show typical siroheme bands in the regions around 400 nm and 540–580 nm and an additional absorption band at 628 nm.[36] On treatment with acetone-HCl, approximately two metal-free sirohydrochlorins can be extracted per molecule of desulfoviridin.[36] Under the same conditions, however, sirohemes are extracted from other sulfite reductases

[35] A. J. Pierik, M. G. Duyvis, J. M. L. M. van Helvoort, R. B. G. Wolbert, and W. R. Hagen, *Eur. J. Biochem.* **205,** 111 (1992).

[36] M. J. Murphy, L. M. Siegel, H. Kamin, D. V. DerVartanian, J. P. Lee, J. LeGall, and H. D. Peck, Jr., *Biochem. Biophys. Res. Commun.* **54,** 82 (1973).

including desulforubidin, desulfofuscidin, P-582, and *E. coli* sulfite reductase.[36,37] Mössbauer spectroscopy has been applied to study the iron centers in desulfoviridin from *D. gigas*[34] and the results are briefly reviewed here.

The Mössbauer spectra of the as-isolated *D. gigas* desulfoviridin shows three distinguishable spectral components: (1) a diamagnetic ($S = 0$) broad quadrupole doublet with parameters that are typical of $[4Fe\text{-}4S]^{2+}$ clusters, (2) a magnetic component that resembles that of an exchange-coupled ferric high-spin siroheme–$[4Fe\text{-}4S]^{2+}$ cluster found in *E. coli* sulfite reductase,[18] and (3) a magnetic spectrum indicative of a mononuclear Fe(III). The relative absorptions of these three components are approximately 80, 10, and 10%, respectively. The most unexpected result from the Mössbauer study was the finding of less than stoichiometric amounts of the exchange-coupled siroheme–[4Fe-4S] cluster in desulfoviridin. Taking the metal determination of 18 iron atoms per molecule into consideration, 10% absorption yields 0.4 siroheme–[4Fe-4S] clusters per reductase. This finding was substantiated by the Mössbauer data of the hydrogenase-reduced desulfoviridin, which shows that only 2% of the total iron absorption is arising from the reduced Fe(II) siroheme (approximately 0.4 siroheme per molecule). Because two sirohydrochlorins were found for each molecule of desulfoviridin, it was concluded that 80% of the desulfoviridin molecules contain metal-free tetrahydroporphyrins.[34]

In its oxidized form, desulfoviridin exhibits EPR signal typical of high-spin ferric heme. In support of the Mössbauer findings, EPR spin quantitation yielded a substoichiometric amount for the heme signal, 0.2–0.4 spin/molecule.[34,38] A resonance Raman investigation also confirmed the presence of metal-free sirohydrochlorins in desulfoviridin.[39] It is unclear whether the existence of metal-free sirohydrochlorin in desulfoviridin is a natural phenomenon or an artifact resulting from damage caused by protein purification. It is, however, interesting to note that neither CO nor CN^- inhibits desulfoviridin activity. Because these molecules generally coordinate to the heme iron and do inhibit the sulfite reduction activity of desulforubidin and *E. coli* sulfite reductase, the enzymatic activity of desulfoviridin may originate from enzymes with metal-free sirohydrochlorin.

More recently, a detailed EPR study on desulfoviridin from *D. vulgaris* (Hildenborough) has been reported by Pierik and Hagen.[38] In addition to

[37] M. J. Murphy and L. M. Siegel, *J. Biol. Chem.* **248,** 6911 (1973).

[38] A. J. Pierik and W. R. Hagen, *Eur. J. Biochem.* **195,** 505 (1991).

[39] K. K. Lai, I. Moura, M. Y. Liu, J. LeGall, and K. T. Yue, *Biochim. Biophys. Acta* **1060,** 25 (1991).

the previously reported high-spin ferric heme EPR signal, signals at g = 17, 15.1, 11.7, 9.4, 9.0, and 4 were detected using intense microwave power (80–200 mW). These signals were interpreted as arising from an $S = 9/2$ system. Even though EPR quantitation yielded only 0.6 spin/molecule, a working hypothesis of two $S = 9/2$ clusters per molecule was proposed for the *D. vulgaris* desulfoviridin. On the basis of such a tentative working hypothesis it has been suggested that the well-established exchange-coupled siroheme–[4Fe-4S] cluster is in question and that the corresponding magnetic Mössbauer spectral component is actually representing two separate magnetic systems: the high-spin ferric siroheme and a novel $S = 9/2$ iron cluster. Although these EPR observations and speculations are interesting and thought provoking, definitive evidence has not been provided. The observation of 80% iron absorption attributable to diamagnetic species in the *D. gigas* desulfoviridin is directly at odds with this working hypothesis, which predicts only 40% of the absorption should be associated with diamagnetic species.

Periplasmic Iron Hydrogenase from Desulfovibrio vulgaris

On the basis of their metal contents, hydrogenases can be grouped into three categories: the [Ni-Fe] hydrogenases, the [Ni-Fe-Se] hydrogenases, and the [Fe] hydrogenases.[40] All these three types of hydrogenases can be found in *D. vulgaris* (Hildenborough) at different locations in the cell.[41–43] The soluble periplasmic hydrogenase belongs to the [Fe] hydrogenase category. The nucleotide sequence of the structural gene indicates that it is composed of two subunits with molecular masses of 45.8 and 13.5 kDa.[44] Amino acid sequencing of the purified hydrogenase, however, indicates that the small subunit lacks a hydrophobic NH_2-terminal peptide having general characteristics of a signal peptide.The molecular mass of the small subunit excluding the signal peptide was estimated to be 10 kDa.[45] Plasma emission studies indicated that the *D. vulgaris* periplasmic hydrogenase contains only iron and no other metals. Owing

[40] G. Fauque, H. D. Peck, Jr., J. J. G. Moura, B. H. Huynh, Y. Berlier, D. V. DerVartanian, M. Teixeira, A. E. Przybyla, P. A. Lespinat, I. Moura, and J. LeGall, *FEMS Microbiol. Rev.* **54,** 299 (1988).

[41] H. M. Van der Westen, S. G. Mayhew, and C. Veeger, *FEBS Lett.* **86,** 122 (1978).

[42] B. H. Huynh, M. H. Czechowski, H.-J. Krüger, D. V. DerVartanian, H. D. Peck, Jr., and J. LeGall, *Proc. Natl. Acad. Sci. U.S.A.* **81,** 3728 (1984).

[43] M. Rohde, U. Fürstenau, F. Mayer, A. E. Przybyla, H. D. Peck, Jr., J. LeGall, E. S. Choi, and N. K. Menon, *Eur. J. Biochem.* **191,** 389 (1990).

[44] G. Voordouw and S. Brenner, *Eur. J. Biochem.* **148,** 515 (1985).

[45] B. C. Prickril, M. H. Czechowski, A. E. Przybyla, H. D. Peck, Jr., and J. LeGall, *J. Bacteriol.* **167,** 722 (1986).

to the difficulties involved in protein determination, the reported iron content varies from 10 to 16 mol of iron per mole of protein.[42,46] A similar range of variation was also reported for labile sulfur content. Sequence analysis revealed that the NH_2-terminal part of the large subunit contains a region that is homologous to the metal-binding region of bacterial eight-iron ferredoxin, suggesting that eight of the iron atoms in *D. vulgaris* hydrogenase may be organized into two ferredoxin-type [4Fe-4S] clusters.[44] The remaining iron atoms were suggested also to form an iron cluster that was believed to be the hydrogen activation site, termed the H cluster.

Reports on Mössbauer investigation of the *D. vulgaris* [Fe] hydrogenase are limited. Presently, most of our knowledge concerning the electronic properties of the iron clusters in the *D. vulgaris* enzyme is from EPR studies. Six characteristic EPR signals have been reported.[47–50] (1) the isotropic 2.02 signal, (2) the complex signal observed for the dithionite-reduced enzyme, (3) the rhombic 2.06 signal (g = 2.06, 1.96, and 1.89), (4) the rhombic 2.10 signal (g = 2.10, 2.04, and 2.00), (5) the axial 2.06 signal (g = 2.06, 2.00, and 2.00), and (6) the high-spin signal (g = 5). The isotropic 2.02 signal was observed in the as-isolated enzyme and quantified to only 0.02–0.2 spin/molecule. The signal is similar to that of a $[3Fe\text{-}4S]^{1+}$ cluster. The presence of a substoichiometric amount of [3Fe-4S] clusters was believed to be an artifact generated during protein purification. The complex signal resembles that of the dipole–dipole interacted $[4Fe\text{-}4S]^{1+}$ clusters in eight-iron ferredoxins and was used as evidence supporting the presence of two [4Fe-4S] clusters in the *D. vulgaris* [Fe] hydrogenase. Other signals were attributed to the H cluster. The rhombic 2.06 signal was detected during reductive titration of the enzyme. It maximizes at about −110 mV with a spin quantitation of 0.7 spin/molecule. The rhombic 2.10 signal was believed to represent the active form of the H cluster. It was detected during reductive titration below −160 mV. Its appearance is in concert with the concomitant disappearance of the rhombic 2.06 signal. On anaerobic reoxidation, only the rhom-

[46] W. R. Hagen, A. van Berkel-Arts, K. M. Krüse-Wolters, G. Voordouw, and C. Veeger, *FEBS Lett.* **203,** 59 (1986).

[47] H. J. Grande, W. R. Dunham, B. Averill, C. van Dijk, and R. H. Sands, *Eur. J. Biochem.* **136,** 201 (1983).

[48] W. R. Hagen, A. van Berkel-Arts, K. M. Krüse-Wolters, W. R. Dunham, and C. Veeger, *FEBS Lett.* **201,** 158 (1986).

[49] D. S. Patil, J. J. G. Moura, S. H. He, M. Teixeira, B. C. Prickril, D. V. DerVartanian, H. D. Peck, Jr., J. LeGall, and B. H. Huynh, *J. Biol. Chem.* **263,** 18732 (1988).

[50] A. J. Pierik, W. R. Hagen, J. S. Redeker, R. B. G. Wolbert, M. Boersma, M. F. J. M. Verhagen, H. J. Grande, C. Veeger, P. H. A. Mutsaers, R. S. Sands, and W. R. Dunham, *Eur. J. Biochem.* **209,** 63 (1992).

bic 2.10 signal was observed.[50] Because aerobically isolated *D. vulgaris* hydrogenase is inactive and requires a reduction process for activation, these EPR observations indicate that the rhombic 2.06 and 2.10 signals may represent, respectively, the inactive and active forms of the H cluster. Exposure of the reduced enzyme to CO generates the axial 2.06 signal in stoichiometric amounts, suggesting that the signal represents CO-bound H cluster.[51] Irradiation of the CO-reacted enzyme with white light at low temperature (below 20 K) transforms the axial 2.06 signal to the rhombic 2.10 signal. Warming the photoirradiated sample to 150 K for 10 min converts the rhombic 2.10 signal back to the axial 2.06 signal.[52] These observations are consistent with the axial 2.06 and rhombic 2.10 signals representing, respectively, the CO-bound and the unligated H cluster. The effect of irradiation is to flash off the bound CO. Both the rhombic 2.10 and axial 2.06 signals have also been observed for the bidirectional [Fe] hydrogenase (hydrogenase I) from *C. pasteurianum* and similar photoinduced interconversion between the two signals have also been observed.[53] The rhombic 2.06 signal was not detected for the *C. pasteurianum* hydrogenase. This is consistent with the suggestion that the rhombic 2.06 signal represents inactive H cluster; the *C. pasteurianum* enzyme is oxygen sensitive and must be purified under strict anaerobic conditions, the as-purified enzyme is active and contains no inactive form of the H cluster.

The 4.2 K Mössbauer spectrum of the as-purified *D. vulgaris* [Fe] hydrogenase recorded with a weak applied field of 50 mT shows a broad quadrupole doublet.[42] The observed parameters ($\Delta E_Q = 1.25$ mm/sec and $\delta = 0.42$ mm/sec) and the general shape of the spectrum resemble those of $[4Fe\text{-}4S]^{2+}$ clusters. Spectra recorded with strong applied fields reveal that the iron centers are diamagnetic. In other words, the Mössbauer data are consistent with the view that the majority of the iron atoms in *D. vulgaris* hydrogenase are organized into two ferredoxin-type [4Fe-4S] clusters and that the H cluster is diamagnetic in the as-purified enzyme. Mössbauer investigation of the H cluster in the rhombic 2.06 inactive form and in the axial 2.06 CO-bound form has been reported.[53a] The paramagnetic spectral components corresponding to the two forms of the H cluster were analyzed using an $S = 1/2$ spin Hamiltonian [Eqs. (2),

[51] D. S. Patil, M. H. Czechowski, B. H. Huynh, J. LeGall, H. D. Peck, Jr., D. V. DerVartanian, *Biochem. Biophys. Res. Commun.* **137,** 1086 (1986).

[52] D. S. Patil, B. H. Huynh, S. H. He, H. D. Peck, Jr., D. V. DerVartanian, and J. LeGall, *J. Am. Chem. Soc.* **110,** 8533 (1988).

[53] A. T. Kowal, M. W. W. Adams, and M. K. Johnson, *J. Biol. Chem.* **264,** 4342 (1989).

[53a] Third International Conference on Molecular Biology of Hydrogenases, Troia, Portugal, July 29–August 1, 1991; N. Ravi, D. S. Patil, M. Y. Liu, J. LeGall, J. J. G. Moura, H. D. Peck, Jr., D. V. DerVartanian, and B. H. Huynh, unpublished results (1991).

TABLE III
MÖSSBAUER PARAMETERS OBTAINED FOR 2.06 RHOMBIC AND 2.06 AXIAL SPECIES IN *DESULFOVIBRIO VULGARIS* IRON HYDROGENASE[a]

Species	$A_{xx}/g_n\beta_n$ (T)	$A_{yy}/g_n\beta_n$ (T)	$A_{zz}/g_n\beta_n$ (T)	δ (mm/sec)	ΔE_Q (mm/sec)	η
2.06 rhombic						
Site 1	−26	−26	−17	0.52	1.3	0.7
Site 2	19	15	6	0.60	2.3	1.0
2.06 axial						
Site 1	−24	−24	−15	0.50	1.3	0.3
Site 2	20	20	10	0.58	1.9	0.0
$[4Fe\text{-}4S]^{1+}$						
Site 1	−23.2	−23.8	−20.4	0.50	1.32	0.78
Site 2	19.3	9.8	6.3	0.58	1.89	0.32

[a] The results for the [4Fe-4S] cluster are from P. Middleton, D. P. E. Dickson, C. E. Johnson, and J. D. Rush, *Eur. J. Biochem.* **88,** 135 (1978).

(5), and (6)]. Each form was found to contain two iron sites, with each site corresponding to a pair of equivalent iron atoms. The obtained hyperfine parameters are listed in Table III and are compared with those of reduced ferredoxin-type $[4Fe\text{-}4S]^{1+}$ clusters.[54] The similarities are striking, suggesting that the inactive H cluster may have a structure similar to that of a [4Fe-4S] cluster. However, on the basis of iron and sulfide content determination (14–16 Fe and 12–14 S^{2-} per molecule) and accounting for the two ferredoxin-type [4Fe-4S] clusters, a structure composed of 6 Fe and 6 S^{2-} was proposed for the H cluster.[48] A similar argument and conclusion have also been made for the H cluster in the *C. pasteurianum* hydrogenase, which was determined to contain 22 iron atoms per molecule; 16 of the iron atoms were proposed to organize into 4 [4Fe-4S] clusters and the remaining 6 atoms were suggested to form the H cluster.[55] A resonance Raman study on three different [Fe] hydrogenases,[56] however, showed evidence for the presence of ferredoxin-type [2Fe-2S] clusters in addition to the [4Fe-4S] clusters in both the *C. pasteurianum* and *Thermotoga maritima* hydrogenases. Consequently, a reevaluation of the iron composition for the H cluster is required for the *C. pasteurianum* hydrogenase. Considering the uncertainties involved in iron determination

[54] P. Middleton, D. P. E. Dickson, C. E. Johnson, and J. D. Rush, *Eur. J. Biochem.* **88,** 135 (1978).

[55] M. W. W. Adams, E. Eccleston, and J. B. Howard, *Proc. Natl. Acad. Sci. U.S.A.* **86,** 4932 (1989).

[56] W. Fu, P. M. Drozdzewski, T. V. Morgan, L. E. Mortenson, A. Juszczak, M. W. W. Adams, S.-H. He, H. D. Peck, Jr., D. V. DerVartanian, J. LeGall, and M. K. Johnson, *Biochemistry* **32,** 4813 (1993).

and the complexity of the iron centers in hydrogenases, iron composition of the H cluster derived solely from metal determination should be viewed with skepticism.

From the above discussion, it is obvious that characteristic and unique spectroscopic properties have been obtained for the H cluster, implicating a novel structure. It is, however, equally obvious that despite the extensive spectroscopic data available, the structure of the H cluster remains elusive.

[39] *In Vivo* Nuclear Magnetic Resonance in Study of Physiology of Sulfate-Reducing Bacteria

By HELENA SANTOS, PAULA FARELEIRA, JEAN LEGALL, and ANTÓNIO V. XAVIER

Introduction

Nuclear magnetic resonance (NMR) spectroscopy is one of the most powerful analytical methods available to chemists, allowing, for example, the qualitative and quantitative characterization of chemical mixtures, the determination of the isotopic distribution within molecules, and the measurement of reaction rates in the steady state. The dramatic improvement achieved in the performance of spectrometers in the last two decades has made it possible for cellular physiologists to benefit from the capabilities of this technique for carrying out useful measurements directly on living systems.[1-5] One of the most attractive aspects of the applications of this method to study cellular physiology derives from its nondestructive and noninvasive characteristics; in fact, not only is it possible to monitor substrate consumption and end-product formation, but it is also possible to follow the changes in the concentrations of intracellular metabolites in the same sample. Furthermore, different aspects of cellular processes can be probed with the different nuclides that can be detected by NMR; for

[1] R. G. Shulman, T. R. Brown, K. Ugurbil, S. Ogawa, S. M. Cohen, and J. A. den Hollander, *Science* **205,** 160 (1979).

[2] J. K. M. Roberts and O. Jardetzky, *Biochim. Biophys. Acta* **639,** 53 (1981).

[3] D. G. Gadian, "Nuclear Magnetic Resonance and Its Applications to Living Systems." Oxford Univ. Press (Clarendon), Oxford, 1982.

[4] M. Bárány and T. Glonek, this series, Vol. 85, p. 624.

[5] R. S. Balaban, *Am. J. Physiol.* **246** (Cell Physiol. 15), C10 (1984).

instance, ^{31}P-NMR allows monitoring of the energetic status of the cell[6] and the transmembrane proton gradient[7,8] and allows measuring of enzyme kinetics *in vivo,*[9] whereas substrate consumption, product formation, and the metabolic fate of individual carbon atoms can be followed by ^{13}C-NMR,[1,10] and cation gradients as well as cation transport over the cell membrane can be measured with ^{23}Na- or ^{39}K-NMR in the presence of shift reagents.[11,12] Provided that suitable probe heads are available, all this complementary information can be obtained for a single sample and in one experiment by interleaved acquisition of the signals from the different nuclei.

The major drawback of *in vivo* NMR techniques is low sensitivity, which forces the use of dense cell suspensions in order to increase the total intracellular space that is accessible to detection. The sensitivity of the NMR measurement is determined largely by the total amount of nuclei present in the detection region and by the field strength of the spectrometer magnet. For useful results with intact cells, spectrometers operating at frequencies higher than 300 MHz for ^{1}H are required and metabolites should be present in intracellular concentrations of at least 0.1 m*M*. However, in many cases these sensitivity limitations are by far overcome by the unique opportunity provided by these measurements to monitor a number of biochemical parameters without disturbing the cellular organization. The method is particularly useful in compartmentation studies[13,14] and in investigating different aspects of metabolic regulation, providing essential information to complement enzymatic studies *in vitro,* in which artifacts due to cell disruption are common.

Keeping Cells under Physiological Conditions for Nuclear Magnetic Resonance

For *in vivo* NMR experiments to yield useful results, the sample must be maintained under defined and homogeneous physiological conditions.

[6] J. K. M. Roberts, *in* "Nuclear Magnetic Resonance" (H. F. Linskens and J. F. Jackson, eds.), Modern Methods of Plant Analysis, New Series, Vol. 2, p. 43. Springer-Verlag, Berlin, 1986.

[7] J. L. Slonczewsky, B. P. Rosen, J. R. Alger, and R. M. Macnab, *Proc. Natl. Acad. Sci. U.S.A.* **78,** 6271 (1981).

[8] K. Nicolay, R. Kaptein, K. J. Hellingwerf, and W. N. Konings, *Eur. J. Biochem.* **116,** 191 (1981).

[9] K. M. Brindle, *Prog. Nucl. Magn. Reson. Spectrosc.* **20,** 257 (1988).

[10] R. E. London, *Prog. Nucl. Magn. Reson. Spectrosc.* **20,** 337 (1988).

[11] R. K. Gupta, P. Gupta, and R. D. Moore, *Annu. Rev. Biophys. Bioeng.* **13,** 221 (1984).

[12] C. S. Springer, Jr., *Annu. Rev. Biophys. Biophys. Chem.* **16,** 375 (1987).

[13] P. S. Belton and R. G. Ratcliffe, *Prog. Nucl. Magn. Reson. Spectrosc.* **17,** 241 (1985).

[14] J. V. Shanks and J. E. Bailey, *Biotechnol. Bioeng.* **35,** 1102 (1990).

Because high densities of cells in the NMR sample are required to maximize signal intensity, significant problems arise with regard to adequate substrate delivery as well as to the removal of waste products and the maintenance of pH. In our studies with cells we employ two different systems.

1. When the cellular metabolic activity does not lead to significant change in the pH, living cells are used as a dense suspension (approximately 40 mg dry mass ml^{-1}) and kept well mixed with a simple airlift system.[15] In this way, efficient delivery of gases (such as oxygen or hydrogen) to the sample is achieved; the sample is kept homogeneous and free of bubbles in the detection zone, with significant improvement in the linewidth of the resonances, and additional gain in sensitivity is obtained under fast pulsing conditions, because the flow brings "fresh" nuclei continuously to the detection region. When anaerobic conditions are required, the system is used with argon, and even strict anaerobes such as methanogenic bacteria can perform well under these conditions (P. Fareleira, H. Santos, and A. V. Xavier, unpublished results, 1990). In every case the impermeability of the tubing to gases is an important requirement.

   To build this flow system, a 5-mm diameter NMR tube (approximately 5 cm long and open at both ends) is inserted concentrically into a standard 10-mm NMR tube by a tight fitting provided by two Teflon disks that had been previously cut in order to allow space for the liquid to flow between the two NMR tubes (see Fig. 1 in Ref. 15). The gas to be delivered is bubbled at about 1 cm below the top of the inner tube at a rate of approximately 10–20 ml min^{-1} and causes movement of liquid upward in the inner tube and downward between the two tubes. This system has proved useful in our hands for NMR studies of sulfate-reducing bacteria for delivery of hydrogen, oxygen, or argon.
2. When considerable changes in the pH of the cell suspension are expected owing, for example, to the production of acids, such as with propionic acid bacteria,[16] cells are immobilized in a solid matrix[17,18] (most frequently κ-carrageenan) and perfused with a large

[15] H. Santos and D. L. Turner, *J. Magn. Reson.* **68,** 345 (1986).

[16] H. Santos, H. Pereira, J. P. S. G. Crespo, M. J. Moura, M. J. T. Carrondo, and A. V. Xavier, *in* "Physiology of Immobilized Cells" (J. A. M. de Bont, J. Visser, B. Mattiasson, and J. Tramper, eds.), p. 685. Elsevier, Amsterdam, 1990.

[17] M. A. Taipa, J. M. S. Cabral, and H. Santos, *Biotechnol. Bioeng.* **41,** 647 (1993).

[18] H. J. Vogel, P. Brodelius, H. Lilja, and E. M. Lohmeier-Vogel, this series, Vol. 135, p. 512.

volume of suitable medium; if necessary, the pH of the circulating medium is kept constant with the help of a pH stat and the system is stable for several hours.

^{31}P Nuclear Magnetic Resonance of Sulfate-Reducing Bacteria

Detection of Unusual Metabolites

Phosphorus-31 is still the most popular nucleus for *in vivo* NMR studies, owing to its high relative sensitivity and the importance of phosphorylated metabolites in energy metabolism. Furthermore, the pH dependence of the chemical shift of the resonance due to inorganic phosphate in the physiological pH range provides a direct and noninvasive method for measuring intracellular pH, an important biochemical parameter.

^{31}P-NMR spectra of cell suspensions of two *Desulfovibrio* strains are shown in Fig. 1. Metabolites occurring in concentrations high enough to be detected in *Desulfovibrio gigas* (trace a in Fig. 1) include nucleotide

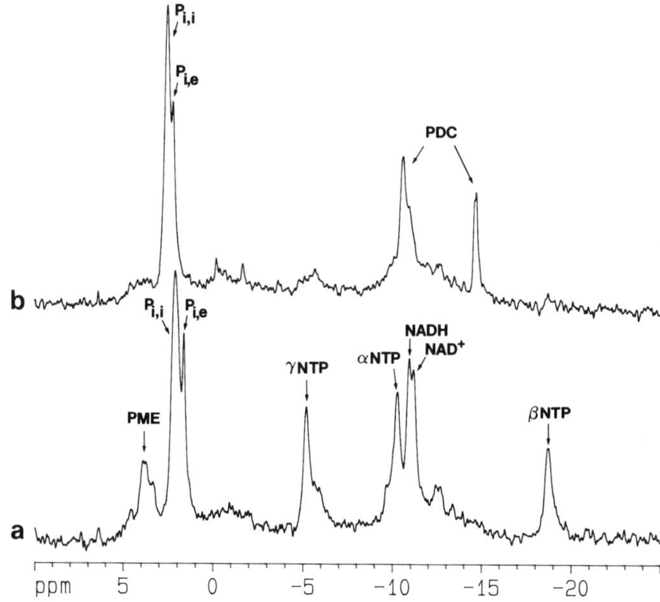

FIG. 1. *In vivo* 202-MHz ^{31}P-NMR spectra of cell suspensions of (a) *D. gigas* and (b) *D. desulfuricans* ATCC 27774 in MOPS buffer at 33°. PME, Phosphomonoesters; $P_{i,i}$, intracellular inorganic phosphate; $P_{i,e}$, external inorganic phosphate; NTP, nucleotide triphosphates; NADH, NAD$^+$, nicotinamide adenine dinucleotides; PDC, phosphoric anhydride diester compound.

triphosphates (mainly ATP), and nucleotide diphosphates, nicotinamine adenine dinucleotides (NADH or NADPH and NAD^+ or $NADP^+$), inorganic phosphate, and several phosphomonoesters that originate a complex broad line positioned to the left of the inorganic phosphate resonance (approximately at 4 ppm). This is a typical ^{31}P-NMR spectrum of an energized living system and immediately denotes the presence of internal reserves in *D. gigas,* because no external substrate has been added. From the chemical shifts of the resonances due to intracellular and extracellular inorganic phosphate, values of 7.1 and 6.6 are inferred for the intracellular and extracellular pH, respectively (see below). The general features in the spectrum of *Desulfovibrio desulfuricans* are significantly different and, in particular, the phosphoric anhydride diester region is dominated by two strong resonances at -10.7 and -14.8 ppm.[19] This latter chemical shift is rather unusual for resonances in the spectra of living systems, and extraction, purification, and NMR characterization of the compound were carried out. The novel metabolite was identified as 3-methyl-1,2,3,4-tetrahydroxybutane-1,3-cyclic bisphosphate,[20] a metabolite also reported to accumulate in *Brevibacterium ammoniagenes* and *Micrococcus luteus* under oxidative stress conditions.[21] Six different strains of *Desulfovibrio* were searched for the presence of the novel phosphodiester compound, but *D. desulfuricans* ATCC 27774 was the only organism in which this phospho compound was present in detectable amounts and, furthermore, only when sulfate (or thiosulfate) was used as an electron acceptor.[19] The physiological role of this compound in *D. desulfuricans* ATCC 27774 has not been elucidated so far, but this provides a good example of the usefulness of NMR to detect and characterize novel metabolites.

Determination of the Transmembrane Proton Gradient by ^{31}P Nuclear Magnetic Resonance

The importance of accurate measurements of intracellular pH is presently well recognized. According to the chemiosmotic hypothesis, a protonmotive force is the essential coupling factor in many membrane-bound energy transducing processes; furthermore, several processes are driven or controlled by the protonmotive force, which is composed of a transmembranal pH gradient (ΔpH) and an electrical potential ($\Delta\psi$).

[19] H. Santos, P. Fareleira, C. Pedregal, J. LeGall, and A. V. Xavier, *Eur. J. Biochem.* **201,** 283 (1991).

[20] D. L. Turner, H. Santos, P. Fareleira, I. Pacheco, J. LeGall, and A. V. Xavier, *Biochem. J.* **285,** 387 (1992).

[21] D. Ostrovsky, I. Shipanova, L. Sibeldina, A. Shashkov, E. Kharatian, I. Malyarova, and G. Tantsyrev, *FEBS Lett.* **298,** 159 (1992).

The principles of NMR techniques for determination of intracellular pH have often been presented in the literature and the validity of the method discussed.[22–24] The principal advantages are that measurements can be made continuously, nondestructively, and without forcing exogenous probes into cells. The method is based on the pH dependence of the position of resonances due to intracellular metabolites; the most commonly used resonance is that due to intracellular inorganic phosphate; however, resonances due to other phosphorus compounds, such as glucose 6-phosphate, have also been used,[25] as have ^{13}C or ^{15}N resonances due to intracellular malate[26] and histidine,[27] respectively. Changes in pH are directly manifested by shifts in the NMR peak positions and absolute values of pH can be determined provided that a reliable calibration curve is available.

The method has been applied to determine the transmembrane proton gradient in resting cells of *D. desulfuricans* CSN[28] and it was concluded that this organism regulates internal pH in a manner similar to other bacteria, such as *Escherichia coli*.[7]

The accuracy of the pH measurement by NMR is largely determined by the validity of the titration curve used to relate chemical shift and pH. In our laboratory the following procedure is followed[8]: calibration of the chemical shift of internal P_i as a function of pH is carried out with a cell suspension of the organism under study in the presence of a 100 μM concentration of the protonophore 3,3′,4′,5′-tetrachlorosalicylanilide (TCS) (Eastman Kodak Co., Rochester, NY) dissolved in *N,N*′-dimethylformamide. To calibrate the chemical shift of the external P_i, a titration curve of the buffer medium used to resuspend the cells is also performed after addition of approximately 1 mM P_i.

The presence of a transmembrane ΔpH in both cell suspensions corresponding to Fig. 1 is easily recognized from the distinct resonances due to extracellular and intracellular P_i. A careful determination of transmembrane proton gradients in *D. gigas* as a function of the extracellular pH

[22] J. K. M. Roberts, N. W. Jardetzky, and O. Jardetzky, *Biochemistry* **20,** 5389 (1981).

[23] R. J. Gillies, J. R. Alger, J. A. den Hollander, and R. G. Shulman, *in* "Intracellular pH: Its Measurement, Regulation, and Utilization in Cellular Functions," p. 79. Alan R. Liss, New York, 1982.

[24] J. K. M. Roberts, *in* "Nuclear Magnetic Resonance" (H. F. Linskens and J. F. Jackson, eds.), Modern Methods of Plant Analysis, New Series, Vol. 2, p. 106. Springer-Verlag, Berlin, 1986.

[25] H. J. Vogel and P. Brodelius, *J. Biotechnol.* **1,** 159 (1984).

[26] K. Chang and J. K. M. Roberts, *Plant Physiol.* **89,** 197 (1989).

[27] T. L. Legerton, K. Kanamori, R. L. Weiss, and J. D. Roberts, *Biochemistry* **22,** 899 (1983).

[28] M. Kroder, P. M. H. Kroneck, and H. Cypionka, *Arch. Microbiol.* **156,** 145 (1991).

is illustrated in Fig. 2, and the corresponding relevant parameters are plotted in Fig. 3. The results show that pH homeostasis in *D. gigas* cells utilizing endogenous energy reserves is good (7.5 ± 0.1) over an external pH range of about 5.8 to 7.8, in which the NTP content is also maintained; for external pH values higher than 7.5, an inverted pH gradient (more acidic inside) is observed and the intracellular pH starts to rise, although more slowly than the external pH. The NTP/NDP ratio decreases at high values of external pH (Fig. 2). It seems that this organism is well capable of pumping protons out efficiently over a wide range of acidic external pH, but is much less capable of preventing leakage of protons when the external pH is more alkaline.

For these measurements, cells are immobilized in κ-carrageenan and perfused in the NMR tube with buffer medium progressively adjusted at discrete pH values over a wide pH range. Excellent signal-to-noise ratios are obtained in this way and the method has the advantage of using a

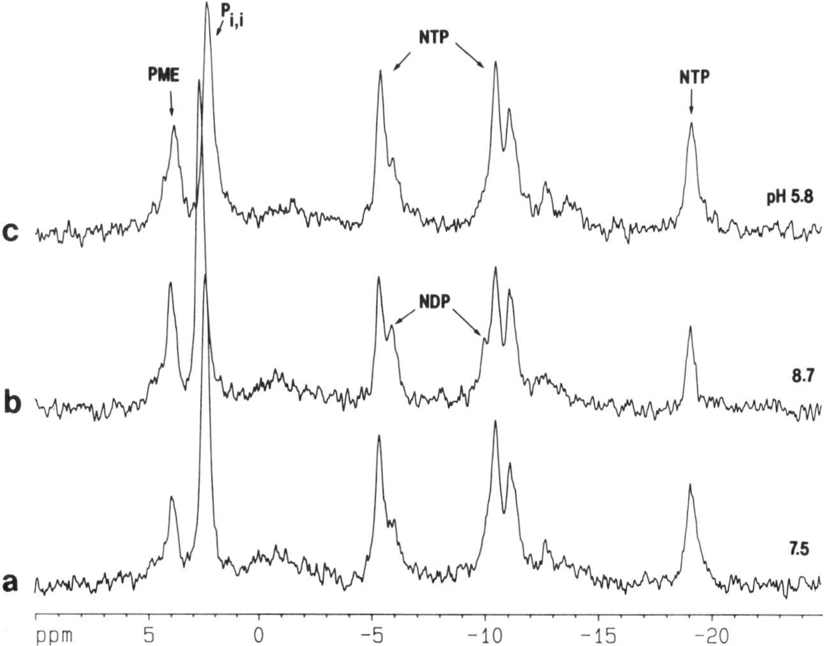

FIG. 2. Determination of intracellular pH and transmembrane proton gradients by *in vivo* ^{31}P-NMR in κ-carrageenan-immobilized cells of *D. gigas,* at different external pH values; (a) 7.5; (b) 8.7; (c) 5.8. PME, Phosphomonoesters; $P_{i,i}$, intracellular inorganic phosphate; NTP, nucleotide triphosphates; NDP, nucleotide diphosphates. Spectra were run in a Bruker (Karlsruhe, Germany) AMX500 spectrometer at 33°.

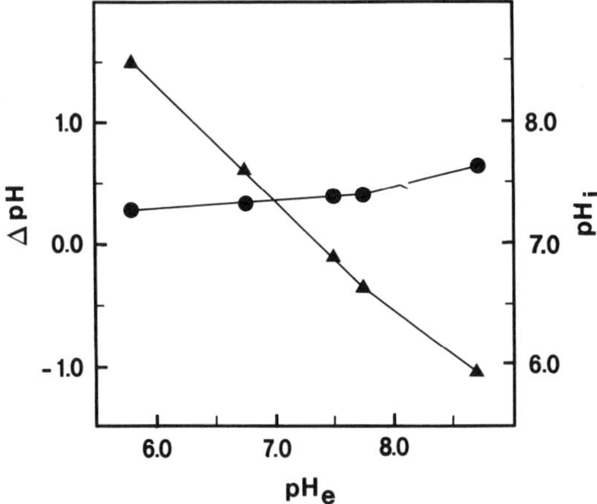

FIG. 3. Intracellular pH (●) and transmembrane ΔpH values (▲) determined by *in vivo* ^{31}P-NMR in *D. gigas* and plotted as a function of the external pH. pH_i, Intracellular pH; pH_e, external pH; ΔpH, transmembrane proton gradient.

single sample for all the measurements without risking cell damage caused by local pH extremes that form on addition of concentrated acid or base directly into cell suspensions; furthermore, a distinct resonance due to intracellular P_i is detected at every pH value, and uncertainties in the measurements of the chemical shifts of very close resonances when ΔpH is small are easily avoided. A deconvolution method that enables determination of the chemical shifts of partially overlapping resonances has been applied to measure intracellular pH in yeast.[14]

Utilization of Internal Reserves by Desulfovibrio gigas: Effect of Inhibitors and Uncouplers

In vivo ^{31}P-NMR spectra of washed cells of *D. gigas* reveal high NTP/NDP ratios (spectrum a in Fig. 4) that are maintained for many hours and result from the utilization of internal endogenous energy reserves. It is known that this organism accumulates polyglucose as an internal carbon reserve[29] and, in fact, resonances due to natural abundance carbon-13 in polyglucose can be observed even in short-time (2.5-min) accumulation spectra (see Fig. 6, below). Cells cultivated as described in Methods (below) contain the equivalent of 0.35 mg of glucose/mg dry mass, as determined by ^{13}C-NMR.

[29] F. J. M. Stams, M. Veenhuis, G. H. Weenk, and T. A. Hansen, *Arch. Microbiol.* **136,** 54 (1983).

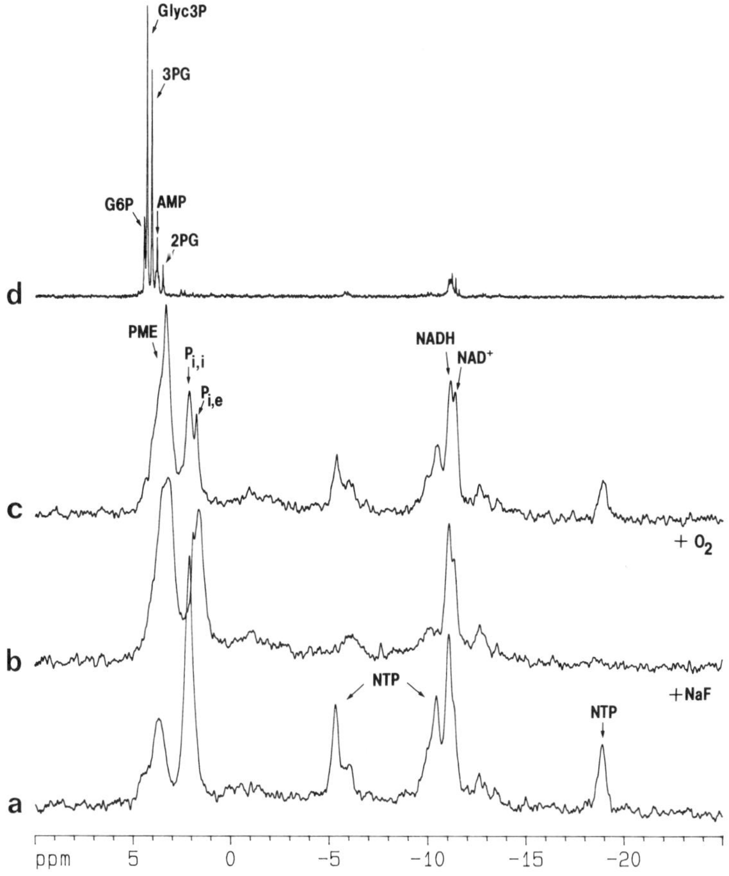

FIG. 4. Effect of sodium fluoride and oxygen on the content of phosphorylated metabolites in *D. gigas* utilizing endogenous polyglucose, as monitored by ^{31}P-NMR. Spectra (a)–(c) were obtained *in vivo* with cells suspended in MOPS buffer (pH 7.5), at 33°: (a) Initial spectrum of washed cells under argon atmosphere; (b) spectrum acquired after addition of 10 m*M* NaF to the cell suspension under argon; (c) gas atmosphere switched to oxygen; (d) perchloric acid extract of a cell suspension treated with sodium fluoride under argon. PME, Phosphomonoesters; $P_{i,i}$, intracellular inorganic phosphate; $P_{i,e}$, external inorganic phosphate; NTP, nucleotide triphosphates; G6P, glucose 6-phosphate; Glyc3P, glycerol 3-phosphate; 3PG, 3-phosphoglycerate; 2PG, 2-phosphoglycerate.

When glycolysis is inhibited with fluoride (10 m*M*) the pool of nucleotide triphosphates decreases drastically and phosphomonoester compounds accumulate (spectrum b in Fig. 4). It is normally difficult to resolve the majority of the phosphomonoester resonances in a spectrum from living cells, because lines are broadened due to intrinsic heterogeneity of the sample and accumulation of paramagnetic ions; cell extracts are useful for the identification of resonances, because factors causing broadening can be controlled and well-resolved spectra can be obtained. Tables with chemical shifts of phosphorylated metabolites commonly detected in ^{31}P-NMR spectra of living systems are available[3,4,18] and provide useful guidance for the assignments. However, firm assignments usually require addition of small amounts of the suspected compounds and, in some cases, analysis of the pH dependence of the chemical shifts of resonances.

Assignment of resonances due to major phosphomonoester compounds accumulated in fluoride-treated *D. gigas* cells was made in perchloric acid cell extracts (Fig. 4, spectrum d). Glycerol 3-phosphate, 3-phosphoglycerate, and adenosine monophosphate are the main components, as expected for enolase inhibition by fluoride. Levels of NTP can be remarkably restored when an electron acceptor, such as oxygen (Fig. 4) or nitrite (spectrum c in Fig. 5A), is provided, a result indicating that reoxidation of NADH formed from polyglucose oxidation in the reaction catalyzed by glyceraldehyde 3-phosphate dehydrogenase is carried out via electron transfer to a terminal acceptor with concomitant formation of NTP. This finding led to the elucidation of the enzymatic equipment that allows electron transfer from NADH to oxygen in *D. gigas*.[30,31] It seems as though this so-called strict anaerobe is able to take advantage of oxygen to produce maintenance energy from polyglucose reserves, and thereby survive in oxic environments.[32]

Some features in the spectra shown in Figs. 4 and 5 deserve further comment. It is interesting to note that changes in the cell NAD(P)H reduction charge are reflected in the ^{31}P-NMR spectra, because at the high magnetic field used, resonances due to NAD(P)$^+$ and NAD(P)H can be partially resolved, NAD(P)$^+$ absorbing at lower frequency (or more to the right in the spectrum) as compared to NAD(P)H. A higher NAD(P)$^+$/NAD(P)H ratio is clearly inferred from the spectra of cell suspensions after addition of an electron acceptor (compare, e.g., spectra a and c in Fig. 4 or 5). Furthermore, changes in the intracellular pH can also be clearly observed in the same figures; on treatment with fluoride, significant acidification of the intracellular space occurs (shift to the right of the P_i resonance), leading to an inverted ΔpH; however, when an electron acceptor is provided, NTP is generated and intracellular pH rises.

[30] L. Chen, M.-Y. Liu, J. LeGall, P. Fareleira, H. Santos, and A. V. Xavier, *Biochem. Biophys. Res. Commun.* **193,** 100 (1993).

[31] L. Chen, M.-Y. Liu, J. LeGall, P. Fareleira, H. Santos, and A. V. Xavier, *Eur. J. Biochem.* **216,** 443 (1993).

[32] H. Santos, P. Fareleira, A. V. Xavier, L. Chen, M.-Y. Liu, and J. LeGall, *Biochem. Biophys. Res. Commun.* **195,** 551 (1993).

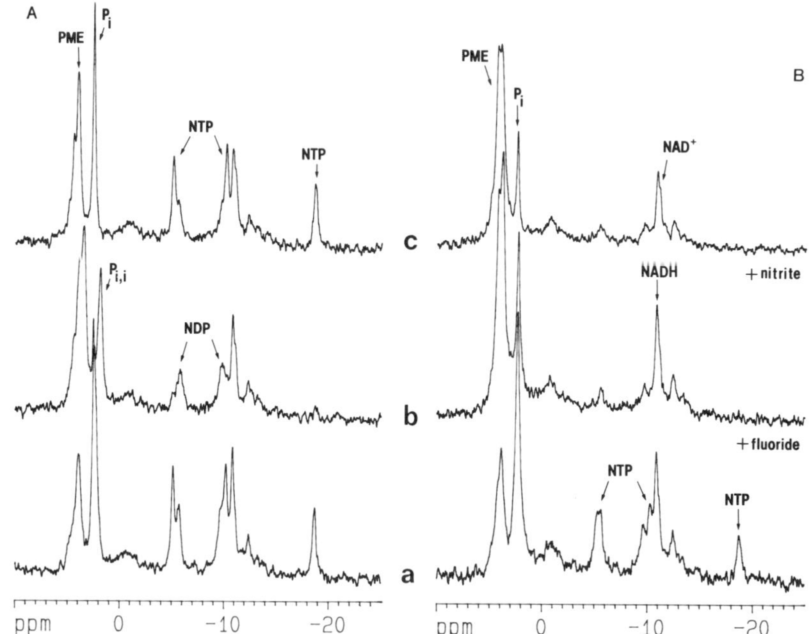

FIG. 5. *In vivo* ^{31}P-NMR spectra of *D. gigas* cells in MOPS buffer in the absence (A) or in the presence (B) of 100 μM TCS. After the acquisition of the initial spectrum (a), 10 mM NaF was added to the cell suspensions (spectrum b); spectrum (c) was obtained with the same cell suspension after addition of 20 mM sodium nitrite. For experiment (B), cells were incubated with TCS for 1 hr at room temperature. TCS was dissolved in dimethylformamide and injected directly into the cell suspension. For experiment (A), cells were treated in a similar way, adding an identical volume of dimethylformamide without TCS. PME, Phosphomonoesters; $P_{i,i}$, intracellular inorganic phosphate; NDP, nucleotide diphosphates; NTP, nucleotide triphosphates.

To investigate the mechanism of NTP synthesis, experiments are carried out in a similar way except that 100 μM TCS is added (Fig. 5B). It is clear that NTP is not synthesized under these conditions when nitrite is the electron acceptor supplied, and this result indicates that, in this case, NTP is formed by oxidative phosphorylation, in agreement with previous reports.[33]

Quantitation of NTP in Desulfovibrio gigas: Determination of Nuclear Magnetic Resonance Detectability

To determine the content of NTP in living cells by *in vivo* NMR, the intensity of the β-NTP resonance is compared with the intensity of the

[33] L. L. Barton, J. LeGall, J. M. Odom, and H. D. Peck, Jr., *J. Bacteriol.* **153,** 867 (1983).

resonance of a phosphorus standard that is added internally to the cell suspension, or contained in a capillary (1-mm diameter) coaxially mounted in the sample tube. The spectrum is acquired under nonsaturating conditions, that is, the repetition delay should be at least five times the longest ^{31}P T_1 value corresponding to the resonances to be integrated. Phosphonate compounds, such as methylphosphonate[34] or methylenediphosphonate,[4,35] are usually used as references because they originate resonances that fall outside the spectral region where common phosphorylated metabolites appear.

Nuclear magnetic resonance permits observation only of freely mobile metabolites, which are considered by some researchers as those that are available to participate in metabolism. For this reason, considerably higher phosphorylation potentials are measured in some systems by *in vivo* NMR than by destructive analytical methods.[36,37] In muscle, the amount of ADP detected by NMR is smaller than by other methods and this is attributed to immobilization of that metabolite by binding to actin.[4] Also, a large proportion of P_i in mitochondria[38] and liver[37,39] is not detected by NMR. To our knowledge the assessment of the NMR detectability of adenine nucleotides in bacterial cells has not been reported in the literature. To evaluate the ATP detectability in *D. gigas,* we compared *in vivo* NMR determinations of NTP in cells metabolizing endogenous reserves and bioluminescence measurements[40] in perchloric acid cell extracts of identical cells. Similar values (7 ± 1 nmol/mg dry mass) were obtained by both methods, indicating that in these cells ATP is fully detected by NMR.

Application of *In Vivo* ^{13}C Nuclear Magnetic Resonance to Study Carbon Metabolism by Sulfate-Reducing Bacteria

Because of the low sensitivity and receptivity of ^{13}C, most of the useful *in vivo* NMR studies involving this nucleus use isotopically enriched substrates that allow tracing of the movement of a specific atom into other

[34] H. Santos, P. Fareleira, R. Toci, J. LeGall, H. D. Peck, Jr., and A. V. Xavier, *Eur. J. Biochem.* **180,** 421 (1989).

[35] C. Roby, J.-B. Martin, R. Bligny, and R. Douce, *J. Biol. Chem.* **262,** 5000 (1987).

[36] J. K. M. Roberts, A. N. Lane, R. A. Clark, and R. H. Nieman, *Arch. Biochem. Biophys.* **240,** 712 (1985).

[37] C. C. Cunningham, C. R. Malloy, and G. K. Radda, *Biochim. Biophys. Acta* **885,** 12 (1986).

[38] S. M. Hutson, G. D. Williams, D. A. Berkich, K. F. LaNoue, and R. W. Briggs, *Biochemistry* **31,** 1322 (1992).

[39] E. Murphy, S. A. Gabel, A. Funk, and R. E. London, *Biochemistry* **27,** 526 (1988).

[40] A. Pradet, *Physiol. Veg.* **5,** 209 (1967).

compounds. The method has been used to follow the fate of the label in [1-^{13}C]ethanol during fermentation by *Desulfobulbus propionicus*.[41] These labeling studies coupled to enzyme measurements demonstrate that in this organism propionate is formed via a succinate pathway involving a transcarboxylase, as is the case in *Propionibacterium* species.

In the experiment illustrated in Fig. 6, pyruvate labeled on the methyl group is supplied to a cell suspension of *D. gigas*, and spectra are acquired consecutively until substrate exhaustion. Pyruvate (and its hydrated form) is consumed at a rate of 20 nmol/(min·mg dry mass); acetate (labeled on the methyl group), succinate (labeled on either one of the methylene groups), and alanine (labeled on the methyl group) are the major end products. The isotopic enrichment of the end products can be evaluated by running ^{1}H spectra of the supernatant solution obtained by centrifugation of the cell suspension once the substrate has been consumed. Significant amounts of unlabeled acetate and succinate are found, which result from the catabolism of polyglucose that is utilized by the cells even when an external substrate is supplied. A high proportion of pyruvate (about one-third) is transformed to succinate, probably through the activities of (*S*)-malate: $NADP^+$ oxidoreductases (EC 1.1.1.40) and (*S*)-malate hydro-lyase (fumarase;EC 4.2.1.2), which are present in this organism[42]; the reducing power required for the reactions leading to the production of succinate can be supplied by the oxidation of internal polyglucose.

Methods

Cultivation of Sulfate-Reducing Bacteria

Sulfate-reducing bacteria are cultivated at 35° in 2-liter glass bottles under argon atmosphere on a Starkey's medium containing the following ingredients: NH_4Cl (1 g/liter), $MgSO_4 \cdot 7H_2O$ (2 g/liter), Na_2SO_4 (4 g/liter), K_2HPO_4 (0.5 g/liter), 5.7 *M* sodium lactate (7.2 ml/liter), yeast extract (1 g/liter), and resazurin (50 μg/liter). This medium is supplemented by the addition of a trace elements solution (1 ml/liter) containing the following: 12.3 *M* HCl (51.3 ml/liter), MgO (10.75 g/liter), $CaCO_3$ (2 g/liter), ferric citrate (0.21 g/liter), $ZnSO_4 \cdot 7H_2O$ (1.44 g/liter), H_3BO_3 (0.06 g/liter), $Na_2MoO_4 \cdot H_2O$ (0.1 g/liter), $NiCl_2 \cdot 6H_2O$ (0.01 g/liter), and Na_2SeO_3 (0.05 g/liter). The pH of the medium is adjusted to 7.2 before sterilization.

[41] A. J. M. Stams, D. R. Kremer, K. Nicolay, G. H. Weenk, and T. A. Hansen, *Arch. Microbiol.* **139,** 167 (1984).

[42] E. C. Hatchikian and J. LeGall, *Ann. Inst. Pasteur* (*Paris*) **118,** 125 (1970).

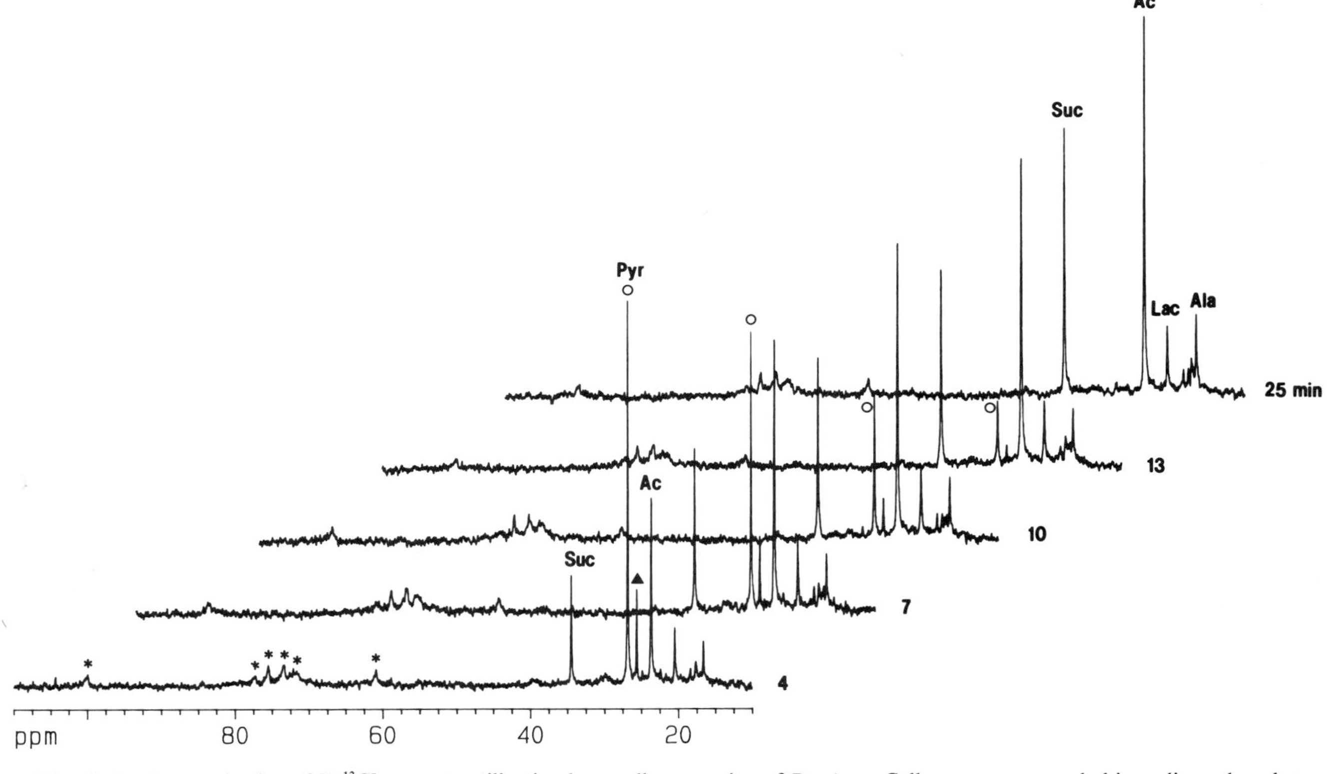

FIG. 6. *In vivo* monitoring of [3-^{13}C]pyruvate utilization by a cell suspension of *D. gigas*. Cells were resuspended in sodium phosphate buffer, pH 7.7, and the experiment was run at 33°. Following the addition of 20 m*M* [3-^{13}C]pyruvate at time zero, consecutive spectra were acquired over a period of about 25 min; each spectrum represents 2.5 min of accumulation and is the average of 100 scans. Pyr, C_3 of pyruvate; ▲, C_3 of pyruvate hydrate; Suc, C_2 and C_3 of succinate; Ac, C_2 of acetate; Ala, C_3 of alanine; Lac, C_3 of lactate (contaminant in the pyruvate supply). *, Peaks due to natural abundance ^{13}C in endogenous polyglucose.

Preparation of Cell Suspensions for Nuclear Magnetic Resonance

Fresh cell suspensions are prepared for each experiment. Cells are harvested at the end of the exponential growth phase by centrifugation at 2000 *g* for 10 min under anaerobic conditions, washed once with an oxygen-free buffer solution [either 50 m*M* morpholinepropanesulfonic acid (MOPS) (sodium salt)–1 m*M* $MgCl_2$ (pH 7.5–7.7) or 20 m*M* sodium phosphate–1 m*M* $MgCl_2$ (pH 7.5–7.7)], resuspended in the same buffer to a total volume of approximately 4 ml, and transferred under argon to a 10 mm NMR tube. ^2H_2O is added to a final concentration of 5% (v/v) to provide a lock signal. The resulting cell suspension contains approximately 40 mg of dry mass/ml.

Gases are delivered directly into the cell suspension in the NMR tube by using an airlift device as described above. To avoid excessive foam, 5 μl of silicone antifoam (SAG 5693; Union Carbide Chemicals and Plastics S.A., Geneva, Switzerland) is added.

Immobilization Procedure

For immobilization, 2 g of freshly prepared cell paste is suspended in 1.5 ml of physiological saline and mixed with 6 ml of 2.2% (w/v) κ-carrageenan (Sigma Chemical Company, St. Louis, MO) under anaerobic conditions at room temperature. The mixture is dropped through a 25-gauge needle into 0.3 *M* KCl solution, forming beads with a diameter of approximately 2 mm. Beads are allowed to cure for 30 min, washed with distilled water, suspended in 50 m*M* MOPS (potassium salt), 1 m*M* $MgCl_2$, pH 7.7, and transferred to a 10-mm NMR tube under argon; a cotton wool filter is fitted on the top of the beads to prevent blockage of the tubing. Buffer solution is circulated through the beads by using two peristaltic pumps and a perfusion device with gas-impermeable Tygon connecting tubes; two glass capillaries are fitted into the NMR tube, one positioned a few milimeters from the bottom (medium inlet) and the other at the top of the region occupied by beads (medium outlet). Argon is bubbled continuously through the buffer solution contained in the reservoir vessel. In addition, anaerobic conditions are assured in the NMR tube by continuous flushing of argon above the cotton wool plug.

Acquisition of Nuclear Magnetic Resonance Spectra of Living Cells

Spectra are obtained using standard 10-mm probe heads. Typical conditions for acquisition of ^{31}P-NMR spectra are as follows: 45° flip angle, 0.9-sec repetition delay, 16K acquisition data points, and 16-kHz spectral width. For quantitative data, spectra are acquired using 30° pulses and a

9-sec recycle time. Chemical shifts are referenced with respect to external 85% H_3PO_4.

In vivo ^{13}C-NMR spectra are obtained with proton broadband decoupling, using a 30° flip angle, 1.5 sec of recycle delay, 32K acquisition data points, and 38-kHz spectral width. For quantitative measurements, 128K data points are acquired, using 30° pulses and a repetition time of 30 sec. Proton broadband decoupling is applied only during acquisition.

Preparation of Perchloric Acid Extracts

To identify unknown resonances originating from small phosphorylated metabolites, perchloric acid cell extracts are performed as follows[43]: the cell suspension is removed from the NMR tube with a syringe and immediately extruded dropwise into liquid nitrogen. The frozen beads are added to 100 ml of ice-cold perchloric acid (2%, w/v) and stirred vigorously for 20 min while maintained in an ice bath. Cellular debris are removed by centrifugation (5000 *g*, 20 min) and the supernatant neutralized with 5 *M* KOH and centrifuged once more to remove the potassium perchlorate. The resulting supernatant is lyophilized and the lyophilisate dissolved in a small volume of H_2O, insoluble materials being removed by centrifugation. Prior to analysis by ^{31}P-NMR, 10 m*M* ethylenediaminetetraacetic acid is added and the pH adjusted to approximately 8.0 or, alternatively, the supernatant is passed through a Chelex-100 (Bio-Rad, Richmond, CA) column to remove polyvalent cations.

Concluding Remarks

This chapter illustrates the potential of *in vivo* NMR, and in particular of ^{31}P-NMR, to probe different features of metabolism in sulfate-reducing bacteria. This technique has an important role in the characterization of many metabolic processes, although most of its potentialities have not yet been fully exploited for all aspects dealing with this group of bacteria. Energy metabolism in sulfate-reducing bacteria is probably the subject for which the most useful results will be obtained in the near future.

Acknowledgments

This work was supported by Junta Nacional de Investigação Científica e Tecnológica, Grant No. PMCT/C/BIO/873/90.

[43] S. Kanodia and M. F. Roberts, *Proc. Natl. Acad. Sci. U.S.A.* **80,** 5217 (1983).

[40] Computational Chemistry and Molecular Modeling of Electron-Transfer Proteins

By JOHN E. WAMPLER

I. Computational Approaches to Structure, Dynamics, and Energy

Details of the structure and dynamic behavior of proteins at atomic resolution are central to understanding mechanisms of catalysis, ligand binding, allosteric modulation, and protein–protein interactions. The methods of computational chemistry and molecular modeling allow us to study such processes with temporal detail and spatial resolution that are difficult or impossible to obtain experimentally. However, this presents a significant problem in validation of the methods, because many of the details are not verifiable. Verification then tends to rest on comparing averaged properties (either spatially or temporally or over some ensemble of atoms or molecules). This chapter describes some of the computational approaches used to investigate these problems, introducing the terminology and a range of useful references to the current literature.

Although considerable effort has been expended on the "protein-folding" problem, at the current start of the art, accurate modeling and simulation usually begin with known structures. For these we rely primarily on the X-ray crystallography, but with increasing information coming from nuclear magnetic resonance (NMR) studies. Both of these sources of structures can be extended by the techniques of homology modeling, using known structures to predict structures of homologous proteins. A major resource of structures for molecular modeling is maintained by Brookhaven National Laboratory (Upton, NY) in the form of a bank of structure files of proteins and other molecules.[1,2] A second large resource of molecular structures, particularly small molecules, is the Cambridge Structure

[1] F. C. Bernstein, T. F. Koetzle, G. J. B. Williams, E. F. Meyer, Jr., M. D. Brice, J. R. Rodgers, O. Kennard, T. Shimanouchi, and M. Tamusi, *J. Mol. Biol.* **112,** 535 (1977).

[2] E. E. Abola, F. C. Bernstein, S. H. Bryant, T. F. Koetzle, and J. Weng, *in* "Crystallographic Databases—Information Content, Software Systems, Scientific Applications" (F. H. Allen, G. Bergerhoff, and R. Sievers, eds.), p. 107. Data Commission of the International Union of Crystallography, Bonn, 1987.

Database[3] available in the United States from the Medical Foundation of Buffalo (Buffalo, NY).

There are two major areas of computational chemistry: quantum mechanical and nonquantum mechanical calculations. For biological macromolecules such as proteins, the quantum mechanical approaches play a limited role, primarily in defining some of the parameters used with nonquantum calculations. However, other roles are emerging, for example, in studying enzyme mechanisms, and there is particular promise in approaches that mix quantum mechanical and nonquantum mechanical methods. For a review of the role of quantum mechanical calculations the reader is referred to Mulholland *et al.*[4] In this chapter the discussion of quantum mechanical calculations is limited to the calculation of parameters for molecular mechanics force fields (Section II,D) and the remainder of the discussion focuses on nonquantum mechanical approaches.

Nonquantum mechanical approaches to the computational description of macromolecules rely on the Born–Oppenheimer approximation, which allows separation of the nuclear and electronic components of the description of a system. Because the motions of electrons are fast relative to nuclear motions, nuclei can then be assumed to respond to average electronic fields. The nuclei, themselves, are then assumed to respond to forces following the laws of Newtonian mechanics. The electronic contribution is approximated by average properties evaluated using either quantum mechanics calculations of small molecules or by estimation from physical measurements.

Many of the programs described in this chapter are complex, containing many user options. In some cases there is little information in the literature on selection between options or it is buried in technical details that the novice does not understand. In either case, it is important to operate in the context of the literature. This is important for two reasons: first, it builds the database of applications of the program aiding future evaluation and, second, the experience of others, some unpublished, is incorporated into one's selections.

This field of study is associated with three-dimensional graphics workstations and multifunction molecular modeling programs (or program suites; see Section I,A) although neither is necessary to carry out many of the calculations and the display of the results can take many forms. The next sections (Sections I,A–C) list the names, sources, and references

[3] F. H. Allen, S. Bellar, M. D. Brice, B. A. Cartwright, A. Doubleday, H. Higgs, T. Hummelink, B. G. Hummeling-Peters, O. Kennard, M. D. S. Motherwell, J. R. Rodgers, and D. G. Watson, *Acta Crystallogr. Sect. B: Struct. Crystallogr. Cryst. Chem.* **B35,** 2331 (1979).

[4] A. J. Mulholland, G. H. Grand, and W. G. Richards, *Protein Eng.* **6,** 113 (1993).

for a number of the major program packages mentioned throughout this chapter. This is not an exhaustive list, nor is citation of any particular program intended as an endorsement. The list is simply intended to give the novice to the field a starting point. Programs discussed in the text that are found in these sections are not referenced repeatedly because their references, often multiple, are given here.

A. General-Purpose Molecular Modeling Programs

Insight/Discover is a commercial package from Biosym (San Diego, CA) that can be purchased with a variety of modules for both quantum mechanical and molecular mechanics calculations

MacroModel[5] is a general-purpose package that contains implementations of the AMBER, AMBER/OPLS, and MM2 force fields. (Source: W. Clark Still, Columbia University, New York, NY)

MIDAS is a noncommercial molecular graphics program available from R. Langridge at the University of California at San Francisco (San Francisco, CA)

QUANTA is a commercial package (see Momany and Rone[6]) available from Molecular Simulations, Inc. (Burlington, MA). It is available with the CHARMM force field and many other computational chemistry modules

SYBYL is a commercial package (see Clark *et al.*[7]) that includes the Tripos force field as well as an implementation of AMBER and interfaces to a number of molecular mechanics and quantum mechanics programs. (Source: Tripos Associates, St. Louis, MO)

Each of these programs makes provisions for molecular mechanics calculations and most include other features and interfaces to other programs.

B. Molecular Mechanics Force Field Programs

Most of the major molecular mechanics force field (see Section II) programs themselves are written in FORTRAN language, and can be compiled to run on a wide range of computers. Although many of them are included in or interfaced to one or more of the general-purpose pro-

[5] F. Mohamadi, N. G. J. Richards, W. C. Guida, R. Liskamp, M. Lipton, C. Caufield, G. Chang, T. Hendrickson, and W. C. Still, *J. Comput. Chem.* **11,** 440 (1990).

[6] F. A. Momany and R. Rone, *J. Comput. Chem.* **13,** 888 (1992).

[7] M. Clark, R. D. Cramer III, and N. van Opdenhosch, *J. Comput. Chem.* **10,** 982 (1989).

grams (Section IA), they can also be run stand-alone without graphics support.

AMBER[8–11] is a suite of molecular mechanics and dynamics programs supplied by P. Kollman and co-workers at the University of California at San Francisco (UCSF)

AMBER/OPLS[12] is a modification of the AMBER force field available from P. Kollman at UCSF

CHARMM,[6,13] like AMBER, is a molecular mechanics and dynamics program available academically from M. Karplus at Harvard (Boston, MA) and commercially from Molecular Simulations, Inc.

ECEPP/2[14–17] is a protein molecular mechanics program available from the Quantum Chemistry Program Exchange (QCPE), Indiana University (Bloomington, IN)

GROMOS[18,19] is a molecular mechanics and molecular dynamics package from W. F. van Gunsteren and H. C. Berendsen, University of Groningen (Groningen, the Netherlands)

MM2[20–23] is a molecular mechanics program available to academic users from the QCPE and commercially through Molecular Design Limited (San Leandro, CA)

[8] P. K. Weiner and P. A. Kollman, *J. Comput. Chem.* **2,** 287 (1981).

[9] S. J. Weiner, P. A. Kollman, D. A. Case, U. C. Singh, C. Ghio, G. Alagona, S. Profeta, Jr., and P. K. Weiner, *J. Am. Chem. Soc.* **106,** 765 (1984).

[10] S. J. Weiner, P. A. Kollman, D. T. Nguyen, and D. A. Case, *J. Comput. Chem.* **7,** 230 (1986).

[11] D. A. Pearlman, D. A. Case, J. C. Caldwell, G. L. Seibel, U. C. Singh, P. Weiner, and P. A. Kollman, "AMBER 4.0." University of California, San Francisco, 1991.

[12] W. L. Jorgensen and J. Tirado-Rives, *J. Am. Chem. Soc.* **110,** 1657 (1988).

[13] B. R. Brooks, R. E. Bruccoleri, B. D. Olafson, D. J. States, S. Swaminathan, and M. Karplus, *J. Comput. Chem.* **4,** 187 (1983).

[14] F. A. Momany, R. F. McGuire, A. W. Burgess, and H. A. Scheraga, *J. Phys. Chem.* **79,** 2361 (1975).

[15] G. Nemethy, M. S. Pottle, and H. A. Scheraga, *J. Phys. Chem.* **87,** 1883 (1983).

[16] M. J. Sippl, G. Nemethy, and H. A. Scheraga, *J. Phys. Chem.* **88,** 6231 (1984).

[17] K. D. Gibson and H. A. Scheraga, *Proc. Natl. Acad. Sci. U.S.A.* **83,** 5649 (1986).

[18] J. Hermans, H. J. C. Berendsen, W. F. van Gunsteren, and J. P. M. Postma, *Biopolymers* **23,** 1513 (1984).

[19] W. F. van Gunsteren, H. J. C. Berendsen, J. Hermans, W. G. J. Hole, and J. P. M. Postma, *Proc. Natl. Acad. Sci. U.S.A.* **80,** 4315 (1983).

[20] N. L. Allinger, *J. Am. Chem. Soc.* **99,** 8127 (1977).

[21] U. Burkert and N. L. Allinger, "Molecular Mechanics." American Chemical Society, Washington, D.C., 1982.

[22] N. L. Allinger, R. A. Kok, and M. R. Imam, *J. Comput. Chem.* **9,** 591 (1988).

[23] J.-H. Lii, S. Gallion, C. Bender, H. Wikstrom, N. L. Allinger, K. M. Flurchick, and M. M. Teeter, *J. Comput. Chem.* **10,** 503 (1989).

MM3[24–27] is the new version of the Allinger molecular mechanics program available from Technical Utilization Corp., Inc. (Powell, OH) and Molecular Design Limited

SPASMS[28] is a molecular dynamics package available from UCSF

TRIPOS 5.2[7] is a molecular mechanics force field available as part of the SYBYL package from Tripos Associates

YETI[29–31] is a molecular mechanics package with specialized code for metalloproteins available from the Swiss Institute for Alternatives to Animal Testing (Ettingen, Switzerland)

C. *Quantum Chemistry Programs*

GAMESS[32–34] is a general-purpose suite with both semiempirical and *ab initio* capabilities. (Source: QCPE and M. Schmidt, Ames Laboratory, Iowa State University, Ames, IA)

GAUSSIAN[35] is a general-purpose suite with both semiempirical and *ab initio* capabilities. (Source: Gaussian, Inc., Pittsburgh, PA)

HONDO[33,36] is a general-purpose quantum mechanical program similar to GAUSSIAN. (Source: IBM Corp., Kingston, NY)

MOPAC[37] is a semiempirical program available from the QCPE

[24] N. L. Allinger, Y. H. Yuh, and J.-H. Lii, *J. Am. Chem. Soc.* **111,** 8551 (1989).

[25] J.-H. Lii and N. L. Allinger, *J. Am. Chem. Soc.* **111,** 8566 (1989).

[26] J.-H. Lii and N. L. Allinger, *J. Am. Chem. Soc.* **111,** 8576 (1989).

[27] J.-H. Lii and N. L. Allinger, *J. Comput. Chem.* **12,** 186 (1991).

[28] D. C. Spellmeyer, W. C. Swope, and E. R. Evensen, "San Francisco Package of Applications for the Simulation of Molecular Systems." Regents of the University of California, San Francisco, 1989.

[29] A. Vedani, M. Dobler, and J. D. Dunitz, *J. Comput. Chem.* **7,** 701 (1986).

[30] A. Vedani, *J. Comput. Chem.* **9,** 269 (1986).

[31] A. Vedani and D. W. Huhta, *J. Am. Chem. Soc.* **112,** 4759 (1990).

[32] M. Dupuis, D. Spangler, and J. J. Wendoloski, "Program QC01." National Resource for Computations in Chemistry Software Catalog, University of California, Berkeley, 1980.

[33] M. Dupuis, P. Mougenot, J. D. Watts, G. J. B. Hurst, and H. O. Villar, *in* "MOTECC: Modern Techniques in Computational Chemistry" (E. Clementi, ed.), p. 306. ESCOM, Leiden, The Netherlands, 1989.

[34] M. W. Schmidt, K. K. Baldridge, J. A. Boatz, J. H. Jensen, S. Koseki, M. S. Gordon, K. A. Nguyen, T. L. Windus, and M. S. Gordon, *QCPE Bull.* **10,** 52 (1990).

[35] M. G. Frisch, G. W. Trucks, M. Head-Gordon, P. M. W. Gill, M. W. Wong, J. B. Foresman, B. G. Johnson, H. B. Schlegel, M. A. Robb, E. S. Replogle, R. Gomperts, J. L. Andres, K. Raghavachari, J. S. Binkley, C. Gonzalez, R. L. Martin, D. J. Fox, D. J. Defrees, J. Baker, J. J. P. Stewart, and J. A. Pople, "Gaussian 92." Gaussian Inc., Pittsburgh, Pennsylvania, 1992.

[36] M. Dupuis, J. D. Watts, H. O. Villar, and G. J. B. Hurst, *Comput. Phys. Commun.* **52,** 415 (1989).

[37] J. J. P. Stewart, *J. Comput. Aided Mol. Des.* **4,** 1 (1990).

MOPAC ESP[37,38] is an adaptation of the MOPAC code to include calculation of electrostatic potential-derived charges. (Source: K. Merz, Pennsylvania State University, University Park, PA)

II. Molecular Mechanics Force Fields

Molecular mechanics describe molecules in terms of their structure, the properties of the atoms, and their interactions. For a given molecule, the structure description consists of the positions of the atoms and their atom type, an abbreviated name, or a number that identifies the atom and its context (hybridization, number and kinds of bonding partners, etc.). The structure description may also contain information such as a list of bonds or names of substructures (amino acid residues).

There are two general types of structure representations: Cartesian coordinates and internal coordinates. With Cartesian coordinates the location of each atom is given separately (x, y, and z). With internal coordinates each atom is placed in relation to previously defined atoms, first by the length of the bond to one, the angle defined by those two atoms and one other, and the dihedral angle defined by those three atoms and yet another. In this case, bonding is explicit (with the exception of loops), but with Cartesian coordinates a list of bonds or connectivities often accompanies the coordinates. Because bond lengths in proteins are usually close to nominal values and the connectivity of each amino acid residue is known beforehand, bonds can be placed on the basis of the x, y, z positions of the atoms and their types without need for a bond list in most cases. The Brookhaven Protein Databank (PDB) uses Cartesian coordinates and a list of connectivities (if needed). Tripos residue and "mol2" files, MM2 and MM3 input files also use Cartesian coordinates. Many of the quantum mechanics programs (GAUSSIAN, MOPAC, etc.) use a form of internal coordinate format called a *Z*-matrix, but the exact order of information and strictures on the sequence of atoms may differ from program to program (see Clark[39]). AMBER uses a different form of internal coordinates for its Prep module input files, but reads PDB files for the starting structure. Most of the commercial programs allow input and output using various formats.

For the purposes of this discussion the combination of atom characteristics and interaction preferences is referred to as the *parameter set*. The parameter set is ideally a general database of information valid for a large class of molecules. The physical properties of the molecule are

[38] B. H. Besler, K. M. Merz, and P. A. Kollman, *J. Comput. Chem.* **11,** 431 (1990).

[39] T. Clark, "A Handbook of Computational Chemistry." Wiley, New York, 1985.

then described by the combination of the atomic characteristics and the correlation between the geometry and the interaction preferences. Molecular mass is the sum of the atomic masses, net charge is the sum of charges, dipole moment is defined by the spatial distribution of these charges, and so on. A molecule in which the geometry positions the atoms near to their preferred bond lengths, angles, and so on is in a low-energy conformational state, one with substantial differences in these parameters from the preferred values is in a high-energy or strained conformation. Quantitation of this conformational energy is the central calculation of molecular mechanics.

Conformational energy is calculated using a force field equation, E_{pot}, which is a function of the geometry and the parameter set. In practice, the energy of a given structure is calculated not only by treating the nuclear and electronic components of force independently, but by further dividing the potential into independent components, bond length energy, bond angle energy, and so on. The total energy is then evaluated by summing the interactions over all appropriate groups of atoms and summing up the individual components [Eq. (1)].

$$E_{pot} = \Sigma V_s + \Sigma V_a + \Sigma V_t + \Sigma V_v + \Sigma V_e + \cdots \quad (1)$$

The differences between molecular mechanics force fields reside in the number of and formulation of these component parts and in the parameter sets that support the calculations. Many early force fields considered the covalent structure of a molecule to be fixed and inflexible, calculating the potential based only on the nonbonded interactions (van der Waals forces and/or electrostatic forces) and rotation of dihedral angles. More recently, these calculations have involved many more terms, including bond length and bond angle terms that allow flexibility even in the bonded structure.

The molecular mechanics force field components used for calculations of macromolecules have simple mathematical formulations (see below). This is a necessity because the large numbers of atoms and interactions involved make even simple calculations time consuming. Indeed, the most time-consuming part of all is calculation of the interaction energy between pairs of atoms that are not bonded to each other. The number of these individual calculations, if all possible interactions are evaluated, increases as the number of atoms squared. Thus, a moderate sized protein of 2000 atoms would involve 4 million nonbonded calculations for a single energy evaluation. As a result many of the choices in formulating a molecular mechanics equation for macromolecules (as compared to small molecule calculations) focus on decreasing the computational load, particularly of the nonbonded component. For example, several programs for protein molecular mechanics offer the option to "fuse" hydrogen atoms with the heavy atom to which they are bound, an extended or united atom approach. Justification for this comes from the following four considerations (after Brooks *et al.*[13]):

1. The X-ray structures, which are the usual starting point for molecular mechanics calculations, do not show the positions of the hydrogens.
2. By including hydrogens as part of their heavy atom partner, the number of atoms modeled typically decreases by half, decreasing the number of nonbonded interactions by as much as fourfold.
3. Motions of hydrogens are fast relative to the heavy atoms, so that, like electrons, it might be justified to include their average effect in the parameter set.
4. When molecular dynamics simulations are being calculated, the lightweight hydrogen demands much smaller step sizes (thus many more calculations) than the other atoms.

This choice is easier to justify for hydrogens on aliphatic carbon atoms than for hydrogens that can participate in hydrogen bonding. As a result, the latter type are often represented explicitly, whereas the aliphatic type are represented by united or extended atoms. A number of tests have shown, however, that with sufficient computer resoures and tractable-sized problems explicit representation of all hydrogens is better, that is, the all-atom approach. Indeed, most of the compromises that save computer time have their cost.

One of the key assumptions of molecular mechanics is the hypothesis that the parameter set associated with a given atom type is transferable, that is, can be used for that same atom type in different structures. This is particularly important for macromolecular problems, in which transferability allows results from small molecules to be used for computations on large molecules. Thus, an evaluation of the average electronic charge around an atom in *N*-acetylglycine amide (NAGA) by quantum mechanics can be used to describe the electronic charge around the corresponding atom in a protein where the size of the system excludes use of quantum mechanics. This hypothesis can, of course, be limited in such a way as to ensure its accuracy. For example, different sets of electronic charges might be calculated for the atoms in NAGA for different conformations of the molecule and then used for protein calculations only for residues with matching conformations. At the limit this subdivision of molecular form and associated parameters would lead to unique parameters for every atom, that is, no transferability. In practice, the approaches used tend to carry transferability in the other direction, ignoring conformational differences and using the same parameters for comparable atoms in a wide range of different chemical moieties. Because these assumptions can be gross approximations to reality, they lead to empirical adjustments of parameters so that they give reasonable predictions of chemical structures, energies, and properties for a range of different molecules. In some cases, the adjustments can be as dramatic as adding a new term to a force field or a pseudoatom to the structure with its own mass, charge, and position. All of these adjustments and additions tend to separate the computations from their theoretical basis.

As shown in Section I,B, there are a number of different programs for molecular mechanics on proteins. Comparative studies[12,40–45] do not clearly support any one of these in preference to the others. Indeed, by adding options to use different parameters sets and different force field components, the newest versions of these programs tend to have overlapping functionality and performance. This section therefore discusses and contrasts four different force fields that have been chosen arbitrarily, with the only condition being that they have been used for molecular mechanics calculations on proteins. They are AMBER as developed by Kollman and colleagues, CHARMM from Karplus and colleagues, MM2 as developed by Allinger and co-workers, and the TRIPOS 5.2 force field from Clark and co-workers as distributed in the SYBYL package. For background information and the foundation of the formulations of these force fields the reader is referred to the monographs of Burkert and Allinger[21] and Clark[39] and the collected literature on these and other such programs cited in Section I,B. MM2 (and its offspring MM3) has been formulated primarily for small molecule calculations and its success in determining geometries and conformation energies rivals experimental approaches (see Engler *et al.*[40]). It is therefore included in this comparison in order to reveal the compromises and simplifications that have been employed to make macromolecular calculations tractable.

A. *Modeling Covalent Interactions*

The covalent structure of a protein molecule is a rather rigid framework at biochemical temperatures. Of the three major variables of covalent structure (bond lengths, bond angles, and torsion angles), only the latter is subject to large variation. This fact allows the formulation of a simple

[40] E. M. Engler, J. D. Andose, and P. von R. Schleyer, *J. Am. Chem. Soc.* **95,** 8005 (1973).
[41] D. Hall and N. Pavitt, *J. Comput. Chem.* **5,** 441 (1984).
[42] M. Whitlow and M. M. Teeter, *J. Am. Chem. Soc.* **108,** 7163 (1986).
[43] I. K. Roterman, K. D. Gibson, and H. A. Scheraga, *J. Biomol. Struct. Dyn.* **7,** 391 (1989).
[44] I. K. Roterman, M. H. Lambert, K. D. Gibson, and H. A. Scheraga, *J. Biomol. Struct. Dyn.* **7,** 421 (1989).
[45] K. Gunderhofte, J. Palm, I. Pattersson, and A. Stamvik, *J. Comput. Chem.* **12,** 200 (1991).

potential energy function to describe bond and angle energetics. Indeed, one successful force field, ECEPP/2, treats both bonds and angles as invariant. However, even if they are allowed to vary, the equations used need fit only the bottom part of the potential well.

1. Bond Length. The shape of the potential energy profile for bond stretching for a wide variety of organic single, double, and triple bonds is well fit[46] by the classic Morse potential [Eq. (2)].

$$V_s = k_s[1 - e^{-A(r-r_0)}]^2 \tag{2}$$

where V_s is the potential, k_s is the well depth, A determines the width, r is the bond length, and r_0 is the preferred or equilibrium bond length. This function is computationally intensive on a digital computer, owing to the exponential. Because the bottom of the well is the only part of the curve of interest in protein molecular mechanics at biological temperatures, it is the only part that must be represented accurately. The simplest approximation and the one used by most programs is a simple harmonic potential [Eq. (3)].

$$V_s = k_s(r - r_0)^2 \tag{3}$$

However, the harmonic function does not fit the shape of the well very accurately, even over a fairly narrow range of distances.[46] In MM2, a cubic function [Eq. (4)] is used for a more accurate fit.

$$V_s = k_s(r - r_0)^2[1 - 2(r - r_0)] \tag{4}$$

This function has the disadvantage of a maximum as well as a minimum, so that if atoms are allowed to get far enough apart, the potential would actually repel them. This has been fixed in MM3 by adding a quartic term.[24] It should also be noted that in MM2, the bond length is coupled to the bond angle (see the next section) with an additional stretch-bend potential.

The initial values of the parameters (k values and r_0) for any given bond are typically derived from vibrational spectra [infrared (IR), Raman, microwave, etc.] or *ab initio* calculations on small molecules. However, these initial values are often used only as a starting point for optimization based on known structures. As a result the value of even the simplest of these parameters varies with the exact formulation of the force field and the structures used to optimize it.

2. Bond Angle. Like bond length, the bond angle term in protein force fields is a simple harmonic function [Eq. (5)].

$$V_a = k_a(\theta - \theta_0)^2 \tag{5}$$

[46] M. Orozco and F. J. Luque, *J. Comput. Chem.* **14,** 881 (1993).

where k_a is the well depth and θ_0 is the preferred or equilibrium angle. Again, MM2 uses a more complex term to refine geometry of small molecules. In this case there is both a bond angle term [Eq. (6)] and a "cross" term that includes both angle and bond length information, the MM2 stretch-bend potential [Eq. (7)].

$$V_a = k_a(\theta - \theta_0)^2[1 - (7 \times 10^{-8})(\theta - \theta_0)^4] \tag{6}$$

$$V_{sa} = k_{sa}(\theta - \theta_0)[(r_1 - R_{0,1}) + (r_2 - r_{0,2})] \tag{7}$$

where the subscripts 1 and 2 stand for the two bonds that make the angle θ.

The derivative of Eq. 5 is ambiguous in practice because the angle is scaled to values between 0 and π. This is a problem because both minimization and force calculations depend on calculation of these derivatives. To overcome this problem, Schlick[47] has suggested a function similar to Eq. (2), in which both angles are substituted by their cosines, and Swope and Ferguson[48] have proposed an alternative approach to derivative calculation for this function that has been incorporated in the molecular dynamics program SPASMS (see Section I,B).

3. Torsion Angle. Unlike bond length and angle potentials, the torsion angle potential usually has multiple minima. For example, that for the planar amide bond in a protein should have a minimum at 0° for the *cis* conformation and 180° for the *trans* conformation. Although in many cases, such as this one, the well depths should be different for different conformers, most protein force fields use symmetrical formulations that do not allow this option. However, to the extent that nonbonding interactions account for the difference in well depths, this asymmetry may be introduced by the other terms.

The torsion angle term for AMBER is the simple cosine function of Eq. (8).

$$V_t = k_t/2[1 + \cos(n\phi - \Delta)] \tag{8}$$

where k_t is the well depth, n is the number of minima, ϕ is the torsion angle, and Δ is the offset or phase angle. The corresponding CHARMM term is given by Eq. (9):

$$V_t\ |k_t| - k_t\cos(n\phi) \tag{9}$$

In this case, the phase information is represented by the sign of k_t. A similar approach is used in the TRIPOS force field [Eq. (10)] except that the sign of n is used to convey the phase information.

$$V_t = k_t[1 + n/|n|\ \cos(|n|(\phi)] \tag{10}$$

[47] T. Schlick, *J. Comput. Chem.* **10,** 951 (1989).

[48] W. C. Swope and D. M. Ferguson, *J. Comput. Chem.* **13,** 585 (1992).

MM2 again represents this potential with a more complex function [Eq. (11)] with three parameters, k_{t1}, k_{t2}, and k_{t3}. Only in this case can the well depths be varied.

$$V_t = k_{t1}/2(1 + \cos \phi) + k_{t2}/2(1 - \cos 2\phi) + k_{t3}/2(1 + \cos 3\phi) \quad (11)$$

These cosine-shaped potential wells do not have steep sides and there are a number of situations that place more constraint on torsion angles, for example, geometry at sp^2-hybridized carbon atoms, ring geometries, asymmetry at optical centers involving united or extended atoms, and so on. Both AMBER and CHARMM use an additional potential, called an improper torsions potential, to constrain these geometries. In AMBER it is in the form of Eq. (8), but the torsion angle ϕ can include four atoms not sequentially bonded, for example, the C_1',C_α-N-H atoms of a peptide backbone. MM2 restricts torsions for the cyclobutane ring by making the value of k_{t3} in Eq. (13) very large compared to the values used for all three k values in other compounds. CHARMM uses a harmonic term [Eq. (5)] to apply the same type of constraint. A similar restriction is achieved with an out-of-plane bending potential in the TRIPOS force field.

4. Other Geometric Constraints. Constraints can be used to preserve geometry about some parts of a molecule during minimization or to correct deficiencies in the force field (as discussed in the case of improper torsion angles, above). In minimization, such constraints are used to allow a local bad geometry to be relieved locally and prevent its effects from disturbing good geometries elsewhere in the molecule. In AMBER and CHARMM additional constraints of atom-to-atom distances and of angles defined by the positions of three atoms are calculated using harmonic potentials [like Eqs. (3) and (5)] with large force constants. Note that it is not necessary for the atoms involved to be bonded to each other. With the TRIPOS force field a similar capability is achieved by allowing an arbitrary group of atoms to be fixed in their geometry, but still used in the potential calculation. As mentioned above, the CHARMM harmonic constraint function is used to define more narrowly some torsional movements as well.

B. Modeling Noncovalent Interactions

Unlike the bonding potentials, which are relatively deep, narrow potential wells, nonbonding potentials, although weak, are generated over fairly large distances. The basic nonbonding interaction potential of most force fields consists of two terms: a term for van der Waals interactions and a term for electrostatic interactions. Because the basis for the potentials used in both cases are the simplest monopole and dipole interactions and proteins contain many elements and groups that have significant higher order interactions (see Burley and Petsko[49] for an excellent review of this matter), it is not surprising that the simple models are not very accurate. As with bonding interactions, the typical molecular mechanics force field uses special potentials and corrections to make up for these deficiencies. The most critical problems that must be corrected in protein molecular mechanics are the representation of hydrogen-bonding interactions, the polarizability of atoms such as nitrogen and sulfur, and proper representation of interactions with aqueous solvent. Each of these topics is discussed below.

1. van der Waals Interactions. The van der Waals interaction term itself consists of at least two parts (for a review see Fitts[50]), an attractive potential due to London dispersion interactions and a repulsion due to overlap between close approaching atoms. The dispersion interactions can be represented by a series of terms for different induced multipole components, but for most purposes the induced dipole–dipole term predominates and depends on the sixth power of the separation distance between the two atoms. The repulsive term can be represented by a switch function when there is no repulsion up to the sum of the van der Waals radii of the atoms and infinite repulsion at any closer distance. A more continuous function is obtained using a high-power (twelfth order) distance dependence for the repulsion along with the sixth order dependence for the attraction component. This so-called Lennard–Jones 6-12 potential function[51,52] can be formulated in a number of different ways.[50] For CHARMM calculations, a table of 406 sets of pairwise interaction parameters, A_{ij} and B_{ij}, is used along with Eq. (12).

$$V_v = (A_{ij}/r^{12} - B_{ij}/r^6) \tag{12}$$

The parameters themselves are calculated from van der Waals radii, polarizability, and so on, using the Slater–Kirkwood formulas.[53] An alternative approach, used in the AMBER and TRIPOS formulations, is to calculate the 6-12 parameters using atom-specific values, for example a well depth k_{ij} calculated as the geometric mean $[k_{ij} = (k_i k_j)^{1/2}]$ of the parameters for the two interacting atoms (k_i and k_j) and the sum of their van der Waals radii, R_{vdw} [Eq. (13)].

$$V_v = k_{ij}[(R_{vdw}/r)^{12} - 2(R_{vdw}/r)^6] \tag{13}$$

[49] S. K. Burley and G. A. Petsko, *Adv. Protein Chem.* **39,** 125 (1988).

[50] D. D. Fitts, *Annu. Rev. Phys. Chem.* **17,** 59 (1966).

[51] J. E. Jones, *Proc. R. Soc. London A.* **106,** 441 (1924).

[52] J. E. Jones, *Proc. R. Soc. London A.* **106,** 463 (1924).

[53] J. C. Slater and J. G. Kirkwood, *Phys. Rev.* **37,** 682 (1931).

In MM2 the twelfth power repulsion is replaced by an exponential in a form referred to by Fitts[50] as an Exp-6 potential [Eq. (14)].

$$V_v = k_{ij}[(2.9 \times 10^5)\exp(-12.5r/R_{vdw}) - 2.25(R_{vdw}/r)^6] \quad (14)$$

When the interacting atoms are very close ($r < \frac{1}{3}R_{vdw}$), this potential fails by going through a maximum and generating an attraction that would tend to fuse the atoms. This is avoided in MM2 by switching to a simple $(R_{vdw}/r)^2$ repulsion for small values of r, a situation that arises only when a poor initial structure involves atomic overlap.

2. *Electrostatic Interactions*. The largest nonbonded interaction component also tends to be the least well formulated (see reviews by Harvey[54] and by Davis and McCammon[55]). In most force fields, electrostatic interactions are represented by simple Coulombic monopole interactions [Eq. (15)].

$$V_e = q_i q_j (4\pi\varepsilon_0 r_{ij}) \quad (15)$$

where q_i and q_j are the atom-centered point charges of the interacting atoms, r_{ij} is the separating distance, and ε_0 is the dielectric constant. One fault of this approach is that the dielectric constant is usually treated as a constant throughout all regions of space and throughout the entire calculation. Second, polarizability and multipole interactions are not accounted for. This approach to calculating electrostatic interactions ignores shielding by intervening charges and is a particular problem when a force field is used to calculate a molecular property for correlation with solution measurements in the absence of an explicit representation of the solvent molecules. The empirical approach to addressing these issues is to make the dielectric constant distance dependent, that is, $\varepsilon_0' = \varepsilon_0 r_{ij}$. Substituting in Eq. (15), the potential becomes

$$V_e = q_i q_j/(4\pi\varepsilon_0 r_{ij}^2) \quad (16)$$

This simple change, even in the presence of explicit solvent, can go a long way in improving the electrostatic component of the energy calculation (see Whitlow and Teeter[42] and Guenot and Kollman[56]).

For MM2 calculations on polar groups, a dipole–dipole interaction term is used rather than a Coulombic potential [Eq. (17)].

$$V_e = (\mu_i \mu_j)/(\varepsilon_0 r_{ij}^3)(\cos\chi - 3\cos\alpha_i \cos\alpha_j) \quad (17)$$

where μ_i and μ_j are bond-centered dipole moments, χ is the angle between them, and the angles α_i and α_j are those between each of the moments and the line connecting their centers, respectively.

[54] S. C. Harvey, *Proteins* **5,** 79 (1989).

[55] M. E. Davis and J. A. McCammon, *Chem. Rev.* **90,** 509 (1990).

[56] J. Guenot and P. A. Kollman, *J. Comput. Chem.* **14,** 295 (1993).

The AMBER, CHARMM, and TRIPOS force fields all have options to use either constant or distance dependent dielectrics with a Coulombic potential [Eqs. (15) and (16)] based on atom-centered point charge parameter sets. CHARMM has a number of additional options including an extended electrostatic interaction potential consisting of a Coulombic term for short-range interactions and a multipole expansion (up to quadrapole) for group-to-group interactions.

3. Hydrogen Bonding. Hydrogen bonds play a fundamental role in protein structure and their importance is supported by a variety of experimental and theoretical evidence (see Burley and Petsko[49] and Rose and Wolfenden[57]). Hydrogen bonds are dipole–dipole interactions with some charge transfer character (see Burley and Petsko[49]). As a consequence, the strength of the interaction depends not only on distance but on the orientation of the dipoles. Polarization of the donor hydrogen bond effectively decreases the van der Waals radius of the hydrogen atom and increases its atom-centered charge. Thus, one approach to represent hydrogen bonds accurately is to use the combined nonbonding potential (6-12 or exp-6 plus Coulombic) but with different parameters for the hydrogen-bonding atoms. This is done with MM2(87),[22] in which the van der Waals radii of donor hydrogens are decreased and the attractive potential parameters are increased. With the TRIPOS force field, the normal nonbonded potentials are used, but the van der Waals radii of the hydrogen partners are given a value of zero. With the AMBER formulation, the nonbonding interactions of hydrogen-bonding pairs are treated differently with a separately parameterized Lennard–Jones 10-12 potential in the place of the normal 6-12 potential. CHARMM uses a different hydrogen bond potential with lower power repulsion r^{m} ($m = 0, -2$, or -4) and attraction r^n ($n = 0$ or -2), terms, and with cosine multipliers for both the donor and acceptor angles.

4. Nonbonding Electron Pairs. The polarizability of atoms due to nonbonding outer shell electrons is a particular problem for molecular mechanics because the charges used are of fixed magnitude, independent of structure or conformation, and are atom centered. Conformational dependence of the electrostatic potential around molecules (and therefore in the atom-centered charges) has been demonstrated by *ab initio* calculations.[58–61] The simplest solution is to place additional charges off atom, thus simulating asymmetry and dipolar character with fixed parameters. Such pseudoatoms may have their own charge, van der Waals radii, and

[57] G. D. Rose and R. Wolfenden, *Annu. Rev. Biophys. Biomol. Struct.* **22,** 381 (1993).

[58] P. Cieplak and P. Kollman, *J. Comput. Chem.* **12,** 1232 (1991).

[59] K. M. Merz, *J. Comput. Chem.* **13,** 749 (1992).

[60] W. A. Sokalsi, D. A. Keller, R. L. Ornstein, and R. Rein, *J. Comput. Chem.* **14,** 970 (1993).

[61] T. R. Stouch and D. E. Williams, *J. Comput. Chem.* **14,** 858 (1993).

TABLE I
GEOMETRY AND PARAMETERS OF SOME WATER MODELS[a]

	SPC	TIP3P	BF	TIPS2	TIP4P	ST2
No. of sites	3	3	4	4	4	5
$r_{O\text{-}H}$	1	0.9572	0.96	0.9572	0.9572	1
H–O–H angle	109.47	104.52	105.7	104.52	104.52	109.47
O--O						
12/6 A $\times 10^{12}$	629.4	582.0	560.4	695.0	600.0	*b*
12/6 B $\times 10^{6}$	625.5	595.0	837.0	600.0	610.0	*b*
O charge	−0.82	−0.834	—	—	—	—
H charge	0.41	0.417	0.49	0.535	0.52	0.2357
m charge	—	—	−0.98	−1.07	−1.04	—
$r_{O\text{-}m}$	—	—	0.15	0.15	0.15	
lp charge	—	—	—	—	—	−0.2357
$r_{O\text{-}lp}$	—	—	—	—	—	0.8
lp–O–lp angle	—	—	—	—	—	109.47

[a] Distances ($r_{O\text{-}H}$, $r_{O\text{-}m}$ + $r_{O\text{-}lp}$) in Å; angles in degrees.
[b] ST2 potential function has Lennard–Jones parameters and a switching function that turns off electrostatics up close.

geometric parameters. This approach is similar to the device used in models of liquid water to correct hydrogen-bonding geometries (see the next section). It is used in AMBER for the sulfur atoms in cysteine and methionine, to which two lone pair pseudoatoms are added. It has also been suggested[60] that off-atom charges might be needed to correct peptide hydrogen-bonding inaccuracies.

5. *Water Models.* Several simple, rigid water models have been described (see Jorgensen *et al.*,[62] Teeter,[63] and Daggett and Levitt[64]) that can be used in molecular mechanics and dynamics calculations with nonbonded terms that involve only Lennard–Jones 12-6 and Coulombic interactions. The simplest of these models involves only three sites, the locations of the three water atoms. The four-site models place the negative charge of the oxygen on the bisector of the H–O–H angle at position *m*. The five-site model (ST2) places the oxygen charge on two lone pair positions with an overall tetrahedral geometry. The geometry and descriptive parameters for some of these models are given in Table I. Obviously, the more sites involved in the model, the more nonbonding terms are involved in any calculation using it. For example, a water dimer calculation involves 9 pairwise interactions with the 3-site models, 10 with the 4-site

[62] W. L. Jorgensen, J. Chandrasekhar, J. D. Madura, R. W. Impey, and M. L. Klein, *J. Chem. Phys.* **79,** 926 (1983).
[63] M. M. Teeter, *Annu. Rev. Biophys. Biophys. Chem.* **20,** 577 (1991).
[64] V. Daggett and M. Levitt, *Annu. Rev. Biophys. Biomol. Struct.* **22,** 353 (1993).

models (only 1 Lennard–Jones calculation is used, the O--O term), and 17 with the 5-site model (16 electrostatic, 1 Lennard–Jones).

The three-site (TIP3P) and four-site (TIP4P) models predict liquid water properties fairly well (densities of 0.982 and 0.999 g/cm^3 compared to the experimental values of 0.997 g/cm^3; heats of vaporization of 10.45 and 10.66 kcal/mol compared to 10.51; compressibility of 18×10^6 and 35×10^6 atm^{-1} compared to 45.8×10^6). The five-site ST2 model more accurately represents the gas-phase water dimer geometry, but offers no improvement in liquid water properties.

Flexible and polarizable water models have also been proposed (see Daggett and Levitt[64]; Corongiu[65]), presenting a more realistic model of solvent. Their disadvantage to protein modeling lies in their computational demands with more complex and/or more numerous terms.

C. *Molecular Mechanics and Metal Centers in Proteins*

The problem of modeling the interactions of metals in proteins is due to their complex coordination chemistry, spin variability, and the polarizability of the outer shell electrons. Attempts to model bound metals using only the potential functions described above often give unrealistically strong nonbonding interactions and distorted metal center geometries. In part these problems may be attributed to parameterization of the force fields, because there are not nearly as many examples of metal center structures for optimization as there are amino acid residue structures in proteins. Indeed, in many cases the parameters for metal centers are little more than educated guesses. Second, quantum mechanical calculations of geometries, charges, and so on, for metal centers are often thwarted by lack of small molecule structure analogs and the computational demands of transition metal calculations (see Salahub and Zerner[66]).

If a nonbonded interaction approach is taken to model the metal–protein complex, unrealistically strong electrostatic attraction and repulsion terms between a highly charged metal ion and other charged groups (either positive or negative) may result in inappropriate metal–protein ligation (e.g., hexacoordinated zinc, when tetravalent coordination is desired) or, in some cases, expulsion of the metal from the protein due to its proximity to other positively charged atoms. it should be noted that the net charge of a typical metal center (the metal ion and its ligating groups) is low and that the effects of each charged atom are reduced

[65] G. Corongiu, *Int. J. Quantum Chem.* **42,** 1209 (1992).

[66] D. R. Salahub and M. C. Zerner (eds.), "The Challenge of d and f Electrons." American Chemical Society, Washington, D.C. 1989.

considerably when a distance-dependent dielectric constant is used. In fact, a distance-dependent dielectric might be a useful first-order correction because the metal ion charge should be masked somewhat from the surrounding atoms by offsetting charges on nearby ligands. However, when solvent is incorporated in molecular models with no distance-dependent reduction in charge, the influence of each atom of such charged centers is more widely felt. Thus, metal ions, in particular, accentuate the limitations of a Coulombic electrostatic model in nonbonded interactions.

The key to this approach is well-developed parameters. Deerfield and co-workers[67,68] have used small molecule (oxygen, dimethyl phosphate, formate, etc.) calcium and magnesium ion complexes minimized on the basis of nonbonded interactions (van der Waals and electrostatics). After optimizing the parameters by comparison with *ab initio* calculations, they compared the effects of the two ions on small cyclic peptides using molecular mechanics minimization (AMBER).

Another way to handle metals using a nonbonded approach is through the use of constraints. The distances and angles between the ligands and metal can then be fixed, thus holding the metal in place.[69] Although this may be artificial, it appears to be acceptable on an ad hoc basis. Merz and Kollman[70] used this approach to study the dynamics of thermolysin, a zinc metalloprotein, constraining residues in the vicinity of the zinc atom so that they could not form bonds with the zinc atom.

There are a number of other ways to handle these problems. Of course, one could simply remove the metal. However, because metal ions play important roles not only in the function of proteins, but in their structure, an apoprotein structure deficient in a major component is not likely to model physiologically important structures. In one study,[71] the active site zinc of carboxypeptidase A was removed before it was minimized and its dynamics simulated. As might be expected, the active site underwent significant structural rearrangements, including *cis* to *trans* isomerization of a peptide bond. Another method, still unacceptable although less drastic, is to reduce the electrostatic contributions of all atoms so that there

[67] A. T. Maynard, M. A. Eastman, T. Darden, D. W. Deerfield II, R. G. Hiskeyl, and L. Pedersen, *Int. J. Pept. Protein Res.* **31,** 137 (1988).

[68] D. W. Deerfield, M. A. Lapadat, L. L. Spremulli, R. G. Hiskey, and L. Pedersen, *J. Biomol. Struct. Dyn.* **6,** 1077 (1989).

[69] M. S. Lee, G. P. Gippert, K. V. Soman, D. A. Case, and P. E. Wright, *Science* **245,** 635 (1989).

[70] K. M. Merz, Jr., and P. A. Kollman, *J. Am. Chem. Soc.* **111,** 5649 (1989).

[71] M. W. Makinen, J. M. Troyer, H. van der Werff, H. J. C. Berendsen, and W. F. van Gunsteren, *J. Mol. Biol.* **209,** 201 (1989).

is less distortion due to these terms.[29–30,72] In the extreme, calculations can be performed without the electrostatic component. However, because electrostatics interaction energies tend to be the largest single component of a force field evaluation and force fields are often parameterized by global optimization, this approach is risky. It can result in a "do-nothing" calculation: a minization that does not change the starting structure or a simulation that samples a biased region of conformational space.

A common approach used with a variety of different metal centers is to bond the ligands to the metal explicitly. A number of examples for electron transport and redox proteins are described below. Once the geometry of the metal center or some small molecule analog has been fixed, it is then parameterized like any other residue in the protein. Although this approach is useful in many cases, the result is essentially to fix the coordination geometry and prevent the possibility of ligand exchange. This is the approach that has been used for heme proteins and iron-sulfur proteins in a number of cases (see Section *D.4* below). Merz[73–75] has detailed an approach to parameterize zinc in order to use the standard AMBER force field for calculations on carbonic anhydrase.

Bernhardt and Combs[76] have modeled a variety of transition metal complexes using a bonded representation with geometrical freedom in the ligand–metal–ligand bond angle. Their force field, bonding terms only, was parameterized using a wide range of crystal structures for carboxylate, amine, and imine complexes of Cd^{3+} and Ni^{2+}; amine and imine complexes of Cr^{3+} and Fe^{3+} (low spin); amines of Zn^{2+} and Rh^{3+}; and amine, imine, carboxylate, and sulfur complexes of Cu^{2+}. Only in the case of copper was the problem of variable geometry and coordination number addressed. Calculations were not extended to biological molecules, however.

Although the approach of Bernhardt and Combs[76] allows some variation in geometry for copper coordination at least, it does not address exchange of ligands or shifts between ligand positions. Sironi[77] describes an addition to MM2 that allows minimization of metal carbonyl clusters with exchange of ligands by calculating the bonding interactions on the basis of the geometric arrangement at each step in minimization.

A more appropriate approach than all of these would be to represent the appropriate ionic bonds of a metal center with a new force field compo-

[72] T. J. Gibson, J. P. M. Postma, R. S. Brown, and P. Argos, *Protein Eng.* **2,** 209 (1988).

[73] K. M. Merz, *J. Am. Chem. Soc.* **113,** 406 (1991).

[74] K. M. Merz, *J. Am. Chem. Soc.* **113,** 3572 (1991).

[75] K. M. Merz, M. A. Murchko, and P. A. Kollman, *J. Am. Chem. Soc.* **113,** 4484 (1991).

[76] P. V. Bernhardt and P. Combs, *Inorg. Chem.* **31,** 2638 (1992).

[77] A. Sironi, *Inorg. Chem.* **31,** 2467 (1992).

nent and the covalent attachments as described above. Ionic bonding to metals, such as hydrogen bonds in CHARMM, could be evaluated with both distance and angle dependence. Vedani and co-workers[29,30] modified AMBER with a specific metal center potential [Eq. (18)] with a Lennard–Jones 10-12 component and a cosine angle dependence:

$$V_{\mathrm{MET}} = \Sigma[(A''/r^{12}) - (C''/r^{10})]\ \pi[\cos^{k}(\phi_{\mathrm{L\text{-}M\text{-}L}} - \phi_0)] \tag{18}$$

where A'' and C'' are the Lennard–Jones parameters and preferred angles, ϕ_0, are defined for all ligands. This force field, called YETI, also uses a reduced electrostatic contribution for protein metal ions and was initially parameterized only for zinc metal centers, but has been expanded.[31]

D. Adding Parameters to the Force Field

Use of molecular mechanics force fields to study electron transfer proteins is currently limited by the availability of parameters for the various prosthetic groups. These in turn are limited by the number of available structures than can be used to optimize the parameters and the size of the molecules that must be treated quantum chemically to derive parameter values.

In determining the parameters for a prosthetic group, there are several considerations. Is it covalently bound to the protein or held in place only by nonbonding interactions? Is its geometry rigorously maintained in different proteins or is it flexible? What is the electronic complexity of atoms involved?

With covalently bound prosthetic groups, one consideration is how many atoms of the linking amino acid residue(s) must be included to model the structure and electronic properties of the center accurately. Because quantum mechanical calculations are often used to obtain starting values for many force field parameters, there is a conflict between the simplicity needed for such calculations and the complexity needed for the derived parameters to be transferable to the macromolecule.

Force field parameters are derived from multiple types of data, vibrational spectra, *ab initio* and semiempirical quantum mechanical calculations, NMR studies, and so on. Examples of procedures used for metal proteins are cited above and in some particular examples detailed below. Atom-centered point charges are a particular problem, yet their values have a significant impact on the evaluation of the nonbonded part of molecular mechanics interactions in proteins. Their determination seems to be a limiting aspect of parameterization of prosthetic groups, particularly of metal centers and groups that are not covalently attached. Another justification for focusing on this issue is the use of these charges in defining interaction sites on proteins and in higher order electrostatics calculations that can now be used to understand binding, redox chemistry, and solvation energetics (see Sections IV,B,3 and IV,B,4). The next section summarizes the approaches. The discussion then focuses on organic components and metal centers separately. Finally, the last section summarizes information from the literature on parameters used for several different cases of both types of prosthetic groups.

1. Point Charge Approaches. It is important to remember that atom-centered point charges are derived from the combination of nuclear charge and the electron charge distribution. Part of the problem in defining point charges lies in defining how the electron density is divided between bonded atoms. Even with an experimental approach (see Coppens[78]), it can be ambiguous. With semiempirical and *ab initio* methods atom-centered charges are not a direct outcome of the calculations and the three approaches to determining them are (1) to sum charges from the electronic population of the atomic orbitals from *ab initio* and semiempirical calculations,[79] referred to as Mulliken populations or Mulliken charges, (2) to integrate or partition the charge density around the atoms based on physical boundaries,[80] referred to here as CD charges, and (3) to fit the charges to an electrostatic potential map of the area in space around the molecule,[81] referred to as electrostatic potential-derived or ESP charges. It is argued that ESP charges are more appropriate for molecular mechanics because they are used, in fact, to calculate electrostatic potential interaction energies. In addition to these methods there are a number of more empirical methods.[82–86]

2. Organic Molecules and Moieties. With organic molecules the correlation between calculated and experimental dipole and quadrapole moments for a variety of small molecules is best with ESP charges[87,88] and

[78] P. Coppens, *Annu. Rev. Phys. Chem.* **43,** 663 (1992).

[79] R. S. Mulliken, *J. Chem. Phys.* **23,** 1833 (1955).

[80] D. A. Case and M. Karplus, *Chem. Phys. Lett.* **39,** 33 (1976); Another example is the Self-Consistent Charge and Configuration (SCCC) calculation charge partitioning of J. J. Alexander and H. B. Gray, *Coord. Chem. Rev.* **2,** 29 (1967).

[81] F. A. Momany, *J. Phys. Chem.* **82,** 592 (1978).

[82] H. Berthod and A. Pullman, *J. Chim. Phys.* **62,** 942 (1965).

[83] L. Dosen-Micovic, D. Jeremic, and N. L. Allinger, *J. Am. Chem. Soc.* **105,** 1716 (1983).

[84] W. J. Mortier, S. K. Ghosh, and S. Shankar, *J. Am. Chem. Soc.* **108,** 4315 (1986).

[85] J. Mullay, *J. Am. Chem. Soc.* **108,** 1770 (1986).

[86] R. J. Abraham and P. E. Smith, *J. Comput. Chem.* **9,** 288 (1987).

[87] S. R. Cox and D. E. Williams, *J. Comput. Chem.* **2,** 304 (1981).

[88] C. Aleman, F. J. Luque, and M. Orozco, *J. Comput. Chem.* **14,** 799 (1993).

a number of methods have been published to evaluate them.[89–92] Selection between empirical, semiempirical, and *ab initio* methods is often determined by the size of the molecular system and availability of high-performance computing. Indeed, it is fairly common to mix charges derived from different procedures in macromolecular calculations. Fortunately, most studies have shown accurate linear correlations between ESP charges derived using different methods (Table II), making it possible to mix charges by using simple scaling factors. Table II summarizes the scaling factors from linear fits where the "standard" is an *ab initio* calculation using a 6-31G* basis set. Choice of this standard is defended below.

On the basis of this standard, then, the methods in Table II can be ordered roughly according the correlation coefficients of the fits to 6-31G* charges, r_{ab}. For the semiempirical methods this order is MNDO > AM1 = PM3. In a detailed study of over 1000 charges for peptides and monosaccharides, Merz[59] found a similar order of correlation to 6-31G*, that is, MNDO (0.96), AM1 (0.81), PM3 (0.70).

With ESP charges the initial requirement is that the potentials themselves be adequately described. Examination of this issue[93,94] seems to support the contention of earlier workers[9,87,95] that *ab initio* calculations with a 6-31G* or 6-31G** basis set are adequate for most organic molecules. Another correlation that seems to support this standard is the comparison with the experimental charges from Coppens,[78] although it is based on limited data (Table II; Su and Coppens[92]). Note, however, that there are significant differences in both sign and magnitude for ESP charges by the method of Su[96] and those from X-ray refinement.

However, when 6-31G* potentially derived charges are used to calculate molecular dipole moments, they consistently overestimate the dipole magnitude (see Table II). Thus the scaling procedure of Weiner *et al.*[9] was to scale up to the 6-31G** level and then correct the charges by the 6-31G** dipole moment correction (F_d = 0.91). Note that this approach is not the same as simply scaling the charges to the dipole moment. The most common comparison in the literature cited here is to charges and

[89] U. C. Singh and P. A. Kollman, *J. Comput. Chem.* **5,** 128 (1984).

[90] R. J. Woods, M. Khalil, W. Pell, S. H. Moffat, and V. H. Smith, Jr., *J. Comput. Chem.* **11,** 297 (1990).

[91] C. M. Breneman and K. B. Wiberg, *J. Comput. Chem.* **11,** 361 (1990).

[92] Z. Su and P. Coppens, *Z. Naturforsch. A: Phys. Sci.* **48,** 85 (1993).

[93] F. J. Luque, F. Illas, and M. Orozco, *J. Comput. Chem.* **11,** 416 (1990).

[94] J. Rodriguez, F. Manaut, and F. Sanz, *J. Comput. Chem.* **14,** 922 (1993).

[95] W. J. Hehre, L. Radom, P. von R. Schleyer, and J. A. Pople, "*Ab initio* Molecular Orbital Theory." Wiley, New York, 1986.

[96] Z. Su, *J. Comput. Chem.* **14,** 1036 (1993).

TABLE II

CORRELATIONS AND SCALING FACTORS FOR ATOM-CENTERED POINT CHARGES FOR ORGANIC MOLECULES FROM DIFFERENT COMPUTATIONAL METHODS

Method	Atoms and atom types[a]	Fab[b]	rab[b]	Fd[c]	rd[c]	Fa[d]	Ref.
Miscellaneous							
X-ray	C, H, N, O, Aro	1.01[e]	—	—	—	0.90	Coppens[f]
CHARGE2	C, H, N, O, P, Br, Cl, S, Aro, Cyc, Hcy	—	—	1.08	0.98	—	Abraham and Smith[g]
EN	C, H, N, O, Hcy	1.17[h]	—	—	—	1.04	Mullay[i]
Semiempirical							
AM1	C, H, N, O, Aro	1.39	0.89	1.06	0.74	1.24	Besler *et al.*[j]
AM1	C, H, N, O, F.Aro, Hcy	1.33	0.88	1.34	0.94	1.18	Aleman *et al.*[k]
CNDO	C, H, N, O, P, Br, Cl, S, Aro, Cyc, Hcy	—	—	1.11	0.96	—	Abraham and Smith[g]
MNDO	C, H, N, O, S, Aro	1.42	0.97	1.26	0.60	1.26	Besler *et al.*[j]
MNDO	C, H, N, O, F.Aro, Hcy	1.31	0.98	1.27	0.92	1.17	Aleman *et al.*[k]
PM3	C, H, N, O, F.Aro, Hcy	1.41	0.88	1.19	0.93	1.25	Aleman *et al.*[k]
Ab initio							
STO-3G	C, H, N, O, B	1.13[l]	—	1.27	—	1.01	Cox and Williams[m]
STO-3G	C, H, N, O, S, Aro	1.10	0.96	—	—	0.98	Besler *et al.*[j]
6-31G	C, H, N, O, B	0.82[l]	—	0.83	—	0.73	Cox and Williams[m]
6-31G*	C, H, N, O, F.Aro, Hcy	—	—	0.89	0.99	0.89	Aleman *et al.*[k]
6-31G**	C, H, N, O, B	—	—	0.91	—	0.91	Cox and Williams[m]

[a] Atoms found in molecules calculated and molecule types: Aro, aromatic; Hcy, heterocyclic; Cyc, simple cyclic.

[b] Scale factor, F_{ab}, to scale the calculated charges, q, from each method to the *ab initio* 6-31G* level. $q_{6\text{-}31G*} = F_{ab}q$. r_{ab} is the linear correlation coefficient.

[c] Scale factor, F_d, to scale the calculated dipole moment, μ_c, from the atom-centered charges to the experimental dipole moments, μ_c. $\mu_c = F_d\mu_c$. r_d is the linear correlation coefficient.

[d] AMBER-like scaling factor, $F_a = 0.89F_{ab}$.

[e] Comparison from this paper and STO-3G results of U. C. Singh and P. A. Kollman, *J. Comput. Chem.* **5,** 128 (1984) on 9-methyl-guanine (15 atoms total) gave a slope of 0.916 and a correlation coefficient of 0.87. Slope was then scaled to 6-31G* by a factor of 1.1.

[f] P. Coppens, *Annu. Rev. Phys. Chem.* **43,** 663 (1992).

[g] R. J. Abraham and P. E. Smith, *Nucleic Acids Res.* **16,** 2639 (1988).

[h] Comparison in this paper to STO-3G scaled to 6-31G* by a factor of 1.1.

[i] J. Mullay, *J. Comput. Chem.* **9,** 399 (1988).

[j] B. H. Besler, K. M. Merz, and P. A. Kollman, *J. Comput. Chem.* **11,** 431 (1990).

[k] C. Aleman, F. J. Luque, and M. Orozco, *J. Comput. Chem.* **14,** 799 (1993).

[l] Scale factor, F_{ab}, is for scaling to 6-31G** level, i.e., $q_{6\text{-}31G**} = F_{ab}q$.

[m] S. R. Cox and D. E. Williams, *J. Comput. Chem.* **2,** 304 (1981).

moments calculated from the 6-31G* basis set rather than 6-31G**. The 6-31G** basis set is identical to 6-31G* with the addition of a set of *p*-type functions added to hydrogen and helium atoms. The comparisons of electrostatic potential by both Luque *et al.*[93] and Rodriguez *et al.*[94] indicate that 6-31G* and 6-31G** calculations on a variety of molecules give similar results. Thus, "AMBER-like" scaling of charges calculated with the other methods of Table II can be carried out using the factors in the column headed "F_a."

As mentioned, the AMBER charges of Weiner et al.[9,10] are a mixture of results from different calculations. For the AMBER protein charges, the individual amino acid residues were treated as the combination of three distinct parts: the main chain atoms N, H, C_1, and O; the R-group spacer Cα and Cβ; and the "chromophore" composed of the rest of the R-group atoms. The *ab initio* calculations for the side chain moieties were performed with the modest STO-3G basis set scaled according to Cox and Williams,[87] as mentioned above. For all residues, the N, H, C_1, and O charges were assigned as 0.28, -0.52, 0.526, and -0.50 (United atom), respectively, leaving -0.246 to be distributed on the bridge atoms (or Cα for glycine) for electroneutrality. These backbone charges came from a more rigorous 6-31G calculation of *N*-methylacetamide scaled by a factor of 0.75.[87] Finally, the charge necessary to achieve the appropriate formal charge on the entire residue (0, 1, or -1) was distributed between the Cα and Cβ carbons according to their Mulliken populations. This approach achieves three purposes: it maintains the appropriate formal charge of each residue, the backbone atoms are the same atom type for all residues, and the ab initio calculations use a minimum number of atoms with the simplest chemical analog representing each side chain. Because the amino acid residues represent, as a group, many of the functional groups and structural moieties found in organic prosthetic groups, the AMBER approach could be used as a template for other organic molecules. However, complex prosthetic groups can simply be too large for high-order calculations. Groups such as hemes, dinucleotides, and flavin nucleotides involve 30 to 60 atoms, including numbers of heteroatoms.

3. Metal Centers and Clusters. A broad and well-balanced analysis of the suitability and applications of semiempirical and ab initio quantum mechanical calculations for transition metals is found in the book edited by Salahub and Zerner,[66] particularly in the overview of Chapter 1. The choice is again between three approaches: conventional *ab initio* methods, density functional methods, and semiempirical methods.

Ab initio methods for systems with *d* or *f* shell electrons generally involve electron correlation effects. Some of the methods involved (see

Hehre *et al.*[95]) are variously known as configuration interaction (CI), limited CI (CIS, CID, CISD), and Moller–Plesset perturbation theory (MP1, MP2, etc.), in which the ground state wavefunction and energy are expanded in a power series (truncated at the MPn term). Other methods of importance to transition metal calculations are the complete active space (CAS), general valence bond (GVB), and modified coupled-pair functional (MCPF) approaches (see Salahub and Zerner[66]; Chong and Langhoff[97]). A simplifying approach to *ab initio* methods is the use of effective core potentials (ECP) to represent the core electrons of the heavy atoms along with a basis set representation of the valence electrons.

With *ab initio* methods it is not clear which basis set performs best for transition metals. Barnes *et al.*[98] report that it is necessary to correlate all of the metal and CO valence electrons in order to evaluate dissociation energies of first- and second-row transition metal carbonyl complexes. Metal$^+$–water-binding energies evaluated with the 10 first-row transition metals (Sc through Zn) using a number of different basis sets and levels of correlation[99] gave only modest correlation with experimental values even in the best case (see Table III). As can be seen from Table III, the various methods gave varying results.

The simplest density functional theory (DFT) approaches are the Xα [Hartree–Flock–Slater (HFS)] or local density approximation (LDA) methods. Developments in these methods for molecular systems have been reviewed by Ziegler.[100] Modifications of the basic Xα and LDA schemes have allowed extension of the methods to include correlation (e.g., GCX) for larger systems and heavier atoms.[100–102] With third-row transition metals relativistic effects must also be included.[103]

In addition to ECP-based calculations, which can also be considered semiempirical, there have been a number of applications of classic semiempirical approaches to metal complex calculations. MNDO, AM1, and PM3 parameters are available variously for zinc, cadmium, and mercury of the transition metals and other metals of biological interest (Na, Mg,

[97] D. P. Chong and S. R. Langhoff, *J. Chem. Phys.* **84,** 5606 (1986).

[98] L. A. Barnes, M. Rosi, and C. W. Bauschlicher, Jr., *J. Chem. Phys.* **93,** 609 (1990).

[99] E. Magnusson and N. W. Moriarty, *J. Comput. Chem.* **14,** 961 (1993).

[100] T. Ziegler, *Chem. Rev.* **91,** 651 (1991).

[101] D. A. Case, *Annu. Rev. Phys. Chem.* **33,** 151 (1982).

[102] A. D. Becke, *in* "The Challenge of *d* and *f* Electrons" (D. R. Salahub and M. C. Zerner, eds.), p. 165. American Chemical Society, Washington, D.C., 1989.

[103] T. Ziegler, J. G. Snijders, and E. J. Baerends, *in* "The Challenge of *d* and *f* Electrons" (D. R. Salahub and M. C. Zerner, eds.), p. 322. American Chemical Society, Washington, D.C., 1989.

TABLE III

CORRELATION COEFFICIENTS, r_c, FOR CALCULATED ENERGIES AND EXPERIMENTAL ENERGIES FOR TRANSITION METAL COMPLEXES

Method	Basis set	Energy calculated	r_c	Ref.
Ab initio methods				
MCPF	Modified Wachters[a]	Metal–acetylene	0.26	Sodupe and Bauschlicher[b]
MCPF	See paper	Metal–noble gas	0.98	Partridge *et al.*[c]
MCPF	Modified Wachters[a]	M^+–benzene	0.87	Bauschlicher *et al.*[d]
CASSCF	Modified Wachters[a]	M^+–methylene	0.91	Bauschlicher *et al.*[c]
CISD[f]	Modified Wachters[a]	M^+–water	0.70	Magnusson and Moriarty[a]
ECP-MP2/3	Double-zeta	Iron–ligand reactions	0.94	McKee[h]
ECP-MP2	Double-zeta	Iron reactions	0.71	Schröder *et al.*[i]
DFT methods				
LDA-GCX	—	Metal–ligand	0.95	Ziegler[j]
Semiempirical methods				
SINDO1	See paper	Various heats of formation (inorganics)	0.99	Li *et al.*[k]

[a] A. J. Wachters, *J. Chem. Phys.* **52,** 1033 (1970) and cited reference.
[b] M. Sodupe and C. W. Bauschlicher, Jr., *J. Phys. Chem.* **95,** 8640 (1991).
[c] H. Partridge, C. W. Bauschlicher, Jr., and S. R. Langhoff, *J. Phys. Chem.* **96,** 5350 (1992).
[d] C. W. Bauschlicher, Jr., H. Partridge, and S. R. Langhoff, *J. Phys. Chem.* **96,** 3273 (1992).
[e] C. W. Bauschlicher, Jr., H. Partridge, J. A. Sheehy, S. R. Langhoff, and M. Rosi, *J. Phys. Chem.* **96,** 6969 (1992).
[f] Quadratic CISD with triples contributions.
[g] E. Magnusson and N. W. Moriarty, *J. Comput. Chem.* **14,** 961 (1993).
[h] M. L. McKee, *J. Phys. Chem.* **96,** 1683 (1992).
[i] D. Schröder, A. Fiedler, J. Hrušák, and H. Schwarz, *J. Am. Chem. Soc.* **114,** 1215 (1992).
[j] Ziegler as reported by A. D. Becke, *in* "The Challenge of *d* and *f* Electrons" (D. R. Salahub and M. C. Zerner, eds.), p. 165. American Chemical Society, Washington, D.C., 1989.
[k] P. Li, P. C. de Mello, and K. Jug, *J. Comput. Chem.* **13,** 85 (1992).

Si, K, Se, and Pb). The INDO-based methods of Zerner and colleagues (ZINDO[104]), and that of Jug and coworkers (SINDO1[105]) have both been parameterized for first-row transition metals.

Other methods, classified as empirical, that have been used for metal

[104] W. D. Edwards and M. C. Zerner, *Theor. Chim. Acta* **72,** 347 (1987).
[105] P. Li, P. C. de Mello, and K. Jug, *J. Comput. Chem.* **13,** 85 (1992).

TABLE IV
QUANTUM MECHANICS CALCULATED CHARGES ON METAL HEXACYANO COMPLEXES

Complex	Atom	Method			
		I[a]	II[b]	III[c]	IV[d]
$Cr(CN)_6^{3-}$	Cr	0.12	0.47	1.21	1.74
	C	−0.36		−0.17	−0.45
	N	−0.16		−0.53	−0.54
$Mn(CN)_6^{3-}$	Mn		0.48	1.13	1.72
	C			−0.14	−0.45
	N			−0.55	−0.34
$Fe(CN)_6^{3-}$	Fe(II)		0.45	0.98	1.66
	C			−0.05	−0.42
	N			−0.61	−0.36
$Co(CN)_6^{3-}$	Co	0.18	0.41	0.84	1.63
	C	−0.21		−0.04	−0.38
	N	−0.32		−0.60	−0.39
$Fe(CN)_6^{4-}$	Fe(III)		0.42	0.83	1.38
	C			−0.06	−0.38
	N			−0.74	−0.52

[a] CD charges from SC calculation; S. Kida, K. Nakamoto, J. Fujita, and R. Tsuchida, *Bull. Chem. Soc. Jpn.* **31,** 79 (1958).
[b] CD charges from SCCC calculation; J. J. Alexander and H. B. Gray, *J. Am. Chem. Soc.* **90,** 4260 (1968).
[c] Mulliken charges from DV-Xα; M. Sano, H. Adachi, and H. Yamatera, *Bull. Chem. Soc. Jpn.* **54,** 2898 (1981).
[d] Mulliken charges from SCF calculation; M. Sano, H. Kashiwagi, and H. Yamatera, *Inorg. Chem.* **21,** 3837 (1982).

center charges are the electronegativity equalization method, EEM,[106] and integrated charge density from X-ray crystal analysis.[107]

As was the case with organic molecules, different approaches give different results in evaluation of point charges. There are few data on the use of potential surface fitting to derive charges involving metal ion complexes,[73,74] but Mulliken population analysis and charge density partitioning results illustrate this situation. Table IV gives the Mulliken charges on some metal hexacyano complexes calculated using of these approaches. In spite of the earlier cautions about using Mulliken charges in molecular

[106] J. Shen, C. F. Wong, S. Subramaniam, T. A. Albright, and J. A. McCammon, *J. Comput. Chem.* **11,** 346 (1990).
[107] N. Li, P. Coppens, and J. Landrum, *Inorg. Chem.* **27,** 482 (1988).

mechanics and about mixing charges from different sources without appropriate scaling, some efforts at molecular dynamics simulations involving metal-containing proteins have used Mulliken charges mixed with potential derived charges.[108,109]

4. Prosthetic Groups in Electron Transfer Proteins. Parameters for prosthetic groups in electron transfer proteins for molecular mechanic, dynamics, and electrostatic potential calculations have been derived from many different sources. Some parameters, particularly metal center charges, are arrived at using little more than an educated guess or by empirical adjustment, such as reducing charges until the geometry is maintained. The following sections summarize some of the available data for charge parameters for a variety of prosthetic groups. Note that other bonded and nonbonded interaction parameters are often given in these same papers.

Cu^{2+} in super oxide dismutase: Superoxide dismutase (SOD) has intrigued a number of workers because of clear evidence that the reaction is driven by electrostatic attraction between the simple substrate, $O_2^{\cdot -}$, and the exposed Cu^{2+} site of the enzyme. Thus, the electrostatic potential field around the Cu^{2+} site has been the focus of electrostatic potential field calculations.[110,111] Although these early studies placed the full formal charge (2+) on the copper, more recent studies (see Shen *et al.*[106]) suggest much reduced values with distribution of charge onto the atoms of the ligands. Such reduced charge models have been used in molecular dynamics simulations.[112–114] In the Banci *et al.* simulation[114], the *ab initio* partial charges of Shen *et al.*[106] were used along with other metal–ligand parameters developed by Merz and colleagues[73,75] for zinc in carbonate dehydratase.

Cu^{2+} in stellacyanin: Fields *et al.*[115] modeled the structure of stellacyanin on the basis of that of cucumber basic protein (CBP), using molecular

[108] L. Banci, I. Bertini, P. Carloni, C. Luchinat, and P. L. Orioli, *J. Am. Chem. Soc.* **114,** 10683 (1992).

[109] L. Banci, I. Bertini, F. Capozzi, P. Carloni, S. Ciurli, C. Luchinat, and M. Piccioli, *J. Am. Chem. Soc.* **115,** 3431 (1993).

[110] E. D. Getzoff, J. A. Tainer, P. K. Weiner, P. A. Kollman, J. S. Richardson, and D. C. Richardson, *Nature (London)* **306,** 287 (1983).

[111] I. Klapper, R. Hagstrom, R. Fine, K. Sharp, and B. Honig, *Proteins* **1,** 47 (1986).

[112] J. Shen, S. Subramaniam, C. F. Wong, and J. A. McCammon, *Biopolymers* **28,** 2085 (1989).

[113] J. Shen and J. A. McCammon, *Chem. Phys.* **158,** 191 (1991).

[114] L. Banci, P. Carloni, G. La Penna, and P. L. Orioli, *J. Am. Chem. Soc.* **114,** 6994 (1992).

[115] B. A. Fields, M. Guss, and H. C. Freeman, *J. Mol. Biol.* **222,** 1053 (1991).

mechanics minimization and molecular dynamics. The charge on the copper atom was empirically reduced to 0.5, copper ligand bonds and angle were defined with low force constants, and an improper torsion was used to maintain the histidine–copper–histidine plane.

FeS_4 in rubredoxin: Molecular mechanics minimizations and dynamics simulations for rubredoxin to date[116–119] have used geometry parameters taken from multiple X-ray structures of the protein and empirically determined force field parameters (see Wampler *et al.*[118]). The charges on the cysteine atoms of the cluster were standard AMBER parameters and that on the iron atom was 1.5. The geometry of the FeS_4 center has been modeled by a number of inorganic complexes.[120,121] ECP[122] and CI[123] *ab initio* and DFT[124] calculations have been reported for model centers, but no populations or charges were reported. Noodleman *et al.*[125] report valence shell populations for a number of different models of iron-sulfur clusters using the SW-Xα method.

Fe_2S_2 in plant-type ferredoxins: DFT Xα and *ab initio* calculations have been carried out on models of the Fe_2S_2 center[126,127] with Mulliken charges on the irons and sulfurs (oxidized Fe = 1.47, S = −0.91; reduced Fe = 1.34, S = −1.04) reported from the latter work. The SW-Xα results of Noodleman et al.[125] also evaluated valence shell populations for this species.

Fe_4S_4 in high-potential iron-sulfur protein and bacterial ferredoxin: Banci *et al.*[108,109] have modeled the molecular dynamics of high-potential iron-sulfur protein (HiPIP) using the AMBER 4.0 force field. The parameters of the iron center were derived from several sources. Nonbonding parameters included atom-centered point charges from the Fe_2S_2 cluster calculations of Carloni and Corongiu[127] and spectroscopic data. The

[116] D. E. Stewart, P. K. Weiner, and J. E. Wampler, *J. Mol. Graphics* **5,** 133 (plate on 145) (1987).

[117] D. E. Stewart and J. E. Wampler, *Proteins* **11,** 142 (1991).

[118] J. E. Wampler, E. A. Bradley, M. W. W. Adams, and D. E. Stewart, *Protein Sci.* **2,** 640 (1993).

[119] E. A. Bradley, D. E. Stewart, M. W. W. Adams, and J. E. Wampler, *Protein Sci.* **2,** 650 (1993).

[120] R. W. Lane, J. A. Ibers, R. B. Frankel, and R. H. Holm, *Proc. Natl. Acad. Sci. U.S.A.* **72,** 2868 (1975).

[121] V. K. Yachandra, J. Hare, I. Moura, and T. G. Spiro, *J. Am. Chem. Soc.* **105,** 6455 (1983).

[122] R. A. Bair and W. A. Goddard III, *J. Am. Chem. Soc.* **99,** 3505 (1977).

[123] R. A. Bair and W. A. Goddard III, *J. Am. Chem. Soc.* **100,** 5669 (1978).

[124] J. G. Norman, Jr., and S. C. Jackels, *J. Am. Chem. Soc.* **97,** 3833 (1975).

[125] L. Noodleman, J. G. Norman, Jr., J. H. Osborne, A. Aizman, and D. A. Case, *J. Am. Chem. Soc.* **107,** 3418 (1985).

[126] J. G. Norman, P. B. Ryan, and L. Noodleman, *J. Am. Chem. Soc.* **102,** 4279 (1980).

[127] P. Carloni and G. Corongiu as reported by Banci *et al.*, Ref. 109.

TABLE V
CHARGES FOR Fe_4S_4 CENTER MODELS

Method	Model	Charges	Ref.
MS-Xα CD charges	$Fe_4S_4^*(SCH_3)_4^{2-}$	Fe, −0.05 S*, −0.27 S, −0.18 C, 0.27 H, −0.09	Aizman and Case[a]
SW-Xα CD charges	$Fe_4S_4^*(SH)_4^{2-}$	Fe, 0.02 S*, −0.35 S, −0.17	Noodleman *et al.*[b]
	$Fe_4S_4^*(SH)_4^{3-}$	Fe, −0.03 S*, −0.44 S, −0.28	
Ab initio Mulliken	$Fe_4S_4^*(SCH)_4^{2-}$	Fe, 1.47 S*, −0.91	Carloni and Corongiu[c]
	$Fe_4S_4^*(SCH)_4^{3-}$	Fe, 1.34 S*, −1.04	

[a] A. Aizman and D. A. Case, *J. Am. Chem. Soc.* **104,** 3269 (1982).
[b] L. Noodleman, J. G. Norman, Jr., J. H. Osborne, A. Aizman, and D. A. Case, *J. Am. Chem. Soc.* **107,** 3418 (1985).
[c] P. Carloni and G. Corongiu as reported by L. Banci, I. Bertini, P. Carloni, C. Luchinat and P. L. Orioli, *J. Am. Chem. Soc.* **114,** 10683 (1992).

Lennard–Jones parameters for iron were taken from values used for myoglobin in early calculations[128] and those for both the cysteine sulfur and the inorganic sulfur were assigned the normal AMBER cysteine sulfur values. Bonds were fixed to the crystal structure values, angle parameters were adjusted empirically, and torsion angle constants were set to zero for the atoms within the cluster. Equilibrium angles were given the crystal structure values. Although the double lone pair representation of cysteine sulfur was retained, the inorganic sulfur atoms were not assigned lone pairs.

Other earlier calculations of charges as well as the ones assigned by Banci *et al.*[108] are listed in Table V. It is clear that the CD charges from DFT calculations and the Mulliken charges from the *ab initio* study do not correlate very well. It is also likely that ESP charges derived even using the same quantum mechanical methods would still be different.[87]

Flavin mononucleotide flavodoxin: Vinayaka and Rao[129] evaluated the conformational energy around the flavin-binding site of flavodoxin, using

[128] D. A. Case and M. Karplus, *J. Mol. Biol.* **132,** 343 (1979).
[129] C. R. Vinayaka and V. S. R. Rao, *J. Biomol. Struct. Dyn.* **2,** 663 (1984).

a three-term force field (van der Waals, electrostatics, and torsion potentials). The van der Waals potential was described by Vasudevan and Rao.[130] A combination of semiempirical charges based on Del Re and Huckel methods[82] were used for the flavin mononucleotide (FMN) atoms. A dielectric constant of 3.5 was used. Torsion potentials (threefold) for atoms along the ribose-phosphate side chain were taken from the literature (see Vinayaka and Rao[129]). The isoalloxazine ring was assumed to be planar. Stewart[131] has carried out molecular dynamics simulations on a protein complex with flavodoxin using a set of AMBER parameters derived from these sources. It is interesting that changes in charges and geometry for one and two electron reductions of lumiflavin have been reported using semiempirical MINDO/3 and *ab initio* 3-21G calculations.[132]

Hemes in cytochromes: Both current versions of AMBER (AMBER 4.0[11]) and CHARMM (CHARMM 2.2[6]) are supplied with parameters for heme groups. Neither set of parameters is documented, but a paper by Lopez and Kollman[133] gives some of the details for AMBER parameters for oxygen and carbon monoxide-binding heme groups and publications describing the CHARMM parameters are in preparation.[134] Parameters for bonding and nonbonding force field terms are also available in papers by McCammon *et al.*,[135] Case and Karplus,[128] and Northrup *et al.*[136] Detailed geometry parameters are also available from X-ray crystal studies of porphyrins.[107]

Partial atomic charges for heme atoms are available from both experimental and theoretical studies. In the case of Lopez and Kollman,[133] the heme charges were developed by combining results from calculations on component parts. The charge distribution of the heme–histidine Prep file supplied with AMBER is more asymmetric (the Lopez and Kollman models are for unsubstituted porphyrin rings), but the charges are of the same magnitude. The CHARMM charges are different and much larger. The heme iron has a full formal charge balanced by equal offsetting negative charges on each of four ring nitrogens. The earlier force field of McCam-

[130] T. K. Vasudevan and V. S. R. Rao, *Int. J. Biol. Macromol.* **4,** 219 (1982).

[131] D. E. Stewart, Ph.D. Dissertation, University of Georgia, Athens (1989).

[132] S. A. Vazquez, J. S. Andrews, C. W. Murray, R. D. Amos, and N. C. Handy, *J. Chem. Soc., Perkin. Trans. 2,* 889 (1992).

[133] M. A. Lopez and P. A. Kollman, *J. Am. Chem. Soc.* **111,** 6212 (1989).

[134] Mackerell *et al.,* manuscripts in preparation for the CHARMm Parameters from the CHARMm 2.2 Documentation, Molecular Simulations Incorporated, Burlington, Massachusetts.

[135] J. A. McCammon, P. G. Wolynes, and M. Karplus, *Biochemistry* **18,** 927 (1979).

[136] S. H. Northrup, M. R. Pear, J. D. Morgan, and J. A. McCammon, *J. Mol. Biol.* **153,** 1087 (1981).

TABLE VI
PARTIAL ATOMIC CHARGES ON Fe^{2+} AND IRON-BOUND ATOMS IN HEMES

Pyrrole nitrogens	Axial atom (type)	Axial ligand	Iron	Ref.
−0.45	−0.23 (N)	Pyridine	0.71	Li *et al.*[a]
	−0.23 (N)	Pyridine		
na[b]	−0.07 (O)	O_2	1.02	Yamamoto and Kashiwagi[c]
	na[b] (N)	Histidine		
−0.18	−0.17 (S)	Methionine	0.24	Northrup *et al.*[d]
	−0.19 (N)	Histidine		
−0.5	−0.10 (S)	Methionine	2.00	CHARMM 2.2[e]
	−0.40 (N)	Histidine		
−0.5	0.0 (O)	O_2	2.00	CHARMM 2.2[e]
	−0.40 (N)	Histidine		
−0.5	0.021 (C)	CO	2.00	CHARMM 2.2[e]
	−0.40 (N)	Histidine		
−0.18	−.21 (O)	O_2	0.27	Lopez and Kollman[f]
	−.12 (N)	2-Methyl imidazole		
−0.18	0.10 (C)	CO	0.20	Lopez and Kollman[f]
	−0.47 (N)	Pyridine		
−0.18	−0.025 (S)	Methionine	0.25	AMBER 4.0[g]
	−0.50 (N)	Histidine		

[a] Experimental charges from electron densities measured by X-ray diffraction; N. Li, P. Coppens, and J. Landrum, *Inorg. Chem.* **27,** 482 (1988).
[b] na, Not available from this paper.
[c] Mulliken population charges from *ab initio* calculations; S. Yamamato and H. Kashiwagi, *Chem. Phys. Lett.* **203,** 306 (1993).
[d] S. H. Northrup, M. R. Pear, J. D. Morgan, and J. A. McCammon, *J. Mol. Biol.* **153,** 1087 (1981).
[e] Parameters supplied with CHARMM 2.2.
[f] M. A. Lopez and P. A. Kollman, *J. Am. Chem. Soc.* **111,** 6212 (1989).
[g] Parameters supplied with AMBER 4.0.

mon et al.[135] was parameterized for molecular dynamics simulations of cytochrome *c* by Northrup *et al.,*[136] and has partial charges on the iron and ring nitrogens that are similar to the AMBER values, but with a more symmetrical distribution on the axial atoms. *Ab initio* calculations[137] and experimental measurements of electron density from high-resolution X-ray crystallography[107] indicate that the partial charges should probably be between these two extremes (Table VI).

[137] S. Yamamato and H. Kashiwagi, *Chem. Phys. Lett.* **203,** 306 (1993).

III. Applications of Protein Force Field

A. Minimization

The variables of the force field equation E_{pot} are the atomic positions. For minimization (evaluation of the minimum potential energy structure) these variables are unknowns to be evaluated on the basis of an initial guess (the starting geometry). The parameter set values (the r_0, ϕ_0, K values, etc., of the equations discussed above) are the constants of the equation. Reviewing these equations, it should be obvious that E_{pot} is a nonlinear function of the unknowns. Thus, minimization becomes a large nonlinear least-squares problem; a 2000-atom minimization involves a solution for 6000 x, y, z coordinate values.

The numerical methods for solution of such problems are well known and widely used in all areas of science, but the size of this problem is unusual. Nevertheless, several of the well-known methods are robust enough to find a minimum energy solution given sufficient computer power and numerical precision. Because the methods used evaluate the derivative(s) of the potential energy function in the local region of conformational space and use iteration to "home in," the minimum found is seldom, if ever, the global minimum. It is simply the nearest minimum to the initial structure. Thus, it will invariably correspond to a structure that is similar to it. For a discussion of different methods of minimization and their advantages and disadvantages the reader is referred to Chapter 3 in Burkett and Allinger,[21] and the appropriate sections of the papers by Brooks *et al.*,[13] Clark *et al.*,[7] and Weiner and Kollman.[8]

The issue of finding the global minimum by a combination of conformational searching and minimization has been discussed by Vajda et al.[138] Comparisons between approaches such as those of Roterman et al.[43,44] must be viewed in light of the information[138] that indicates that the force field minimum can be considerably different from the X-ray structure. This calls into question the formulation of the force fields and the form of representation of environmental effects during such attempts.

B. Minimization of Known Structures

There are three reasons to minimize a known X-ray or NMR structure: to test the molecular mechanics force field, to refine the structure, and to prepare the structure for molecular dynamics simulation. This section focuses on the first of these. Although force field minimization is often

[138] S. Vajda, M. S. Jafri, O. U. Sezerman, and C. DeLisi, *Biopolymers* **33,** 173 (1993).

used for structure refinement (e.g., Warme and Scheraga[139]), the force field and/or parameter sets used are often not the same as those used for other purposes. Minimization in preparation for molecular dynamics simulation, on the other hand, is subject to the same concerns discussed here.

One of the tests of molecular mechanics force fields that has been used by a number of workers is to evaluate how little change occurs when a high-quality X-ray crystal structure is minimized.[7,23,42] One caution should, however, be kept in mind concerning such tests. According to this criterion of a good force field, a computer program that reads the input structure and writes it to the output file carries out the best possible molecular mechanics minimization! Similarly, a test that excludes the long-range interactions of nonbonded electrostatics[7] is suspect unless the force field is parameterized to operate without this component. Another variable of such tests is the environment of the protein used for the minimization. Strictly speaking, minimization of an X-ray structure with the goal of minimal change should be done in the X-ray environment with atoms from neighboring molecules place according to crystallographic packing. It is well documented that minimization of X-ray structures in explicit solvent environments or in the absence of a molecular surrounding can change the structure from as little as a few tenths of an angstrom to over 1 Å root-mean-squared over all atoms.[7,23,42,118,140] Comparisons between solution NMR structures and X-ray structures and between multiple X-ray structures of the same protein show similar differences.[118,141]

The approach to minimization can be confusing with the large number of variables involved. Selection of force field type (All-Atom or United), nonbonded cutoff, options of the electrostatic calculation (or even to not calculate electrostatic potentials), minimization environment, minimization method, minimization in stages versus one step, and so on, will all have a bearing on the minimization results. Kini and Evans[140,142] have tested the AMBER force field as implemented in the commercial package SYBYL, concluding that X-ray coordinate positions are best maintained by minimization with an All-Atom force field with a single procedure for unrestricted conjugate gradient minimization. Whitlow and Teeter[42] investigated a number of different combinations of electrostatic selections, environment, and cutoff for AMBER United-Atom minimizations of cram-

[139] P. K. Warme and H. A. Scheraga, *Biochemistry* **13,** 757 (1974).

[140] R. M. Kini and H. J. Evans, *J. Biomol. Struct. Dyn.* **10,** 265 (1992).

[141] I. Chandrasekhar, G. M. Clore, A. Szabo, A. M. Gronenborn, and B. R. Brooks, *J. Mol. Biol.* **226,** 239 (1992).

[142] R. M. Kini and H. J. Evans, *J. Biomol. Struct. Dyn.* **9,** 475 (1991).

bin. The best results (defined by minimal change in the X-ray structure) are obtained, as might be expected, with minimizations in a crystal environment.[23,42] It should be noted that selections for minimal change in an X-ray structure may not be best for other uses of minimization, such as homology modeling. Indeed, in contrast to the recommendation of Kini and Evans,[140] Stewart and co-workers[118] found that multiple stages of minimization are needed in this application.

C. Homology Modeling

Another test proposed for comparing force fields and minimization methods is to mutate one known structure into another and minimize the resulting model.[143] With both structures known, this approach allows direct comparison in a case in which the "do-nothing" result is not acceptable. Different environments may be tested by minimizing both the model and original structures under the same conditions before comparison. Given appropriate test structures, this method can be extended with conformational searching procedures, simulated annealing, or any other methodology designed to find the global minimum structure. The ideal test structures would be small, but rigid, monomeric proteins lacking prosthetic groups. Unfortunately, in the current database of structures most of the multiple structures for small, rigid proteins contain prosthetic groups. Indeed, most such examples involve proteins that are the inspiration for this chapter! In our previous tests of the AMBER force field in homology modeling,[116,118,131] prosthetic groups were included along with the accompanying imprecisions in their parameters (see Section II,D).

Although there is a large database of X-ray structures, there will always be many more proteins of known amino acid sequence being studied experimentally and a much larger body of spectroscopic and enzymatic data pertaining to them. Therefore an important role for computational methods is to extend the structure database for sequenced proteins by predictions based on the known structures of homologs. For reviews of homology modeling approaches see Blundell *et al.*[144] and Sali *et al.*[145]

[143] J. E. Wampler, D. E. Stewart, and S. L. Gallion, *in* "Computer Simulation Studies in Condensed Matter Physics II" (D. P. Landau, K. K. Mon, and H.-B. Schuttler, eds.), p. 68. Springer-Verlag, Heidelberg, 1990.

[144] T. L. Blundell, B. L. Sibanda, J. E. Sternberg, and J. M. Thornton, *Nature* (*London*) **326,** 347 (1987).

[145] A. Sali, J. P. Overington, M. S. Johnson, and T. L. Blundell, *Trends Biochem. Sci.* **15,** 235 (1990).

More recent work includes a number of new methods.[146–149] Homology modeling takes advantage of two important facts of protein chemistry: (1) closely related proteins have similar tertiary structures,[150–153] and (2) single amino acid substitutions are generally compensated locally.[154,155]

There are three steps to homology modeling: sequence alignment, substitution or mutation of the differing residues, and refinement of the structure with the substituted residues. Because structure is often preserved to a greater degree than sequence,[144,151] a major challenge in homology modeling is to find the proper alignment. One method to improve the alignment process is to score the alignments of different amino acids on the basis of chemical properties[156] or genetics relatedness.[157] The actual alignment calculation and scoring can be accomplished using any number of programs[158–160] based on the Needleman–Wunsch algorithm.[161] Multiple sequence alignment algorithms have also been described[162–164] and selections of gap penalties and scoring methods can be weighted on the basis of the structural information.[159]

The substitution or mutation of the aligned residues can be carried out as simply as by editing the text version of the structure file (e.g., Stewart et al.[113]) or by using the mutation or change feature of one of the graphics-based protein modeling programs (see Section I,A). A more sophisticated approach is to place the mutated side chain atoms in a low-energy confor-

[146] C. A. Schiffer, J. W. Caldwell, P. A. Kollman, and R. M. Stroud, *Proteins* **8,** 30 (1990).

[147] A. P. Heiner, H. J. C. Berendsen, and W. F. van Gunsteren, *Protein Eng.* **6,** 397 (1993).

[148] S. H. Bryant and C. E. Lawrence, *Proteins* **16,** 92 (1993).

[149] N. Srinivasan and T. L. Blundell, *Protein. Eng.* **6,** 501 (1993).

[150] R. M. Sweet, *Biopolymers* **15,** 1565 (1986).

[151] C. Chothia and A. M. Lesk, *EMBO J.* **5,** 823 (1986).

[152] G. Vriend and C. Sander, *Proteins* **11,** 52 (1991).

[153] D. P. Yee and K. A. Dill, *Protein Sci.* **2,** 884 (1993).

[154] A. M. Lesk and C. H. Chothia, *Philos. Trans. R. Soc. London A* **317,** 345 (1986).

[155] T. Alber, S. Dao-pin, K. Wilson, J. A. Wozniak, S. P. Cook, and B. W. Matthews, *Nature (London)* **330,** 41 (1987).

[156] L. Kelly and L. A. Holladay, *Protein Eng.* **1,** 137 (1987).

[157] R. M. Schwartz and M. O. Dayhoff, *in* "Atlas of Protein Sequence and Structure Supplement 3" (Dayhoff, M. O., ed.), p. 353. National Biomedical Research Foundation, Washington, D.C., 1978.

[158] G. J. Barton and M. J. E. Sternberg, *Protein Eng.* **1,** 83 (1987).

[159] A. M. Lesk, M. Levitt, and C. Chothia, *Protein Eng.* **1,** 77 (1986).

[160] K. Nishikawa, H. Nakashima, M. Kanehisa, and T. Ooi, *Protein Sequence Data Anal.* **1,** 107 (1987).

[161] S. B. Needleman and C. D. Wunsch, *J. Mol. Biol.* **48,** 443 (1970).

[162] M. Murata, J. S. Richardson, and J. L. Sussman, *Proc. Natl. Acad. Sci. U.S.A.* **82,** 3073 (1985).

[163] M. S. Johnson and R. F. Doolittle, *J. Mol. Evol.* **23,** 267 (1986).

[164] W. R. Taylor, *J. Mol. Evol.* **28,** 161 (1988).

mation, scoring different conformations by a force field calculation.[146,165,166] One particularly difficult problem for this step in homology modeling is insertion or deletion of residues where a gap in sequence is found either in the known structure or in the unknown sequence, respectively. These changes usually occur in surface regions of proteins, areas where the structure is often variable even for very closely related proteins. One approach is to use the technology of secondary structure prediction (see Boscott *et al.*[167]) to develop a model of the inserted or deleted region. With deletions there is often an obvious solution that can be implemented by interactive modeling such as "clipping" off a turn of a helix, truncating an external loop, or shortening an antiparallel β-sheet. Dudek and Scheraga[168] have described an approach to global minimization of surface loops using a step-wise search sampling both backbone and sidechain conformations.

The final step, refinement, can be carried out using many of the same methods used to refine X-ray crystal or NMR structures: energy minimization with a step-wise[113,118,169] or one-step approach[142] or molecular dynamics and simulated annealing.[144,170]

Several more complex approaches to homology modeling have been proposed. Methods using structural elements abstracted from multiple proteins[145,150] are now included in several commercial software packages (e.g., the COMPOSER module of SYBYL and the Homology module of Insight/Discover). A more sophisticated, and more computationally intensive, approach to homology modeling has recently been described by Heiner *et al.*[147] In this case single site mutations in a protein are accommodated by continuous change during molecular dynamics simulation. The procedures are analogous to the slow growth approach used in free energy perturbation calculations (see below).

D. Molecular Dynamics

Molecular dynamics calculations are based on the solution of Newton's equations of motion, where the atom (mass $= m$) at position r_i (position vector associated with location x_1, y_1, z_1) moves due to the combination of its current velocity, $\mathbf{V}_{\text{init}}$, and that due to acceleration provided by the force, $\mathbf{F}$, exerted by the potential function, E_{pot}. The force is calculated

[165] F. Eisenmenger, P. Argos, and R. Abagyan, *J. Mol. Biol.* **231,** 849 (1993).
[166] C. Wilson, L. M. Gregoret, and D. A. Agard, *J. Mol. Biol.* **229,** 996 (1993).
[167] P. E. Boscott, G. J. Barton, and W. G. Richards, *Protein Eng.* **6,** 261 (1993).
[168] M. J. Dudek and H. A. Scheraga, *J. Comput. Chem.* **11,** 121 (1990).
[169] M. Whitlow and M. M. Teeter, *J. Biomol. Struct. Dyn.* **2,** 831 (1985).
[170] A. T. Brunger and M. Karplus, *Acc. Chem. Res.* **24,** 54 (1991).

by evaluating the spatial derivatives of the potential at $r_i(t)$. The new position of the atom, $r_i(t + dt)$, is then determined by the vector sum of two velocity components, $\mathbf{V}_{\text{init}}$, and the action of the acceleration ($\mathbf{F}/m$) over some small time step dt.

$$r_i(t + dt) = r_i(t) + \mathbf{V}_{\text{init}}\, dt + \mathbf{F}/m\, dt^2 \tag{19}$$

Providing that dt is small, the trajectory of movement of an atom can be mapped by calculating successive steps of this type.

Equation (19) is strictly true only if the time step is infinitesimally small, because the force **F** changes continuously with position. If this equation is used with a time step that is too large, the new position of the atom will not be correct, but if the step is too small the computational time increases. A number of numerical integration methods have been tried to improve prediction of the new position with larger time steps.[171-173] All of the major molecular dynamics programs use one of these approaches.

The accuracy of the trajectory can be accessed by evaluating energy conservation. At any point in the trajectory with no external forces applied, the sum of the potential energy and the kinetic energy should be constant.

$$\tfrac{1}{2}\Sigma m_i v_i^2 + E_{\text{pot}} = \text{constant} \tag{20}$$

where the sum is over all atoms.

1. Simplifications for Computational Speed

1. One simplification used by most protein force fields is to use lists of interacting pairs of atoms or groups for the nonbonded calculation and then to update the list infrequently. In combination with a nonbonding cutoff (see below), this approach saves computations of distances between all pairs of atoms except when the lists are updated. Of course, this means that some atoms may move within the cutoff distance and not be accounted for and some will move outside and still have their interactions summed.

2. Because the nonbonded interactions fall off as $1/r^n$, where n is at least 1, it can be argued that contributions outside of some distance r will be negligible when computing the interaction energy of any particular atom. However, as r increases the volume of included atoms increases

[171] W. F. van Gunsteren, *Protein Eng.* **2,** 5 (1988).

[172] J. A. McCammon and S. C. Harvey, "Dynamics of Proteins and Nucleic Acids." Cambridge Univ. Press, Cambridge, 1987.

[173] Appendix 1 of, C. L. Brooks, M. Karplus, and B. M. Pettitt, "Proteins: A Theoretical Perspective of Dynamics, Structure, and Thermodynamics," p. 52. Wiley, New York, 1988.

as the cube. In conjunction with infrequently updated lists of interacting partners (see above), the use of a nonbonded cutoff, typically ≥8 Å, can save considerable computational time. Use of a cutoff introduces a discontinuity in the force and lack of energy conservation.[56,174] This has led to a number of methods for smoothing the calculation of forces across the cutoff boundary. Different methods have been reviewed.[56,174,175] Guenot and Kollman[56] have discussed the selection of cutoff method in conjunction with treatment of dielectric constant and the method of temperature conservation in solvated simulations.

3. Large time steps can allow atoms, particularly lightweight hydrogens, to move distances that result in unusual bond lengths. By constraining bond distances during simulation, larger steps can be taken. One method is the SHAKE algorithm,[176] in which motions are constrained so that bond lengths do not exceed some preset threshold. The constraint is applied as adjustments to positions rather than as adjustments to step size. The SHAKE constraints reposition the lightest atoms the most, and the solution of the new positions is iterative; one atom is adjusted at a time until all constraints are satisfied. SHAKE typically allows time steps to be doubled. An alternative approach is the RATTLE algorithm,[177] in which velocities are constrained. The RATTLE algorithm has been used in two important tests of cutoff effects in protein simulations.[56,174]

4. *In vacuo* simulations or simulations without the explicit representation of solvent molecules have been used to investigate protein dynamics with fewer atomic loci. Typically solvation of a protein will increase the number of atoms by a factor of two or more. The distance-dependent dielectric constant (see above, Section II,B,2) has been used to represent an aqueous environment, but in direct comparisons between solvated and unsolvated simulations, this does not appear to be an adequate representation.[116,174,178–183] One positive result of unsolvated simulations is that a larger region of conformational space is investigated in a short simula-

[174] D. B. Kitchen, F. Hirata, J. D. Westbrook, R. Levy, D. Kofke, and M. Yarmush, *J. Comput. Chem.* **11,** 1169 (1990).

[175] R. J. Loncharich and B. R. Brooks, *Proteins* **6,** 32 (1989).

[176] W. F. van Gunsteren and H. J. C. Berendsen, *Mol. Phys.* **34,** 1311 (1977).

[177] H. C. Andersen, *J. Comput. Phys.* **52,** 24 (1983).

[178] S. Swaminathan, T. Ichiye, W. van Gunsteren, and M. Karplus, *Biochemistry* **21,** 5230 (1982).

[179] J. Wendoloski, J. B. Matthew, P. C. Weber, and F. R. Salemme, *Science* **238,** 794 (1987).

[180] M. Levitt and R. Sharon, *Proc. Nat. Acad. Sci. U.S.A.* **85,** 7557 (1988).

[181] L. X.-Q. Chen, R. A. Engh, A. T. Brunger, D. T. Nguyen, M. Karplus, and G. R. Fleming, *Biochemistry* **27,** 6908 (1988).

[182] C. L. Brooks and M. Karplus, *J. Mol. Biol.* **208,** 159 (1989).

[183] V. Daggett and M. Levitt, *J. Mol. Biol.* **223,** 1121 (1992).

tion[116,178,179] and in some cases the movements seen are directly correlated with those seen in longer time solvated simulations.[179,181] As indicated above, full solvation of a protein with periodic boundary conditions[180] often increases the number of atoms by a factor of two and the calculation time by as much as a factor of four. Guenot and Kollman[184] have investigated the use of less complete solvation along with a distance-dependent dielectric constant. They found that accurate simulations were obtained with as little water at that localized in the crystal structure, using a distance-dependent dielectric constant.

As with all simplifications used to allow minimization or dynamics calculations with limited computer resources, these selections have their costs and certain combinations are not compatible with different methods of temperature maintenance (see Guenot and Kollman[56]).

2. *Different Types of MD Runs*. MD simulations are characterized by the control exerted on the state functions. When total energy and volume are controlled and the number of atoms in the simulation is kept constant (nVE = constant), the collection of structures at each different time step can be considered a sample from a microcanonical ensemble. In a solvated simulation, this type of control requires some boundary condition where molecules of solvent moving outside of the box of atoms are replaced, typically by a like molecule moving into the box along the same translational vector. The protein is not allowed to translate, so that as long as the box is big enough, the shape changes in the protein are accommodated. Translational and rotational energy components for the protein itself are typically removed and replaced by increasing all of the atomic velocities by an appropriately small, proportional amount.

Constant temperature simulations (nVT = constant) can also be carried out to generate a canonical ensemble of structures. Two approaches have been used to maintain constant temperature. In that of Andersen[185] the atoms in the simulation have their velocities randomized by "collision" with atoms from a temperature bath of "solvent" molecules. The simpler approach of Berendsen *et al.*[186] is to rescale velocities to maintain temperature.

Constant pressure simulations (nPH = constant) to generate an isobaric/isoenthalpic ensemble can also be achieved by scaling. Both Berendsen et al.[186] and Andersen[185] offer the solution where both positions and boundaries are scaled.

[184] J. Guenot and P. A. Kollman, *Protein Sci.* **1,** 1185 (1992).

[185] H. C. Andersen, *J. Chem. Phys.* **72,** 2384 (1980).

[186] H. J. C. Berendsen, J. P. M. Postma, W. F. van Gunsteren, A. DiNola, and J. R. Haak, *J. Chem. Phys.* **81,** 3684 (1984).

E. Free Energy Perturbation

The collection of structures sampled during a molecular dynamics simulation at constant temperature can be treated as a canonical ensemble for the purposes of a statistical thermodynamics evaluation of the properties of the system. Unfortunately, the configurations sampled over any reasonable simulation time are unlikely to include a sufficient sample of the low-probability, high-energy regions of conformational space that would be necessary to define the proper ensemble average for direct calculation of thermodynamic parameters.[187] However, several methods have been devised to use thse methods for calculation of thermodynamic differences (for reviews see van Gunsteren[171] and Straatsma and McCammon[188]). Even so, relatively long simulation times are required for stable results.[189–191]

There are two basic approaches to free energy calculations using molecular dynamics simulations: thermodynamic perturbation methods and thermodynamic integration methods. Variations of these methods and some additional approaches are discussed in the review by Straatsma and McCammon.[188] Both methods depend on the capability of the molecular mechanics method to calculate a potential energy and force that can be composed of two fractional contributions. Consider the simplest mutation of changing a single atom, the oxygen of serine into the cysteine sulfur. In practice this change would be complicated in some force fields by the addition of two pseudoatom lone pairs to the sulfur, but for simplicity consider only the O to S change. The force field parameter set will contain parameters for both atoms (bond lengths, angles, dihedrals, force constants, van der Waals terms, charges, etc.). Now consider the hypothetical protein containing a hybrid residue made up of half-cysteine and half-serine, $\lambda = 0.5$. The force field potential can then be calculated for any particular x, y, z location of the atom centers on the basis of the full sum of potential contributions, using half of the potential calculated using the oxygen parameters and half of that using the sulfur parameters. Obviously, a similar calculation could be carried out for any value of λ, the coupling parameter. When $\lambda = 0$ the protein contains serine; when $\lambda = 1$ the protein contains cysteine.

[187] J. P. Valleau and G. M. Torrie, *in* "Statistical Mechanics Part A. Equilibrium Techniques" (B. J. Berne, ed.), p. 169. Plenum, New York, 1977.

[188] T. P. Straatsma and J. A. McCammon, *Annu. Rev. Phys. Chem.* **43,** 407 (1992).

[189] M. J. Mitchell and J. A. McCammon, *J. Comput. Chem.* **12,** 271 (1991).

[190] J. Hermans, R. H. Yun, and A. G. Anderson, *J. Comput. Chem.* **13,** 429 (1992).

[191] A. Hodel, T. Simonson, R. O. Fox, and A. T. Brunger, *J. Phys. Chem.* **97,** 3409 (1993).

With thermodynamic perturbation calculations, separate simulations are carried out with different values of λ and the differences in ensemble averages are used to evaluate stepwise free energy changes with relatively few discrete λ values. The overall change in free energy is then the sum of these components. For this approach to be accurate there must be enough different values of λ to assure that equivalent samples of the two populations are obtained.

With thermodynamic integration calculations the value of λ is changed continuously during the course of a single simulation. In this case for our example, the simulation would start with the protein containing serine and end with the mutant protein containing cysteine with integration to determine free energy carried out over the course of this change. This approach is also referred to as a "slow growth" calculation. Both methods should be tested by carrying out the reverse calculation with comparison of the forward and reverse calculations use to test microscopic reversibility and precision.[188]

The utility of such calculations arises when they are used to evaluate two sides of a thermodynamic cycle. Consider the mutation mentioned above. The relative stability of the cysteine mutant, $\Delta\Delta G_{s,c}$, is determined by the difference between the free energies of denaturation of the proteins containing the two residues, that is, $\Delta\Delta G_{s,c} = \Delta G_{denat,\ Cys} - \Delta G_{denat,\ Ser}$:

$$\begin{array}{ccc} \text{folded native (serine)} & \xrightleftharpoons{\Delta G_{folded,mut}} & \text{folded mutant (cysteine)} \\ \Big\downarrow\Big\uparrow \ \Delta G_{denat,ser} & & \Big\downarrow\Big\uparrow \ \Delta G_{denat,cys} \\ \text{unfolded native (serine)} & \xrightleftharpoons{\Delta G_{unfolded,mut}} & \text{unfolded mutant (cysteine)} \end{array}$$

Although these values ($\Delta G_{denat,\ Cys}$ and $\Delta G_{denat,\ Ser}$) cannot be calculated by simulation, conservation of energy demands that $\Delta\Delta G_{s,c}$ can also be calculated as the difference between free energy perturbation or integration calculations on the folded and unfolded proteins, that is, $\Delta\Delta G_{s,c} = \Delta G_{folded,\ mut} - \Delta G_{unfolded,\ mut}$. Similar energy cycles can be used to evaluate binding differences between different chemical analogs to a protein, solvation energy differences, relative redox potentials, and so on.

A great deal of attention has been paid to using these methods to evaluate relative stabilities of mutant proteins.[191–195] In all of these calculations, the unfolded state of the protein is represented by a short peptide

[192] L. X. Dang, K. M. Merz, and P. A. Kollman, *J. Am. Chem. Soc.* **111,** 8505 (1989).
[193] M. Akke and S. Forsen, *Proteins* **8,** 23 (1990).
[194] B. Tidor and M. Karplus, *Biochemistry* **30,** 3217 (1991).
[195] S. F. Sneddon and D. J. Tobias, *Biochemistry* **31,** 2482 (1992).

segment in the vicinity of the mutation. As pointed out by Yun-yu *et al.*,[196] the assumptions of this representation are that long-range interactions of the unfolded chain have negligible differential effect, that the "unfolded" chain adopts an extended conformation, and that residues in the "unfolded" state are fully solvent exposed. Thus, accurate evaluation of $\Delta G_{\text{unfolded,mut}}$, which demands adequate sampling of the ensemble of the unfolded state, has been called into question.[196] This sampling problem is also illustrated by the study of a single mutation in staphylococcal nuclease.[191]

IV. Other Molecular Modeling and Computational Chemistry Programs

Molecular modeling and computational chemistry calculations for protein molecules can be categorized according to how much information is required. Some calculations, particularly those used in the three-dimensional graphics display of molecular structure, require only the coordinates that describe the geometry of the system; some use only this information and miscellaneous data on atomic characteristics. Some, such as molecular mechanics minimizations and dynamics, use a full set of parameters, geometry, characteristics, and preferences. In additions to the components of the molecular mechanics description of a protein molecule, a number of quantum chemistry programs also have application particularly in the area of defining parameters as discussed in Section II,D.

The most useful programs for molecular modeling are actually suites of programs that carry out most of the needed calculations coupled to graphics display "engines" that allow convenient, three-dimensional display of the information obtained. There are a number of commercial and noncommercial packages of this type for molecular modeling listed in Section I.

A. Calculations from Structure Description

The structure description is the basis of graphical display. Wire frame models can be displayed, rotated, and translated with color used to identify different atom types. The display projections can be calculated. Bond distances, angles, and dihedrals can be obtained and analysis of structure motifs can be carried out. In this latter case, depending on the algorithm, some of the characteristic information about atom types may be needed from the parameter set.

[196] S. Yun-yu, A. E. Mark, W. Cun-xin, H. Fuhua, H. J. C. Berendsen, and W. F. van Gunsteren, *Protein Eng.* **6**, 289 (1993).

The geometric transformations of molecular structures, and measurements of distances, angles, and dihedrals are essentially the same as those used in displaying any three-dimensional object on a two-dimensional surface. A useful source of program subroutines for these types of calculations is the appropriate subroutines from any one of the commonly available molecular mechanics programs discussed above.

Analysis of protein secondary structure motifs can be carried out by examination of torsion angles along the peptide backbone. Repetitive values for the C1′-N-CA-C1 (the ϕ angle) and the N-CA-C2-N′ (the ψ angle) dihedral angles in the vicinity of -57 and $-47°$, respectively, are characteristic of an α helix whereas repetition of -119 to -139 and 113 to 135 ϕ, ψ values indicate β-sheet conformations.[197,198] More detailed analysis of structural motifs requires both the structure description and information from the parameter set to identify hydrogen bonding partners (see below).

B. Other Structure-Based Calculations

When a given structure is analyzed in conjunction with a database of parameters, a range of useful information can be calculated: hydrogen bonding patterns, electrostatic fields, surface exposure of chemical groups, and so on. Comparisons between structures can be used to gain insight into structure–function relationships or effects of group substitutions and mutations.

1. Hydrogen Bonding Patterns and Structural Motifs. Hydrogen bond identification requires atom types, distances (and angles in some algorithms), and information from a parameter set concerning hydrogen-bonding partners, preferred bond lengths, and so on. This analysis is typically provided by major programs; however, the program DSSP[199] give detailed information about H bonding and secondary structure in a very useful format.

2. Molecular Surface Calculations. Richards[200] defined two useful molecular surface envelopes for examination when one is concerned with interactions between molecules in drug design, protein–protein interac-

[197] G. N. Ramachandran and V. Sasisekharan, *Adv. Protein Chem.* **23,** 283 (1968); Note change in angle standard, p. 438.

[198] The nomenclature for phi, psi, and omega angles along the protein backbone was defined by the IUPAC-IUB Commission on Biochemical Nomenclature and published in *Biochemistry* **9,** 3471 (1970). For a note and explanation of the change in nomenclature relative to that previously recommended, see L. B. Anfinsen, Jr., M. L. Anson, J. T. Edsal, and F. M. Richards, eds., *Adv. Protein Chem.* **23,** 438 (1968).

[199] W. Kabsch and C. Sander, *Biopolymers* **22,** 2577 (1983).

[200] F. M. Richards, *Annu. Rev. Biophys. Bioeng.* **6,** 151 (1977).

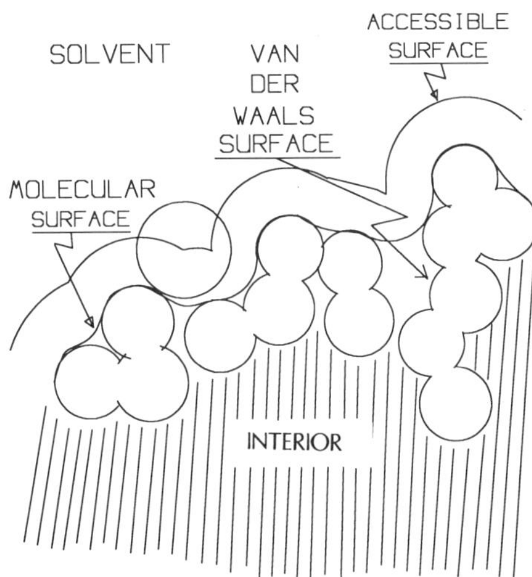

FIG. 1. Illustration of surface definitions of F. M. Richards, *Annu. Rev. Biophys. Bioeng.* **6,** 151–176 (1977).

tions, and so on. Figure 1 shows three defined surfaces. The van der Waals surface, typical of space-filling physical models, excludes some volumes that are not accessible to the smallest solvent or solute species. Thus, a more important representation of the surface of a molecule is the "molecular surface," the boundary that is accessible for van der Waals contact and occupation by a solute molecule. Although it may seem obvious that this surface represents the closest approach of a solute, in a molecular mechanics representation of molecules nonbonding interaction energies are calculated between atomic centers. The surface of closest approach of the atomic centers is defined as the "accessible surface."

For real-time graphics display, molecular surfaces require large amounts of computer time when three-dimensional transforms are calculated. Max[201] has reviewed a number of algorithms for calculation and display of molecular surfaces. New algorithms and new implementations of older algorithms have been described.[202,203] The most computationally efficient surface for interactive graphics display is one that is represented by an envelope of dots rather than solids. This approach has been used

[201] N. L. Max, *J. Mol. Graphics* **2,** 8 (1984).

[202] B. S. Duncan and A. J. Olson, *Biopolymers* **33,** 219 (1993).

[203] M. L. Connolly, *J. Mol. Graphics* **11,** 139 (1993).

for rapid display of the van der Waals surface[204] and the molecular surface.[205] The Connolly Molecular Surface program (MS), available from the QCPE, generates dot coordinates for representation of the solvent accessible surface very efficiently[205,206] and presents the contact and reentrant parts separately. Because the dots used are uniformly spaced, surface area exposure may be calculated by summing the dots from the contact part of the molecular surface. Newly coded versions of several of surface programs written in C language on a Macintosh PC are available directly from Connolly.[203]

3. Electrostatic Potential and Potential Gradients. One of the more useful and easily calculated displays of information that can be obtained from a structure is to map some chemical or physical property displayed on the molecular surface using color coding. This can be as simple as color by residue type (acidic, basic, neutral, etc.) or a color-coded display of electrostatic properties. For electrostatic properties the simplest of calculations[207] is the surface electrostatic potential calculated using Coulomb's law (Fig. 2). The potential is typically calculated at points above the surface (where the center of the solute atom would be) and then mapped back onto the molecular surface, color coded for display.

Another useful display of electrostatic information is the gradient of potential near to the surface of a proteins (Fig. 3). Two approaches are often used to display this information: potential vectors and isopotential contours. In the first case, the gradient is calculated at each surface dot on a number of concentric surfaces so that the "flow" of the field is easily seen, and the vectors are drawn of uniform length with field magnitude encoded in the color of the vector.[110] Contour plots of electrostatic potential are available in a number of the commercial molecular modeling programs.

As pointed out above, the Coulombic potential with a fixed dielectric constant is not an accurate representation because of the complexity of the dielectric environment, the polarizability of atoms of both the protein and solvent, and the screening of charge due to specific interactions such

[204] P. A. Bash, N. Pattabiraman, C. Huang, T. E. Ferrin, and R. L. Langridge, *Science* **222,** 1325 (1983).

[205] M. L. Connolly, *Science* **221,** 709 (1983).

[206] The molecular surface drawn by the Connolly algorithm is often referred to as the solvent accessible surface when according to the definitions of Richards it is the combination of the contact and reentrant surfaces. The two are easily distinguished as shown in Fig. 1. The accessible surface has no concave features characteristic of the reentrant parts of the molecular surface. These distinctions are carefully made in the documentation provided with the Connolly program.

[207] P. K. Weiner, R. Langridge, J. M. Blaney, R. Schaefer, and P. A. Kollman, *Proc. Natl. Acad. Sci. U.S.A.* **79,** 3754 (1982).

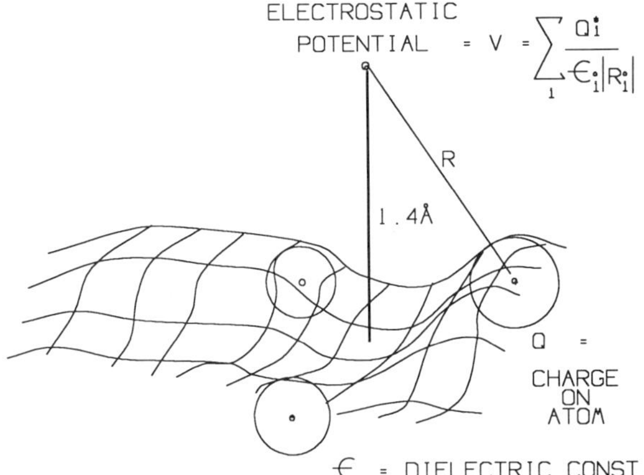

FIG. 2. Electrostatic potential calculated using Coulomb's law at a point above the molecular surface.

as ionic bonds. Two reviews[54,55] give detailed examinations of the problem of calculating electrostatic potential in this complex system. As Harvey points out, the calculation of electrostatic potential on a static structure is even more likely to be in error than a calculation using the same functionality on dynamic structures. This is because dynamic simulations sample an ensemble of conformations and arrangements that, if of sufficient size, should give representation of the effects of polarization and screening.

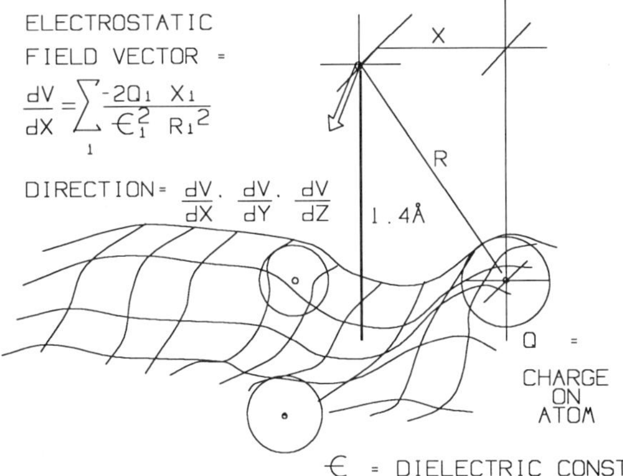

FIG. 3. The electrostatic potential gradient calculated at a point above the molecular surface.

However, there is some question as to whether the rigid water models used to represent solvent explicitly and the accessible time periods of current simulations are sufficient even in this case. Two programs that seem to offer reasonable solutions to the problem of calculating a more accurate electrostatic potential are DelPhi[208] and POLARIS.[209] DelPhi is based on a numeric, iterative, finite-difference solution of the Poisson–Boltzmann equation for the system using a grid of points and interconnections. POLARIS uses the PDLD algorithm[210] representing the solvent as a grid of dipoles and expanding the charge description of the protein atoms by addition of point dipoles at each atom. The orientation and magnitude of the point dipoles are evaluated iteratively on the basis of the calculated electrostatic field and polarizability at each atom. Xu *et al.*[211] obtained improved performance with the PDLD method by substituting the charges from GROMOS and calculating individual atomic polarizabilities.

4. Energy Difference Calculations from Electrostatics. The more accurate representation of electrostatic potential given by both DelPhi and POLARIS allows evaluation of the energy of solvation (see Gilson and Honig[212]). Free energy difference calculations can then be used to evaluate binding, protein conformation changes, the effect of mutations, and so on. Lee *et al.*[213] have compared free energy difference calculations by free energy perturbation (Section III,E) with PDLD results and some experimental data, finding reasonable agreement between the methods.

V. Summary

The methods of computational chemistry and molecular modeling are becoming more and more accessible to biochemists with the advent of fast, inexpensive graphics workstations and well-tested computer programs. The state of the art in small molecules allows chemists to use these programs as "black boxes" and be confident of the results at an amazingly high level of precision. This is not the case, however, for biological macromolecules at this time. Therefore, it is necessary before using the programs listed in Section I that we familiarize ourselves with their theoretical basis and limitations. It is also important that they be used in the context of the accumulated literature on their use. The survey given in this chapter

[208] M. K. Gilson, K. A. Sharp, and B. H. Honig, *J. Comput. Chem.* **9,** 327 (1987).

[209] A. Warshel and S. Creighton, *in* "Computer Simulation of Biomolecular Systems" (W. F. van Gunsteren and P. K. Weiner, eds.), p. 120. ESCOM, Leiden, The Netherlands, 1989.

[210] A. Warshel and A. Aqvist, *Annu. Rev. Biophys. Biophys. Chem.* **20,** 267 (1991).

[211] Y. W. Xu, C. X. Wang, and Y. Y. Shi, *J. Comput. Chem.* **13,** 1109 (1992).

[212] M. K. Gilson and B. H. Honig, *Proteins* **4,** 7 (1988).

[213] F. S. Lee, Z. T. Chu, and A. Warshel, *J. Comput. Chem.* **14,** 161 (1993).

is intended as an introduction to these tools and as a source for initiating the discovery process with the literature cited.

Acknowledgments

This work was funded in part by grants from the National Institutes of Health (Grant No. GM41482) and the North Atlantic Treaty Organization (Grant No. 0404/88). Special thanks are due to Dr. Walter McRae, Assistant Vice President for Computing, for support funds and computer time on the high-performance machines.

[41] Immunoassay of Sulfate-Reducing Bacteria in Environmental Samples

By J. MARTIN ODOM and RICHARD C. EBERSOLE

Introduction

A method for immunoassay detection of sulfate-reducing bacteria has been developed specifically in response to the need for real-time estimations of sulfate-reducing bacteria in surface and subsurface oil and refining operations. This method involves modification and optimization of standard immunoassay principles for efficient capture and detection of an internal molecular marker for sulfate-reducing bacteria. Methods were also develoepd to obviate immunoassay interferences present in environmental samples. The procedures and devices described give order-of-magnitude estimates of sulfate-reducing bacteria in environmental samples. Although this method was developed specifically for sulfate reducers, the strategy and methodologies can, in principle, be adapted for detection of other microorganisms from a variety of clinical or industrial environments.

The need for a field test for sulfate-reducing bacteria can be found in the historical association of sulfate-reducing bacteria with oil souring, corrosion of underground metal and concrete structures, and plugging of oil-containing formations associated with oil exploration since its beginning.[1] Although these consequences of microbial activity have been recognized for at least a century, the technology for detection and control of sulfate-reducing bacteria has not advanced since the early days of oil exploration. At that time, culture methods were developed that have

[1] J. M. Odom, *in* "The Sulfate-Reducing Bacteria: Contemporary Perspectives" (J. M. Odom and R. S. Singleton, Jr., eds.), p. 189. Springer-Verlag, New York, 1993.

remained in use within the oil industry until the present time.[2] Culture methods for bacterial detection, particularly detection of sulfate reducers, suffer from the problems of incubation time (days to weeks) and uncertain specificity associated with a particular nutritional condition. Early detection and real-time monitoring of sulfate reducers for the purpose of assessing microbicide effectiveness, and sources and extent of bacterial contamination, are the principal driving forces behind development of this method.

General Considerations for Immunoassay of Environmental Samples

Choice of the analyte for detection is critical to test specificity. Immunoassays for particular groups of related bacteria require a conserved antigen unique to and present in all members of that group. For detection of a particular species, a species-specific antigen or epitope must be isolated. Taxonomic makers for immunoassay are ideally extracellular and thus freely accessible for antigen–antibody reaction. Antigenic properties of extracellular cytochromes c_3 and cell surface antigens from *Desulfovibrio* spp. and *Desulfotomaculum* spp. have been investigated by others with the conclusion that species or genus specificity prevented useful application of these components as markers for sulfate-reducing bacteria as a whole.[3–5] Species-specific antigens appear to be more common than broadly cross-reactive antigens. This is probably due to the deep taxonomic divisions within this grouping with the result that the only trait many sulfate reducers have in common is the ability to reduce sulfate to hydrogen sulfide.[6]

The enzymatic pathway of respiratory sulfate reduction consists of, at a minimum, the ATP-sulfurylase, adenosine-5′-phosphosulfate (APS) reductase, and bisulfite reductase. We chose APS reductase, an obligatory enzyme for sulfate reduction, as a likely taxonomic marker protein for all sulfate-reducing bacteria because of its restricted occurrence and high cellular concentration.[7] A complicating factor is the presence of the enzyme in some sulfide-oxidizing bacteria. Thus, one question to be addressed is whether sufficient antigenic dissimilarity exists beween APS reductases from sulfate reducers and sulfide oxidizers for the enzyme to be useful as a taxonomic marker for sulfate reducers.

[2] American Petroleum Institute (API), "Recommended Practice for Biological Analysis of Subsurface Injection Waters," Second Ed. API, Dallas Division, 1982.

[3] A. Norqvist and R. Roffey, *Appl. Environ. Microbiol.* **50,** 31 (1985).

[4] R. Singleton, Jr., J. Denis, and L. L. Campbell, *Arch. Microbiol.* **139,** 91 (1984).

[5] R. Singleton, Jr., J. Denis, and L. L. Campbell, *Arch. Microbiol.* **141,** 195 (1985).

[6] R. Devereaux and D. A. Stahl, *in* "The Sulfate-Reducing Bacteria: Contemporary Perspectives" (J. M. Odom and R. S. singleton, Jr., eds.), p. 131. Springer-Verlag, New York, 1993.

[7] R. N. Bramlett and H. D. Peck, Jr., *J. Biol. Chem.* **250,** 195 (1975).

A number of operational constraints are also inherent in development of an immunoassay for field use. Definitive and relevant criteria for detection limit, incubation times, and specificity are crucial and should be targeted at the outset of the development process. Limits on test sensitivity and specificity are properties of the antibody reagents but test design and sample pretreatment may enhance these parameters significantly over that observed with reagents in a standard enzyme-linked immunosorbent assay (ELISA) microtiter plate format. Another constraint is the nature of the sample to be assayed. Detection of an antigen contained within a toxic or inhibitory environmental milieu poses a unique challenge in terms of sample clean-up and pretreatment.

Field detection also imposes additional constraints on the immunoassay test format. A field test format is generally constrained to the use of nonradioactive procedures relying on visual or colorimetric readout. Furthermore, field detection requires simple sample processing, which enables rapid and efficient analyte capture.

Field Immunoassay for Sulfate-Reducing Bacteria

Field immunoassay components consist of the basic antibody reagents and the hardware for sample processing and performance of the analyses. Antibody reagents for field use can also be utilized in the laboratory microtiter plate format using standard ELISA procedures.[8] A formal description of the microtiter plate format is minimal and focus is on components, methods, and problems unique to the field test.

Overview: Immunoassay Components and Test Procedure

Sample Pretreatment. Purification of bacteria from the environmental sample and removal of interfering substances are achieved by entrapment of bacteria in a diatomaceous earth (DE) matrix (Fig. 1). The diatomaceous earth and sample mixture are collected in a filter, washed to remove solutes, and then transferred to a reagent vial and resuspended in a buffer containing anti-APS reductase–alkaline phosphatase conjugate.

Cell Lysis. Reaction between antigen (APS reductase) and anti-APS reductase–alkaline phosphatase conjugate occurs simultaneously on sonication of the diatomaceous earth–bacteria–conjugate mixture.

Capture. Immobilization (Fig. 2) of the antigen–antibody conjugate complex is attained by passing the filtered, sonicated suspension through the capture bead. The capture bead consists of a glycidyl methacrylate-

[8] P. Tijssen, *in* "Practice and Theory of Enzyme Immunoassays" (R. H. Burdon and P. H. van Knippenberg, eds.), p. 221. Elsevier, Amsterdam, and New York, 1985.

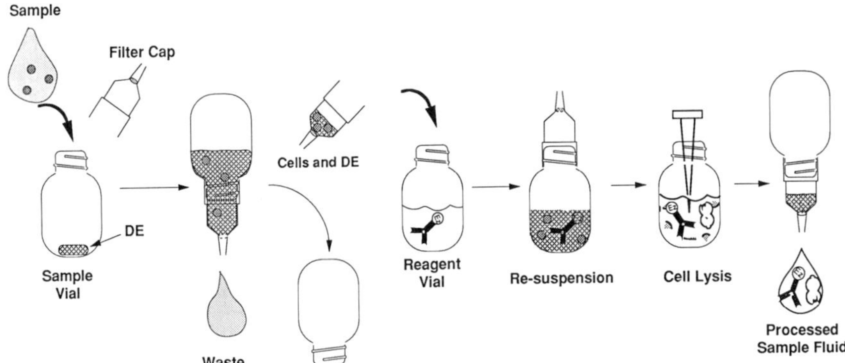

FIG. 1. Sample pretreatment. Diatomaceous earth (DE) is mixed with a volume of the sample to be assayed. The bacteria are entrapped in the DE matrix when the sample is expelled through a porous filter housed within the filter cap. The bacteria–DE matrix is then added to a reagent vial containing anti-APS reductase–alkaline phosphatase conjugate in test buffer. The cells are lysed by sonication and the soluble APS reductase–conjugate complex is expelled through a second filter cap for subsequent assay.

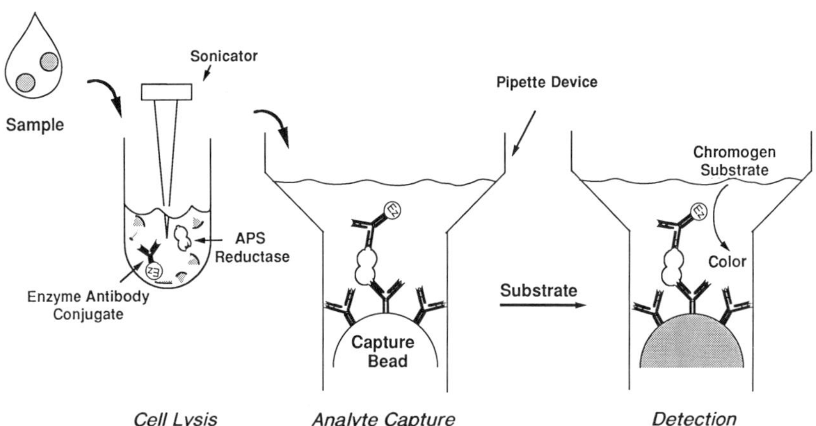

FIG. 2. Field immunoassay for sulfate-reducing bacteria. The environmental sample (pretreated or untreated) is lysed by sonication in the presence of anti-APS reductase–alkaline phosphatase conjugate. The APS reductase–conjugate complex is then captured by passage over a porous capture bead. The capture bead can be housed within a plastic pipette. After washing off excess conjugate,the amount of bound APS reductase–conjugate is visualized by addition of a chromogenic substrate,which imparts a color to the capture bead proportional to the amount of APS reductase in the sample.

grafted microporous plastic bead to which affinity-purified anti-APS reductase antibody has been covalently attached.

Detection. Visualization of bound antigen is achieved by washing excess, unbound conjugate off of the capture bead with a wash buffer followed by exposure to a chromogenic alkaline phosphatase substrate (5-bromo-4-chloroindolyl phosphate). Color intensity, generated over a standardized time interval, will be proportional to the amount of bound APS reductase on the bead.

Estimations of Cell Numbers. Quantitative estimations of sulfate-reducing bacteria present in the sample are derived by visual comparison of the color generated by the test sample with a standard curve relating color intensity to cell numbers (determined by microscopic counts) of a standard sulfate-reducing organism.

Preparation of Immunoassay Components

Growth of Bacteria

Desulfovibrio desulfuricans G100A, a strain originally isolated from oil well production water, was used as a source of the APS reductase.[9] Adenosine-5′-phosphosulfate reductases from the following bacterial strains were required for specificity studies: *Desulfomicrobium baculatum, Desulfovibrio salexigens, Desulfosarcina variabilis, Desulfovibrio gigas, Desulfovibrio vulgaris, D. desulfuricans* 27774, *D. desulfuricans* 13541, *D. desulfuricans* API, *Desulfovibrio multispirans, Desulfotomaculum nigrificans, Desulfotomaculum orientis, Desulfotomaculum ruminis* and *Desulfobulbus propionicus, Thiobacillus denitrificans, Chromatium vinosum, Chlorobium thiosulfatophilum, Escherichia coli,* and *Streptomyces lividans*. All organisms and cell extracts were grown and prepared as previously described.[10]

Adenosine-5′-phosphosulfate Reductase Purification

Antibody purification involves affinity chromatography using covalently bound APS reductase to extract purified antibody from crude antisera. It is critically important, particularly for a large antigen such as APS reductase (190 kDa), that its native configuration be conserved and

[9] P. J. Weimer, M. J. van Kavelaar, C. B. Michel, and T. K. Ng, *Appl. Environ. Microbiol.* **54,** 386 (1988).

[10] J. M. Odom, K. Jessie, E. Knodel, and M. Emptage, *Appl. Environ. Microbiol.* **57,** 727 (1991).

denaturation be limited. Therefore a rapid, high-yield purification procedure is required to satisfy the aforementioned criteria.

A cell-free crude extract is prepared from a suspension of *D. desulfuricans* G100A (50 g of frozen cell paste) cells in 100 m*M* phosphate buffer (1 : 2.5, w/v). This suspension is lysed by passage through an SLM-Aminco French pressure cell (SLM-Aminco, Urbana, IL) at 20,000 lb/in^2. Cell debris and unbroken cells are removed by centrifugation at 108,000 *g* for 1 hr. The supernatant from step 1 is adjusted to 30% of saturation with ammonium sulfate (grade III; Sigma Chemical Co., St. Louis, MO) and then centrifuged at 20,000 *g* for 20 min and the pellet discarded. The supernatant from this step is increased to 60% of saturation with ammonium sulfate. The pellet from the 60% saturation supernatant contains most of the APS reductase and is redissolved in 50 m*M* Tris [Tris(hydroxymethyl)aminomethane] (Sigma Chemical Co.) buffer, pH 7. This extract is then applied to a phenyl-Sepharose CL-4B (Sigma Chemical Co.) column (15 × 2.5 cm) equilibrated with 700 m*M* ammonium sulfate in 50 m*M* Tris, pH 7. The column is washed with 10 m*M* ammonium sulfate in 50 m*M* Tris, pH 7, and then the APS reductase elutes with a 50% (v/v) mixture of glycerol and 50 m*M* Tris, pH 7. DE-52 (Whatman Biosystems, Ltd., Maidstone, Kent, England) ion-exchange chromatography is used to remove the glycerol and concentrate the enzyme for gel filtration.

The phenyl-Sepharose CL-4B eluate is applied directly to a Whatman DE-52 column (5 × 1.5 cm) followed by washes of 40 m*M* potassium phosphate buffer, pH 7, and then eluted with 100 m*M* potassium phosphate buffer, pH 7. Gel filtration on Sephacryl S-300 (2.5 × 30 cm) at 0.5 ml/min in 50 m*M* Tris buffer, pH 7, results in an electrophoretically pure enzyme. The purification scheme is shown in Table I. The same purifica-

TABLE I
PURIFICATION OF ADENOSINE-5′-PHOSPHOSULFATE REDUCTASE FROM *Desulfovibrio desulfuricans* G100A[a]

Step	Volume (ml)	Total units[b]	Units/ml	Protein (mg/ml)	Specific activity (units/mg)
Crude extract	120	1590	13	37	0.36
Ammonium sulfate	63	1123	18	25	0.71
Phenyl-Sepharose	50	712	14	7	2.0
DE-52	12	438	36	18	2.0
Sephacryl-S300	30	360	12	3	4.0

[a] Courtesy *American Society for Microbiology*.
[b] Micromoles of ferricyanide reduced per minute.

tion scheme was used for APS reductases from other sulfate-reducing and sulfide-oxidizing bacteria.

The key step in the purification is the high affinity of all APS reductases for phenyl-Sepharose CL-4B. The purified enzyme from *D. desulfuricans* G100A is composed of M_r 70,000 and 20,000 subunits with a total molecular weight of 190,000. Subunit stoichiometry is unknown but the enzyme contains 0.74 mol of FAD per mole of enzyme, and 6.3 mol of nonheme iron and 6 mol of acid-labile sulfide per mole of enzyme. The enzyme was assayed spectrophotometrically by measuring AMP and sulfite-dependent ferricyanide reduction.[7]

Antibody Production

Antiserum to *D. desulfuricans* G100A APS reductase is generated by initial intradermal immunization of New Zealand White rabbits with 250 μg each of enzyme in Freund's complete adjuvant. Subsequently the rabbits receive subcutaneous booster injections of 50 μg of enzyme every 30 days. At biweekly intervals, 20 ml of blood can be obtained from each rabbit for antisera production. Hazleton Research Products, Inc. (Denver, PA) performed all rabbit maintenance, immunizations, and bleeds.

Antibody titer determinations are performed on antisera using standard microtiter plate ELISA methodology.[8] Immulon II microtiter plates (Dynatech Laboratories, Alexandria, VA) are coated with APS reductase (1 mg/ml) in buffer composed of 50 m*M* Tris, pH 7.5, and 100 m*M* sodium chloride overnight at 4°. Blocking of the plates against nonspecific adsorption is achieved by 2-hr incubation with the same buffer containing bovine serum albumin (1 mg/ml) and 0.1% (v/v) Tween 20 (Sigma Chemical Co.) (blocking buffer). Antisera or purified antibody dilutions are added to the plates in blocking buffer and incubated for 2 hr followed by a blocking buffer rinse. Goat anti-rabbit immunoglobulin G–alkaline phosphatase conjugate (Boehringer Mannheim, Indianapolis, IN) at 3000-fold dilution in blocking buffer is applied at 200 μl/well and incubated for 2 hr. Finally the plates are washed four times with blocking buffer and color development initiated with 200 μl/well of *p*-nitrophenol phosphate [1 mg/ml Sigma 221 buffer (Sigma Chemical Co.)]. Color development proceeds for approximately 5 min, at which time color intensity is determined by absorbance at 410 nm. Microtiter plate ELISAs of antisera or antibody for specificity studies are carried out in a similar fashion as previously described.[10]

Antibody Reagent Preparation

The multisubunit nature of APS reductase presents special problems for using the enzyme in affinity purification of the antibody. Typically a covalently bound antigen is used to adsorb the specific antibody from crude antisera and then the antibody is eluted with chaotropic agents or low pH. Adenosine-5′-phosphosulfate reductase must first be dispersed into individual subunits before covalent binding to an affinity column matrix to preclude the possibility that subunit complexes may contribute a high percentage of noncovalently bound material to the column matrix. This noncovalent material will then be present as an antigen contaminant in the eluted antibody material.

In our procedure the enzyme is pretreated by dispersing APS reductase (0.2 mg/ml) in a buffer consisting of 100 m*M* sodium bicarbonate and 100 m*M* *n*-octyl-β-D-glucopyranoside (Sigma Chemical Co.), pH 8.3, for 1 hr at 37°. Cyanogen bromide-activated Sepharose 4B (Pharmacia, Piscataway, NJ) is then added to the detergent-dispersed enzyme at a ratio of 2.7 mg of enzyme per gram of gel (dry weight) and allowed to incubate at 4° overnight with gentle mixing. The gel is then placed in a column and washed (overnight at 5 ml/min) with 10 m*M* sodium phosphate, 10 m*M* NaCl, and 20 m*M* *n*-octyl-β-D-glucopyranoside (pH 7) buffer or until all traces of ELISA-detectable antigen are removed from the column effluent. This step is critically important for removal of noncovalently bound enzyme. For immunoadsorption of specific APS reductase antibody, titered antiserum is passed over the column at 5 ml/min. The column is then washed with 10 m*M* sodium phosphate, 10 m*M* sodium chloride (pH 7) buffer until all the residual nonadhering protein is removed as determined by a stable optical density at 280 nm. A wash buffer composed of 100 m*M* glycine (pH 4) and 100 m*M* NaCl is used to elute more tightly bound contaminating material from the column. The anti-APS reductase antibody is then eluted with 100 m*M* glycine, pH 2.5, and 100 m*M* NaCl; the pH of the eluate is immediately adjusted to pH 7 with an excess of 1 *M* potassium phosphate buffer, pH 7, on collection.

Anti-APS reductase–alkaline phosphatase conjugate preparation is by a one-step glutaraldehyde conjugation procedure linking antibody to alkaline phosphatase.[11] The procedure utilizes alkaline phosphatase (Boehringer Mannheim) and antibody in a 2 : 1 (w/w) ratio. Purification of the conjugate away from unconjugated antibody or enzyme is performed by preparative-scale high-performance liquid chromatography using a Bio-Sil SEC 250-5 (300 × 7.8 mm) column (Bio-Rad Laboratories, Richmond CA) preparative column in 100 m*M* potassium phosphate buffer. Conjugate titer is determined by ELISA using Immulon II microtiter plates coated with APS reductase (1 mg/ml, 200 μl/well) and blocked in blocking buffer.

[11] P. Tijssen, *in* "Practice and Theory of Enzyme Immunoassays" (R. H. Burdon and P. H. van Knippenberg, eds.), p. 221. Elsevier, Amsterdam, New York, 1985.

Dilutions of the conjugate are then plated directly on the antigen-coated plate and incubated for 2 hr. Color development is performed as described above for antibody titer. Conjugate synthesis is a process that is difficult to control precisely, thus the product may vary in its activity and final concentration. Optimization of the amount of this product to be used in the field assay is an empirical process that entails matching conjugate dilution in test buffer with the antibody capture bead such that a minimal blank and maximal sensitivity are obtained.

Antibody Solid-Phase Support: "Capture Bead"

The surface properties of the porous plastic supports, used in the field test device for analyte capture, are important to facilitate antibody attachment and to minimize assay interferences that can result from oil fouling and nonspecific adsorption of the enzyme conjugate reagents to the surface of the solid support. We have employed electron beam graft polymerization to modify the surface properties of the plastic supports.[12] The grafting process enables different types of polymers to be chemically bonded to the surface of the plastic support. In this way a means is provided to alter the surface properties of the solid-phase support.

Glycidyl methacrylate (GMA) is an attractive monomer for this application because the resulting graft GMA polymer provides reactive epoxide groups facilitating direct antibody coupling to the surface of the support via reactive amino groups on the antibody itself. Furthermore, the treatment both increases the quantity of antibody that can be attached to the support and suppresses binding of interfering substances.

Grafting is accomplished by irradiating the porous plastic supports with high-energy electrons (3 MeV). This produces radicals throughout the plastic support. The radicals provide sites for initiating polymerization when the irradiated support is placed in contact with monomer such as glycidyl methacrylate.

Surface Activation of Supports

Bullet-shaped porous polyethylene beads (0.169-in. diameter, 0.162-in. length) are purchased from Porex, Inc. (Fairburn, GA). The supports are constructed of interconnecting pores ranging in size from 20 to 40 μm. Glycidyl methacrylate (>97% pure), the monomer used for surface modification, is obtained from Aldrich (Milwaukee, WI). Prior to use, a grafting monomer solution is prepared by dissolving 15% (v/v) GMA in reagent grade tertiary butanol at 40°.

[12] H. R. Allcock and F. W. Lampe, *in* "Contemporary Polymer Chemistry," p. 203. Prentice-Hall, Englewood Cliffs, New Jersey, 1981.

Surface grafting is accomplished by placing the porous beads (100 gm) in a polyethylene bag containing 200 ml of the GMA grafting monomer solution. Air is then removed from the bag by introduction of argon gas. Excess argon is removed from the bag and the top closed by heat sealing. For irradiation, the bag is placed in a level tray and the beads spread out in a uniform layer over the interior of the bag. The bead–monomer mixture is then irradiated with a 4-Mrad dose of electrons (2.8 MeV) at a dose rate of 1.2 Mrads/sec. Following irradiation the bag is shaken and the polymerization reaction is allowed to proceed at room temperature for 4 hr at room temperature. To free the beads of homopolymer and residual monomer, methylene chloride (750 ml) is then introduced and allowed to equilibrate for 10 min. The beads are then washed three times with equal portions of methylene chloride. The grafted beads are then dried to constant weight under vacuum, using a slow stream of nitrogen. Analysis of the dried supports should indicate about 5 to 20% weight increase over untreated beads.

Antibody Attachment to Activated Capture Bead

The treated bead supports (100) from the polymerization step are placed in a 25-ml container fitted with a serum stopper, which is then evacuated to remove trapped air in the capture bead. An antibody solution (5 ml) containing affinity-purified anti-APS reductase (100 μg/ml) in 10 m*M* potassium phosphate buffer (pH 7)–10 m*M* sodium chloride (PBS buffer) is injected and the vial adjusted to atmospheric pressure.

Evacuation and repressurization is repeated four times in order to promote contact of the interior surfaces of the bead with the antibody solution. This process facilitates contact between the antibody and the internal pores of the capture bead. The container is then rotated (10 revolutions/min) for 24 hr at 4°. Following equilibration, excess antibody is removed by aspiration. A blocking solution (10 ml) containing 0.1% bovine serum albumin (BSA) (Grade IV; Sigma Chemical Co.) in PBS buffer is introduced and the supports equilibrated with rotation at 4° in 15 min. The supports are washed four times with cold PBS buffer and stored in the cold in PBS containing 0.1% (w/v) sodium azide. The antibody-coated support beads can also be maintained in the dried state following freeze drying.

Pipette-Capture Bead Device

The antibody-coated support beads are inserted into a pipette (3 ml, Cat. No. 233-9525; Bio-Rad, Inc.) by first removing the pipette tip and then inserting the bead to a point 0.5 cm from the neck of the pipette. By drawing the solution to be tested into the bulb of the pipette, and then expelling it, the pipette–bead assembly provides a convenient means of moving fluid through the bead. This format maximizes contact between the capture antibody on the surface of the porous bead and the solution to be tested.

Assay Buffers

Samples to be assayed in the pipette–capture bead device should be in test buffer, which consists of Tris (50 m*M*, pH 7.5), sodium chloride (75 m*M*). SL-18 detergent (Olin Chemical Co., Stamford, CN), BSA (0.1%, w/v), and azide (0.02%, w/v). After the sample–conjugate mixture in test buffer has been cycled through the pipette–bead device, excess conjugate is washed out of the device with wash fluid [Tris (50 m*M*, pH 7.4) and azide (0.05%, w/v)]. The pH of the solution is adjusted to pH 7.4. For visualizing the amount of conjugate bound, a chromogenic substrate reagent consisting of 2.3 m*M* 5-bromo-4-chloroindolyl phosphate (Sigma Chemical Co.) in a 10% (v/v) 2-amino-2-methyl-1-propanol buffer at pH 10.2 containing 0.01% (w/v) $MgCl_2$ (Sigma Chemical Co.) is drawn into the pipette–bead device.

Diatomaceous Earth Sample Processing Step

This step relies on diatomaceous earth (DE) or a similar cell-entrapping matrix to facilitate collecting and washing the bacteria free of interfering solutes. In this process 10 ml of sample fluid is added to a sample collection vial containing a small amount of diatomaceous earth filtration medium (40–80 mg/ml sample). A filtration cap, containing a porous plastic frit (average pore size <10 μm) is then attached. The mixture is suspended by shaking the vial. The vial is then inverted and the suspended solids are allowed to settle into the filtration cap. The bottle is then squeezed to expel sample liquid to waste. During filtration, bacteria are entrapped in the filtration matrix and collected in the filtration cap. The bacteria are washed by reattaching the cap to a bottle containing 3.0 ml of wash fluid and squeezing the wash bottle. Cell lysis is achieved by 60-sec sonication cycles using a Branson model 450 sonifier equipped with a 1-mm probe tip.

Celite (Celite Corp., Lompoc, CA) has been used successfully although effective cell recovery can be obtained with a number of different sorbent materials. Diatomaceous earths such as Kenite (200 or 700) (Witco Chemical Co., Chicago, IL), Celatom FW60 (Eagle Picher Co., Reno, NV), and Cuno M901 (AMF/Cuno Microfiltration Products, Meriden, CN) also give adequate recovery of sulfate-reducing bacteria.

Test Performance

Specificity of Antibody Reagent

Microtiter plate ELISA response of anti-*D. desulfuricans* G100A APS reductase crude antisera to crude extracts from sulfate-reducing bacteria, sulfide-oxidizing bacteria, as well as to control bacteria (which should not contain the enzyme) is shown in Table II. There appear to be three group-

TABLE II
CROSS-REACTIVITIES OF CRUDE EXTRACTS OF SULFATE-REDUCING BACTERIA AND SULFIDE-OXIDIZING BACTERIA WITH ANTI-ADENOSINE-5′-PHOSPHOSULFATE REDUCTASE ANTIBODIES[a]

Organism	Strain G100A ELISA[b] response (%)
Desulfovibrio desulfuricans	
G100A	100
ATCC 27774	92 ± 16
API	100 ± 15
ATCC 13541	84 ± 10
Desulfomicrobium baculatum	33 ± 12
Desulfovibrio vulgaris	63 ± 10
Desulfovibrio gigas	92 ± 11
Desulfovibrio salexigens	76 ± 9
Desulfovibrio multispirans	69 ± 15
Desulfotomaculum orientis	48 ± 6
Desulfotomaculum ruminis	12 ± 3
Desulfotomaculum nigrificans	22 ± 4
Desulfobulbus propionicus	26 ± 6
Desulfosarcina variabilis	21 ± 4
Thiobacillus denitrificans	8 ± 2
Chromatium vinosum	4 ± 2
Chlorobium thiosulfatophilum	0
Escherichia coli	0
Streptomyces lividans	3 ± 3

[a] Courtesy *American Society for Microbiology*.

[b] The ELISA response at an optical density of 410 nm as an average percentage of the *Desulfovibrio desulfuricans* G100A response over a range of from 6 to 600 ng of crude extract protein per well for each organism.

ings of reactions, the most strongly cross-reactive (33–100% of G100A response) being mostly *Desulfovibrio* spp., including marine and freshwater strains. The second grouping consists of significantly less cross-reactive (12–48% of G100A response) *Desulfotomaculum, Desulfosarcina,* and *Desulfobulbus* species. Cross-reactivity with *Desulfotomaculum orientis* is surprisingly high in light of the response with the other *Desulfotomaculum* species investigated. The third grouping consists of bacteria that lack APS reductase (*E. coli* and *S. lividans*), photosynthetic sulfide oxidizers (*Chl. thiosulfatophilum* and *C. vinosum*), and *T. denitrificans*. Of the non-sulfate-reducing bacteria investigated, the strongest response was consistently observed with *T. denitrificans*. It is anticipated from these results that antibody specificity will allow reasonable order-of-magnitude estimations of many *Desulfovibrio* spp., based on calibration of the response with cell suspensions of *D. desulfuricans* G100A, but will significantly underestimate the distantly related *Desulfosarcina, Desulfobulbus,* and *Desulfotomaculum* spp. A remedy for this would be the inclusion of antibodies to one or more of these organisms into the antibody capture and detection reagents.

To see if these differences were actually due to the antigenic variation in the enzymes or to some other factor such as cellular enzyme content, enzyme extractability, or enzyme stability, enzymes from *D. desulfuricans* API. *D. desulfuricans* 27774, and *Desulfomicrobium baculatum, D. vulgaris, Desulfotomaculum orientis,* and *T. denitrificans* were purified to comparable specific activity and assayed by standard ELISA microtiter plate methodology (Table III). The data show that the purified enzyme reactivities paralleled crude cell lysate reactivities with the exception of *Desulfotomaculum orientis,* which gave essentially 100% of *D. desulfuricans* G100A response, suggesting that variations in response are due to antigenic differences in the enzymes themselves.

Sample Pretreatment

Environmental samples may contain high concentrations of solid sediments, colloids, emulsions, soluble and insoluble salts, and industrial chemicals that can prevent or invalidate the immunoassay. To remove these potential interferences, sample processing is required prior to immunoassay. This is useful both to concentrate the small numbers of organisms found in field sample materials and to remove interfering materials that prevent analysis or cause inaccurate test results. Generally, sample treatment processes can involve centrifugation, membrane filtration or chemical precipitation; however, these procedures require laboratory facilities

TABLE III
SPECIFIC ACTIVITIES AND CROSS-REACTIVITIES OF ENZYMES TO ANTI-ADENOSINE-5′-PHOSPHOSULFATE REDUCTASE POLYCLONAL ANTIBODIES[a]

Organism	Strain G100A ELISA response (%)[b]	Specific activity[c]
Desulfovibrio desulfuricans		
G100A	100	3.5
API	84	2.0
ATCC 27774	57	2.0
Desulfomicrobium baculatum	38	0.8
Desulfovibrio vulgaris	76	1.8
Desulfotomaculum orientis	100	3.0
Thiobacillus denitrificans	8	2.3

[a] Courtesy *American Society for Microbiology*.
[b] Purified protein, 240 ng per well.
[c] Micromoles of ferricyanide reduced per minute per milligram of protein.

and are not readily accomplished in the field. We have found that bacteria can be efficiently entrapped in and recovered from small filters containing a diatomaceous earth (DE) as the entrapping matrix.

Microorganisms can be effectively recovered over a wide range of cell concentrations typically encountered in environmental samples as shown in Table IV. In this example, production water from an oil well (courtesy of Conoco, Inc.) is seeded with varying concentrations of *D. desulfuricans*

TABLE IV
SAMPLE PRETREATMENT CELL RECOVERY EFFICIENCY VERSUS SAMPLE CELL DENSITY OF *Desulfovibrio desulfuricans* G100A

	Response (OD_{410})		
Cells/ml (added)	Tris buffer[a]	DE[b]	Filtration[c]
0.5×10^7	1.70	1.80	1.60
0.5×10^6	0.80	0.80	0.70
0.5×10^5	0.16	0.20	0.12
0.5×10^4	0.05	0.08	0.04
0	0.04	0.07	0.08

[a] Control, no treatment.
[b] Diatomaceous earth (Celite) at 80 mg/ml sample fluid.
[c] Membrane filtration.

G100A up to 0.5×10^6 cells/ml. Cells are recovered from 1.5-ml samples at each of the different cell concentrations. Cell recovery by means of the sterilizing membrane filters is determined by drawing 1.5 ml of each sample through a 0.45-μm sterilizing membrane filter (Gelman Sciences, Inc., Ann Arbor, MI). The recovered cells are then washed by drawing 2.0 ml of wash fluid through the filter. They are then removed from the filter for testing by using a strong back flush of 1.5 ml of sample buffer, using a 10-ml syringe to move the fluids through the filter disk.

Using these samples, cell recovery by the DE method (recovery device containing an optimized ratio of 80 mg of DE per milliliter of sample fluid) is compared with recovery by filtration through a 0.45-μm membrane filter (Acrodisc; Gelman). Release of intracellular APS reductase from the bacterial cells is accomplished by attaching the filter cap containing the washed cells to a bottle containing 1.5 ml of test buffer containing anti-APS reductase–alkaline phosphatase conjugate. The cells and solids are transferred from the cap into the test buffer and sonicated for 1 min. Diatomaceous earth is removed by filtration through a second filter and test fluid collected for assay in the pipette–bead device.

Buffer (50 m*M* Tris, pH 7) samples seeded with the same cell concentrations as the oil well samples were used as controls. All the samples are sonicated for 90 sec, then assayed for APS reducase by means of the APS reductase microtiter plate immunoassay. All aliquots (1.5 ml) from the Tris buffer controls are analyzed directly, without any treatment. The levels of APS reductase determined for the control samples are considered to represent 100% of detectable APS enzyme contained at each concentration. The results shown in Table IV indicate cell recovery by the DE method is efficient at all cell concentrations tested. Recovery approached that achieved by membrane filtration and nearly 100% of the microtiter plate immunoassay-detectable APS reductase in the samples.

Field Immunoassay Specificity and Sensitivity

Standard ELISA microtiter plate test results are performed with typical incubation time couses of at least 2 hr for equilibrium between antigen, antibody, and antigen–antibody complex. In the rapid test format, incubation times for antigen–antibody complex formation are on the order of 1–2 min. Conditions are such that an overwhelming excess of antibody is present on the microporous bead for enhanced antigen capture.

Three strains were chosen for determination of field test sensitivity. The strains against which antibodies were raised included *D. desulfuricans* G100A, *D. desulfuricans* API, and *Desulfomicrobium baculatum*. Calibrated cell suspensions of these organisms are prepared by direct counting using a Petroff-Hauser cell counter such that dilutions up to 10^8 cells/ml are available for testing. Figure 3 shows the field immunoassay response to standard cell dilutions of *D. desulfuricans* G100A. The following steps and incubation times were employed.

1. Cell suspensions (10 ml) are collected by diatomaceous earth pretreatment and washed free of residual medium with wash fluid.

2. The diatomaceous earth–bacteria mixture is then mixed with 1.5 ml of test buffer containing conjugate and sonicated for 30 sec. The soluble APS reductase–conjugate complex generated in this step is removed from the diatomaceous earth matrix by filtration.

3. The filtrate from step 2 is passed through the antibody-coated microporous bead–pipette device four times. The bead–pipette device is rinsed four times with wash fluid.

4. For the purpose of colorimetric quantitation, the beads are then removed from the pipette and placed in individual microtiter plate wells containing *p*-nitrophenol phosphate at 1 mg/ml in Sigma 221 buffer and allowed to incubate for 15 min before quantitation of the color formed. In an actual field assay a solution of 5-bromo-4-chloroindolyl phosphate is drawn directly into the bead–pipette device and cell density is estimated visually by comparison of bead color intensity with a calibrated color

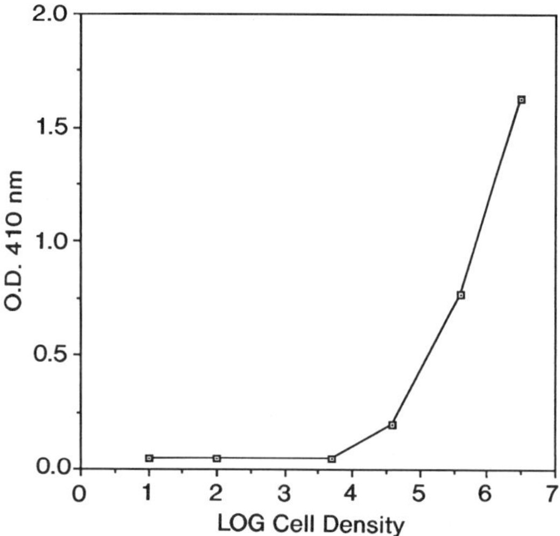

FIG. 3. Response of field immunoassay assay to different concentrations of *D. desulfuricans* G100A. Response was numerically quantitated by removal of the capture bead from the pipette for color development in a microtiter plate well. (Courtesy National Association of Corrosion Engineers.)

chart relating color intensity to cell density of *D. desulfuricans* G100A. The curve in Fig. 3 relates cell numbers to optical density at 410 nm; note that the limit of detection is close to 10^4 cells/ml.

To determine the extent of underestimation of poorly cross-reactive strains as well as cross-reactivity with a strong responding strain, *D. desulfuricans* API (100% of *D. desulfuricans* G100A response) and *Desulfomicrobium baculatum* (~30–40% of *D. desulfuricans* G100A response) are tested to check the correlation between actual cell counts and those determined by comparison with the colorimetric response by *D. desulfuricans* G100A. Field immunoassays were performed as described above for the *D. desulfuricans* G100A strain. Table V shows almost a 100-fold underestimation for cell numbers for the *Desulfomicrobium baculatum* strain even though the crude extract and purified enzyme ELISA data indicated only a 2- to 3-fold difference in response under equilibrium conditions. There is generally much less than an order of magnitude difference in cell numbers for the *D. desulfuricans* API strain. More importantly, the detection limit appears to be between 10^3 and 10^4 for *D. desulfuricans* API and 10^4 and 10^5 for *Desulfomicrobium baculatum*. It is important to emphasize that the field immunoassay is a short time course nonequilibrium assay as opposed to the microtiter plate assay, which is essentially

TABLE V

RESPONSE OF FIELD TEST FORMAT TO CELLS OF *Desulfovibrio desulfuricans* API AND *Desulfomicrobium baculatum* AS DETERMINED BY COLORIMETRIC RESPONSE RELATIVE TO *Desulfovibrio desulfuricans* G100A

	Cell density by immunoassay[b]	
Actual cell density[a]	*D. desulfuricans* API	*Dsm. baculatum*
0	0	0
10^2	0	0
10^3	0	0
10^4	2×10^3	6×10^2
10^5	2×10^5	6×10^2
10^6	9×10^6	4×10^4
10^7	2×10^7	9×10^4
10^8	5×10^7	4×10^6

[a] Cell density as determined by microscopic counts using a Petroff-Hauser counter.

[b] Cell density as determined by rapid immunoassay and comparison of color response with *D.desulfuricans* G100A standard response curve.

an equilibrium assay. The nonequilibrium format apepars to be more sensitive to differences in affinity between antibody and the different APS reductases. However, we have found, using the identical reagents in the microtiter plate format, that the detection limit for *D. desulfuricans* G100A is 10^5 to 10^6 cells/ml after a total incubation time of 2 hr as opposed to 2–3 min for the capture-bead format.

Concluding Remarks

We describe a method for estimating cell numbers of sulfate-reducing bacteria in crude environmental samples. There are two unique and key elements to this method: (1) the sample pretreatment step, which isolates the bacteria and removes soluble interferences, and (2) the antibody-coated microporous capture-bead device, which enables rapid and efficient capture of the antigen from the sample. At present antibody specificity restricts the sensitivity of the test to *Desulfovibrio* spp., however, inclusion of antibodies to APS reductases from other genera into the antibody reagent should expand the overall specificity of the test.

Author Index

Number in parentheses are footnote reference numbers and indicate that an author's work is referred to although the name is not cited in the text.

A

B

C

D

G

H

I

J

K

L

M

P

Q

R

S

T

U

V

W

X

Y

Z

Subject Index

A

B

D

E

I

K

L

N

O

P

Q

R

S

T

U

V

W

ISBN 0-12-182144-7